# CITIES' IDENTITY THROUGH ARCHITECTURE AND ARTS

PROCEEDINGS OF THE 1ST INTERNATIONAL CONFERENCE ON CITIES' IDENTITY THROUGH ARCHITECTURE & ARTS, CAIRO, EGYPT, 11–13 MAY 2017

# Cities' Identity Through Architecture and Arts


*Editors*

Anna Catalani

*College of Arts, University of Lincoln, Lincoln, UK*

Zeinab Nour

*Associate Professor in Painting at the Faculty of Fine Arts, Helwan University, Egypt*

Antonella Versaci

*Assistant Professor at the Faculty of Engineering and Architecture, University "Kore" of Enna, Italy*

Dean Hawkes

*Architect and Award-Winning Academic, Cambridge, UK*

Hocine Bougdah

*Professor of Architectural Technology, Canterbury School of Architecture, University for the Creative Arts, UK*

Professor Adolf Sotoca

*Chair Professor in Architecture at Luleå University of Technology, Sweden*

Professor Mahmoud Ghoneem

*Associate Professor in Architecture Department, Faculty of Fine Arts, Helwan University, Egypt*

Professor Ferdinando Trapani

*Professor of Urban Planning, Architecture, Palermo University, Italy*



Routledge
Taylor & Francis Group
LONDON AND NEW YORK

Published 2018 by Routledge
2 Park Square, Milton Park, Abingdon, Oxon OX14 4RN
605 Third Avenue, New York, NY 10017

First issued in paperback 2020

*Routledge is an imprint of the Taylor & Francis Group, an informa business*



Typeset by V Publishing Solutions Pvt Ltd., Chennai, India

ISBN 13: 978-0-367-73460-2 (pbk)
ISBN 13: 978-1-138-05409-7 (hbk)

Published by:
CRCPress/Balkema
P.O. Box 447, 2300 AK Leiden, The Netherlands
e-mail: Pub.NL@taylorandfrancis.com
www.crcpress.com – www.taylorandfrancis.com

*Cities' Identity Through Architecture and Arts – Catalani et al. (Eds)*
*© 2018 Taylor & Francis Group, London, ISBN 978-1-138-05409-7*

# Table of contents

## *Planning and approaching the city*

### *Architecture*



### *History and arts*

*Cities' Identity Through Architecture and Arts – Catalani et al. (Eds)*
*© 2018 Taylor & Francis Group, London, ISBN 978-1-138-05409-7*

# Preface

Intended to be a guide for academics, scholars, and interested leaders, this book was designed to critically assess issues related to architectural identity, the city as a scene, the city as an organism, the city as a subject, and the planning or rather approaching of one.

A pressing issue for many researchers in the field, the book discusses the negative repercussions resulting from globalization. Studies have indicated that globalization, despite all the positive effects, has resulted in a loss of identity within a city. As a city develops over time, its identity is evolving as well and may even be lost due to rapid and constant changes it is subjected to. Discussed as well are examples and tendencies in dealing with urban identities as well as the transformation of cities and urban cultures mentioned in terms of form, identity, and art.

This book is a combination of innovative research submitted to a conference on Cities' Identity Through Architecture and Arts (CITAA) whereas scholars from all over the world gather in one venue to discuss cultural, historical, and economic issues of the city. Thus, the book offers a collective and global solution that is applicable on a universal level.

The research presented in this book was conducted by authors, or rather participants of the conference from, three different continents of the world and organized by IEREK. It was a distinct opportunity for them to share their thoughts with leading scholars and professionals in the field of Architecture, Arts, and Planning.

The research and materials in this book are directed at those who are actively engaged in the decision-making processes and to a heterogeneous audience who has an interest to critically examine all the new literature available in the field.

A special word of thanks should be made to the editors of this book and to all the authors and co-authors of the chapters who collectively provided the academic community with unique and increasingly valuable literature.

*Cities' Identity Through Architecture and Arts – Catalani et al. (Eds)*
*© 2018 Taylor & Francis Group, London, ISBN 978-1-138-05409-7*

# Acknowledgements

IEREK would like to express its appreciation to all the members of the staff, scientific committee, chairpersons, and editors for contributing to the tremendous growth of this institution and for making the Cities' Identity Through Architecture and Arts (CITAA) conference what it is today. IEREK would also like to thank conference chairpersons, Professor Safia El-Kabany, who is the core reason as to why this conference was transformed from a mere vision into an on ground success. This institution is greatly indebted to the conference guest editors: your efforts and continuous support is what led to our success.

IEREK takes distinct pride in being an institution that amasses a highly qualified and competent team who restlessly worked for months to make this conference what it is today. With regards to the success of this conference, any step forward towards the ultimate goal of creating a well-rounded society was made possible by the highly reputable scientific committee that worked competently to prepare for and revise research papers. It would also like to thank all the members of the Scientific Committee who made it their duty to help this institution spread knowledge to the masses.

*Cities' Identity Through Architecture and Arts – Catalani et al. (Eds)*
*© 2018 Taylor & Francis Group, London, ISBN 978-1-138-05409-7*

# Organization

## SCIENTIFIC COMMITTEE OF REVIEWERS



## ORGANIZING COMMITTEE

*Cities' Identity Through Architecture and Arts – Catalani et al. (Eds)*
*© 2018 Taylor & Francis Group, London, ISBN 978-1-138-05409-7*

# Organizing institutions

Helwan University of Egypt

Faculty of Fine Arts, Helwan University, Egypt

Egyptian Architects Association

International Experts for Research Enrichment and Knowledge Exchange

 *ISBN 978-1-138-05409-7*

# Sponsoring organizations

Beta Egypt for Urban Development

Scib Paints

*Cities' Identity Through Architecture and Arts – Catalani et al. (Eds)*
*© 2018 Taylor & Francis Group, London, ISBN 978-1-138-05409-7*

# Keynote lecturers

Professor Nabil Khelifi
*Managing Editor in the Middle East and North Africa, Springer, Heidelberg, Germany*

Professor Hocine Bougdah
*Professor of Architectural Technology, Canterbury School of Architecture, University for the Creative Arts, UK*

Professor Adolf Sotoca
*Chair Professor in Architecture at Luleå University of Technology, Sweden*

Architect Rasem Badran
*Architect and Recipient of the Aga Khan Award for Architecture for the Great Mosque of Riyadh and the Redevelopment of the Old City Centre, Riyadh, Saudi Arabia*

Professor Chris Watts
*Professor Emeritus, Department of Fine Arts, Washington State University, Pullman, USA*

*Architectural identity and globalization*

*Architecture*

*Cities' Identity Through Architecture and Arts – Catalani et al. (Eds)*
*© 2018 Taylor & Francis Group, London, ISBN 978-1-138-05409-7*

# A great Chinese 'rural' metropolis—the unity and contradictions in Beijing's urban identity

Xie Li
*ICOMOS CHINA (Chinese National Committee for the International Council on Monuments and Sites), Beijing, China*

ABSTRACT: A historic city, with its complex webs of relations and stories past, is often a bafflingly complicated system, and as such any attempt to summarise its identity becomes questionable. However, such efforts offer us an opportunity to better understand the role cities play in shaping our lives through their urban landscapes.

The author has been working on the practical solutions for Beijing's urban conservation and regeneration. This paper attempts to step back and look at the city in its historical and spatial totality in order to expand our vision of the possibilities for the conservation and rehabilitation of Beijing's unique architectural heritage. By looking at the city as a complex, contradictory organism, this study seeks to redefine our general approach to urban conservation and development.

> 'How noble this city must be! For we have spent four hundred years trying to finish it off, and we still have not succeeded.'
>
> – Doctor Juvenal Urbino, *Love in the Time of Cholera*

## 1 INTRODUCTION

The old city of Beijing has experienced greater changes in the past seven decades of rapid development than in all its earlier existence of the previous eight centuries. As modern China embraced new ideologies from the 1950s onward, the capital faced the radical demolition of historic landmarks such as old city walls, gates, and decorated archways. Many noble mansions and large courtyard houses were filled with new populations of working-class owners, without much effort to adapt their design. With the real estate boom and large-scale urban construction projects of the past 30 years, have come even further erasures of traditional houses in favour of wider roads, high-rise buildings and commercial developments. An awareness and practical knowledge of cultural heritage protection and urban conservation have grown rapidly in China, but never fast enough to catch up with the speed of the changes made in the name of 'modernisation'. Preservation of the historic city of Beijing has evolved via a similar path to that followed by the international community, moving from the protection of individual monuments to larger historic districts, and now to historic urban landscapes. Regrettably, however, the damage of previous actions has left irreversible marks on the city.

As the Beijing of today becomes more and more fragmented and its distinctive history less and less visible, is it still possible to reconnect with that urban past and its creative spirit? In this paper, I attempt to delve into the genes that make the capital unique to provide one possible answer.

## 2 RETHINKING OLD BEIJING'S IDENTITY

### 2.1 *Beijing was built in a day*

The old saying goes that 'Rome was not built in a day'. We take it almost as an inherent truth that great cities do not spring into being fully formed, but are developed incrementally over time. Beijing, however, presents a different story.

Beijing differs from other ancient capital cities around the world, such as Athens, Rome, or Istanbul, whose premodern forms primarily grew out of the natural evolution of urban space according to changes in population, economic wealth and political power. As the capital of a vast empire, historic Beijing was unique because it came into being as a 'planned entity' (Liang, 1986).

#### *Born to be old: An 800-year-old city carrying a 3,000-year-old tradition*

The actual construction that made Beijing the capital city it is today took around 20 years in the 13th Century, but its planning strictly adhered to principles laid out 3,000 years ago in the ancient Chinese Confucianism classic the *Zhou Li*, or the *Rites of Zhou.*

> 'Here, where Heaven and Earth are in perfect accord,
> where the four seasons come together,
> where the winds and the rains gather,
> where the forces of yin and yang are harmonised,
> One builds a royal capital.'

Though compiled later in the Spring and Autumn period, which began in the 7th century BC, the Chapter text 'Record of Construction' of the *Zhou Li* already set forth the conditions for the site selection of a capital city as early as the 11th century BC. The above poetic description was no random vision, but an expression of a rich geomantic system that explained ideals of place. Mountains and rivers, sunlight and wind—all elements of the natural environment were carefully observed and recorded in order to calculate the exact place where a city representing the heart of the empire would be able to prosper.

The book also specified the proper layout of a city, with guidelines illustrating in great detail what a good capital city should look like. It stated that such a capital should be laid out in a square so as to align with the four directions, north, south, east and west. Three city gates should be built on each side and interconnected via gridded streets. According to the *Zhou Li*, the inner organisation of a city should reflect the order of the universe itself, with every social function having its proper place. The court was to be placed in the centre, facing south, with the market behind, and an altar for paying respects to ancestors on the left and another for worshipping the gods on the right. It sets forth plans for the layout of main streets, with access designed hierarchically to lead from the most public to the most private spaces.

The actual design and construction of Beijing took place during the Yuan dynasty (1271–1368). It was built to be a strategic and ideal city, in the shape of a square, perfectly situated on the geomantic point representing the central place of power on earth. The guidelines set forth in the *Zhou Li*'s Records of Construction were materialised in almost every scale and dimension. Everything was planned, from the capital's placement within the empire's geographic environment—with attention paid to the major mountains and rivers to locate the city's most symbolically central point—to the width of residential lanes, so as to guarantee the proper social hierarchies, good circulation of traffic, and sufficient sunlight and ventilation.

In the Ming dynasty, Beijing was enlarged from its original square shape to a 凸 shape by including the Temple of Heaven to the south, and the urban waterway system was reorganised. The Qing dynasty inherited the city of the Ming, adding its own aesthetic preferences and decorative details. In terms of architecture and urban design, the same ancient Chinese principles prevailed.

Typically, other older Chinese capitals were destroyed when a new dynasty came to power. But Beijing, an 'unparalleled masterpiece of urban design' (Wheatley, 1971), stood solid and intact through 800 years. By crystallising Chinese philosophy and cosmology in its physical manifestation, it is the ultimate example of ancient Chinese urban planning, and perhaps the largest single work of art in the world.

If we use the modern language of urban design to summarise, some of the main features of the historic city of Beijing would be as follows:

1. Site selection according to geomantic omens, a perfect location with all natural environmental elements well-considered, balancing the forces of Yin and Yang
2. The north-south central axis of 7.8 kilometres, the spine of the city, which dictates the capital's spatial order and symmetry, and gives it a clear orientation
3. The city walls of the Yuan and Ming dynasties, the big square in which every house will identify itself with the grid of the city and the earthly world
4. The consideration of natural landscape elements in its design, which serves practical functions such as providing water for urban usage, but also shows how Daoist aesthetics balanced and completed the Confucian order of Beijing's urban planning, softening severe, square rigidity with natural shapes and curving lines
5. The chessboard street patterns, with fishbone lanes dividing and connecting neighbourhoods used for commerce and social congregation, just as much as for transportation
6. An absolutely horizontal composition, with only one type of building (courtyard house) of one-storey high, forming a uniquely open and graceful skyline
7. Unity and variety in form, colour and decorative details, with a hierarchy achieved through a play of symbols, where their accurate organisation creates a kind of visual symphony in space.

All of the features listed above were part of a holistic urban planning vision that reflected the integral, complete philosophy of what ancient Chinese people thought made up the ideal human habitat. Theirs was a philosophy of order and harmony. As an entirely designed space, the old city of Beijing was the best illustration of this ancient philosophy.

It is common in urbanism today to put more emphasis on the idea of an urban fabric before thinking about individual architectural elements. But from the very beginning, those planning the old city of Beijing were already thinking about the capital in such a way, coming up with an open yet well-structured plan which succeeded in flexibly adapting and accommodating actual on-the-ground realities as time went on and dynasties alternated. Beijing has for centuries functioned as what Richard Sennett terms an 'open city' (Sennett, 2006).

### 2.2 *A city of nomads too*

Situated at a crossroads of the diverse nationalities and cultures of the north China plain, Beijing has been home to rulers from a wide variety of backgrounds, hailing from very different agricultural, herding and hunting cultures. The urban planning and design principles of the old city derive from a purely Han Chinese philosophy developed by the settled agricultural civilisation of China's central plain, but two very different nomadic peoples—the Mongols of the Yuan and the Manchu of the Qing—also made distinctive contributions in shaping its identity.

When the Mongolian emperor Kublai Khan decided to build the great capital Dadu in the Yuan dynasty according to Han Chinese principles, he had not long been separated from his previous nomadic life with horses on the grasslands. The nomadic Mongols, herders from China's north, were used to moving around in search of places 'where there is abundant water and grass'." It is little wonder that when Kublai Khan overthrew the Jin dynasty, he first chose to settle not in their old palace, but rather in an imperial garden in a suburb to the north-east, where its huge natural lake must have better recalled the nature he was accustomed to.

Ancient Chinese capitals would classically have had their palace and city centre placed on dry, flat land with water to the south, flanked by mountains. But the Mongolian emperor made an exception. He made the wide, open expanse of the natural lake the very centre of his city, and built the palace complex around it, enveloped by the rest of the city. This bold choice gave Beijing an extraordinary city centre—one in which classical urban planning was infused with the free spirit and imagination of nomadic traditions. Emotionally still attached to the green landscape of the steppes, Kublai Khan ordered grass and vegetation be brought from the wild north into what today is the Beihai Imperial Park, to remind his descendants of their origins. This man-made hill in the middle of the imperial garden amazed Marco Polo, who declared it a 'green mountain', stating admiringly that here gathered 'the most beautiful trees and plants of the world'. Nomadic culture thus brought a touch of wilderness to the highly symbolic social and spatial order of Han Chinese city planning, making the old city of Beijing a green place with what is probably the earliest 'central park' in the world.

The Manchu emperors, too, attached great importance to their nomadic origins, while still appreciating the culture of the Han. Up until the reign of Emperor Qianlong (1711–1799), a plethora of imperial gardens, summer palaces and mountain resorts were built, turning Beijing into a true garden city.

### 2.3 *A metropolis of diverse cultures*

Beijing was the capital of one of the world's most powerful empires in the Yuan dynasty, and enjoyed long periods of extreme economic prosperity during the Ming and Qing dynasties. As such, it was a place where different cultures and ideas met and merged. Several religions such as Buddhism, Christianity, Islamism, Judaism, and shamanism all left numerous historic monuments and sites in Beijing, each with their own distinct artistic styles and architectural elements. The white Buddhist pagoda, built by the Yuan court architect Araniko (1244–1306) from Nepal, remains a distinctive landmark of the city, as does the Tibetan Buddhist white pagoda built during the later Qing dynasty on the hill of the Beihai Imperial Park, which remains to this day one of the most notable and beautiful elements of the historic city's skyline, perched high on the park's man-made hill.

### 2.4 *A city that stays rural*

The two most important components of the old city of Beijing are the *siheyuans* (traditional courtyard houses) and the *hutongs* (alleys between the lanes of courtyard houses). They are traditional architectural forms that originated 3,000 years ago, and are common throughout China, where they are still very much present in many villages and small towns today.

A siheyuan has the basic form of a courtyard surrounded by timber houses on four sides, typically with only one gate or door opening on the south-east corner. Elaborations can be made to this basic form, for instance via the addition of courtyards. Noble mansions and other such important buildings were often large courtyard complexes with their own gardens. With the door closed, people living in a siheyuan are enveloped in a peaceful space of their own, even in the middle of a buzzing market. This design allows one to be both integrated in and detached from city life.

In Mandarin Chinese, the words for home are 家园 or 家庭, with the first character literally denoting the house where one lives, and the second meaning garden or yard. This suggests that the ideal living space is one consisting of a built structure, plus nature—specifically, to have nature inside your home. The Beijing-style courtyard house realises this Chinese ideal of 'home'. This unusually pastoral image of the ideal home makes Beijing a very special metropolis—one that has retained dreams of an agricultural past in the heart of the city, and in each courtyard house.

The old city of Beijing is unique for being of only one storey high. It is a city of courtyard houses, with the Forbidden City as the biggest courtyard of all at its centre. Ancient Chinese urban planners chose this model to impose a sense of order and grandness in the capital, organising space to evoke a sense of hierarchy as residents move through it. Courtyard houses not only sheltered men, but also gods. Large courtyard houses could also serve as temples and public buildings, allowing them to be used by people of varied social statuses and income levels.

The old city of Beijing is thus a collection of modular, inward-looking spaces. Each courtyard house unit is independent and self-sufficient, and represents a rich tradition of philosophical thought. In a way, such homes are almost anti-urban, yet together they form a kind of alternative urban harmony, maintaining a touch of rural living without compromising the conveniences of urban life.

Hutongs perform a similar function. The name *hutong* is said to be derived from the Mongolian word for a well (Wu, 1999), likely because of the old tradition of planning blocks and houses around the city's numerous wells. Ever since the city began construction in the Yuan dynasty, there were clear and detailed regulations on the size and standards of streets, with hutongs as the smallest unit of the urban grid. If the pattern of growth in European cities centred around *squares* (*Piazzas/agora*), in China cities grew more out of a network-based pattern of planning. Without social gathering spaces such as the *agora* in the west, streets and alleys served as more ambiguous and flexible nodes for congregation. Streets and hutongs make up this network.

At a neighbourhood level, hutongs were not simply there for transportation. Though not spacious, they became a key public space for life, essentially fulfilling people's need for a 'public outdoor room' which is communal yet not as formal as a typical public space, put forward by architect Alexander Christopher (*Christopher, 1977, 69*), as a pleasant social space that modern cities are lack of. However, in Beijing, hutongs naturally blurred the boundaries between public and private in a pleasant way. They are places where the elderly can sit and play chess while children can roam in a relatively safe environment, at once an informal nursing home for seniors and an open 'kindergarten'.

Hutong residents often say that 'far-off relatives are not as good as close neighbours' – a phrase that used to apply to rural societies. Hutongs are just wide enough for small services such as open-air barbershops or fresh food markets, with farmers or vendors lined up along one side of the hutong to sell vegetables directly from their carts three times a day, as regularly as in a village.

This unique vernacular architecture—the courtyard house, and the multifunctional historic hutong space—added a relaxed, pleasant touch to the bustling metropolitan atmosphere of a city directly under strict imperial rule, keeping the capital in close touch with nature and folk culture.

These are just a few of the characteristics that shape old Beijing's urban identity. Contradictions abound in the old capital. Confucian Han Chinese cultural roots run deep throughout its entire urban history, yet Beijing has also always been a place where other cultures merge and thrive. Despite being the country's most developed urban centre for 800 years, Beijing has nevertheless retained characteristics more often found in rural areas. Harmonious yet not monotonous, Beijing, like many other historic cities, is a place whose streets and structures are rich with meaning.

## 3 CONCLUSION

### 3.1 *A different path for a different city?*

When it comes to urban development and renewal, or historic preservation and regeneration (or the lack thereof), modern Beijing has primarily followed the trends and examples set by Western cities. In a somewhat regrettable manner, it is more prone to following the trends of North America than of Europe. But given the particularities of Beijing expressed in the second part of this paper, one should question whether it is a good idea to apply conservation tactics that have worked for historic Western cities to China's capital.

There are certainly overlapping problems and challenges when it comes to conserving and developing Beijing versus other historic cities, and technical solutions used elsewhere may very well prove effective in China. Nonetheless, in this paper I have tried to stress some of the unique qualities of historic Beijing which might inspire a fundamentally different approach to looking at the same issues.

In a Chinese cultural context, it is worth rethinking from scratch all the seemingly unquestionable dichotomies of 'rural vs urban' or 'unity vs diversity', and even the concept of 'city' itself. What makes old Beijing special is not any single monument, or building, or even district, but rather the ways order and variations in its urban planning come together to create a coherent and harmonious whole. The city is thus a single, indivisible entity, best considered holistically. To apply generally accepted modern urban planning concepts such as the use of zoning or 'protected areas' to such a place may itself be questionable.

Beijing's unique melding of rural with urban spaces may also be an inspiration for urbanisation in a new Internet era where we rely less and less on fixed physical spaces and conventional grand scale urban infrastructures. Do we have to continue with the same urbanisation processes that have come before in our future cities, now that 'rural' areas are less disadvantaged for residents who still seek modern services? Perhaps the old Beijing mind set, in which the rural coexists with the urban in harmony, could suggest an interesting alternative path forward. Could 'rural' be our new 'urban'?

### 3.2 *Reconnect through creation?*

An old Chinese saying states: 'Once you have grasped the essence, forget about the form'. It was typically used to describe a certain stage in the practice of calligraphy, a highly abstract aesthetic pursuit that looks beyond form to try and capture the essence of a subject.

As urban conservationists, we try to understand and analyse urban space and delve into how cities were built up and constructed—essentially, we investigate the static results of a rich, dynamic past.

When it is no longer possible to recover or reconstruct the physical form of architectural heritage, perhaps it would be possible to try and capture the fluid spirit of Beijing, a place loyal to its origin yet open and accommodating in its evolution. It is the special disposition of the city, rich in time and meaning, that provides for us today a solid yet mobile ground to walk ahead and continue to live creatively. It is this living spirit that we might find most valuable to the people in the present and of the time to come, because we could only imagine to live wholly, with the ability to connect with it. When reconstructing the physical form of architectural heritage is not possible, Beijing might suggest that we should rather try to grasp the flowing of the spirit, and live as one with it, and without the pressure of being overwhelmed. One way to achieve this end for me is through contemporary art; it might not succeed automatically in rebuilding such connection, but it will be an authentic attempt and effort in getting closer to it; and in these pieces of genuine creative processes of the present, we might find ourselves meeting halfway with the spirit of integrity, a past we have never lost, and a future of becoming from the very beginning.

## REFERENCES


Alexander, C. (1977), *A Pattern Language*. New York: Oxford University Press.

Hou, R. (2009). *The living memory of Beijing City*. Beijing: Joint Publishing.

Liang, S. (1986). *Selected Works of Liang Sicheng* (vol. 4). Beijing: China Architecture and Building Press.

Sennett, R. (2006). *The open city*. Urban age (2006): 1–5.

Wheatley, P. (1971). *The pivot of the four quarters: A preliminary enquiry into the origins and character of the ancient Chinese city*. Chicago: Aldine Pub. Co.

Wu, L. (1999). *Rehabilitating the old city of Beijing*. Vancouver: UBC Press.

Zhang, J. (2012). *The cultural gene of ancient Chinese space*. Beijing: Qinghua University Press.

*Cities' Identity Through Architecture and Arts – Catalani et al. (Eds)*
*© 2018 Taylor & Francis Group, London, ISBN 978-1-138-05409-7*

# Architectural ornaments in the twenty-first century: An analytical study

Ghada Mohammad Elrayies
*Faculty of Engineering, Port Said University, Port Said, Egypt*

ABSTRACT: More than 100 years ago, the use of ornaments in architecture was described as a crime. In recent years, ornaments have become ecological devices that share to support the entire performance of the buildings amongst their built context. Between the two judgements of ornaments as being a crime or a function, and between the two states of rejection or reinvention, this paper aims to investigate the validity of the recent claim and trace the reasons for this shift, through an analytical study of a number of contemporary ornamental buildings in which the use of the ornament's characteristic is emerging. The paper's methodology consists of two sections: the first deals with a literature review of the discussions and different judgements of ornamentation in architecture from antiquity to contemporary, and the second deals with an analysis of 35 contemporary ornamented buildings in the last decade (of the years 2006 to 2016). The study concludes that the ornament is a translator and a bridge that connects the building to the environment and connects people with history and culture; thus, ornamentation is an instrument for the local identity of the building, the environment, and the entire urban context within the frame of sustainability.

*Keywords*: ornaments; ornamentation; patterns; contemporary architecture; twenty-first century

## 1 INTRODUCTION

The famous manifesto of Adolf Loos in which he titled the ornament as being a crime, concurrent with the beginning of the twentieth century, led to the disappearance of the ornaments for almost a century (Mitrache, 2012). Since the beginning of the twenty-first century, it could be said that ornamentation strongly announces its return to the architectural scene (Mitrache, 2012; Picon, 2014). Ornaments have attracted attention in the last decade, almost since 2005, that is clearly demonstrated in art exhibitions, journals, and books (Balik & Allmer, 2016). The architectural ornaments study is an uninterrupted source of architectural research, and the discussions about nature, use, social and cultural determinants of decoration are still ongoing (Mitrache, 2012). Over the history of architecture, the issue of ornaments' revival had been raised three times, and each time there was a different reason for the resurfacing. The first time was in the early twentieth century, where the transition was from handicraft to industrialisation and mass production; the second time was in the second half of the twentieth century with the transition to the post-structural processes of postmodernism, and the third time was at the transition from the mechanical to the digital age (Gleiter, 2009).

Before initiating the theoretical review, let us first cast our eyes on ornaments over the architectural history and hold a narrative of ornamentation throughout the history of architecture. Figure 1 and Table 1 together draw the timeline of the evolution of ornaments chronologically.

Figure 1. The evolution of architectural ornaments from antiquity to the contemporary.

Table 1. Ornaments—A walk through time.

| No. | Architectural age | Time period[*] | The most featured ornamentation | Epitome (Figure 1) |
|---|---|---|---|---|
| a) | Ancient Egypt | 3,050 BC to 900 BC | Murals, painted carvings, symbolic motifs, and botanical motifs, for decorative purposes as well as to record historic events inspired from nature | Temple of Medinet Habu, Luxor, Egypt |
| b) | Roman | 850 BC to 476 AD | Geometric and plant-based ornaments | Pantheon, Rome, Italy |
| c) | Byzantine | 527 to 565 AD | Geometric rich use of mosaics and polychrome marble veneers | St Mark's Basilica, Italy |
| d) | Islamic | 622 to 1600 AD | Geometric, calligraphic, and vegetal decorations | Alhambra Palace, Spain |
| e) | Romanesque | 800 to 1200 AD | Zigzag, spiral and animal head motifs, and painted murals | Pisa Cathedral, Italy |
| f) | Gothic | 1100 to 1450 AD | Huge windows with ornamental stone openwork filled with stained glass | Amiens Cathedral of Notre Dame |
| g) | Renaissance | 1400 to 1600 AD | Cylindrical columns, Corinthian capitals, entablatures, and semi-circular arches | St Peter's Basilica, Italy |
| h) | Baroque | 1600 to 1830 AD | A lavishly ornate decoration, called Churrigueresque | Santa Maria della Salute, Italy |
| i) | Rococo | 1650 to 1790 AD | Ornate sculptured cartouches, and natural organic objects that have been described as being an organised chaos and playfulness | Dresden Frauenkirche, Germany |
| j) | Neoclassicism | 1730 to 1925 AD | Antiquity's decorations revival | Hungarian National Museum |
| k) | Art Nouveau | 1890 to 1914 AD | Leaves, tendril motifs, flowers, intertwined organic forms, lavish birds, insects, and women statues (metaphorical-based) | Casa Milà |
| l) | Art Deco | 1925 to 1937 AD | Motifs were inspired by the architecture of ancient Egypt | National Diet Building, Tokyo |
| m) | Modernism | 1920 to 1960 AD | Completely devoid of applied ornamentation | Weissenhof Museum |
| n) | Postmodernism | 1970 to 1980 AD | Ornaments have returned, replacing the unornamented modern styles | 1000 de la Gauchetière, Montreal |
| o) | Deconstructivism | 1980s onwards | The geometry of collage | Ray and Maria Stata Center 2004 |
| p) | Contemporary | | 2000 onwards | John Lewis Department Store |

Sources: (Wikipedia3, Craven, 2016; Wikipedia1, Majewski, 2012; Wikipedia2, Kharazmi & Sarhangi, 2016; Khairi, 2011; ArtEncyclopedia, Bothireddy, 2007; SaylorAcademy, 2011; Moussavi & Kubo, 2006).
[*] Time periods are approximate, since they vary slightly from one source to another.

## 2 METHODOLOGY

This paper aims to investigate the judgements of the ornaments, whether traditional or contemporary, in terms of rejection or renovation, and to explore reasons for their revival through an analytical study of a number of contemporary ornate buildings. The paper methodology consists of two parts. The first part deals with a literature review of discussions and various judgements of ornamentation in architecture from ancient and contemporary times. The second part deals with an analysis of 35 ornate contemporary buildings (particularly from the years 2006 to 2016) to determine the contemporary ornament's identity, its role in the twenty-first century, and the ornaments' contribution towards lending the cities identities.

## 3 A LITERATURE REVIEW

### 3.1 *Exposition of ornament*

Herein, various definitions of ornament will be reviewed. Several scholars agree that the traditional definition of the term 'ornament' refers to the decoration utilised to embellish objects of the buildings (Scranton, 1922; Paner, 2013; Mitrache, 2012). As the contemporary ornament has no particular definition yet, due to its multifaceted nature (Balik & Allmer, 2016; Bothireddy, 2007), the section below will try to raise a complete picture of the concept of the contemporary ornament.

Ornamentation comes from the innate human need for visual pleasure (Riisberg & Munch, 2015). It is a source of aesthetic pleasures rooted in our small and big world. It is observed in the quilt surrounding our bodies and the carpet under our feet (Massey, 2013). It imparts joy to life and adds charm and fun to the buildings (McNicholas, 2006). As ornamentation is the process of adding items to a work of art in order to enhance the aesthetic characteristics and the depth and the clarity of its symbolic implications, it is used by every culture as an integral part of artwork (Mitrache, 2012). However, it should be noted that the public artwork inherently is not just concerned with aesthetic values; it extends to involve social, cultural and political aspects (Sandle, 2000), as will be discussed later.

Ornaments link people to things and make their relationship more intimate with their surroundings, compared to the blank surfaces (McNicholas, 2006). Ornaments are the main sources of human visual cultures; hence, they are related to the individual identity, and accordingly, the community's identity (McNicholas, 2006). They indicate the social status by marking affiliations and distinctions. They distinguish the society classes, as well as the multi-religious and sectarian denominations (Massey, 2013).

Ornamentation indicates the historical and cultural values of the society in a certain period of time. It is a way of architectural communication through an organised narrative process in order to provide the identity of the building and the community as a whole (Bothireddy, 2007). In the same way, Riisberg and Munch define ornaments as being communication devices. They state that ornaments produce resonance (Riisberg & Munch, 2015).

Siwalatri et al. (2012) stated that ornaments hold a meaning that is conveyed to the observer either directly or in a metaphorical way. The purpose here could have been educational for the community. It educates on the history of humans and links them to the past (Siwalatri et al., 2012; McNicholas, 2006). In the same way, Bothireddy (2007) has defined ornamentation as being the action of remembering, and it is related to history as remembering begins when history ends (Bothireddy, 2007).

Ornamentation is a way to express the meaning within all its types: the symbolic, historical, aesthetic, and metaphorical. Most architects agree on one issue: that the ornamentation is necessary for architectural expression, and is provided to maintain at least one of its fundamental functional or symbolic roles (Siwalatri et al., 2012; Balik & Allmer, 2016; Mitrache, 2012), and according to Balik and Allmer (2016) the symbolic ornament expresses the building functionally and aesthetically together. The symbolic aspect of ornaments makes the building an icon and a landmark (Balik & Allmer, 2016), and according to Robert Levit

(2008), symbolic ornamentations require a certain level of cultural familiarity or erudition to ensure fair access to the code. According to Levit (2008), that is Moussavi's argument too. The consent and recognition of the public are significant to avoid alienating forms (Levit, 2008). To underline the argument of articulation, one should review Lilac and Diez's argument. They claim that the ornament is not a truth: it is never a real object; it is an articulation derived from nature. The distribution of real things for adorning purposes is the role of *décor*. That is the difference between ornamentation and decoration. It prefers simulation, deception, and illusion in an enjoyable and pleasing manner to be accepted by the eye and the mind (Lilic & Diez, 2009).

When we talk about ornament, there is no room for the omission of the political aspect. As politics relate to the management of wealth, and the ornament was an economic problem, there is a relationship between politics and ornamentation. The argument is to prohibit unnecessary expenses that consequently could adversely affect the state. However, Picon pointed out that the ornament can be an investment (Picon, 2014). Mitrache has indicated economic benefits from the use of ornaments, reasoning her claim on the flourishing of the construction market and mass production those are dependent on the production of full façade systems integrated with ornamental effects (Mitrache, 2012).

Through the above contemporary reviews of the different concepts of ornaments in architecture, it is found that the contemporary ornament is associated with several important matters: aesthetics, culture, religion, history, society, and politics.

## 3.2 *Incarnation of ornament*

### 3.2.1 *Materialisation*

The ornament is a detail that is used to attract the attention of the observer, whether it is a form, colour or texture (McGraw-Hill, 2003). In architecture, every detail represented in the shape, texture, or colour is employed to attract the observer, and since the ornaments are related to detailing, they are an integral part of architecture (Bothireddy, 2007).

### 3.2.2 *Motifs of ornamentation*

Various motifs used in the ornamentation are geometric, botanical, and animal motifs. Geometric motifs are made up of abstract shapes, which include all geometric shapes from the point to the polygon. Botanical motifs are made up of leaves, flowers, and fruit. Examples of these are lotus and papyrus, palm leaves and acanthus. Animal motifs consist of real or imaginary animals as well as human figures. These motifs could be put together in combinations such as the integration between geometric and animal motifs, or geometric and floral, such as arabesques (TGSE, n.d.).

### 3.2.3 *Types of ornamentation*

Ornaments in architecture are divided into integral and applied ornaments. The integral ornament is an integral part of the framework of the building, whether it is façades, ceilings, roofs, or the entire building. According to Riisberg and Munch (2015), the most appropriate ornament is that stemming from the building's structure and materials, and which is not masking the real building. The contemporary ornaments disclose the architecture of the building with various materials, innovative technologies, and structure (Riisberg & Munch, 2015). If a contemporary building lacks unifying space and surface, the ornaments become like wallpaper, and structural ornaments free ornaments from being an applied decoration (Balik & Allmer, 2016). The applied ornament is an architectural detail, on a small scale compared to the scale of the building to which they are attached; they often include carvings, sculptures, paintings, mosaics and inlay (Encyclopedia, n.d.).

## 3.3 *A literature of the traditional ornaments*

Before reviewing judgements about traditional ornamentation, it is worthwhile to show the opposite trends and arguments. The judgements of modernists against ornamentation are well known to those interested in this field. The modernists believed that ornamentation falsifies the pureness of the form (Siwalatri et al., 2012). They conceived the ornament as being 'unsuitable' in terms of function, materiality, and construction (Riisberg & Munch, 2015). In his manifesto, *Ornament and Crime* (1908), Adolf Loos argued that ornaments consume manpower and health (Gleiniger & Vrachliotis, 2009). He bragged that he turned the meaning of ornament from the synonym of beauty to be that of inferior and superfluity (Massey, 2013).

Interestingly, although modernists decried ornaments, they had the means to express the visual pleasure somehow. Modernists advocated standardised solutions, naked structures, white walls, and geometric forms. They adopted the visual pleasure of the veining in wood and stone (see Figure 2(a)), sparkling chrome, the colour and reflectivity of glass, and the flatness and whiteness of paint (Massey, 2013). Moreover, according to Bothireddy, works of modernists were not in one way or another devoid of ornamentation. It has been represented at the urban level in the geometric pattern on Le Corbusier's *City for Three Million Plan*, where ornamentation was derived from manipulating the structure and construction processes. That was named 'ornament structuralized', a term coined by Thomas Beeby on the basis of the relationship between ornament and structure. Another structural-related ornament in modernism is in Mies van der Rohe's Lakeshore Drive Apartments. The façade's I columns were used to generate ornamental effect. This type of structural-related ornament is called 'structure ornamentalized'. In Frank Lloyd Wright's Unity Temple (see Figures 2(b), 2(c)), the structure of the ornamentation design was derived from the structural module of the design. That is so named 'ornament constructed'. Ennis-Brown's house (1924) is as good an example of this type of ornamentation (see Figure 2(d)). Bothireddy stated that Loos aimed not to eliminate ornaments, but instead the so-called 'ornate structure'. The type of ornaments related to this category is applied ornaments or surfaced ornaments, which can be seen, for instance, in Art Nouveau (Bothireddy, 2007).

Besides the foregoing, it should be noted that, akin to Wright, who abstracted plant forms to geometric shapes, Claude Bragdon created a Projective Ornament system (1915) to generate decorative motifs and geometric patterns to be a universal language to replace the historical patterns. These modernists reinvented the ornament rather than cancelling or eliminating it. From the late 1960s to the 1980s, postmodernists reshaped ornaments by abstraction and popular art. They developed iconographic motifs in which ornaments were evoked from antique to the 1920s modernism. Critics decried postmodernism for its nostalgia and commercial trend (Massey, 2013).

In classical architecture, ornaments were either separate small scale architectural forms derived from the main form of the building, or they represented stemming parts of the structure, such as Ionic, Doric, and Corinthian columns (Bothireddy, 2007). The purpose of ornaments from the Renaissance to the beginning of the twentieth century, according to Picon (2014), is not merely to please the vision. It is further to convey information about the nature and the purpose of the building, the level of the owner, the nature of occupants, and the rank of the firm it hosted, as an expression of the social values. In other words, it acts a

Figure 2. (a) The veining in the stone: the decorative instrument by the modernist, Mies van der Rohe, Barcelona Pavilion (1928); (b), (c) Frank Lloyd Wright's Unity Temple, USA (1908); (d) Wright's Ennis-Brown house (1924).

communicator (Picon, 2014). Along the same line, Siwalatri et al. stated that architecture in classical times managed to use ornaments more than at any other time, not only because the ornament is an aesthetic element, but rather for its symbolic role which holds an explanation, and for its role as a means of communication between the building and the users (Siwalatri et al., 2012).

Massey (2013) stated that ornaments in antique architecture conveyed the building's purpose, status, and character from the ornamentation's order, the proportion, the details of mouldings, and motif configurations that stem from mythology, history, and the military. Massey believes that, from the Renaissance to the nineteenth century, classicism unfolded clearly through ornaments. This classical order contributed to modularity, which consequently aided in subdividing surfaces and gaining the sense of proportions. Massey found that ornaments also reflected the economic status at this time; furthermore, the style of ornaments also reflected the imperialism (Massey, 2013). Industrialisation also affected the buildings' decorations in terms of production speed and scale, and the development of new materials such as stainless steel (Massey, 2013). Chrysler Building (1928) here testifies to this argument.

### 3.4 *A literature of the contemporary ornaments*

Recent publications have shown a clear shift in attitudes towards the ornaments in architecture, where they are now encouraged and supported. According to Fairhurst, it is now functional (Fairhurst, 2007). According to Opincariu (2011), ornaments in contemporary architecture go beyond decorating, to be further tools for expression and cultural reflectance. Ornaments demonstrate themselves as communication tools while acting as façades layers. She stated that the technology and digital revolution contribute to linking structure with the aesthetic values. She defines modern ornaments as the mirrors that reflect the new materiality and the technical logic (Opincariu, 2011).

Picon (2014) asserted that the subjectivity of the present-day ornament is not obvious. This ornament's subjectivity just has an abstract character that hardly takes us to a specific visage. On the other hand, it greatly articulates materiality as a witness to the development of the material and computer sciences. He pointed out that this technological development draws attention to new designs and manufacturing. From his viewpoint, materiality is supposed to be a tool, not an objective. Contrary to materiality, immateriality gives further contact with the physical world. So, materiality is closely constrained to specific physical and social conditions (Picon, 2014).

According to Balik and Allmer (2016), Jörg H. Gleiter argued that what causes ornaments to ignite again is the digital technology more than styles (Balik & Allmer, 2016). The return of ornaments in contemporary architecture is strongly attributed to the advanced technology (Mitrache, 2012; Pantazi, 2008; Balik & Allmer, 2016). According to Massey, while Le Corbusier rejected decorations in architecture as a hinder to societal progress, decoration has become today a witness to the technological progress (Massey, 2013).

The advanced technology-attributed programs, such as Computer-Aided Design (CAD), and Computer-Aided Manufacturing (CAM), have promoted architects to process complicated motifs through introducing patterns, complicated textures, and colours into their designs. Parametric design and its tools involve computer numerical control (CNC) milling, laser-cutting, robotic layering, 3D printing and water jets. Structural innovative approaches along with the production techniques resulted in the employment of ornaments through repetition, pattern, light, and numerical similarity to achieve structural and construction performance and environmental capabilities in tandem with aesthetic values (Massey, 2013; Balik & Allmer, 2016; Picon, 2014; Moussavi & Kubo, 2006).

In parametric architecture, patterns are considered the strongest architectural expression tools that produce a high-performance dynamic ornamentation (Schumacher, 2009). According to Robert Levit (2008), pattern is one of the chief incarnations of ornaments, including patterned-colours, materials, structures, and assemblies. He added information also to the pattern makers (Levit, 2008). Amongst this technological stream, digital ornamental façades combine between materiality and immateriality, and permanence and temporality. Within the contemporary age of visual communication and the domination of images on architecture that attract public attention, the current culture is constituted. It could be said that the ornament in contemporary architecture contributes to image-making and trade marketing (Balik & Allmer, 2016).

Ironically, with the more state-of-the-art technology, there is a heavier return to classic ornaments. This can be found in the case of 3D-printed buildings technology, an example of which is the 3D-printed villa by WinSun, China (Kerns, 2016). The villa represents a strong return to the classics of architecture by the means of advanced technology.

## 4 THE ANALYTICAL STUDY

The second part of this paper deals with an analysis of 35 ornamented buildings. Since there is no specific definition of the ornate building, the buildings that have been selected are the product of the author's search over the year preceding the publication of this paper in the framework of the research topic.

Based on the foregoing literature review, and in order to draw the analysis structure, a map has been drawn to conceive the process of making the contemporary ornament (see Figure 3). Based on the literature, the physical aspect of ornamentation can be incarnated in one, or more of five visible aspects: structural, cladding, pattern, surface, and colour. These materiality features handle one of the three main motifs of ornament: geometric, floral, or faunal, or a combination of any two of them. The product, as an invisible aspect of materiality, could be served as a historical, functional, political, cultural, and/or aesthetic value.

In accordance with the foregoing review, the analysis structure will comprise of the building name, an illustration figure, the motifs of ornaments, the year of construction, architect/s, building type, morphology, materials, the origin of genesis and synthesis of ornaments, other aspects (access to daylight, ventilation, and other design statuses), and ultimately, the invisible aspect of materiality according to the designed map. Most of the information that was compiled about the buildings is based on the architectural pages on the Internet such as ArchDaily, Divisare, Archilovers, Arch20, Designboom, and ArchiTravel. The analysis is represented in the seven Tables that follow.

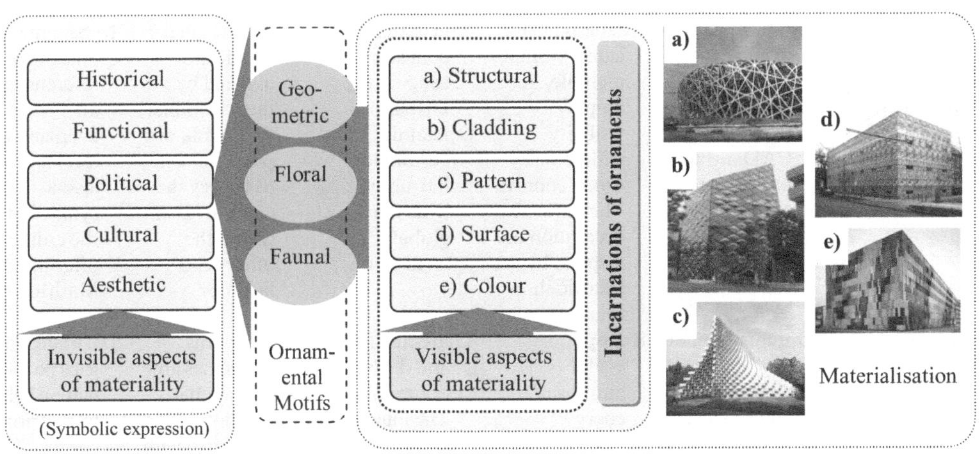


Figure 3. Ornamental incarnations in contemporary architecture.
Source: Author based on (Moussavi & Kubo, 2006; Riisberg & Munch, 2015). (a) Structure: Beijing National Stadium, Beijing, China, 2008 (Balik & Allmer, 2016); (b) Cladding: Prada Aoyama Store 2003; (c) Pattern: Serpentine pavilion 2016; (d) Surface: Eberswalde library 1999 (Moussavi & Kubo, 2006); (e) Colour: Pharmacological research laboratories, Germany, 2002 (Picon, 2014).

Table 2. A comparative analysis of ornamented buildings in contemporary architecture.

| | (1) | (2) | (3) | (4) | (5) |
|---|---|---|---|---|---|
| Name | Dorobanti Tower Bucharest, Romania | ILUMA building, Singapore | Contemporary Art Centre, Cordoba, Spain | Polish Pavilion at Shanghai Expo 2010, China | ABC museum, Madrid, Spain |
| Illustration |  |  |  |  |  |
| Motifs of ornament | 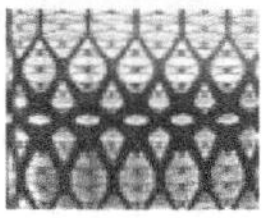 | 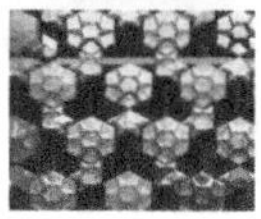 | 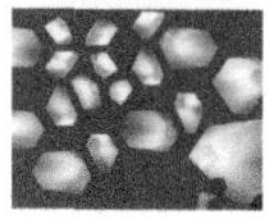 |  |  |
| | Floral | Floral/ Geometric | Geometric | Floral/ Geometric | Geometric |
| Year | 2013 | 2009 | 2013 | 2010 | 2010 |
| Architect | Zaha Hadid | WOHA Architects | Nieto Sobejano, Realities: United | Polish studio WWAA | Aranguren & Gallegos |
| Building type | Residential | Entertainment and retail | Museum | Pavilion | Museum |
| Morphology | Meandering structural lattice, with chamfered diamond-like structure | The tessellated pattern crystal mesh media façade (interactive façade) | Media skin façade (the varying-size-lit-bowls represent pixels of a large display system) | Folk art paper cut-out motifs | Tessellated triangular glass and steel tiles façade |
| Material | Stainless steel filled with concrete | Crystal | GRC prefabricated panels | CNC-cut plywood | Glass and steel tiles |
| The origin of genesis and synthesis of ornaments | Establishing an iconic presence, structural requirements, CAD, urban parameters, and site constraints | Establishing an iconic vibrant nightlife, amplify visibility within the urban context. The intricate decoration are inspired by the historic shop houses | The pattern comes as an echo of the Hispano-Islamic culture in harmony with the global civilisation | The perforated façade are inspired by the traditional Polish folk art paper cut-outs, to convey the cultural idiom within the other expo facilities | To become an artistic reference at an international level and also a symbol of the cultural offer in Madrid |
| Other aspects | Optimisation of daylighting, view outside, and seismic resistant structure | It amplifies visibility, and saves energy | Glare treatment, and daylight optimisation. The interactive façade links the space with the public | The patterns' design has an educational function associated with the main theme of the expo | The triangular gaps provide natural light to interior spaces |
| Invisible aspect of materiality | Structural-based | Cultural/ symbolic-based | Cultural/ symbolic-based | Cultural/ functional-based | Cultural/ symbolic-based |

Table 3. A comparative analysis of ornamented buildings in contemporary architecture.

| | (6) | (7) | (8) | (9) | (10) |
|---|---|---|---|---|---|
| Name | Torre de Especialidades, Mexico | Gantenbein Winery, Fläsch, Switzerland | Airspace Tokyo, Japan | 0–14 Tower, Dubai, UAE | Argos, Cali, Colombia |
| Illustration | 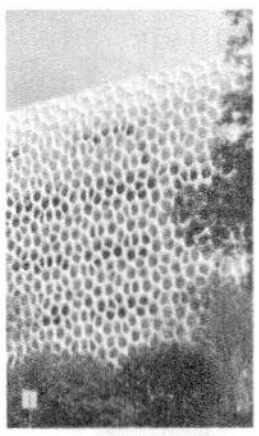 |  |  |  | 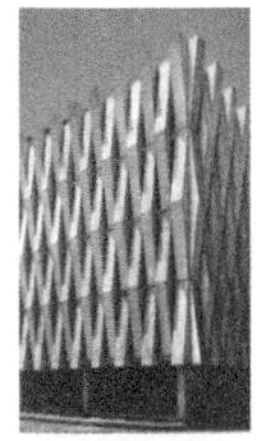 |
| Motifs of ornament | 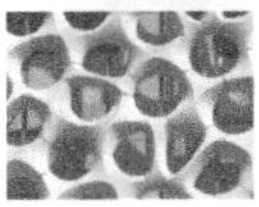<br>Geometric | <br>Geometric | 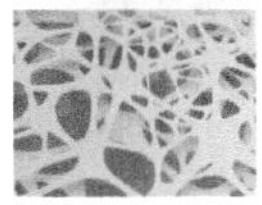<br>Floral | 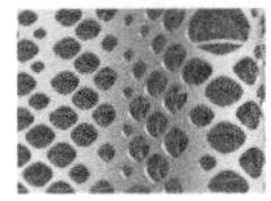<br>Geometric | 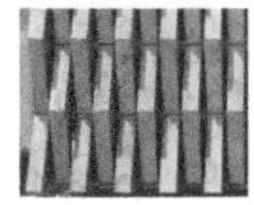<br>Geometric |
| Year | 2013 | 2006 | 2007 | 2011 | 2008 |
| Architect | The Berlin-based architects | Bearth & Deplazes, Gromazio & Kohler | Faulders Studio | Reiser + Umemoto | Felipe Gonzalez-Pacheco |
| Building type | Hospital | Vineyard | Mixed use (residential/ commercial) | Offices/ commercial | Electrical generator at cement factory |
| Morphology | Quasicrystal façade air-purifying textiles | Pixels bricks façade | Foliage-like façade | Lace-like thick concrete exoskeleton shell | A woven concrete bamboo-like façade |
| Material | Prosolve370e coated with a superfine ($TiO_2$) | Bricks | Aluminium Composite Materials (ACM) | Concrete | Prefabricated concrete |
| The origin of genesis and synthesis of ornaments | Decorative architectural modules (inspired by fractals in nature) with photocatalytic pollution-fighting technology are used to clear Mexico City atmosphere | Robotic production method alongside a simulation-generating process is utilised to find a façade structure system that meets building function as a vineyard | The cellular design and double-layering screen simulate the lush vegetation in the site to provide a level of privacy | It is modulated depending on structural requirements, outside vision, and environmental considerations (sun exposure, and luminosity) | Simulates the textures made by the locals in fibres to fabricate artisanal objects. Conveying an image of the building function and product |
| Other aspects | Achieves a synergy between design form and molecular technology. | Controlling light & air penetration and buffering temperature. | The airspace screen acts as a buffer zone that provides privacy, and isolates noise. | The design creates a chimney effect, cooling the building, control sunlight & luminosity. | Making a reference for the local people and generates a play of light and shadow. |
| Invisible aspect of materiality | Environmental-based value | Functional/ structural-based value | Functional/ symbolic-based value | Structural/ environmental-based value | Symbolic-based values with technical solution |

Table 4. A comparative analysis of ornamented buildings in contemporary architecture.

| | (11) | (12) | (13) | (14) | (15) |
|---|---|---|---|---|---|
| Name | (LIMS) Melbourne, Australia | RMIT Swanston Academic Building, Melbourne | Lille Métropole Musée extension, France | Nantong Urban Planning Museum Nantong, China | John Lewis department store & cineplex, UK |
| Illustration |  |  |  |  | 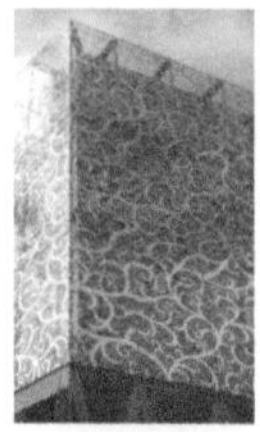 |
| Motifs of ornament |  Geometric |  Geometric | 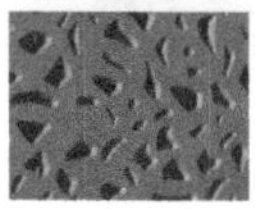 Floral |  Geometric | 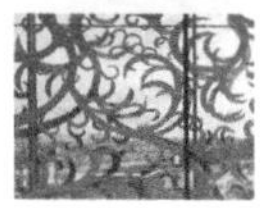 Floral |
| Year | 2013 | 2013 | 2010 | 2012 | 2008 |
| Architect | Lyons | Lyons | Manuelle Gautrand | HENN architects | Foreign Office architects |
| Building type | Educational (university) | Educational (university) | Modern art museum | Museum | Retail |
| Morphology | Mosaic geometry-based tessellated façade | A chameleon and a mirror façade | Openwork screens (modern Mashrabiya) | Diamond-Shaped diagrid | Lace-like pattern façade (vegetal) |
| Material | Wood | Anodised aluminium panels | Ultra-High-Performance Concrete (UHPC) | Reticulated metal structure | Glass and ceramic |
| The origin of genesis and synthesis of ornaments | It is inspired by the hexagonal geometry of the molecular structure to express the research nature undertaken within the building. | It derives its identity from its surroundings as a chameleon and a mirror. | As an extension of the historical context. The irregular pattern used as a light filter and rain screen, while affording outside vision. | Fluid gradient fashion of the façade came from the need to optimise the requirements of lighting interiors. | To reflect the city's cultural and historical traditions (textiles and weaving), privacy considerations, and admitting daylight. |
| Other aspects | The organisation of low and high windows maximising natural light. The cellular concept creates meeting spaces and canopies. | The façade elements provide degrees of shading and optimise thermal performance. The façade creates a dialogue with the surrounding community. | The irregular pattern restricts light levels within the galleries and protects art works. | The diamond-shaped diagrid controls and regulates sunlight and allows for varying degrees of opening from 9%–60%. | Filtering sunlight, and achieving interiors privacy |
| Invisible aspect of materiality | Functional/ symbolic-based value | Culture-Based value | Heritage/ environmental-based value | Structural/ environmental | Functional/ environmental-based value |

Table 5. A comparative analysis of ornamented buildings in contemporary architecture.

| | (16) | (17) | (18) | (19) | (20) |
|---|---|---|---|---|---|
| Name | Ravensbourne College of Design & Communication, UK | Italian Trade Center 'I Principi d'Italia', China | Ibermutuamur Building, Murcia, Spain | St Teresa's Academy Windmoor Center, Kansas, USA | A simple factory building, Singapore |
| Illustration |  |  |  |  |  |
| Motifs of ornament | 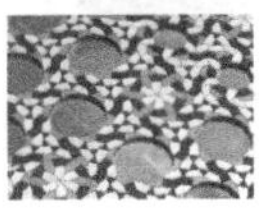 Geometric/Floral | 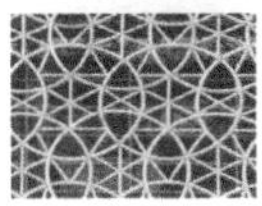 Geometric | 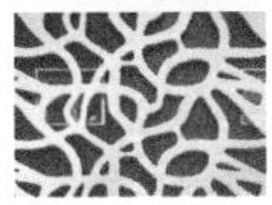 Floral | 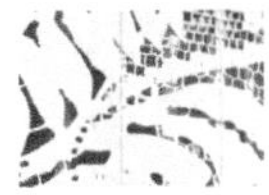 Floral | 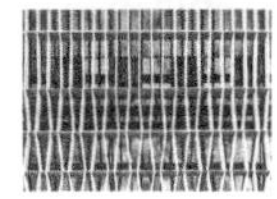 Geometric |
| Year | 2010 | 2010 | 2010 | 2012 | 2012 |
| Architect | Foreign Office | MDU Architects | Roberto Gómez | Gould Evans | Pencil Office |
| Building type | Educational (college) | Trade, office, residential | Hospital | College for women | Factory |
| Morphology | Non-periodic tiling pattern | Lattice façade (double skin) | Lattice façade (double skin) | Lace façade | Optical façade |
| Material | Anodised aluminium tiles | Bronze glass façade, steel frame | Aluminium (Cortizo/ Duralmond) | Aluminium lace fabrication | EIFS screen and bronze window wall |
| The origin of genesis and synthesis of ornaments | It is inspired by the Gothic rose windows and flower patterns alongside the articulation of technology and novel materials | The traditional Chinese curving latticework and the Venetian medieval stone and wood frames form the lace skin to express the commercial and cultural tie between China and Italy | To show emblematic building (an icon of popular ideas) without more costs | It is inspired from the narrative of St Teresa, the patron saint of lace makers. The façade combines the school's history (the core values of the school) and the vision of modern education | To mitigate the tropical solar radiation, and achieve openness, outside vision, and transparency |
| Other aspects | Patterns were produced as an abstract construction. The façade configuration supports both structure and daylighting penetration | The façade is at once light-emitting and shading. | Achieving direct solar load reduction on the building that creates a small pocket of air waves through the building. | Worship spaces are flooded with soft light and connected to nature. Patterns merged both the sacred and secular elements. | It acts as a climatic engine. The veil protects the interiors from the harsh sun, allows views outside, amplifies daylight, and natural ventilation |
| Invisible aspect of materiality | Structural/ symbolic-based value | Symbolic-based value | Environmental/ symbolic-based | Religious symbolic-based value | Environmental-based value |

Table 6. A comparative analysis of ornamented buildings in contemporary architecture.

| | (21) | (22) | (23) | (24) | (25) |
|---|---|---|---|---|---|
| Name | Central Mosque of Pristina competition entry, Kosovo | Louis Vuitton Matsuya Ginza, Japan | French Ministry of Culture & Communication, Paris, France | Podčetrtek Sports Hall, Podčetrtek, Slovenia | The House of Chutes-Lavie, Marseille, France |
| Illustration |  |  |  |  |  |
| Motifs of ornament |  |  | 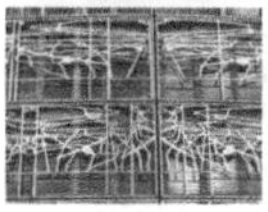 | 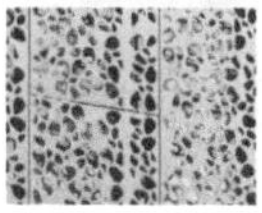 |  |
| | Geometric/Floral | Geometric | Floral | Geometric/Floral | Geometric |
| Year | 2013 | 2013 | 2006 | 2010 | 2012 |
| Architect | Maden Group | Jun Aoki & Associates | Soler Francis, Druot Frederic | Enota | Damien Fluchaire & Julien Cogne |
| Building type | Masjed | Retail | Public | Sports hall | Sports hall |
| Morphology | Arabesque cladding façade | Louis Vuitton's damier | Organic-Weave mesh of steel lace façade | Flower pattern perforated cladding | Flower and map-like pattern perforated cladding |
| Material | NA | Steel | Stainless steel | Metal | Stainless steel cladding |
| The origin of genesis and synthesis of ornaments | Using arabesque to keep the tradition and to make a city landmark. There is a combination between geometry and flower patterns inspired by Islamic art | The highly abstract and stylised geometric pattern inspired by both the history of Ginza with its art deco design, and by Louis Vuitton's damier | The continuous mesh is used to regroup the different architectural styles (modern and classic) of the two existing buildings into a single unit | To highlight the main access that directs visitors to the main hall entrance, and to break the intensity of the serious heavy façade | The vegetable patterns inspired from pine trees in the middle of Marseille. It represents a mystical code as it interprets the pattern of the neighbourhood plan |
| Other aspects | Arabesque openings maximise the usage of natural light, insulation, ventilation and air conditioning. | The façade pattern with its bulges and dents, with LED lights, reveals various appearances through both day and night. | The network of steel lace brings natural light into the buildings and allows outside views. | Perforated cladding covering a large window prevents the disturbing, direct impact of sun on the playgrounds. | It protects the building from solar gains and allows privacy. |
| Invisible aspect of materiality | Religious/ symbolic-based value | Trade marketing/ historical based value | Functional-based value | Functional-based value | Social/ symbolic-based value |

Table 7. A comparative analysis of ornamented buildings in contemporary architecture.

| | (26) | (27) | (28) | (29) | (30) |
|---|---|---|---|---|---|
| Name | Princess Nora bint Abdulrahman University, KSA | Suzhou Science and Cultural Arts Centre, Suzhou, China | CIB Biomedical Research Center, Pamplona, Spain | Parking Structure Art Façade, USA | John Curtin College of the Arts, Fremantle, Australia |
| Illustration |  | 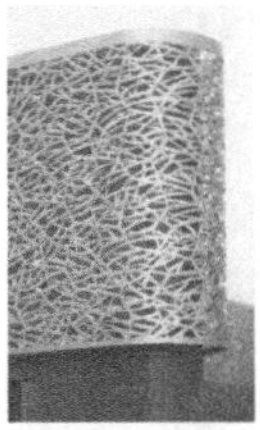 |  |  |  |
| Motifs of ornament | 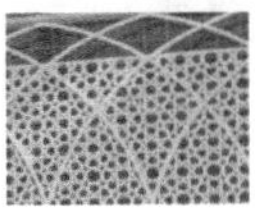 | 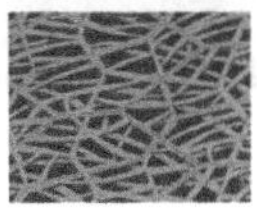 |  | 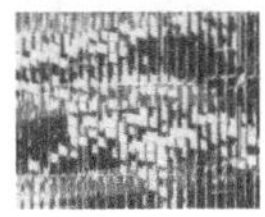 |  |
| | Geometric/Floral | Geometric | Geometric | Various | Geometric |
| Year | 2011 | 2006 | 2011 | 2014 | 2015 |
| Architect | Perkins+Will, and Dar Al-Handasah | Paul Andreu & ECADI | Vaíllo & Irigaray | Rob Ley Studio | JCY Architects & Urban Designers |
| Building type | University for women | Science & cultural arts centre | Research centre | Services (parking structure) | Educational (school) |
| Morphology | Latticework façade | Hexagonal matrix façade | Origami façade | Art dynamic façade | Sculptural veil façade |
| Material of ornament | GFRC and aluminium | Aluminium | Perforated aluminium plates | Folded metal panels (steel) | Aluminium |
| The origin of genesis and synthesis of ornaments | It is inspired by Islamic geometrical pattern. It symbolises regional architecture and cultural tradition. It exploits innovative techniques as a single architectural language. | The overlayered hexagons form the matrix that is inspired from the classical Suzhou timber window. Science and arts (the main activities of the building) form a continuous dialogue within the façade pattern. | The origami façade simulates the biotype of leaves to reflect the building's intrinsic functionality (linking the façade biomimicry-based concept with the biomimicry research nature of the building). | The active camouflage façade seeks to create a dynamic interactive façade for the viewers in the urban context. The visual screen masks the scene that could be seen in ordinary parking structures. | The 3D Mobius elements, developed from the simple geometry of the golden right-angled triangle, contributes to enriching the simple structure components. |
| Other aspects | It acts as shading screens, enhances daylight distribution, and allows privacy and visibility. | The ornamental metal screen acts as sun shading. | The façade geometry generates folds to always keep the sun and allow vision. | It allows natural ventilation with regard to pattern design. | The sculptural screen provides passive climate control (sun shading and heat filtration). |
| Invisible aspect of materiality | Environmental/ cultural-based value | Environmental/ heritage-based value | Functional/ symbolic-based value | Functional/ aesthetic-based value | Environmental/ aesthetic-based value |

Table 8. A comparative analysis of ornamented buildings in contemporary architecture.

| | (31) | (32) | (33) | (34) | (35) |
|---|---|---|---|---|---|
| Name | Rose Museum, Beijing, China | The Newtown School, India | Cultural and Leisure Centre, Chaville, France | Ecuador Pavilion, Milan, Italy | Shanghai Natural History Museum |
| Illustration |  |  |  |  |  |
| Motifs of ornament |  Floral |  Geometric (text) | 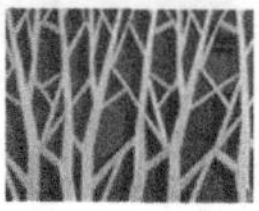 Floral |  Geometric | 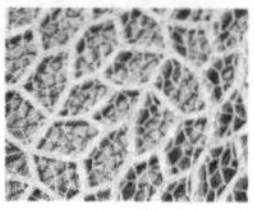 Geometric/ Faunal |
| Year | 2016 | 2015 | 2015 | 2015 | 2015 |
| Architect | NEXT Architects | Abin Design Studio | Laraqui Bringer Architecture | Zorrozua & Associates | Perkins+Will |
| Building type | Museum | School | Cultural Centre | Exhibition | Museum |
| Morphology | Rose-shaped perforated pattern façade | Bespoke stencil screen | Lace-like woodwork skin | A graphic façade | A striking cellular glass wall façade |
| Material of Ornament | Stainless steel | Fibre-Reinforced Plastic (FRP) | Wood (raw larch) | Anodised aluminium curtains | Glass and concrete |
| The origin of genesis and synthesis of ornaments | Rose shape and Chinese walled-off courtyard stem from Chinese culture. The intention was to create a new version of traditional Chinese culture based on the rose-like shapes and the semi-transparent walls surrounding the courtyards. | The theme of the façade, the alphabets, numbers and symbols, stems from the type of the building as a school. At the same time, it was intended to distinguish the building within its urban context | It was inspired by the neighbouring forests to stand out as a remarkable building. The building, used as a cultural centre, uses its ornamental latticework to teach the people to use natural materials to be in harmony with nature. | It was inspired by the multi-coloured traditional fabrics of Ecuador to convey the environment and the cultural heritage. The effect of the sun and the wind raises the sense of movement and subsequently attracts the visitors. | The atrium façade was inspired by the cellular structure of plants and animals. The whole building expresses the harmony of humans with nature as an abstraction of the basic elements of traditional Chinese art. |
| Other aspects | The enclosed spaces provide protection and privacy, and identify the boundary between outside and inside | The perforated façade filters the harsh sunlight, acting as a sun shading device | The façade connects the inside with the outside landscape while admitting the warmth sunlight | It highlights the biodiversity of the agri-food wealth of Ecuador and its role in sustainable agriculture. | The façade maximises daylight and minimises solar gain. |
| Invisible materiality | Cultural-based | Functional/ environmental | Functional/ environmental | Cultural-based | Cultural/ environmental |

## 5 RESULTS AND DISCUSSION

Undoubtedly, the traditional ornaments have contributed to the beauty of the buildings, and have conceived and embodied the culture and history of the city. They are still a destination for tourists, and sources of inspiration for artists, art historians, scholars, and writers. Despite those considerable aesthetic, cultural, and historical roles, they are not classified as a function of architecture. Regardless of support or opposition, the ornament has announced its resurgence strongly in contemporary architecture. But can we state that contemporary ornaments beat and surpassed the old ones as being both aesthetic edifices and ecological toolkits? The answer of this question will be provided, based on the results of both the analytical study and the literature review. The results could be summarised as follows:

The development of ornaments in architecture throughout history confirms the human evolution over the centuries. As the human in ancient times relied on his craftsmanship, consequently, that was incarnated clearly through the rich ornamentation that emphasises his handicraft skills. As humans began to progress with the help of industry and technology, this was obviously reflected in architecture, including the ornamentation. It could be said that ornaments have never died. The changes that have occurred are the technological progress and the society's culture.

Many of the modernists strongly decried the ornament, or in particular, the decoration, or the real elements that are affixed to buildings without any purpose. However, they still found different ways to adorn their buildings. Maybe even in terms of that simplicity, with all its forms, leads to visual pleasure as well. Other modernists reinvented modern decorative models to replace the old ones. Many scholars refuted the common fact that classic ornaments are merely lavish decorative pieces that are affixed to façades. They have indicated to them as originating from the building's structure, and society's traditions, culture, and history.

The results of the analysis of this paper are consistent with the recent judgements and provisions that have been reviewed. The case studies clearly announce that modern ornaments originate as a reflection of:

### 5.1 *Function*

- Architectural functions (as a response to the structure system, providing privacy, or hiding defects).
- Environmental functions as ecological adaptive devices that result as a response to environmental issues. Since the climate has become an influential element in urban development, architecture has begun referring to the ecosystem, and ornaments have started to be combined with buildings' façades to allow the urban environment to take advantage of this synergy. Contemporary ornamentation is an intermediary for communication between the building and its urban context. Therefore, it has become an integral part of this ecosystem as ecologically passive and/or active devices that respond and interact with climate change to improve the performance of the building.
- Trade marketing functions (as brand attracting items), politics, and/or economic.

### 5.2 *Articulation or expression (symbolism)*

The contemporary ornament is linked to the cultural-social-historical surrounding contexts; it is a mirror that reflects the close correlation between the history of the nation on the one hand, and art, culture, tradition, society, and religion on the other. Mostly, the expression of this relationship occurs in an expressive or symbolic way.

### 5.3 *Aesthetic*

It stems from the formation of different typologies as pattern, symmetry, colour, and/or 2D or 3D incarnation (see Figure 4). Further, the results of this paper are consistent with the same judgements which attribute the ornamentation resurgence to the technological progress,

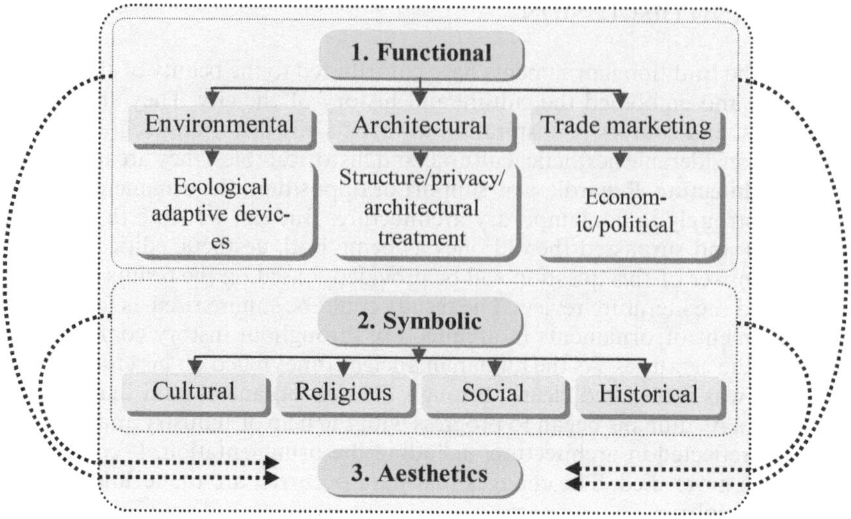


Figure 4. Ornament formation in contemporary architecture.

as technology provides mass production of ornamented façade systems, and the advanced materials that allow various designs, flexible formation, and eco-friendly performance compared to the potentials of old plaster.

Hence, it could be said that there are no radical differences in the causes for the advent of the ornament over the architectural eras. Forms of ornamentation may differ, while the reasons have remained the same. The ornamentation was, and has become, an expressive tool that testifies to the technology and science, the civilisation of people, their evolution, their cultures and beliefs, their social and economic circumstances, and their environmental awareness, all of which are the factors which draw the identities of people, cities and nations.

## REFERENCES

Artencyclopedia. *Art Nouveau: Characteristics, history of curvilinear design movement*. [Online]. Encyclopedia of Art Education. Available: http://www.visual-arts-cork.com/history-of-art/art-nouveau.htm [Accessed October 20 2016].

Balik, D. & Allmer, A. (2016). A critical review of ornament in contemporary architectural theory and practice. *ITUJFA, 18*, 157–169.

Bothireddy, H. (2007). Syntactic and semantic role of ornament in architecture. Master, University of Cincinnati.

Craven, J. (2016). *Architecture timeline – Historic periods and styles of the West*. [Online]. Available: http://architecture.about.com/cs/historicperiods/a/timeline.htm [Accessed October 13 2016].

Encyclopedia. (n.d.). *Ornament*. [Online]. Available: http://encyclopedia2.thefreedictionary.com/ornament [Accessed November 14 2016].

Fairhurst, E. (2007). *Rejection or reinvention – The decriminalization of ornament*. [Online]. Newcastle University Newcastle-Upon-Tyne UK. Available: http://www.presidentsmedals.com/Entry-11130 [Accessed 4 November 2016].

Gleiniger, A. & Vrachliotis, G. (2009). *Pattern: Ornament, Structure, and Behavior*. Birkhäuser.

Gleiter, J.H. (2009). *Ornament: The battleground theory*. Abitare.

Kerns, J. (2016). *Bringing 3D printing home – To actual homes*. [Online]. Machine Design. Available: http://machinedesign.com/blog/bringing-3d-printing-home-actual-homes [Accessed October 20 2016].

Khairi, S. (2011). *History of Islamic architecture*. [Online]. SlideShare. Available: http://www.slideshare.net/shahrilkhairi/history-of-islamic-architecture [Accessed 5 November 2016].

Kharazmi, M. & Sarhangi, R. (2016). An analytical study of the methods of design and geometric constructions in architectural ornaments of the Friday Mosque of Forumad. *Nexus Network Journal, 18*, 275–310.

Levit, R. (2008). Contemporary ‘ornament’: The return of the symbolic repressed. *Harvard Design Magazine*, *28*, 70–85.
Lilic, R. & Diez, M.A. (2009). *The quest for ornament.* [Online]. LA76. Available: http://blog.la76.com/2009/06/the-quest-for-ornament/ [Accessed 7 November 2016].
Majewski, M. (2012). *Geometric ornament in art and architecture of Western cultures* [Online]. Available: http://atcm.mathandtech.org/ep2012/invited_papers/3472012_19700.pdf [Accessed October 9 2016].
Massey, J. (2013). Ornament and decoration. In G. Brooker & L. Weinthal (Eds.), *The Handbook of Interior Architecture and Design.* A&C Black.
McGraw-Hill. (2003). *Ornament.* [Online]. McGraw-Hill Dictionary of Architecture and Construction. Available: http://encyclopedia2.thefreedictionary.com/ornament [Accessed November 13 2016].
McNicholas, M.T. (2006). The relevance and transcendence of ornament: A new public high school for the South Side of Chicago. Master, University of Notre Dame, Indiana.
Mitrache, A. (2012). Ornamental art and architectural decoration. *Procedia-Social and Behavioral Sciences*, *51*, 567–572.
Moussavi, F. & Kubo, M. (2006). *The function of ornament.* Actar.
Opincariu, D. (2011). Structure and Building Facades, the new concept of ornament. *Acta Technica Napocensis, 54*, 193–203.
Paner, C.M. (2013). *The evolution of ornaments.* [Online]. Slide Share. Available: http://www.slideshare.net/crisenciopaner/the-evolution-of-ornaments [Accessed October 15 2016].
Pantazi, M. (2008). *Actualizing ornament with shape grammars.* [Online]. Available: http://the-t-machine.blogspot.com.eg/2008/09/actualizing-ornament-with-shape.html [Accessed 6 November 2016].
Picon, A. (2014). *Ornament: The politics of architecture and subjectivity.* John Wiley & Sons.
Riisberg, V. & Munch, A. (2015). Decoration and durability: Ornaments and their ‘appropriateness’ from fashion and design to architecture. *Artifact*, *3*, 5–1–5.13.
Sandle, D. (2000). Public art and city identity. Political and cultural issues in the development of public art in the UK city of Leeds. *On the w@terfront*, 91–103.
Sayloracademy. (2011). *Postmodern architecture.* [Online]. Saylor Academy. Available: http://www.saylor.org/site/wp-content/uploads/2011/05/Postmodern-architecture.pdf [Accessed October 15 2016].
Schumacher, P. (2009). Parametric pattern. *Architectural Design*, *79*, 28–41.
Scranton. (1922). *History of Architecture and Ornament.* Scranton, International Textbook Company.
Siwalatri, N.K.A., Prijotomo, J. & Setijanti, P. (2012). Meaning of ornament in Balinese traditional architecture. *Journal of Basic and Applied Scientific Research*, 7121–7127.
TGSE. *Ornament.* [Online]. *The Great Soviet Encyclopedia* (3rd ed.). (1970–1979). Available: http://encyclopedia2.thefreedictionary.com/ornament [Accessed November 13 2016].
Wikipedia1. *Ancient Egyptian architecture.* [Online]. Available: https://en.wikipedia.org/wiki/Ancient_Egyptian_architecture#cite_ref-7 [Accessed October 12 2016].
Wikipedia2. *Architecture of cathedrals and great churches.* [Online]. Available: https://en.wikipedia.org/wiki/Architecture_of_cathedrals_and_great_churches [Accessed October 13 2016].
Wikipedia3. *History of architecture.* [Online]. Available: https://en.wikipedia.org/wiki/History_of_architecture [Accessed October 20 2016].

*Cities' Identity Through Architecture and Arts – Catalani et al. (Eds)*
 *ISBN 978-1-138-05409-7*

# Conservation strategies to revive the imageability of the Kumbakonam historic town

K. Thirumaran & K. Kiruthiga
*Department of Architecture, National Institute of Technology, Tiruchirappalli, India*

ABSTRACT: Kumbakonam, a historical town, addresses new challenges and changes. The visibility and significance of the town and its heritage settings are decaying due to the process of physical transformation. Thus, this paper intends to highlight the imageability of the Kumbakonam heritage settings that still survive in the town and reflect its characteristic elements. The findings reveal that the town of Kumbakonam is being confronted with the pressure of development and changes to its skyline and its urban fabric/façade elements. Moreover, this confrontation is accompanied by meagre changes in how towns conserve and promote their heritage settings against the haphazard development. The findings suggest that the heritage settings of the town contribute to the imageability of places. In this regard, the conservation strategies should be introduced to revive the imageability of the town, as the image of the town is distinct only when its heritage settings are conserved.

*Keywords*: Kumbakonam; historical town; heritage decay; heritage conservation

## 1 INTRODUCTION

Conceived out of a profoundly established legacy, the noteworthy towns throughout the world have turned into immortal spaces. Historical towns are the legacy of our ancestors which narrate their picturesque culture and traditions that have been passed on over the centuries (Gast, 2007). The characteristic elements of the towns add to the legacy; also favors to the history of the place and the sense of belonging to the residents. Throughout the years, these towns with their bountiful history evolve their character and create a wonderful image. This manifested image of the towns is, for the most part, a sample of imperial architecture.

Since the 20th century, historical towns have been addressing new challenges and changes. The historic buildings, temple precincts, processional routes and the public squares all enriched with moonstones, remained engulfed in obscurity (Kiruthiga & Thirumaran, 2017). The respectable image of the town and its heritage settings are decaying due to a process of physical transformation (Feilden & Jokilehto, 1998). Kumbakonam, the historical town, is no exception to this process. The town, undoubtedly incorporates colourful gopurams directed towards the sky from its temple. Additionally, the town was once an honour of medieval South Indian power (Nanda, 1996). The imageability of the Kumbakonam town is engraved in its characteristic elements.

In recent decades, the Kumbakonam town has been facing new developments and changes in architecture, that hasve led to faulty and inconsiderate additions to the built heritage. This disintegration of the built heritage slowly, but surely, ruins the characteristic elements of the Kumbakonam town. This paper intends to highlight the imageability of the Kumbakonam heritage settings that still survive in the town of Kumbakonam, and to reflect on its characteristic elements. The main objectives of this paper are (1) to identify the characteristic elements that manifest and portray the image of the town, and (2) to study the problem and look for potential in the heritage settings of the Kumbakonam town. The primary data was gathered through direct observation to examine the characteristics of the constituent elements. Questionnaire surveys were carried out by a group of architecture students on the residents about their experiences in the town. The findings reveal that the town of Kumbakonam is facing pressure of development and changes to the skyline, as well as its urban fabric and façade elements. This is accompanied by meagre changes in the conservation and promotion of its heritage settings against the haphazard development. The findings suggest that the heritage settings of the town contribute to the imageability of places. Efforts to conserve the built heritage of historic towns are weak and in need of proper strategy enforcement. In this regard, the conservation strategies should be introduced to revive the imageability of the town, as the image of the town is distinct only with an intact heritage setting.

## 2 LITERATURE REVIEW

### 2.1 *The context of imageability*

The imageability of a place is not limited to prominence. The unique characteristics of a place may catch an observer's attention for a limited amount of time, while its imageability is eternal. The integrated and intact physical settings of the town produce a vivid image through visual exposure. The physical attributes such as the skyline, the streetscape, the buildings with architectural elements, and the temple tanks and its precincts, create a fabulous picturesque portrayal of the heritage settings of the historic town. Thus, the visual experience of the urban elements within the heritage settings provide a sturdy image.

Kevin Lynch, a well-known icon in the field of urban design and city planning, was committed through his work to visual elements and cognitive concepts of the urban environment. Lynch (1960) additionally focuses his empirical study on imageability. He argues that the visual exposure of the urban elements heightened its image. He further categorises the physical elements of the townscape into five imaginable elements that include paths, edges, districts, landmarks and nodes. These elements play a vital role in evoking vivid images of the place. Moreover, Gordon Cullen (1961) states that the characteristic visual theme contributes to a cohesive sense of place, where he relates the imageability to a 'sense of place'. Silva (2006) also states that the character of the town is often perceived via visual exposure and fully experiencing the town.

### 2.2 *The components of imageability*

Once the historic town is classified as significant by its heritage settings, the study of its imageability becomes equally as significant. Evidently, the characteristic elements of the town become highly imageable due to the living heritage settings of the town. Each town has constituent physical attributes in its settlement pattern which changes over time. The physical attributes of the town, such as the traditional buildings, temples or religious buildings, have streetscapes and skyline demonstrating the architectural and morphological elements that are unique to the heritage settings of the town. The need for conservation is then understood. Moreover, Ewing and Clemente (2000) provided the voluminous empirical literature on the perceptual qualities of the urban environment, in which imageability perceived urban qualities with more significance. This study presents a review on the residents' perception with regards to imageability in the historic town through their reminiscent memories, hence explaining the existence of physical attributes and the residents' perceptions that strongly define the imageability of the town. These inferences of imageability are applicable as conservation strategies for new developments and architectural interventions in the historic town. This would maintain the town's consistency and harmony in the face of its historic fragmentation.

## 3 METHODOLOGY

The comprehensive review of the literature provides knowledge on how the physical attributes influence the imageability of the place. Based on the theoretical premises, imageability of the

Kumbakonam town has been analysed by and noted for its physical attributes. The study furnished much about the visual character and the physical attributes of the existing town.

A multilayer methodology was used for the study. At first, the study started with the division of the data into categories. The study concluded with a reintegration of the data to form a whole. The primary data was gathered through direct observation to examine the characteristics of the constituting elements. The data, such as the Kumbakonam town planning area map and land use map from secondary sources, has been collected from the Kumbakonam town planning department.

Furthermore, the questionnaire surveys were carried out by a group of architecture students on the residents of the Kumbakonam town to collect feedback on their perceptions and experiences. The questionnaire is composed of interviews as verbal descriptions to provide supporting information and substantiate the research data. The traditional core of the town is composed of the target residents of this study. Each resident was allowed 10–15 minutes to answer the questionnaire. Consequently, the survey data has been interpreted and analysed into graphs and chart diagrams with the help of MS-Excel. Finally, the study produced the comparative analysis of the qualitative data gathered from the primary and secondary data.

## 4 KUMBAKONAM: THE HISTORIC TOWN

### 4.1 *An overview*

Kumbakonam, the South Indian historic town is located at 10.97°N and 79.42°E in the Thanjavur district in the state of Tamilnadu, India. Kumbakonam is one of the river edge settlements in Tamilnadu, bounded by the two rivers known as *Cauvery* and *Arasalar* (see Figure 1). The rulers of Kumbakonam, especially Cholas, acknowledged the significance of these rivers and pioneered the production of paddies in this region. The region was also known as the 'rice bowl of India'. The town possesses strong and eventful history of the region (Hunter, 1908). The evolution of the town took place around the third century and received prominence during the seventh century (see Figure 2). Since then, the town has led to the prosperity of the Hindu mythology through its religious architecture (Ayyar, 1920).

Moreover, more than a hundred temples in the town date back to the early centuries, awarding it the title of the 'Temple town' (TNUIFSL, 2007). The town has become a popular religious centre over the past era. It is esteemed for its famous Mahamaham Festival, that takes place once every twelve years and guarantees the attendance of Hindu pilgrims from different parts of the world (Bansal, 2008). The urban fabric of the town has made Kumbakonam one of the best surviving ancient Tamil towns (Nanda, 1996).

Figure 1. (a) Location of Kumbakonam town in Tamilnadu State; (b) Location of Kumbakonam in Thanjavur District; (c) Kumbakonam Town map showing 1. River Cauvery 2. River Arasalar 3. Mahamaham Water tank 4. Porthamarai Water tank.

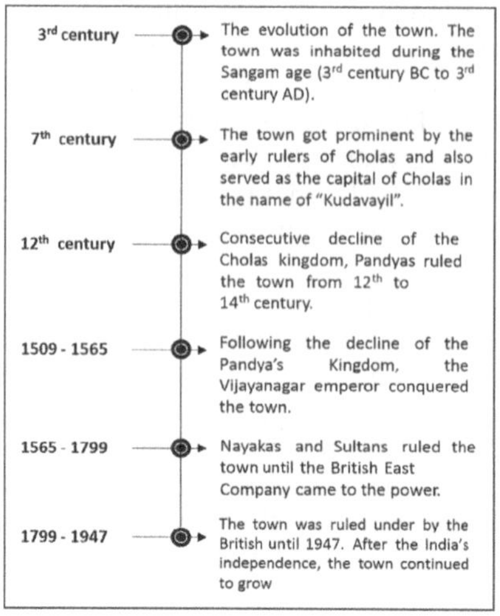


Figure 2. Kumbakonam timeline.

### 4.2 *Physical components of imageability*

The character of the South Indian historic town is influenced by its physical attributes, such as the traditional settlements, river edges and water tanks. Buildings with architectural values, and the skyline, dominate the temple shrines. According to the town of Kumbakonam's mythology, the emergence of the built environment was centred at the temple complex. The ritual topography of the town comprises of the temple complexes and the sacred tanks of Mahamaham Tank (famous for its Mahamaham Festival) and Porthamarai theertham.

As shown in Figure 3, the processional route connects the two sacred tanks and provides accessibility to the main temple complexes. The sacred royal tank has significant physical attributes of the town, along with the residential and commercial areas. This processional route represents the ritual importance, both in the everyday life and in the celebratory dimension, and creates deeper a connection to the image of the place. It can also be said that the processional route has been the historic centre as well as the traditional core of the town. Most of the historical special temples are located along the traditional core of the town with only a few in other locations. These prominent temples evoke the magnificent and ritual experiences to the viewers. Some of these temples have been in the town for centuries and exhibit a rich background that enhances their imageability. Historical associations are powerful example, and include the Adhi Kumbeshwaran Temple, the Sarangapani Temple, the Nageshwaran Temple, the Ramasamy Koil and the Kasi Viswanathar Swamy Temple.

The famous Mahamaham Festival of Kumbakonam held on the Mahamaham Tank is shown in Figure 4. The Mahamaham Tank is effectively recognisable to the stranger by its spatial, shape that is accompanied by functionality and deliverance of a vivid image. Its strength as a landmark is felt by its religious status and its cultural heritage. Once a historical background, a myth, or a meaning is associated with an element, its image and significance is then heightened and strengthened. Some of the historical buildings established a reputation as recognisable landmarks, such as the Porter Town Hall building, the Srinivasa Ramanujan Centre and the clock tower.

The residential buildings in the precincts of Mahamaham Tank and the Porthamarai theertham exhibit the architectural elements of previous decades, as shown in Figure 5. The architectural elements such as arched windows, pilasters, cornices, lean-to roof, pot-tiled roofing, ornamental parapets and entrance door or gateways, surely impact the imageability and the built heritage of the town (Kiruthiga & Thirumaran, 2017). Only a few buildings in the central commercial areas are dominated by the contemporary façade elements in their built fabric. However, the town's skyline is still dominated by the temple towers as the previous manifestations have turned them into principal elements for the whole town, as shown in Figure 6. The temple towers are also known as the long-distance landmarks, due to their symbolic visual significance.

The town's heritage exists in the physical attributes of the traditional settlements, such as its water tanks, buildings with architectural values, temples from the previous centuries, and its skyline that prevail in the midst of all the temple shrines. The modern interventions of new architectural styles and new materials are experienced by the residents. Still, a certain amount of the physical attributes maintains the strong image of the historic town in the heritage settings of Kumbakonam town. Now the physical attributes of Kumbakonam town have been

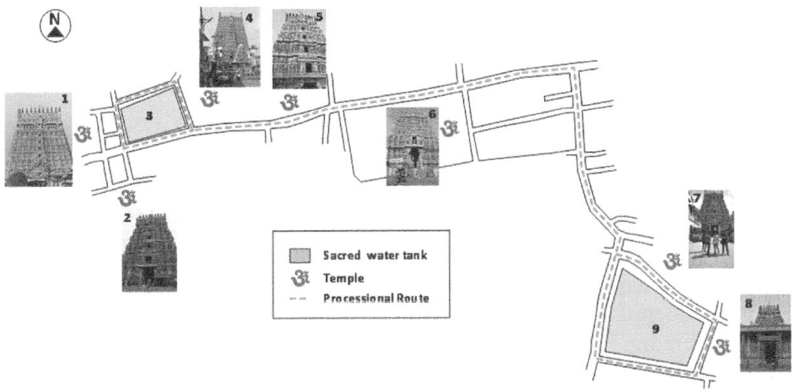


Figure 3. Map showing the location of the principal temples and sacred water tanks in the Kumbakonam town: (1) Adi Kumbeswarar Temple; (2) Ramasamy Temple; (3) Pottramarai water tank; (4) Sarangapani Temple; (5) Somesar Temple; (6) Nageshwara Temple; (7) Kasi Viswanathar Temple; (8) Abhimukhewara Temple; (9) Mahamaham water tank.

Figure 4. More than 40 lakh devotees took holy dip in the Mahamaham Tank during the Mahamaham Festival in February 2016. (Source: Deccan Chronicle).

Figure 5. Images of the building show the traditional architectural style on the processional route.

Figure 6. The temple tower along with the commercial buildings on the processional route.

changed, and from now, there is without having the proper conservation strategies for the development in the city, the character and the image of the town would lose.

### 4.3 *Perceptual viewpoint of the residents*

MEMORY AND IDENTITY: The significance of theories substantiates the stable connections between the time and memory to promote the identity of the town and its history (Ferdous & Nilufar, 2008). A place's identity has significance in towns with historical backgrounds and creates memories that evoke strong images of the town. The questionnaire survey points out the memorable spaces in Kumbakonam town.

As outlined in Figure 7, the Kumbakonam town attained a unique position which reflects the historical significance of the town. Furthermore, the role of heritage settings instilled the image and memory of the town. Additionally, the results show that the residents' perception of identity can be conceived by the physical elements. Likewise, the Mahamaham Tank, Temples of Kumbakonam, Cauvery River, Commercial streets such as TSR Big Street, Bazaar Street, and Nageshwara Koil Street endured as the most significant spaces of the Kumbakonam town by the residents' perception.

COHERENCE AND LEGIBILITY: The residents or observers need a visible space related to the physical elements of the town del Rio6 (Del Rio, 2016). The consistency of the place refers to an 'immediate understanding', whereas the legibility of the place refers to 'inferred understanding'. Consequently, the proper understanding of the physical elements by the residents provides high coherence and legibility for the place. The questionnaire survey reveals the understanding of spaces in the Kumbakonam town.

Shown in Figure 8 are the results' unveiling of the understanding level of the town residents. The Mahamaham Tank precincts and the temple complexes of the town are highly legible and coherent. In Kumbakonam, the streets and precincts associated with the temples and tanks become profoundly meaningful places and fosters the residents' mental map.' The legibility and coherence of the physical elements of the Kumbakonam town are affected by its traditional core settlements and the natural edges of its rivers, the Cauvery and Arasalar Rivers.

VISUAL APPEAL: The physical elements in the heritage settings intensify the visual qualities and the image of the town. The visual appeal of the town depends on the perception of the viewer, who could regard the physical form or pattern, quality of the space, or circumstance during which it has been seen. The questionnaire survey proved the visually appealing elements in Kumbakonam town.

Figure 9 shows how the Mahamaham Tank in Kumbakonam town sustains its vitality in the long run, thus attracting the residents, pilgrims and the viewers. The skyline of the town is filled with life, from the mixture of residences and commercial buildings. Out of all the existent structures, the temple towers are implicit in the skyline. The streets in the traditional residential settlements of Kumbakonam town are lined with buildings, whose traditional typology is punctuated by their windows and doors, columns and their style, generating a beautiful manifestation. The visual experience of space receding diagonally along the Cauvery and Arasalar river banks with agricultural land is reminiscent of the Kumbakonam town.

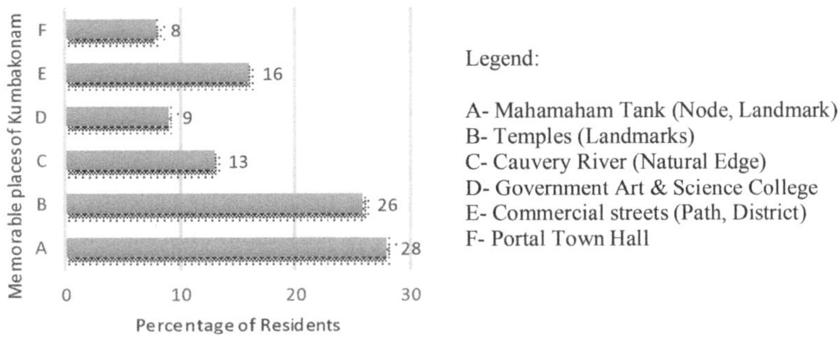


Figure 7. Graph showing the memorable places of Kumbakonam town from the perception of its residents.

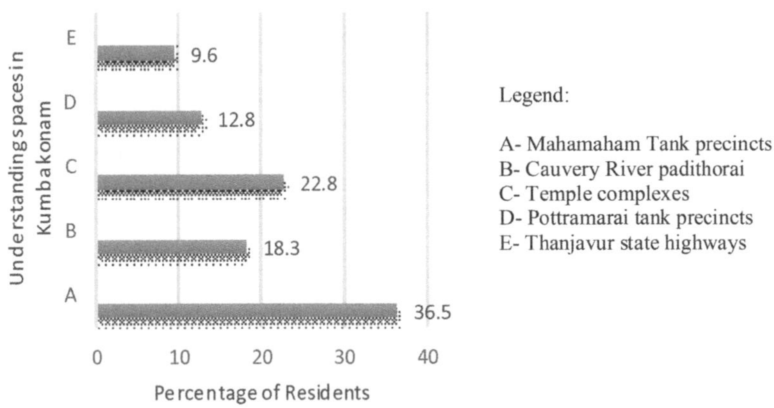


Figure 8. Graph showing the understanding of places in Kumbakonam town by its residents.

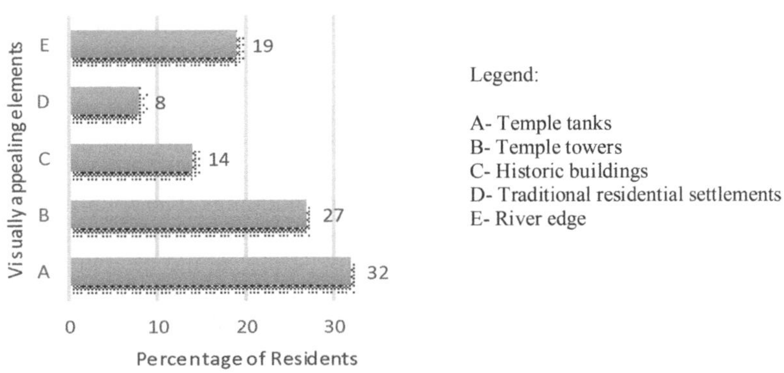


Figure 9. Graph showing the visually appealing elements of Kumbakonam town by its residents.

## 5 FINDINGS

The physical attributes of the town differ from region to region. The character of the South Indian historic town is influenced by its physical attributes such as the traditional settlements, river edges and water tanks, buildings with architectural values, dominating skyline, and the temple shrines. Though the traditional settlement has undergone a large number of changes in land use from residential to commercial, some of the buildings in the precincts of Mahamaham Tank are still in residential use. The increase in commercial buildings leads to a rising in property value. Due to this higher property value, the town faces such problems as traffic congestion, lack of infrastructural facilities, and reduction of open spaces in the processional route.

On the other hand, the demand for commercial use has meant a flood of high-rise structures, resulting in a major change in the skyline of the town. The temple shrines were once the dominant element of the town's skyline but have now been replaced by commercial contemporary buildings. The buildings in the traditional settlement area hold the heritage values of the town (Silva, 2001). Due to the commercial pressure and urbanisation, the traditional buildings are changing their architectural elements into contemporary ones. This interference of a new style of architecture and materials is completely irrelevant to the context and rather damages the town's image. The endowment of the historic town's streetscape is then completely ruined in this process.

The Kumbakonam town differs from the other ancient historic towns of India due to its unique urban and ritual topography. The physical elements of the town contribute to all sorts of manifestations. From the gathered literature and site survey, the physical attributes of the Kumbakonam town comprised of imageability in its heritage settings. The results of the questionnaire survey highlighted the spaces by the residents' perception, consisting of historically significant spaces. Hence, the results of physical and perceptual components of the Kumbakonam town played the key role of evoking a sturdy image of the town's heritage settings. On the other hand, the uncontrolled development of the town is threatening the heritage setting which gradually causes a decline in the general image of the town. Therefore, the town is in need of immediate attention and conservation. The new development and growth should not destroy or ever interfere with the town's physical attributes.

## 6 SUGGESTED RECOMMENDATIONS

The development plan of the town should consider the historic centre a conservation zone that comprises of the processional route and the sacred royal water precincts. Most of the historically significant buildings and temples in this area reflect the character of the town. The following conservation guidelines are general for the existing and new buildings in the historical centre of the town:

1. Earlier decade buildings reflect the historic architectural features, such as pot-tile roofing, arched windows, cornices, pilasters, ornamental parapets, that should be retained and preserved. These building features are incorporated in the new buildings to contribute to the visual integrity of the existing historic buildings.
2. The existing historic buildings should not be demolished. Instead, they could be preserved with some alterations that should merge with their original style.
3. Maintaining the original materials used in historic buildings. The colours should be chosen to complement the traditional architectural style existing on the overall streetscape.
4. In the commercial areas, the signboards of the buildings should be appropriately sized in a manner respectful of the historic fabric of the street. It is also necessary to avoid huge banners alongside historical buildings as they cause visual disorientation.
5. The land use of the buildings in the Mahamaham Tank should be retained as a residential zone so that the cultural activities like *kolam**, *bajanai***, and social gatherings, could take place in the traditional sense. (**kolam* is a handmade drawing, using coloured powder or rice powder on the house front in the streets; ***bajanai* is a group of people singing a song with a religious or spiritual theme in their own language).
6. In the processional route, the height of the buildings is generally single storey or double storey. It is necessary to avoid extensions or an increase in floors of the buildings, so that they will not differ from the surrounding buildings by way of their actual height. The height restriction should be implemented so that the temple could be maintained as the dominant element in the skyline.

7. The sacred royal water tanks of Mahamaham Tank and Porthamarai theertham should be cleaned periodically and their spirituality retained.

## 7 CONCLUSION

More results have been gathered to illustrate that the changing physical elements of the Kumbakonam town have the potential to affect the heritage settings of the town. The heritage settings of the town can play a major role in enhancing the imageability as the town strengthens its physical elements. Kumbakonam fluvial myths, transferred down to the present day, elevate it to the status of the celestial threshold, one that is as primordial and the beginning of creation 1996 (Nanda, 1996). The historic town of Kumbakonam is an indisputable part of our ancestors' heritage and pride, but the action to revive these towns from declining lacks cohesive efforts. The Kumbakonam town has the potential of being revived to its past glory and to adapting to the current development growth through the integrated conservation strategy that will enhance its imageability. In short, this study reveals that the Kumbakonam town is in need of conservation to avoid the mentioned issues. Therefore, the implementation of the conservation strategy is recommended as an essential means to promote the town's image.

## REFERENCES


Ayyar, P.V.J. (1993). *South Indian shrines illustrated.* New Delhi: Asian Educational Services. ISBN 81-206-0151-3.

Bansal, S.P. (2008). *Hindu pilgrimage: A Journey through the holy places of Hindus all over India.* Delhi: Hindology Books. ISBN 978-81-223-0997-3.

Conversion of City Corporate Plan to Business Plan for Kumbakonam municipality-TNUIFSL (PDF) (Report). *Commissioner of Municipal Administration, Government of Tamil Nadu.* (2007). Retrieved 2012-11-16.

Cullen, G. (1961). *The concise townscape.* London: Reed Educational and Professional Publishing.

Del Rio, V. (2016). Urbanity, the flâneur, and the visual qualities of urban design: A walk in Lisbon, Portugal. *Focus: Urban, community and Regional Planning commons, 12*(1), 16.

Ewing, R. & Clemente, O. (2013). *Measuring urban design: Metrics for livable places.* Washington: Island Press.

Feilden, B.M., & Jokilehto, J., (1998). *Management guidelines for world cultural heritage sites.* Rome: ICCROM.

Ferdous, F. & Nilufar, F. (2008). Cultural space – A conceptual deliberation and characterization as urban space. *Protibesh, Journal of the Department of Architecture.* BUET, ISSN: 1812-8068, 12–1(29–36).

Gast, K.P. (2007). *Modern traditions: Contemporary architecture in India.* Basel: Birkhauser. ISBN 3764382988, 9783764382988.

Hunter, Sir W.W. (1908). *The imperial gazetteer of India 1908 Vol. 16.* Clarendon Press.

Kiruthiga, K. & Thirumaran. K. (2017). Visual perception on the architectural elements of the built heritage of a historic temple town: A case study of Kumbakonam, India. *Frontiers of Architectural Research, 6*(1) http://doi.org/10.1016/j.foar.2016.10.002.

Lynch, K. (1960). *The image of the city.* Cambridge, MA: The MIT Press.

Nanda, V. (1996). Cambridge Kumbakonam project: The ritual topography of Kumbakonam. *Architectural Design*, 124, 30–34.

Silva, K.D. (2001). Preserving historic urban districts: Issues of culture change, stress, and culture-supportiveness. In M. Edge (Ed.), *Old World – New Ideas: Environmental and Cultural Change and Tradition in a Shrinking World* (pp. 15–21). Edmond, OK: Environmental Design Research Association.

Silva, K.D. (2006). Conserving Asian urban heritage: The need for re-conceptualizing the value-based management approach. In S. Silapacharanan & J. Campbell (Eds.), *Asian Approaches to Conservation* (pp. 16–23). Bangkok, Thailand: Chulalongkorn University.

*Cities' Identity Through Architecture and Arts – Catalani et al. (Eds)*
*© 2018 Taylor & Francis Group, London, ISBN 978-1-138-05409-7*

# Envisioning the unseen: Interdisciplinary approach between painting and architecture

Eman Abdou, Doaa Abouelmagd & Rania Elhelw
*Faculty of Fine Arts, Helwan University, Egypt*

ABSTRACT: The comprehensive development of historical neighbourhoods in Cairo has been a major concern lately. This paper relates to the previous notion while researching and investigating the collaboration between painting, social practice and architecture. It is a case study of an interdisciplinary pedagogy approach between painting and architecture students, and how it can initiate a difference in the design process of the development of historical and informal areas, through collaboration, aiming to include artists for innovative solutions and futuristic visions.

It approaches a holistic vision through an interdisciplinary methodology that included interviews with stakeholders, sketching, photography and painting. The paper examines and analyses the workshop *Envisioning the Unseen*, initiated by the authors in 2015 during the first Egyptian Urban Forum (EUF), in order to question the transformative potential and the pedagogical partnership between painters and architects in the development of historical deteriorated urban areas, with a special focus on Al-Darb al-Ahmar area in Egypt.

*Keywords*: Egyptian Urban Forum; AL-Darb al-Ahmar; painting and architecture; social practice painting

## 1 INTRODUCTION

This paper contributes to the ongoing discussion regarding interdisciplinary approaches in the pedagogy of art and architecture, especially painting. The collaboration between artists and architects had been discussed extensively under the topic of Public Art. However, this paper experiments widely in different trajectories aiming to reach new forms of collaborations, support communities and aid their development; this is mainly regarding the use of social practice methodologies in painting and architecture, and the possibility of including artists in the innovative design process for architects. Due to the increased rapid change in our time, there is a strong need for a holistic understanding of the world, to look beyond discipline boundaries and conventional models of organisation, and to develop more flexible transdisciplinary study models for architecture and art students (Franz & Lehmann, 2004).

Interdisciplinary learning had been defined by Newell and Green as the inquiries that critically draw upon two or more disciplines and lead to an integration of disciplinary insights (Haynes, 2002). Furthermore, there had been four models of interdisciplinary learning identified by Lattuca, Voight and Fath: informed disciplinary, synthetic disciplinary, transdisciplinary, and conceptual interdisciplinary. A synthetic disciplinary approach is when instructors combine theories' concepts and even research methods from different disciplines, while the contributing discipline remains clearly identifiable (Lattuca et al., 2004). This model was approached in the initial planning of the case study workshop *Envisioning the Unseen*, although the implementations opened the doors to further integrations, raised questions and stimulated innovative insights. Among these questions: painters are known to depend on their personal information and interpretation as resource for their work; can a social practice methodology be used to include the stakeholder's aspirations and dreams as a resource in painting a fictional reality (solution) of an existing problem? On the other hand, is it possible to include painters in the innovative stage of designing, to aid architects in their planning for the conservation or the sustainability plan of an urban area that needs development? Furthermore, can the visions that were introduced by the artist inspire the design of the development plans of the architects?

The word *Utopia* was used by Thomas More, as a title for his book that was written in 1516 (More, 2010). Since then, the contrast between the social and the urban ideal presented by utopia had been responsible for the definition of the concept behind the word; this ideal is not possible to concretise on proposal, but can be feasible some years later (Fernandes & Silvia, 2014). Art students in the *Envisioning the Unseen* workshop (July 2015) visited the sites, envisioned the future and created artworks that included the social dream of the inhabitants of the Al-Drab al-ahmar area, even if utopia is a far aspiration, but these visions can open the gates in time to feasible manifestations. The book *Imaginary Cities*, by Darran Anderson, demonstrates that Marco Polo was not a liar and that each city dreamt by an artist, a writer or an architect has an equivalent life. Concluding that imaginary cities do not just simply exist in fiction or the mind, but they manifest and we actually live in them (Anderson, 2015). Later, Anderson presented his findings in the *Imagining the Future* debate, *Venice Architecture Biennale* 2016, in the lecture series programme titled *Encounters on Optimism: Utopia in a Finite World*, which strength this paper's argument on the importance of creating wild fictive imaginary future cities (Donald, 2016).

## 2 METHODOLOGY

The pedagogy methodology executed was introduced to the art students in a series of lectures, covering architectural information about a deteriorated historical area of Al-Darb al-Ahmar and informal settlements in Egypt, how artists envision the future, utopia and examples of collaboration between artists and architects, social practice methodology in investigation, and design in cityscapes. Followed by a site visit, we applied an interdisciplinary approach by involving students from the architecture department to guide the art students during the field visit. The students viewed the site and asked the residents about their dreams and aspirations, using the social practice methods. Students took a black and white photograph of a location, and created an artwork that included a more factious reality per the vision and the received information during the workshop. The execution of the artwork took two weeks, and was followed by an exhibition at the first Egyptian Urban Forum (EUF). In a third stage, further investigation took place by including architecture students in a site visit and a survey to evaluate and discuss the artwork, which is a developed stage of the interdisciplinary approach.

## 3 CASE STUDY SELECTION

The choice for using the Al-Darb al-Ahmar district for the study was based on many qualifications: the district is part of the medieval Cairo and it has many rich historical Islamic monuments; the district passed through a regeneration project sponsored by the Aga Khan Trust for Culture, in which the Azhar Park was developed, and environmental rehabilitation to cultural restoration and socioeconomic development for the residents took place. Historical monuments, and parts of the public spaces and houses were developed, renovated or rehabilitated. On the other hand, the district has faced the deterioration of many buildings and the spread of informal buildings in the area, especially after Egypt's 2011 revolution.

### 3.1 *The workshop*

The *Envisioning the Unseen* workshop was designed to be delivered by the authors to 11 students from the third year of the painting department, Faculty of Fine Arts, Helwan

University, with the attendance of three students from the architecture department from the same faculty. It was funded by the UN-Habitat and the Faculty of Fine Arts, Helwan University, Egypt.

The workshop was executed over two stages that started with three sessions held on 3 June 2015. The first session was titled 'The first Egyptian urban forum—Rethinking informality'. The lecture was prepared for an audience of art students with a limited background in the topic. It started by introducing the share of responsibility between the public and private sector in providing housing in Egypt, the definition and classifications of informal areas in Egypt, and the percentage of formal and informal housing in Egypt. Then the lecture moved to the first EUF and introduced its aims to the students. After that, the students were introduced to the Al-Drab al-Ahmar district and the Aga Khan urban regeneration project. The lecture showed different examples and case studies, supported by images, maps and graphs. The lecture was followed by an open discussion from the students to answer their inquiries about the topic.

The second session 'Envisioning the unseen, the future, art and concept' presented the futuristic approaches in visualising the future, or fictive realities, in contemporary artwork, and major events that present the ongoing dialogue between art and architecture, especially exhibitions that presented city strategic planning ideologies in opposition to that encountered in everyday living. A team activity was introduced, aimed at showing the importance of including the stakeholder's ideas. Students were given photographs of the Fine Arts campus, and were asked to present futuristic plans for the specific locations of the shots. At the end, students presented their fictive plans for their campus. Social practice methodology of investigating the needs of the community was introduced, especially the part where the artist interacts with the community, asks questions and tries to understand and analyse the problems within a neighbourhood. The process of asking questions about problems, analysing community problems and identifying their reasons was introduced. Students were asked to develop their personal set of questions, to ask the inhabitants of the Al-Darb al-Ahmar area during their visit the following day. They were also directed towards asking them about their future aspirations and visions, and thus including the stakeholder's hopes in their artwork, rather than creating a cityscape from an aligned vision of an artist who is interested only in a historic cityscape.

Subsequently, a third session 'Art, design and architecture' explained to the students that design is a universal visual language that shares many principles between different forms of art. Moreover, it is found to play a great role in intersecting the interdisciplinary fields both visually and socially, which was important to the students in the execution of their projects. Furthermore, examples of artwork were presented, which are deeply involved in the urban scenery and their impact on the viewer who normally might be passing it on daily basis. The discussed selected artworks were paintings with different cityscapes, identifying through them design styles applied in planning the cityscape, besides other examples of mural paintings that suggested visual solutions for some disfigured scenes.

On the second day, the workshop went on a guided site visit of Azhar Park and Al-Darb al-Ahmar area to conduct their social research and to choose the location of their black and white photos, with the assistance of three students from the architecture department. The Azhar Park was the starting point. Then the group moved towards the Ayubi wall, crossing it to the Aga Khan successor 'Mazala Foundation', then moving to Darb Shaalan path, Alslam mosque and Al-Darb al-Ahmar St. towards Abou Hureiba mosque. Then coming back through El-Tabanaa St., visiting El-Merdany Mosque, Beit Madkour and the blue mosque, and going back to the Azhar Park. Through social interviews, the students discussed with the people inhabiting the places their needs and their visions of a better place, while trying to imagine how their artwork would affect the community and their inhabiting environment, visually and socially. Each student would finally choose a single photo of his/her own to make their artwork that projects both their view and that of the community, envisioning the unseen potentials each place holds, and how it could be transformed in the future. Following that, the second stage of the workshop (10-day period) included studio follow-ups, where students were guided through one-to-one sessions, where their working steps were supervised, starting with visualising the concept required to be delivered, sketching and painting. Two students asked to present their work as a photo-collage rather than as painting, and it was approved. Later the artwork was exhibited on 14th June 2015, on the second day of the EUF at the Azhar Park venue. Upon exhibition, the audiences were so highly appreciative of the interdisciplinary workshop and the artwork, that the forum organisers requested the work be exhibited once again on the third day of the forum at the Marriott hotel in Cairo.

Art students submitted a text regarding their experience through the site visit and upon questioning the inhabitants of the community. The following are some examples of the students' reflections regarding the process. Dina Hany (Figures 3 and 4) commented that at first people were somehow afraid that the students were government inspectors or journalists, but in time, when they knew that they were only fine art students, they started to open up and tell them about their thoughts. Hany elaborated:

> *The majority of them complained about the garbage that was everywhere, and the 'tuk-tuks' that keeps passing by with their noises and bad driving. Also, they wished to have a playground for their children, which I find realistic and could be fulfilled but I focused in my painting more around their complaints. I highlighted their problems in a surreal exaggerated way by using vibrant colours and contrast between the old building and the garbage surrounding it, creating flying 'tuk-tuks' and polka dots garbage bags.*

As for Amgad Elsharkawi (Figure 5), he mentioned that he was in search for expanding the known, from a humanitarian view; he highlighted a problem and tried to present a fictive solution for it within the place. Another student had also concluded that the people were aware of the historical importance of their location but they had basic needs, and with her belief that colours can have a positive impact on people's lives, she represented the scene colourfully, aiding the inhabitants to peruse their life easily in a happy manner (Figure 4). Also, Fatma al-Zahraa mentioned:

> *I've noticed through the visit that even though the main colour of the buildings and streets in the neighbourhood was grey, when the people repainted the indoors of their shops and cafes, they painted them with a blue, turquoise colour, I thought it was as if they needed the sea, an optimistic scenery, so I repainted the scene in Al-Darb al-Ahmar area with blue, aiming to fulfil their subconscious dreams.*

Later, upon the success of the students' exhibition at the UN-Habitat Forum, and the high level of engagement by the audience, who were mainly distinguished architects and engineers in the Egyptian architectural scene, a third stage proposed itself for further research investigation.

### 3.2 *The third stage of the workshop: Al-Darb al-Ahmar site visit*

Twelves students from the third and fourth years in the architecture department, Faculty of Fine Arts, Helwan University were selected to visit Al-Darb al-Ahmar district. Eleven students had not visited the area before; only one student was a resident from the area and was nominated to join the group to help during the visit. Prior to the visit that took place on 9 October 2015, the students were introduced to the Agha Khan regeneration project in a lecture held in the Faculty of Fine Arts. A similar route was adopted during the EUF workshop in May 2015. At the end of the visit students filled in a survey. They were asked about the role and importance of art in the urban renewal and upgrading projects, their opinion about the paintings and art works of their colleagues from the painting department, and the possible cooperation between artists and architects. The main results of the survey can be concluded as follows.

#### 3.2.1 *The role and importance of art in the urban renewal and upgrading projects*

The students highlighted the importance of art in the urban development of any area. They agreed that the cooperation with artists can create new ideas and bring about innovative solutions. Mentioning that artists use their imagination, their work reflects the people's needs, beliefs and daily lives. The following quotes highlight the students' opinion: 'Imagination creates different options for development, this can be the role of artists' (Aisha Mahmoud, students survey).

> *Artists and architects have different visions; the artist see what the architect does not see, I thought that the role of the art in the urban upgrading is limited to mural painting, after seeing the paintings of my colleagues I discovered new ideas that we as architects can use to develop the area.*
>
> *Nada Elsherif, students survey*

3.2.2 *The most inspiring painting and implication*

The students were asked about the most inspiring paintings and their vision to apply it. Figures 2 and 5 shared the first rank with five choices each; Figure 1 got two choices. Students who chose Figure 5 agreed that the centre of the work was the human being; the work reflects how the environment affects the society and how the society is the core of any real development and urban transformation. Ideas like creating cultural centres, reviving the traditional vocations, and founding children cultural centres were suggested by the students. K. Sami clarified by saying, 'Figure 2 is the most inspiring as it opens the area to the Azhar Park and integrates the greens with the district'. Students suggested adding more greens to the public spaces, roof gardens, balconies and colouring the facades of the buildings. Students who chose Figure 1 suggested adding more restrictions for the construction of new buildings, and cleaning and organising the area. Other students stressed on the importance to renovate historical buildings and bring back their original vital image.

3.2.3 *Paintings inspiring architects to have creative ideas for upgrading*

The students were asked to name the paintings that can give them ideas to upgrade the buildings and the chosen urban area, clarifying their choices with explanation. Figure 5 received two choices, but those who chose it commented that although it inspires them with ideas, because it focuses on humans, it lacks solutions. Figure 2 got five choices; again, students suggested the use of soft-scape elements and establishing children play areas for urban development. Meanwhile, Figure 4 received five choices. Students agreed that the work carries different notions of sarcasm, highlighting the current deteriorated situation of a historical building in the area, in addition to the problems of noise and the garbage; the work brings imaginative solutions and integrative designs. Students who chose Figure 1 agreed on the importance of the colours and the artistic touch to bring beauty, vitality and identity to the area. Again, the unity of the colours, cleanness and landscape elements are the main reasons of the students' choice. Other paintings (photos not included) received four choices for ensuring the importance of the details to bring beauty, and for the use of the colours to distinguish the buildings. Two paintings did not receive any choice.

Figure 1. Above: Fatma Al-Zahra, oil painting and mixed media on original size, 35 × 50 cm canvas, June 2015.

Figure 2. Below: Passant Elshafei, oil painting on canvas 35 × 50 cm, June 2015.

Figure 3. Above, Dina Hany, Madkour's house, photo of the house.

Figure 4. Below: Dina Hany, acrylic painting and collage on canvas, June 2015.

3.2.4 *The cooperation between architects and artists in the urban development*

All the students highlighted the importance of including artists in the process of the urban development. They agreed that artists can add an artistic vision to the architecture and urban project, and they can also work with the residents to develop their talents and revive the intangible cultural heritage and the traditional vocations. Elaborating that artists can also work with

Figure 5. Amgad Elsharkawy, photo collage and color manipulation, 35 × 50 cm, June 2015.

the residents in the process of social and economic development, Mostafa Khaled mentioned: 'Maybe what we are missing is the absence of the artists in the development process'. Meanwhile, Hantash clarified: 'The diversity of the development team will bring creative ideas; the secret can be in using the imagination of the artist and the creativity of the architect'.

## 4 CONCLUSION

In the traditional norms of painting pedagogy, students in the painting department are trained to paint *plein-air* cityscapes in historical Cairo, but they rarely engage with the community; they just paint the landscape in front of them. In this case study a new dimension was added, where the stakeholders' dreams and daily living habits were taken into consideration. For example, the use of blue colour in the interior of the shops was detected by the painting students and reused to create a dream vision of the future of the place inspired visually by the inhabitants' social interaction. These visions were considered by the artists, sensed with their intuition, and translated by their brushes and colours. We argue that it makes the social dreams more visible to the architect. Roy Ascott mentions that the initial part of the design process of the artists depends on the trained intuition:

> *The artist is no analyst or statistician. We work by intuition, psychic apprehension. We're a lot nearer to the shaman than to the scientist. Paradoxically, at the very time when art is embracing high technology, the shamanic way is the only way forward for the artist (Ascott, 1997).*

In conclusion, there is an increased tendency towards creating a social interaction in the everyday build environment, including organising stakeholders' workshops prior to the designing of public buildings, and applying bottom-up strategies in the designing process (Salama, 2016; Salama et al., 2017). Therefore, we recommend the inclusion of artists trained in social practice investigation in these workshops, and to participate in the design process following a methodology similar to the one done in the paper. Thus, opening the doors to comprehensive integration and creative innovative site-specific visions for developments of deteriorated or informal parts of the city, it can also have a leading role in distinguishing the city's identity and adding an artistic value. It is also worth mentioning that the interdisciplinary collaboration between the students was highly appreciated; most of them wanted to collaborate with the other field in the future. Furthermore, it is recommended to include an advanced module that incorporates teams from the architectural and the art departments, for creating visions of the future and envisioning the unseen.

## ACKNOWLEDGEMENT

The authors express their thanks to the students from the painting and architecture departments at Faculty of Fine Arts, Helwan University. Their input and feedback were essential to the completing of this paper and the success of the different in-class and out-campus exercises.

## REFERENCES

Anderson, D. (2015). *Imaginary cities*. London: Influx Press.

Ascott, R. (1997). *The connective state: Europe's technoetic dimension*. Telepolis, viewed 2 January 2017. https://www.heise.de/tp/features/The-Connective-State-Europe-s-technoetic-dimension-3445997.html.

Donald, A. (2016). *La biennale di Venezia: August events*. British Council, viewed 2 January 2017. http://design.britishcouncil.org/blog/2016/aug/02/la-biennale-di-venezia-august-events/.

Fernandes, E. & Silvia, A.C. (2014). *From Moore to Calvino, The invisible cities of 20th century planning.* 20th Century New Towns. Archetypes and Uncertainties, Oporto, May 22–24, 2014.

Franz, J.M. & Lehmann, S. (2004). *Side-by-side: A pedagogical basis for (design)*.

Haynes, C. (2002). *Innovations in interdisciplinary teaching*, West Port, CT: American Council on Education ORYX Press.

Lattuca, L.R., Voight, L.J. & Fath, K.Q. (2004). Does interdisciplinary promote learning? Theoretical support and researchable questions. *The Review of Higher Education*, Fall 2004, *28*(1), 23–48.

More, T. (2010). *Utopia: Thomas More*. United States: Feather Trail Press.

Salama, A. (2016). *Understanding society and the everyday environment.*

Salama, A.M., Azzali, S. & Wiedmann, F. (2017). The everyday urban environment of migrant laborers in Gulf cities: The case of the old center of Doha, Qatar. *City, Territory and Architecture*, *4*(1), 5. doi: 10.1186/s40410-017-0061-5.

*Cities' Identity Through Architecture and Arts – Catalani et al. (Eds)*
*© 2018 Taylor & Francis Group, London, ISBN 978-1-138-05409-7*

# Heritage management: Investigating current practices in sustainable retrofitting of built heritage, methodologies, tools and approaches

HabibaTallah Omar ElShabrawy
*Department of Architecture, Faculty of Engineering, British University in Egypt, Cairo, Egypt*

Laila Khodeir
*Department of Architecture, Faculty of Engineering, British University in Egypt, Cairo, Egypt*
*Ain Shams University in Egypt, Egypt*

ABSTRACT: It is commonly acknowledged that the conservation of the heritage of buildings is crucial to any culture, especially since conservation is essential to understanding a country's background through preservation and control. However, most of the current practices regarding conservation and management of heritage buildings, with special reference to Egypt, are not guided by specific policies or strategic management plans. Consequently, the main objective of this paper is to analyse the current practices that are applied in sustainable retrofitting of heritage buildings, whether nationally or internationally, in order to extract lessons learned for the management of heritage building conservation. A qualitative analysis, through a literature review, was implemented in order to document the nature of reused heritage buildings. This was followed by a comprehensive comparative analysis between a numbers of case studies for heritage buildings, either nationally or internationally. The findings have been extracted from the literature review and case studies.

*Keywords*: Conservation; Heritage Buildings; Cultural Revitalisation; Heritage Buildings Management; Strategic Management Plans; Sustainable Retrofitting of Heritage Building

## 1 INTRODUCTION

Historical buildings, unlike those currently constructed, are built using a variety of normal materials that includes different types of wood and stone. These materials are appreciated because of the magnificence of the outdated designs (Martins & Carlos, 2013). Heritage management consists of development techniques and approaches that encompass an investigation of historic and prehistoric remnants for retrofitting (Nemaheni, August 2003). Furthermore, heritage management is a matter of rising importance, with the intention of identifying and preserving cultural heritage in the communal interest. Such interest was conveyed, according to an American source, back in the nineties when 50% of construction funds were spent on the conservation of buildings (Mitropoulos & Howell, 2002).

Value management is a methodical function of an approved plan that operates through recognising and classifying project functions. This is executed using a selective approach to make the best use of the overall performance (SAVE International, 2007). Retrofitting and renovating an existing building may, in many cases, turn out to be more cost-effective when compared to creating a new centre (Gultekin, 2009). In 1983, the United Countries' Brundtland Fee defined sustainable development as 'a development that complies with the needs of today without compromising the power of future generations to meet their own', which also results in a pleasurable experience explaining the need for its heritage. Generally, tradition conservation will therefore be provided by any truly ecological development that fits the definition. Sometimes, sustainability objectives will inevitably clash with the needs of an individual and compromises will need to be considered (Akrson, 2015).

### 1.1 *Research problem*

Sustainable retrofitting in heritage building, as a subset of conservation, has been adopted in a number of countries as it represents an opportunity for heritage building reuse. In Egypt, for instance, some heritage buildings, currently in a deplorable state, were once reused for something other than their original purpose. In order to maintain the idea of sustainable retrofitting, heritage buildings ought to have developed a plan that fulfils the requirements of both the present and the future. In several studies, the title of conservation has suffered from exclusion and neglect, (Giancola, 2014). However, most of the current practices regarding conservation and management of heritage buildings, with special reference to Egypt, are not guided by specific policies or strategic management plans.

It is evident that the conditions in Egypt require international and national assemblies of policy makers and leading authorities, with regards to the existing performance and future guidelines intended for the continued existence of the antiquities of Egypt (Elnaggar, 2014). Furthermore, in a number of studies, the Ministry of Antiquities has undertaken a feasibility study to restore and document the historical buildings, to produce methods or techniques for conservation (Aref, 2012). This study was also done on the Baron Empain Palace in Heliopolis, Cairo, which highlighted a lack of guidelines and policies regarding sustainable retrofitting implementation as a type of conservation to the palace (Aref, 2012).

### 1.2 *Research objectives*

The main objective of this study is to analyse the current practices that are applied in sustainable retrofitting of heritage buildings (either nationally or internationally), in order to enhance education on the management of heritage buildings and conservation, and produce policies and guidelines for Egypt. The aims of this study also include promoting environmental and sustainable performance measures for existing heritage buildings without impacting their cultural heritage, increasing sustainability awareness, and encouraging the contribution towards conserving our heritage. The actions capable of accomplishing these discussed aims and objectives include:

- identifying and discussing the sustainable retrofitting nature of the national and international projects.
- studying diverse types and conditions of national and international case studies for retrofitting.
- identifying the sustainable retrofitting applications that are applicable to heritage buildings.
- investigating the practices of sustainable retrofitting for heritage buildings that are usable in Egypt.

### 1.3 *Research methodology*

This paper examines sustainable retrofitting implementations in heritage buildings. A qualitative analysis, through a literature review, was implemented in order to document recent practices of sustainable retrofitting and was used when needed to follow the right procedures. Together, the panels of experts, selected according to field relevance, were always prompt to cooperate with the strategies upon request. Different studies are analysed, and then re-adapted by retrofitting heritage buildings.

The method of this paper follows a technique that should be implemented to accomplish the purpose of this study and the objective supported by the subsequent points, as shown in Figure 1.

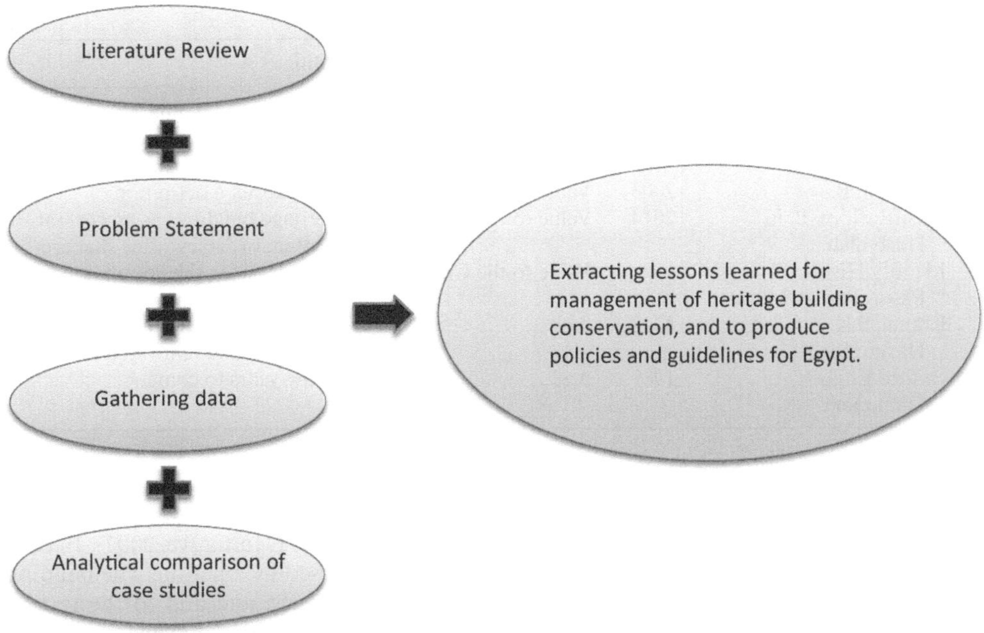


Figure 1. The outline of the methodology.

- Explain the basic terminologies supported by academic sources and integrate them.
- Declare the problem statement and analyse a manageable objective.
- Gather data to identify the various types, methods, climate zones and project sizes of the retrofitting projects in different case studies nationally and internationally.
- Analyse the case studies and examples of sustainable retrofitting projects.

### 1.4 *Value of research*

This paper is crucial as there exists a lack of management plans for sustainable retrofitting of heritage building and their implementation in Egypt. Thus, further studies need to be considered and implemented, similar to as in other countries. It should also include the possibility of its success and failure, as well as further plans suitable for implementation. Public awareness of project conservation of heritage buildings takes into account the future sustainable techniques used to renovate such types of building. Furthermore, this research is essential to the 2030 plan and vision explained, as it is concerned with the sustainable development, protection, and conservation of heritage buildings. Finally, if sustainable retrofitting is used in conservation of heritage building projects, it will assist in accomplishing an improvement policy plan.

## 2 LITERATURE REVIEW

This section of the dissertation covers the literature review, identifies the issues, while assessing the literature review sources. This section explains the meaning of Cultural heritage Management, Value of Heritage Management, Conservation, Sustainable Retrofitting and Heritage Buildings Conservation, and the multidisciplinary role of Sustainable Retrofitting.

### 2.1 *Nature of heritage*

Historical places may be composed of one or more buildings, regions, landscapes or archaeological sites that have previously been acknowledged and declared to be historical regions.

Table 1. Nature of heritage findings.

| Author | Year | Focus on | Findings |
|---|---|---|---|
| Laila Khodeir Shaimaa Tarek Dalia Ali | 2016 | Age | These authors summarise briefly the identification and definitions of heritage buildings and extract a definition of heritage buildings as assets that contain ordinary value that are conserved for the population, and as being important properties that need to be protected and renovated for many years to come. |
| Feinlden, B. | 2003 | Value to the Culture | |
| HasifRafidee, B. in Hasbollah | 2014 | Value to the Culture | |
| Identify Historical Places Part 1 (Initiate) | | Value to the Culture | |
| Kamal, S.K. Harun, S.N. | 2002 | Age | |
| Oxford English Dictionary | 1989 | Age | |

Such identification and declaration assists in defining the identity of a society and presents unusual insights into its culture, traditions and background (SAVE Initiative, 2007). In other words, heritage buildings provide symbols to be carried and used by a nation. The buildings facilitate the creation of the nation's character and social values, and add to the tourism industry (Hasbolla, 2014). The Egyptian law no. 119, of 2008, considers the historical buildings to be important properties that agree with, and rather encourage, the need for renovation and maintenance (Khodeir et al., 2016).

Egypt is loaded with the multiplicity of existing heritages that, unfortunately, suffer from several issues. There are countless resources concerning the terminology of 'built heritage', that has been defined as 'a historic building that gives us a sense of wonder and makes us appreciate culture and our heritage' (Feilden, 2003). Moreover, there exist several identifications and definitions of 'heritage' from several sources, such as the Oxford English (1989), that state and characterise heritage as being an 'asset' that could be acquired. Table 1, the illustrated sources and terminologies are gathered to identify, as selective criteria, the age and value, the findings of the statistics, and the scarcity of data.

### 2.2 *The Philosophy of value of heritage buildings and its classification*

The current practices consider the value of built heritage to be an important factor in the conservation field. Values classify the importance of buildings and categorise some places and objects as historical (Avrami, 2000). The Welsh government classified the values of the existing built heritage into several types, as shown in Table 2. Cadw's main beliefs of conservation were represented by a system for surveying the importance of every noteworthy resource in view, and categorised such values into four parts (Cadw, 2011).

Four models of order were contemplated through the previous review. Every model has at least one particular aspect of arrangement. The grouping by Cadw was categorised by evidential and recorded qualities, in which the work in itself could act as proof of a particular piece of history or act as an extensive part of a verifiable period. Social qualities, for instance, were grouped under what was called mutual esteem (Khodeir et al., 2016). In Table 2 below the classification of values, illustrated from several sources, were used to extract the definitions and the relationship between value and heritage buildings.

### 2.3 *Cultural heritage management*

The sources represented in this section of the literature review have been used in defining the terminology of cultural heritage as being 'the result of interactions between people through a period of time and it contains all features of the surrounding environment as a reflection of rituals, values and traditions'. These results have been collected into a group of resources

Table 2. Basic terminologies for value of heritage management.

| Model type classification | Defined basic terminologies |
|---|---|
| | Evidential Value<br>• Value yielded from the capability of a place to harvest new confirmation about past human movement (Heritage, 2008).<br>• Value obtained from historical benefits which provide verification for human movement and were proved with evidence in several cases (Cadw, 2011). |
| Cultural Value → | Historical Value<br>• The benefits of historical association of notable movements of an event, person or a family that exemplify a particular feature of life in the past (Heritage, 2008).<br>• Values extracted from the routes where individuals from the past, occasions, and parts of life can be associated into a present place (Cadw, 2011). |
| Cultural Value → | Aesthetic Value<br>• Values invented by people with an intellectual stimulation from history (Heritage, 2008).<br>• Values yielded from the courses in which individuals draw tangible and scholarly from a place (Cadw, 2011).<br>Communal Value<br>• Values that are concerned with the collective experience or memories people derive from historical benefits passed onto them (Heritage, 2008).<br>• Values acquired from the implications of a place, for a general population that identifies with it, (Cadw, 2011). |
| Usage Value → | Social Value<br>• Values that usually are the core for building conservation. It is one that is connected to an object or a place as it embraces a definite significance for people in the society, Torre, 2002).<br>Economical Value<br>• Values very powerful in recognition, revision and decision-making in the relation of things (De La Torre, 2002). |
| Age Value → | Comparative Value<br>• Values that people connect to assets containing economic, symbolic, spiritual and comparable values (Cadw, 2011). |
| Risk Value → | Vulnerability Value<br>• Identified by the risk level that the building draws (Cadw, 2011). |

inherited from the past (Gnedovsky, 2013). On the other hand, the relation between values and cultural heritage is described in economic and social sectors. Some of the authors and sources focus on the type of cultural heritage as built heritage, existing heritage, and variable and archaeological heritage, while others focus on the general part of cultural heritage management (Gnedovsky, 2013).

Alongside various verifiable periods, people presented an assortment of cultural heritage that kept their recordings of heritage throughout time. Cultural heritage is characterised by the Council of Europe's Framework convention as a gathering of assets acquired from the past which individuals recognise, autonomously of proprietorship, as a reflection and articulation of their always advancing qualities, convictions, information and conventions. It incorporates all parts of the earth coming about because of the connection amongst individuals and places through time' (Khodeir et al., 2016). Fabricated legacy is an essential kind of cultural heritage as it can exceptionally address various parts of a general public and its advancement throughout history (Khodeir et al., 2016).

The research study sources found that during the collection of data for the basic terminologies of the value of heritage management, the same definitions were agreed upon, as shown above in Table 2. The tangible along with permanent values are classified as cultural

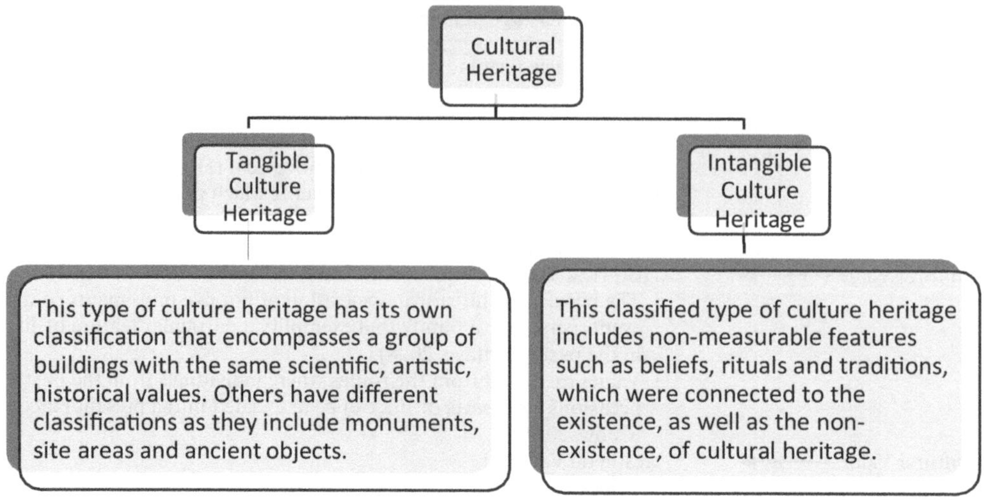


Figure 2. Culture heritage classification types (Hasbolla, 2014).

values which include only four forms, such as aesthetic value, historical value, evidential value and comparative value, from a classification of use that includes economic, symbolic, social and spiritual values. Communal values that are classified under the title of cultural values, in addition to vulnerabilities classified under risk values, are more contingent and reliant on situations (Cadw, 2011).

In Figure 2 below the classifications of tangible and intangible for cultural heritage, which illustrate basic terminologies and descriptive meanings from reliable sources, are presented to extract the classification nature of cultural heritage.

According to this classification, heritage building is considered and described as a tangible classification of cultural heritage. Tangible is defined as a physical feature that can be measured, while intangible is a non-physical feature that cannot be measured. The basic terminology of tangible cultural heritage, according to UNESCO, describes the physical objects that were extracted, conserved and broadcast throughout the community. Meanwhile, intangible cultural heritage is considered to be a foundation and resource of various cultures and acts as a motivation for sustainable development (UNESCO, 2003).

### 2.4 *Nature of conservation*

The conservation in this part of the literature review focuses on the conservation of heritage as the scope of the paper and as per its sequence. The heritage conservation theory is universally accepted, adapted, and practised. An important aspect of its process is to restore the aesthetic features of a heritage building. The most significant characteristic of its background is the ethical feature rather than the aesthetic feature (Heba, 2011). Consequently, the harmony and the beauty of a building are then deemed unacceptable, considering measurements, when restoring it to its original state (Harun, 2005).

The cautious management of alteration is conservation as it protects the historical buildings, qualities, and features to secure it in its original state for present and future generations (Heba, 2011). The conservation of a historical building will be achieved by understanding its importance to be able to accomplish the following aspects (Heritage, 2008):

- the classification to the terminology of heritage value as it is exposed to modification
- classifying the constraints to achieve heritage value
- focusing on the authenticity and future value of the building by a decision-making model.

The act of preservation generally encompasses two main aspects that include caring and protection from being decimated without cautious arrangements (Harun, 2005).

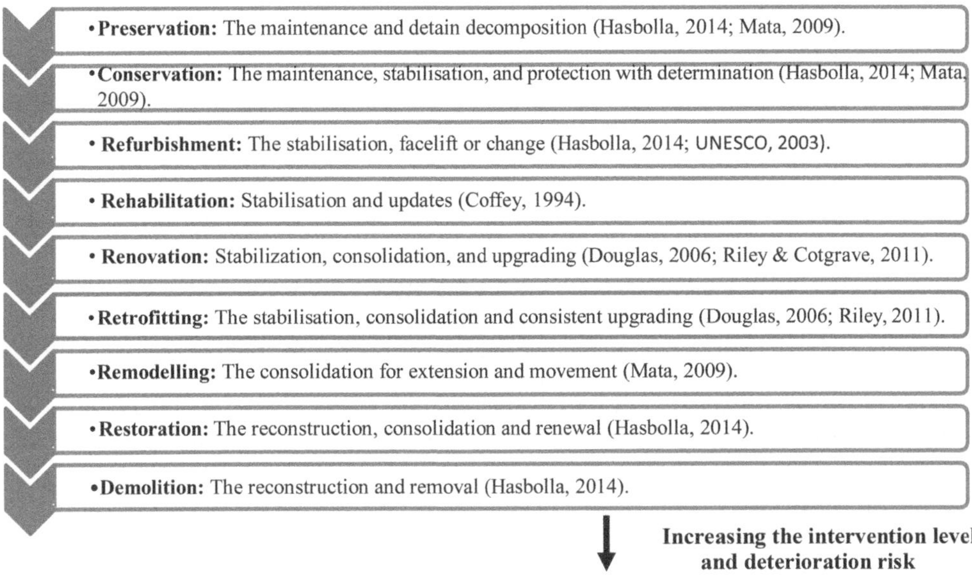


Figure 3. The basic terminologies of conservation approaches.

Table 3. Terminologies of retrofitting and its relation to heritage management.

| Authors | Year | Basic terminologies of sustainable retrofitting |
|---|---|---|
| Almedia, Sara Lucia Gonclaves de | 2014 | The term 'Retrofit' means to afford, in a sustainable manner, something while constructing. |
| Riley, M; Contgrave, A. | 2011 | Retrofitting is fitting new and more current frameworks into an existing building. |
| Douglas, J | 2006 | 'Retrofitting' is generally connected to building administrations, regardless of the building structure and texture being extensively longer than that of the introduced administrations, as indicated by Douglas. |
| Wilkinson, S. | 2017 | 'Retrofit' is 'any work to a working far beyond upkeep to change its ability, capacity or execution'. It is 'any mediation to alter, reuse or update a working to suit new conditions or prerequisites. |
| Mansifield | 2002 | Retrofitting can be replaced by many terms such as upgrading, renovating, conversing and refurbishing. |
| Traykova, Maryna; Charkadova, Tanya | 2014 | The existing heritage building retrofitting represents a sustainable approach to demolish the new building construction. |
| Oget N.COCEN | 2013 | The heritage buildings conserved for a period of sustainable features without alteration in its contextual characteristics. |

As indicated by a study, 'saving will prompt to draw out the life and the social property for its use until further notice and later on' (Hui, 2004).

### 2.5 *Identifying sustainable retrofitting and its relation to heritage building*

Although sustainable retrofitting as a concept was introduced back in 2002 through the publication of sources for 15 years, different authors have made different attempts to redefine it.

These many definitions are represented in Table 3 and are illustrated in the findings section of the literature review.

Based on a discussion of several definitions and terminologies of sustainable retrofitting, it could be concluded that the most comprehensive definition according to the scope of this research study is the one that merges between retrofitting and heritage buildings. It concludes the terminology finding by summarising all the sources represented above into one definition, stating that retrofitting is a procedure of making conceivable utilisation of a property through repair, changes, and increases, while safeguarding its authentic, social or engineering values.

## 3 COMPARATIVE ANALYSIS OF CASE STUDIES

In this section of the research, a qualitative analysis of some case studies is presented. These cases are selected according to a number of principles that encompass conservation, heritage building, value management, and integration between heritage building management and sustainable retrofitting for conservation. The factors and selection criteria that the case studies are based on include:

- varied climatic zones
- different sustainability aspects
- varied building use
- renovation of existing building or heritage building.

### 3.1 *Case studies*

This particular section of the case studies is meant to assess and discuss the cases of heritage buildings' residential and sustainable retrofitting with unique reference to Egyptian palaces

Table 4. Sustainable retrofitting and renovation of heritage building cases.

| Classified Data | Case Study 1 | Case Study 2 | Case Study 3 |
|---|---|---|---|
| Name | Deniz Palace | Hong Kong University | Alexandria National Museum |
| National/International | International | International | National |
| Age | 1920 | 1930 | 1931 |
| Location | Turkey | China | Alexandria, Egypt |
| Original Use | Residential Building | One floor residential and two for commercial | Residential Building |
| Recent Use | Administrative Building | University Building | National Museum |
| Relation with the classification of Cultural heritage:<br>• Cultural Value<br>• Use Value<br>• Age Value<br>• Risk Value | Use Value<br>Age Value | Cultural Value<br>Use Value | Cultural Value<br>Use Value |
| Aim of Reuse | Additional loads to the building lead to Structural Element Retrofitting | Changing the use without interfering with value and building preservation | Enhancing the circulation and preservation of the main villa |
| Added Element and Retrofitted Part | • Reconstruction of foundations, columns, beams<br>• Adding an elevator and fire escape staircase<br>• Preserving the façade | • New mechanical systems added<br>• Preservation of façade<br>• Facilities for disabled<br>• New glass windows, staircases & fire safety measures | • New mechanical systems added<br>• Adding safety measures and techniques<br>• New glass windows |

Table 5. Existing building retrofitting.

| Classified data | Case Study 1 | Case Study 2 | Case Study 3 |
|---|---|---|---|
| Name | Passive House | Clarence House | Telus Building |
| National/International | National | International | International |
| Age | 1940 | 1825 | 1940 |
| Location | Cairo, Egypt | England | Vancouver |
| Original Use | Farmhouse | Residential | Office Building |
| Recent Use | Farmhouse | Residential | Office Building |
| Relation with the classification of Cultural heritage: • Cultural Value • Use Value • Age Value • Risk Value | Use Value Age Value Risk Value | Cultural Value Age Value | Use Value Age Value |
| Aim of Reuse | Active solar retrofitting in order to reduce dependency on non-renewable energy | To reduce carbon emission of $CO_2$ and to preserve heritage assets | Existing buildings should be reused and green strategies should be incorporated |
| Added Element and Retrofitted section/ finding | Providing an examination and vision for retrofits that may be implemented in the future on a large scale in Egypt | • Installing 32 solar panels on the roof • Raising the profile of new technology for sustainable retrofitting reuse in old residential assets | • Providing pre-heating in winter conditions • Using double facades to prevent demolition • Reusing existing systems |

and houses. Analysts have handled the examined target through exploring the favourable circumstances and weaknesses of sustainable retrofitting in writing throughout the period of 2008 to 2016. The analysis of the cases results in a description of the integral relationship between sustainable retrofitting and heritage buildings.

The illustrations written below in Tables 4 and 5 are compared with the literature review on cultural heritage classification and components of retrofitting. The investigation additionally incorporates the correlation between reuse objectives and the components that were added to meet these objectives.

## 4 FINDINGS

The findings of this paper are extracted from the literature review after defining terminologies and focusing on the scope of topic that is sustainable retrofitting. It also produces a qualitative comparative analysis of the case studies previously presented.

### 4.1 *Findings of literature review*

The scope of work focuses on the sequence and flow constructed in the literature review, as shown in Figure 4. The statistics show that there is a scarcity in data with regards to sustainable retrofitting and its integration with heritage management, as is evident by its lowest percentage of the variable.

a. Heritage Building Management has had six resources indicated within the literature review that present the definition from an age and value point of view. The author summarises the differences and controversies surrounding this terminology. The extracted terminology is an asset that contains ordinary value conserved for the population; one that requires protection and renovation during upcoming years.

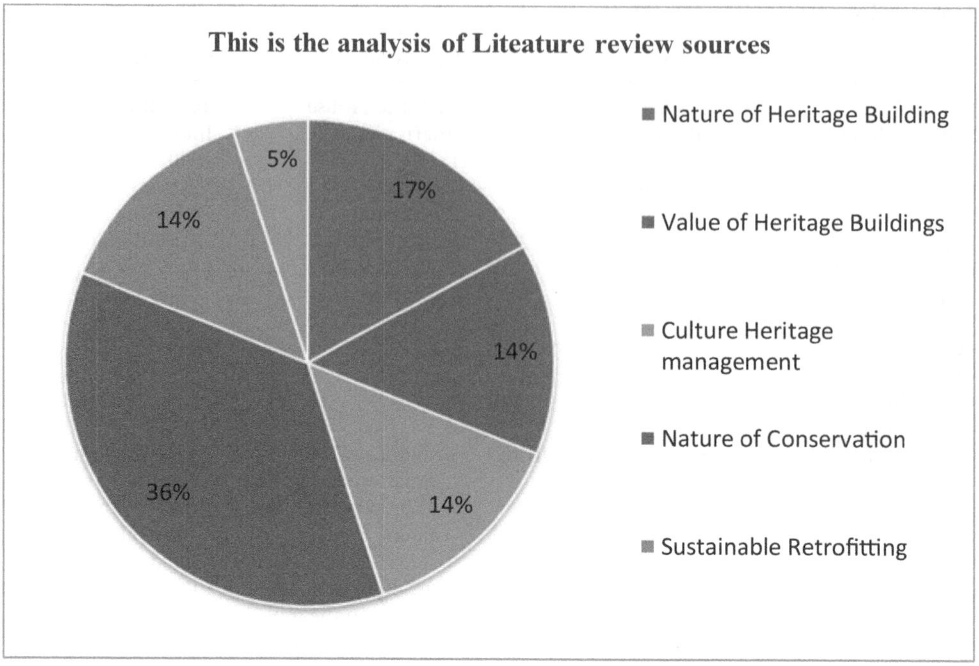


Figure 4. The literature review sequence of extracted data percentage.

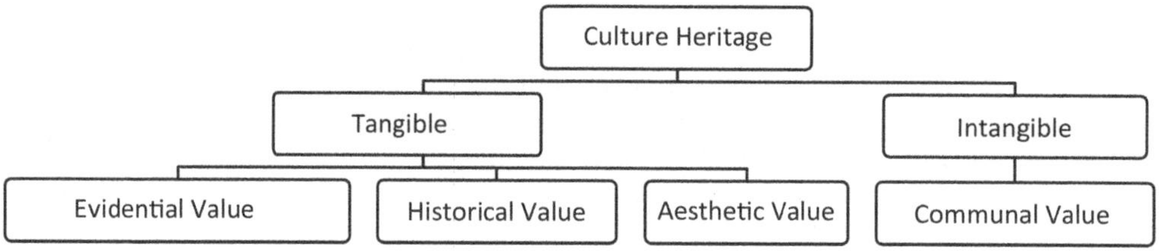


Figure 5. Classification of value and culture heritage.

b. The Value of Heritage Management includes four selected values which are the Evidential Value, Historical Value, Aesthetic Value and Communal Value. The illustration shown below is just a collection of the sources to identify the terminology findings that support the flow of this study.

- Evidential value is value acquired from the capability of a place to gather information on past human movement.
- Historical value is one acquired from the routes where individuals from the past can be associated with those of the present.
- Aesthetic values are those attained from courses through which individuals draw tangible and scholarly views from a place.
- Communal value is value yielded from a place for the general population that identifies it.

c. Cultural Heritage is the result of interactions between people through a period of time and contains all features of the surrounding environment as a reflection of rituals, values and traditions. These results were collected from a group of resources inherited from the past.

Tangible means a physical feature aspect that can be measured; however, Intangible is a non-physical feature that cannot be measured.

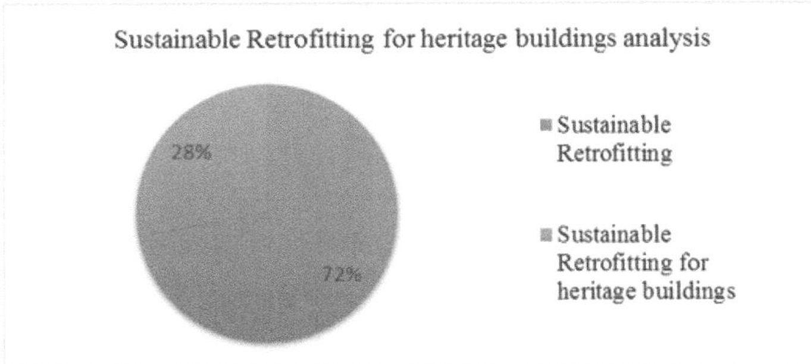


Figure 6. Sustainable Retrofitting Analysis.

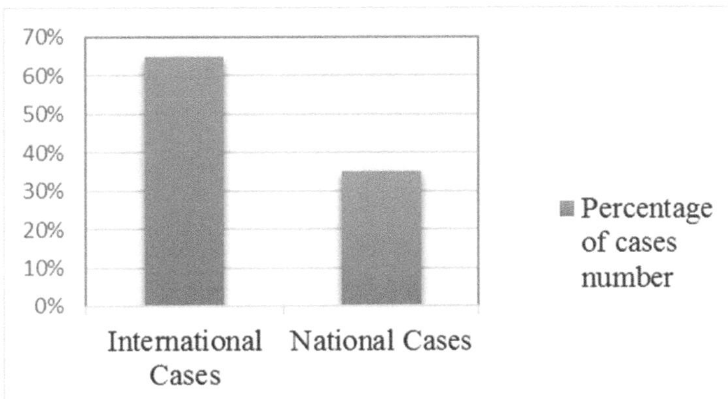


Figure 7. Analysis of the available case studies.

d. Conservation, according to the sources represented above, is defined concisely in the 'Nature of Conservation' section of the literature review. It is stated that conservation is an action taken to draw out the life and existence of important structures over time. In Figure 5 conservation approaches and a statistical analysis of the intervention level and deterioration risk are represented. (Douglas, 2006; Riley & Cotgrave, 2011; Mata, 2009).

The most comprehensive definition, according to the scope of this research, states that retrofitting is an upgrade of a building without major alterations. It has proven to be the most suitable for heritage buildings to be upgraded for future generations, and to protect and save the value of buildings that are in a deplorable state and have been closed without any use.

e. Sustainable Retrofitting was summed up by seven sources represented through an analysis of its terminology throughout the last 15 years. Seventy-two per cent of the sources focus on the sustainable retrofitting of a new building or a part of a building, while 28 per cent focus on sustainable retrofitting's integration with heritage building. The results in Figure 6 shows the analysis that indicate that there exists a scarcity in the integration of sustainable retrofitting and heritage buildings which represents the topic of this research.

### 4.2 *Findings of case studies*

The literature review extracted that the sustainable retrofitting of heritage building has the lowest percentage, which was the focus and scope of the selected case studies. The statistics show that there exists a scarcity in data with regards to the number of cases in Egypt, while the percentage for the international cases was proven to be higher.

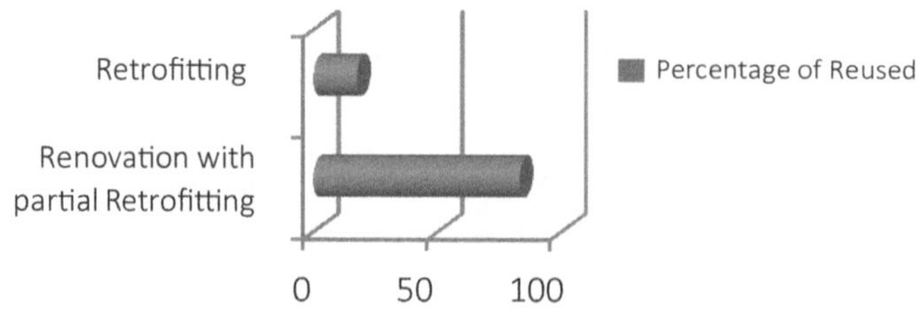


Figure 8. Percentage of retrofitting and renovation in the building.

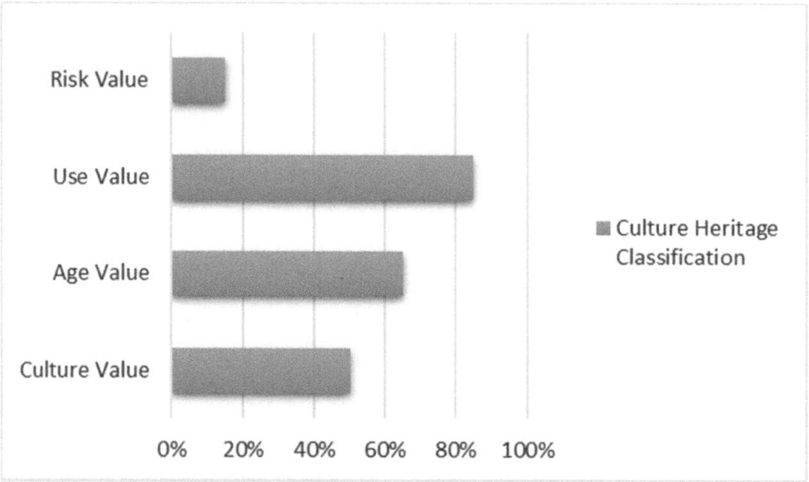


Figure 9. The cultural heritage value classification.

With regards to the case studies that have been used and compared, the majority of the results show that parts of the building have been retrofitted with renovation plans for the whole structure.

The statistics show that there is a scarcity in data of the risk value integrated with heritage building, and the majority of building focuses on the use value, and then the age. The case studies that were selected concluded that half of the cases were renovated to be reused for a different process, so the renovation plan is more important than the retrofitted part in the building to adapt the new uses of the building, whereas the other cases using retrofitting kept their original use. This lead to the fact that the use value of the building controls the plan of conservation in the building. Figure 9 show that it is the higher percentage.

## 5 CONCLUSION

This paper has sufficiently fulfilled and presented the basic terminologies related to heritage management and sustainable retrofitting in the form of tubular data and statistical analysis. Researchers agreed that heritage buildings are defined by age and value through the literature review results and the analysis extracted. Consequently, it has been declared that cultural heritage management is divided into tangible and intangible, thus guiding the process of case studies selection. One of the types of conservation presented and analysed in the literature review is sustainable retrofitting as the main scope of the comparative analysis of the case studies.

Furthermore, the outcome of the comparative analysis is extracted by two methods of renovation and retrofitting. The retrofitting aspect in the buildings analysed was only partially applied. On the other hand, the renovation was applicable to the whole building. Finally, based on an investigation of Egyptian cases, it was clear that sustainable retrofitting is already being used. However, the framework, guidelines and lessons learned from the implementation require further studies by following the codes of conservation plans in Egypt and facilitating an availability of resources for implementation models. The case studies have achieved a basic analysis, but still require further integration for Egypt's codes and policies, in order to reach a lessons learned phase.

## REFERENCES


Aref, N. (2012). Baron Palace to be put back on Egypt's tourism map.

Akrson, P. (2015). *The challenge of retrofitting London's heritage buildings.*

Avrami, E. (2000). Values and heritage conservation. *Conservation: The Getty Conservation Institute Newsletter.*

Cadw. (2011). *Criteria for assessing historical buildings for the purpose of grant Assistance.* Welsh Government.

De La Torre, M. (2002). *Assessing the values of cultural heritage.* Los Angeles: The Getty Conservation Institute.

Douglas, J. (2006). Building Retrofit Butterworth Heinemann.

Elnaggar. (2014). *A Storage of Egyptian heritage: Risk assessment, conservation needs and policy planning* (p. 9). Egypt: Faculty of Archaeology, Fayoum University, Egypt.

Feilden, B. (2003). *Conservation of historic buildings* (3rd ed.). Great Britain: Architectural Press.

Gnedovsky, C.D. (2013). The *social and economic value.*

Gultekin, S. (2009). *Retrofit design of a sustainable eco-office from an existing building.*

Harun, S. (2005). Amalan Pumelihaan Bangunan di Malyasia (Historic Building Conservation Practices). Unpublished PhD thesis.

Hasbolla, H.R. (2014). *A theoretical framework for conserving cultural values of heritage buildings in Malaysia from the perspective of facility management.*

Heba, K. (2011). Preserving architectural heritage in historical cities.

Heritage, E. (2008). Conservation principles: Policies and guidelines. Retrieved 2010, from www.English.Heritage.org.uk.

Hui, S. (2004, 11 16). Sustainable building service for historic buildings. In *Proceedings of the Joint Symposium.* Retrieved 4 16, 2012, from Service Dense Built Environment: www.hku.hk.

Intiative, C.H. *Identify historical places part 1.* Heritage Survey Codes.

Kamal, S.K. & Harun, S.N. (2002). Building research methodology in the conservation of the historic buildings in Malaysia. In *Proceedings of the International Symposium Building Research and the Sustainability of the Built Environment in the Tropics.*

Khodeir, L.M., Aly, D. & Tarek, S. (2016). Integrating HBIM (Heritage Building Information Modelling) tools in the application of sustainable retrofitting of heritage buildings in Egypt.

Martins, A.M.T. & Carlos, J.S. (2013, 5 10). The retrofitting of the Bernardas' Convent in Lisbon. Energy and Buildings.

Mata, R.L. (2009). Applying scientific method in architecture the Lingayen Capitol Building (1918). Architecture, Landscape Architecture and the Designed Environment.

Mitropoulos, P. & Howell, G.A. (2002). Renovation projects: Design process problems and improvements mechanisms.

Nemaheni, T.I. (August 2003). *A culture heritage resources management plan for Thulamela heritage.* University of Pretoria.

Riley, M. & Cotgrave, A. (2011). *Construction technology 3: The technology and refurbishment and maintenance.*

Giancola, E. (2014). Energy efficiency and buildings with heritage values: reflection, conflicts and solutions, Proceedings of the International Conference on Energy Efficiency and Historic Buildings.

SAVE International. (2007). Value standard and body of knowledge.

UNESCO. (2003). Convention for the safeguarding of the intangible Cultural Heritage.

*Cities' Identity Through Architecture and Arts – Catalani et al. (Eds)*
*© 2018 Taylor & Francis Group, London, ISBN 978-1-138-05409-7*

# New urbanism and its reflection on residential interior design in Egypt

Hebatalla Sherin Nazmy & Suk-Kyung Kim
*School of Planning, Design and Construction, Michigan State University, East Lansing, Michigan, USA*

ABSTRACT: Sustainable development has been increasingly embraced as a worldwide concept. Consequently, urban sustainability is gaining conscious attention in many cities. Along with urban sustainability, design principles in new urbanism were referred to by many architects and urban designers. But, little research was done which aligned with the interior spaces through the lens of new urbanism. This research aims to understand how developed countries established design criteria for neighbourhood settlements based on the new urbanism. Consequently, it provides insights for developing countries to adapt appropriate design criteria for new cities. Accordingly, the principles of new urbanism and how they were transformed into strategies were explored through the Leadership in Energy and Environmental Design for Neighborhood Development rating system. Even though new urbanism has basic principles that can be applied all over the world, there will still be some principles that need to be adjusted according to different cultural, ecological, and economic circumstances. Egypt has a rapidly growing population and this has led to an expanding sprawl in some areas. A case study was conducted to identify which new urbanism principles are successfully applied and which need to be adopted within a new Egyptian settlement. This study explains how urban design, landscape architecture, architecture, and interior design share common features and elements that shape cities' identities. Furthermore, it highlights areas where new urbanism principles and interior design are mutually influenced. Eventually, it is expected that this exploration of a developed nation's design criteria will provide valuable insight for an interior designer's contribution to the successful urban sustainability models and guide future research.

*Keywords*: New Urbanism; Egyptian Neighbourhoods; Residential Developments; Interior Design

## 1 INTRODUCTION

Globally, increasing population is leading to urban growth which is associated with multiple challenges. One such challenge is in providing adequate housing (Smith, 2014). Industrialisation, technological advancement, and car dependence have changed the way traditional cities were planned, designed, and built. The suburban communities in the United States led to increased infrastructure costs and environmental degradation. In addition, the building footprint increased, housing costs rose, and natural ecosystems services were disrupted. Roads and highways were extended due to the over-reliance on cars for commuting, leading to an exhaustion of natural resources, traffic congestion and air pollution. People became more isolated and lacked a healthy amount of social interaction (Grant & Tsenkova, 2012). A declining quality of life is associated with the poorly planned urban growth.

Several movements arose as a reaction to the urban spaces that were characterised by over-reliance on passenger cars as a means of transportation and the slums that were created during the era of industrialisation. One example is 'The Garden Cities of Tomorrow' which aimed to integrate the economic and social advantages of a traditional town with the privacy and calmness of a country life. 'The City Beautiful Movement' is another reform concept that was driven by planners' efforts. It aimed at providing a better design for a more beautiful city, which highlights civic virtues. The focus of these movements was to provide urban dwellers with homogeneous mixed-use, prosperous, healthy, connected places (Smith, 2014). These concepts were advanced by new urbanism which incorporated the features of the neo-traditional planning strategy with the concept of transit-oriented design (Grant & Tsenkova, 2012; Smith, 2014).

The United States Green Building Council (USGBC) (2017) took a step towards sustainability through incorporating the leading principles of green building, new urbanism, and smart growth, and released the Leadership in Energy and Environmental Design for Neighborhood Development (LEED-ND) rating system, created as a collaborative effort by the organisations leading each model. LEED-ND encourages designers, planners, engineers, and developers to work collaboratively, creating new development that is expected to 'revitalise existing urban areas, reduce land consumption, reduce automobile dependence, promote pedestrian activity, improve air quality, decrease polluted storm water runoff, and build more liveable, sustainable, communities for people of all income levels' (USGBC, 2014). Moreover, there are several worldwide attempts to develop sustainable urban places. One example is the Building Research Establishment Environmental Assessment Method (BREEAM) of assessing, rating, and certifying buildings and communities in the United Kingdom. Another example, Comprehensive Assessment System for Built Environment Efficiency (CASBEE), is the green building management, rating, and certifying system in Japan (Smith, 2014).

In Egypt, new neighbourhoods were developed outside city centres, due to the increase in population, and the internal migration of peasants from rural to urban areas in search of better job opportunities. The planning of these new cities did not follow the traditional planning principles for the desert environment. They were not designed according to predefined customised guidelines that fit with the Egyptian context. Instead, strategies were borrowed from Western countries and adapted to the Egyptian governmental regulations (Shokry, n.d.). The non-efficient development standards that were adopted for the new cities ended up creating a chaotic mixture of aesthetic and cultural values (Shafik & Bayar, n.d.). This has led to a lack of identity and historical background as well as social exclusion. Consequently, this results in the development of communities that do not sustain the environment nor meet the needs of their residents (Hegazy & Moustafa, 2013; Ismail & Fattah, 2008; Shafik & Bayar, n.d.).

## 2 LITERATURE REVIEW

The United Nations Environment Programme (UNEP) declares that buildings consume about 40% of global energy, 25% of global water, 40% of global resources, and contribute to emitting up to 30% of global greenhouse gas emissions. (United Nations, 2009). Developing the built environment in a sustainable way can have a positive impact on ecosystems that are intensely managed and modified by humans. In this sense, ecosystem services are defined as being the benefits that people attain from ecosystems (Tzoulas et al., 2007). These can be in the form of provisioning, regulating, supporting, and cultural services. These services directly affect the constituents of human well-being, including security, basic material for good life, health, good social relations, and freedom of choice and action, and thus affect the quality of life (Mooney et al., 2005). Ecosystem services provide an organising concept to make environmental, economic, and social objectives explicit and measurable, as addressed by Windhager et al. (2010). Linking performance-based design goals to ecosystem services allows for an assessment of design decisions. Such an evaluation contributes to the achievement of a more sustainable culture, according to Windhager et al. (2010).

Green building, new urbanism and smart growth are three recent and well-recognised movements. They share common principles and strategies which aim to improve the impact from the built environment on the natural environment and human beings. However, each of them works on a different scale. Green building mainly focuses on the environmental impact of an individual building considering energy, water, lighting, and so on. Because green building is mainly limited to the scale of a single building, it is hard for this model to influence the overall sustainability of a given location. New urbanism adopts a holistic planning and design strategy for creating better communities which offer a harmony between buildings and neighbourhoods. Smart growth supports the promotion of building and maintaining sustainable towns and cities within a regional scale in which urban, suburban, and rural communities are designed and follow transit-oriented forms (Smith, 2014). Each of these efforts has faced the obstacle of lacking practical design guidelines, which results in the limitation of their implementation (Smith, 2014).

Built upon the foundation of principles developed previously by green building, new urbanism, and smart growth, in partnership with the Congress for the New Urbanism and the Natural Resources Defense Council, the USGBC released a version of the LEED-ND Rating System (Grant & Tsenkova, 2012; Smith, 2014). It attempts to tie credit attainment with building performance, ecosystem services production, and human well-being (Windhager et al., 2010). Accordingly, the LEED-ND rating system gave encouragement to achieve specific credits within five categories; the main prerequisites and credits are categorised as Smart Location and Linkage (SLL), which is directly related to: Smart Growth, Neighborhood Pattern and Design (NPD), which is an output of new urbanism principles; and Green Infrastructure and Buildings (GIB), which is a descendant of green building concept. In addition, the rating system includes the Innovation and Design Process, and Regional Priority (Smith, 2014).

The LEED-ND rating system embodies physical features and psychological associations to create thriving and healthy developments. Two of the LEED-ND strategies are directly related to the field of interior design. The first is the NPD category which is worth up to 44 points, and the second is the GIB category which is worth up to 29 points. These two categories entail guidelines that are relevant to assessing the ecosystem health and resident well-being. LEED for NPD has been adopted as a guideline to measure the 'liveability' of cities (Boeing et al., 2014). In the USGBC's handbook it is emphasised that the character of the neighbourhood, which is shaped by the design of different building types together with the streets and open space, affects the quality of life (USGBC, 2014). However, it is observed that some of its criteria are based on aesthetics that express diverse values and preferences (Boeing et al., 2014). The concept of GIB introduced in LEED-ND upgrades urban green systems, which are a compromise of natural, semi-natural, and artificial networks of the

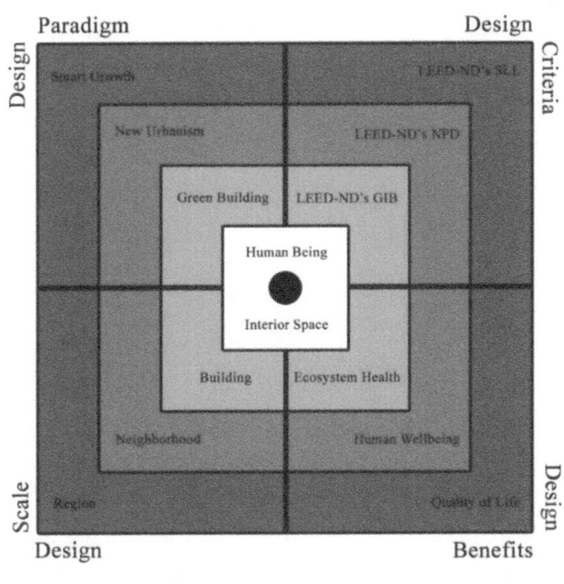


Figure 1. Diagram that shows the relation between built environment, natural environment and the human being in terms of design paradigm, scale, criteria, and benefits. Source: created by the authors.

ecological system. GIB is an approach that if planned, developed and maintained properly, the neighbourhood will offer opportunities for integration between urban development, ecosystem health, and human well-being (Tzoulas et al., 2007).

## 3 METHODOLOGY

First, an exploratory methodology was applied to introduce the basic principles of new urbanism, smart growth, and green building, and explain how they are related to the LEED-ND rating system. In addition, the two criteria NPD and GIB were utilised as supplemental information to understand how built environment contributes to the ecosystem health and human well-being. Second, a case study was conducted to examine the applicability of new urbanism principles within Egyptian neighbourhoods and how they are related to the design of interior spaces. Finally, a recommendation to adapt these principles into the Egyptian local urban conditions was proposed.

## 4 NEW URBANISM

New urbanism principles were initially implemented in projects designed by Andres Duany and Elizabeth Plater-Zyberk (Grant, 2012; Shafik & Bayar, n.d.; Smith, 2014). This movement emerged to react against urban sprawl and integrate design with planning at different scales. It encourages the development of the communities that promote mixed-use and high-density within a community, and also the promotion of connected, compact, walkable, transit-oriented, and diverse neighbourhoods. The sustainability concept occurred as a reaction to the overconsumption of environmental assets. It advocates meeting the human needs while protecting natural ecosystems, regulating the use of natural resources, and eliminating the waste outputs. Consequently, the Congress of New Urbanism was formed in 1993. Its aim

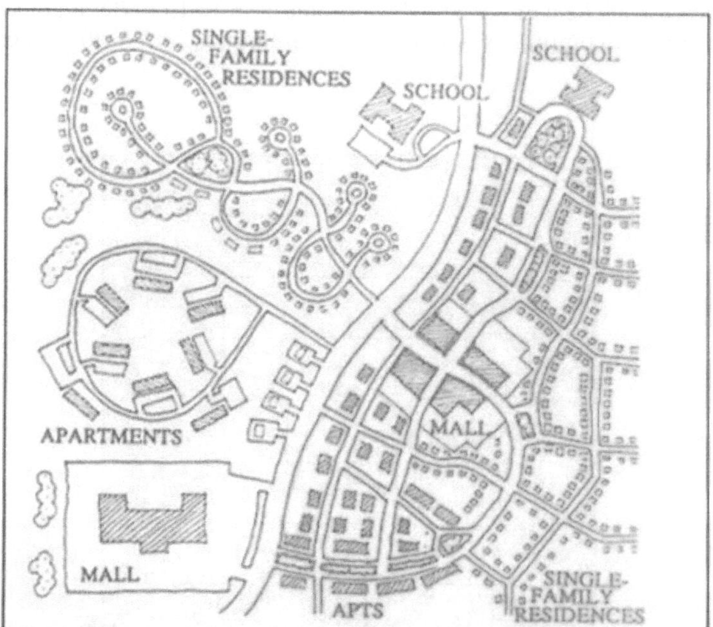


Figure 2. Sketch of suburban sprawl versus traditional neighbourhood design.
Source: http://msue.anr.msu.edu/news/good_urban_form_promotes_walkability_and_physical_health_part_2.

was to show how cities can expand in a way that makes growth attractive to the public and profitable to developers. As an effort towards this goal, new urbanism designers and planners developed a set of design principles, known as the Charter of New Urbanism (CNU), to guide their strategies (Grant, 2012).

The CNU guides public policy, development practice, and design (Shafik & Bayar, n.d.). The charter consists of 27 principles under three main categories. The first category is the 'Metropolis, City, and Town', that is concerned with the region scale. The second is the 'District and the Corridor', which focuses on the neighbourhood scale. The third is the 'Block, the Street and the Building', which is narrowed to the block scale (The Cooperation Council, 2012). This paper was guided by the new urbanism principles focusing on the block scale. Within the block, street, and building scale, the CNU established a preference to expand the thought of architectural design of buildings so that buildings are addressed as a part of the whole, and thus define street and public space profiles. This will result in a perimeter block development which emphasises the importance of the design of spaces between the building, considering the scale of the buildings and activities that will take place within this space. Successful residential environments tend to maintain a clear defined structure in which housing units, open spaces and other activities are within a clearly structured framework of interconnected routes (Tiesdell, 2010).

The CNU also states that individual architectural projects should be designed in integration with their surroundings (The Cooperation Council, 2012; Tiesdell, 2010). In addition, it argues that the design of streets and buildings should reinforce safe and secure environments without compromising the accessibility and openness. Moreover, the CNU emphasises the design of pedestrian-dominant but car-accessible environments in which the design of buildings and spaces meets the need of people through integrating traffic with other activities. Accordingly, a frequently promoted strategy of new urbanism is to provide service allies which allow the cars to park at the rear of the block, minimising their visual dominance within the streetscape. Consequently, having an active frontage and properly configured streets, squares, and pavements will provide a safe, comfortable, and interesting pedestrian-friendly environment in which residents can feel encouraged to walk, know each other, and protect their communities (The Cooperation Council, 2012; Tiesdell, 2010).

Within new urbanism principles, the architectural idiom is considered an important factor in emphasising the sense of place. The CNU encourages the developments responding to the local and regional identity in which architectural and landscape designs should arise from local climate, topography, history, and building practices. Moreover, civic buildings and public gathering places are important elements of the new urbanism as they provide a physical and social focus of the neighbourhood and thus develop the sense of community. The charter principles address the environmental aspects within the building and the occupant's connection to the surrounding environment in terms of location, weather, and time. Finally, the CNU expresses the importance of preserving historic assets as an integral part of the planning system (The Cooperation Council, 2012; Tiesdell, 2010).

## 5 DISCUSSION

### 5.1 *Combined design principles based on the new urbanism and green building*

Table 1 below summarises how design principles on a basis of the green building, new urbanism, and smart growth are linked to the architecture, landscape architecture, and interior design disciplines. It also shows the applicable strategies. The table was created in order to enlist the design elements at the block scale that were mentioned in the CNU. The design elements were classified according to the understanding of the NPD and the GIB criteria. The reason behind this classification is to link the exterior and interior design aspects of the buildings with their surrounding environment. Thus, the scope of the design could be identified from the table.

Table 1 presents how new urbanism principles are connected to the elements in landscape architecture, architecture, and interior design.

Table 1. The new urbanism principles that affect the block scale, and which scope of design it influences such as landscape architecture, architecture, or interior design.

| 3Es | New Urbanism Principles / Factors Affecting Block Scale | Scope of Design: L.A. | Arch. | I.D. |
|---|---|---|---|---|
| Environment | **Improve Land Use Pattern** | | | |
| | Preserve Open Space | Direct | Direct | Indirect |
| | Infill Development | Direct | Direct | Indirect |
| | Rehabilitation Opportunities | Direct | Direct | Direct |
| | **Link to the Surrounding** | | | |
| | Location | Direct | Direct | Direct |
| | Weather | Direct | Direct | Direct |
| | Time | Direct | Direct | Direct |
| | **Respect Local Character** | | | |
| | Topography | Direct | Direct | Direct |
| | Resources | Direct | Direct | Direct |
| | Building Practices | Direct | Direct | Direct |
| Equity | **Create Sense of Place** | | | |
| | History | Direct | Direct | Direct |
| | Culture | Direct | Direct | Direct |
| | Style | Direct | Direct | Direct |
| | **Transform Community Identity** | | | |
| | Safe | Direct | Direct | Direct |
| | Walkable | Direct | Indirect | Indirect |
| | Pleasant | Direct | Direct | Direct |
| | **Reinforce Environment** | | | |
| | Accessability/Connectivity | Direct | Indirect | Indirect |
| | Compactness | Direct | Direct | Direct |
| | Openness | Direct | Indirect | Indirect |
| Economic | **Create Mixed Land Use** | | | |
| | Dense/Diverse Population | Direct | Direct | Direct |
| | Range of Housing Choices | Direct | Direct | Direct |
| | Variety of Building Types | Direct | Direct | Direct |
| | **Define Shared Places** | | | |
| | Street/Square/Pavement | Direct | Indirect | No |
| | Parking Lots | Direct | Indirect | No |
| | Public Spaces | Direct | Direct | Direct |
| | **Accommodate Smart Transportation** | | | |
| | Public Transit | Direct | Indirect | No |
| | Private Cars | Direct | Indirect | No |
| | Bikes | Direct | Indirect | No |

Direct Relationship

Indirect Relationship

No Relationship

## 5.2 *Environment aspects*

The first part of the table consists of environmental design aspects which are influenced by the GIB criteria in combination with new urbanism principles. One aspect is the need of land use pattern improvement. This can be achieved through preserving open space, infill development, and rehabilitation opportunities. The rehabilitation opportunities are one of the aspects that is directly related to the field of interior design. Another aspect is how the built environment is related to the surrounding environment with a conscious sense of location, weather, and time. The interior designer should have a clear understanding of these aspects to create efficient spaces. The last set of design elements of this part is dedicated to the urge for respect of the local character. This entails the local topography, resources, and building practices. Interior designers can contribute to the reduction of resources use and benefit from the local building practices.

### 5.3 *Equity aspects*

The second part of the table shows the equity design aspects. This section is derived from the NPD in the light of the Charter of the New Urbanism. Creating a sense of place is one of the aspects that the interior designer could greatly influence. Together with the projects team members, the interior designer can imply history, culture and style within the design. The second aspect is the community identity. This aspect is in an indirect relationship with the field of interior design. However, interior design can play a role in presenting the building façades. For instance, they can reflect the features of interior space design and the building façades should also contribute to the creation of attractive places for pedestrians. The interior layout of the building would also determine how it is connected to its surrounding environment and promote walkability. Similarly, interior design would indirectly affect accessibility, connectivity, compactness, and openness of a building within its surrounding environment.

### 5.4 *Economic aspects*

The last part of the table is dedicated to the economic aspect. This aspect is more related to the neighbourhood scale; however, some of them have direct relations to the interior design. Creating a mixed land use with a range of housing options, different building types, and the dense and diverse population, will have a direct impact on the way interior spaces are designed to accommodate these purposes. Accommodating smart transportation can be of indirect relation to the interior design as they may be asked to provide some storage spaces for bikes, for example, and waiting spaces to support people using public transit. The last two design elements (i.e. shared places and smart transportation) were added in respect of the economic aspect because achieving them should contribute to the decrease of grey infrastructure cost and ultimately lead to cost reduction. The first set of design aspects covers the issues of preserving natural resources, creating high performance buildings, and the use of green infrastructure which eventually contributes to the ecosystem health. The second set of design aspects are related to the development of liveable communities which contribute to human well-being. Eventually, a thriving economy is expected to arise as a result of achieving the last set of design aspects. Moreover, ecosystem health and human well-being lead to a high quality of life which is the main goal for many developments.

## 6 CASE STUDY

After reviewing the principles of new urbanism and green building, and how it is implemented within the United States with the guide of LEED for Neighborhood Development, a table was created to show the factors that might influence building and neighbourhood design at the block scale. This case study explored these factors within an Egyptian neighbourhood. This study focused on the city of Sodic West. Sodic West City is the largest residential development, located in El Sheikh Zayed, off the Cairo-Alexandria Desert Road. In 2001, over 3,000 single family homes, townhouses, and apartments were delivered. Nowadays, Sodic West City has expanded to include a large number of different residential units, besides schools, clinics, banks, clubhouses, restaurants, office buildings, and shopping malls. A case was analysed based on a site visit, literature search, and an observation of physical environmental characteristics.

## 7 MAJOR FEATURES IN BUILDING AND NEIGHBOURHOOD DESIGN

Sodic West City has been recently planned and developed on desert land. Hence there are no infill development or rehabilitation opportunities yet. Its buildings showed a good link to the surrounding environment.

From an interior design perspective, this was achieved through different types of building openings such as windows, terraces, roofs, and private gardens. Although pictures from

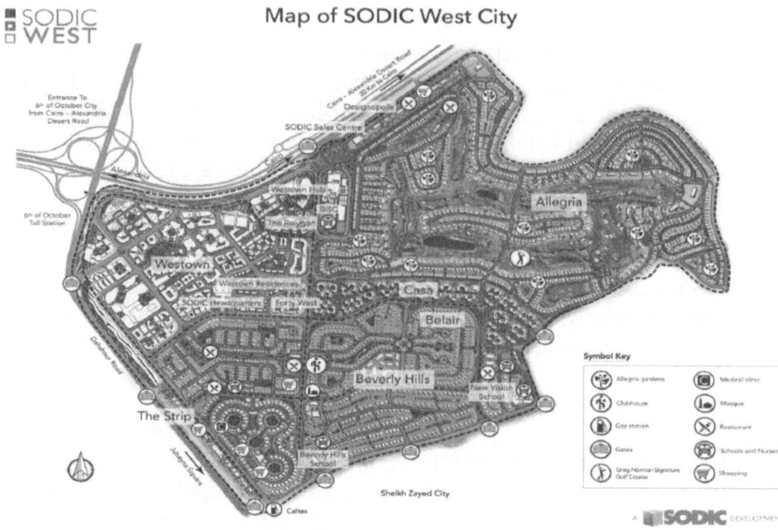


Figure 4. Mass plan of Sodic West City.
Source: https://plus.google.com/photos/+tamergroop/albums/5983808916225460993/5983810974054592978.

Figure 5. (Top left) Interior space of an apartment on the ground floor that shows how the space is linked to the surroundings through windows.
Source: http://www.mlseg.om/en/egypt/96015/penthouse-casa-beverly-hills-october-city-for.

Figure 6. (Top right) Roof of a building in which residents can gather.
Source: https://www.linkedin.com/pulse/beverly-hills-apt-sale-tarek-salem.

Figure 7. (Bottom left) Terrace within an apartment.
Source: https://villaforrentorsale.wordpress.com/tag/beverly-hills/.

Figure 8. (Bottom right) Private garden surrounding an apartment on the ground level.
Source: http://property-advisors.net/property/villa-sale-beverly-hills/.

different residential buildings show the connection of interior spaces to exterior surroundings, it was noticed that environmental aspects are not on top of the designer's criteria list. For instance, there was no specific guideline or building code that designers could follow while designing the building envelope. Some buildings had large windows and others had small ones, which did not reflect a consistent use of energy consumption considerations. It was also noticed that roofs were mainly covered with tiles and green roofs were not well considered yet. It was observed that residents sometimes covered a portion of the private gardens with tiles for the ease of use and maintenance. Planning and design of the city showed some respect to the local character. A good example for respecting the topography was that some streets were inclined so that they led to elevated parts of the land. Egyptian marble was installed as interior finishes in some buildings. In addition to the use of palm trees and some other types of plants that were adapted to the Egyptian climate, less water was needed for irrigation.

Sodic West City is a gated community with nine gates located around its perimeter. It has been developed in several stages, starting with the area referred to as the 'Beverly Hills' area which contains several multistorey residential buildings with an average area of 200 $m^2$ and stand-alone villas with an average area of 400 $m^2$. These types of buildings are surrounded with green open space and a network of streets that accommodate vehicle and pedestrian-friendly pavements. The façades of the buildings were designed referring to European styles.

They feature tilting roofs that are not suitable for the Egyptian warm and dry climate. Most of the façades are painted in white, which is not convenient for the Egyptian dusty weather, requiring more maintenance efforts for the buildings' exteriors. The design of the façades is reflected in the interior spaces. The windows sizes were chosen only in consideration for the design of the façade, and ignoring the interior function and design. For example,

Figure 9. (Top left) Beverly Hills phase one multistorey residential buildings.
Source: http://6october4realestate.blogspot.com/2016/06/beverly-hills-villa.html.

Figure 10. (Top right) Bel Air townhouses.
Source: http://www.gardencity-eg.com/Developments.htm.

Figure 11. (Bottom left) Casa multistorey residential buildings.
Source: http://www.ahram.org/en/egypt/40265/pent-house-for-sale-casa-compound-beverly.

Figure 12. (Bottom right) Alegria stand-alone villas.
Source: https://egypt.aqarmap.com/en/for-rent/villa/cairo/el-sheikh-zayed-city/650639.

Figure 13. (Top left) The British International School in Cairo, located in Sodic West City.
Source: http://wassefdesigngroup.com/british-international-school-in-cairo-bisc/.

Figure 14. (Top right) Westtown Hub.
Source: https://www.behance.net/gallery/26340775/Westown-Hub-Architectural-photography-Egypt.

Figure 15. (Bottom left) Beverly Hills main square and streets.
Source: http://sodic.com/our-developments/beverly-hills/.

Figure 16. (Bottom right) Public transportation provided by Sodic West City to its residents and employees.
Source: http://www.beverlyhillsegypt.com/.

it is very common to have very small window openings in kitchens and bathrooms, which can cause an increased dependency on artificial lighting in those spaces. In the later phases, such as Alegria, Casa, and Bel Air, some of these aspects were improved. The network of streets and pavements were expanded. In addition, some water features and urban furniture were added which promotes walkability. Modern design is the main theme of these areas, with earth colour schemes which are more compatible with their surroundings. Moreover, compact design was noticed in the new phases within both the multistorey residential buildings as well as for the townhouses and stand-alone villas.

The residential development was the primary purpose of Sodic West City. It thus has a diverse variety of housing choices, ranging in sizes and prices, starting with the compact apartments up to the spacious stand-alone villas. In addition to the residential buildings, there is a zone dedicated to mixed-use facilities. It consists of multistorey residential buildings where the ground floors are for commercial use.

Moreover, Sodic West City hosts two schools, one of which is the British International School in Cairo, and the other Beverly Hills School. One of the distinctive features of Sodic West City is that they created the Westtown Hub, The Strip, The Walk, and The Polygon. Westtown Hub is a place dedicated for restaurants and cafes where the landscape and buildings design are well-integrated, offering a variety of indoor and outdoor spaces. This place also is where social events take place. These features make it a live spot of the city, both day and night. The Strip is located at the edge of the city, and it has several banks, a couple of restaurants, and a supermarket. The Walk is a pedestrian-only street that has shops and cafes on both sides. Finally, The Polygon is a multistorey office building. These diverse activities

Table 2. The new urbanism principles that are applied in Sodic West City at the block scale, and whether they are applied sufficiently to enhance the interior space design.

| 3Es | New Urbanism Principles / Factors Affecting Block Scale | Scope of Design: I.D. | Scope of Design: SWC |
|---|---|---|---|
| Environment | **Improve Land Use Pattern** | | |
| | Preserve Open Space | Indirect | / |
| | Infill Development | Indirect | - |
| | Rehabilitation Opportunities | Direct | - |
| | **Link to the Surrounding** | | |
| | Location | Direct | / |
| | Weather | Direct | / |
| | Time | Direct | √ |
| | **Respect Local Character** | | |
| | Topography | Direct | √ |
| | Resources | Direct | / |
| | Building Practices | Direct | / |
| Equity | **Create Sense of Place** | | |
| | History | Direct | / |
| | Culture | Direct | / |
| | Style | Direct | / |
| | **Transform Community Identity** | | |
| | Safe | Direct | / |
| | Walkable | Indirect | / |
| | Pleasant | Direct | / |
| | **Reinforce Environment** | | |
| | Accessability/Connectivity | Indirect | √ |
| | Compactness | Direct | / |
| | Openness | Indirect | √ |
| Economic | **Create mixed land use** | | |
| | Dense/Diverse Population | Direct | √ |
| | Range of Housing Choices | Direct | / |
| | Variety of Building Types | Direct | √ |
| | **Define Shared Places** | | |
| | Street/Square/Pavement | | x |
| | Parking Lots | | x |
| | Public Spaces | Direct | √ |
| | **Accommodate Smart Transportation** | | |
| | Public Transit | | x |
| | Private Cars | | x |
| | Bikes | | x |

Direct Relationship
Indirect Relationship
No Relationship
√ Available
/ Needs Improvement
- Unavailable
x Not Applicable

bring a lot of people to the city besides its residents. Sodic West City has an interconnected structure of streets and squares that connects different areas. Most of the streets have pavements on both sides; however, sometimes there is a long distance between one place and another, so this discourages walking, especially during sunny days. Cars are allowed to park at both sides of the streets, except for in the Casa area which has underground parking. In addition to the private cars, Sodic West City has its own buses for the use of the residents and employees. Moreover, it provides bike storage in different areas.

Table 2 presents how new urbanism principles are connected to the elements in interior design and an assessment of their application within Sodic West City. It was developed based on the new urbanism principles and the Sodic West City case study. It highlights factors and capacity in which they affect interior design within residential buildings. It shows how far new urbanism criteria could meet within three levels: 'available', 'needs improvement', and 'unavailable'. Although Sodic West City was not planned on a basis of new urbanism concepts, it was found that most of it is covered but needs improvement.

On the environmental side, Sodic West City is a relatively new development which does not have an opportunity for rehabilitation or infill development for the time being. However, preserving open space in the future will be highly beneficial for the residents as this will give them a better outdoor view. Despite that, most buildings have sufficient link to the surroundings through windows, terraces, roofs, or private gardens. The size, location, shape, and materials of the openings and spaces are not optimised in terms of energy efficiency and the use of local materials. Introducing green roofs is highly suggested as it will have direct benefits in energy efficiency and provide the residents with a link to the nature and gathering places.

On the equity side, the European and modern design of the exterior of the buildings is reflected in the interior spaces. They do not reflect the Egyptian history, culture, and style; thus, it is not expected to create a high level of attachment and sense of place for its occupants. Sodic West City provides a safe, walkable, and pleasant environment for its inhabitants. Nevertheless, there is still room for the improvement, especially in allocating trees on the streets and its relationship to the buildings. This provides potential energy savings and better outdoor views for the residents. In addition, privacy needs to be addressed properly within the design. The city was successfully planned to be easily accessible and connected. However, reinforcement for the compactness and openness needs to be addressed. Designing more compact and opened buildings will have a direct impact on the shape, size, and orientation of the buildings, which in turn will affect the interior spaces.

On the economic side, Sodic West City is regarded as one the most thriving new cities in Egypt. This is due to the mixed land use that offers a wide range of building types and increases diversity within the city. Although there is a variety of housing choices within the city, affordable housing options are insufficient. Despite the benefits of mixed-use, defining shared spaces and private spaces is recommended to maintain the residents' perception of safety and maintain the economic values of the properties. Sodic West City offers a variety of transportation services. There are several buses designated for employee and resident use. These factors are not directly related to interior design, but they increase the market value of the owned properties.

## 8 CONCLUSION

After reviewing the new urbanism principles and testing them on an Egyptian example of residential neighbourhood, it was found that most of them will fit within the Egyptian context. It is encouraged that the government plays a role in managing growth through long-term planning strategies to ensure a sustainable future. This can be achieved by developing a collaborative effort. To establish a qualitative and quantitative connection between the design goals and the ecosystem services and human well-being, an Egyptian rating system needs to consider the current financial resources and social assets. Interdisciplinary work between Egyptian urban planners, landscape architects, architects, and interior designers is encouraged to incorporate the best of the principles of new urbanism to develop an Egyptian model with a flexible set of guidelines. These guidelines will be locally tailored for the Egyptian environment and culture, being committed to sustainability, positively influencing the character, and emphasising the identity of the city.

Since implementing sustainable urban development is associated with multiple challenges, learning from the experiences in developed countries could be beneficial for developing ones. It is noticed that creating a certification system is acknowledged by government officials, occupants, developers, and investors. It also promotes sustainable practices. Inspired by the USGBC, environmental designers, including architects, landscape architects, and interior designers, should collaborate to develop performance-based guidelines that incorporate ecosystem services into design goals within the built environment. Some strategies that directly influence the interior design of residential spaces is the design of the building exteriors. Floors, walls, windows, and roofs of the building, which are the components of the interior environment, would be connected from the exterior of the building.

First, the floors are affected by the land use pattern strategies, surrounding environment, and local character. The compactness of the building will have a direct impact on the area of the interior spaces. On the other hand, the success of compact developments is largely dependent on the success of the design of interior spaces to accommodate the occupants' needs. Second, wall materials and how walls are designed to minimise the solar gain in summer and maximise it in winter have significant impacts on the energy efficiency. From a broader perspective, exterior wall design is important in creating a sense of place, enhancing the community identity, and reinforcing the environment. Third, the windows have a complementary role to the walls. Moreover, they play a vital role in providing the occupants with enough daylight and outdoor views. They can thus improve the occupants' satisfaction and performance. The placement of windows on the façades reflects the signs of occupancy which contribute to the essential elements of safety and being pedestrian-friendly. Finally, the roof materials and the means of insulation affect indoor temperatures of a building. Moreover, green roofs provide a shared space for residents to gather and foster social interactions.

To implement these considerations for interior spaces, this paper finally suggests a more collaborative multidisciplinary team should be established to pursue synergetic outcomes based on the clear goals and performance plans for residential and neighbourhood designs.

## REFERENCES


Album—Google. Retrieved February 05, 2017, from https://plus.google.com/photos/tamergroop/albums/5983808916225460993/5983810974054592978.

Aqarmap—Real Estate Search Engine. Retrieved February 05, 2017, from https://egypt.aqarmap.com/en/for-rent/villa/cairo/el-sheikh-zayed-city/650639.

Beverly. Retrieved February 05, 2017, from http://www.beverlyhillsegypt.com/.

Beverly Hills Apt for Sale. Retrieved February 05, 2017, from https://www.linkedin.com/pulse/beverly-hills-apt-sale-tarek-salem.

Beverly Hills Villa for Sale In—Properties for sale in October, Sheikh Zayed. Retrieved February 05, 2017, from http://property-advisors.net/property/villa-sale-beverly-hills/.

Beverly Hills Villa from Sodic Projects. Retrieved February 05, 2017, from http://6october4realestate.blogspot.com/2016/06/beverly-hills-villa.html.

Boeing, G., Church, D., Hubbard, H., Mickens, J., & Rudis, L. (2014). LEED-ND and livability revisited. *Berkeley Planning Journal*, *27*(1), 31–55.

Grant, J.L. & Tsenkova, S. (2012). New urbanism and smart growth movements. *International Encyclopedia of Housing and Home* (vol. 5). Elsevier Ltd. https://doi.org/10.1016/B978–0–08–047163–1.00494-X.

Hegazy, I.R. & Moustafa, W.S. (2013). Toward revitalization of new towns in Egypt case study: Sixth of October. *International Journal of Sustainable Built Environment*, *2*(1), 10–18. https://doi.org/10.1016/j.ijsbe.2013.07.002.

Ismail, A. & Fattah, A. (2008). *Appropriating the new urbanism principles to the Egyptian context.*

Mooney, H.A., Cropper, A., Capistrano, D., Carpenter, S., Chopra, K. & Dasgupta, P. (2005). *Ecosystems and human well-being*. https://doi.org/http://dx.doi.org/10.1016/B978–0-12–409548–9.09206-X.

Shafik, S. & Bayar, S. El. (n.d.). New urbanism and sense of community in new Egyptian settlements: Case study—El Sherouq City, Egypt, *173*, 311–322. https://doi.org/10.2495/SDP1302.

Shokry, H.M. (n.d.). *Is a sustainable urbanism possible in 21st century Egypt? A future direction for walkable settlements.*

Smith, R.M. (2014). Planning for urban sustainability: The geography of LEED®–Neighborhood Development™ (LEED®–ND™) projects in the United States. *International Journal of Urban Sustainable Development*, *0*(0), 1–18. https://doi.org/10.1080/19463138.2014.971802.

The Cooperation Council. (2012). The Charter, 1–4. Retrieved from http://www.gcc-sg.org/eng/indexfc7a.html?action=Sec-Show&ID=1.

Tiesdell, S. (2010). The new urbanism and English residential design guidance: A review. *Journal of Urban Design*, *4809* (October 2011), 37–41. https://doi.org/10.1080/13574800220000393 8.

Tzoulas, K., Korpela, K., Venn, S., Yli-Pelkonen, V., Kaźmierczak, A., Niemela, J. & James, P. (2007). Promoting ecosystem and human health in urban areas using green infrastructure: A literature review. *Landscape and Urban Planning*, *81*(3), 167–178. https://doi.org/10.1016/j.landurbplan.2007.02.001.
United Nations. (2009). *Buildings and Climate Change: Summary for Decision-Makers*, 1–62. https://doi.org/10.1127/0941–2948/2006/0130.
US Green Building Council. (2014). LEED 2009 for neighborhood development rating system. *Development*, (July 2014), 149. Retrieved from http://www.usgbc.org/leed.
US Green Building Council. (2017). LEED v4 Neighbourhood Development Guide. Retrieved from https://www.usgbc.org/guide/nd.
Windhager, S., Steiner, F., Simmons, M.T. & Heymann, D. (2010). Toward ecosystem services as a basis for design. *Landscape Journal*, *29*(2), 107–123. https://doi.org/10.3368/lj.29.2.107.

*Cities' Identity Through Architecture and Arts – Catalani et al. (Eds)*
*© 2018 Taylor & Francis Group, London, ISBN 978-1-138-05409-7*

# Place identity/place making in the built environment—towards a methodological perspective

Eman El Nachar
*Professor of Architecture, Fine Arts, Helwan University, Cairo, Egypt*

Aleya Abdel-Hadi
*Emeritus Professor of Interior Architecture, MSA University, Giza, Egypt*

ABSTRACT: Most research cases on place making discuss aesthetic perspective of the built-environment. Beyond the instinctive approaches of aesthetic qualities, this research contends the question of how the identity of places could be examined in respect to design purposes.

The research aims to introduce an integrative agenda to examine characteristics of spaces and buildings of the built-environment in accordance to place and residents' identity needs. Methodology is based on the analyses of theories explaining components relevant to the processes of place identity in order to develop a conceptual model in which characteristics of buildings and spaces together with identity aspects contribute to identity of places holistically perceived. The interpretation of results initiates an integrated agenda for housing policies, decision-making, and design implementation.

*Keywords*: Place; identity; integrative agenda

## 1 INTRODUCTION

Research in the topic of place-making witnessed remarkable works devoted to discuss aesthetic perspective of the built-environment. Arguments focus on the problem of imported western design and its fit and/or misfit in non-western cities. Instinctive visions together with descriptive interpretive criticism are usually followed to debate local versus global designs or vernacular versus international building styles. In most architecture and urban research outcome emphases are mostly placed on visual qualities of buildings and spaces; hence, socio-culture dimensions are rarely incorporated although their inclusion is essential to reach holistic view of a place (Salama, 2012, Devine, 1994 and Graumann, 1983). Beyond the inherent approaches of aesthetic qualities, this research contends the question of how the identity of places could be examined in respect to design purposes.

Based on environment-behaviour perspective, this research aims to introduce an integrative agenda to examine characteristics of spaces and buildings in accordance to the place identity needs, with the intention to develop a conceptual model in which place characteristics and identity aspects are holistically perceived.

The research follows a theoretical analytic methodology to explore the speculative nature of the concept of place identity and its reflection on empirical test. Three main constituents are traced:

*A theoretical overview* tackling the elaboration of the research basic terms: place, sense of place and place-identity. In accordance, places, in this study, are perceived from the 'transactional' approach as socio-physical entities that inter-correlate with all aspects of identity needs.

*A theoretical stance* to develop tangible understanding to the contribution of physical aspects of places in achieving identity needs in any living environment. Three main theories are debated: place identity theory, social identity theory and identity process theory.

*A methodological standpoint* to establish a core set of routes that guide investigations of place-identity in concern of design determination of the built-environment. It represents a key approach devoted to interpret how the identity of any place is constructed and transformed over time.

The previously mentioned theoretical analyses reveal two main overlapping directions in the construct of place-identity: the first is 'motivations' that guide actions related to identity needs and the second is 'mechanisms' by which those needs are achieved. All properties related to motivation and mechanisms of place-identity correlate with aspects of the built-environment.

Given the above, a conceptual model is developed to exemplify by a multi-dimensional matrix, the inter-connected relations among components and processes involved in the construct of place-identity and their bond with physical aspects of any living environment. In respect to the design field, the matrix introduces an integrative research agenda to study place identity; it advocates several methodological procedures and tools to examine characteristics of buildings and spaces and suggests their contribution to identity needs.

## 2 PLACE/SENSE OF PLACE/PLACE IDENTITY

This part elaborates the three basic terms concerning the aim of this study as follows:

### 2.1 *Place*

Several environment behaviour researchers have tried to create terms defining place in their empirical research, especially when employing qualitative research methods. Among those is (Barker 1968) who used the term 'Behaviour Setting' to describe bounded standing patterns of human activity, "that entail corresponding systems images of social groups and individual, as well as their behaviour and the social and physical environments in which their behaviour unfolds" (Popov & Chompalov 2012). Barker's theory has been elaborated by Wicker (1979) who took into account the internal reality of the individual and introduced into the analysis considerations about the experiential world of the individual; he explained 'Social Settings' as social design evolved over time. On the other hand, Canter (1977–79), has been inspired by both theories and developed his 'Psychology of Place', claiming that place is seen as a product of physical attributes, human conceptions and activities.

Based on Canter's theory, (Stokols, & Altman, 1987) have developed the theory of 'Transactional Settings' which explains the relationship between a place and a person as an interdependent process. In parallel, the term 'place' has been used since the 1970s in geography as a 'location', and then it elaborated as a 'place to dwell'. In 1976, Norberg-Schultz defined that spaces where life occurs, are places and that the 'genius loci' meaning 'spirit of place' has been recognised since ancient times. In this sense, 'Place' can delineate the distinctive features both tangible and intangible as well as the associations and feelings that we have when we see or hear about a particular place (Elnachar, 2011).

### 2.2 *Sense of place*

The feeling of perception held by people describes their sense of place; it is a symbol that makes a place exclusive; it is the way people experience, express, imagine and know that place. It could be interpreted as attachment to a place and alludes to the complex relationship between humans and their environment. This kind of relationship includes both the impact of the natural environment on humans as well as the development human activities have placed on the surrounding built-environment; this viewpoint explains how people are identified by places in which they inhabit. Purposely, places are intended to function in multiple ways that provide a sense of belonging, construct meaning, foster attachments and mediate change.

A place is not only about getting used to it but it is also about creating and developing a strong relationship with that place. "The sense of place is an experience created by the setting combined with what a person brings to it" (Cross, 2001); thus relations between people and places are transactional and place becomes part of who we are and shapes our identity.

### 2.3 *Place identity*

Strong debates in the fields of psychology and human geography have taken place in the 1990s to situate identity; is it geared more towards the 'social' or towards the 'place'? (Lalli, 1992; Bonnes & Sechiarolli, 1995). For example, social psychologist Irwin Altman and anthropologist Setha Low (1992) addressed the concept of 'place attachment' to define the ways in which people connect to various places, and the effects of such bonds on identity development, place-making, perception and practice. On the other hand, to some environmental psychologists, place is related to identity through the term 'place identity', a construct promoted by Proshansky who proposes that place identity is another aspect of identity comparable to social identity that describes the person's socialisation with the physical world (Proshansky et al., 1983; 1987). The term is defined as a mixture of cognitive, emotional and perceptual processes formed through individuals' transactions with natural and built-environment. In this reasoning, it suggests that such environments do not simply serve as settings for individuals' activities, actions or behaviour but are instead vigorously 'incorporated' as part of the self (Krupat, 1983; Proshansky, Fabian & Kaminoff, 1983).

The concept explains why people feel at home, as well as why displacement—forced or voluntary—can be so distressing for individuals and groups. In this sense, it is argued that all aspects of identity have place related implications (Twigger Ross 1996). In accordance, places' influences on identity could be perceived as a result of a holistic and reciprocal interaction between people and their built environment; people affect places, and in turn, the way places are affected influence how people see themselves (Hauge, 2007).

Purposely for the built-environment, designers' understanding of the integrative socio-physical unity of place identity is essential in order to design for identity needs. In this concern, examining how physical aspects of any living environments are interrelated with the construct of people's identity needs, calls for an interdisciplinary model to study place identity. Subsequently, integrative knowledge base for designing recognisable built-environments could be reached. Following is a comprehensive theoretical foundation steered to conceptualise a framework to study place identity in concern to the design field.

## 3 PLACE IDENTITY AND THE BUILT-ENVIRONMENT—A THEORETICAL STANCE

Place identity—as observed previously—is mostly debated by theorists to address whether or not identity was more 'social' or more 'place' (Lalli, 1992; Bonnes & Secchiarolli, 1995). Evidently, there has been a neglect of the physical aspects of the environment by self-theorists (Twigger-Ross, 1996), whilst for environmental psychologists, the construct promoted by Proshansky et al. (1983, 1987) called for a more radical evaluation of the concept of identity.

In this concern, three main theories debate how the identity of places is constructed: place identity theory, social identity theory and process identity theory.

### 3.1 *Place identity theory*

Place identity theory indicates how individuals' sense of self arises in part through their transactions with the physical environments. It suggests that such environments do not simply serve as settings for individuals' activities, actions or behaviour but instead are actively incorporated as part of the self. Inside this perspective, Krupat (1983) and Proshansky et al. (1983) rationalised that place identity constructed through emotional and perceptual processes is formed through individuals' transactions with natural and built-environments.

People memories, conceptions, interpretations, ideas and related feelings about specific settings, as well as types of settings assembled their place-identity (Hauge, 2007; Qazim, 2014).

Place identity is composed of processes of observations and interpretations regarding the living environments and changed throughout a person's lifetime (Proshansky & Fabian, 1987). Framed by the accomplishment of five central functions: (recognition, meaning, expressive-requirement, mediating change, anxiety and defence function), place identity develops a cognitive database against which every physical setting is experienced (Proshansky, Fabian and Kaminoff, 1983).

The place identity theory recognised characteristics of spaces and buildings, as referred to 'schemata' that Neisser (1976) and Piaget (1954) portrayed to indicate perceptions and ideas that concern the physical environments. Nevertheless, the theory has been criticised by environmental psychologists for not specifying much detail regarding the structure and the process involved in the construct of place identity (Twigger-Ross, Bonaiuto & Breakwell, 2003). Meanwhile, from experiential perspectives, the place identity theory was criticised for the scarcity of empirical work and for the lack of adequate instruments for measuring the concept in the study of urban identities (Lalli, 1992).

### 3.2 *Social identity theory*

Social identity theory is mostly limited by the dominance of 'self-concept' as the only principle of identity (Twigger-Ross, 1996 and Hauge, 2007). The individual's knowledge of belonging to certain social groups, as well as the delivered personal emotions and values, construct his/her social identity. It all depends on the quality of groups or entities people belong to or have—as a positive reference—such as nationality, culture, religion, family, neighbourhood and/or others (Hauge, 2007). Accordingly, group characteristics and behaviour might be produced by members of a social group who perceive themselves positively.

In this sense, self-esteem is central to the construct of social identity, while place is often associated with a certain group of people and their lifestyle and social status (Qazim, 2014). In accordance, physical aspects of places are related to social identity in their impact on self-esteem. Characteristics of the built-environment work as symbols, they maintain and enhance positive self-esteem for a specific group of people.

Although elements of the physical aspects are neglected when testing and developing the theory of social identity, nonetheless in urban studies the social identity concepts are used in different contexts to explain the symbolic meaning of spaces and buildings, attitudes towards environmental sustainability and identification of place (Uzzell et al., 2002).

### 3.3 *Identity process theory*

Breakwell's identity process model conceptualises identity as a biological organism moving through time and growing through processes of accommodation, assimilation and evaluation of the social world. The selection of information to be accommodated, assimilated and evaluated is governed by four principles: distinctiveness (uniqueness of a person), self-esteem (feeling of personal or social value), self-efficacy (person's perception of self-effectiveness) and continuity across time and situation (Breakwell, 1986; 1992 and 1993).

In this sense, identity is perceived by several theoreticians, as a social product that has both structure and processes brought by the collaboration of abilities for memory, consciousness and organised construal; whereas aspects of places are perceived as symbols that convey meanings representing both personal and shared memories (Twigger-Ross et al., 2003).

Therefore, regarding the built-environment studies, physical aspects of places are components of different sub-identity categories and have important influences as sources of identity elements. Yet, aspects of the physical environment have been mostly simplified within the studies of identity in social and psychological research in a way that could not facilitate the construction of a database for design purpose.

## 4 INVESTIGATING PLACE IDENTITY—METHODOLOGICAL STANCES

It is well observed that previous theories explaining the concept of place identity have implications that are strongly related to the design fields. Despite the revealed conceptual and methodological tensions, a core of set routes could be deduced to guide investigations of place identity concerning design decisions of the built-environment. In literature, a key approach brought by studies devoted to interpret how the identity of any place is constructed and changed over time (Twigger-Ross, 1996; Jorgensen & Stedman, 2001; Dixon, 2004).

Two main overlapping steered are revealed: motivations and mechanisms (Fig. 1).

They both correlate with aspects of the built-environment and expressed in several literature as processes and components of place identity as follows:

### 4.1 *Components of place identity*

In respect to design studies, four mechanisms reveal the intensity of accomplished identity needs in any living environment:

*Familiarity* refers to, a sense of bodily, sensuous, social and autobiographic 'frequencies' (Rowles, 1983); it is the result of people habituation to their physical surroundings.

*Attachment* refers to a sense of emotional belonging to particular physical aspects of the environment. It takes the form of a "psychological investment with a setting that has developed over time" (Vaske & Corbin, 2001 p.17) and, that is captured in the everyday phrases such as feeling 'at home' or having a 'sense of place'.

*Self-expression* refers to the process of personalisation and concerns about how aspects of the built-environment are employed to express or symbolise the self. This dimension of place-identity can operate at a collective as well as an individual level and at socio-spatial scales (Twigger-Ross & Uzzell, 1996 and Dixon, 2004).

*Self-regulation* refers to how people actively and imaginatively 'appropriate' their physical surroundings to create environments where the goals of self-coherence, self-worth and self-expression can be pursued (Korpela, 1989).

### 4.2 *Place-identity processes*

Four key motivations are revealed by Breakwell's identity process model and have the following physical aspects implications:

*Distinctiveness* is the need to maintain personal uniqueness. Aspects of place related self-referent are used by people to display dissimilarity from others. The built-environment,

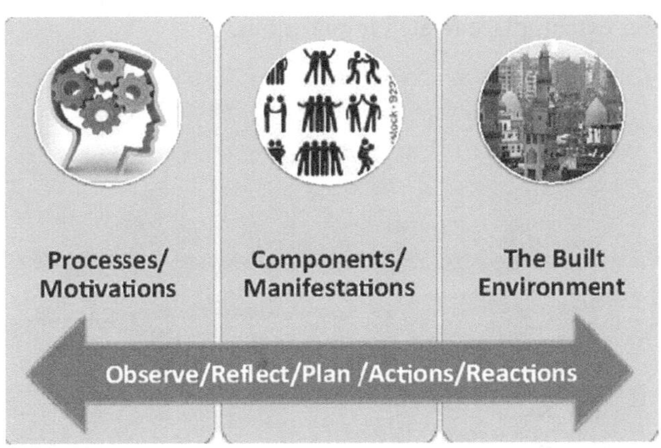


Figure 1. Construct of place identity – Copyright the authors.

then, summarises a lifestyle and establishes a specific type of relationship with the living environment which is clearly distinct from any other type of relationship (Feldman, 1990 and Hummon, 1990).

*Continuity* is the need to preserve continuously self-concept over time. It has two types: place-referent continuity and place congruent continuity. The first refers to the maintenance of continuity via specific places that have emotional significance for a person whilst the latter refers to maintenance of continuity via characteristics of places which are common and transferable from one place to another (Devine, 1994 and Graumann, 1983).

*Self-esteem* is the need to uphold a positive conception of oneself. It has been regarded as a central human within social identity theory (Hogg & Abrams, 1988; Abrams, 1992). With respect to the built-environment, favourite physical aspects of any living environment can support self-esteem and enhance senses of pride by associations (Korpela, 1989; Lalli, 1992; Uzzell, 1995). This differs from simply evaluating a place positively as its qualities would boost a person's self-esteem; therefore, it may be possible to evaluate a place positively but this may not impact upon one's self-esteem, though the two may be related.

*Self-efficacy* is the individuals' belief in their self-competence to meet their demands. With regard to the built-environment, feelings of self-efficacy are maintained if features of places facilitate or, at least, do or hinder person's actions related to everyday lifestyle. In residential communities, it is expressed by the concept of manageable home environments (Winkel 1981). It indicates that residents are able to organise information from their socio-physical environment in such a way that can predict whether a setting supports their purposes to carry out their chosen activities (Breakwell, 1986).

## 5 THE PLACE IDENTITY MULTI-DIMENSIONAL STUDY MATRIX

Based on the above, the concept of place-identity is debated in terms of how 'Identity' is manifested and achieved through familiarity, attachment, self-expression and self-regulation. The four components are not only involved in the contrast of place identity, but they also work as indicatives for examining the dominance of place-identity in specific area or place. In course of everyday life, the four components reflect the bond between person and place while motivated by processes of distinctiveness, continuity, self-esteem and self-efficacy. As a result, in the achievement of place identity, relevant actions are enabled; components and processes involved in the construct of place identity could be exemplified by a grid matrix in which each component could be inter-related with every process that motivates place identity needs (Fig. 2).

In respect to the built-environment, both physical and social aspects of places are important sources of identity since they are symbols that convey meaning to us. Meanwhile, the development of identity in any environment, with both its mechanisms and motivations has to a greater or lesser extent, place related implications.

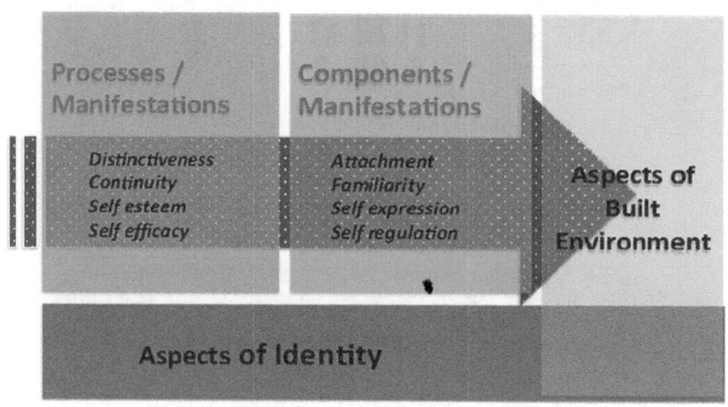


Figure 2. The place identity construct model – Copyright the authors.

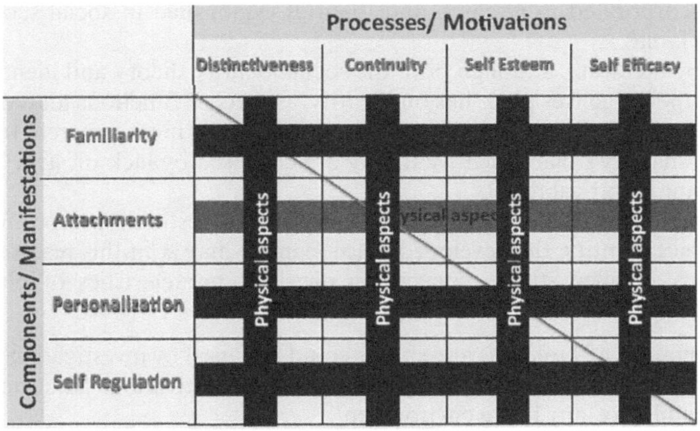


Figure 3. The place identity grid matrix – Copyright the authors.

Thus, physical characteristics, together with social aspects of places are intermediary milieus for the construct of identity. Such integrative bonds could be represented through horizontal and vertical layers crossing the components/motivations identity grid matrix (Fig. 3).

For design purposes, the matrix establishes an integrative framework for the study of place identity in several levels. For example, physical features of residential environments could be examined in their impact on the relation between people's attachment and familiarity with their built environment and motivations that foster their identity needs like their desire to preserve continuity of the self.

The matrix draws several methodological positions in the study of place-identity as follows:

- It inspires multidimensional perspectives for studying settings in the built-environment, since layers of place identity construct could be understood in their mutual relation to each other. The matrix raises questions of how each action related to identity needs is brought by specific motivation and how features of the built-environment are merged within this relation. In accordance, rich hypothetical exploration could be empirically tested in research on place and identity.
- The inquiries of place identity are framed by wide-ranging knowledge correlated to each other. In this sense, developing comprehensive understanding to place identity examines physical features of the environment together with social dimensions associated to psychological impulses and functional desires. Thus, interdisciplinary research procedures are encouraged in respect to data gathering tools and analyses techniques.
- Physical aspects of place are not a separate set of identity next to gender, social class or family, but all aspects of identity have to greater or lesser extent, place related implications (Twigger-Ross & Uzzell 1996). Thus, aspects of attachments/familiarity in any living environment and their relative physical characteristics of spaces and building combined with human motivations towards identity needs could be measured to create an integrative knowledge base for transdisciplinary research.

## 6 CONCLUSION

Along this research, the study of place is approached based upon transactional theories of the people-place relationship. Meanwhile, models that have been found in identity theories and tested in social and psychological disciplines, are employed to illuminate the relation between place and identity. Evidently, it was addressed that there is no social identity that is not also place-related and object-related. From this perspective, fields of design are beneficially incorporated in research and theories established in social sciences to understand identity issues.

As argued by literature, although both the social identity theory and identity process theory explained the influences place has on identity, aspects of functions and manageability of places driven by motivations of identity needs, are left out in most research. In the meantime, studies on Proshansky's place identity theory are criticised by lack of structured tools provided to examine empirical results.

Consequently, in respect to design fields, behind the aesthetic perspectives dominating research on place identity, the developed place identity matrix in this research, allows more interdisciplinary detailed studies concerning physical characteristics of places relative to identity aspects as follows:

- Issues of functions of buildings and spaces could be raised by investigating people's desires to achieve self-esteem and self-efficacy, and how such motivations guide actions related to the manageability of any living environment.
- Within previous perspective, it could be argued that actions pertaining to the expressions of people's identification needs in their living environments are resourceful constituents in respect to creating urban and community identity. For instance, in the case of designing for low-income areas, understanding previously stated actions has informative significance and impact on both housing policies and designs' decisions.

To sum up, this research which aimed at developing a comprehensive innovative approach for investigating place identity within the perspective of how identity theories and their underlying concepts can be merged into an 'integrative examination agenda'. The approach incorporates novel ideas to the study of place identity, in respect to the built-environment, where issues deriving from social and psychological disciplines are integrated.

In significance, the proposed approach involves number of correlations that aim to reach reliable results among variables of identity aspects and the associated physical characteristics. In addition, the invention of integrative research questionnaires to examine assorted variables is recommended. The value of such an approach lies in the value of how planning and architectural aspects, social and cultural issues, all incorporated into one mechanism towards a comprehensive inquiry on the identity of communities and cities.

Finally, this paper introduces the study of the first phase of a research project initiated by the authors two years ago to place theoretical and epistemological foundations for the second phase: an empirical examination. In progress, an integrative questionnaire is designed, guided by the place identity grid matrix to explore aspects of place identity in low income residential areas in Cairo.

## REFERENCES

Abrams, D. 1992. Processes of Social Identification. In G. M. Breakwell, Ed., *Social Psychology of Identity and the Self-concept*. Surrey: Surrey University Press, UK.

Altman, I. & Low, S. Eds, 1992. *Place Attachment*. New York Plenum Press, USA.

Barker, R. 1968. Ecological Psychology: Concept and methods for studying the environment of human behavior. Stanford, CA: Stanford University Press, USA.

Bonnes & Sechiarolli, 1995.

Breakwell, G. M. 1993. Integrating paradigms: methodological implications. In G. M. Breakwell & D. V. Canter, Eds., *Empirical Approaches to Social Representations*. Oxford: Clarendon Press, UK.

Breakwell, G. M. 1992. Processes of self-evaluation: efficacy and estrangement. In G. M. Breakwell Eds., *Social Psychology of Identity and the Self-concept*. Surrey: Surrey University Press, UK.

Breakwell, G. M. 1986. Coping with Threatened Identity. London: Methuen, UK.

Devine, P. 1994. A time and a place for the Irish? – an explanatory analysis of significance of Irish historical places for maintaining Irish National Identity. Unpublished MSc. Thesis, Department of Psychology, University of Surrey.

Dixon, J. & Durrheim, K. 2004. Dislocating identity: Desegregation and the transformation of place. Journal of Environmental Psychology 24, 455–473.

Elnachar, E. 2011. The concept of place as a data generating tool in architecture education. INTD 2011, Valencia, Spain, proceeding cd, ISBN: 978-84-614-7422-6.
Feldman, R. M. 1990. Settlement identity: Psychological bonds with home places in a mobile society: *Environment and Behavior*, 22, 183–229.
Graumann, C. 1983. On multiple identities. *International Social Science Journal*, 35, 309–321.
Hauge, A. L. 2007. Identity and Place: a critical comparison of three identity theories. *Architectural Science Review. March 1, 2007*. HighBeam Research.
Hogg, M. A. & Abrams, D. (1988). Social Identifications. London: Routledge, UK.
Hummon, D. 1992. Community attachment, local sentiment and sense of place in I. Altman and S. Low (Eds), *Place attachment* (pp. 253–278) New York: Plenum Press.
Jorgensen, B. S. & Stedman, R. C. 2001. Sense of place as an attitude: Lakeshore owners' attitudes toward their properties. *Journal of Environmental Psychology*, 21(3), 233–248.
Korpela, K. M. 1989. Place identity as a product of environmental self-regulation. *Journal of Environmental Psychology*, 9, 241–256.
Krupat, E. 1983. A place for place identity. *Identity of Environmental Psychology*, 3, 343–344.
Lalli, M. 1992. Urban related identity: theory, measurement and empirical findings. *Journal of Environmental Psychology*, 12, 285–305.
Neisser, U. 1976. Cognition and Reality. San Francisco: W. H. Freeman and Company, USA.
Norberg-Schultz, Ch. 1976. Genius Loci—Towards a Phenomenology of Architecture. Rizzoli, New York. Edinburg College of Art Library.
Piaget, J. 1954. The Construction of Reality in the Child. New York: Basic Books, USA.
Popov, L. & Chompalov, I. 2012. Crossing Over: The Interdisciplinary Meaning of Behavior Setting Theory. International Journal of Humanities and Social Science. Vol. 2, 19.
Proshansky, H. M. & Fabian, A. 1987. The development of place identity in the child. In C. S. Weinstein & T. G. David, Eds., *Spaces for Children*. New York: Plenum Press.
Proshansky, H. M., Fabian, A. K. & Kaminoff, R. (1983). Place identity: physical world socialisation of the self. *Journal of Environmental Psychology*, 3, 57–83.
Qazimi, S. 2014. Sense of place and place identity. European Journal of Social sciences Education and Research, Vol. 1, No. 1.
Salama, A.M. 2012. "Architectural Identity Demystified: Visual Voices from the Arab World," in P. Emmons, J. Lomholt, and J.S. Hendrix, eds., The Cultural Role of Architecture: Contemporary and Historical Perspectives (London: Routledge, 2012), pp. 175–84.
Speller, G. M., Lyons, E. & Twigger-Ross, C. L. 2002. A community in transition: The relationship between spatial change and identity process. Social Psychological Review, 4(2), 39–58.
Stokols, D. & Altman, I. 1987. Handbook of Environmental Psychology, John Wileg & Sons, New York.
Twigger-Ross, C. L., Bonaiuto, M. & Breakwell, G. M. 2003. Identity theories and environmental psychology. In M. Bonnes, T. Lee & M. Bonaiuto (Eds.), Psychological theories for environmental issues (pp. 203–234). Aldershot, Ashgate Publishing, UK.
Twigger-Ross, C. & Uzzell, D. 1996. Place and Identity Processes, in Journal of environmental Psychology, 16, 205–220. Academic Press Ltd.
Uzzell, D. L., Pol, E. & Badenas, D. 2002. Place identification, social cohesion, and environmental sustainability. Environment and Behavior, 34, 26–53.
Uzzell, D. L. 1995. 'Creating Place Identity Through Heritage Interpretation.' The International Journal of Heritage Studies, 1(4), 219–228.
Wicker, A. W. 1979. An Introduction to Ecological Psychology. Monterey, CA. Brooks/Cole.
Winkel, G. H. 1981. The perception of Neighbourhood Change. In J. Harvey (Ed.), Cognition, Social Behavior, and the Environment. Hillsdale, NJ: Lawrence Erlbaum, pp. 489–512.

*Cities' Identity Through Architecture and Arts – Catalani et al. (Eds)*
*© 2018 Taylor & Francis Group, London, ISBN 978-1-138-05409-7*

# Sustainable urban heritage conservation strategies—case study of historic Jeddah districts

Samaa Badawy
*Department of Architecture, College of Engineering, Mansoura University, Egypt*
*Department of Architecture, College of Architecture and Design, Effat University, Saudi Arabia*

Ahmed M. Shehata
*Department of Islamic Architecture, College of Engineering and Islamic Architecture, Umm Al-Qura University, Saudi Arabia*

ABSTRACT: Due to the economic boom in the Saudi kingdom during the seventies, the urban area of the economic capital of the kingdom and its historic main port were subject to a lot of modifications and expansions. Later, during the eighties, the city expanded dramatically to the north and the old city was deserted. During the nineties, preservation and conservation strategies were implemented to maintain the city's culture and heritage. Since then, governmental and civil society organisations have been engaged in maintaining and conserving the historic once-gated old city, known these days as Al-Balad District. Moreover, successful efforts have been made by the Saudi authorities to register the vast majority of the remains of this historic area as human race heritage through the United Nations Educational, Scientific and Culture Organization (UNICCO).

Through this, practitioners and researchers spent a lot of effort in developing and adopting strategies and policies to revive and preserve this heritage urban area. This research, as part of these efforts, aims to evaluate the applied urban conservation strategies and policies for Al-Balad District, to test their success in preserving the heritage while maintaining socio-physical, socio-cultural, and environmental aspects of the local community. To do this, benchmark case studies for developing and conserving historic urban areas were analysed and compared to the applied strategies. The research concludes with recommended policies to achieve the most sustainable conservation strategies in this valuable area that will fulfil the recent needs of the local community.

*Keywords*: Old Jeddah; Conservation; Sustainability; Socio-Cultural

## 1 INTRODUCTION

During the last few decades, most cities all over the world have been facing the problem of deterioration of their historic districts caused by their rapid socio-economic development. Despite the economic benefits of this rapid development, it has affected the historic parts of the cities in a negative way. It has torn down the old urban fabric and character of the city, (Chohan, 2006) and directed the development towards new districts while neglecting the inner parts of the city, which has resulted in poor services and infrastructure in those parts. As a result, the residents have migrated to new districts, leaving their buildings abandoned, or they have tried to modernise their buildings with unplanned initiatives to cope with the recent requirements. Such initiatives tend to harm the historic districts. In the end, the result is the loss of identity, history and the collective memory of the city. (UNEP, 2016).

It is the permanent contest between new development and heritage conservation which has been the subject of research and study for a long time, in order to find the best solutions to create a balance between the changing needs and conserving the past.

The previous scenario happened in the Saudi cities, especially in Jeddah city, the main port and the second capital of the kingdom, due to the economic boom during the seventies. A lot of modifications and expansions happened to the city in ignorance of the historic Al-Balad District centre, which resulted in the loss of many important historic buildings.

Since then, governmental and civil society organisations have been engaged in maintaining and conserving the old city or Al-Balad District. Moreover, an application to the United Nations Educational, Scientific and Culture Organization resulted in recognising the vast majority of the remains of the historic area as human race heritage.

Through this, practitioners and researchers spent a lot of effort developing and adopting strategies and policies to revive and preserve this heritage urban area. This research, as a part of these efforts, aims to evaluate the applied urban conservation strategies and policies for Al-Balad District, to test their success in preserving the heritage while maintaining socio-physical, socio-cultural, and environmental aspects of the local community. To do this, benchmark case studies for developing and conserving historic urban areas will be analysed and compared to the applied strategies. The research concludes with recommended policies to achieve the most sustainable conservation strategies for this valuable area that will fulfil the recent needs of the local community.

### 1.1 *Research problem*

The research is directed to answer the following questions:

- What is the relation between historic conservation and sustainable development?
- What are the strategies of sustainable historic conservation?
- Are the conservation strategies used in Al-Balad District sustainable?

### 1.2 *Research objectives*

The main aim of the research is to develop a sustainable strategy for historic conservation to guide the development in Al-Balad District in Jeddah city. This is achieved through a combination of sub-objectives as follows:

- Clarify the relation between historic conservation and sustainable development.
- Identify the strategies of sustainable historic conservation.
- Evaluate the impact of the applied strategies on the development of the historic area.
- Formulate a set of recommendations for developing Al-Balad District to achieve a sustainable development.

### 1.3 *Research methodology*

The methodology is based on three approaches: theoretical, analytical and applied approach. The theoretical study aims to identify the concept of sustainable historic conservation. This is followed by an analytical study for benchmark case studies of historic conservation to understand the different strategies used in sustainable historic conservation. Then the applied study of Al-Balad District in Jeddah, KSA, and the evaluation of the applied urban conservation strategies and policies for Al-Balad District, to test their success in preserving the heritage while maintaining socio-physical, socio-cultural, and environmental aspects of the local community. Finally, the paper concludes with the recommended policies to achieve the most sustainable conservation strategies for this valuable area that will fulfil the recent needs of the local community.

## 2 HISTORIC CONSERVATION AND SUSTAINABILITY CONCEPT

The general definition of conservation is the protection from loss and depletion for tangible and intangible elements, so historic conservation helps to extend places and values of the past into the present (Alison, 2007). The concept of historic conservation has existed since the 1960s (Elnokaly, 2013) and has become the core of cultural identity, and ownership of the past discourse over the last few decades (Alison, 2007).

The sustainability concept emerged in 1970 as a significant mode of thought in nearly every field, insisting that environmental quality, economic growth, and social and cultural values must be considered for viable long-term development strategies. Since then historic conservation became an important part of the sustainability agenda (Alison, 2007). Although their roots are different, conservation and sustainability share common ground (Elnokaly, 2013). Sustainability in this context means ensuring the continuing contribution of heritage to the present through the thoughtful management of change, responsive to the historic environment and to the social and cultural processes that created it. By creating the balance between change and continuity, between the old and the new, and when history is viewed as continuous change, it can be expanded into culturally responsive strategies that provide an alternative to imported solutions that do not relate to, or grow out of, the existing cultural context. Conservation can contribute to the sustainable environment (Alison, 2007).

### 2.1 *Historic conservation within the three-dimensional sustainability*

The most popular definition of sustainable development is the World Commission on Environment and Development (WCED) definition 1987, which defines sustainable development as '*the development that meets the needs of the present without compromising the ability of future generations to meet their own needs*' (Chohan, 2005). In this context, it was made clear that sustainable development can only be achieved through the integration between the three main dimensions: environmental, economic and social (United Nation Environment Program, 2016). Historic conservation can lead a way to sustainable development through achieving environmental upgrading, economic viability and social coherence, by the comprehensive policy for sustainable historic conservation (Chohan, 2005), as will be discussed.

#### 2.1.1 *Historic conservation and environmental sustainability*

The main focus of the environmental dimension is the reduction of harmful emissions and in reducing the consumption of non-renewable resources (Yung, 2012). So, environmental sustainability depends on the three Rs: 'reducing-reusing-recycling'. In order to make historic conservation environmentally sustainable, one should consider the reuse of the historic resources. Reusing historic buildings is a significant way to practise the three Rs. On a much bigger scale, a new life breathed into old buildings helps to reduce urban sprawl, conserve energy and promote sustainable communities. According to the Life Cycle Assessment (LCA), which examines the energy and material usage during the entire life of the building, it indicates that retaining and rehabilitating buildings is more environmentally friendly than new construction; it diverts waste, and reduces the consumption of materials, energy, and land for new development.

Aspects of traditional designs which take into account natural ventilation, heat retention and cooling treatments in historic buildings (Allison, 2003) can be expanded into environmentally responsive strategies, which provide an alternative to imported solutions that do not relate to, or grow out of, the existing cultural context (Allison, 2011). The same concept can be applied at the scale of urban design. A historic neighbourhood is by default a sustainable neighbourhood. The layout of these neighbourhoods placed stores, schools, jobs, and recreation in close proximity to one another. Encouraging people to walk reduces the dependency on the car and reduces the carbon footprint (Gilderbloom, 2009).

Therefore, the traditional strategies of designing buildings and neighbourhoods can help in developing local design strategies.

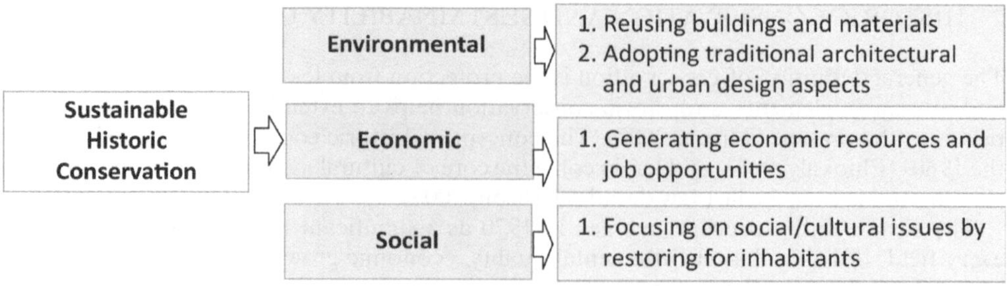


Figure 1. Sustainable historic conservation strategies.

2.1.2 *Historic conservation and economic sustainability*
Economic sustainability seeks to meet the service needs of the general public, particularly the poor, while enhancing the naturalness of the urban environment (Basiago, 1999). Historic conservation can be economically sustainable when it contributes in generating economic resources, and this can be achieved by providing job opportunities. The studies indicated that historic conservation results in more job creation than most other kinds of investments, whether in the restoration process or through the different activities which will exist in the area after development (Gilderbloom, 2009).

2.1.3 *Historic conservation and social sustainability*
Social sustainability can be defined as a life-enhancing condition within communities, and a process within communities that can achieve that condition (McKenzie, 2004). The town centres, where socio-economic pressure is high, become a driving force, whose victims are working class residents. In order to make sustainable historic conservation, a specific focus has to be given to social perspectives for the renewal of inner cities. Since 1996, within the United Nations Educational, Scientific and Cultural Organisation's (UNESCO) Management of Social Transformations (MOST) programme, many studies and initiatives have been made to improve socio-economic and cultural issues while improving the physical environment and buildings in historic districts. The main aim for conservation is how to restore and preserve historic districts as habitable areas where residents can enjoy the benefits of the revitalisation (UN-Habitat, 2016). Figure 1 summarises the strategies of sustainable historic conservation.

## 3 BENCHMARK CASE STUDIES

### 3.1 *Palestine rehabilitation of historic centres*

3.1.1 *Project background*
During the period from 1991 until 2001, there was an increase in the rate of destruction and the loss of architectural heritage in Palestinian cities. In 1991 Riwaq was established to save Palestine's heritage and restore historic buildings. Riwaq started its programme by restoring chosen single buildings. By 2000, the programme recognised the need to develop its strategies to include the three dimensions of sustainability. The new focus was to make historic conservation an income generator and a source for jobs. From 2001, all renovation projects had large job opportunities, local workers were being trained and then employed in the restoration works, and up to 60% of the construction costs went to the employment of local labourers. The programme aimed to save a threatened cultural heritage in many ways, especially in the case of Israeli occupation. A generation is coming of age that no longer remembers the geography of historical Palestine: they know only a fragmented landscape, where villages are cut off from their rural hinterland by political zones, checkpoints and barriers (Lamprako, 2013).

Figure 2. The location of Birzeit town (Lamprako, 2013).

### 3.1.2 *Birzeit rehabilitation of historic centres*

Birzeit, located ten kilometres north of Ramallah, is one of the largest towns in the Ramallah area and is significant for its history. The historic centre of Birzeit lies on a small hill on the town's eastern side. The area of the site is about four hectares (ten acres). The historic centre encompasses 108 historic buildings. Most of these buildings date back to the Ottoman era (RIWAQ, 2016).

After the 1967 war, the people of Birzeit abandoned the historic centre and moved to newly developed areas near the new regional road linking Ramallah and the northern village. In 1980 Birzeit University relocated its campus out of the historic centre. In turn, Birzeit lost its central role and had fell prey to rapid urbanisation. Several new housing projects, neighbourhoods, and cities were proposed in the region, causing a reshape of the area. Streets and public spaces were neglected (RIWAQ, 2016).

Riwaq started a five-year project to revive the decaying town of Birzeit (ARCHNET, 2016). Birzeit was chosen to be a pilot project for the revival of 50 other projects for the following reasons: proximity to Ramallah, the seat of the Palestinian authority; proximity to, and historical links with, Birzeit University; the presence of a supportive municipality and a number of active local NGOs (Non-Governmental Organizations), including Rozana School; a diverse, but socially and economically disadvantaged, population (Greek Orthodox, Roman Catholics, and Muslims); a largely intact historic fabric, which was compromised by neglect and degradation; and the presence of diverse resources around the town, including archaeological sites, natural springs and olive groves (Lamprako, 2013).

### 3.1.3 *The project vision and principles*

The project aimed to rehabilitate the historic fabric of the town to create a sense of value, and use this as a basis for revitalisation and development. The goal was to keep people living in the old town, and to create new interest and investment, making it a desirable place to live, work, do business, and pursue leisure and entertainment (Lamprako, 2013). The project vision was therefore to create a lively centre which is attractive to its residents, investors and visitors, within a mixed-land use context, and which is respectful of its identity and cultural heritage (Aga Khan Award for Architecture, 2013), Riwaq's conservation work involves: a) conservation/upgrading of the public realm; and b) restoration of selected buildings with public/community functions to act as catalysts for further interest and development (Lamprako, 2013).

The development principles included: improving the quality of life for the inhabitants and the users of the centre, the protection of the tangible cultural heritage as a space for

Figure 3. Map for the historic centre of Birzeit (State of Conservation Report, 2015).

socio-economic development, encouraging investment through the creation of key projects to make the place suitable for contemporary life, and promoting the historic centre as a cultural and touristic centre. The objectives of the project are obtained through a wide involvement of the community, which was encouraged from the start, including local NGOs, the private sector, owners, tenants and users (Saudi Commission for Tourism and National Heritage).

### 3.1.4 *The development projects*

The development projects targeted mainly the real needs of the population. Those projects included: upgrading infrastructure for the existing water system and the provision of a future sewer system; paving streets and the provision of street naming and signage project (those names were based on local memory and narratives); the creation of a heritage walkways and a map; planting trees; the rehabilitation of public spaces and the creation of new ones (streets, courtyards, gardens); the conservation of stone facades and walls lining the public space; and the restoration and reuse of selected buildings, such as for several restaurants, an Internet café, and a science museum devoted to children's science and environmental education as a way to communicate to children the links between past and future, and between built and natural environments. The Circus School has occupied a former university building; it offers training in modern circus techniques to young boys and girls from across Palestine; also, the Birzeit Guest House, which is used as a small hotel. The old town is slowly becoming a destination for residents, university students, and local and foreign tourists (Lamprako, 2013 & NG Architects 2016).

### 3.1.5 *Local architectural character, structure, materials, and technology*

The development projects respect the architectural and material characteristics of the historic centre. The material used is generally local limestone, lime mortar and lime plaster. Most buildings in the historic centre are one storey in height. The traditional house consists of one or more square modules, each covered by a dome (*qubba*). In addition to houses, the old town includes a number of religious structures, including the Greek Orthodox Church, the

Figure 4. Birzeit Guest House before and after development (Aga Khan Award for Architecture, 2013).

Figure 5. Birzeit streets before and after development (Lamprako, 2013).

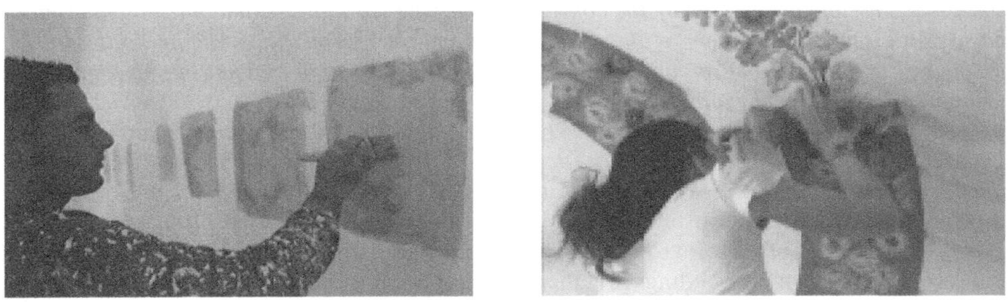

Figure 6. Local labourers are trained, and worked on the project (Aga Khan Award for Architecture, 2013).

Latin Catholic Church and a mosque. The project focused on providing training workshops for local craftsmen; 90% of the labour was local labourers (Lamprako, 2013 & NG Architects 2016).

### 3.1.6 *Sustainability issues*

The projects demonstrate several sustainable dimensions: environmental, social, and economic. The dimensions table below summarises the different strategies and the achieved sustainability objectives.

## 3.2 *Conclusion of regional experience*

After analysing four urban conservation projects (which were awarded some prestigious awards like the Aga Khan and other national and international awards in four different countries),

Table 1. The conservation of historic areas. Response to historic area issues using three sustainability axes.

| Sustainability dimensions | Sustainability strategies | Birzeit sustainability strategies |
|---|---|---|
| Environmental | 1. Reusing buildings and materials.<br>2. Adopt traditional architectural and urban design aspects | Saudi Commission for Tourism and National Heritage. (2012). Submitted report to UNICCO property value, *Historic Jeddah, the gate to Makkah*. State of Conservation Report, November 2015. Available online: http://archnet.org/system/publications/contents/2570/original/FLS3269.pdf?1384764408. Last accessed June 2016. |
| Economic | 1. Generating economic resources and job opportunities | • Provision of job opportunities through training the local labourers and promoting the traditional crafts<br>• Provision of variety of activities which provide job opportunities<br>• Encourage local trades |
| Social/Cultural | 1. Focusing on social/cultural issues by restoring for inhabitants | • Saving and restoring historic centre to promote awareness of the old social and cultural bonds<br>• Creation of series of open spaces that act as socio-cultural hubs<br>• Using cultural and historic values in naming the streets<br>• Promoting the traditional crafts helps in saving culture |

Enokaly and Elseragy were able to conclude that conservation success factors are: '*Public participation and awareness at early stages of the project; an implicit approach of integrated urban conservation; employing many of the concepts underlying cultural sustainability; preserving the basic pattern and morphology of the urban fabric and the use of space; minimal interventions following a scrutinized analysis and assessment of the monuments or buildings involved; a bottom up approach and involving all stakeholders from the beginning of the projects; restoring buildings to their original form using traditional decorative features and motifs. Another key feature that led to success of most of these revitalisation projects is high impact propaganda that helped in mobilizing and selling the idea of rehabilitation to the residents*' (Elnokaly, 2013). From analysing the Birzeit rehabilitation project and the literature review analysed projects, sustainable urban conservation of historic towns and cities follows a comprehensive approach, that combines conserving historic buildings and their physical urban context along with the their socio-economic environment. This means utilising policies and programmes to mitigate the local community's needs, such as reducing poverty, increasing employment, and upgrading local communal services. Enabling the local community to positively participate in the planning and management of their area guarantees a sustainable conservation.

In all reviewed previous projects, community participation, training of local professionals and local institution-building, public awareness campaigns were the key factors in achieving sustainable conservation.

## 4 THE CASE STUDY—AL-BALAD AREA

### 4.1 *Relative importance of the case study*

Jeddah, with its three million population, is the second largest Saudi city. It belongs to what can be called the Red Sea civilisation. Historic Jeddah consists of urban property extending over an area of seventeen hectares, composing the heart of nowadays Jeddah city. Its architectural style is characterised by the tower houses, decorated by large wooden *Roshans* that were built in the late 19th century by the city's mercantile elite. Figure 7 shows some of these

Figure 7. The architecture style of the area where buildings are decorated by large wooden *Roshans*.

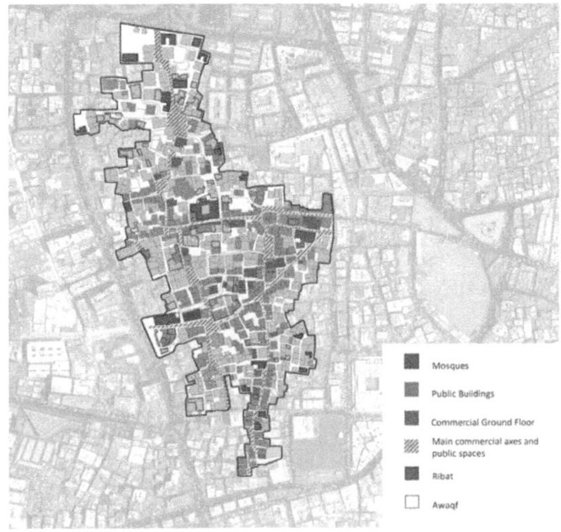


Figure 8. Public spaces and commercial activities within the area.

houses with their *Roshans*. The outstanding universal value of the area relates to its unique development of the Red Sea architectural style and to its preserved urban fabric. It can be considered as being the last surviving urban centre of this cultural region that still preserves its original urban fabric. An extraordinary pre-modern urban environment is shown in the map in Figure 8, where tower houses, lower coral stone houses, mosques, Rabats, souks and small public squares compose a vibrant space, inhabited by a multicultural population that still plays a major symbolic and economic role in the life of the modern metropolis.

### 4.2 *Development of the historical area*

As mentioned before, the historic area that was once a whole gated city, is now only a small district within the city itself. This was the result of several factors affecting the area through the past fifty years. After the old city walls were demolished in 1948, due to the creation of the modern port of Jeddah on reclaimed nearby land, the airport, the new road axis and the high-rise buildings were constructed. Dahab street in the heart of the old city was opened, and later in the 1970s, new types of modern buildings like shopping malls and office buildings were constructed. The old city of Jeddah has lost entire sectors of its original urban fabric, mostly replaced by the new high-rise structures in the Bahr Quarter, next to the original coastline. It should be noted that large areas of the Yemen, Mazloum and Sham quarters are still preserved to their overall original structure. Figure 9 illustrates the rapid growth of Jeddah city and its modern metropolis through the last fifty years.

The historic area quarters shown in Figure 10 preserve a very evident coherence at urban, architectural, economic and social levels. Therefore, it can be said that the existing historic area, though covering only part of the original walled-in city, contains the ensemble of the attributes that convey its outstanding universal value.

### 4.3 *Urban and architecture characteristics*

Historic Jeddah is a living urban environment primarily hosting residential and wholesale commercial activities, with some mosques. In the immediate surroundings of the area some traditional and cheap hotels and accommodations for pilgrims can be found. The traditional infrastructures created for the reception of pilgrims, the *ribats*, and the ground floors of

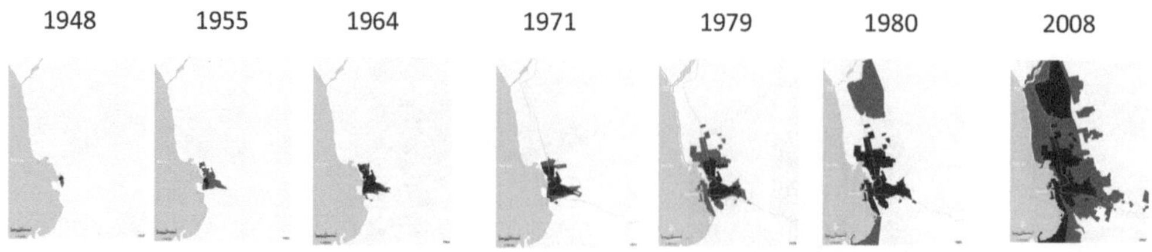


Figure 9. Set of maps show the rapid growth of the metropolitan area of modern Jeddah city through the last fifty years.

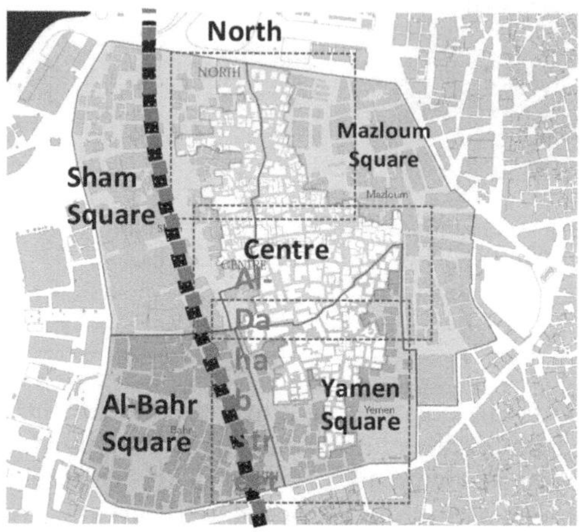


Figure 10. Historic Jeddah quarters.

private houses, that used to be rented out to pilgrims, are now mostly empty or replaced by other commercial activities. The historic mosques within the area perimeter have maintained their function and role for the community and most of their original features.

### 4.4 *Administrative structure*

Since the 1980s the four quarters have been reduced to two districts, but the traditional *Umdahs* continue to exist and to play an important role for the residents. Though contemporary Saudi society has also developed other management administrative structures, both at the local and central levels, this traditional management system is still active and has even been revitalised in the recent past.

### 4.5 *Cultural and economic life*

It is a traditional urban environment, where there are still concentrated the headquarters of century-old economic enterprises, retail shops, traditional souks, small cafés, popular restaurants, street food sellers and semi-legal fruit market stalls. A surprisingly rich human environment exists where Yemeni, Sudanese, Somali, Pakistani and Indian immigrant workers buy and sell their products to Saudi and non-Saudi clients in crowded 'traditional' souks, as shown in Figure 11. This an area where even the food shows multiple influences: Indian spiced rice and Red Sea colourful fishes, as in the past, constitute the menu of many traditional restaurants where old immigrants, newcomers and Saudi clients convene in the hot evenings.

Figure 11. Historic Jeddah crowded 'traditional' souks.

### 4.6 *Social character*

The major transformation that occurred in the last 50 years in Jeddah, is that poor newcomers have progressively replaced the rich local families who themselves do not reside anymore in their traditional family houses, but who look for modern comfort in the rich suburbs. Though this phenomenon is extremely common in many other cities, there are two specificities that distinguish historic Jeddah:

1. Most of the original merchant families that abandoned their houses have often kept their headquarters of their commercial companies in their old properties.
2. The poorer newcomers that have progressively replaced the merchant families are not coming from the local rural areas, as in other countries, but from neighbouring Islamic poor countries.

These two specificities created the city identity, a merchant and multicultural city, at the social and economic levels. The functional and social identity of the area is therefore fully authentic, even though the city, like any living urban environment, has continued to evolve and change, mirroring the evolution of the Saudi society (Saudi Commission for Tourism and National Heritage, 2012).

## 5 ANALYSIS OF PRESENT STATUS OF THE AREA

In its present state, the urban environment is no longer able to provide the comfort required by the Saudi families. The replacement of its original residents that moved to the rich modern suburbs has gradually transformed the heart of the city into a low-class ghetto, mostly rented out to poor foreign immigrants. Favouring in turn the accelerated deterioration and the collapse of many of its traditional coral masonry houses.

The area has deteriorated with the passage of time, by the transformation of their inhabitants, and with the fragility of their structure system. Pictures in Figure 12 show the condition of some of the buildings of the historic area. Some of the buildings have been changed by modern additions and some in-depth transformations that affected their form and substance, as shown in Figure 13.

## 6 CONSERVATION EFFORTS

The historic area of Jeddah city was subject to many studies, development and conservation projects. Being listed on the World Heritage List inspired even more efforts from the private

Figure 12. Deteriorated buildings and urban fabric in historic Jeddah.

Figure 13. Buildings facing modifications that affect their condition and style.

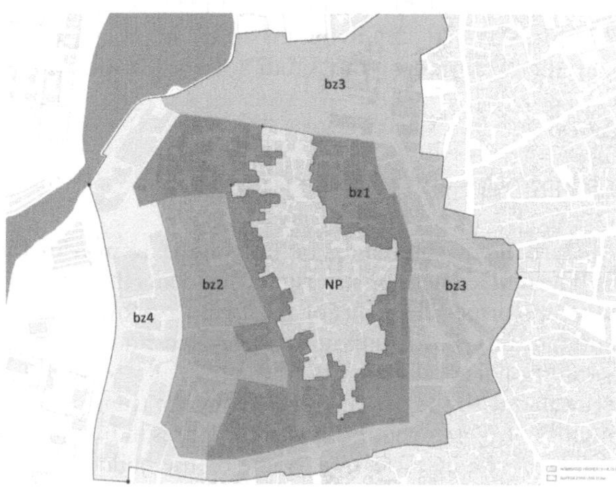


Figure 14. Subdivision of historic area according to urban regulations.

sector, municipality and Saudi Commission of Tourism and Antiquities (Saudi Commission for Tourism and National Heritage, 2012).

- The urban regulations approved by the municipality in 2012 set a new, much needed legislative framework, providing the legal tools permitting the control of the speculative moves that have caused the loss of many historic buildings in the past. The map in Figure 14 presents the preserved area, in addition to its surrounding buffer that was divided into four different categories. Each of these buffer areas has its own building regulations.

- Cultural and seasonal activities managed by the SCTA to raise the awareness of the importance of the cultural value of the area. Traditional shows are performed and local street

Figure 15. The annual festival of traditional activities in the historic area.

Figure 16. Restoration of Al Shafee' mosque.

Figure 17. Improving built environment conditions to help conservation efforts.

food is served to native and foreign visitors. Plazas and famous buildings are lit and opened to give a festival atmosphere. Figure 15 illustrates the annual festival atmosphere.

- The renewal efforts of the SCTA and the municipality of Jeddah for the conservation and restoration projects of main houses and mosques, under the joint control of the municipality and the SCTA, are actively countering the decay processes. The picture in Figure 16 shows Al Shafee' mosque during its restoration process.
- The urban fabric of the listed area is in relatively good condition and the deterioration impact is being tackled by the concerned authorities, as per the preservation strategy proposed in the management plan of the area. Lighting was replaced with decorative lighting, road pavement was replaced by traditional tiles and street furniture was replaced with designed ones.

## 7 ASSESSING CONSERVATION EFFORTS OF THE AREA IN THE SHADOW OF SUSTAINABILITY

- Working sites, directed by Jeddah professionals and municipality engineers with the support of foreign consultants, have used compatible traditional materials and techniques, and have been an opportunity to redevelop traditional constructive skills.
- Existing commercial activities did not enrich the social life of the area.
- Facilitating an economic base will help in achieving sustainable conservation and regeneration of the Jeddah historic area.
- Lack of adequate services and failing to accommodate modern communication and transportation is still preventing native families from returning permanently.
- Cultural activities increase the sense of belonging and ownership for Jeddah's inhabitants.
- Municipality efforts in regenerating the urban environment along with the conservation efforts is very useful to maintain a sustainable conservation.

## 8 CONCLUSION AND RECOMMENDATIONS

Regional successful interventions in conserving historic centres started from an overall vision, but they did not proceed as per the planned schemes. Intervention efforts should be based on incremental development as a response to tangible requirements, perceived from real site experience. The decision-making process should be based on the assessment of the local needs and requirements. An overall strategic vision concerned with heritage conservation and urban regeneration can result in achieving sustainable development.

Heritage buildings conservation and restoration should be integrated with their urban context sustainable regeneration, combining interactively many different disciplines. Conserving heritage is a good tool for sustainable tourism. Utilising cultural activities would enhance the spirt of belonging and ownership of the inhabitants. The conservation initiative gives to local inhabitants a strong sense of belonging and ownership of their social, cultural and historical values.

Involving the local authorities, local community and different interest groups is very important for the success of heritage conservation. The community involvement in any urban heritage conservation is a general expression which needs comprehensive strategies for preservation. Despite legal and institutional framework importance, local community and local authorities play the key role in achieving sustainable results in the urban development. Authorities should spend effort in involving the community and motivating other stakeholders, community professionals and volunteers to work for sustaining their heritage that could be a symbol of pride and give a sense of belonging and ownership to the coming generations.

## REFERENCES

Aga Khan Award for Architecture (2016). *Revitalization of Birzeit historic centre*. Birzeit, Palestine. Available online: http://www.akdn.org/architecture. Last accessed June 2016.

Allison, E.W. (2011). *Historic preservation and the livable city*. Hoboken, New Jersey: Wiley & Sons.
ARCHNET. (2016). *RIWAQ centre for architectural conservation*. Available online: http://archnet.org/authorities/484/publications/2569. Last accessed June 2016.
Basiago. (1999). Economic, social, and environmental sustainability in development theory and urban planning practice. *The Environmentalist*, *19*(2), 145–161_1999. Boston: Kluwer Academic Publishers. Manufactured in the Netherlands.
Chohan, A.Y. & Ki, P.W. (2005). Heritage conservation a tool for sustainable urban regeneration. *41st ISoCaRP Congress*.
Elnokaly, A. & Elseragy, A. (2013). Sustainable heritage development: Learning from urban conservation of heritage projects in non-western contexts. *European Journal of Sustainable Development*, *2*, 31–54.
Gilderbloom, J.I., Hanka, M.J. & Ambrosius, J.D. (2009). Historic preservation's impact on job creation, property values, and environmental sustainability. *Journal of Urbanism, Routledge,* 2(2), 83–101.
Heritage Branch. (2001). *How does historic conservation contribute to sustainable development?* (2001). Available online: www.for.gov.bc.ca/heritage. Last accessed May 2016.
Lamprako, M. (2013). *Revitalization of Birzeit historic center. On site review report*. 2013.
McKenzie, S. (2004). *Social sustainability: Towards some definitions*. Hawke Research Institute, University of South Australia, Magill, South Australia.
NG Architect. (2016). *The regeneration of the historic centre of Birzeit*. Available online: http://www.ngarchitects.co.uk/#!birzeit/c1wdy. Last accessed June 2016.
RIWAQ. (2016). *Birzeit*. Available online: http://www.riwaq.org/content/birzeit#the_village. Last accessed June 2016.
Rodwell, D. (2007). *Conservation and sustainability in historic cities*. Oxford, UK: Blackwell Publishing Ltd.
Saudi Commission for Tourism and National Heritage. (2012). Submitted report to UNICCO property value, *Historic Jeddah, the gate to Makkah*.
State of Conservation Report, November 2015. Available online: http://archnet.org/system/publications/contents/2570/original/FLS3269.pdf?1384764408. Last accessed June 2016.
UN-Habitat. (2016). *Best practices on social sustainability in historic districts*. Available online: http://www.bestpractices.org. Last accessed May 2016.
United Nations Environment Program (UNEP). (2016). *The three dimensions of sustainable development*. Available online: http://web.unep.org/ourplanet/march-2015/unep-work/three-dimensions-sustainable-development. Last accessed May 2016.
Yung, E.H.K. & Chan, E.H.W. (2012). Implementation challenges to the adaptive reuse of heritage buildings: Towards the goals of sustainable, low carbon cities. *Habitat International, Elsevier Ltd. 36*(3), 352–361.

*Cities' Identity Through Architecture and Arts – Catalani et al. (Eds)*
*© 2018 Taylor & Francis Group, London, ISBN 978-1-138-05409-7*

# The correlation between art and architecture to promote social interaction in public space

Maher Mk. Dawoud
*Department of Mural Painting, Faculty of Fine Arts, Helwan University, Giza, Egypt*

Ebtesam M. Elgizawy
*Department of Architecture, Faculty of Fine Arts, Helwan University, Giza, Egypt*

ABSTRACT: Public Space gradually becomes an absolutely necessary resource to a successful community in the new cities. It can provide lots of opportunities for people to meet and be exposed to their neighbours. This agrees with evidence of modern sociologists, who have proved that the strong social interaction between residents leads to a healthy community. However, these social connections and neighbour meetings often take place by coincidence or with active organising. In addition, the way of promoting the interaction among people in public spaces has been mainly ignored in many communities.

More recently, modern artists and architects reveal that arts and culture strategies can help to enhance the social form of the community by shaping the scene of its public space. Public art administrations, institutions, and cultural centres can play an important role in designing, managing, and planning these public spaces. Nowadays, artists are cooperating with landscape architects and city planners to design creative public spaces.

From this point of view, the study will display the meaning, the value, and the main characteristics of that new approach of designing to be part of the community's collective identity and promote the community social interaction. Moreover, the result confirms a supposed relationship between a creative community vision of designing their public space and their social interaction in these places.

*Keywords*: Social Interaction; Public Spaces; Visual Art; Contemporary Art; Landscape architecture; New Communities

## 1 INTRODUCTION

Community engagement is an interaction connection between residents and a place that can provide them a satisfaction, loyalty, and passion; a place where they can connect and socialise. They then feel a bond to their community which is stronger than merely being satisfied about where they live. Public space is the most attached form of place to the people who want to interact and share their interests together. Moreover, it plays an important role in identifying the community image and culture scene. In addition, public art such as memorials, sculptures, murals, and other aesthetic elements, can be involved with landscape architecture to personalise that public space and enhance the social connection between the community residents.

In that way, this paper will use the descriptive methodology which is generally used for humanities and social science research to highlight the crucial value of the correlation between art and architecture in the public space, to promote the social ties between residents which leads to a liveable and healthy community.

## 2 PUBLIC SPACES FOR CREATIVE COMMUNITIES

In the last decades, public space at first glance has been taken as meaning a common name for a public park. However, that meaning has been transformed these days to meet other new aspects and connotations. For instance, there is a prevalent vision around the world to see recreation centres, marketplaces, and gallerias as the modern kinds of public spaces. In addition, these spaces are considered to be crucial ingredients in every successful community. It can draw and infer an identity between the society and the whole city.

Public spaces can provide many chances for residents to gather and be exposed to various sorts of neighbours. These gatherings most probably take place by coincidence. However, they also can be organized by active associations or come through creative organising. Despite this, the art of enhancing and boosting the good vibes of interaction among people in the public spaces has been almost forgotten in many communities. Most of the urban planners, architects, and landscape architects have focused more on designing aesthetic places which provide various spaces to accommodate any kind of public activities, rather than creating places that promote social interaction between the community residents.

More recent studies in Chicago (Earls & Carlson 2001) have shown a very surprising result that the most significant factor of health from one community to the next was not wealth or easy access to healthcare, or any of the expected factors. It is simply the capacity of people who interact with each other on matters of common interests which makes a huge difference in health and well-being for individuals and neighbourhoods.

Another study was conducted by William H. Whyte, a famous American urbanist and sociologist writer, who emphasised that crowded, pedestrian-friendly, and active public spaces are more likely conducive to healthy civic communities than secured and controlled public spaces (Whyte, 1988). In addition, they are deemed to be safer and economically productive according to other sociologists who have asserted that strong social interactions are crucial ingredients of economic success.

'What attracts people most, it would appear, is other people' by this evidence, we now have a clear vision of, to build a creative community, you should create spaces where people can cooperate, connect, and share their goals.

## 3 INVOLVING PUBLIC ART IN PUBLIC SPACE

For Penny B. Bach, a seasoned executive of the Association for Public Art, public art is the easiest way for collecting people together and encouraging them to communicate well. 'It's free. There are no tickets. People don't have to dress up. You can view it alone or in groups (Figure 1). It's open to everyone' (Bach, 1992).

Furthermore, according also to Bach, many studies have indicated that the economic benefits of art has been increasing recently when public art is involved with public space. A survey of 43,000 residents in 43 new cities (John & James 2010) showed that viewing public art was the second most desirable activity in the public spaces, ranking above jogging, hiking, and biking.

In that way, it seems obvious that public space is the combined work of many design and artistic disciplines, predicting the prospering future of public art. Public art organisers, visual

Figure 1. People watch art work in different contexts as individuals or groups.

Figure 2. Various parades in public spaces.

artists, and cultural administrators can play a significant role in designing, organising, and programming the public space.

Recently, the awareness of how art can be an added value for any public space leads to involving visual artists with architects, landscape architects, and city planners in designing and creating these spaces with their unique facilities. Increasingly, there is a strong belief that as important as the space, pieces of art, or annual events, is the process by which they are created. For instance, you can organise a puppet parade involving only a group of dancers marching in the street, or it could be the result of a lengthy, community-wide process contributing many residents who create unique themes, paint the puppets, conduct the activities, and march together with their families in the neighbourhoods and public spaces (Figure 2).

### 3.1 *Promoting collective identity through public art*

While the design of public space influences its visitors, a public art event could form the collective identity of the community. By now, the meaning of public art as a piece of art work in an open space has been changed to include music, performance, and ceremonial show, besides the usual fine arts of sculpture, painting, and mosaic. In order to called the above, contemporary art which became a big hunt nowadays.

Innovating the kind of interaction between people that leads to collective identity is daring for any urban planner, organiser, or community administrator. Annual public art events or contemporary art exhibitions can play an important role. They boost the self-image of the community beyond the aspects that have been shown by them. Visitors and local residents now come from other communities to attend these events. They have become a secret formula as well as essential ingredient for any creative and liveable society.

As well as art associations, public galleries, or art institutions making the change to contribute in forming the community's collective identity, individual artists can also make a distinction. Artist Barnaby Evans, who works in many media including sculpture installations, photography, landscape design, and architectural projects, encouraged hundreds of volunteers and supporters to create a public art event in Providence, Rhode Island, USA. It was called Water Fire and its mission was to inspire Providence city and its residents by restoring the urban experience, boosting community interaction, and innovatively transforming the image of the whole city, by presenting the Water Fire public art event for all to enjoy (Frenchman, 2004).

The event engages music, performances, and sparkling bonfires. Filling the air with the fragrant scent of firewood, flickering firelight and enchanting music from all over the world involves all the senses and emotions of over ten million visitors, who have been captured by that kind of art which brings life to that public place and revives the connection between people every time it happens (Figure 3).

### 3.2 *Public art to restore the urban and social landscape*

Once you see an art work in a public landscape area, your movement through the space is slowed down. None of us can deny that one public art work can have a great effect on viewers. They gather up around it, take photos of themselves with it, or debate the underlying message behind it. Public art works attract people and lead to an interesting controversial

Figure 3. Water Fire public art event, Providence, USA.

Figure 4. Rainbow, Culver city, California.

talk. Public art could be provocative and joyous. However, it could also be annoying. It does not matter what feelings and emotions it raises up; an art is a sudden stop in pedestrian life.

For this case, artist Tony Tassel has restored the urban and social landscape experience of Culver city, California through his creative public contemporary art work of a 94-foot rainbow (Figure 4). Many visitors now are welcomed, to be excited and socialised with each other around a wide neighbourhood area watching that unique kind of art. Actually, the rainbow is showable from miles around that people could believe that it is real!! Back then, during the rainbow's first inauguration ceremony, the visitors and the community residents were feeling dizzy and talking about how it has really changed the image of their city (Andrew, 2012).

Eventually, where the art work hits the ground, an exciting conversation starts between the people and the landscape.

## 4 THE CORELATION BETWEEN ART AND ARCHITECURE

Throughout time since the ancient ages to the modern history and then by now, the correlation between art and architecture is deemed to be as an inseparable symbiotic relationship. The aesthetic norms of art have been always engaged with the synthesis of architecture to produce a very successful product of the two in any specific field. However, public art is seen to be unappreciated much like landscape architecture is. But then, by now, great efforts have been made to indicate the role that landscape architects play in helping public art work.

In this instance, there are many good examples of the combination between artists and landscape architects to make a difference. Janet Echelman the famous sculptor and artist who graduated from Harvard University and was named an Architectural Digest Innovator in 2012 for changing the very essence of urban spaces (Echelman, 2016), always says that the landscape architect can play a leadership role in innovating a space for her art work. She is always the person who is charge. Echelman Studio explores the cutting edge of sculpture, public art, and urban revitalization.

The design team usually collaborates with creative architects, landscape architects, mechanical engineers and lighting designers. Furthermore, their design often focuses on creating a large scale public art work of contemporary sculpture. These sculptures embody collective identity and attract residents to form a personal and dynamic relationship with the art and place. In the Vancouver Project, Canada, Janet Echelman said that Phillips Smallenberg, the seasoned landscape architect was the leader of her work. He designed a creative landscape architecture product for the Vancouver convention centre, which is covered with a Six-Acre green roof to make a lovely space for her art work. Echelman and Smallenberg collaborated and integrated the art ideas and the design concepts into the landscape. Especially in that project, Smallenberg redesigned the idea several times to adjust the water garden so that Echelman's art work became remediating.

Art critic Mary Louise Schumacher describes this project as in 'A social space that is simultaneously physical and virtual'. Many visitors came from all over Canada to see that huge and magnificent art work and engaged with each other in the surrounding space around (Figure 5). Since that time, the Vancouver convention centre has held annual public art events to please their visitors and raise the prosperity of the city.

Another exciting combination between public art and landscape architecture has begun with Echelman and Weiler to produce a contemporary project in Dilworth Plaza, Philadelphia (Figure 6). The both qualified artist and architect have added to the detailed work of the city hall's historic architecture with a creative virtual Rothko painting in the landscape area. The art work innovates with layers of coloured light moving in water mist, that physically and psychologically lead the people on a path through the underground lines below and traces them above the ground in real time (Harris, 2012).

Enthusiastically, Egypt is on the track. More and more new communities now believe in public art annual and seasonal events in their public spaces, to attract people to connect and socialise. It is clear findings, that events play an important role to promote the economic development and the human well-being for these communities. In addition to, they help to draw up the evidence of how public art could be a great element in place making and contribute well to landscape design and urban restoration.

In this instance, Sodic Residences, one of the greatest real estate companies in Egypt, has a missioned to that offering a variety of open spaces, overlooking green open areas and recreation plazas as the best way to sell a liveable community and a sustainable and maintain neighbourhood (Sodic website). According to their mission, the Sodic design team has created West Town Hub Project in one of their successful communities, Sodic West, Cairo-Alexandria desert road, to be a place where people can share and cooperate (Figure 7).

In the beginning, the project suffered a lot to make a foot print on the way to success, because it would depend only on the architecture landscape elements of recreation centres such as restaurants, cafes, market places, plazas, and outdoor areas. However, it has achieved a great boon after involving public art performances and events to these ingredient landscape areas.

Figure 5. Jelly Fish, Vancouver Centre, Canada.

Figure 6. Contemporary public art in Dilworth, Philadelphia.

Figure 7. West Town Hub, Sodic West, Egypt.

In that way, West Town Hub became the famous place that always opens a room for the individuals and institutions of art, to produce their innovative ideas in contemporary art with a lovely environment for residents and visitors.

## 5 DISCUSSION AND RECOMMENDATION

The information drawn from the study indicated that people gravitate towards other people. Besides the proof that public spaces are the places which can best provide them a healthy and happy community to interact with each other, these spaces are the life-blood of a successful society and they have become an essential component of any community.

Urban designers, landscape architects, and city planners have been devoted more on creating social spaces that encourage residents to interact and find common interests. Moreover, enormous efforts have been made to develop these public spaces in order to improve the community civic participation and healthy living.

As another way to promote the social connections in public spaces, new cities, stakeholders, planners, and organisers recognise that public art, along with landscape architecture, is one of the crucial tools the community can use to build strong meaningful interaction between people and places.

Through the development of creative community design, seasoned artists are commissioned to innovate interesting products of public art works that are helping to enhance the scene of the place and shape the city.

In this respect, the study has discussed many examples of public places in various communities which depend on public art work to attract lots of visitors and help residents to connect and socialise in open spaces. Meanwhile, these art works contribute to raise up the civic participation and urban revitalisation in those communities.

Therefore, it is obvious to suggest the involvement of public art work with the landscape design of public spaces to promote the social interaction between people. This paper can provoke an interesting question about the correlation between art and architecture to enhance the collective identity of the creative community.

## 6 CONCLUSIONS

Through the investment and the development of successful community, public art can contribute to urban planning and landscape design to create a serene environment where people can make a social interaction. That increases the community attachment which engages people with their place.

Today, many new communities have recognised that involving public art in master plans increases the opportunities to build a creative society. Public art is ideally matched and attuned to its social and environmental context.

Public art investment and integration can assist in overcoming of economic, and physical challenges by increasing the social interaction in public space. The relationship between the economic health of a community and the quality of its connection has been increasingly strengthened.

This paper has concluded that public space has become one of the main ingredients in any successful community. It is the place where the community residents actually live and attach to their community. As well as this, public art has been considered nowadays as one of these ingredients needed to create a liveable community from which it draws its identity. Furthermore, public art administrators and cultural planners are being tapped to collaborate with urban planners and landscape architects in designing innovative public places which encourage the community residents to socialise and connect.

## REFERENCES


Andrew, K. (2012). Huge rainbow art piece aims to bring Culver City oohs and Oz. *Los Angeles Times*. Articles.latimes.com. Access Feb. 2017.

Bach, Penny B. (1992). *Public art in Philadelphia*. Pennsylvania, USA: Temple University Press.

Earls, F. & Carlson, M. (2001). The social ecology of child health and well-being. *Annual Review of Public Health,* 22(1), 143–166.

Echelman, J. (2016). *Every beating second sculpture*. USA: Janet Echelman, Inc. Available online at http://www.echelman.com.

Frenchman, D. (2004). Events-Places in North America: City meaning and making. *Places*, 16(3), 35–49.

Harris, M. (2012). Philadelphia's water supply: how human capital and collaboration can overcome significant man-made challenges. *Journal of Organizational Dynamics*. 2(1): 21.

John, S. & James, L. (2010). *Grand Forks*. North Dakota, USA: Gallup, Inc.

Sodic Projects. Sodic human development at http://sodic.com.

Whyte, William H. (1988). The social life of small urban spaces (documentary film). Available online at http://vimeo.com/21556697.

*Cities' Identity Through Architecture and Arts – Catalani et al. (Eds)*
 *ISBN 978-1-138-05409-7*

# The miniatures of Lahore Darbar

Ayesha Kamal Bhatti
*University of Gujrat, Punjab, Pakistan*

ABSTRACT: This research analyses the court miniatures painted during the reign of Maharaja Ranjit Singh, otherwise known as The Lahore Darbar Miniatures. Furthermore, the essay highlights the main characters of the Darbar depicted in these paintings, which helps us to understand the medium in layers as well as the industry which was formed to aid this school of paintings. The city of Lahore plays a vital role in the identity of these paintings, as not only did the city work together to produce these paintings, but the paintings have become evidence of the city's history and its changing hands of power.

Throughout India and Persia, numerous styles and schools of painting were practised. When Maharaja Ranjit Singh came to power and claimed his throne in the capital of Punjab, Lahore, he decided to bring painters from all over India and Persia and settle them in Lahore to form ateliers. The amalgamation of artists, schools, material, techniques and ideals of beauty, power, authority and the holy come together to form what would be called the Lahore School of Miniatures or the Miniatures of the Lahore Darbar.

Working with two major museums of Lahore, the Lahore Museum and the Faqir Khana Museum, this research studies the material and the ideological aspects of these paintings and shows how the painting itself becomes evidence of a city's artistic triumph.

Using a few case studies representing a microcosm, the story of the painting and the city in the story are revealed.

*Keywords*: Court Miniatures, Maharaja Ranjit Singh, Lahore Darbar Miniatures, Darbar, Lahore School of Miniatures

## 1 INTRODUCTION

### 1.1 *The formation and development of the Lahore Darbar Miniatures*

Painting and sculpture prevailed in the city of Lahore for a long time. Specifically about painting, we know that during the reign of the Mughals in the 14th century when Akbar moved his court along with his artists and ateliers to Lahore, there was an established style followed by the painters in Lahore. The evidence of this lies in the famous project of Hamzanama which was an illustrative depiction of the tales of Amir Hamza from the 'Tilism Hosh Ruba'. This was a big project, and though headed by the Persian painters, local painters from Lahore were employed. The paintings produced by the artists from Lahore had a distinct style where individuality of faces of characters are apparent, unlike the traditional Persian style where the faces are almost identical. This style later became a part of the Mughal Style of Miniatures.

By the 19th century, this particular 'Lahori' style of miniatures reached its peak under the reign of Maharaja Ranjit Singh.

Punjab was divided into 12 misls or estates, owned by their respective leaders or sardars. Lahore, which was in a miserable condition and under the tyrannical rule of the Bhangi Sardars (Lehna Singh, Gujjar Singh and Shobha Singh), reached out to Ranjit Singh to be its saviour. Maharaja Ranjit Singh was a young sardar and was known for his gallantry, wit and leadership qualities. He marched into Lahore and with least resistance took the reins of the city in his hands. On 12 April 1801 he was crowned the Maharaja of Punjab at the venue of Lahore Fort. He ruled until his death in 1849.

Initially called Darbar Khalsa, its name later changed to Lahore Darbar as the Maharaja stressed on the secularity of his court. The 48-year rule of Maharaja Ranjit Singh is considered the golden period of Punjab. Lahore and its environs were peaceful, safe and prosperous. Maharaja Ranjit Singh, in an attempt to update his methods of administration, employed and imported experts from all trades. He employed European army officers who had served in the French army to train his men. Likewise, he also sent for artists from various adjoining areas. In an interview with Faqir Syed Saif ud Din, he mentions that Maharaja Ranjit Singh issued a farman or official orders to Faqir Noor ud Din who was the governor of Lahore, to bring skilled experts and technicians from various areas and settle them in Lahore and Amritsar.

Artists were settled in specific localities in Lahore and Amritsar. In Amritsar, there was a 'Gali Musawwaran' (Artists' street), while in Lahore these artists set their ateliers or studios in 'Gumti Bazaar' ('Gumti', the name of the market, originated from the fancy cloth which was sold in bulk there) which was surrounded by the residences of the well-heeled patrons. Also, studios were set up at 'Tehsil Bazaar' (inside Mochi Gate) which received generous patronage by the Maharaja. There were over 120 functioning art schools/studios and more than 60 ustad (teachers of calligraphy) during this time, some of these art schools predating the Sikh rule.

Some prolific artists, like Perkhu, travelled with their patrons like Raja Sansur Chand of Jammu, who moved to Lahore for political reasons.

When these artists from the schools of Kangra, Guler, Chamba, Bhasoli, Jammu, Lucknow, Delhi, Agra, and Rajastan, with their distinct styles and a diverse visual vocabulary, came together and mixed with the already prevailing traditions in Lahore, they produced paintings which showed the blending of style and visual understanding. This became the signature style of the Lahore Darbar School of Miniatures.

## 2 ARTISTS, ATELIERS AND TECHNIQUES

Whenever establishing the definition of the Lahore Darbar Miniatures, we see a combined contribution of the artists, ateliers, techniques and a chain of historic events which leads us to the development of this school.

Artists who had been imported from different areas and schools around India and Persia came with their sensibility and ideals of beauty. When these artists eventually settled in Lahore, the interesting features began to show. Suddenly, not only do we see the regional styles of the artists showing in the paintings, but also certain characteristics that define the style that becomes the signature of all paintings made in Lahore.

This Lahori Ang, or Lahore Style, had to be integrated for the favour of the patron and became a formula for commercial gains. The integration of multiple schools made the Lahore School of Miniatures an interesting mélange of different styles and yet remained the paintings of Lahore Darbar.

With the different schools of miniatures coming together in one school, the flavour of work became individual, multi-regional yet remained Lahori. What follows now are the characteristics of the Lahore School of Miniatures and the factors that form its distinctive style.

### 2.1 *Colour*

A definite colour palette is used in the Lahore School of Miniatures. Bright and vibrant colours were selected for the paintings. Bright red, blue and yellow were used abundantly and reflected upon the opulence of the patron. Since gems were used in the making of pigments, they were colour-fast and did not lose most of their brilliance over years. It is not that these colours or pigments were not used before; they were not used in such abundance and not in such saturated amounts.

The characteristic colours of the Lahore Darbar Miniatures are bright blue, taken from ground lapis lazuli, bright red, yellow taken from a stone locally known as gattu, and with generous use of gold and silver leaf. White, which was usually used as a base and commonly known as safaida, was made from ground pearls. Even today, we can see the brilliance of these pigments.

### 2.2 *Portraiture*

Portraiture of Lahore Darbar paintings did not adhere to the idealised Chinese features which were fondly made by the Persian miniaturists. They also did not stylise the face, as did their other regional Indian counterparts, but made individual and accurate features. This was partly because most of these paintings were made for historiographical reasons and all those present in the paintings were well recorded and accounted for. As a matter of fact, names were also recorded on the frame of the paintings to retain the record of the characters present in the painting. This tradition can also be seen in the sculpture tradition of Lahore where accurate features were made.

In Figure 1 the faces of the Imaedin religiou scholars) show individual features along with distinctive attire. Notice the variations of skin tone, face, and hair types along with the character shown on their faces. Along with the faces, these paintings show the individual dress sense observed by the members within a group.

Along with the accuracy of features, the character study of a person is also evident through the paintings in subtle and sometimes vivid details. This is a unique feature of Lahore Darbar paintings, and contributed greatly in the narrative which was created in the paintings. Narrative was given more importance than the adherence to the realistic depiction of the scene, yet the realistic depiction in portraiture was demanded by the patrons. This left very little room

Figure 1. Individuality and portraiture: Syed Ghulam Noshasani 'Faces of the Imaedin' (Religious Scholars).
Detail: DarbarFaqir Syed Ghulam Moheyeddin Noshasani, Faqir Khana Museum, Lahore, Pakistan.

Figure 2. Character study of the courtiers involved in the murder of Maharaja Sher Singh.
Detail of Darbar Sher Singh, Artist: Kher Singh, Faqir Khana Museum, Lahore.

to manoeuvre around stylising the characters enough to tell the story. In other Indian schools of miniatures which are heavily stylised, the archetypal characters are given stylised features, establishing their role in the story.

Still, the painters of the Miniatures of Lahore Darbar possess the unique quality of showing character traits through postures and detailed facial features.

Notice the faces of the members of Sher Singh's court. Distinctive individual attire is noticeable as well as the facial features. Individual hand gestures can also be seen in this painting. The most interesting part of the details of this painting is that these characters were involved in plotting the murder of Maharaja Sher Singh. The entire painting told the story of the murder being plotted in the court of Sher Singh. The names of the persons present in the painting are written on the border of the painting. This was an important record-keeping technique which allowed the painting to preserve the historic moment.

Taking a closer look, you may notice the flared nostrils and focused gaze of the characters.

Notice the sneer on But Singh's face as his nostril is flared and the curve of his moustache arches sharply over his mouth. This shows the ferociousness of the character who is waiting to pounce on his prey. But Singh and his brother Lehna Singh were instrumental in the assassination of Sher Singh.

Characteristic physical deformity is also shown here in Figure 6, as Raja Dehan Singh's sixth finger is carefully rendered as it oddly springs out. The artist delicately comments on the unnecessary presence of the villain who sticks out like an ugly sixth finger; though his hands are closed together to show respect, his devilish plot cannot be ignored, just like his sixth finger. Raja Dehan Singh was the mastermind behind the assassination plot against Sher Singh.

Figure 3. Detail of But Singh's face from Darbar Sher Singh.
Artist: Kher Singh, Faqir Khana Museum, Lahore.

Figure 4. Detail of Raja Dehan Singh and his six-fingered hand from Darbar Sher Singh.
Artist: Kher Singh, Faqir Khana Museum, Lahore.

Figure 5. Detail of Faqir Ghulam Moheyeddin Noshasani.
Artist: Perkhu, Faqir Khana Museum, Lahore.

Figure 6. Detail of Bhai Ram Singh's clothes from Darbar Maharaja Sher Singh,
Faqir Khana Museum, Lahore.

Iconography from various cultures and religions is evident in the image-making as well as establishing character traits of the person. European cupids are presented as angels, and halos around revered persons are shown even in the Mughal School of Miniatures, but the presence of a Muslim saintly figure in the posture and mien of Baba Guru Nanak, the founder of Sikh religion, is something unique. Using the local/regional religion of Sikhism as a point of reference for the holy, shows the amalgam of the school of thought and ideals of divine reverence.

Figure 5 shows the Faqir Ghulam Moheyeddin Noshasani seated in a position and represented with a likeness of Guru Nanak. Faqir Ghulam Moheyeddin Noshasani was a Sufi scholar and a poet, and was revered by people of all faiths. His legs are showed uncovered up to the calf, which matches the representation of Guru Nanak. Compositionally, it appears he is sitting under a tree which is a common archetype for representing teachers and people of faith in the Indian subcontinent. This Sikh iconography is then mixed with the European Christian tradition of placing a halo behind the head of the revered person. This halo is also present in the Mughal miniatures from the time of Akbar. When the Europeans illustrated bibles in India, they brought their iconography which was picked up by the artists of the Mughal court. Following the same tradition are the painters of the Lahore Darbar. Another European iconographic tra-

Figure 7. Detail of embossed pearl rendering on Sher Singh's jewellery, Darbar Sher Singh. Artist: Kher Singh, Faqir Khana Museum, Lahore.

dition can be seen in the background in the form of an angel who is showering petals and gold. The angels had also appeared in Mughal miniatures before. The act of showering petals and gold was a common practice in India to show reverence, respect and the showering of blessings.

The secular court of Maharaja Ranjit Singh had created an example of tolerance and moderation, and so it is evident in these paintings how cross-cultural imagery comes together in one composition.

Another trend of portrait within a portrait is exercised in the painting of Maharaja Sher Singh.

## 2.3 *Three-dimensionality of the painting*

A strong sense of three-dimensionality is also found in the paintings, which is rare in this particular school of miniatures, where otherwise flat washes or heavily textured fabric was used to show the flatness of the figure. This created a three-dimensional study of the body which would be further rendered with the design of the cloth that followed the curves of the body beneath it.

The three-dimensionality is also evident in the treatment of canvas. Finger nails were used to slightly emboss or sink the details on the paper. This tradition of treating the paper is believed to have come from the calligraphy done through the same technique known as Khat e Nakhoun. Calligraphers were a revered part of the ateliers and through the introduction of this technique calligraphers of Khat e Nakhoun became a part of the image-making process.

Jewellery, like a string of pearls, would be painted using real ground pearls and applied thickly to appear embossed.

As seen in Figure 7, embossing with thick layers of pearl pigment to make pearl necklaces in the painting created a slight relief. When observed very closely, these thick layers of white paint appear like rounded pearls.

These kinds of techniques have not been practised in any other school of miniatures and Lahore Darbar School stands unique in producing such interesting works.

## 2.4 *Perspective*

The intention of the painting was to tell the entire story or the happening in one visual. This required certain sacrifices. Real perspective was one of them.

The perspective in these paintings is not actual, but as perceived by the artist. The interior and exterior view is sometimes required in the same painting. Though this perceived perspective was heavily gauged by the patron, the artist did not lose his lyricism and logic. The compositions still balance out harmoniously and do not lose focus of the most important characters.

In many paintings, the interior view of court is juxtaposed with the exterior view through the window. The perspective is almost the same, whereas multiple perspectives are shown inside the court. Likewise, in some paintings, piled-up perspective is paired with the one-point

Figure 8. Perspective flow, Darbar Maharaja Ranjit Singh, Lahore Museum.

perspective to highlight the important areas. The dimensions of the characters vary according to their rank and importance.

## 3 INDIVIDUAL PICTURE BREAKDOWN

### 3.1 *Maharaja Ranjit Singh in Shikarpur, Lahore Museum collection*

This painting depicts the historic event which took place at Shikarpur, Sindh. Maharaja Ranjit Singh planned to extend his empire towards Sindh. On advancing towards Shikarpur he was stopped by the British, who sent their Wakeel (lawyer) or representative to stop him in his tracks and to request him to go back.

On hearing this, Maharaja brought out all the contracts and told him there was no law stopping him from this invasion, but he really lost his temper when the Wakeel, Basharat Khan, suggested that the Maharaja's greed was getting the better of him. Ranjit Singh pulled out his sword and told the Wakeel off by saying that he had a birthright to his land, that the British had no right to claim what was not their motherland, and that it was indeed the British whose greed was getting the better of them.

The following are the numbered details of the painting:

1. Maharaja Ranjit Singh in his winter apparel sits with his sword out
2. Basharat Khan—Wakeel (lawyer) of the British
   2.1. Smudged feet show the artist's mistake in depicting shoes, whereas no one was allowed to wear shoes inside the court.
3. Phula Singh Akali—Trusted and respected general of the Akalis
4. Akali Army—The diehard Sikh army, loyal to Ranjit Singh, who used to fight with religious zeal. They were led by Phula Singh Akali. (3)
5. Divan Ratan Chand—Minister of the court, who was added later by the artist
6. Quilted curtains depicting that it was wintertime
7. Clouds made in Rajasthani style of miniatures
8. Floor design is typical to that of Lahore

This miniature shows the time of the year by showing elements of weather. Heavy clothing is worn by the courtiers, while the windows are covered with quilted curtains (6).

Figure 9. Numbered details of the painting, Maharaja Ranjit Singh in Shikarpur, Lahore Museum, Lahore.

### 3.2 *Darbar Maharaja Sher Singh, Kakir Khana Museum collection*

This painting is phenomenal in its subject. The artist Kher Singh paints the court of Maharaja Sher Singh before his assassination. As the plot against the king develops, the attendees of the court are brewing up plans for the assassination. The right half of the painting consists of the villains, while in the left half are the victims. Kher Singh draws himself as one of the victims who foresees the situation. A careful study of the faces of these villains shows intense hatred as Kher Singh renders his focused gaze on Maharaja Sher Singh with flared nostrils.

Showing the view from the north through the Naulakha pavilion at the Lahore Fort, we can see the River Ravi, Kamran's Baradari, Jehangir and Asif Jah's mausoleum.

On the top left corner of the painting smoke rises from the direction of the city. Tiny angels are being sucked into this smoke showing that even the king's guardian angels cannot help him anymore.

The interesting combination of one-point perspective and piled-up perspective adds to the narrative when the interior of the Naulakha Pavilion opens the view into the placement of the Old Ravi River, mausoleums of Asif Jah and Jehangir (Mughal rulers), and at the same time comes back to the city of Lahore which smoulders at the thought of the assassination of the rightful king. This is a very strong propaganda painting which legitimises the rule of the Sher Singh, and then attaches it to the death of the former great rulers who ruled from the same fort Sher Singh and his attendees are sitting in.

1. Maharaja Sher Singh—Son and successor of Maharaja Ranjit Singh.
2. Bhai Ram Singh.
3. Raja Dhian Singh—Mastermind behind the assassination of Sher Singh. Famed to be six-fingered; a closer look at his hands shows his sixth finger branching out of his thumb.
4. Wahbian Singh.
5. Raja Heera Singh.
6. Attar Singh Kalianwaley.
7. Sardar But Singh Sandhianwala—Active participant in the assassination plot.
8. Sardar Lehna Singh Sandhianwala—Brother of But Singh and actively involved in the plot.
9. Kher Singh—One of the leading court painters of the time and artist's self-portrait.
    9.1. Kher Singh's Album on lap where he is pointing towards another miniature of Sher Singh riding his favourite horse Dhollu (this detailed miniature exists separately).

Figure 10. Numbered details of Darbar Maharaja Sher Singh, Faqir Khana Museum, Lahore.

10. The Naulakha Pavilion of the Lahore Fort. Opposite Sheesh Mahal.
11. River Ravi, which flowed right next to the Lahore Fort in those times but later changed its course.
12. Kamran's Baradari—Here shown as double-storied building. It now stands with one storey and in the middle of the River Ravi.
13. The Minarets of Jehangir's mausoleum, seen through the trees of Shahdara. Next to it is Asif Jah's Mausoleum (can be seen very faintly).
14. Rising of a smoke cloud from the side of the walled city of Lahore with small angels, showering flower petals, being sucked into the smoke cloud.

## 4 CONCLUSION

The Miniatures of the Lahore Darbar established the industry of book arts in Lahore like none before. These also become one of the true first-hand representations of the Sikh court of Punjab. A Maharaja's vision to unify his people and bring tolerance for all cultures and styles is evident in how he encourages the embracing of multiple philosophies and schools of Indian Miniatures, and how in doing so, records his feats, his companions, his enemies and the city he loved the most. Lahore shines brilliantly in these paintings along with her ruler.

## REFERENCES

Darbar e Khas Hazrat Faqir Syed Ghulam Mohiuddin Shah Sahab Nosha Sani. (n.d.). [gouache on wasli] Lahore: Faqir Khana Museum.

Darbar of Maharaja Ranjit Singh. (n.d.). [gouache on wasli] Lahore: Lahore Museum.

Din, Faqir Syed Saif ud, interview by Ayesha Kamal Bhatti. *The Miniatures of the Lahore Darbar* (December 10, 2011).

Khalid, D. (2011). *The Miniatures of Lahore Darbar*.
Khalid, Prof. Kanwal, interview by Ayesha Kamal Bhatti. *The Miniatures of Lahore Darbar* (December 10, 2011).
Maharaja Ranjit Singh in Shikarpur, Sindh. (n.d.). [Gouache on Wasli] Lahore: Lahore Museum.
Saifuddin, F. (2011). *The Miniatures of Lahore Darbar*.
Singh, K. (n.d.). *Darbar Sher Singh*. [Gouache on Wasli] Lahore: Faqir Khana Museum.
Singh, K. (n.d.). *Sher Singh on a Horse*. [Gouache on Wasli] Lahore: Faqir Khana Museum.

*Cities' Identity Through Architecture and Arts – Catalani et al. (Eds)*
*© 2018 Taylor & Francis Group, London, ISBN 978-1-138-05409-7*

# The potential of the eco-passive construction technique for the Western Desert in Egypt

Mohamed A. Mahdy & Magdy M. Mousa
*Department of Architecture, Faculty of Fine Arts, Alexandria University, Egypt*

Ahmed R. Abdin
*Department of Architectural Engineering, Faculty of Engineering, Cairo University, Egypt*

ABSTRACT: A proposal for the eco-passive construction technique has been discussed to maintain thermal comfort in hot, arid climates. The study focuses on the case of Kharga, Egypt – climate classification BWh—for its strategic location, as well as for the availability of data and accurate weather data files for the Western Desert Region in Egypt. The proposal discusses an integrated multi-layered wall using eco-friendly insulation (rice husk), thermal storage materials, and Bio Phase-Changing Materials (Bio PCM), with a total thickness of 24.6 cm (one-brick), using a computerised simulation for three prototypes of multi-layered walls applied to three rooms. According to their geometrical roof shape (flat, domed, vaulted), the simulation and discussion are consequently run across three stages, which leads to successfully achieving comfort according to the Adaptive Comfort Model in American Society of Heating, Refrigerating and Air-Conditioning Engineers (ASHRAE) Standard 55-2010 for the Kharga climate (19.6°C–30.7°C) with 100% comfort operative temperatures in summer and 91% in winter.

*Keywords*: Eco-Passive Construction; Thermal Comfort; Eco-Friendly Insulation; Thermal Storage Materials; Bio Phase-Changing Materials; Bio PCM; ASHRAE

## 1 INTRODUCTION

Many aspects of globalisation have a direct impact on the built environment in general and specifically on the control of the architectural environment. Aspects of the globalisation process

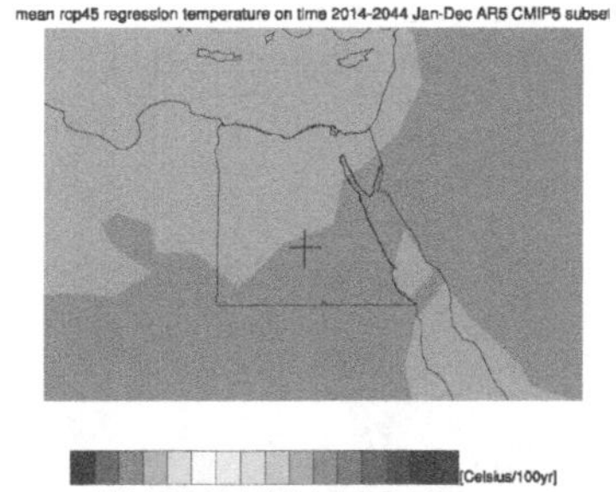


Figure 1. Prediction for the next 30 years of the rate of climate change in Egypt. The map illustrates that the Kharga oases zone, where the study takes place, is under a change rate of 4°C/100 years. The prediction was simulated by the KNMI Climate Explorer web-based tool. The cross refers to the location of Kharga (KNMI, 2014).

affect, and are affected by, factors like business and work organisations, economics, sociocultural resources and the natural environment (Babones, 2007). Environmental challenges such as global warming, climate change and air pollution, and overfishing of the ocean are linked to globalisation (Bridge, 2002). Climate change is impacting on our economy, health and communities in diverse ways. Over the past 50 years the average global temperature has increased at the fastest rate in recorded history (NRDC, 2011) (see Figure 1). In addition, it is impacting on the architectural identities of nations that document their history and shape their unique character. The Kharga Oasis in the Western Desert Region in Egypt is lacking the required number of people to settle due to the hot arid climate which characterises this territory. The climate is characterised by high and wide diurnal temperature fluctuation. At certain points around the year, the temperature difference can reach up to 20°C between day and night. This leads to insufficient thermal comfort in the territory, and it is interrupting the governmental plans to encourage people to settle in this territory, with fears that they may abandon it because of insufficient thermal comfort. Furthermore, this wide diurnal fluctuation affects human health with chronic kidney disease (Barsoum, 2013, Kalaitzidis et al., 2014).

## 2 RESEARCH HYPOTHESIS

The construction materials and the thickness of walls and roofs are important to consider in terms of their thermal resistivity. The architect Hassan Fathy presented a project proposal to New Bariz in the 1960s for settling people in the Kharga territory, representing the folkloric style that was accumulated from the Pharaonic, Christian and Islamic times. This would be carried out using a passive design approach. However, due to the economic and political issues at the time, this planned development was not completed (see Figure 2). During that time, Fathy's strategy to design passive buildings for hot, arid climates depended on the use of local materials such as mud bricks, and in Kharga to have wall thicknesses of up to 80 cm, to maintain thermal comfort with the appropriate fabric inertia. In addition, he used the geometrical roof structure of domes and vaults to diffuse solar rays in order to maintain his initial aim of reaching indoor comfort zones. The experiment of Hassan Fathy in Cairo (Fathy, 1986) revealed that a wall thickness of two mud bricks can act as a thermal energy storage system, a thermal lag for heat dissipation, and a thermal insulator for heat dissipation at night (see Figure 3). The experiment also revealed that this technique is more beneficial in controlling temperature fluctuation than that of prefabricated concrete slabs with a thickness of 10 cm. This was an example of ingenious urban development in the Western Desert Region in Egypt. However, at the moment, mud-brick structures alone may not be the most durable and applicable for the new century's urban development in Kharga. According to Harper (Harper, 2011), questions have been raised about mud brick's long term durability and susceptibility to water damage. In addition, it is commonly known that the overuse of mud brick as a core construction material in Kharga affects the sustainability of the agricultural cultivation in this location (*Attia* and *Raslan*, 2011).

Figure 2. Architect Hassan Fathy used mud brick as a local construction material in the project New Baris Village, Kharga, Egypt. His design represents an ideal expression for the architectural identity for this particular location. (Archnet, 2014).

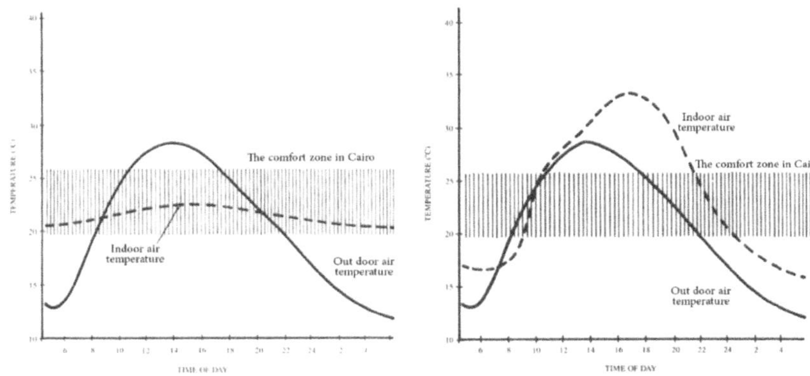


Figure 3. An illustration of a comparison between heat flux rate over 24 hours, both indoors and outdoors, by using a 50 cm mud brick (right), and a concrete wall (left). The graphs show the comfort zone in Cairo (Fathy, 1986).

## 3 RESEARCH BACKGROUND

Several researches on this concept have been reviewed. For example, Tyagi and Buddhi (2007), Pasupathy et al. (2008), and Whiffen and Riffat (2012, 2013) discussed the different forms of thermal energy storage integrated with building an envelope, whether it is for the targeted climate zone in this paper (hot, arid climate) or as a general approach for other climates. By giving examples without any limitations, Tyagi and Buddhi (2007) discussed passive solar and heating/cooling systems which can work through three main functions: solar energy collection (wall external surface), storage (Thermal Energy Storage (TES) of multi-layered materials), and distribution (internal surface material). Storage can be maintained through the multiple layers of a wall, as the construction materials such as brick, stone and concrete can be utilised as sensible forms of heat storage, or PCM (Phase Change Material) as latent heat storage. Passive storage systems for heating can be used through two strategies with wall construction. The first one is 'Direct Gain' and the second one is 'Indirect Gain'. The Direct Gain concept can be explained as receiving heat through a single-layer external wall surface (solid–solid phase change method), while Indirect Gain can be maintained by utilising the thermal mass storage wall between direct solar radiation and the living space, whether with sensible heat storage using bulk materials or latent heat storage through a layer of PCMs (solid-liquid phase change method). The literature review related to this topic assisted in conducting the research methodology below to fulfil the required research aim.

## 4 RESEARCH AIM AND METHODOLOGY

This research investigates the feasibility of replacing the very old and thick construction wall that uses earth materials for the vernacular architecture context in Kharga with an equivalent thin lightweight multi-layered wall, to be used as an internal and external wall for multi-storey buildings in the Western Desert Region in Egypt (see Figure 4). The research identified organic non-paraffinic PCMs (called bio-based PCMs), which are nano-engineered (Yu et al., 2014, Jeong et al., 2013, Jeong et al., 2014) and derived from natural resources such as palm oils. They are expected to be stable for thousands of phase-changing cycles. Bio-based PCMs can absorb, store and release large amounts of latent heat, like general paraffinic PCMs. They can be manufactured to have various melting points varying between −22.7°C to 78.33°C, so they can be used in different applications and in different climate zones (Phase Change Energy Solutions, 2014). However, they still has low thermal conductivity compared to the conventional organic PCMs. The research used the climatic computer

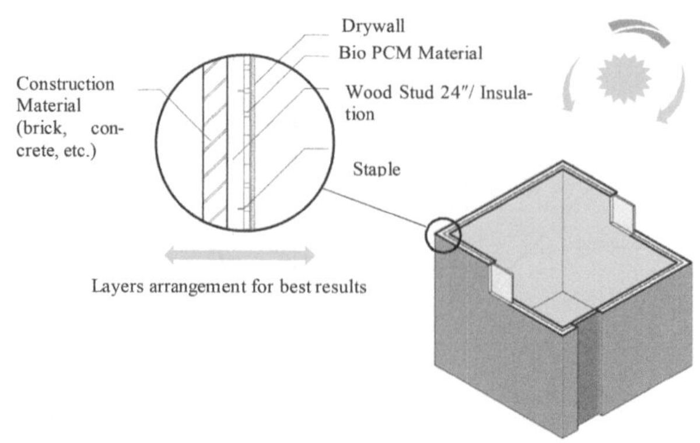


Figure 4. This illustration for the concept being studied shows the proposed wall construction layers. Details are redeveloped by the researcher based on (Phase Change Energy Solutions, 2013f).

simulation method to introduce an application of simulation that data sheet information of the defined Nano PCM layer product which is integrated with mud-brick walls to investigate operative indoor temperatures for a thinner integrated multilayer wall, and to compare with those temperatures produced by using the conventional 50 cm walls. The performance was studied in all seasonal conditions using the latest typical meteorological year data (Energy, 2003) for exterior boundary conditions in Kharga, Egypt. Comparisons were made between different cases of layer arrangements and parameters to investigate certain arrangements of layers. To continue with these various variable parameters, some parameters were aimed to be fixed, such as the rooms' boundary conditions (represented in the floor area), opening areas, and air volume and flow rate. The comparison is hypothesised to be limited and represents a precise thermal simulation for the multi-layered proposals.

## 5 COMPUTER MODEL

Three prototypes of multi-layered walls are proposed to be applied to three different rooms (see Table 1). Several computerised simulations were carried out and led to the final three proposals in Table 1 for discussion. With reference to the room's geometrical shape of their roof (flat, domed, vaulted), see Table 2, the proposal discusses the integrated mud-brick multi-layered walls and roof using eco-friendly insulation (rice husk), and the thermal storage materials (Bio PCMs); the exception is for the flat roof room, which is reinforced using concrete. The investigation took place using fixed and variable parameters and was then modelled and simulated through computerised simulation. The aim of designing these three experimental multi-layered surfaces is to investigate their impact on comfort temperatures, besides the impact of using different roof shapes with respect to the orientation of the sun. The fixed and variable parameters approach narrows the investigation to let the study target only the effect of the integrated proposed multi-layered construction materials and the shape of the roof.

### 5.1 *Parameters (Fixed and Variable)*

The study considered that all rooms have the same internal area of 7.840 $m^2$ and that they have the same internal volume of 26.6 $m^3$ as well. According to the advised dimensions by ECP 306-2005 (2006), the opening dimension has been designed to consider that the sum of all opening areas for each room is 3.48 $m^2$ (including doors and roof vents). The opening orientation was also considered as a fixed parameter. Based on climatic analysis for the location of Kharga, the windows and doors are positioned in all rooms to face the prevailing winds from the north-west direction in Kharga. Two windows are placed in the position to maintain cross-ventilation. All domes maintained stack effect ventilation using a crown on the

Table 1. Surfaces prototypes, illustrated by a detail in the roof of the domed room. Bricks are used in Egypt with dimensions of 25 × 12 × 6.5 cm (Quality, 2005). Mud bricks used to be used with standard dimensions of 24 × 12 × 9 cm in Egypt (Correas-Amador, 2012). According to (Sheweka, 2011), mud brick has a lot of economical, eco-friendly and thermal benefits as a core construction material in the location of study.

SURFACES PROTOTYPES

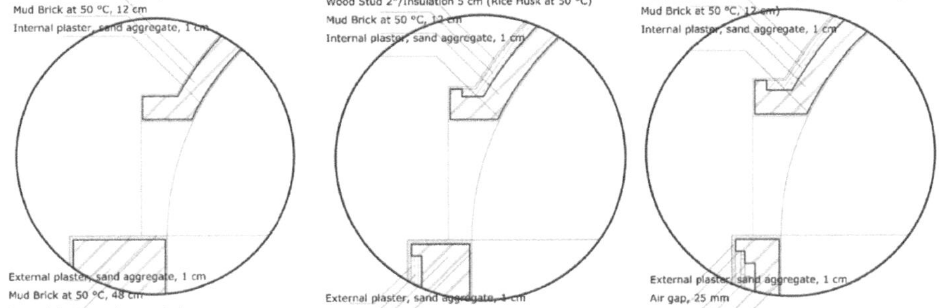

| | | | |
|---|---|---|---|
| WORKSHOP DETAIL (roof) | External plaster, sand aggregate, 1 cm<br>Mud Brick at 50 °C, 12 cm<br>Internal plaster, sand aggregate, 1 cm | External plaster, sand aggregate, 1 cm<br>Wood Stud 2"/Insulation 5 cm (Rice Husk at 50 °C)<br>Mud Brick at 50 °C, 12 cm<br>Internal plaster, sand aggregate, 1 cm | External plaster, sand aggregate, 1 cm<br>Wood Stud 2"/Insulation 5 cm (Rice Husk at 50 °C)<br>Mud Brick at 50 °C, 12 cm)<br>Internal plaster, sand aggregate, 1 cm |
| WORKSHOP DETAIL (wall) | External plaster, sand aggregate, 1 cm<br>Mud Brick at 50 °C, 48 cm<br>Internal plaster, sand aggregate, 1 cm | External plaster, sand aggregate, 1 cm<br>Wood Stud 2"/Insulation 5 cm (Rice Husk at 50 °C)<br>Mud Brick at 50 °C, 24 cm<br>Internal plaster, sand aggregate, 1 cm | External plaster, sand aggregate, 1 cm<br>Air gap, 25 mm<br>BioPCM® M51/Q29, 21 mm<br>Wood Stud 2"/Insulation 5 cm (Rice Husk at 50 °C)<br>Mud Brick at 50 °C, 12 cm)<br>Internal plaster, sand aggregate, 1 cm |
| THICKNESS (CM) | Total Roof thickness = 14 cm<br>Total Wall thickness = 51 cm | Total Roof thickness = 19 cm<br>Total Wall thickness = 32 cm | Total Roof thickness = 19 cm<br>Total Wall thickness = 24.6 cm |
| U-VALUE ($W/M^2.K$) | Roof = 5.423<br>Wall = 1.505 | Roof = 0.859<br>Wall = 0.748 | Roof = 0.859<br>Wall = 0.684 |

Table 2. Three rooms with three different roof geometrical shapes.

| FLAT ROOF ROOM ( REINFORCED CONCRETE WITH 2% STEEL) | DOMED ROOF ROOM | VAULTED ROOF ROOM |
|---|---|---|

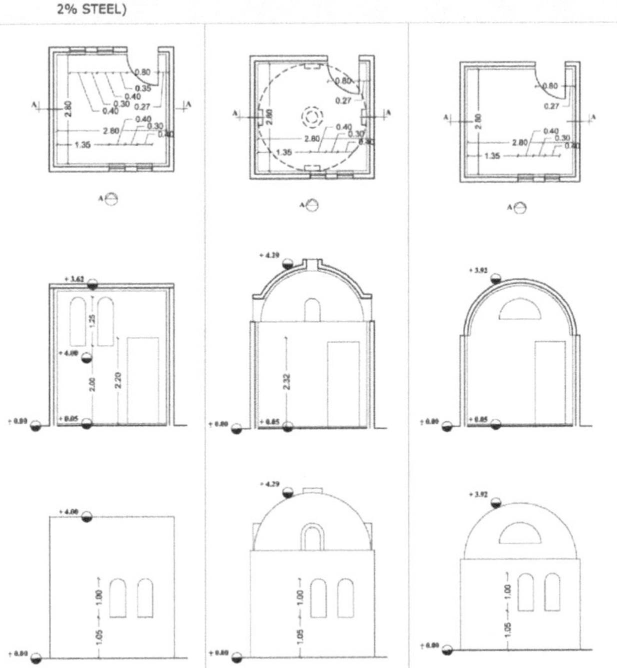

top of the dome and four vents positioned at the dome's axes. In the case of the vault, natural ventilation is maintained via two vents at the two flat surfaces of the vault. The vaulted roof is orientated to face the prevailing wind to maintain stack effect as well. The windows have a vertical rectangular shape to maintain the longest possible shading. All rooms have a floor made up of sandstone tiles (3 cm) fixed with mortar (2 cm). The occupation density differs from one place to another across Egypt, where it is related to the typological characteristics. The average number of family members in the rural areas in Upper Egypt is 4.37 capita and the residential space per capita is 13.7 $m^2$. The typological studies for the location of the study states that occupation differs across the day. The average occupation density in the location of study is calculated to be 0.072 people/$m^2$. The study considers an activity of eating/drinking, with metabolic factor 0.90, winter clothing has insulation effect of 1 clo, and the summer clothing 0.50 clo. A medium crack template for air tightness was considered with a constant rate of 0.700 ac/h. On the other hand, the study considered the following variable parameters:

- defining surface multi-layers
- shape of the roof
- surface layers orientation
- opening scheduling (night purging).

The study in this research has been discussed through a set of climatic simulation studies to utilise certain PCM material with its available data sheets, using Design Builder software v4.2.0.054 with the Energy Plus simulation engine of the accredited and the most up-to-date weather data file for Kharga. The software is approved by the building regulations bodies of various countries.

## 6 RESULTS AND DISCUSSION

The study investigated using numerous computerised simulations to produce the final three proposals, which are presented in Table 1 for results and discussion. The discussion will address the results through three stages. Each stage addresses each variable parameter that contributes to the particular stage.

### 6.1 *Stage 1*

This stage discusses the variable parameter defining multi-layered surface which is defines the layers of each proposed multi-layered case. At this stage, the discussion addresses the parameter's effect on a room's thermal comfort. Reviewing Bio PCM products, they are produced in different shapes, such as a mat of pouches or sticks. The mat is ideal for this study, as it is possible to be affixed onto curved surfaces or embedded into walls. The utilisation of the appropriate Bio PCM depends on the peak melting temperatures and its thermal energy storage capacity. Bio PCM is placed facing the maximum radiant, conductive and convective energy exchange. To determine how much Bio PCM is needed, first the cooling load is determined using loads software or a delta T over area calculation. Secondly, the total Bio PCM material that matches the thermal energy storage capacity cooling size for the space is specified. More PCMs than the design-cooling load is not typically required but can be used for increasing the thermal energy storage and peak demand time lag. Thirdly, the external surface exposure to solar gains is evaluated. Fourthly, the Bio PCM capacity is divided among surfaces, plus consideration is given to using materials of the highest capacity to face the highest exposure to heat gains (Phase Change Energy Solutions, 2013). According to the design heating load equation (Szokolay, 2014):

$$Q = A \times U \times \Delta T \quad (1)$$

where

- Q design-cooling load of building (W)
- U U-value surface to surface (W/$m^2$ K)

– A  area of surface (m²)
– ΔT  design temperature degrees difference

Cooling load is an hourly rate that refers to the capacity of equipment required to account for such load. It may seem logical to define that space heat gain is equivalent to space cooling load, but in practice 'Heat gain ≠ cooling load' (Bhatia, 2012). Thus, considering the maximum wall area of a flat room,

$$Q = 6.25 \times 12.752 \times 21 = 1673.7(W) = 5710.9014512 \frac{Btu}{h}$$

Then, the calculated average Btu/h value is 41.60 Btu/h per feet or M41.60 (according to the manufacturer's specifications) for the sum of the surface area. According to the communication with the manufacturer (Crossett, 2014a), the mentioned calculation above is valid as long as the cooling load is considered with the minimum value, which can be considered as being M51 among the available prototypes by the manufacturer (Phase Change Energy Solutions, 2013a). In a further communication with the manufacturer (Crossett, 2014b), this prototype has been confirmed for simulation for this study. More Bio PCM than the design-cooling load is not typically required but can be used to provide additional benefits by providing additional Btu/hr thermal energy storage and peak demand time lag. The average life cycle of the used bio-based PCM is 85 years. The study did not consider thermal bridging. A simulation was run to test hourly inside surface temperature in the middle day of both the summer design week (01 AUG) and the winter design week (25 JAN). Therefore, the study can stand on the feasibility of the three prototypes, compared with the outside Dry-Bulb Temperature (DBT). Results in Figure 5 confirm the fact that the temperature fluctuation in Kharga is 17–20°C. While the three proposed prototypes record swing in temperature that varies between only 2–4°C.

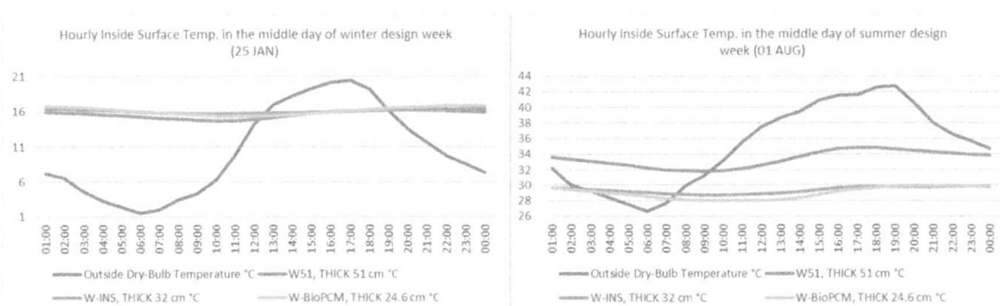


Figure 5. Results of hourly inside surface temperatures for the three proposed multi-layers prototypes. The data is for the middle day of the design week in summer and winter in Kharga.

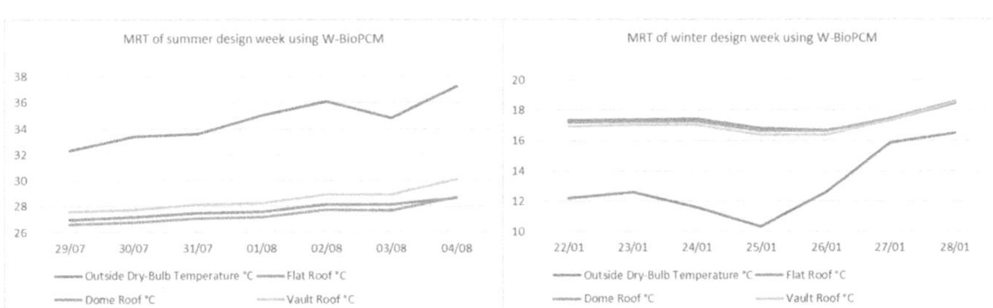


Figure 6. Result of comparing MRT for flat, domed and vaulted rooms in summer and winter design weeks, using prototype W-BioPCM.

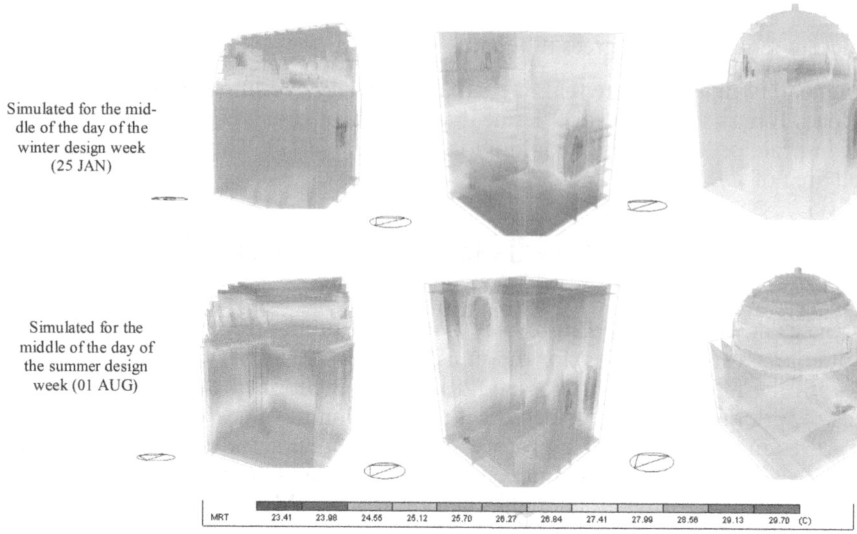


Figure 7. CFD simulation for contour temperature distribution for the three rooms.

### 6.2 *Stage 2*

Stage 2 discusses the variable parameter shape of the roof (flat, domed, vaulted) with the prototype W-BioPCM 24.6 cm thick, using the simulation method of solution algorithm finite difference. Figure 6 shows results for the summer design week. The average Mean Radiant Temperature (MRT) for the three rooms is lower than the DBT by 5–10°C, while the difference among the three rooms is AV 1°C. The domed roof room recorded the lowest MRT among the three rooms, while the vaulted roof in this particular room design and the windows orientation recorded the highest MRT. In the winter design week, the variation between DBT and MRT is 2–6°C; the difference in MRT for the three rooms is notably small.

Figure 7 illustrates a Computational Fluid Dynamic (CFD) simulation for the three rooms according to the temperature distribution of each room. During the summer, the MRT for the three rooms are more likely to be increased with heat transfer by convection of exposed surfaces. In winter, the MRT is more likely to be stimulated by the stored earth heat capacity.

### 6.3 *Stage 3*

Stage 3 discusses the thermal comfort of a domed room through its operative temperature according to wall orientation. Numerous simulations optimise several alternatives, and final alternatives are discussed through two proposed levels, Level 1 and Level 2. Each level will be discussed separately in order to interpret the produced results.

Level 1 (Figure 8) discusses comfort through daily mean operative temperatures according to walls W-BioPCM orientation and the use of insulation in the non-exposed direction(s) in summer and winter. The variables key for Level 1 results (Figure 8) is explained as follows:

- All Directions: refers to using W-BioPCM in all four directions.
- West & South + Ins. Non-exposed: refers to using W-BioPCM at both west and south only, and using insulation for non-exposed surfaces.
- West + Ins. All directions: refers to using W-BioPCM at west orientation only and using insulation in all other directions.

The simulation results in Figure 8 show that in summer, all the alternatives presented lower temperatures than DBT (AV 4°C), and they were all in the comfort zone range of Kharga, Egypt; ASHRAE climate zone BWh, according to Adaptive Comfort Model (ACM) in ASHRAE Standard 55–2010 for Kharga Climate (19.6°C–30.7°C). With regards to the win-

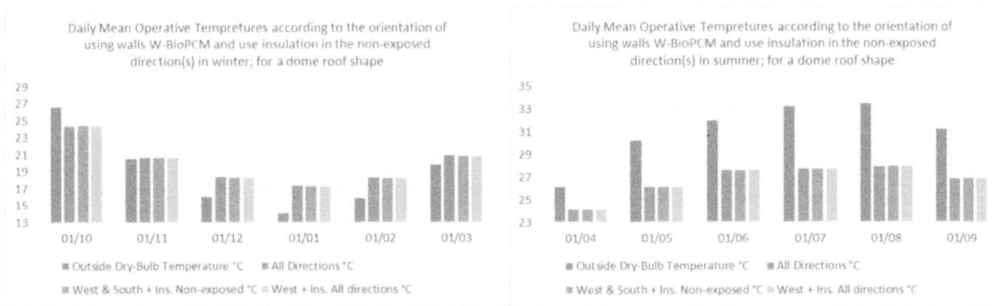


Figure 8. Level 1 daily mean operative temperatures according to the orientation of using walls W-BioPCM and use insulation in the non-exposed direction(s).

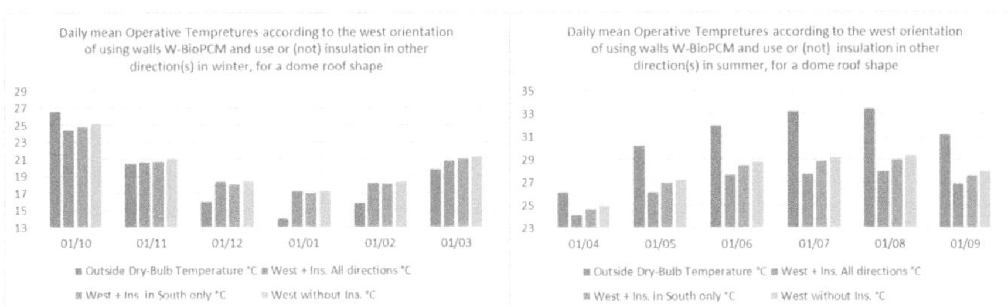


Figure 9. Level 2 daily mean operative temperatures according to the west orientation of using walls W-BioPCM and the use (or not) of insulation in the other direction(s), for a domed roof shape.

ter results, the alternatives resulted in operative temperatures above DBT with AV 4°C. Both summer and winter results at this level show that using W-BioPCM in the west and rice husk insulation in all other directions represents the reasonable choice in terms of cost for this level.

Level 2 discussion is as a result of Level 1, which led to the alternative of using W-BioPCM only in the west and using insulation for the other three directions. Level 2 discusses operative temperatures according to the use of wall W-BioPCM in the west orientation (most exposed), and whether or not using insulation in other direction(s) in summer and winter. The variables key in Level 2 results (Figure 9) is explained as follows:

- West + Ins. All directions: refers to using W-BioPCM in the west and using insulation in all other directions.
- West + Ins. In the South only: refers to using W-BioPCM in the west and using insulation in the south only.
- West without Ins.: refers to using just W-BioPCM in the west direction.

Results in Figure 9 show that in summer simulations, proposed alternatives can decrease the DBT down by 4–7°C. Using W-BioPCM in the west orientation and insulation in all other directions records the maximum difference in temperature degrees between DBT and operative temperatures, but using W-BioPCM in the west orientation without any insulation applied to other directions records the minimum result. On the other hand, the winter results showed the opposite. Proposed variables record higher operative temperature degrees than the recorded DBT with AV 4°C.

Using W-BioPCM in the west orientation and insulation in all the other directions recorded a higher difference between DBT and operative temperatures than when using W-BioPCM in the west orientation without any applied insulation to other directions. Using W-BioPCM in the west and insulation in the south recorded reasonable results with respect to the comfort range.

It can be concluded from the simulations discussed above for Level 1 and Level 2 that the case of using W-BioPCM in the west and rice husk insulation boards in all the other directions, and the case of using W-BioPCM in the west, and rice husk insulation boards

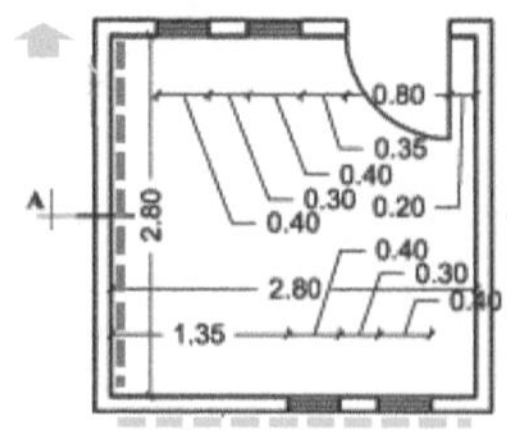

Figure 10. The proposed solution using W-BioPCM at west and rice husk insulation at south.

in the south, both reach the comfort zone in Khraga according to ACM. Comfort has been achieved by 100% in the summer as well as in winter within the ACM defined acceptability limit, amounting to 91%. Despite using W-BioPCM in the west and insulation in the south direction, it achieves less difference in operative temperatures than when using W-BioPCM in the west orientation and insulation in all other directions. As illustrated in Figure 10, using less insulation is preferable when it comes to cost as long as temperature differences are still within the comfort range.

## 7 CONCLUSION

It can be concluded from the aforementioned results and discussion for the three stages of simulation that the prototype W-BioPCM 24.6 cm thick has the economic advantage to introduce an equivalent thermal control performance to the conventional bulk materials walls (mud-brick walls with 51 cm thickness). Discussing the prototype W-BioPCM 24.6 cm thickness combined with the shape of the roof (flat, domed, vaulted), the domed roof room recorded the lowest MRT among the three rooms. This has been verified from the conducted CFD simulations. Discussing the aimed thermal comfort range of the domed room through its operative temperature according to wall orientation, the results have been discussed through two levels. It can be concluded from the discussed simulations for Level 1 and Level 2 that using both W-BioPCM in the west and rice husk insulation boards in all the other directions and using W-BioPCM in the west, and rice husk insulation boards in the south reaches the comfort zone in Khraga according to ACM. Comfort of 100% has been achieved in summer, as well as in winter within the ACM defined acceptability limit amounting to 91%. The study shows the potential benefit of using a proposal of multi-layered lightweight construction technique which has thermal capabilities similar or could be equivalent to the thermal characteristics of those buildings constructed with heavyweight construction materials such as bearing-walls buildings. This opens the scope towards thermally comfort multi-storey building with bigger capacity rather than the conventional bearing-walls technique which is limited in height with limited capacity. The research opens the scope for considering urban planning development in Kharga with flexible multi-storey skeleton concrete structures, rather than by using the conventional 50 cm bearing-walls construction that is limited in height.

## REFERENCES


A. Bhatia, B. E. 2012. HVAC Made Easy: A Guide to Heating & Cooling Load Estimation. Available: http://www.pdhonline.org/courses/m196/m196content.pdf [Accessed 21/12/2014].

ARCHNET. 2014. *New Baris Village, Kharga, Egypt* [Online]. Aga Khan Trust for Culture and the Aga Khan Documentation Center. Available: http://archnet.org/collections/11/sites/2560 [Accessed 19/10/2014].

*Attia, K. & Raslan, Y.* 2011. Cause and Combat of Agriculture Land Desertification in Egypt Related to River Processes. *International Journal of Water Resources and Arid Environments.*

Babones, S. 2007. Studying globalization: Methodological issues. *The Blackwell companion to globalization.* Oxford: Blackwell Publishers.

Barsoum, R. S. 2013. Burden of chronic kidney disease: North Africa. Available: http://www.nature.com/kisup/journal/v3/n2/pdf/kisup20135a.pdf.
Bridge, G. 2002. Grounding Globalization: The Prospects and Perils of Linking Economic Processes of Globalization to Environmental Outcomes. *Economic Geography,* 78, 361–386.
Correas-Amador, M. 2012. A survey of the mud-brick buildings of Qena. England: The Egypt Exploration Society.
Crossett, J. D. 20/12/2014 2014a. *RE: Re: Website Submission.* Type to MAHDY, M.
Crossett, J. D. 22/12/2014 2014b. *RE: Re: Website Submission.* Type to MAHDY, M.
Energy, U. S. D. O. 2003. Weather Data, Egypt. *In:* ENERGY, U. S. D. O. (ed.). U.S.
Fathy, H. 1986. Natural Energy and Vernacular Architecture, Principles and Examples with Reference to Hot Arid Climates. Chicago and London: The University of Chicago Press.
Harper, D. 2011. Alternative Methods of Stabilisation for Unfired Mud Bricks. Available: http://www.ewb-uk.org/filestore/20110514-Mud%20brick%20construction.pdf.
Hbrc, H. A. B. N. R. C. 2006. ECP 306–2005. *Egyptian code to improve energy efficiency in residential buildings*. Cairo: HBRC.
Jeong, S.-G., Chung, O., Yu, S., Kim, S. & Kim, S. 2013. Improvement of thethermal properties of Bio-based PCM using exfoliated graphite nanoplatelets. *Solar Energy Materials and Solar Cells, ELSEVIER,* 117, 87–92.
Jeong, S.-G., Lee, J.-H., Seo, J. & Kim, S. 2014. Thermal performance evaluation of Bio-based shape stabilized PCM with boron nitride for energy saving. *international Journal fo Low-Carbon Technologies,* 71, 245–250.
Kalaitzidis, R. G., Damigos, D. & Siamopoulos, K. C. 2014. Environmental and stressful factors affecting the occurrence of kidney stones and the kidney colic. *INTERNATIONAL UROLOGY AND NEPHROLOGY.*
KNMI. 2014. *mean rcp45 regression temperature on time 2014–2044.*
NRDC. 2011. *An Introduction to Climate Change* [Online]. Natural Resources Defense Council. Available: http://www.nrdc.org/globalwarming/climatebasics.asp [Accessed 10 April 2014].
Pasupathy, A., Velraj, R. & Seeniraj, R. V. 2008. Phase change material-based building architecture for thermal management in residential and commercial establishments. *Renewable and sustainable energy reviews, ELSEVIER.*
PHASE CHANGE ENERGY SOLUTIONS, I. 2013. *How to determine how much bioPCMTM to place in a building?* [Online]. Phase Change Energy Solutions, Inc. Available: http://www.phasechange.com/index.php/en/standard-products/biopcmat [Accessed 19/10/2014].
PHASE CHANGE ENERGY SOLUTIONS, I. 2013a. *biopcm* [Online]. Phase Change Energy Solutions, Inc. Available: http://www.phasechange.com/index.php/en/ [Accessed 31/05/2014].
PHASE CHANGE ENERGY SOLUTIONS, I. 2013f. *Architectural Details* [Online]. Phase Change Energy Solutions, Inc. Available: http://www.phasechange.com/index.php/en/architectural-library [Accessed 06/10/2014].
PHASE CHANGE ENERGY SOLUTIONS, I. 2014. *Phase Change Energy Solutions* [Online]. Phase Change Energy Solutions, Inc. Available: http://www.phasechange.com.au/biopcm [Accessed 31/05/2014].
Quality, E. O. F. S. A. 2005. ES: 451 – 1 / 2005. *Refractory Bricks—Dimensions.* Cairo: Egyptian Organization for standardization and Quality.
Sheweka, S. 2011. Using Mud Bricks as a Temporary Solution for Gaza Reconstruction. *Energy Procedia, Elsevier*, 236–240.
Szokolay, S. V. 2014. *Introduction to Architectural Sciences: The basics of sustainable Design*, Oxon, UK, Routledge.
Tyagi, V. V. & Buddhi, D. 2007. PCM thermal storage in buildings: A state of art. *Renewable and sustainable energy reviews, ELSEVIER,* 11, 1146–1166.
Whiffen, T. R. & Riffat, S. B. 2012. A review of PCM technology for thermal energy storage in the built environment: Part 2. *International Journal fo Low-Carbon Technologies*.
Whiffen, T. R. & Riffat, S. B. 2013. A review of PCM technology for thermal energy storage in the built environment: Part I. *International Journal fo Low-Carbon Technologies,* 8.
Yu, S., Jeong, S.-G., Chung, O. & Kim, S. 2014. Bio-based PCM/carbon nanomaterials composites with enhanced thermal conductivity. 120, 549–554.

*Cities' Identity Through Architecture and Arts – Catalani et al. (Eds)*
*© 2018 Taylor & Francis Group, London, ISBN 978-1-138-05409-7*

# Towards public identity and climate awareness architecture

Ahmed Kaihoul
*Department of Architecture, College of Earth Sciences and Architecture, Larbi Ben M'hidi University—OEB, Oum El-Bouaghi, Algeria*

Leila Sriti
*Department of Architecture, College of Sciences and Technology, Mohamed Khidher University, Biskra, Algeria*

ABSTRACT: Of all of the arts and technology produced by man, architecture is the most conspicuous of them all. The search for an identity is a difficult endeavour and for a multicultural nation the effort is even greater. To search for one's identity is also something of a peculiar endeavour, since it implies that one has either lost their identity or does not have a clue as to who one is. The Aga Khan Organisation gives awards to architectural projects every three years and there is a large selection criteria; but the most important ones concern the identity issue data. So, his highness the Aga Khan is interested in the heritage and identity of Islamic architecture. This study focuses on the views of architects contributing to the production of cultural identity in public architecture and analyses examples of their projects. For the purpose of this study, samples of architecture\projects that won the Aga Khan Award for Architecture (AKAA) were selected from a group of international architects expressing Islamic cultural identity in their projects. The methods used for data collection included an analytical framework using standardised thematical axes and the analysis of examples of contemporary international public architecture that represent the expression of cultural identity. The aim of this work and research is to clarify the duality of Identity/Climate awareness in public architecture and to define the influence of different variables and indicators on the architectural design of the public projects, such as architectural identity variables and climate data variables.

*Keywords*: Public Identity Architecture; Climate Awareness; Aga Khan Award for Architecture

## 1 INTRODUCTION

### 1.1 *General introduction*

The search for a public architecture identity seems to be a must for countries that are either newly independent or those with a leadership that stresses that certain groups or races are 'better' than others (Baper & Hassan, 2010). In the last few decades, we have often heard about identity in architecture, and this is now a daily concern for architects, designers and deciders. The context of the Aga Khan Award takes identity as a major factor that needs to be respected by the participating projects. 'We can consider that there are no less than three (3) different identities in any nation's architectural works; a natural identity, a forced identity and a manufactured one' (Tajuddin & Rasdi, 2005).

This study discusses the concept of identity and the role of modernity in the shift towards globalisation, which became a crucial study in identifying the factors that influence the concepts of change and continuity in architectural identity. In this short essay, we will attempt to classify the various approaches towards answering the problem of a national architectural identity (Zarzar & Guney, 2008). The literature study covers definitions of the keywords, which are: identity and architecture, Aga Khan Award for Architecture, climate awareness. The definitions include descriptions and issues related to the key words, 'case studies:

Aga Khan awarded projects' and checking whether there could be a climate awareness issue in addition to that of identity representation.

### 1.2 *Public architecture and identity*

The Aga Khan committee tries as hard as possible during the selection process to select projects that fulfil the standard of 'Search for Meaning', which is 'considered the particular context in which each project has evolved, as well as the unique social, economic, environmental and technical factors to which it responds'. The quality is assured by the 'appropriate and creative utilization of available resources in meeting functional and cultural needs, as well as the higher potential in each project to set a standard for the future'. Gracía Lorca: "*time, not man, makes architecture*", one of the best justifications for the specialised skill of the architect is that he or she makes possible the continuation into the future of valuable qualities of the environment, form, texture and materials, and also of details and decorations that would otherwise disappear. By signalling buildings and towns for special care, the architect distinguishes them from their fellows, emphasising their potential to serve the ends Lorca proposes for them. But this skill also has dangers. Since the action of time on buildings is judged to be an important factor, it does not do to rebuild them, to create pristine forms and details, to replace the patina of age with spanking new materials and textures or to put newly carved decoration in the place of old (Lewcock, 1989).

Societies are organised to ensure their own continuity and thereby to serve the function of preserving something from the past. Society's awareness of the past is, in fact, society's awareness of its own continuity. The concern here is to reinterpret the past in a way that is useful and suitable for the present: that is, in a way that will re-establish a sense of continuity and eliminate the rupture and sense of alienation being voiced as a result of the introduction of a contemporary environment. This concern implies that the past has a certain value for the present (Al-Hathloul, 1998).

### 1.3 *Public architecture and the climate awareness issue*

The designer architect is interested in those aspects of climate that affect human comfort and the use of buildings. They include averages, changes and extremes of temperature, the temperature differences between day and night, humidity, sky conditions, incoming and outgoing radiation, rainfall and its distribution, air movements and special features such as trade winds, thunder storms, dust storms and hurricanes (Pradeepa, 2013).

Shelter is the main instrument for fulfilling the requirements of comfort. It modifies the natural environment in order to approach optimum conditions of liveability. The architect's problem is to create an environment that will not place undue stress upon the body's heat-compensation mechanism. It is the task of the architect to make the utmost use of the natural means available in order to produce a more healthful and liveable building, while also achieving a savings cost by keeping to a minimum the use of mechanical aids for climate control (Boake, 2008).

'We must begin by taking note of the countries and climates' in which homes are to be built if our designs for them are to be correct. One type of house seems appropriate for Egypt, another for Spain...one still different for Rome...It is obvious that design for homes ought to conform to diversities of climate' (Vitruvius, 1st century B.C., pp. 170). What we as architects are aiming for is to take the climate-motivated, environmentally sustainable, valid ideas and practices, from both indigenous and vernacular buildings, and to incorporate them into current architecture that clearly responds to the issues of climate (and comfort) in the design of the building.

'When the full power of a human imagination is backed by the weight of a living tradition, the resulting work is far greater than any that an artist can achieve when he has no tradition to work in or when he wilfully abandons its tradition' (Fathy, 1988). This is because the best vernacular construction has always been primarily answerable to natural forces. As Fathy himself once said in a lecture at Dar al Islam: 'If the architect does not respect the God-made environment, he commits a sin against God. The God-made environment is the landscape; the atmosphere, the flora, the fauna, and the human beings who live in this environment. In this God-made environment there is nothing that is inharmonious. If we become one with nature, beauty is defined as it is. Beauty, then, is obtained when form considers the forces that are working on it. It is only when man has ignored the environment and has been cut off from nature that problems arise. We must not distort any of the forces in nature' (Steele, 1992).

## 2 THE ANALYTICAL STUDY

### 2.1 *Methodology*

In this study, we will try to show that, while we are searching to apply an identity in design, inevitably, we are affected by the environment and climate data, and by giving climatic design solutions. To verify this hypothesis, the approach method carried out was a qualitative analysis based on a case study. The adopted approach relies on selecting and analysing a set of public buildings that represent the expression of cultural identity. The architectural analysis was based on design characteristics (criteria) expressing identity and climate adaptation. The criteria used throughout the study of the existing literature were presented in the form of two matrices using, respectively, an 'identity design' criteria and a 'climatic design' criteria as standardised thematic categories. The matrix is a synthetic visual representation of the data at an intermediate stage of the analysis, which will then be used during the interpretations. It presents the information in a compact and orderly form, allowing the researcher to visualise a set of variables together and draw conclusions accordingly. Thereby, each of the seven projects selected was analysed according to the matrix criteria. The outcome matrices indicate the different criteria fulfilled by each project. By comparing the two matrices, we aim to clarify the duality of Identity/Climate awareness and define the extent to which these two factors influence each other.

### 2.2 *Case study of 'AKAA' project winners*

With reference to the analytical framework, the selected examples had to be public buildings and to express cultural identity. The Aga Khan Organisation rewards architectural projects every three years. The selection criteria are many; but the most important one concerns the identity issue data. For the purpose of this study, a sample of seven architects\projects that won the Aga Khan Award for Architecture (AKAA) was selected from a group of international architects who expressed Islamic cultural identity in their projects. Our study focuses on the views of the architects contributing to the production of cultural identity in public architecture, and analyses examples of their projects. The seven AKAA projects were selected according to two main criteria: 'public architecture and identity awareness'. The selected projects (based on the author's architectural analysis by referring to the AKAA official sites and books) are: Madinat al-Zahra museum, Spain; Arab World Institute, France; Nubian museum, Egypt; Great mosque, Saudi Arabia; National museum, Qatar; Alexandria library, Egypt; and the Islamic cemetery, Austria.

- Madinat al-Zahra museum, Spain:

It is concluded that the museum has managed to create an image of the authenticity of ancient times (Madinat al-Zahra) with regards to their architecture, culture, history and quality of life. Visitors/tourists to the region who discover this archaeological site will visit the museum to see the first modern simulation traces of the Islamic era. The former environmental strategies of the Islamic architecture of Madinat al-Zahra (patio, underground, compact size, opacity and massiveness of the walls) are used in the museum. The museum remains a landmark of this architecture for other generations, as shown in Figure 1.

- Arab World Institute, France:

In this project, the Islamic traditional concept of Mochrabeiyah has used technologically 'the diaphragm of each window opens and closes due to the sunlight', also, the central courtyard has been directed to a valuable cultural building in the city 'Notre dame de Paris'. The institute creates a communicating space 'extension' between the institute and the Sorbonne university 'a grand plaza', as shown in Figure 2.

- Nubian museum, Egypt:

The edifice is surrounded by a natural botanical garden, which contains a large variety of Egyptian flora. All of the museum's construction materials are local and natural, and it was

Figure 1. Madinat al-Zahra museum. The source: http://www.akdn.org/architecture/project/madinat-al-zahra-museum.

Figure 2. Arab World Institute. The source: http://www.akdn.org/architecture/project/institut-du-monde-arabe.

Figure 3. Nubian museum. The source: http://www.akdn.org/architecture/project/nubian-museum.

Figure 4. Great mosque. The source: http://www.akdn.org/ru/architecture/project/great-mosque.

all built of stone and adobe. The museum's orientation is towards the Nile, the same as local identity houses. Nubians have questioned whether or not the edifice represents the local identity, and they have confirmed that it does (Figure 3).

- Great mosque, Saudi Arabia:

The great mosque of Riyadh, along with the urban development of nearby public squares, is one of the master plans for revitalising the Qasr al Hokm district, which is the old centre of Riyadh. Rasem Badran has recreated and transformed the spatial character of the local Najdi architectural idiom without directly copying it. The complex is a group of buildings behind walls, punctuated by such traditional elements as gates and towers. Within, columns, courtyards and narrow passageways recall the traditional uses of space. Mosque components, including 'courtyards, arcades, and the flat-roofed prayer hall' are ordered and articulated in the traditional way, as shown in Figure 4.

- National museum, Qatar:

Qatar's old Amiri Palace was reconstructed to form the nucleus of the museum. The Palace complex consists of three courtyard houses, two reception halls and various service quarters. Built in 1918, a two-story arcaded structure at the centre of the compound dominates the site (Figure 5).

- Alexandria library, Egypt:

The Alexandria library is a revival of the legendary ancient library that was built in classical Greek times. The rebuilding of the library has returned Alexandria to its former status as a centre for learning and exchange and has provided the city with a landmark building.

Figure 5. National museum. The source: http://www.akdn.org/architecture/project/national-museum.

Figure 6. Alexandria library. The source: http://www.akdn.org/tg/architecture/project/bibliotheca-alexandrina.

Figure 7. Islamic cemetery. The source: http://www.akdn.org/fr/architecture/project/islamic-cemetery.

The column and structure design is based on and inspired by the Egyptian flower, which applies an identity to the structure and external façade. This is also seen by the half sun disc shape. In contrast, the architect ensured that the buildings modernity was clearly shown on the selective façade 'entrance' (Figure 6).

- Islamic cemetery, Austria:

This finds inspiration in the primordial garden, and is delineated by roseate concrete walls in an alpine setting. It consists of five staggered, rectangular grave-site enclosures, and a structure housing assembly and prayer rooms. The prayer room on the far side of the courtyard reprises the lattice-work theme with Kufic calligraphy in metal mesh on the qibla wall. Mochrabeiyah-ornaments have been used to assure and profess the Islamic identity as a symbol in a minimalistic way (Figure 7).

### 2.3 *Defining thematic design categories: The matrix criteria*

Through the study of previous theoretical concepts, we focused on elements embodied in the design and took into account both the environment and the question of identity, and on this basis we selected examples of both traditional buildings and contemporary projects (AKAA selected project winners). We aimed to highlight their role in giving a clear idea of the identity of Islamic architectural areas, as well as in achieving climatic efficiency, which was reflected in the design parameters and their components, in order to achieve an integrated architectural environment and preserve their identity and privacy.

#### 2.3.1 *Analytical pattern—projects of (AKAA)*

The review of the projects and the selection of the award recipients is the responsibility of an independent master jury, which is specially appointed for each award cycle. Each jury is multidisciplinary, bringing together specialists in such fields as history, philosophy, art, engineering and architectural preservation, in addition to practising architects, landscape architects and urban planners.

The reviewers are required to comment on a detailed set of criteria in their written reports, and they must also respond to specific concerns and questions prepared by the master jury for each project. To ensure maximum objectivity, reviewers report on projects located outside of their native countries.

2.3.2 *The 'identity design' thematic category*
The analysis contained several axes covering different enquiries, including the views and intent of architects when producing this architecture, in order to understand how they perceive what they are doing, what they use for their references, and how they are influenced by the client, the context, the climate and the history of the country.

- Results
- Identity design 'Aga Khan's criteria', see Table 1.

Table 1. Identity axe results collection.

| Criteria/Projects | Madinat al-Zahra Museum, Spain | Arab World Institute, France | Nubian Museum, Egypt | Great Mosque, Saudi Arabia | National Museum, Qatar | Alexandria Library, Egypt | Islamic Cemetery, Austria |
|---|---|---|---|---|---|---|---|
| 1/ Projects that respond to people's physical, social and economic needs | ■ | ■ | ■ | ■ | ■ | ■ | ■ |
| 2/ Projects that stimulate and express people's cultural expectations | ■ | ■ | ■ | ■ | ■ | ■ | ■ |
| 3/ Building schemes using local resources and appropriate technology in innovative ways | | ■ | ■ | ■ | | | |
| 4/ Set new standards of excellence in architecture, planning practices, historic preservation and landscape architecture | ■ | ■ | ■ | ■ | ■ | ■ | ■ |
| 5/ Building projects that affect today's environment | | ■ | ■ | ■ | | | |
| 6/ Architecture that reflects the pluralism that has characterised Muslim societies and communities | ■ | | | ■ | ■ | | ■ |

• Interpretations:
1/ We find that the Aga Khan seeks to give more awards to projects for identity, idiom and culture than for anything else (1,2,4);
2/ It appears that the environment and use of local resources are not one of the main reasons why a master jury gives awards to projects (3,5);
3/ It is also considered that the projects reflect the pluralism of Islam (6);
4/ From this way of thinking, we were forced to ask a question regarding the relationship between climate and identity to verify whether or not they are connected, which we did by reanalysing the chosen projects and checking the texts and plans of them (see Table 2).

Collected by referring to the texts containing the views of architects and specialists (Aga Khan Award for Architecture—cycle's books and official site).

2.3.3 *The 'climatic design' thematic category*

- Climatic design 'Aga Khan's criteria', see Table 2.

Collected by referring to the texts containing the views of architects and specialists (Aga Khan Award for Architecture—cycle's books and project's books).

Finally, by comparing the two matrices, it appears that when cultural identity is expressed in a building, it leads to a climate responsive project. So, by validating the idiom and identity criteria's elements in the design, we are, at the same time, assuring the climate adaptability of the architectural product.

Table 2. Climatic axe results collection.

| Criteria/Projects | Madinat al-Zahra Museum, Spain | Arab World Institute, France | Nubian Museum, Egypt | Great Mosque, Saudi Arabia | National Museum, Qatar | Alexandria Library, Egypt | Islamic Cemetery, Austria |
|---|---|---|---|---|---|---|---|
| *Architectural level* | | | | | | | |
| 1/ Exterior morphology | ■ | ■ | ■ | ■ | ■ | ■ | ■ |
| 2/ Spatial organisation | ■ | ■ | | ■ | ■ | | ■ |
| 3.A/ Façades and their treatment | ■ | ■ | ■ | ■ | ■ | ■ | ■ |
| 3.B/ Building materials | | | ■ | | ■ | | |
| 3.C/ Responsive building elements | ■ | ■ | ■ | | ■ | | ■ |
| *Urban level* | | | | | | | |
| 4/ Buildings position and implementation | ■ | ■ | ■ | ■ | ■ | ■ | ■ |
| 5/ Orientations and guidance | ■ | ■ | ■ | ■ | ■ | ■ | ■ |
| 6/ Urban structure (grid, density, …) | | | | ■ | ■ | ■ | |
| 7/ Shape and create urban spaces (courtyards, gardens, …) | ■ | ■ | ■ | ■ | ■ | | ■ |

- Interpretations:

1/ As a general overview, we notice that the projects are validating the climate awareness assumption, by applying most of them;

2/ With reference to the architectural level, we find that the exterior morphology and façades treatments are applied in all of the projects (1, 3.A). Climatic elements are also important (2, 3.C);

3/ With reference to the urban level, we find that the implementation and orientation are always taken into account (4, 5). The projects also reveal a variety of urban and green spaces, which play a decisive role in climate performance (7).

## 3 CONCLUSIONS

Projects awarded by the 'AKAA', which show a high quality in expressing idiom, culture and excellence in innovative architecture, are also sustainable and climate-efficient buildings. So, what is cultural identity in public architecture?

It is a process, and not a 'found' object. It may be likened to the trail left by civilisation as it moves through history. The trail is the culture, or identity, of that civilisation. As it is a process, identity cannot be fabricated. We develop our identity by tackling what we perceive to be our real problems (climate, for example).

In this paper, we focused on identity and climate as design criteria and we attempted to define the extent to which these two factors influence each other. We found that the way that we are designing and expressing our identity is itself the way that we are solving climate issues; it is a harmonious duality.

## REFERENCES


Al-Hathloul, S. (1998). *Aga Khan Award for Architecture, 1996–1998 cycle's book (Legacies for the future, contemporary architecture in Islamic societies)* (pp. 18–22), edited by: C. Davidson Cynthia, published by: Thames & Hudson Ltd, London—publication date: 1998, ISBN: 0-500-28087-8, printed and bound in Italy by Conti Tipocolor.

Baper, S.Y. & Hassan, A.S. (2010). The influence of modernity on Kurdish architectural identity. p. 12, the journal: *American J. of Engineering and Applied Sciences 3* (3): 552–559, 2010 ISSN 1941-7020 © 2010 Science Publications.

Boake, T.M. (2008). *Arch 125: Introduction to environmental design course* (p. 6). School of Architecture, University of Waterloo.

Fathy, H. (1988). Natural energy in vernacular architecture: Principles and examples with reference to hot arid climates, edited by: Walter Shearer and Abd-el-rahman Ahmed Sultan, published by: The United Nations University by The University of Chicago Press Chicago and London—publication date: 1986, ISBN-10: 0226239179, ISBN-13: 978-0226239170, printed in the United States of America.

Lewcock, R. (1989). *Aga Khan Award for Architecture, 1987–1989 cycle's book (Architecture for Islamic societies today)* (pp. 18–21), edited by: James Steele, published by: Academy Group Ltd—publication, date: 1994, ISBN 1 85490 207 5, printed and bound in Singapore.

Pradeepa, C. (2013). *Climate responsive architecture course* (p. 4). University of SRM.

Steele, J. (1992). *Aga Khan Award for Architecture, 1990–1992 cycle's book (Architecture for a changing world)* (pp. 51–52), edited by: James Steele, published by: Academy Editions—publication date: 1992, ISBN 1 85490 173 7, printed and bound in Singapore.

Tajuddin, M. & Rasdi, M. (2005). Malaysian Architecture, Crisis Within, Chapter: The question of identity in Malaysian architecture, p. 5, published by: Utusan Publications & Distributors Sdn Bhd—publication date: 2005, ISBN 967-61-1728-5, printed by YEOHPRINCO SDN. BHD.

Vitruvius, (1st century B.C.), The Ten Books on Architecture, Book VI, Chapter 1 On Climate as Determining the style of the House, (pp. 170).

Zarzar, K.M. & Guney, A. (2008). Understanding Meaningful Environments: Architectural precedents and the question of identity in creative design, p. 9, published by: IOS Press—publication date: 2008, ISBN-10: 158603913X, ISBN-13: 978-1586039134.

- Aga Khan Award for Architecture (cycle's books):

AKAA. (1980). *Architecture and community: Building in the Islamic world today, 1978–1980 cycle's book*, p. 171, edited by: Renata Holod with Darl Rastorfer, published by: Aperture, New York—publication date: 1983, ISBN: 089381-123-8.

AKAA. (1989). *Architecture for Islamic societies today, 1987–1989 cycle's book*, p. 147, edited by: James Steele, published by: Academy Group Ltd—publication date: 1994, ISBN 1 85490 207 5, printed and bound in Singapore.

AKAA. (1995). *Architecture beyond architecture, 1993–1995 cycle's book*, pp. 85–88, edited by: C. Davidson Cynthia with Ismaïl Serageldin, published by: Academy Editions—publication date: 1995, ISBN 1 85490 443 4, printed and bound in Singapore.

AKAA. (2001). *Modernity and community: Architecture in the Islamic world, 1999–2001 cycle's book*, p. 102, edited by: Kenneth Frampton, Charles Correa and David Robson, published by: Thames &

Hudson Ltd, London—publication date: 2001, ISBN 0-500-34185-0, ISBN 0-500-28330-3, printed and bound in the United Kingdom by Butler and tanner.
AKAA. (2004). *Architecture and polyphony: Building in the Islamic world today, 2002–2004 cycle's book*, p. 17, edited by: Philippa Baker, published by: Thames & Hudson Ltd, London—publication date: 2004, ISBN 0-500-28533-0.
AKAA. (2010). *2007–2010 cycle's book*, p. 58, edited by: Melissa Vaughn, published by: Lars Müller Publisher, Switzerland—publication date: 2011, ISBN 978-3-03778-242-2.
AKAA. (2013). *Architecture is life, 2011–2013 cycle's book*, p. 204, edited by: Mohsen Mostafavi, published by: Lars Müller Publisher, Switzerland—publication date: 2013, ISBN 978-3-03778-378-8, printed in Germany.

- Aga Khan Award for Architecture (project's books):

AKAA. (1989). *Institut du Monde Arab's book*, pp. 53–55, the publisher is the Aga Khan organisation (http://www.akdn.org/sites/akdn/files/media/documents/AKAA%20press%20kits/1989%20AKAA/Institut%20du%20Monde%20Arabe%20-%20France.pdf).
AKAA. (1995). *Great Mosque and redevelopment of the old city center's book*, pp. 9–11, the publisher is the Aga Khan organisation (http://www.akdn.org/sites/akdn/files/media/documents/AKAA%20press%20kits/1995%20AKAA/Great%20Mosque%20-%20Saudi%20Arabia.pdf).
AKAA. (2001). *Nubian Museum's book*, pp. 5–7, The publisher is the Aga Khan organisation (http://www.akdn.org/sites/akdn/files/media/documents/AKAA%20press%20kits/1980%20AKAA/National%20Museum%20-%20Qatar.pdf).

- Aga Khan Award for Architecture (official site):

Madinat al-Zahra – http://www.akdn.org/architecture/project/madinat-al-zahra-museum.
Arab World Institute – http://www.akdn.org/architecture/project/institut-du-monde-arabe.
Nubiam Museum – http://www.akdn.org/architecture/project/nubian-museum.
Great Mosque – http://www.akdn.org/architecture/project/great-mosque.
National Museum – http://www.akdn.org/architecture/project/national-museum.
Alexandria Library – http://www.akdn.org/architecture/project/bibliotheca-alexandrina.
Islamic Cemetery – http://www.akdn.org/architecture/project/islamic-cemetery.
AKAA official site – http://www.akdn.org/our-agencies/aga-khan-trust-culture/aga-khan-award-architecture/about.
AKAA official site – http://www.akdn.org/our-agencies/aga-khan-trust-culture/aga-khan-awardarchitetture/review-selection-procedures.
AKAA official site – http://www.akdn.org/our-agencies/aga-khan-trust-culture/aga-khan-award.

*Cities' Identity Through Architecture and Arts – Catalani et al. (Eds)*
*© 2018 Taylor & Francis Group, London, ISBN 978-1-138-05409-7*

# Vernacular culture and its contribution to the identity of Auresian houses in Algeria

Mohamed Cherif Ammari & Noureddine Zemmouri
*Mohamed Khider University, Biskra, Algeria*

ABSTRACT: The Berber house in the Aures region of Algeria is known for being an example of austere architecture, defined by the construction materials available on site. What makes it different from other traditional settlements in North Africa is the fact that it is influenced by the vernacular culture of the Shawia[1] community. This is mainly the result of a process of adaptation to the common lifestyle of the region, which is the main reason why Shawia house typologies and urban configurations look so similar in their general aspect. However, diversity can be observed from one village to another, resulting in a rich urban fabric. The aim of this research is to investigate and diagnose, through a comparative analysis, the cultural influences on the architecture of two examples of Shawia houses situated in two different villages, the first from the Oued-Abdi valley in the western part of the Aures, and the second from the eastern part of the Oued-Labiod valley.

*Keywords*: Berber house; austere architecture; vernacular culture; urban configurations; Shawia houses

## 1 INTRODUCTION

The Auresian house is one of the few remaining footprints of the North African indigenous civilisations. It is a reflection of the Shawia lifestyle through their architecture, an adaptation of an austere and autonomous society in which culture and attachment to land is a main feature. In this paper, a comparative analysis is conducted after an investigation within two different villages of the Aures region, aiming to build an hypothesis on the cultural reasons for the differences that were noticed.

## 2 GEOGRAPHICAL CONTEXT

The name 'Aures' is given to the wide mountainous area that lies in the east of Algeria. Starting from the trough from which the Kantara River flows, it follows the roman track from Lambiridi to Biskra. Its north east borders start from Batna, to Khenchela through the highway. And goes down from the east to the south through Oued El Arab Valley, arriving to the Ziban. Aures is surrounded by the quadrilateral: Batna, Biskra, Khanget Sidi Nadji, and Khenchela (R. De Latrigue, 1904) (Figure 1).

The Aures region covers about 9,000 Km Sq. and is situated in the north east of Algeria at a latitude of 35° North, and a longitude of between 6° and 7° East. In the south west, Djebel Metlili separates Aures from Ziban. From the north east, the links to the high plains of Constantine shape a border, as also does the Seggana synclinal oriented to the east and the synclinal break of the Ain Touta that separates Aures from the Belezma Mountains (R. De Latrigue, 1904).

1. Native inhabitants of Aures mountains

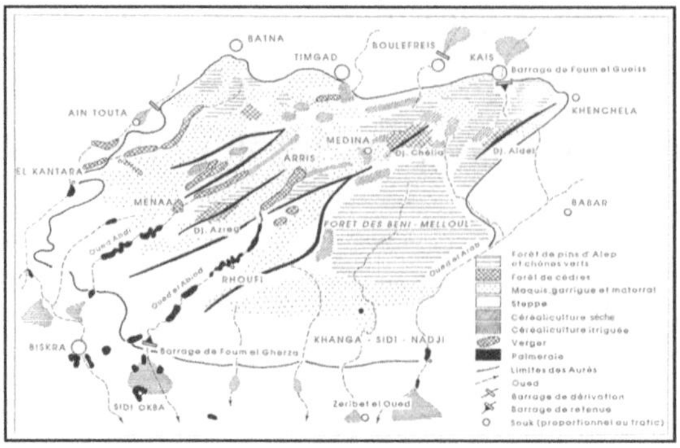


Figure 1. Economical map showing limits of Aures (Côte, 1983).

The climate in Aures differs from one place to another due to the rich geographical landforms. It is categorised by three different ranges of microclimates, depending on height (Benbouaziz, 2011):

- Above 1,400 m high there is a cold winter, with 60 snow days per year and precipitations of about 1000 ml/Year. The climate is sub-humid.
- Under 1400 m the climatic conditions change: 24°c in summer, 5°c in winter and 15 snow days per year. The climate is semi-dry.
- In the southern piedmont, the climate is hot and dry.

## 3 SOCIAL STRUCTURE

Fanny Colonna defines Aures as a group of tribes united by history and conflicts, based on identical social and cultural practices, and on the exchange of properties, goods, men and saintliness. Despite disparities due to certain differences in their lifestyles, they are also united on identical principals of social organisation (Colonna, 1995). The Shawia settlement is an independent entity, both economically and socially; the villages are the property of the tribe 'El Arch'. The tribe is composed of numerous fractions called in local language "Harfikth", and each fraction is divided to a number of families. The fraction is the most common scale of division; it is generally named after the agnatic ancestor's name. (Benbouaziz, 2011) Shawia have always lived in a closed traditional economical entity, based on self-sufficiency, on products that are as varied as possible and on a basis of controlled consumption. Together, all members of a family play a major role in economic life; both men and women participate in agricultural activities.

## 4 SHAWIA SETTLEMENTS

Settlements of North African tribes are found in different configurations, they often lay on a complex fragment of space. They are distributed on the hillsides, and most of the time on the least accessible summits dominating a valley. Even in the lower relief of the Saharan oases, settlements are found in hardly reachable positions. This is a characteristic of the setting of the Berber settlement (Bellal & Brown, 2001).

The Shawia settlement in the Aures demonstrates a moderate degree of savoir faire. It consists of a limited number of clusters grouped here and there. Most of the time the stones are used as a back wall. The cluster has fortified granaries and underground storage rooms where grain, crops and water are stored. In most of the Shawia settlements, the granary is the focal point of the community (Bellal & Brown, 2001).

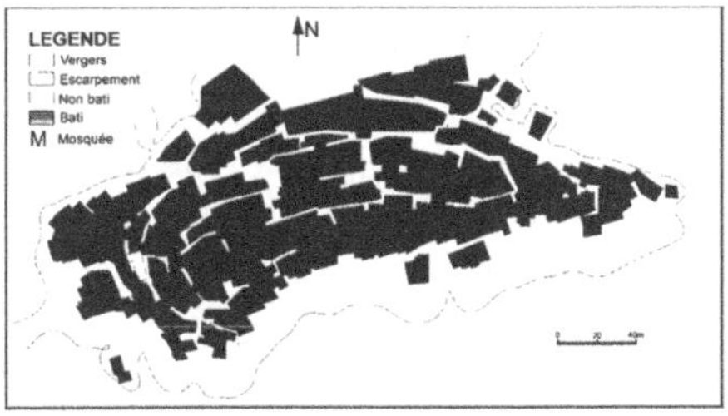


Figure 2. Urban plan schema of Menâa—regrouped settlement (Benbouaziz, 2011).

Figure 3. Perspective general view of Menaa village (inumiden.com, 2016).

Figure 4. Postcard showing a view of Tifelfel in 1930–dispersed Auresian settlement (declamp.net, 2016).

Figure 5. Houses of Himsounin around the courtyard (Flickr, 2012).

The idea of the Berber house makes sense, but it is still not easy to define. It has been mostly described as poor architecture: modest and ephemeral. Its main components are mud, stones and palm trunks. But it is far from being commonplace; many sophisticated techniques are found in the different types of buildings, and also many original and significant shapes. These buildings are most commonly used for residential purposes, but granaries and castles can also be found (Sekkour, 2011).

It is claimed in many studies that the typologies of Auresian houses change in relation to the climate, local environment, available materials and the building's function. These parameters are the same in most of the Shawia vernacular architecture settlements. (Benbouaziz, 2011) Accordingly, these settlements are classified in three types (Adjali, 2002):

- A dispersed settlement covering part of the vast area of the Constantine high plains and north piedmont of Aures.
- A regrouped settlement, more structured and more compact, often situated in mountain crests or in the depth of a valley; they are called 'Dechra' of the Aures region (Figures 2 and 3).
- A settlement of houses that gives premises a Saharian typology without undergoing its constraints; the southern piedmont house of Aures. It is defined by a group of properties around a courtyard, belonging to one fraction (Figure 5).

## 5 CONSTRUCTION RELATED CULTURE

Besides the influences of a physical nature, such as context, climate and construction materials, which had an impact on the architecture of the Shawia houses (Sekkour, 2011), there were other elements related to the culture of the local society that made house typologies distinguishable and different from one village to another, or more obviously from one region to another. Numerous daily life practices, or lifestyles, may affect the configuration of the houses. For example, in some regions in Aures, a form of builder-architects existed in each village, which the natives called 'Azurkaw' or plural 'Izurkawen'. They had a certain authority over construction related decisions affecting the shape and the configuration of houses, according to their experience and vernacular knowledge. Twiza (Tawiza or Thawiza) is also among the most influential traditions on the architectural layout and typology of Berber houses, as it is a common tradition in North African communities; it is a type of solidarity action that provides materials and physical assistance to persons in need. Community members organise Twiza to mobilise labour and assistance, in order to build new houses. It is often orchestrated by one of the builders of the village.

In the villages used for the case studies, many houses have been modified due to changes in social needs, some because of joint families becoming nuclear ones, and others due to the lack of adaptation to the requirements of modern life. However, there are still houses that are in the same condition as when they were first built, which are the ones that are the subject of this study.

## 6 COMPARATIVE ANALYSIS

### 6.1 *Case studies*

The case studies constitute representative samples in the Oued-Labiod valley, in the village of Tifelfel **(House 1)**, and the Oued-Abdi valley, in the Dechra of Menâa **(House 2)**. The two villages are situated parallel to each other between the two valleys; a mountainous region separates the sites of the two villages, giving them identical climatic characteristics. Tifelfel is home to a nomadic society that used to live an unstable mobile lifestyle. The year is divided into two long seasons; a cold season, during which the inhabitants settle in the lower valley beside the river to work on the land, harvest the crops and also as a refuge from the cold in the upland, and a warm season, which they spend in the heights of the Buyeman mountain at the other side of the river, at a higher level than the valley, where the settlement is dispersed (Figure 4), seeking for a better airflow. Meanwhile in Menâa, the village is more elaborate on an urban level (Figures 2 and 3); it has gained a structured appearance over time, giving it the prominent configuration of an Auresian village and resulting in a stable, settled society.

The information in the table below (Table 1) has been gathered after an investigation and measurement conducted by the author. It gives a description of the studied houses from an architectural point of view.

### 6.2 *Analysis*

#### 6.2.1 *House 1*

The first house represents a traditional house built between 1860 and 1870. It has not undergone any modification. It reflects the Auresian vernacular type of constructive culture in the Oued-Labiod valley. It was built with stonewalls—with a thickness of between 50 cm and 54 cm—and a supportive structure of cedar trunks carrying lightweight flat ceilings composed of juniper beams.

The house contains five main rooms with an accessible intermediate terrace. It also has two fireplaces in two separate rooms, indicating the presence of two families living in this house. The space within these rooms is used for multiple purposes, such as traditional textile fabric manufacturing, cooking and families gathering for meals. It is noticed that the two rooms open on to an intermediate terrace, allowing light and air to penetrate and having the auxiliary use of drying both food and washed rags and clothes. The guest room has two doors; one that leads directly to the outside and another going into the interior of the house. The barn is situated in the depth of the house, which means that the animals had to pass through the living room to reach their shelter. It is a spatial incarnation of the aptitude of nomadic societies to live together with their cattle (see floor plans row in Table 1).

Storage spaces are relatively limited; except for the small built-in wall closets in the entry, the terrace and the family rooms, there are no other storage spaces in the house, which is probably due to the presence of the granary in the village, where the inhabitants store most of their stocks (Figure 6).

The façade is 6 m large and oriented to the west. It has small openings in the shape of triangles, with no other form of opening except for the doors (See Figure 6 and Table 1).

Table 1. Case studies morphological presentation (Author).

| | House 1 – Tifelfel | House 2 – Menâa |
|---|---|---|
| Date | 1860 to 1870 | 1880s |
| Urban context | Dispersed; most of the houses are built beside a grove. With the presence of a focal point in the village: the granary. | Structured urban configuration; gradual range of streets. A vernacular urban fabric. See Figures 2 and 3. |
| Surface | Ground 87 m$^{2;}$ First floor 87 m$^{2}$; Cumulative surface 176 m$^{2}$ | Ground 89 m$^{2;}$ First floor 109 m$^{2}$; Cumulative surface 198 m$^{2}$ |
| Ground floor plan | 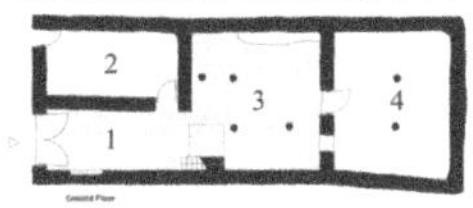 | 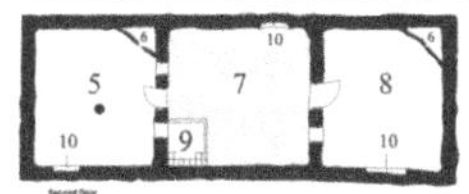 |
| Second floor plan |  |  |
| | 1 Entry 'Tasqift' 2 Guest room 3 Living room 4 Barn 'Tafrakt' 5 'Family room' 'Taddart n-Laamarth' 6 Fireplace 'isli' 7 Terrace 8 Second family room 9 Void<br>10 Storage closet 11 Bachelor room 12 Storage room 13 Ladder | |
| Façades | 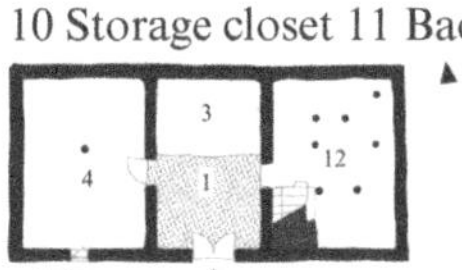 | 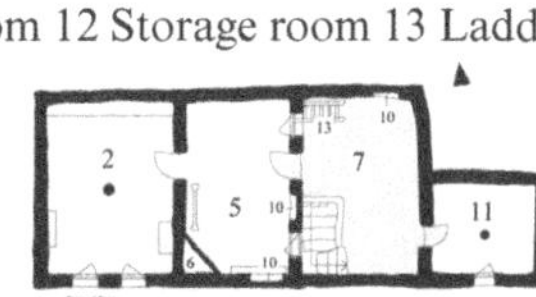 |
| Sections |  | 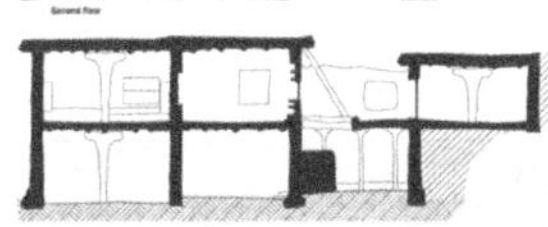 |

Figure 6. Exterior perspective on a Tifelfel house—case study 1 – (Author).

### 6.2.2 *House 2*

The second case study is a family house situated in the village of Menâa in the Oued-Abdi valley. It has recently been transformed into a museum for tourists. Apart from decoration, there has been no modification. The owner claims that the house was built around 1880, simultaneously with the eastern adjoining house, which explains the overlapping (see the façade in Table 1). Stone is used for the ground floor and the boundary walls on the second floor. The interior walls on the second floor are earthen built and they are about 40 cm thick.

This second case study shows a different layout on the plan level; the entry door opens directly on to the visually exposed exterior space, which is used as a living room or a guest room. The entry also separates the barn from the storage room. In this case, the barn is open with a small window for aeration on the exterior. At the other side, a storage room of 24 $m^2$ is used to stock reserves, tools and different life items. It has an opening onto the stairs, which leads to an intermediate terrace between the family room and a second smaller room with no fireplace. This space has smaller dimensions than the other rooms—including height—and the owner states that it is a bachelor room. The guest room is only accessible through the family room, and has two windows facing on to the street. The storage spaces inside the rooms are numerous (Figure 7). Shelves are installed within window wall breaks (Figure 8).

## 6.3 *Comparative study*

By taking a general view of this comparison, a lot of fundamental similarities are noticed between the two case studies of the Shawia houses, however, there are also a number of differences. For example, the entrances are made to visually protect the interior of the houses from the exterior passengers, through an articulation making the access indirect "The chicane". Even though that the example in case study 2 opens more to the exterior, since the external door also leads to a small living room. In the Oued-Labiod case study, there is an additional door that opens on to the guest room from the street. Families occupy private spaces in the village of Menâa in the Oued-Abdi valley; the streets are divided gradually from the largest to the tightest until they reach dead ends. This might be the reason behind the less intimate living space at the entry.

In the Oued-Labiod case study, the village is composed of dispersed settlements. Also, there is the presence of additional spaces due to different family compositions and different comfort range requirements. As can be seen in the Oued-Labiod valley case study, two fireplaces indicate the presence of two separate families, living together in a floor surface of 87 $m^2$, which indicates the degree of austerity and the low level of consideration given to

Figure 7. Interior view main family room—Menaa House (Author).

Figure 8. Storage space within window—Menaa House (Author).

comfort. On the other side of the hill, in the Oued-Abdi valley, the degree of comfort can be seen by the single-family house built on a surface of 109 $m^2$ and comprising of three rooms and a terrace (see Table 1). The materials used are identical, except for a difference noticed in the second floor walls in case study 2; they are earthen built, which facilitates the excavation of storage spaces in the walls, unlike with the stonewall, where the process is more complicated. The orientation of the facades is different in the case studies; case study 1 shows a tight façade of only 6 m long, oriented to the west, with no openings except for the two doors and triangular aeration holes.

Meanwhile in case study 2, the façade is oriented to the south, with openings on to the street from most of the spaces. The physical austerity aspect of both case studies is noticed from both the exterior and the interior, from the irregular stone used in the walls and the lack of any form of decoration. This comparison helps to build a hypothesis regarding the effect of cultural influences and traditional patterns on Shawia houses in the studied two regions. The plural identity of Shawia houses appears in the differences between the valleys.

## 7 CONCLUSION

The Berber settlement configurations in North Africa are the result of many outstanding traditions and lifestyles. It is generally an adaptation to specificities of the context. From the present research, it is clear that Shawia houses differ from one region to another. These differences are the result of numerous social and environmental practices.

In the Oued-Labiod valley, the nomadic lifestyle of the tribes makes the quest for comfort less important than issues of mobility. This results in austere dwellings, with few comfort features. This is shown in the case study by the lack of storage spaces and the relatively reduced floor surface. In addition to this, the family composition helps in the collective daily life of this society, which is why this house is a home for two families.

The village of Menâa in the Oued-Abdi valley shows more elaborate urban structures, following a logic process through its dense fabric, allowing its inhabitants to settle in a single portion of the mountain and develop a stable lifestyle, which has led to the adaptation of the houses in this static rhythm. As can be seen from images of this adaptation, more storage spaces are present in this case (Figures 7 and 8). The standards of comfort are higher in this case study, which is probably due to the community exposure.

Through the exploration of these examples from two faces of the Aures region, it is obvious that vernacular culture is one of the essential components of the Auresian Shawia identity, and that it acts not only as a major influence on the local architecture, but also on the urban scale.

Throughout the previous examples and many others, we can see that adaptation and sustainability is what this native community mastered the most. The vernacular architecture is the solid foundation of Auresian villages. The house acts like a singular piece, not only as a shelter for families, but also as a manufacturing and working place, playing a main role in the economic circle of the local community and allowing it to put up with climatic and geographical constraints and drawing out of it a singular vernacular frame, which is a result of a historical process, making the present identity of the Aures mountain villages.

## REFERENCES


Adjali, S. (2001). Traditional housing in Aures: Case of Oued Abdi Valley. In Baduel, P.R. (Ed.), Housing, State, society in Maghreb. CNRS Editions. doi:10.4000/books.editionscnrs.817.

Bellal, T. & Brown, F. (2001). Comparative analysis of M'zabite and other Berber domestic space. *Proceedings of the 3rd Space Syntax Symposium*. Atlanta.

Benbouaziz, A. (2011). *Architectural and morphological transformations of traditional housing in Aures: Case of Menâa*. (Masters thesis). Mohamed Khider University, Biskra.

Bourdieu, P. (2001). *Sociology of Algeria* (1st ed.). Paris: Presses universitaires de France.

Colonna, F. (1995). *The verses of invincibility: permanence and religious changes in contemporary Algeria*. Paris: Presses de Sciences Po.
Côte, M. (1983). "*Algeria*, Classes of 6th Year, National institute of pedagogy", Algiers.
De Lartigue, R. (1904). From the 3° Zouaves: "Monography of Aures", Constantine. Digital file. P. 4.
Hilton-Simpson, M. (1921). *Among the hill-folk of Algeria* (1st ed.). New York: Dodd, Mead and Co.
Sekkour, I. (2011). *A system of the Auresian Architecture, a genetico-syntaxic study* (Masters thesis). Mohamed Khider University, Biskra.

## WEBOGRAPHY

declamp.net. (2016). *Algerian centenaire en 1930,aures tifelfel*. Retrieved from https://www.delcampe.net/en_GB/collectables/postcards/algeria-other-cities/algerie-centenaire-en-1930-aures-tifelfel-ref-49512–407591589.html.
Encyclopédie Berbère. (2011). *Guelaa* (Taq'liath) *of Benian, Aures (photo M. Bovis)*. Retrieved from http://encyclopedieberbere.revues.org/1780.
Flickr.com. (2012). *La maison familiale de Si El Haoues à Mchounech*. Retrived from https://www.flickr.com/photos/qatkkhal/16943637306.
inumiden.com. (2016). *Dechra of Menaa*. Retrieved from http://www.inumiden.com/la-dechra-de-menaa-un-monument-de-plus-de-10-siecles/10.

*Cities' Identity Through Architecture and Arts – Catalani et al. (Eds)*
*© 2018 Taylor & Francis Group, London, ISBN 978-1-138-05409-7*

# The historical symbolizing of Istanbul city through its iconic buildings

H. Coskun
*Department of Architecture, MSGSU, Mimar Sinan Fine Arts University, Istanbul, Turkey*

ABSTRACT: Istanbul city has been historically symbolised through its iconic buildings during three different historical periods. The city that would later become a Byzantine city was formerly known as a Roman city in AD 196 and also as Constantinople, the new capital of the East Roman Empire. Finally, after 1453, during the Ottoman era, it was renamed as Istanbul which it still is today under the Turkish Republic.

According to J. Ebersolt, Constantinople's previous cityscape had been determined by its structures that were interspersed along the antique seven-hills, rising above the city walls where the blue waters of the Golden Horn ended. (Ebersolt, 1918) Thus, as an ancient features of Istanbul throughout its history, the antique hills evolved and were emphasised by newly constructed forums or buildings.

These buildings gave the city of Istanbul its identity and they were generally placed according to their importance along the antique axis of the ancient city hills. Although no longer wishing to regulate a city panorama sprinkled over the antique hills in a similar way to the Roman planning during the Ottoman period, the city still reflects the iconic structure emphasising the axis of those antique hills. (Petruccioli, 1991).

After the Byzantine period, Istanbul city was transformed from a Roman-Byzantine city in to an Ottoman city with the newly built Ottoman mosques replacing the antique Roman forums and with their columns emphasising the antique hills axis.

## 1 ISTANBUL CITY'S THREE HISTORICAL PERIODS SYMBOLISED THROUGH ICONIC BUILDINGS

### 1.1 *Istanbul city's three historical periods and constructing of its iconic buildings*

Istanbul city has been symbolized by its unique buildings during three different historical periods.The first and oldest one was the aqueduct of Valens which strongly symbolised the

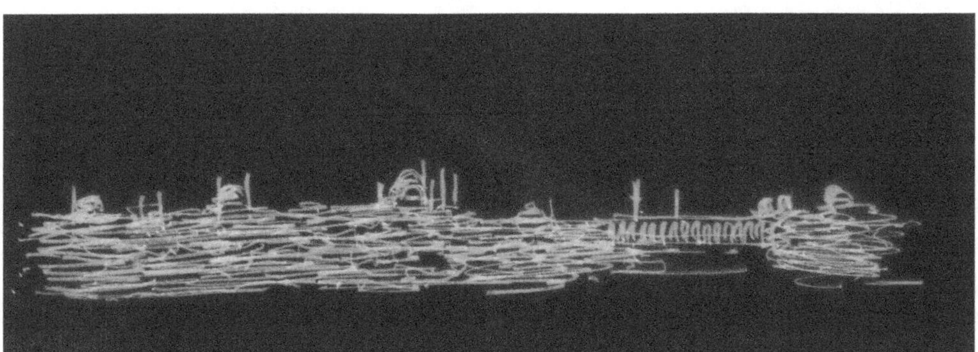

Figure 1. Istanbul city, from the character of a Roman-Byzantine city to an Ottoman city with mosques replacing the Roman forums and their columns. Suleymaniye Mosque is situated in the centre of the city and right, Roman Valens Aqueduct connecting through the axis of the hills. (Coskun).

antique Roman era of Constantinople by giving it the characteristics of a Roman city. The iconic structure of the aqueduct of Valens was first located in the silhouette of the city of Istanbul during the Roman period. The iconic aqueduct of Valens has been situated on Istanbul's antique city axis on the hills from the Roman period until the present day.

The second landmark was Saint Sophia which symbolised the antique Byzantine era of the city of Constantinople, after the Roman era by giving the city a unique Byzantine character that would later be called Byzantium.

## 2 ISTANBUL'S HISTORICAL ICONIC BUILDINGS

### 2.1 *Roman period and Valens Aqueduct*

According to periodical order aqueduct Valens had been seen in the silhouette of Constantinople as a first iconic structure during the antique Roman period.

Completed by the Roman Emperor Valens in the late 4th century AD, it was maintained and used by the Byzantines and later by the Ottomans. It still remains as one of the most important landmarks of the city. The exact date of its construction is still uncertain, but it is said to have been completed in 368 AD during the reign of the Roman Emperor Valens, whose name it bears.

Valens aqueduct extended along the antique axis of the seven hills, emphasising the third and fourth hills of the city. This spectacular multi arched bridge was laid along the valley between the third and fourth hills of Constantinople, occupied respectively at that time by the Capitolium and the Church of the Holy Apostles (Muller, 1977). Near the east end of the aqueduct there was a distribution plant, and another has been laid near Hagia Sophia's water feeds the zone of the imperial palace (Mamboury, 1953).

This iconic Roman arch between the antique hills connects Byzantine era cathedrals and the later Ottoman mosques. This connection specified due to later transformation from the Roman-Byzantine city to Ottoman city character through the newly constructed mosques replaced the old Roman forums columns with their slender minarets emphasising the antique city's axis on the hills.

### 2.2 *Byzance city's iconic Cathedral of Saint Sophia*

Cathedral of St. Sophia is located on the first hill of antique Istanbul known as early era Constantinople which is the most important of the antique axis of hills, and is a unique structure symbolising the Byzantine character of the city.

Figure 2. St. Sophia, iconic building of the Byzantine city's most important first hill. Bizans Yuruyus Yolu, (2008).

Figure 3. St. Sophia, Byzantine cathedral in the Ottoman period. View from Sultanahmet Square. Anonymous.

Constantinople was turned into an unique Byzantine city, which later would be called Byzantium. Mainly emphasised by its topographic character Constantinople, constructed by ancient Roman genius loci on its antique axis of hills with the columns at the center of the Roman forums on the eternal hills.

St. Sophia, as a preliminary starting point, was placed on the first hill of the Constantinople ancient hills sequences. So, being the most important building in the Byzantine era St. Sophia, was seen in the silhouette of the outer city from the Marmara Sea and in the silhouette of the inner city from the Golden Horn, with the perception of being three dimensional. It was was also the city's most important place, symbolising the city's aesthetic, strategic and religious values.

St. Sophia, importance as the an iconic Byzantine cathedral meant that it was also the city's most important structure during the Ottoman era, similar to its status during the Byzantine period.

## 2.3 *Ottoman city character and Suleymaniye Mosque*

The Ottomans lived in the Istanbul city did not ignore the antique axis of the hills deriving from the ancient Roman period. On the contrary, they chose to emphasise the ancient hills axes by constructing their own buildings, especially the Ottoman mosques.

Suleymaniye Mosque, placed on the third hill of the Istanbul city was the most important mosque giving the city a new definitive character after the Sultanahmet Mosque on the first hill. During the Ottoman period, Istanbul was adorned with by the axis of the ancient Roman period. Suleymaniye mosque, mostly perceived as panoramic, emphasising the silhouette of the city, and by its newly built buildings in the ancient hills that stressed the new identity of the city.

Actually, Suleymaniye Mosque was built on the third hill of the old, antique Istanbul city's axis of hills where one of the old Roman forums called Forum Theodosius took place. Thus newly built Ottoman mosques replaced of the antique Roman forums, as they were the only empty spaces in the city during the Ottoman period.

Antique Istanbul city's panoramic view with newly built Ottoman mosques were changed slowly and replaced of old Roman columns. So, the antique axis of Istanbul gained a more Ottoman character with Ottoman mosques. Fatih Mosque was the first mosque built in the city place of the demolished old Byzantine Church of the Holy Apostles, and then Sultanahmet was built on the first hill of the city next to St. Sophia and Sehzade Mosque was placed on the antique axis of the hills.

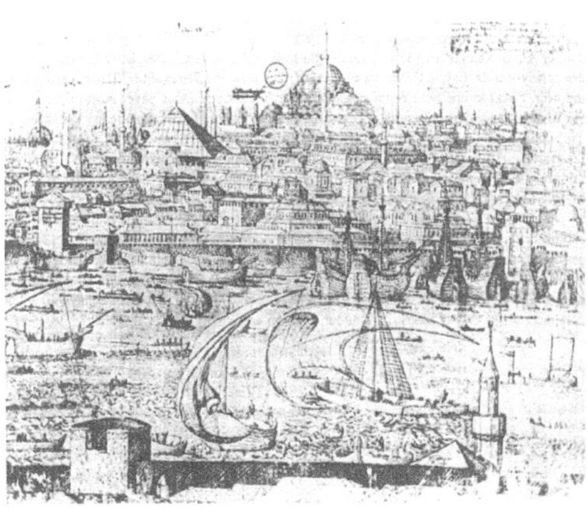

Figure 4. Istanbul, Suleymaniye Mosque Ottoman period. 1474, Gravure of Melchior Lorics.

Figure 5. Istanbul, Suleymaniye Mosque, beginning of the 20th century. IFA Archives.

However, located and integrated at the top of the centre hill of the Istanbul city's third hill, the old Roman Forum Theodosius was replaced by Suleymaniye Mosque. This mosque is still specified as the most important place within the city centre.

## 3 CONTINUITY OF ISTANBUL CITY'S IDENTITY WITH ITS ICONIC BUILDINGS

### 3.1 *Roman, Byzantine and Ottoman planning methods as continuity of a city's identity*

According to Borie and Pinon, small Ottoman domes and minarets replaced the antique obelisks remaining from the Byzantine era where their old places still could be perceived clearly in the silhouette of the city today (Borie, Pinon, 1988).

It is possible to say that since, the ancient Roman period the silhouette of the ancient Istanbul city has been determined by the demolished cathedrals that are marked as focal points. They have always been determinants and guides to the new constructions in the city. Whereas in the relatively new western quarters near Galata the imperial foundries had been such an entity (Gönen,1999).

The rules of structuring the Istanbul city hills, which were not visible like the antique seven-hills city of ancient Rome, had been pioneered by the relative continuity of the city. According to Borie, the Ottoman architects also pointed to the traditional system, saying that the architects of the time had adopted the construction rules arrangements of the buildings towards the Golden Horn, the hilltop skirts or the silhouette of the city (Borie, 1993).

These construction rules were a continuation of the Roman genius loci in the Istanbul city they were only seen in the city's historical peninsula section which had a Roman substructure. But, in such quarters as Üsküdar which was also constructed in the Ottoman period, the mosques were positioned in the lower part of city which was close to the seaside and far from monumentality unlike the old Ottoman rules.

The continuity of the Istanbul city after the Roman and the Byzantine eras, with their iconic, monumental and aristocratic constructions, had been constituted as the main reason for the continuity of these hilly structures. According to Guidoni, the fact that the construction of the Istanbul city was carried out with more monumental elements, had been seen similarly in the many Islamic cities but, their reflections of the monumentality was more symbolic than visual. (Guidoni, 1991).

Figure 6. Istanbul City, beginning of the 20th century view from the Golden Horn. On the left St. Sophia emphasising the Byzantine character and on the right Sultanahmet Mosque shows the Ottoman character of the city. IFA Archives.

## 4 CONCLUSION

The most radical changes on the historical peninsula with regards to the identity of Istanbul city were seen in during the Ottoman period. The Byzantine character of the city of Istanbul was gradually transformed into the character of an Ottoman city, with the dominant appearance of the newly built Ottoman mosques replacing the destroyed Byzantine cathedrals and also the Roman antique forums and its their columns.

Thus, Istanbul's Byzantine-Ottoman character was almost completed with the transformations during the late Ottoman period the newly built Ottoman mosques in the city's antique hills. But, in the late Ottoman period, although it had gained its predominant character from the Ottoman mosques as a continuation of antique Roman forums and Byzantine cathedrals which placed in the ancient axis of the hills where could still visible in the Istanbul city's silhouette.

Even today, as a meaning of the same idea of urban continuity with the urban identity. Istanbul city axis still depriveted from fundamental old Roman period city planning rules determining the constructing new structures in the city as long as Roman genius-loci.

As a result of the structures being built according to the eternal construction of the antique hills determined by the silhouette of the Roman city, whether it be a Byzantine cathedral or an Ottoman mosque the buildings in the Istanbul city have continued with their iconic urban fabric from antiquity until the present day.

According to S. Kostof, in Istanbul, likewise the seven hills complex topography of the Rome we also called Ottoman period Constantinople's topography with seven hills because they wanted to realise a seven-hilled city vision like Rome precisely placed with domed mosques (Kostof, 1991).

Perhaps the Ottoman structures in the Istanbul city which emphasised the antique seven-hills urban structure with its their monumental forms outlined this ancient cityscape to a grater extend than did the structures of the Roman period. It is fact that in Istanbul city, the Ottoman hills structures were more accomplished than the Roman city structures due to the continuity of the rules transposed from the antique Roman period, as S. Kostof stated (Kostof, 1991).

As emphasised by Kostof, from the antique Roman genius-loci of the topography of Istanbul city and the construction of the hills with the eternal rules lead to preserving the city's antique identity and its character also remain unchanged and have been sustained even to the present day (Kostof, 1991).

This unchanged ancient structure has been reflected by a unique character of Istanbul city that was enriched and crowned by the Ottoman mosques during the Ottoman period, along with the Byzantine cathedrals and some antique Roman city elements, such as the Roman forums and their columns that underline the antique hills and Roman aqueducts.

By the beginning of the 20th century, Istanbul city had not changed its Byzantine-Ottoman city character. However, during this period some buildings and even some Ottoman mosques would be destroyed due to the city's new planning regulations, which were more in the context of a modernisation project rather than reconstruction. New 20th century arrangements for transportation and modernisation had been purposefully conducted to render the city similar to modern European capitals. Large streets were implemented and even some important buildings were destroyed.

From the 1950's, onwards, the Istanbul city's historical peninsula was affected by the intense and massive internal migration to the city. New and modern streets had been opened in the Istanbul city, and on the edges of these streets were constructed new multi-storey buildings that would transform the area into a modern city. It was only in the historical peninsula of Istanbul city's that green areas and gardens were still seen, and the city's panoramic view has been replaced by multi-storey apartment-style residences, which has affected the city's iconic Byzantine-Ottoman character and silhouette.

## REFERENCES

Borie, A., (1993), "Sinan's Kulliye's Architectural Compositions", Mimar Sinan the Urban Vision, *Environmental Design*, Istanbul, Carucci Editore, (p. 121).

Borie, A., Pinon P., Stéphane Yerasimos Atilla Yücel, Nur Altıntıldız, (1988), *Mimar, N. 114, Journal with Institut Français d'Architecture in Collaboration,* English translation, K.Hylton, (p. 23–27).

Costenec, J. (2008), Bizans Yuruyus Yolu, Istanbul, Zero Books.

Ebersolt, J., (1918), Constantinople et les Voyageurs du Levant, (p. 17), Paris, E. Leroux.

Gönen, E., (1999), *Sur La Réhabition et l'Avenir de la Fondarie Impérialed' Istanbul,* Anatolie Moderne Yeni Anadolu VIII, Librarie d'Amerique et d' Orient, Adrien Maisonneuve Bibliothéque De L'Institut Française d'Etudes Anatoliennes-George Dumézil XLIV, Paris, (p. 221–236).

Gravure of Melchior Lorics (1474).

Guidoni, E., (1991), "Sinan construction of the urban panorama", Mimar Sinan the urban vision, *Environmental Design,* Istanbul, Carucci Editore, (p. 20).

IFA Archives, Paris.

Kostof, S., (1991), The citys shaped, (p. 288), London, T&H.

Mamboury, E., (1953), The tourist Istanbul, (p. 57). Istanbul, Çituri Biraderler Basımevi.

Müller-Wiener, W., (1977), Bildlexikon zur Topographie Istanbuls: Byzantion, Konstantinupolis, Istanbul bis zum Beginn *d. 17 Jh* (in German), (p. 78), Tübingen: Wasmuth.

Petruccioli, A., (1991), Mimar Sinan The Urban Vision, *Environmental Design*, Istanbul, Carucci, Editore, (p. 8–9).

*Interior design*

*Cities' Identity Through Architecture and Arts – Catalani et al. (Eds)*


# Anticipating possible future visions in interior architecture

A. Abdel-Hadi & A. Harb
*October University for Modern Science and Arts (MSA), Giza, Egypt*

ABSTRACT: An in-depth analysis of today's design practices and discourses can help in forming a vision about the possible future of a city's identity. This paper focuses on the hypothesis that new visions for future trends in interior architecture could be anticipated through the choices of the current concepts expressed in the graduation projects of the Interior Design students at October University for Modern Sciences and Arts (MSA). The interpretation of the results shows a major tendency towards global design trends, with some local cultural influences, according to each student's contextual affiliation.

*Keywords*: interior architecture—future vision—pattern metamorphosis

## 1 INTRODUCTION

The city is a reflection of constant changes, both tangible and intangible. The process of its dynamics is due to the fusion of a multiplicity of factors: historical, cultural and the degree of exposure to the global trends at large in all facets of life. An in-depth analysis of today's design practices and discourses can help in forming a vision about the possible future of a city's identity.

This paper examines the question of what the possible design directions and trends could be in the upcoming years. It focuses on the hypothesis that new visions for future trends of interior architecture could be anticipated through the choices of the current concepts expressed in the projects of Interior Design students.

Interior Design students at MSA University are required to submit a bachelor thesis on a chosen topic for their graduation project and to apply research findings in the design phase. In the year 2015/2016, the students were encouraged to thoroughly explore, analyse and study a pattern of their choice in response to the theme, 'Pattern Metamorphosis: Historical, Natural, Social or Invented' which was given to them by the teaching staff. Suggesting that the students research their chosen topic within a broad theme has encouraged them to form a well-structured exploratory research, thus helping them to obtain insights for their final design projects and to consolidate coherent philosophical design concepts. The students' design outcomes adopted different approaches, ranging from bio-mimicry to historical and geometrical visual analysis, which inspired their process of form generation, use of materials and the creation of their projects' visual identity.

The aim of this paper is to verify whether the visions of the students, through their work give an indication of the future identity of the city's interior architecture and whether the outcome is more geared to local or global trends.

The methodology of anticipation follows a systematic enquiry using participant observant techniques during the research and design phases followed by a comprehensive analysis of each design project. The students' cultural identity and the contextual factors are also thoroughly investigated and analysed. The study made up on 35 students (28 females/7 males) from the 2015/2016 interior GRADS.

## 2 RESEARCH AND DESIGN

### 2.1 *Research design*

As stated above, the Interior Design students for the graduation year 2015/2016 were asked to work on a design research project that should encompass two phases: the first was to investigate and explore the requirements and specifications of a particular design project and the second was to explore and analyse a chosen pattern (historical, natural, social or invented) and accordingly form a solid outcome combining both the functional and the philosophical aesthetical aspects of a specific interior design project.

Their research was intended to reinforce the students' knowledge and skills by providing them with the practical expertise needed in all phases of the design process. Its main goals were to qualify the students to identify and apply their knowledge in formulating research issues and the aims pertaining to the topic of their project, and, also, to effectively write the literature review, the research methodology and the results of their studies. Moreover, it teaches students how to use effective communication aids in presenting their research outcomes and leads them to further develop their project's design programme and concept throughout the design phase.

### 2.2 *Thematic issues*

'Pattern Metamorphosis' was the theme given to the students, in which they were greatly encouraged to thoroughly explore, analyse and study a pattern of their choice. The types of patterns are explained in terms of historical, natural, social and invented:

*Historical patterns*: to explore the visual identity of a specific period of time, its contextual factors and transformation through time, as well as the reasons for its change and its logic.

*Natural patterns*: to discern patterns in cosmological phenomena; in natural resources, such as rivers and topography; in environmental factors and forces, such as water ripples and earthquakes' visual consequences; and in biomimetic which imitates any biological occurrences.

*Social patterns*: to observe and document behaviour and activities whether seasonal, self—generated, economically driven and/or cultural personifications.

*Invented patterns*: to analyse a variety of design directions, such as parametric design and digital fabrication, glitch art or abstract design.

The thorough exploration of a particular pattern led each student to deeply understand the meaning and philosophy behind the chosen pattern, and to be able to apply this knowledge in the following design phase. For the students' research projects, it was important to note that, despite efforts to define the design's research approach and to establish it more clearly as a research discipline in its own right (Friedman, 2008, cited in Lockton, 2012), there remains little consensus regarding the boundaries of what counts as 'design methodology' in an academic sense (Kimbell, 2011).

This justifies the suggestion of a mixed academic research with a design project, as design is thought to be a mixture of many other disciplines, such as sociology, human factors, physiology and architecture, and this concept is particularly relevant to the design approach of interior design, as it is concerned with understanding users in terms of culture, behaviour and characteristics. According to Pontis (2010) 'research through design involves both understanding the process of design itself and developing new design actions, artefacts or methods'. It is also noteworthy to mention that 'research is systematic enquiry whose goal is communicable knowledge; systematic because it is pursued according to some plan; an enquiry because it seeks to find answers to questions; goal-directed because the objects of the enquiry are posed by the task description; knowledge-directed because the findings of the enquiry must go beyond providing mere information; and communicable because the findings must be intelligible to, and located within some framework of understanding for, an appropriate audience' (Archer, 1995). Students were encouraged to follow the above-mentioned criteria in their graduation research project, and they have been able to establish and consolidate a solid research outcome that helped them to inform their final design project.

## 3 DESIGN CONCEPTS DEVELOPMENT

In this section, an introduction about concept design formulation has been given, along with an explanation of the projects chosen by the students and their design concepts. Design concepts development is the backbone of any design project, as it is the process by which designers and innovators approach various design solutions with regards to a specific problem. A concept development process involves using a set of creative processes in order to achieve the desired goals; these processes include brainstorming sessions, visual analysis researches, affinity mapping and sketching ideas. The existence of a conceptual framework helps designers to think in context and to always be focused on specific design goals and intents (Kotsopoulos, 2007).

Interior Design students are required to learn how to envision spatial solutions as well as to create functional and aesthetic values; the nature of interior design requires the ability to represent the contextual relationship between humans and the surrounding environment (McAuliffe, 2007).

### 3.1 *Current design discourse*

In recent years, design has been inclining towards the recent developments and innovations in mathematics, physics and biology, and this has led to a strong relationship with the form generation process. Although these developments appear as if they are concerned with the spatial values, programmatic solutions and building structures, the fact is that the main issue that directs these developments is the problem of form generation. For instance, fractal geometry, non-linear geometry and topological geometry are used in architecture design as a tool for form generation (İnceköse, 2007). Other developments such as parametric design, biomimetic and non-Euclidian geometry, have also been used as tools for form generation. These trials and tools are made to present distinct and novel forms and they have already had a significant effect on the form generation process.

The current discourse reading architecture and interior design has highlighted that architects and interior designers are always attracted by the new geometrical concepts and theories, such as the afore mentioned developments, in order to produce complex, dynamic and novel forms and structures. This notion is accentuated by the fact that several of these developments and attempts have been adopted by a number of giant design firms and consultancies, such as Foster + Partners, Greg Lynn Form, Zaha Hadid Architects and many more. These developments are thought to show the current direction of solving form problems through various methodological processes of visual research, form analysis and algorithmic mathematical form generation developments. The geometric analysis of patterns and forms has made a great contribution to the process of form generation and spatial design. Through the analysis of the structural aspects behind each chosen pattern, designers go through a process of the visual exploration of patterns, trying to reveal the system behind the formation of this pattern and thus develop this system and integrate it into their conceptual design process (İnceköse, 2007).

### 3.2 *Concept and form generation*

The afore mentioned theme was given to students as a methodological way of introducing them to these concepts and techniques, thus allowing them to provide their design project with the outcome of an extensive visual, conceptual and philosophical research. Students were advised to follow a complete concept development process through the analysis of a specific pattern of their choice. Likewise, guidance was offered to them regarding the selection of the relevant pattern for each design project, as well as directing them through the visual analysis and form generation processes and mentoring them on the techniques and methodologies of 'concept and form' generation.

Students were introduced to the double diamond process developed by the British Design Council (2007) so that they could work within its framework, which consists of four distinct design phases: discover, define, develop and deliver (Figure 1). This framework maps the

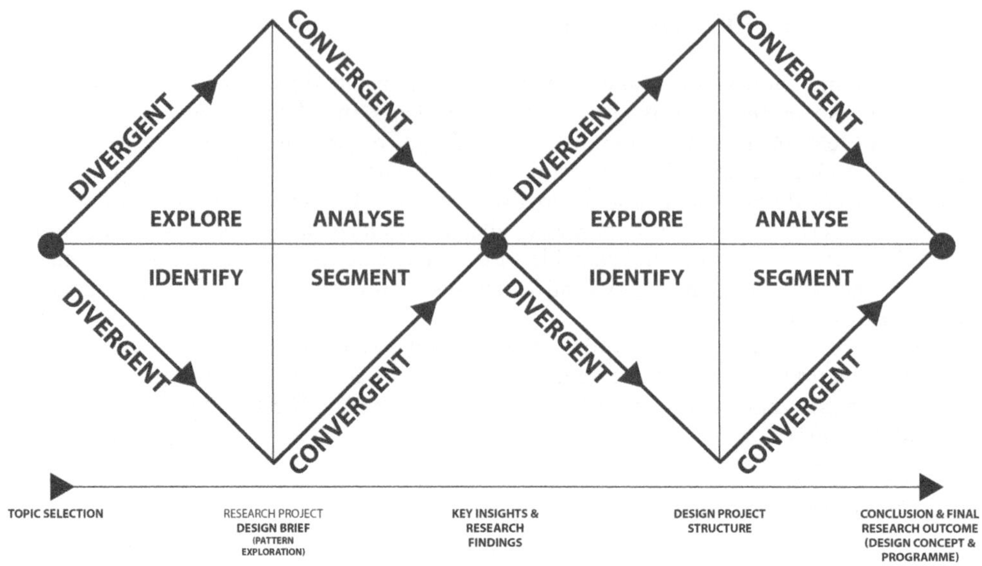


Figure 1. The double diamond process model. Copyright Ammer Harb.

divergent and convergent stages of the design process, thus showing the different modes of thinking that designers use.

Considering the above, the students went through a comprehensive design process from design research to a complete design project; this included design problem identification, design thinking, verbal communication, data gathering and a literature review. Visual aspects of the design were explored and reflected through sketches and 3D models. Throughout the process, there was a clear transformation (metamorphosis) in their ideas during the research and concept development process. It has to be noted that visual representations in the students' projects, such as sketches and 3D models, are seen as an aesthetic problem-solving method, in which the conceptual ideas of the design projects are consolidated in visual form.

## 4 DESIGN OUTCOMES

### 4.1 *Theme implemented*

A total number of 35 students have completed their design projects, and the results have shown a tendency towards adopting the natural patterns as a core element for the form generation process; the number of students who adopted natural patterns was 17, historical patterns were implemented by 9 students, invented patterns were implemented for 7 projects and, lastly social patterns were used in 2 out of 35 projects.

Within the scope of the natural patterns, the students' selections have shown a tendency towards both the patterns created from natural resources and biomimetic inspired patterns. Students also used the historical patterns as a source of inspiration and found in them a significant opportunity to approach a design style that is both global and local. This was achieved by analysing a very local pattern yet applying different geometric transformation methods and techniques on them, thus approaching this mixture of a traditional inspired form but in an international design language. Both the invented patterns and the social patterns were not of great interest to the students; this low uptake might be explained by the difficulties in adopting those directions. These difficulties arise from the fact that a social and/or invented pattern would need a defined in-depth analysis and numerous observation sessions, in order to reach an expressive and relevant pattern. This might take a longer time and greater efforts than a straightforward pre-created pattern that can be found in nature or history. It is also worth noting that one student had a mixture of two approaches, which were the glitch art and the parametric design directions; this mixture would have been approved based on the philosophical point of view of the student that the project would need to implement both directions in order to achieve her vision.

The projects taken are segmented into nine categories: Adaptive Reuse, Community Service, Culture Centres, Educational, Healing Environment, Renovation, Roads' Facilities, Sports Centres and Tourism.

Within the natural patterns selections, almost 89% of the project categories have been tackled however, for the historic patterns, only 55% of the categories have been tackled, which is the same as for the invented patterns; with the social patterns scoring only 22%. It has to be noted that some of the project categories have been tackled by more than one project and some of them have a very limited uptake; while the community service project category scored the biggest number with 11 projects out of 35, the roads facilities category had the lowest uptake, scoring only one project out of the 35 projects. However, the results have shown the diversity in the projects' selections and this has been proven by the fact that almost every category has at least one project selection (Figure 2).

Table 1. The numerical relationship between the selected patterns and the selected projects.

| Types of patterns | | Adaptive resuse | Community service | Culture centres | Educational | Healing environment | Renovation | Roads facilities | Sports centres | Tourism | Total |
|---|---|---|---|---|---|---|---|---|---|---|---|
| Historical patterns | | 2 | 3 | 2 | 1 | | | 1 | | | 9 |
| Natural patterns | cosmological phenomena | | 1 | | | | | | | | 17 |
| | natural resources | | 2 | 1 | 2 | 1 | | | | 1 | |
| | environmental factors | | | 1 | 1 | | 1 | | 1 | | |
| | biomemtics | 2 | 2 | | | | | | 1 | | |
| Social patterns | seasonal | | | | | | | | | | 2 |
| | self generated | | | | | | | | | | |
| | economic driven | | | | | | | | | | |
| | cultural personification | | 1 | | | | 1 | | | | |
| Invented patterns | parametric design | | | | 0.5* | | | | | 1 | 7 |
| | digital fabrication | | 1 | | | | | | | | |
| | glitch art | | | | 0.5* | | | | | | |
| | abstract design | 1 | 1 | | 1 | | | | 1 | | |

(*) The 0.5 indicates that one case has combined both types of patterns: the parametric and the glitch art.

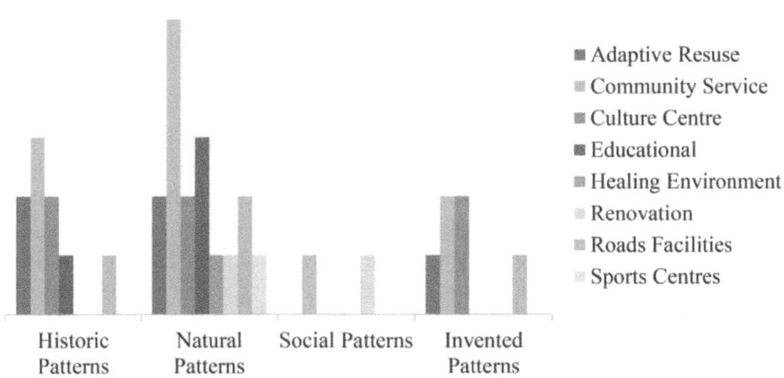


Figure 2. Bar chart showing the proportional relationship between selected patterns and selected projects.

Table 2. The results of pattern implementation in the projects.

| | Pattern implementation | Fully implemented | | | Partially implemented | | | Non-implemented | Total number of projects | Results |
|---|---|---|---|---|---|---|---|---|---|---|
| Types of Patterns | | 2D&3D | 2D | 3D | 2D&3D | 2D | 3D | | | |
| Historical patterns | | 1 | 3 | 0 | 0 | 1 | 0 | 4 | 9 | 56% of the students have implemented the historical patterns in their design projects and 44% have not. |
| Natural patterns | | 5 | 1 | 0 | 0 | 2 | 0 | 9 | 17 | 47% of the students have implemented the natural patterns in their design projects and 53 % have not. |
| Social patterns | | 0 | 0 | 0 | 0 | 2 | 1 | 0 | 2 | 100% of the students have implemented the social patterns in their design projects. |
| Invented patterns | | 4 | 2 | 0 | 0 | 0 | 0 | 1 | 7 | 86% of the students have implemented the invented patterns in their design projects and 14% have not. |

The students adopted various design approaches that can be segmented as follows: geometric transformation; metamorphosis and evolutionary; morphological progress; biomimetic; and, finally, abstraction and symbolic representation of spatial aspects.

These categories can be seen clearly in the students' works; in their visual analysis research, exploration methodologies and their design projects. The choice of patterns, visual analysis methodologies and the selection of projects show how the background, cultural identity and personal experience of each student affect the design process; in each particular category of projects and with the same adopted pattern, each student came up with a totally different design project and direction, thus fulfilling its unique vision.

Sixty percent (60%) of the students implemented the patterns in their final design projects, yet 40% of the students were not able to implement the pattern. This relationship is shown in Table 2 as percentages for each type of pattern. This uptake indicated that some of the students' choices were not suitable for the types of projects selected and, thus, the implementation of such patterns would not serve their intended design goals. It is worth noting that, although the teaching staff were already aware of this, they allowed the students to freely choose what they thought was suitable in order to let them learn from the full experience.

### 3.2 *Student projects*

In the following section, a sample project for each type of patterns has been briefly explained and illustrated in order to show the development (metamorphosis) of the patterns to the final design projects.

*Historical patterns*: Zienab Salam has chosen to work on an Islamic Cultural Centre in Spain. She has implemented the Islamic geometric patterns in her design concept, as well as working on a complete pattern analysis process to reflect the modern and unique identity of Islamic Spain. The objectives of the cultural centre were to provide a gathering point where Muslims can perform their prayers, and to accommodate a wide spectrum of educational and cultural activities, such as lectures, seminars and exhibitions. Through these activities the centre helps to provide the non-Muslim community with a clear and better understanding of Islam (Figure 3).

*Natural patterns*: Neveen Hadi's project is an educational and health care centre for rural villages in Egypt. Her concept is inspired by the process that ants use to build their colonies, and this process has fed the form generation process and the design concept in general. She has analysed the colonies' building systems and then imitated these efficient systems in the design of the centre. Her project is a sustainable model and system for educational and health centres, which can be used in any of the rural villages across Egypt; she has also chosen 'Nazlet Khater', an Egyptian rural village, as a case study for her project (Figure 4).

*Social patterns*: Youssef Sherif's project was renovating the current 'El-Sawi Cultural Wheel'. Youssef' s design concept was inspired by a very special social pattern, which is the 'Tanoura' dancing pattern; he has linked the concept of the dancing to his project and this link was made on the understanding that both the cultural activities and the Sufism activities have a lot of common values. One of the values on which he built his concept is the 'Catharsis' philosophy by Aristotle; it emphasises the meanings of emotions' purification and soul isolation through the art of dancing. He analysed and de-constructed the movement pattern of the 'Tanoura' dancers to get several forms of inspiration, which he then turned into spatial forms (Figure 5).

*Invented patterns*: Farah Khaled's project was to design a TV channel building and training centre. She has worked on the concept of 'cellular automata' along with the 'glitch art' in order to formulate her design concept. Cellular automata is a mathematical model in computability theory, physics and microstructure models; it is a pure unexpected behaviour and

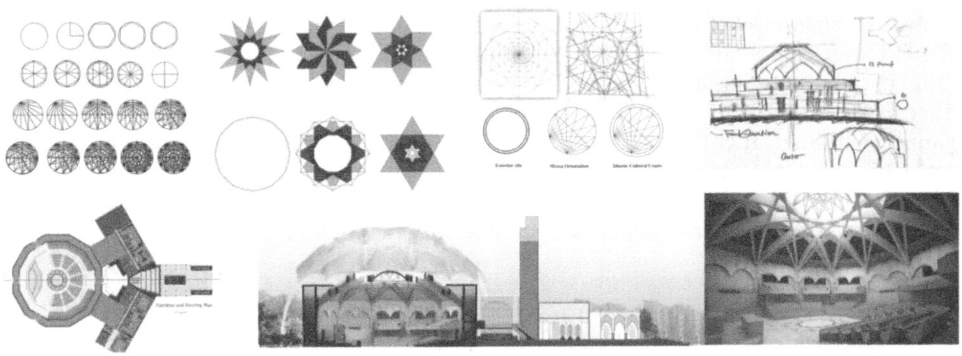

Figure 3. Development and implementation of historical geometric patterns in the design of the Islamic Cultural Centre in Spain—Copyright the designer—Zienab Salam—used with permission.

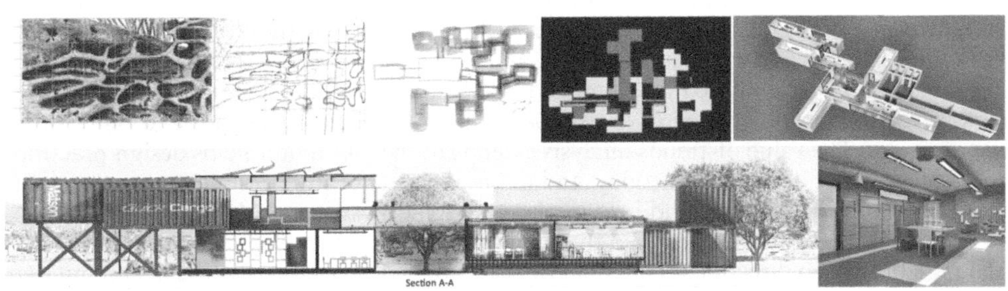


Figure 4. Development and implementation of natural patterns in the design of an Educational and Health Centre in Egypt—Copyright the designer—Neveen Hadi—used with permission.

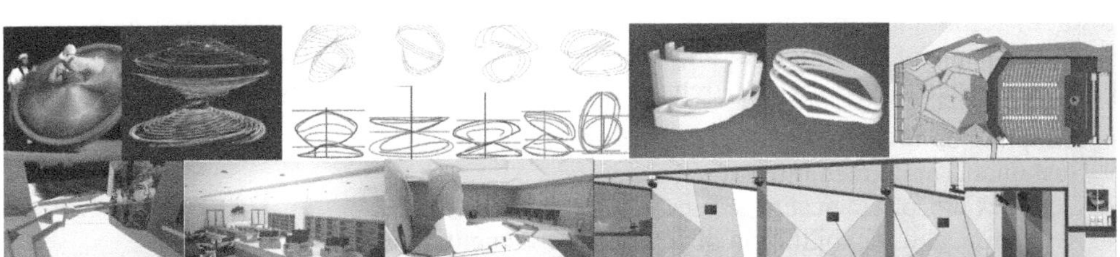

Figure 5. Development and implementation of social patterns in the renovation of 'El-Sawi Cultural Wheel' in Cairo. Copyright the designer—Youssef Sherif—used with permission.

Figure 6. Development and implementation of invented patterns in the design of a TV channel in Cairo. Copyright the designer—Farah Khaled—used with permission.

an abstract environment pattern formation of cells of an infinite number of states (on and off). However, 'glitch art' is the deliberate introduction of digital interference artefacts into digital photographs; it is an image marked by all kinds of 'errors' in its structure. Errors occur either in the form of undefined colours, or as pixelated parts, or with vertical stripes offset against each other. It could also be the combinations of all three types of 'errors'. In order to mix the cellular automata with the glitch art, Farah has analysed the 'Conway's Game of Life', which is a very common application for the cellular automata, and then she applied the rules of the game in order to generate spatial forms (Figure 6).

## 5 ANTICIPATING FUTURE VISIONS

Trends analysis is a method frequently used in design to analyse current and possible future trends. It might have different terms used to describe the same meaning; however, it essentially highlights the knowledge and awareness of current and potential tendencies in design directions. A trend is the general direction in which something is developing or changing (Evans 2005). Lindgren & Bandhold (2003) argue that a trend is considered to be some act that affects a deeper change than a 'vogue'. The existence of a trend means that it already has some tendency and inclination; that is, any direction could be a trend, yet the uptake of this direction, either a limited or extended uptake, would indicate its importance. A trend is discovered rather than created (Cornish, 2004); it should be easy to spot and segment in a clear manner. The value of trends analysis extends to the fact that it helps design practitioners to make better decisions for today, and to anticipate the future consequences in the form of opportunities and threats and help in planning to address them (Malhotra et al., 2014).

The researchers have performed a trend analysis of the students' projects and have identified possible design directions that could be the leading design trends in the upcoming years. Analysing the students' projects has accentuated many possible future directions in interior design and in design thinking in general. The analysis also indicated the affinity of the students towards the global and international design directions, and the clear effect of globalisation on the design thinking and design decisions taken throughout the process. This effect might be relevant to the hypothesis that the world is going through a fusion of cultures as a result of technology and the ease of accessing the cultures of others. However, and it has to be noted, some of the students, although a very limited number, have adopted some ethnic and local design directions. The trend analysis has shown a tendency towards the following:

- Adopting biomimetic approaches in interior architecture;
- Sustainable and green interior architecture;
- Complex morphological processes in interior architecture;
- Reforming and rehabilitation of informal settlements;
- A limited uptake for ethnic or traditional design directions.

Last but not least, this research represents a case study assessment with results that are indicative but could not be generalised; hence, further studies in this domain are needed together with comparative evaluation between similar cases in different cultures. Hereafter, the authors are looking forward to pursuing other research in the area of 'Global vs. Local' design approaches from an educational point of view.

## ACKNOWLEDGEMENTS

*The researchers would like to thank all of the teaching staff—Instructors and Tutors—who taught and tutored the students throughout this course of the graduation project I & II (2015/2016).*

- *Prof. Dr. Hassan Metwalli, Professor of Interior Architecture & Head of Décor Department, Fine Arts Cairo, HU.*
- *Dr. Diaa Tantawy, Associate Professor of Interior Design, Faculty of Applied Arts, HU.*
- *Dr. Hoda Gad El Rab, Lecturer of Interior Design, Faculty of Arts & Design at MSA.*

*Together with the Lecturer & Teacher Assistants in the Interior Design Department, Faculty of Arts and Design, MSA: Heba Eissa, Noha Essam, Ammer Harb, Sara Sabahy, Asmaa Awad and Laila Alaa.*

## REFERENCES

Archer, B. (1995). The nature of research, *Co-Design, Interdisciplinary Journal of Design*, 1 (2): 6–13.

British Design Council (2007). Eleven lessons: Managing design in eleven global brands, a study of the design process. Retrieved from: http://www.designcouncil.org.uk/resources/report/11-lessons-managing-design-global-brand (accessed on 15/01/2017).

Cornish, E. (2004). Futuring: The exploration of the future. USA. World Futurist Society.

Evans, M. (2005). *I-SPY: Utilising forecasting and scenario planning for design futures*. European Academy of Design Conference, University of the Arts, Bremen, Germany. Conference paper: DOI: 10.13140/RG.2.1.4800.9763.

İnceköse, Ü. (2007). Geometrical Transformation: A Method for the Creation of Form in Contemporary Architecture. *Paper presented at the Tenth Annual Bridges Conference, Bridges: Mathematical Connections in Art, Music, and Science*. School of Architecture, The University of the Basque Country, San Sebastian (Donostia), Spain. Retrieved from: http://archive.bridgesmathart.org/2007/bridges2007-415.pdf.

Kimbell, L. (2011). *Designing for service as one way of designing services, International Journal of Design*, Vol 5, No 2 (2011).

Kotsopoulos, S. (2007). Design concepts in architecture The porosity paradigm. *Proceedings of The First International Workshop on Semantic Web and Web 2.0 in Architectural, Product and Engineering Design*. Busan, Korea.

Lindgren, M & Bandhold, H. (2003). *Scenario planning*. New York, USA: Palgrave. Macmillan.

Lockton, D. (2012). *Design with intent, A design pattern toolkit for environmental & social behaviour change*, (Ph.D.). School of Engineering & Design, Brunel University, London, UK.

Malhotra, S., Das, L.K. & Chariar, V.M. (2014). Design research methods for future mapping. *International Conferences on Educational Technologies and Sustainability, Technology and Education 2014*. New Tapei City, Taiwan.

McAuliffe, M. (2007). Considering the role of presence in the conceptual design of interior architectural environments. *10th Annual International Workshop on Presence*. Barcelona, Spain.

Ponis, S. (2010). Types and approaches of (design) research. Mapping complex in formation blog. Retrieved from: http://sheilapontis.wordpress.com/2010/11/05/types-and-approaches-of-design-research/ (accessed on 20th of December 2016).

*Cities' Identity Through Architecture and Arts – Catalani et al. (Eds)*
 *ISBN 978-1-138-05409-7*

# Contemporary Egyptian theatre and heritage

Omneya Ahmed Yehia
*Faculty of Fine Arts, Cairo, Egypt*

ABSTRACT: There are a number of different types of spectacle and paratheatrical forms that constitute popular Egyptian heritage. By the end of the twentieth century, and the beginning of the twenty first century, new and diverse theatrical attempts had appeared, which tried to invest these forms in order to promote Egyptian theatre and give it a local identity through the use of themes and characters derived from heritage and traditional sight relationships. These forms renewed the relationship with other arts and forms of celebration and spectacle. The idea of gaining inspiration from heritage merged from the desire to reach the minds of the citizens and to modify the present through the revival of ancient folk traditions. Some of the findings of the research are that the theatre is a social phenomenon that affects society and is influenced by its culture and heritage. Moreover, the contemporary theatre uses new theatrical spaces and performances that are inspired by heritage and are trying to preserve the Egyptian identity.

*Keywords*: paratheatrical forms; Egyptian heritage; Egyptian theatre; contemporary theatre, theatrical spaces

## 1 INTRODUCTION

'The Egyptian history in the Middle ages till the 18th century interested the orientalists during the 19th century, especially after the release of the book 'Description of Egypt' written by the scientists accompanying Napoleon's campaign to Egypt. They draw many paintings, and wrote many books discussing this period, and after the foundation of Cairo University in 1925, many thesis analyses its main features' (Paul, 2015, P15). From these books and theses we could recognise the different types of spectacle and paratheatrical forms that existed in popular Egyptian heritage. Among these forms we can list the storyteller, the narrator, the imitator, the shadow show, the puppet (*Al Aragoz*) and the rural Samer. The research studies the new and diverse theatrical attempts that tried to invest these forms in order to promote Egyptian theatre and give it a local identity through the use of themes and characters derived from heritage and traditional sight relationships. The research discusses the main ideas and movements that have tried to develop the modern and contemporary Egyptian theatre by returning to its roots. Moreover, it studies the new theatrical spaces, and analyses plays and performances that are inspired by heritage and are trying to preserve the Egyptian identity.

For the extrapolation of the future of the Egyptian theatre, we have to contemplate the people's nature, their signs, indications, culture, heritage, origins, and also their present. Also, we have to contemplate how much the Egyptian theatre was influenced by western theatre, with its openness to the prevailing theatrical currents in the world, in order to reach a clear plan enabling us to develop theatre as a cultural institution in an attempt to upgrade the theatrical culture of the Egyptian audience and enable it to catch up with the movement of civilisation and culture. Meanwhile, a critical awareness of heritage is needed, and this is the first step on the way to finding our true cultural identity. True heritage awareness cannot be completely achieved in a frame of cultural isolation. The correct perspective cannot be achieved, nor can the real aspects of communications be unfolded, amid the different cultural varieties without comparing our heritage with the harvest of human experiences in different cultures. In the last few years, most of the creative people in Egypt have headed towards heritage elements for inspiration, in order to establish the Egyptian heritage theatrical forms and derive their creative entities from heritage.

In the second half of the twentieth century, especially since the beginning of the sixties—which witnessed the peak of Egyptian theatre's glory—Egyptian theatre witnessed many serious experiences that dealt with the most important trends and distinguishing features of the contemporary theatre, such as inspiring elements from heritage, image theatre, the use of untraditional theatrical spaces, audience participation in the theatrical game, touring theatre, people's theatre, provincial theatre and mixing cultures. The Egyptian theatre knew the epic theatre through watching Bertold Brecht's works. The recognition of epic theatre coincided with the orientation of Arab theatre to use elements derived from local heritage, such as the storyteller '*Al hakawati*' and the narrator '*Al rawi*'. Brecht's theatre was also one of the things that drew the attention of the Arab playwrights to the possibility of returning to their heritage and seeking local elements in it. Brecht's theatre was one of many local and international branches that were found in the Egyptian theatre. The accompaniment of these theatrical currents, with the conflict between their philosophies, had the greatest effect in enriching the theatrical renaissance.

This evolution in perspective coincided with the multiple attempts that had taken place since the beginning of the twentieth century to revive forms of theatrical performances from the past and from different cultures. It was also crystallised as a result of the evolution of human sciences, such as anthropology, sociology, psychology and semiology, which were concerned with theatrical performance, and by studying the celebration phenomena that preceded the appearance of the theatre in historical societies. 'The attempts of the Western theater to return to theater origins became a distinct orientation when it was associated with the desire to search for new formulas for the reviving of the theatrical phenomenon, through returning to the old forms that accompanied the birth of theater, and through trying to benefit from the rituals and the ceremonies that are still practiced in their primitive forms in other cultures' (Elias & Kassab, 1997).

'During the journey of theater in Egypt, there were attempts to found a theater, through the mating between the public celebrative manifestations with their dramatic roots, and the techniques of the western theater, Yacub Sanu initiated them when he benefited from some dramatic manifestations in Farce actors 'Al Mohabeziin', shadow play 'khayal Al zel' and the Puppet 'Al Aragoz', on both form and subject levels. These attempts widened with the increased interest in folklore academically and technically' (Hussein, 1999, P246).

The playwright Alfred Farag stressed two basic ideas, those of social thinking and identity instilling. He says: 'The concept of identity was developed in the fifties, and firmly urged the young creative to translate it into a creative reality; our job is not just taking a story from One Thousand and One Nights to consolidate our theater with, but to do so in a way that reflects the Egyptian reality'. (Alra'i, 1997, P95) Yehia Haqqi explained that in order to establish an Egyptian identity for theatre, we have to turn to folklore and resort to folk storage. Due to this thinking, Yusuf Idris probably headed towards the Samer Theatre. (Alra'i, 1997, P97) Tawfik Al-Hakim also wanted to give his theatre a folk dye. The most important things with regards to consolidating and identity instilling were not only returning to folklore but, essentially, raising social issues through these frameworks. An orientation emerged that searched for the secrets of the strength and greatness of the Egyptian people through searching and studying their heritage, which led to inquires that preoccupied the playwrights. The motivation for these inquires was the expansion of national sentiment and the increasing role of the people. 'The two inquires that the revolution had inspired the playwrights' consciences with were: was this theater closely related to the heritage of the country, or was it an imported translated one, and why shouldn't we have our own Egyptian theater? Also, was the presented theater a real folk theater, or was it just for the intellectuals?'. (Alra'i, 1997, P101). 'The attempts to answer those two inquires have led to the search for a new theatrical form expressing the People in a language and techniques that were close to their folklore, the folk spectacle that ruptured the playwright's sense of being Egyptian, and also imitating the multiple attempts of socialist countries to benefit from heritage as an inspiration' (Hussein, 1999, P217). Benefiting from heritage was not only limited to folk literature, but it also surpassed it to benefit other forms of spectacle.

We can separate the methods of trying and benefiting from literature, heritage and folk art in the search for an Egyptian theatre into two routes:

First: trying to benefit from the different forms of folk literature, such as myths, biographies and tales, to create dramatic themes that are used to project ideas on to contemporary reality, or to create a mating between heritage (originality) and reality (contemporaneity), just as Naguib Sorour did when creating the plot of his play From where to bring people '*Mnen agib nass*' where he coupled between Isis and Osiris myth with its Pharaonic origin, and the folk tale of Hasan and Na'ima and edited them into a texture of contemporary reality of the pre-revolution Egypt. 'And from folk tales with social impact, Naguib Sorour inspired his anecdotal material of Yassin and Baheya, he also resorted to other sources like myths and folk Mawwal from contemporary reality' (Hussein, 1992, P215). Thus the playwrights surpassed One Thousand and One Nights as a playwriting source to other forms, thus Shawqi Abdel hakim wrote Shafiqa and Metwali based on the fictional Mawwal known with the same name, Yosry Al Gendy wrote Ali Al Zeibaq and Ya Antara, but Alfred Farag was committed to One Thousand and One Nights in The Barber of Baghdad.

Second: the attempt to find a new form of theatrical spaces, or to benefit from the folk spectacle forms in creating a theatrical performance with an Egyptian identity and character.

The theoretical callings adopted by Yusuf Idris, Tawfik Al-Hakim and Ali Alra'i began to demand the search for an Egyptian theatrical form in 1964.

Yusuf Idris began with a series of articles calling for the search for an Arab theatre, where he said that he believed that in order to have our own theatre we have to take account of all forms of celebrations and folk spectacle, in order to create a theatre with new content and new heroes. He confirmed that the theatre that we currently presented depended on an imported formula that was strange to us, and it was our duty to search for an Arabic form of theatre that was already known and accepted by our audiences, which was the rural Samer Theatre. (Hussein, 1992, P227).

He invited the playwrights to participate in the search for a theatrical form stemming from heritage. Yusuf Idris' theory is summed up in the definition of the concept of theatre, and he believes in the need to provide group co-operation between the performer and the recipient in order to be able to call the theatrical phenomenon a play. He says: theatre is neither the place nor the gathering where we watch something, our people had invented for this the words: spectacle, or watching and seeing, but the theatre is a meeting that each one of the attendants must participate in. In all of these theatrical forms, two elements should be found. First: the group and the group attendance; and second: the whole group action. From this perspective, Yusuf Idris believes that theatre is not proprietorship to a specific people or gathering than another, because of its innate nature that is associated with the human presence regardless of place and time. From this perspective, our people created their own theatrical manifestations.

Yusuf Idris selected some of the gatherings or theatrical forms that are frequently seen in our daily lives, in the occasions and celebrations that people created. 'The solution that Yusuf Idris believes in returning to the heritage forms in folk spectacles like the Samer that was crystallized for the vast majority of the folk masses in the country side and cites, and was held as a mass ceremony in special occasions and markets, it mainly depended on singing and dancing, and some chapters and improvised novels to some extent which depended upon the spontaneous character of the farfour' (Hussein, 1992, P228).

Yusuf Idris presented an example of this heritage theatre in his play *Al farafeer* in 1965, into which he put characters and sketches taken from national and international folk comedy. This play is considered an application to his theory. He wrote in its introduction: 'this novel is written on the basis of the audiences' participation with the actors in presenting the theatrical work. And the perfect theater for performing this novel is not the traditional theater (the Italian box set) but the arena theater or the ring formed due to the gathering of a group people. And there is no need for entry and exit doors, as the actor can penetrate the rows on his way to the circular stage in the middle' (Hussein, 1992, P229).

Thus Idris defined the most important feature of his theatre, which is a theatre that depends on group participation between the actors and the audiences, unrestricted by a specific type of architecture or building; however, it is a ring or a place where the audiences gather, the only lighting existing is what illuminate the theatrical events. 'Despite the fierce arguments raised by these opinions between supporters and opponents, they refreshed the theater movement, and provoked the playwrights to take attention of the local folk acting phenomena and benefiting from their fertile potentials' (Moses, 1996, P770).

'Tawfik Al-Hakim believed that the Samer was not the right expressing form of the Egyptian spirit, and that if we wanted to reach the theatrical roots that expresses us through searching in our heritage, and in an attempt to search for the form or the suitable mold, we have to return to the Pre-Samer, and to stop specifically at the imitator, and the storyteller' (Moses, 1996, P770). 'He called in his book Our Theatrical Mold in 1967 to use the folk theater method based on imitation through the storyteller—the imitating narrator—and not the acting; thus, it is preferred to present a play by a narrator and an imitator, both depending on the performance that doesn't aim at the merging of the actor in his role, nor seeking to convince the spectator that what he is watching is events occurring in reality and not acting' (Hussein, 1992, P231).

'Al-Hakim believed that we should be returning to the stage when we only knew the storytellers, and the imitators, and people found a great pleasure in listening to the storyteller telling the biographies and epics, and the imitators mimicking people, without any settings, costumes, or theatrical stage. He believed that in order to call it a real mold, it should be valid for all different types of plays to be cast in: national, international, old and contemporary' (Moses, 1996, P770).

He depended on full communication with the audiences, which is the essence of theatre that is based on the live communication between man and art. 'And though Yusuf Idris has provided a technically feasible model, Al-Hakim call and experience didn't find approval nor was it implemented. This call had come after he tried to present some folk arts, as in The Piper 'Al Zamar' which he wrote in 1930 inspired by the rural Samer' (Hussein, 1992. P231). 'Then he wrote The Deal 'Al Safqa' in 1956 trying to emerge the folk arts as dancing, 'Tahtib' and singing in the frame of the play, where it all took place in the outdoors, the barn or in front of the mastaba 'terrace'. In 1962, there was another attempt to link our old folk features with the most recent manifestations of the contemporary art in O' Tree Climber 'Ya Tale' Al Shagara" (Al-Hakim, 1967, P11).

Ali Alra'i resorted to looking at a stage a little bit closer to the current time, which was the stage of the improvised theatre. In his book The Improvised Comedy in the Egyptian Theater, he called for benefits to be taken from the nature of the improvised theatre that was introduced to Egypt by the Syrian artist George Dakhol at the end of the nineteenth century, which depended on an improvised dialogue between the actor and the audience, and which he kept presenting in the coffee shops and the improvised folk theatres until the twenties of the twentieth century.

His goal was to create the actual participation of the audience, which was the main feature of the theatrical phenomenon, and the search was still ongoing for it in order to create an Egyptian theatre based on heritage. Because of the dependence of the folk shows on a great deal of abstraction, there were no settings, no special costumes and no written scripts, just an inherited idea around which the improvised performance took place. This is how it was in the shadow play, the Samer, the shows of the storyteller, and the narrators of biographies, where the role of the audience went beyond watching to participating.

However, the reality of the Egyptian theatre showed that practical experience had preceded theorising. The operetta of O' night O' eye '*Ya Leil Ya Ein*', written by Zakaria Hijjawi in 1958, surpassed the employment of the heritage of folk literature to the manifestations of folk spectacle. The experience was presented to show and embody folk art in a theatrical form and in folk dance.

Also from these earlier performances came the play The Barber of Baghdad by Alfred Farag. These theatrical calls were accompanied by shows that were more penetrating in the search for heritage, as in Shafiqa and Metwali by Shawqi Abdel Hakim; many trials and experimental attempts appeared to arrive to theatrical forms for dramatic shows that depended on some folk spectacle manifestations.

Karam Metawea had initiated these when he directed Yassin and Baheya in 1964, which provoked critics on two levels; the new dramatic phrasing of the known fictional Mawwal written by Naguib Sorour, and the form choose by Metawea depending on the storyteller, the poet and the narrator, creating a sort of mating between the folklore in the folk spectacle forms and the imported from the epic theory in both form and content. This form still exists in our contemporary life in the form of folk performers, who narrate folk biographies and epics, and some of these are still scattered in the village markets and *Mawlid*. 'Karam Metawea used the stage of the Pocket theater close to the audience, and free from coulisse and traditional curtains to keep his show away from the traditional theater architecture that was not suitable with this folk form which depended on the intimate relationship between the actor and the recipient' (Hussein, 1992, P238).

The call to use the Samer formula in theatre was accompanied by the desire to create a new theatre away from the prevalent city culture, by the attempt to discover folk spectacle forms in the rural societies, and by the desire to be inspired by the well-known topics in the folk memory that were known in the rural theatre. 'The playwright Mahmoud Diab gave the folk Samer a more specific definition and successfully used it in his play The Harvest's Nights 'Layali Al Hassad' in 1967 and The Worthless 'Al Halafit' in 1969, where he found his goal in the rural Samer that he saw in one of the villages, when he was watching a group of peasants chatting in a ring mimicking each other, and through this mimic the imitator was expressing his opinion about the character he was mimicking, showing what is hidden in it. Only then, an image of a genuine Egyptian theater was revealed to Diab. This was the Samer definition developed by Diab, which was different from Yusuf Idris' definition that he sought in the comic act or the improvised comedy. However, improvisation and folk humor remained a common feature between the two formulas or the two Samers' (Moses, 1997, P770).

'The efforts succeeded for the detection and clarification of the character and feature, trying to link the art of theater new to us to some old artistic manifestations in our folk societies' (Al-Hakim, 1967, P11). The enthusiasts of this movement had called for the foundation of a national theatre with a folk character, which considered the rural theatre to be a representation of the true face of the unknown people on the literary side. Within this orientation, the play The Deal by Tawfik Al-Hakim is considered to be the first play about the countryside in the Egyptian theatre. Other plays were written afterwards, such as The Guarded '*Al Mahroussa*' by Saad El din Wahba and Naguib Sorour's play Full of Night '*Ghazir Al Leil*', which was presented by the *Al Warsha* troupe in 1993, and used the Samer formula and all the dimensions of the spectacle forms in the countryside and dramatically employed them.

Both modern and contemporary Egyptian theatres have tried to benefit from their heritage, while keeping pace with the international theatre. In the show, The City of Dreams, the scenographer and director Nagy Shaker used the image theatre method together with some folk spectacle elements, such as the shadow play and the puppet. The show was presented by Cairo Puppet Theatre in 1965 and gave plastic art the full opportunity to be the main tool of expression on the theatrical stage; this was done by merging the puppets and the settings in a simple artistic form, varying the existing relationships on the stage from one artistic form to another and being accentuated and deepened by the music. The idea was taken from a poem written by the poet Fouad Qa'oud, which was inspired by an old folk tale. The designer was impressed by the new content that reverberated inside the old folk form. The show portrayed the ethics falling behind the social and political achievements and the need to change the internal configuration of the individual to fit the reality of the society. There was no dialogue except for some verses of poetry that were said separately, and the communicating language between the form and the recipient became the movement, the music, the abstraction and the symbol. Nagy Shaker chose the black theatre and the shadow play as two methods to express

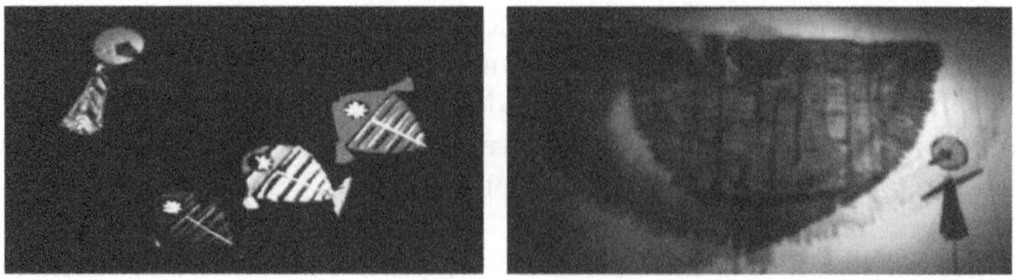

Figure 1. The black theatre and the shadow show in The City of Dreams show. Private archive Nagy Shaker.

the content of his ideas; thus the black theatre expressed the society controlled by matter, while the coloured transparent shadow play expressed the new ideal society.

The City of Dreams was preceded and prefaced by folk selections that were called The City of Dreams' selections, which were presented in a chromatic and dynamic form that confirmed that our folklore was a rich source from which we could take different methods of artistic expression from music to the spoken word, singing and plastic arts. In addition to the folklore art, there were obvious influences from the old Egyptian art, and the music was Egyptian, based on sundries of folklore. The City of Dreams travelled in an artistic tour to Europe, where it impressed all who saw it. Some of the critics' comments in Europe were: 'The artists coming from Cairo force us to shyly think about our arts and folk traditions', and 'Despite keeping the modern outlines, the design doesn't forget the Egyptian folklore, which gives the show a special glamour'.

In the fifties and sixties, some performances were political and presented 'the features of the severe crisis suffered by the Egyptian mind, while moving from the walls of the bourgeois society and agricultural prospects to the industrial progress and scientific mentality. The Egyptian mind was unable to balance between the intellectual affiliation and practical situations' (Eissa, 2015). After the war of 1967, the Egyptian theatre presented some plays that represented the defeat in a symbolic way, using elements and tales from heritage. Right after the war the National Theatre presented the play *AL-Zeer Salem*, written by Alfred Farag and directed by Hamdy Gheith, and with the settings and costumes by Nagy Shaker.

Alfred Farag used a historical folk tale and discussed contemporary issues through it. The show contained multiple disparate scenes, and the designer used an abstract symbolic setting, a multi–levelled theatrical ground and multiple curtains made of gauzes that divided the stage into different levels of depth and created different abstract forms that changed the theatre space depending on the interaction between the gauzes and the light, which resulted in a network of shadows that scattered on the theatrical stage and its components in different levels in time and place and the historical dimension of a script that went beyond 40 scenes. Those curtains did not prevent the audience from seeing the preparations for the next scene. Thus, the contemporary perspective was fulfilled in the set design. This was consistent with Brecht's method, which considered that what was happening on the stage did not aim to delude or merge. This was why the designer did without the curtain, thus leaving the theatre open to the auditorium throughout the whole time of the show. In order to achieve a historical atmosphere, the designer depended on the bright costumes with contrasting colours that carried an obvious symbolic significance for each character.

'Few of the shows presented in the seventies and early eighties were inspired of heritage and folk, on both form and content levels, such as Abdel Aziz Hamouda being inspired by Isis and Osiris myth in his play People in Thebes in 1981, where the playwright changed a lot in the axes and symbols of the myth in order to project contemporary significant. Yosry El-Gendy, was committed to casting away the folk biography of Antara Bin Shadad during the adaptation of his play O' Antara presented by Avant-garde theatre in1977. He succeeded in keeping the biography core, employing it to explain his contemporary reality' (Elias & Kassab, 1997, P456). 'The folk Coffee Shops were places for meeting and watching, where the shadow play show and the storyteller sittings took place, and the coffee shop theater with

Figure 2. Scenes from *Al-Zeer Salem* play.
The national centre of theatre.

Figure 3. Scenes from All in One, showing the change in settings.
Private archive Nagy Shaker.

its western meaning wasn't known then, except one attempt: The Coffee Shop Theater Troup in 1970' (Bakir, 2000, P27).

Among the serious theatrical models in the eighties was the play All in One, directed by Mamdouh Tantawy, with settings by Nagy Shaker and costumes by Abdel Ghani Abu El-Enein. The mass culture and the folk performance arts sectors both produced the play. It tried to link Tawfik Al-Hakim's biography, life, thinking and theatre to the historical, cultural, social and artistic conditions in Egypt, through using some scenes from his works along with dancing, singing, performing and the supernumeraries' movement. The show also combined narration with monologue. The director based his directing on the fact that all elements are key elements; meaning that the actor, the settings, the lighting and all the other elements integrate to create a piece of art. The show revealed an era in Egypt's history, and also revealed the struggle and unity of the Egyptian people against colonialism throughout history. The set design was the framework that allowed the presentation of multiple atmospheres for the different plays of Al-Hakim and for the different performing and acting forms, using a fixed setting and very limited added elements (curtains, tent fabric and flags), so that the change would depend mainly on lighting.

The idea of establishing The Touring Theatre appeared in the eighties, and aimed to create theatrical troupes that toured with their shows throughout Egypt's provinces, which were lacking serious theatre on a technical level. The Touring Theatre achieved its experimental goal, which was an attempt to rescue the Egyptian theatre from its peccadilloes with young elements in writing, directing, acting and settings, besides taking an interest in public issues through paying attention to the political theatre. These shows appeared, in their experiments, to be closer to the techniques of the political theatre in the west. Also, its shows were experimenting with heritage.

Among the important topics that were raised in the theatrical field during the eighties and the nineties was that of the provincial theatre. Dr. Nehad Seliha talked about it, saying that the theatrical activities' centre had moved from the main cities to the provinces, and this appeared in the shows of the theatrical festival held by the mass culture for one hundred nights on the Samer Theatre. 'The repertoire concept endangering the Egyptian theatrical heritage was applicable through the provincial theater. The theater was experimenting the most recent multiple methods in the international theater, in an attempt to select the most suitable ones for the Egyptian spectator. Thus, the message of the provincial theater was that theater was a script, an actor and a place, not settings, stars, and high costs, as some shows depended on a simple stage' (Seliha, 1986, P171). When Ali Alra'i wrote some studies about the people's theatre, he aimed to send a message to the playwrights to help them to discover the hidden treasures of people's culture and arts that people produce in their daily lives, creating the spectacle complemented by thinking. The folk actors voraciously took ideas from the heritage of the folk tales, myths and the everyday conflicts, their memory storage and what was accumulated from their collective and individual awareness.

The director, Ahmed Ismail, tried to develop the people's theatre based on the folk drama in his village, Shobra bakhom. He returned to the Egyptian countryside in an attempt to seek a simple, spontaneous folk theatre through the peasants. Ahmed Ismail presented his shows with his theatrical troupe with the aim of being inspired by folklore in both form and content, but developing it to be closer to our contemporary reality. One of the national artistic ambitions and dreams was to achieve an Egyptian theatre that differed in its imprints from the foreign theatres, a theatre whose vocabularies were all Egyptian: including the themes, the characters, the writing style and the portraying. This theatrical troupe was—perhaps—the only one that had a clear artistic method, which could be summarised as the search for a theatre with an Egyptian privacy, and they had a large number of shows based on this method. After the initial success of the troupe, the experience was generalised in other cultural locations. The members of the troupe were from different classes from the same village or neighbouring villages and they were practically trained in the theatrical arts. This experience is worth studying, not only because it presented its shows in a theatre built by the personal efforts of the village, but also because it represented a great success in the revolutionary dreams of the coalescence of art with the people in its attempt to make a real difference. The troupe has presented many shows, some short acting sketches and spectacular songs. The troupe in its initial stage presented its shows in a limited courtyard in the village on a small stage, and the number of spectators from beginning to end ranged between 300 and 500. The crew presented a series of shows, such as A Point of View about the Third Sex by Yusuf Idris in 1983, in which the director presented an experimental show built on a dramatic structure of heritage, folk, audio and visual vocabularies.

Ahmed Ismail presented the trilogy A Rural Evening for ten years, using the trees and the village houses as a natural open theatre, where the actors and the audience, both from the same village, coalesce on their land and social reality. The experience reverberated among both the audience and the critics and the success encouraged the director to go on, presenting *Al Shatter Hasan*, which was performed for four years both inside and outside the village,

Figure 4. A Rural Evening show and the village children helping with the theatre preparation and décor. Private archive Ahmed Ismail.

and, many years later, he also presented The Harvest's Nights show. A Rural Evening was an improvised celebrative group experience, which represented an important step on the way to creating an Egyptian folk theatre. The first part was presented in the courtyard of the village youth centre in 1982 and was followed, until 1986, by two other parts. For the presentation of the first part, the theatrical stage was a high hill on one of the corners of the playground of the club, while the spectators sat in the playground. This high hill had a background of rural houses that were employed in the show scenes as décor, adding some simple elements to the scenery, in addition to a large number of benches from the neighbouring houses. The audience represented the core of the experience. The number of spectators increased in 1984 for the presentation of the second part by approximately 50%. In 1986 the number of spectators reached almost 6,000. Thus the idea of a new space and a permanent theatre for the shows emerged. The troupe found that the summer theatre was the most appropriate one, and that the simple form that is consistent with nature was the most appropriate form; the theatre stage was built in the form of a ring surrounded three quarters of the way round by the auditorium, and the background of the stage was a high rectangle. This stage achieved two things; first, the ring form in the markets while watching the magicians and the praisers. Second, the rectangle in the background acted as a semi-traditional theatrical stage, thus, the theatre became a part of the village's reality and geography. The stage offered flexibility for different show forms. The theatre's stage and hall were intertwined and the auditorium was simple and made of clay. The setting was some simple accessories of the village's remains that the children used in the embodiment of places in a satirical way. The show was enhanced by the folk songs of the village children, which were entangled with the events.

Ahmed Ismail said: 'I wasn't isolated from the rising spirit in the Egyptian theater of the sixties, the storytelling theater by Roget Assaf in Lebanon, and the celebrative theater by Al-Tayeb Al Sidiki in Morocco, or the theatrical experiences in the Egyptian villages. Also the artistic and cultural movement especially those writings that linked our renaissance with the cultures and the achievements of the developed world. Yehia Haggy and I—the two designers of this theater—are villagers, who had seen how the village people gathered around its chanters and storytellers in the Mawlid, and its praisers and magicians in the markets. We had a great deal of a clear vision while constructing a simple theater that was consistent with the privacy of the village'. The troupe re-performed the trilogy A Rural Evening. The show focused on the triangular relationship between the performer, the receiver and the theatrical space; nonetheless it could not be isolated from the rest of the elements that were entangled with it. The actors had done a field gathering of the events from the village reality, also the village people attended to narrate and personate what actually happened and, with a small amount of guidance from the director, life was presented in an artistic way inspired of the reality that the theatre artistically rephrased, and was consistent with the shape of the rural gathering in the sprees. The critic Nahed Ezz Al-Arab said: 'Some provincial theater attempts searched for a theatrical show that was consistent with the folk spectacle and its heritage'. (Seliha, 1986, P173).

Arab theatre has seen many attempts to search for a new theatrical form that originated from the folklore, trying to present a theatre that is different from the well-known European theatre. Among the important experiences in this field was the show presented by Ahmed Ismail in '*Wekalet Al Ghoury*' in 1984. The show was a model of a folk theatre that was inspired by the awareness of our folklore. Dr. Sabry Hafez emphasised that the show interpreted the folk tale in a contemporary manner, and that it was a totally modern theatrical presentation for the Egyptian theatre that was rich with its own artistic traditions and capable of presenting some of the contemporary reality issues. He wrote: 'The show belongs to the series of Avant-garde attempts that the director went through depending on the actor and the conventions of what we might call the ascetic theater based on a good script and an actor with various abilities'. (Seliha, 1986, P180) Ahmed Ismail's theatrical attempt started with *Al Shatter Hasan* from the moment he chose the script written by Fouad Haddad and Metwali Abdel Latif, which was a contemporary adaptation of the popular folk tale *Al Shatter Hasan.* The director was able to portray it as a drama, where the poetic picture mixed with the dramatic events, the myth and the spirit of the folklore, in a framework that absorbed the audience, and in a theatre that was specially designed for the show and which looked like a semi-circle surrounded by the auditorium.

Ezz El din Naguib wrote: 'Ahmed Ismail participated with a selection of new creative playwrights in creating an artistic current that was serious in its search for a genuine and contemporary identity for the Egyptian theater. The director went through a difficult challenge by directing this work; as it wasn't a theatrical show in the conventional meaning, but rather a form of folk biography and spectacle told by the narrator on the tunes of the Rebec, as it had neither dialogue nor characters, except within the context recited by the narrator using the past tense'. (Seliha, 1986, P182). The narrator and the rebec poet portrayed a character and spoke with its tongue. The anecdotal material did not represent a dramatic event, just a great number of novels revolving around a superhero. Nabil Badran wrote: 'the narrator recites and narrates, then portrays the situation he was narrating, or the character he was talking about, then goes back to the narrator's role once again', which reminds us of Jerzy Grotowski's theatre. (Seliha, 1986, P183) The show's participants emphasise the essential role of the storyteller or the narrator in the folk theatre, who recites with the rebec a tale from the folklore, only this time he performs by using pointing and gestures and, with the ability of expression of the human body, he also adds contemporary indications to the folk tale. The text is mixed between the common prose language and the dialect poetry, creating a literary linguistic texture that expressed a holistic vision where the reality, the fiction and the myth blended. Dr. Nehad Seliha believed that the director created a theatrical experience from the folk narrative text that was both new and genuine. She added that the new thing about the contemporary vision of the story was the redefinition of *Al Shatter Hasan's* identity, where he was transformed from a legendary hero to a symbol of the Egyptian man. (Seliha, 1986, P186). After the traditional beginning that pulled us into the atmosphere of the inherited folk tale, prince Hasan began his transformation journey from a legendary character into a realistic one, from a symbol of an intellectual heritage pattern based on one hero, into a modern pattern where work is the new standard value. The two poets were keen to strengthen the characteristic of heroic courage, which was one of the most prominent characteristics of the heroes in the folk biographies and tales. Nabil Badran wrote: 'the spectator is eager to find out the new image that the heritage folk tale hero will appear in. This communication is doubled by the director's insistence on keeping the spectator's mind away from the idea of watching a traditional performing show. As the folk tale narrators drew close to the audience to the point of direct contact, the narrator sits amid the spectators for a little while then goes back to stage continuing the narration of the tale'. (Seliha, 1986, P183).

The folk tale was characterised by the flexibility that made its contents scalable and editable by the new narrator according to his attitude or the conditions of his social environment. The myths and folk tales were often used to portray social issues and reality. The new thing that Ahmed Ismail brought with the dramatising of the text and the folk tale was the creation of a new dramatic/narrative formula that combined the actor and the narrator. Mixing the narration and the novel with the acting and portraying was a common thing in the modern theatre, especially in Brecht's epic theatre. Thus, Ahmed Ismail achieved what Brecht was aiming for, which was the objectivity of the personating actor, where he made him both a narrator and a performer at the same time. He mixed the choral chanting clips with the individual portraying and narrating clips, and he embodied the paradox in the scenography of

Figure 5. Scenes from *Al Shatter Hasan* show directed by Ahmed Ismail in *Wekalet El Ghoury.* Private archive Ahmed Ismail.

the show, as the place he chose embodied the historical paradox. The director successes in creating a communication between the narrators, the singers, the musicians and the audience through the poetic. Ezz El din Naguib believed that the director was trying—through dramatising this tale—to revive the folk art and the natural climate for the folk biography; the rebec poet on the wooden bench and the spontaneous folk artists, both singers and musicians, with their rural gowns on their primitive folk instruments. (Seliha, 1986, P190) He did not resort to professional actors, instead he hired a very limited number of amateurs who were fluent in acting and singing; each one of them played the role of the narrator of one of the characters. The director borrowed the genuine folklore melodies and used the contradiction between stability and movement on the two levels of the stage; the upper level where the rebec poets sat in a theatrical setting inspired by the village, and the lower level where the spectators circled around the singers and actors while they were moving, eliminating completely the barrier between the actors and the audience. The director chose the folk Samer form in a simple scene that consisted of a set of benches and some folk motifs that could be easily moved to any location, barn or folk courtyard. Ezz El din Naguib added that the simple inspiring décor done by Salah Bissar played an essential role in directing the spectators' imagination towards the rural environment, through using real peasant tools, such as the thresher, ornate straw mats, horse saddles and colourful flags, which are usually seen in the Mawlid festivals and wedding nights. (Seliha, 1986, P190) The director used the elements and the tools inspired by the folk environment; the background made of straw mat and the stage covered with sackcloth, as if the spectators were in a village barn waiting for a folk troupe to present chanting, biographies, tales, Mawwal or rural songs. The setting became the natural cradle in which the story turned into a living theatre. The director used the lighting to move from one scene to the other, and was keen not to use the colourful lighting. He mixed the stage with the audience seats, he also engaged the audience in the singing and acting of some scenes, and also involved them with the actors and chanters in the performance of some climaxes.

Fouad Dawara wrote: 'The spontaneity took over the performing method and singing in the Harvest's night play, so that it seemed to the spectator that the show had no playwright, director or composer, and that they were just a group of peasants gathered to entertain and listen to the tale. The director left a limited space for improvising and heading towards the audience in an attempt to integrate and involve them in the show'. (Seliha, 1986, P192).

Some serious attempts appeared, such as The Puppets' Act '*Shoghl Aragozat*' show, written by Mohsen Meselhi, directed by Ahmed Ismail, and on a stage designed by Nagy Shaker at The Ghoury Agency Theatre in 1993. The theatrical space suggested an atmosphere of folk festivals, specifically the Mawlid, and of the touring troupes that built their marquee in the districts to present their shows. The performance, the singing, the dancing, the acrobats, the costumes, the puppets and the music were all elements that worked on the fine line between the professional theatre and the state of the touring troupes in the Mawlids and the folk festivals. The show was considered to be an experimental folk spectacle, a new theatrical form and a special experience of searching for a theatre inspired by the folklore, depending on the folk spectacle form represented by the puppet. The main idea that the show presented was a criticism of the negative side of the Egyptian character through the puppet (*Aragoz*), who appeared in the show in the same way that it was known in the folklore; as a sharp tongue that criticised the human and social contradiction. Throughout these events, the dramatic structure of the work stemmed from the folklore structure in a sophisticated manner, despite encountering the modern currents in the contemporary theatre, while retaining the Egyptian privacy. The show is based on the relationship between the actor and the puppet, and in the relationship between them a third level is produced: the level of fiction and fantasy. Thus the method became neither the total delusion, according to the Aristotelian theatre, nor the breaking of the delusion (breaking the fourth wall), according to the epic theatre. The exclusiveness in the show was the core of the folklore based on the fine line between fiction and reality, in the frame of a philosophy based on a particular perspective of the Egyptian cultural evolution that accommodates the means of the modern times and reacts with them. The wooden puppet had a philosophy and a history, which is why it was the most capable—on the folk side—of criticism and it was seen as the voice of the simple people and the nation's conscience.

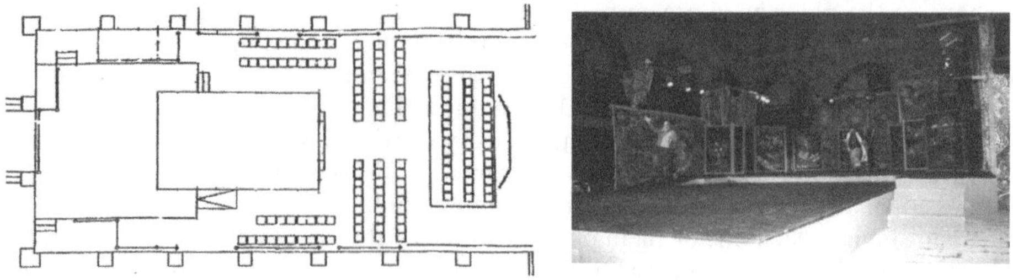

Figure 6. Plan and image showing the relationship between the stage and the auditorium. The theatre takes the form of a thrust theatre where the audiences surround the stage from three sides.
Private archive Nagy Shaker.

Figure 7. The wooden cutters used to show the puppeteer and his dolls from behind it.
Private archive Nagy Shaker.

Many ideas of the contemporary theatre were presented in the play Al-Kohl pillow '*Mekhadet Al kohl*' in 1998, such as the inspiration of heritage and the audio theatre. Moreover, the idea of mixing cultures, raised by Peter Brook, clearly appeared in the play, where the Eastern, Bedouin, Sudanese and African cultures mixed, and performers and chanters from different nationalities with multiple instruments and percussions participated in this work. In the play, the director, Intisar Abdel Fatah, presented the world of the oriental woman with all its charm and mystery.

The show is a journey inside the secret world of the woman, illustrating the restraints imposed upon her, the suppression from which she suffers and the condemnation she would receive from society if she wanted to rebel. The kohl pillow, where the kohl containers are held, is one of the characteristics of the oriental woman; where the kohl container symbolises the customs and traditions. The show begins with an oriental folk theme, which is the entrance to the woman's world with its privacies, vocabularies and contradictions, and also an entrance to the inherited folk celebration of the journey of searching for a theatrical privacy and oriental Arab identity for the theatre. The play is considered to be a new vision in form, content, directing, acting, expressive means and techniques; it raised an issue that had been in evidence for a long time in a totally new way, and the director tried to seek solutions for it through using the new expressive means. The body performance, acting, dancing, singing and musical methods were able to achieve a persuasion that was not achieved by the prevalent methods. Intisar Abdel Fatah introduced the spectator into a special world in order to enable them to receive the details of the story that was being told to them in words and songs, accompanied sometimes by music. Music and songs exchanged places in the symphony led by the director.

*Mekhadet Al kohl*, in its reliance on the folklore and the oriental musical formulation, is a development of the director's efforts and musical research that was previously handled in his initial shows, using the same folk, oriental and African musical instruments and the tools that stemmed from the environment. As for the singing, it was at times oriental, sometimes gulf and at other times African, thus conjuring all the places where the woman suffers the same concerns. The singing forms varied in the show; there were carols, folk Mawwal and Arab hymns that were all driven from the heritage and the old folklore. The folk songs merged with the percussions and the Nubian musical instruments. Abla Al-Reweni wrote:

'the woman of the show is an ornamental woman, coming out of the oriels and One thousand and one night's books. (Theatrical magazine, 1999, P45).

That aesthetic profiling of the woman and this tunica of charm and mystery that was portrayed by the patronizing of the female characters whether gathering or moving as symbols was emphasized by the scenography designed by Naima Agami through converting the entire theatrical hall into a brass bed with its decorative poles divided into six parts, with eight girls sitting inside it, they were a female choir reflecting the woman's world, and through a white transparent tunica we see the girls wearing their makeup'.

The grandmother sat in the centre of the hall, amid them all, starting the sewing machine. It is the entrance to the woman's world that the scenography design was aiming for, where the deliberate impression was given to the audience that they were not entering a theatrical hall but a woman's room. The show depended on body language rather than on verbal dialogue. It abandoned the actor and the actress after treating them as a dancer or a décor set or simply as a colour or a material that is reshaped inside the theatrical space. The black curtains form the girl's room, where her brass bed and her red kohl pillow appear, while the girl is sitting in front of it in awe, wearing her black clothes, as if sitting in a sanctuary, and on the ground there are the glowing candles and the folk mat, which gives a charming mysterious atmosphere to the secret room of the girls. The theatrical space is a rectangular hall where the audience sits on three of its sides, while the show is presented on the fourth one in the heart of the hall.

It is a small hall with no stage, where the spectators sit beside the actors on a platform approximately 50 cm high on the theatrical sides. Only one set design is used throughout the show, made of a black platform in the forefront and in the background appears a two-level bed with brass poles and handicrafts inspired from the folk art. The presence of cushions extending in to the recipient space gives the spectator the feeling that he has entered the world of a woman, and the shape of the bed with its brass material gives a feeling of oldness that suggests the steadiness of these ideas in society. The bed contains all the elements of the show, including the recipient, and there is only one man sitting on the edge of the bed, symbolising the oriental man with all the thoughts that control his look to the woman. The set designer used the ceiling to add a horizontal curtain with Bedouin fringes hanging down from it, in addition to the lighting and the small hanging lanterns. Also among the setting elements used were the vocabularies of the oriental environment, such as the Bedouin rugs hanging on the walls, the Islamic oriel that was used as a symbol of the woman and her position in the oriental society, and some elements representing the Egyptian countryside were also found. The costumes were simple, expressing the environment and the time, using only an Arab gown that had centred on its chest a red square with old folk engravings and symbols. All the girls wore black gowns.

The costumes were generally inspired by the folk Bedouin style, and the anklet was used as one of the accessories of this style. The grandmother's costumes were limited to a black gown

Figure 8. Scenes from *Mekhadet Al Kohl*; the theatrical space and the shadow show scene. Photos by the researcher.

embroidered with folk symbols, mostly in red, and an embroidered scarf with yellow copper rings hanging from it covering her head. The musicians wore wide trousers and a vest that was embroidered with Bedouin engravings. One of the oriental traditions was that the mature girl must hide her face with a veil. The veil used in the show was red, while the white veil was used to hide the grandmother's face, expressing her death. The lighting played an important role in expressing the situations and emotions. In one of the scenes, a white curtain was dropped over the bed, highlighted by a blue light from behind, showing the shadows of the girls, which added an atmosphere of charm. The show won the Cairo Festival of the Experimental Theatre award in 1998 for all its hardworking attempts in seeking a theatrical privacy.

The Life Box '*Sandoq El Donia*' show in 2000 was a collaboration between the director, Hanaa Abdel Fatah, and the Polish Witkacy theatre troupe. The show was an Egyptian/Polish dual production, where the director was Egyptian while the playwright, the scenography and the actors were Polish. The combined play is based on the scenes that we see in the small districts of the city and in the Egyptian villages, which are inspired by the Arab customs and traditions. The show is also based on the Arab Drama, the theatrical shows, the shadow play and Alfred Farag's theatrical adaptation of the fiction tales taken from One Thousand and One Nights. The melodies and monologues are overflowing with stories about life in Egypt, which are presented to the theatre through the storyteller. Dances are also performed, such as the candlestick dance where the candlestick is carried over the heads of the actresses. Hanaa Abdel Fatah also employed the condolences phenomenon and a part from the play Al Hussein a revolutionary and a martyr by Abdel Rahman Al Sharkawi, trying to make both of the Egyptian and the Polish teams reach a stage of an extreme Sufism. The show overflows with a lot of things that are unknown about the Arab theatre in Poland. The Polish actors discovered the varied culture without losing their entity and identity. Thanks to these actors, the director was able to enrich his artistic character while keeping his own culture. This theatrical mating resulted in both troupes discovering themselves. Hanaa Abdel Fatah tried to introduce the Polish actors to the Egyptian heritage, customs, traditions and folklore forms, and he also tried to find the common values between the Arabs and the Polish, presenting them to the theatre. In the Salam Theatre the stage took the shape of an axial thrust theatre and the audience sat on the theatrical stage surrounding the acting space from three sides, and on the fourth side there was a low platform with a shadow play screen on it, which was installed in front of the audience during the show. Breaking the fourth wall and using a different theatrical space aimed to create an intimate relationship between the actor and the spectacle in a game that looks like the folk life box. The setting consisted of some theatrical stuff and the box that the actors carried and entered with in the beginning of the show, then opened it bringing out the theatrical stuff.

The play The Harvest's Nights by Mahmoud Diab was presented for two years (1999–2000) at the Shobra Bakhom Theatre. The script was one of the most difficult scripts that Ahmed Ismail directed. This was because the dramatic structure was unconventional and mainly based on a rural evening and the village's Samer in The Harvest's Nights. The peasants attended to their amusement and entertainment by listening to the harvest songs while mimicking each other, and through this game they introduced their problems, thoughts and dreams. The theatrical space was consistent with the show in most of its elements, and the dramatic structure was based on narration and imitation. Ahmed Ismail presented the show in a big courtyard that summed up the Egyptian rural features; its nature, buildings, residents and the people's enthusiasm for the Samer theatrical game. Thus the show achieved many vocabularies on the technical and audience sides that can be disclosed by several points; the position of The Harvest's Nights in the Egyptian theatrical heritage, the relationship of the script and the show with the ideas that were introduced in the sixties about seeking a new theatrical mould and the cultural inspiration of folk in theatre. The critic Nahed Ezz Al-Arab wrote: 'Mahmoud Diab wrote The Harvest's Nights in response to the calls of Yusuf Idris, Tawfik Al-Hakim and Ali Alra'i for searching in the folk forms seeking an Egyptian theatrical identity'. (Radio and Television Magazine, 1999) The script was one of the most prominent works to benefit from the folk Samer.

The mixture between reality, narration, imitation and fiction reflected the unique value of the script and led to the search for a relationship that links its artistic structure with the

Figure 9. The stage and the auditorium in The Harvest's Nights.
Private archive Ahmed Ismail.

ideas of the theatre leaders, who asked for the search for an Egyptian theatrical mould that was inspired by the folk forms; the Samer and the dramatisation state could be found in the artistic structure of the play, the storyteller, and the imitator in Al-Hakim's call could be found in the structure of the script through a narrator who controls the theatrical game and different types of peasants who alternate on the imitation game, moreover, the script also achieved a very important element mentioned by Ali Alra'i which is the possibility of acting in the minimal equipped places and with minimum costs. Nahed Ezz Al-Arab emphasised that the experiment gave hope of achieving a theatre with an Egyptian identity that was well connected to its audience. (Radio and Television Magazine, 1999) In addition, Dr. Thanaa Mounir believed that the show was characterised by the theatrical experiment's exclusiveness in dealing with the audience and the surrounding environment. The theatre was established in the village's courtyard among the trees, which represented a natural scenographic scope. (Theatrical Magazine, 1999).

The play was presented in this setting with its simple technical facilities, deep in its audio-visual values. The critic Ahmed Abdel Hamid believed that Diab's script had found the right place to be presented geographically, intellectually and theatrically, when the show was presented in a village, in an open theatre in the folk Samer form, and thus the artistic shape of the experience accomplished the basic element that Mahmoud Diab was aspiring to while writing the play. It was inspired by the folk Samer form, with its spirit stemming from spontaneity, where the events took place with no preparation and self-developed during the show, which was characteristic of the Egyptian folk Samer. (Gomhouria Journal, 1999).

The directors first mission was to delude the recipient that it was a non-composed incident, and that it was a real spontaneous show. The script was inspired by the artistic formula of the traditions and techniques of the village Samer and based on the core idea of the theatre, which is that of being an entertainment that relies on the rebec poet and folk biographies. Ahmed Abdel Hamid added: 'The experience transforms the narrating drama, the imitation and the Samer with its old folk form into a contemporary living theater. The main difficulty lied in the psychological makeup of the roles and the dimensions, where every character in the show played more than one role' (Gomhouria Journal, 1999). Mahmoud Diab mixed entertainment, narration and epic drama in one artistic mould. The narration was undertaken by narrators who had multiple roles, sometimes coming in to contact with the chorus role in the classical theatre, while at other times achieving Brecht's method of breaking the delusion. The epic style of the script was represented by the style of dramatisation and the entry and exit of the actors in the show. The performing and portraying of multiple characters was an artistic method that aimed to emphasise alienation and the lack of the full integration of the spectator with the dramatic act. The Harvest's Nights was a poetic, philosophical and social play, as well as a political one, which was not free from symbolism and projected on the war of 1967. The show was considered to be an important experience, as it tried to create a theatrical mould that had the mechanism of communicating with the spectators in the village. It also presented realistic solutions to the problem of providing theatre on a small financial budget, was deep and influential in its ideas and artistic structure, and accentuated the theatre's values and its role in leading the audience into a better life. These shows were sponsored by The Cultural Palaces Organisation, while The Central Organisation of Provinces and Artistic Matters strongly supported the experience. The show was experimental in inspiring the Egyptian folk dramatic phenomena to present contemporary theatre with a theatrical privacy.

In the twenty first century there are many organisations, associations, and cultural and creativity centres in Egypt that are trying to present plays or other kinds of performances inspired by heritage and folk, using different historical sites to preserve the Egyptian identity.

The Artistic Creativity Centre in Al Ghoury Dome, with its outstanding location, is treated as an international lighthouse that embraces all the heritage forms, such as music, theatre and plastic art. Its main purpose is reviving the Egyptian artistic heritage in a new and sophisticated way that conveys the language of modern times, and giving people the opportunity to be introduced to the arts of different people and civilisations through opening fields of conversation between the Egyptian creatives and creatives from all over the world. Among the plays performed in Ghoury Palace was The Mevlavi Phantoms '*Atyaf Al Mawlaweya*', which is an outstanding theatrical show directed by Intisar Abdel Fatah. It was presented in 2009 within the international experimental theatre festival. The show discussed the Egyptian's relationship with God; Sufi songs were used with Coptic melodies and selections from ancient Egyptian heritage. The show was mixed between myth and reality, and stressed the uniqueness of the Egyptian character and the depth of the Egyptian culture based on references from their heritage, the proverbs and the parables in ancient Egyptian literature. The play showed that the main spring from which Egypt took its religions always aimed to the elevation of the spirit.

In 2015 the Prince Taz Palace in Cairo presented The Prince '*Al Amir*' show within the theatre. Many projects were sponsored and organised by the cultural developmental fund, with its historic creativity centres. This project was based on showing the charm of the archaeological and heritage sites by providing a dramatic simulation of the historical events that took place in the yards of the palaces and houses in the historic Cairo region. The Prince show displayed the biography of one of the princes in the Bahri Mamluks sultanate era. In 2016 the Prince Taz Palace hosted the theatrical show 'Theatrical Miniature' by Saad Allah Wanous, directed by Tarek Ghaleb. The show was produced by the public association of cultural palaces. The events of the show took place in Damascus when it suffered from the Tatar's attack, offering a panoramic view of the community in a defining moment in the history of the Arab nation.

In the first decades of the twenty first century, some troupes and centres tried to preserve folk music, songs, religious chanting, such as '*tawashih*' and rare '*maqams*', and paratheatrical forms inspired from heritage. For example, The Drums' dialogue for the Peace and the Egyptian Nubian Drums troupe is interested in the ancient Nubian heritage, in order to keep the singing and percussion heritage that stemmed from the Nile River. The troupe now contains the most important set of different folk musical instruments chosen from all the provinces of Egypt and which touched the spirit of the Egyptian character through its folk arts. Among the cultural centres in Cairo is 'Makan Centre for arts and culture', which is keen to present the heritage arts in their traditional form, and also folk music in its various forms with the Sufi songs and the Zar parties. Makan is a wide space with a theatre in the centre and the audience intertwined with the theatre. The details in the place are few and are all based on the very ancient heritage forms. Great musicians among the heritage keepers of Egypt were participating in this creative musical experience. It is considered an attempt on the way of renovation that stands on the rich bases of the folk music in order to build new forms, sounds, melodies and various structures. The Makan centre hosts '*Mazaher*', one of the folk troupes, to present the folk Zar arts, in order to revive one of the Egyptian heritage colours, presenting it to the audience with its traditional meaning and all of its details in a civilised way.

## 2 CONCLUSION

The theatre is a social phenomenon that affects society, and is influenced by its culture and heritage. The human being is a social creature by nature and, as the theatre depends mainly on human gatherings, thus the theatre is a celebrative ritual that fulfils this need and is the essence of the concept that the theatre has been built on since the beginning of time, extending in to the depths of history, insuring its survival and continuity for many centuries to come. The theatre is an anthropological phenomenon too, and is derived from all of the spectacle and paratheatrical forms of different cultures.

The critical awareness of heritage is the first step on the way to understanding our true cultural identity. The Egyptian theatre aimed to revive the theatre through using folk spectacle forms that were mostly based on narration and storytelling, such as the shows of the storyteller '*Al hakawati*', the narrator '*Al rawi*' and the Samer. It tried to benefit from some dramatic manifestations of Farce actors '*Al Mohabeziin*', shadow plays '*khayal Al zel*' and the puppet '*Al Arajoz*', from both form and subject levels. The Arab theatre orientation was to use elements derived from the local heritage, where the role of the audience went beyond watching to participating.

True heritage awareness cannot be completely achieved in a frame of cultural isolation and without comparing our heritage to the harvest of human experiences in different cultures.

Both modern and contemporary Egyptian theatres tried to benefit from their heritage, while at the same time keeping pace with the international theatre. It witnessed many serious experiences that dealt with the most important trends and distinguishing features of the contemporary theatre, such as image theatre, audience participation in the theatrical game, touring theatre, people's theatre, provincial theatre, mixing cultures and the use of untraditional theatrical spaces.

The contemporary theatre uses new theatrical spaces that are different from the traditional theatrical architecture, such as the Thrust and Arena stage to fit the performances, and also forms inspired from heritage, to make the relationship more intimate between the performer and the recipient.

Many Egyptian sectors, organisations and cultural centres, such as the mass culture sector, the folk performance arts sector, The Cultural Palaces Organisation, The Central Organisation of Provinces and Artistic Matters and the cultural developmental fund, help contemporary theatre to survive through funding and organising performances inspiring elements from heritage and ancient folk, and through performing in historical sites and so trying to preserve the Egyptian identity.

## REFERENCES


Abdel Hamid, A (1999), *Gomhouria Journal, 24/10/1999*, Cairo, Egypt.

Al-Hakim, T. (1967). *Our theatrical mold*. Cairo: Library of Egypt.

Alra'i, A. (1997), *Theater in the Arab world*. Kuwait: World of Knowledge.

Al-Reweni A, (1999) *Theatrical Magazine, Jan, Feb*, Cairo, Egypt.

Bakir, A. (1998). *Theatre school of people*. Cairo: Supreme Council of Culture.

Bakir, A. (2000). *Folk biographies in the Egyptian theater*. Cairo: The Egyptian General Book Organization.

Eissa, S. (2015). *Tales from home book*. Cairo: The Egyptian General Book Organization.

Elias, M. & Kassab, H. (1997). *The theatrical dictionary: Terminology of theatre and performance arts*. Lebanon Library Publishers. Beirut, Lebanon.

Ezz Al Arab, N (1999) *Radio and Television Magazine, 19/11/1999* Cairo, Egypt.

Hussein, K.E. (1992). *The theater and the social change in Egypt*. Egyptian Lebanese House. Cairo, Egypt.

Hussein, K.E. (1993). *Heritage in modern Egyptian theatre* (1st ed.). Egyptian Lebanese House. Cairo, Egypt.

Hussein, K.E. (1999). *The Egyptian theatre and its audience*. The National Center for Social and Criminal Research.

Moses, F. (1996). *The dictionary of theater*. Cairo: The Egyptian General Book Organization.

Mounir, T, (1999), *Theatrical Magazine, October 1999*, Cairo, Egypt.

Paul, S. (2015). *A history of Egypt in the Middle ages* (3rd ed.) (S. Ahmed, Trans). Cairo: The Egyptian Lebanese House.

Seliha, N. (1986), *Theatre between thought and art*. Cairo: The Egyptian General Book Organization.

*Cities' Identity Through Architecture and Arts – Catalani et al. (Eds)*
*© 2018 Taylor & Francis Group, London, ISBN 978-1-138-05409-7*

# Design ideology through architectural identity: A hybrid dynamic potential

Mai M. Youssef
*Faculty of Fine Arts, Alexandria University, Egypt*

ABSTRACT: A city's identity is an act of interplayed synthesis of different patterns through arts and architecture to form its basic structure and planning. This shows not only in a direct link but in a collaborative, shared relationship between architecture, arts and the preserving of cultural identity. During the life cycle of a building, a lot of changes happen to the structure and the facades, which reflect on the interior divisions of the space and its ability to meet the user's requirements. This paper focuses on representing the present cultural identity through smart materials and digital fabrication techniques in a dynamic potential that reflects the diversity of internal spaces to integrate with the form of the architectural building. It relies on smart materials and untraditional patterns in interior spaces, structure, and interactive responsive properties without pulverising or neglecting old values. It signifies a specific hybrid revival concept underlying the visual aspects in architecture to deploy this metamorphic ability characterised by diversity; transformation through modern technology.

## 1 ARCHITECTURAL IDENTITY BETWEEN PRESERVATION AND NEGLECTION

### 1.1 *What is architectural identity*

Cities are mainly identified through their architectural features and arts in several ways: as an expression of culture, as an aspect of development and as a functional liaison between the people and the environment. Emergent trends in arts now possess the power of integrated patterns, structural form and the responsive ability to be coherent and work in accordance with the architectural identity derived from within. Architecture has through the ages, been a transcended way to preserve this identity and adding more values and perspective. It translates the advancement and development of the city into structures, buildings, urban planning and interior spaces; therefore, architecture shapes and clarifies the main characterized aspects of the city and creates a distinctive identity. The interplay of these functions serves as a dynamic force to insert the technological development to preserve a city's identity.

The identity of a city represents a special curriculum in all aspects of life and the most important of the style of the building, but today with modern technology and urban growth, the structure became one solid and irregular liaison between variant patterns and a mingling of cultures and ideas, unable to maintain a well-defined concept. Therefore, this has paved the way for the collapse of architectural identity through engagement with other cultures, and there is no doubt that architectural identity is fading (Heynen, 1999).

The importance of each individual nation is having a distinctive identity; it was mainly seen from the architectural perspective exposed only to the advantages of global integration. This merger covers important elements correlated to the culture, such as the planning, the belonging to the land, the conservation of the existing and how to modernize it with extensions, specific materials, and defined patterns. The similarity between many architectural features led to canceling distinctive characters to make copied constructions without thinking or looking at the other side factors and their impact on every nation culturally, socially and morally. Numerous of the historical architectural eras and ancient eras have many similarities between them, but there are also many features of each era that are different from the others, and this consolidates the importance of each nation and every era from the other. In the past, it was quite easy to identify each places's own architectural identity from a vision of the buildings, but nowadays, especially in the suburbs, it is hard to distinguish the identity of the architectural structures, nor its history and the style of any used of pattern in the existing buildings (Conway & Roenisch, 2005).

There are several factors that have contributed to the disappearance of architectural identity, most notably the lack of awareness of the importance of a present identity portfolio of their history. Also, certain laws that have been enacted by municipalities are responsible for the movement of construction hat has disturbed the retention of traditional architectural identity (Baudrillard & Proto, 2003). The Arab, influenced by the world's great admiration for Western culture, soon became open to their customer, the use of excess material, globalization and neutral principles that have become the basic concepts of the present era. These factors contributed to the loss of our Arab identity and disappearance of its distinctive architectural features.

### 1.2 *Architecture and arts defined as cultural heritage*

Architectural heritage is considered to be one of the most important aspects of human development in all periods of history. The urban environment was influenced and changed by the needs of each stage of human evolution and thus produced what we see today, 'architectural heritage' that was part of life in the past: similarly, our needs today produce the buildings and facilities that we use in our daily lives. These translate the risk that lies in wait for our Arab nation, as it has long been characterized by cultural and religious character and distinctive intellectual wealth. It manifested itself throughout the ages, as it was respected and admired for its progress and development by the different civilisations of the world. Then, Western culture imposed due to sources of globalisation (with its economic, military and technical global influence). It promoted itself as the best single culture and the extension of cultural hegemony was able to replace Arab culture. There is no doubt that the Arab architecture was affected dramatically, as it was exposed to many factors that aimed to undermine its distinctiveness. When we wonder about the architectural style used in residential buildings, we do not find an answer to this question because of the absence of the atmosphere and method that characterised Islamic architecture, such as a unique style, wide circulation, geometric patterns, internal partitions, ornamented facades and absolute privacy (Petruccioli & Pirani, 2013). Now, we are

Figure 1. Represents a photograph taken of two different buildings in Cairo, Egypt. It demonstrates the contradiction and the peripheral cursory external façade, thus the loss of architectural identity.
Figures 2 and 3. Represent Masdar institute residential facades in Abu Dhabi to demonstrate the preservation of identity through architecture using Islamic geometric patterns and modern technology techniques.

heading towards a blind identity that is not tied to our history or the roots of our civilization and culture that we rarely see them. We find ourselves with the loss of almost all of our full cultural identity, unable to cope with the rapid development of Western culture, faced with a problem that is not easily solved without data and national perspectives.

## 1.3 *Significance of the study*

This paper aims at promoting Arab cultural identity concerning interior design, and distinctive features, regardless of the style used.

Each culture has its own identity and characteristic features that refer to the dimensions of its existential forms, advantages, proprieties, atmosphere, indications, customs and inevitable privileges. This special identity is the ability to see the difference between cultures and recognise their nature, even in the absence of any connection or relationship.

Because architecture and the arts are the foundation of cultures and the way that they are identified, architectural identity varies mainly depending on the self-variation of its people through time, the geographical area and from one civilization to another. Some civilisations still trigger a specific recognisable identity through the art, structures and architectural features that they practised thousands of years ago.

Certainly there may be things in common in the architectural language of many civilisations, but those that are still considered to have an inherited identity are those characterised by a certain distinctive aspect in their architecture and the history of their construction. These architectural features of certain cultures are the main determinants and the essential foundations for so-called 'architectural identity' and this particular identity is one of the main pillars of preservation in order to document the history of peoples and their existence, and the interests of individuals and civilisations by emphasising their own identity in their planning, urban environment, arts and architecture. We can restore the heritage features in the form of the vocabulary that we feed into the modern models that are compatible with modern needs and contemporary dimensions which are known as identity. Bourassa says 'The identities of cultural groups can be achieved symbolically, there is no culture without a system of symbols to represent this culture, and culture does not seek to only confirm the figurative but seeks to maintain itself through these forms' (Hubbard, 1993). Human nature is based on building strong relationships with the surrounding figures, and expressing the attributes linked to beliefs, values and norms over time.

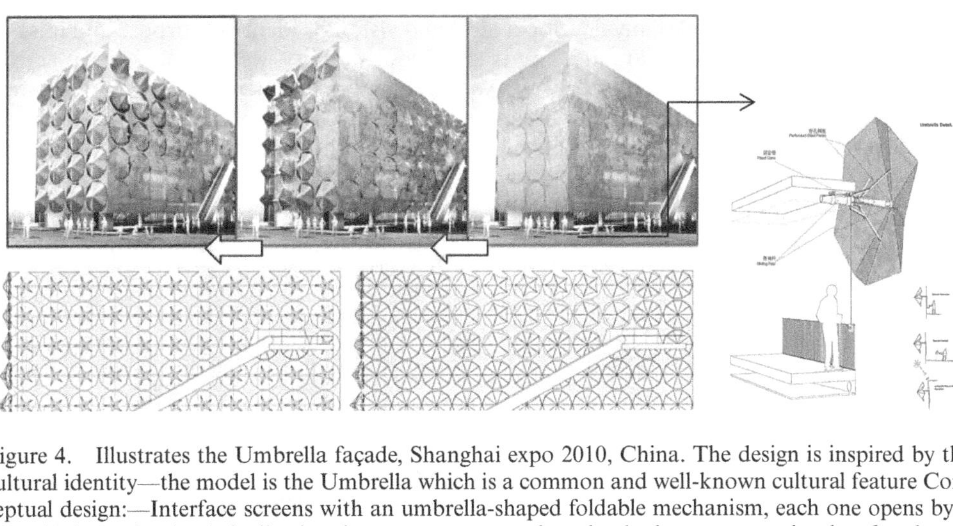

Figure 4. Illustrates the Umbrella façade, Shanghai expo 2010, China. The design is inspired by the cultural identity—the model is the Umbrella which is a common and well-known cultural feature Conceptual design:—Interface screens with an umbrella-shaped foldable mechanism, each one opens by a central joint spring-loaded, allowing the users to open or close the shades to protect the glass façade and the interior from the sun in the summer and allow more light in during the winter. The idea is to make every umbrella controlled by the rotating movement of the mechanism to allow people to engage and respond with the building and the surrounding (Youssef, 2017).

### 1.4 *Historical backgrounds*

When the Muslims first entered Egypt, they found blocks and architectural features of different styles from different eras. Pharaonic models, Ptolemaic remains and two features of Christian architecture (the dawn of Christianity, Constantinople) were the inspirations for the Islamic architecture. They took the Constantinople-style dome, the columns of the Pharaonic and roman models and the idea of the minaret to inspire of the building of the church towers. The Egyptian Muslim artists started shaping the Islam Egyptian culture, which was characterised with specific features, for example, they were not concerned that Egypt had consistently limited the ornamentation of the walls of temples. Within the prohibition of Islam, they replaced them with specific ornaments, carvings and plant-line mutations and excelled in them; they created systems and a vocabulary consistent with being a Muslim «Egyptian» (Petruccioli & Pirani, 2013).

Other models that continued in Egypt were the one seen during the Mamluks era because they adopted a high proportion of the Egyptian cultural heritage, and the Mosque of Mohammed Ali pattern from Hagia Sophia. One of the most striking examples was the Byzantine ornamentation in the Ottoman Architecture, as well as the minaret of Ahmed Ibn Tulun mosque taken from the twisted Samarra in Iraq. Pharaonic heritage had an impact on Islamic culture and was reflected in the Islamic architectural character, and we find evidence on this in the military architecture gates of Cairo, which are very similar to the stated Pharaonic temple and in the similarity of the Islamic content within the contexts of architecture and expression. The prosperity of the Islamic state has influenced world civilisation, as it purely resembles the architectural elements, different vocabularies of expression, ratios, calligraphy and geometric patterns that have produced the parametric patterns and algorithms seen in architecture nowadays.

## 2 THE ART OF ARCHITECTURE AS AN EXPRESSIVE INDENTITY

### 2.1 *Towards a new dimension in architecture*

Nowadays, what we see is a state of architectural chaos and clear evidence of the loss of the features and determinants of personal and national culture; the city's identity has not yet been determined. The buildings became architecturally deaf and lacked both the defined pattern and the means of interaction with the user. The Egyptian character faces more deformity and loss of features and many develop a distorted vision of identity, purpose and mission. Although the basic intent of architecture is to achieve stability, we plan shelters to actively seek balance by using behaviours that involve a frequent adjustment in order to accommodate changeable settings. Smart materials are the hybrid mechanisms used as adjustable interactive factors: they are inherently designed to become an integral part of life and activities (Jaskiewicz, 2013).

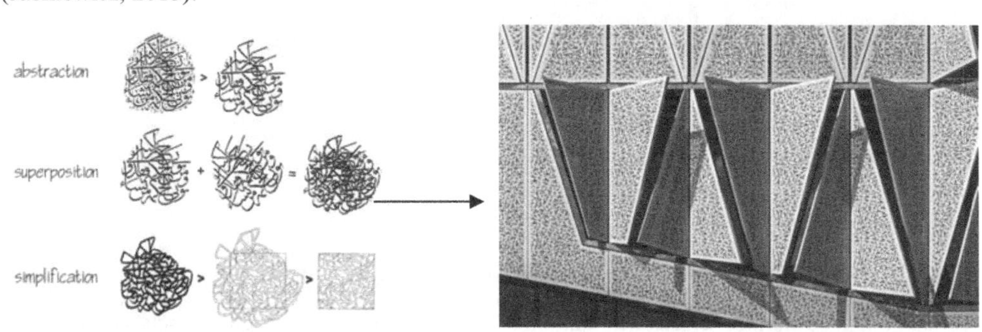


Figures 5 and 6. Demonstrate how to engender complex patterns that differ in size, form and direction using Arabic Islamic calligraphy and how to use these generative patterns to characterise architectural facades.

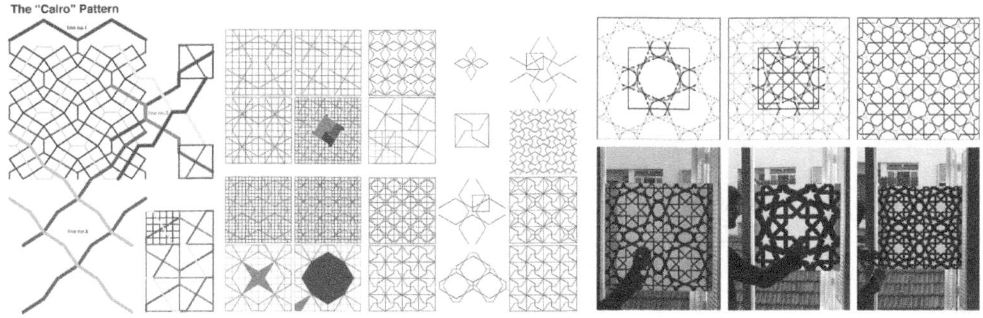


Figure 7. By analysing the urban radial planning of Cairo, many linear or curved patterns are engendered. By simplifying the lines, the planning of the city turns into intersected patterns, therefore, following simple methods of addition and subtraction, we can create new generative forms.

Linear lines are combined and interrelated to form a matrix of Islamic patterns, as Cairo was last influenced by the Islamic culture. Underlining the concept of preserving this particular identity, these patterns are inserted as decorative ornaments and digitally fabricated to enhance the design aesthetic peculiarities in both the external facades and internal wall claddings and partitions.

Computer embedded systems have replaced static structures with active shape-changing ones that reflect the design process. They have helped to engender intelligent, compatible and distinctive functions and interactivity. This blurs the distinction between the simulated and the generated structures and opens up the possibilities of developing space associations based on independent and group behaviours. This framework starts initially with a basic unit that has the ability to undertake distinctive form arrangements through an open finished combinatorial procedure, with novel locomotion and self-structuring qualities (Oxman, 2008). The unit emphasises its kinetic mechanism, joint system, self-governance and the unification of the entire system. The kinetic function is an essential feature of dynamic structures that depends on the division of the wooden surface into modular repetitive units. Responsive design appeared as a developmental movement altering the functional dimension of interior spaces. It is a response to the spatial adjustment and the interaction with the surrounding bi-folding movement.

The metamorphic design is an emerged pattern, a multifunctional concept created to perform any given task to the implanted design. It is an adaptive structure that contributes synergetic arrangements from models generated from multiple inspirations; it creates spaces parametrically by following mathematical equations to produce a complex structure that is automated and efficient (Kim & Maher, 2008). It is an interactive form that is continually changing to create balance, perspective, symmetry, and consistency. It uses a computational knowledge of the design principles, patterns, proportions, ratios, and mechanisms. It provides the necessary guidelines as a balanced shaping force to meet any demands. It affects the function of interior spaces, the performance of its contents, their responsive behaviour, thought structure, kinetic movements, generated models, human interactivity, sustainable energy and other ontologism. Also, through their proximity and implementation in every function and performance, these materials play a potent part when it comes to symbolic and multifunctional expression.

### 2.2 *The design's identity*

The study of architectural identity has a significant impact on the ingenious process of the geometry of space. Architecture is a form of space order. The plan dialectic in architecture and design reflects the way we think, interact, and learn. A developing level of understanding of interior spaces and the environment is considered the next step in the design evolution. This space information template responds to the environment or other sources and possible factors that affect the design by turning them into equations or graphs and then applying them to the

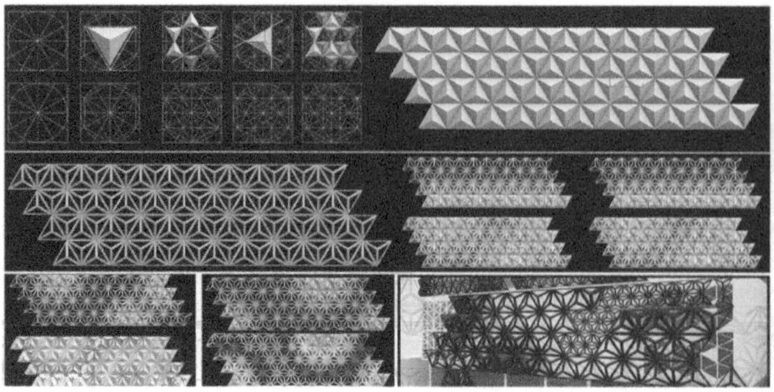

Figures 8 and 9. Illustrate the application of different Islamic patterns in interior spaces. These spaces have the capacity to self-reconfigure to respond to human stimuli-will as they tend to adapt individually.

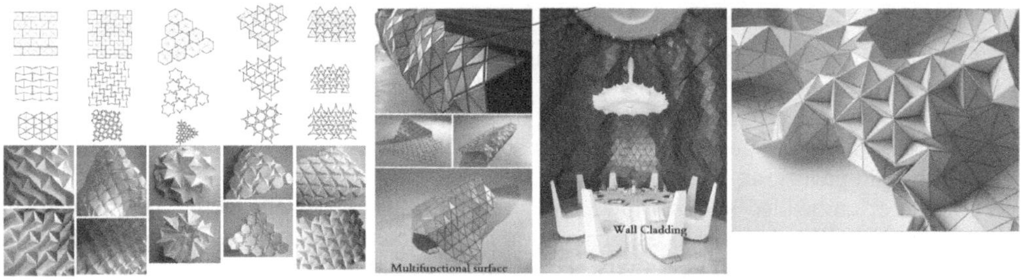


Figure 10. A series of forms inspired from the Islamic patterns of the urban planning of Cairo to create a new dimension of foldable surfaces using a sustainable material as a multifunctional surface (wood).

shape to change the design (Lunenfeld, 2000). All of this has led to the birth of new styles of architecture that did not previously exist, such as folding, parametric, and digital. From different cultural identities and architectural features, it engendered modern design trends as a result of the merging and integration of computer power and computational architectural technology and translated it into design configurations, such as the facades of buildings, structures and interior partitions that adapt environmentally known by their diversity and shape shifting through a combination of simulation technology, arithmetic design technology and digital manufacturing techniques. They allow space to be continuously shaped by being exposed to different conditions, such as environmental surroundings, temperature and pressure, to answer the requirements of the function of the space and its users.

Another pattern was inspired by the distinctive architecture of Hassan Fathy, creating a style of non-static architecture to ensure a self-efficient building and eco-friendly material compatible with the environmental surroundings. It preserves its defined identity while providing a source of natural ventilation. This material is hygroscopic, so it responds to the degree of the humidity and weather conditions, as demonstrated in Figures 12 and 13.

This kinetic movement is duplicated digitally to explore and understand the methods of convergence between the different units and the system. The transformation of these assemblies from one state to another will allow the emergence of a system that is capable of adaptation and dynamic behaviours that will give the structure the autonomy to interact with other objects as well as responding to its interior spaces and the needs of its inhabitants. Another distinctive identity is manifested in the distinctive structural pattern of the Cairo tower, which is, in turn, influenced by the shape of the Lotus Pharaonic column. This pattern can be re-executed parametrically using modern technology and computer programs. This form of a combination of tectonic units can be used to construct a variety of algorithmic structures as an envelope of architectural buildings, towers, skyscrapers or in interior spaces.

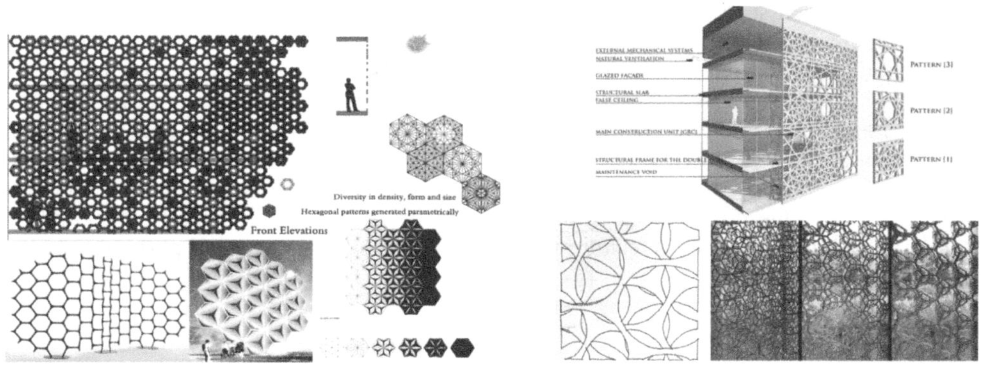


Figure 11. These structures prompt a wide range of distinctive spatial aggregations that are continuously reconfigurable and a variation of coloured glass that reflects during daylight in interior spaces.

Figures 12 and 13. Each component is designed so that it could differentiate itself through morphological behaviour to become part of the structure, then acts as a whole system and is driven by its single parts.

Figures 14 and 15. Demonstrate the implementations of the pattern of the Cairo tower translated using digital tools into determinants of the interior spaces that inherit its structure and identity.

## 3 CONCLUSIONS

As a conclusion, the forms of architecture, arts and design, as images of culture and civilisation, have changed in Egypt, not only with the change of belief, but in the expression of the terms of ideology as a symbol of the transition of the features of cultural identity. This was expressed in the various arts, particularly in architecture, which monitored how the Egyptians could preserve their cultural identity through adapting designs to suit the renewable data within the different stages of evolution during the disparate eras that have passed. Accordingly, these changes did not cover it up nor replace the primary structure, but it was originally drafted and clearly formulated in various forms of human expression and cultural interaction by the ability of modern technology. Smart materials are deployed to utter what has not been expressed and represented in a hybrid untraditional pattern; they are more engaged with the environmental surroundings and responsive to the knowledge of cultural identity. In construction, the data represented in the 3D digital space has generated new ways to analyse and construct complex forms. For instance, by using code we are able to deconstruct the form and link the model with new fabrication techniques, such as CNC milling, laser and plasma cutters.

## REFERENCES


Baudrillard, J. & Proto, F. (2003). Mass, identity, architecture: Architectural writings of Jean Baudrillard, Chichester, West Sussex Wiley-Academy.

Conway, H. & Roenisch, R. 2005. *Understanding architecture: An introduction to architecture and architectural history*, Psychology Press, Routledge francis and taylor group.

Heynen, H. (1999). *Architecture and modernity: A critique*, MIT press, library of Congress cataloging-in-publication data.

Hubbard, P. (1993).The value of conservation: a critical review of behavioural research. *Town Planning Review,* 64, 359.

Jaskiewicz, T. (2013). *Towards a methodology for complex adaptive interactive architecture,* TU Delft, Delft University of Technology.

Kim, M.J. & Maher, M.L. (2008). The impact of tangible user interfaces on spatial cognition during collaborative design. *Design Studies,* 29, 222–253.

Lunenfeld, P. (2000). *The digital dialectic: New essays on new media*, MIT Press, from Leonardo Book series.

Oxman, R. (2008). Digital architecture as a challenge for design pedagogy: Theory, knowledge, models and medium. *Design Studies,* 29, 99–120.

Petruccioli, A. & Pirani, K.K. (2013). *Understanding Islamic architecture*. London and NY: Routledge Curzon.

Youssef, M.M. (2017). Kinetic behavior, the dynamic potential through architecture and design, *International Journal of Computational Methods and Experimental Measurements,* 5 (4), 607–618.

*Cities' Identity Through Architecture and Arts – Catalani et al. (Eds)*
*© 2018 Taylor & Francis Group, London, ISBN 978-1-138-05409-7*

# The mutual influence of the Mamluk interior architecture and urban planning in Damascus

Karam Abdallah
*Department of Interior Design, Faculty of Art and Design, October University of Modern Science and Arts, Cairo, Egypt*

ABSTRACT: During the Mamluk era, Damascus was again the capital of the Levant area and the main passage for the Hajj convoys that travelled from all of the countries in the east to Mecca. This research establishes a study of the relationship between Mamluk political and religious concepts concerning the distribution of the main Islamic buildings in Damascus, the impact of this distribution on the urban planning of the city and, finally, the impact of all these factors on its interior design.

## 1 INTRODUCTION

The Mamluk sultans were new to embrace Islam, and so their main reason for strengthening their rule over the Islamic region was to obtain legitimacy, through taking over the status as the defenders of Islam and its sacred places. To achieve this aim they applied several methods, some of which had a significant impact on the architecture and urban planning for the cities. The methods they used were as follows:

**Firstly:** They launched their first mission to fight and expel the Crusaders from Jerusalem.

**Secondly:** They protected the pilgrimage route to Mecca in all of the countries under their control, which led to the re-emergence of the pilgrimage route from Damascus to the south.

**Thirdly:** The Mamluk policy to consolidate their rule also involved getting closer to the Islamic nations through other social and intellectual aspects, such as spreading fads that were related to religion and exaggerating the celebration of religious ceremonies. The most important of these was the pilgrimage howdah, which was invented by El-Zaher Baybars in Damascus at the beginning of his reign in order to prove his sovereignty over the region. The pilgrimage howdah involved travelling through the pilgrimage route, and was launched from one of Damascus' mosques with grand celebrations. The previous two factors had a great impact on the urban planning of the city, as we will see, since the concept of pilgrimage rituals was connected to the need to establish grand spacious mosques that were located in places that were suitable for accommodating a large crowd to practice the rituals. They also led to the necessity to place these mosques on the route of the pilgrimage, starting from the gate of the citadel to the south that was in alignment to the western city wall.

**Fourthly:** Trying to obtain blessing by getting closer to the buildings of their predecessors, who owned the real legitimacy to rule the Islamic State. Since the Mamluks took great care in choosing their buildings sites, the location of the building was seen to be as important as the architectural significance, and possibly more. In his book, Rabbat Nasser Rabbat mentioned that the historians of the Mamluk era did not care about the aesthetic values of the buildings mentioned in their books, but that the Waqf documents were describing the distribution of the internal spaces without describing the ornaments or decorative elements in detail. Besides that, they paid great attention to announcing the location of the building and they mentioned all of the important buildings in the neighbourhood, which clarifies the importance of choosing the buildings' locations.

Damascus was considered to be one of the oldest continuously inhabited cities; the archaeological studies on the old city of Damascus, carried out by Zack, Dorothee, showed that it had been based on perpendicular street planning since the Hellenistic era. This planning started to disappear with the emergence of Islam in Damascus, until it finally disappeared in the Mamluk era and the city became made up of twisted alleys that may end with blocked lanes. Also, the urban architecture extended outside the city walls to several new important places: Sook Saroja in the north, Senjekdar district in the west and Al-Midan in the south. These are the regions that are included in this survey. We are going to study the distribution of the Islamic buildings of mosques and schools in these areas, including the buildings that mentioned in Sauvaget and Zak, D. researches about Damascus including the buildings that reportedly confined at Sauvagetand Dorothee in Damascus. Some of these buildings are still there and will be within the scope of the field study, others have ceased to exist and their descriptions come from books and references.

The field study shows a great diversity in the interior architecture design style; we are going to make a comparison between the distribution of the buildings and the interior architectural style. Thus, we will use examples of the most important buildings, as shown in the attached table (Table 1). These examples were selected because they were built by the sultans, their representatives or their high ranking officials in Damascus. Therefore we could attain results that reflected the thoughts prevailing during the Mamluk era.

On the map (Figure 2), we notice the distribution of mosques and schools in the old city of Damascus, both inside and outside the walls, and the increase in the buildings' density in the north of the Umayyad Mosque area, the area between the castles and the straight street that was named Sook Al-Kmh, as well as on the pilgrimage route, starting from the locality of Al-Senjekdar in the alignment of the city wall and then turning towards the south. We also observe similar density on both sides of Sook Saroja, which runs from west to north-east of the city walls. The attached diagram, (Figure 1), shows the distribution in numbers and its explanation is as follows:

Table 1. Table shows the sample of the most important buildings in the old city of Damascus in chronological order. By the author.

| No. | Building | Function | Date | Plan | Site | Main entrance |
|---|---|---|---|---|---|---|
| 1 | Al-Zaherya | School | 676 HA/1277 AD | Complex | North Umayyad mosq. | West |
| 2 | Tinkez | Mosque | 718 HA/1318 AD | Arcades | West of the old city | South |
| 3 | Al-Kremy | Mosque | 718 HA/1318 AD | Arcades | Pilgrimage route | East-west |
| 4 | Al-Tankezya | School | 739 HA | Iwan | South of the citadel | North |
| 5 | Yalbuga | Mosque | 747 HA/1347 AD | Arcades | West of the old city | E – W – N |
| 6 | Al-Afridonya | School | 749 HA/1348 AD | Iwan | Pilgrimage route | East |
| 7 | Senjekdar | School | 749 HA/1348 AD | Complex | Pilgrimage route | East |
| 8 | Manjek 1 | Mosque | 763 HA | Arcades | Pilgrimage route | East-north |
| 9 | Moaayad | Mosque | 802 HA/1399 AD | Arcades | North of the walls | west |
| 10 | Al-Jawzah | Mosque | 1401 HA/804 | Arcades | Sook Sarouja | East-west |
| 11 | Al-Aqsab | Mosque | 811 HA/1411 AD | Arcades | Sook Sarouja | North |
| 12 | Al-Thekafe | Mosque | 811 HA/1411 AD | Complex | East of the walls | West |
| 13 | Al-Ekhnaeya | School | 820 HA/1417 AD | Complex | North Umayyad mosq. | North |
| 14 | Jaqmaqyah | School | 824 HA/1422 AD | Complex | North Umayyad mosq. | North |
| 15 | Al-Tawrizy | Mosque | 825 HA | Complex | West of the old city | North |
| 16 | Al-Ward | Mosque | 830 HA/1426 AD | Arcades | Sook Sarouja | South |
| 17 | Hesham | Mosque | 831 HA/1427 AD | – | The Straight street | – |
| 18 | Manjek 2 | Mosque | 835 HA | Arcades | Pilgrimage route | South, north |
| 19 | Belban | Mosque | 840 HA/1441 AD | – | Sook Sarouja | South |
| 20 | Shazebkya | School | 857 HA | Iwan | West of the old city | East |
| 21 | Al-Moaalaq | Mosque | 860 HA | Complex | North of the walls | North |
| 22 | Sabounya | School | 1457 AD | Complex | Pilgrimage route | East |
| 23 | Al-Qaley | Mosque | Late 9th century | – | The Straight street | – |
| 24 | Saybaeya | School | 921 ha/1415 ad | Complex | Pilgrimage route | East |

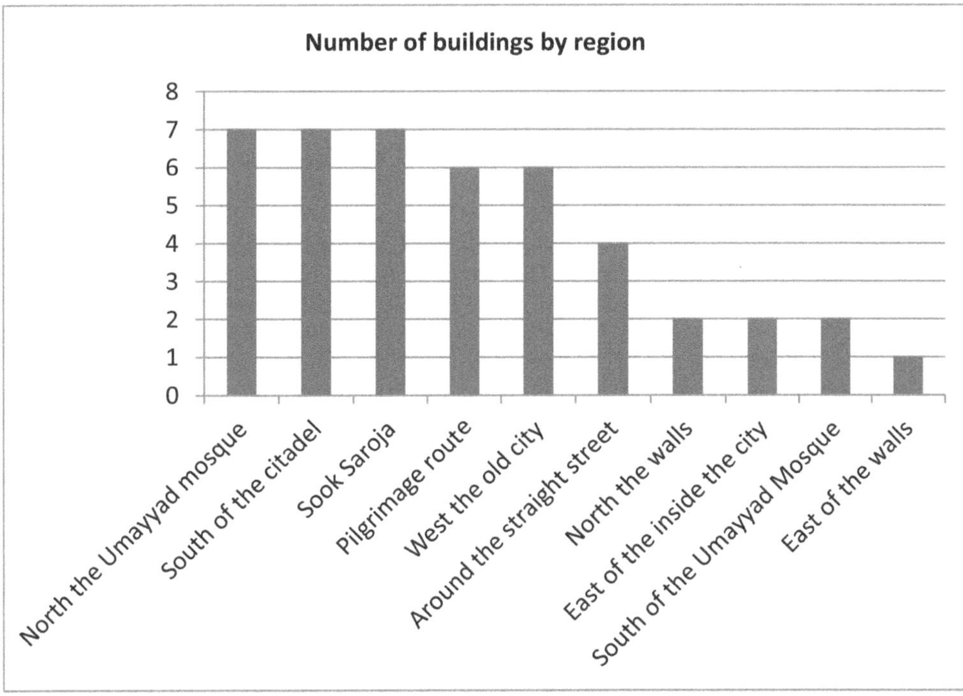


Figure 1. Diagram shows the number of religious buildings according to the regions in the old city of Damascus and the places surrounding it.

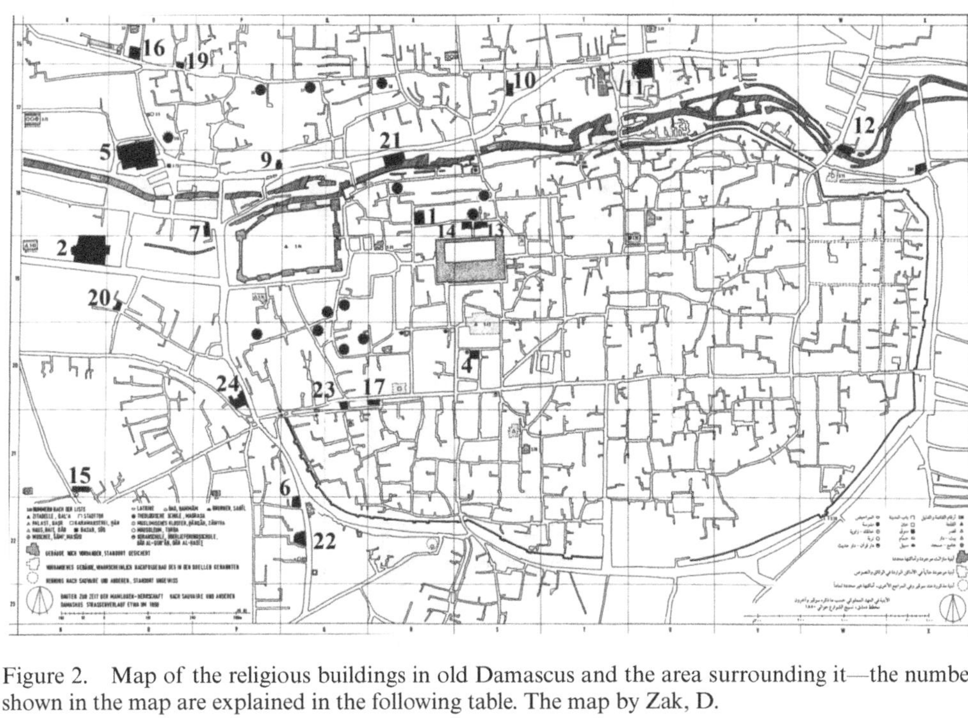


Figure 2. Map of the religious buildings in old Damascus and the area surrounding it—the numbers shown in the map are explained in the following table. The map by Zak, D.

**North Umayyad Mosque area:** Tracking the construction dates of the most important architectural structures in the attached table, we note that the first thing that was done by Sultan Baybars upon the receipt of power was to create Al-Zaheria, which is a mausoleum and a school, and he chose the site facing in the face of Al-Adelya school because of its religious and political fame, making them complement each other and become counterpart twin buildings, as if they belonged to one architectural unit in which the shrines' domes and entrances corresponded to each other. The southern façade is an extension of the southern façade of Al-Adelya school. Here we should mention that the Ayyubids had set up a number of important buildings adjacent to the north of the Umayyad Mosque, notably the Saladin mausoleum and Al-Rashaeya school. This explains the architectural density in this area, which affirms the previously mentioned desire to have neighbouring buildings next to the buildings of the former legitimate kings. So they built Al-Ekhnaeya School (Figure 4) in the place of the Al-Rashaeya, directly followed by the Al-Jaqmaqyah School (Figure 3) in the place of an old mausoleum. The concept of the twin buildings was repeated once again, as these two schools constituted one architectural unit where the domes and the great entrances of both buildings faced each other.

**West of the city:** When we trace the historical sequence of buildings in the Mamluk era, we see that the great rulers in the early Mamluk era gave their attention to the construction of great mosques that were miniature copies of the Great Umayyad Mosque. This interest in the tradition of the Umayyad Mosque was not only due to the importance of its architectural and aesthetic values, but also because it represents a symbol of the Sunni group that the Ayyubids and then the Mamluks tried to revive and spread but also for its religious to the Sunni group of Islam that the Ayyubids first and then Mamluks tried to retrieve its control on the Islamic religion and to be related to its Umayyads creators, the previous legitimate rulers. This was achieved by choosing appropriate sites that provided space for such great mosques in the western part of the old city, looking over the branches of Barada's River. In other words, selecting sites, depending on the availability of the necessary landscape area, where the Great Mosques of Tinkyz and Yalbuga and the Shazebkya School (Figure 5) were built.

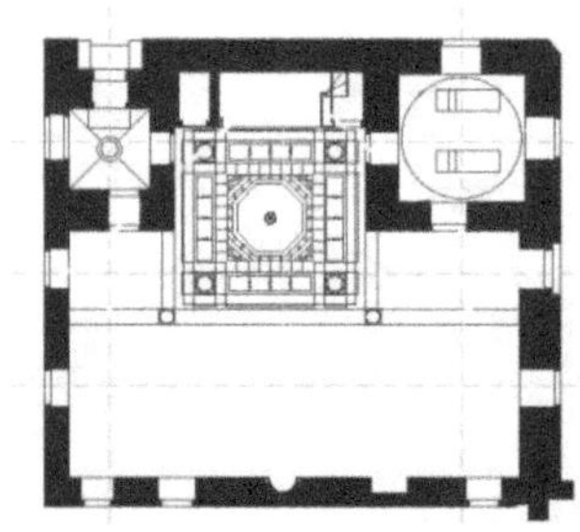

Figure 3. Al-Jaqmaqyah School plan.

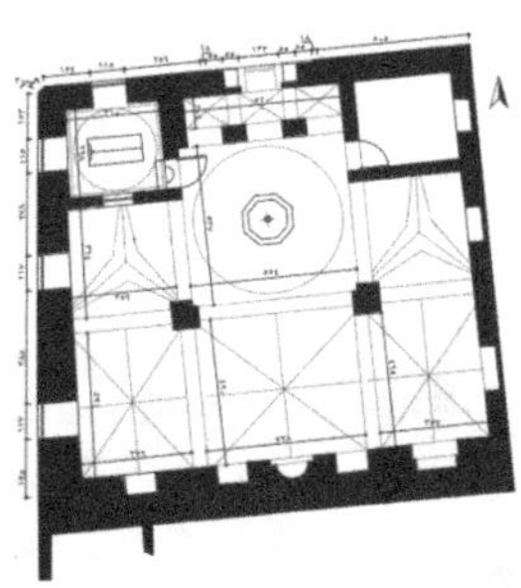

Figure 4. Al-Ekhnaeya School plan.

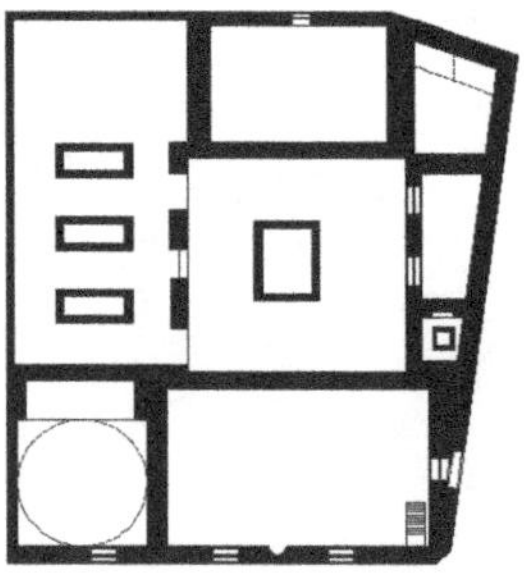

Figure 5. Al-Shazebkya School plan.

**Pilgrimage route:** Meanwhile, a new and important reason appeared for choosing a site on which to build mosques and schools, and that was the pilgrimage route. Through knowing the ceremonial rituals organised by the Mamluks before the pilgrimage season, we consider that the pilgrimage route started from the Senjekdar district, opposite the castle. It then stretched in alignment to the city wall, where the Afridonia and Alsabonia schools were situated, and then it turned south where the two important mosques, Al-Kremy and Manjek, were built.

**The area south of the Castle and north of Sook Al-kamh:** We note the density of school buildings in a relatively small space, since this region is one of the most vital areas in the old city of Damascus and is close to the surrounding public markets, such as Sook Al-Kamh in the south and Al-Bzoryah in the east. Most of the schools were built in distant streets, far from the main public streets. This proves that this area was considered to be a residential area and that it contained some of the most important residential buildings, such as the hospital of Al Nouri, the palace of Alsadaa, which was south-west of the castle, and the palace of Dar-Althahab, which was south-east of the Umayyad Mosque. Therefore, many of the rulers and princes had lived in this region reign since the Ayyubid period. Many historical references, such as Al-Nuaimi Al-Noaymy's mention of schools in Damascus, pointed out that many of the rulers and princes had transferred their houses into religious schools after their death. The most famous of which were the Athrawyah School 580 AH and the inner Al-Shamya School 612 AH. As a result, many of the rulers and high ranking men did the same thing during the Mamluk period. Yet, most of these buildings were destroyed and there is no trace left whatsoever, and even their locations were not determined except in the historical researches of Sauvaget and Zack Dorothee.

**The area north of the walls (Souk Saroja):** When the Burjy Mamluks gained power, they changed the basis on which they chose the sites of the religious buildings. They paid attention to the areas located north of the walls, Sook Saroja and its eastern extension. This was because these regions had reached their constructional completeness in that era and also due to the existence of some significant Ayyubid religious buildings, such as the external Al-Shamya School, Al-Tawba and Al-Aqsab mosques. The former two mosques were built in the style of the Umayyad Mosque; even the remains of Al-Shamya School's arcade shows that this great school was also built in the Umayyad Mosque style. Thus, the remainder of the buildings in this region were an attractive factor in encouraging the residential completion of the Saroja region. As a result, the new religious buildings, such as the mosques of Moaayad, Al-Jawzah, Al-Ward, Belban and Al-Moaalaq, were built on the sides of Saroja's main street.

## 2 THE IMPACT OF THE URBAN PLANNING ON THE INTERIOR ARCHITECTURE

We conclude from the previous sections that political and religious thoughts were the main inspiration for construction and architectural style in the Mamluk era, in order to attain their political aims, and that this had a great impact on the interior architecture.

**Arcades style:** In the early part of their rule, the Mamluks decided to build mosques, and they tried to imitate the Umayyad Mosque. They tended to choose wide spaces, far from the crowded old city, and that allowed them to use the style of the opened courtyard surrounded by arcades.

The Mamluks used several ways to imitate the Umayyad Mosque; the most important of these was by using the same plan, thus the mosques were built with an open courtyard, surrounded by four arcades, with the southern one being the largest because it was in the qibla direction. These arcades overlooked the open courtyard from semicircular arches, topped with another row of arched windows. Also they added a nave to the prayer hall of the Al-Kremy and Al-Ward mosques and the minaret was built in the middle of the northern arcade in Tenkiz and Al-Kremy. In addition to that, all the mosques had several entrances, in the north, east and west, which were in the typical style of the Umayyad Mosque.

In this era, it was preferred that schools, which had no less religious or political importance than mosques, should be built near the residential areas for daily use. For this reason, these schools were built both inside and outside the city, to the north and west of the walls.

## 3 THE IMPACT OF THE LOCATION UPON THE INTERIOR SPACES OF THE SCHOOLS

**Firstly, schools of the iwan style:** This was the typical plan for schools, which had been adopted since the Ayyubid period; as the qibla direction faced the south, these buildings tended to extend from north to south to the northern south direction. The chosen site should provide enough space to make the qibla iwan the biggest one. We see this in Al-Afridonya (Figure 6) and Shathebkya (Figure 5), where we note that the entrances were put in the eastern façade in both buildings, which is the main façade of the building. As for the location of the mausoleum, this may vary according to the street location; it should be put in one of the corners of the building, so that it could overlook the street. Thus, in Shathebkya, the entrance was put in the south-west corner, as the façade was single and not attached to other buildings. But as Al-Afridonya had only one single façade, the mausoleum was put in the north-east corner so that it could overlook the street. However, the Tenkyz School (Figure 7) was the only remaining one that was built to the iwan plan, located inside the walls of the old city. It seems that the powerful of Tinkes enabled him to build his school in such important site facing Dar al-Thahab palace and inside the walls of the city using the typical plan for schools. It seems, from the early date of the building and the builder Tinkez, Demascus' most powerful ruler-who built his school southern of Dar al-Thahab Palace in the place of the Othman palace of Al-Azem, separated only by a street—it seems that the importance of the school builder helped to offer enough space to be used in the iwan plan inside the walls of the city. The entrance was made to face the north, which is the main façade, in order to overlook the street. This important change in the planning of this building occurred as a result of its site. Also, this school did not have a mausoleum, which was considered to be rare. Other schools had the mausoleums of their builders, unless they were transferred from houses to schools.

**Secondly, the complex plan:** The most important example is Al-Zaheria School, which was built opposite facing the Adelya School in the Ayyubid period. It contained four iwans, but they were not symmetrical around one axis. In addition, the main entrance opened on to a transverse arcade, and we can clearly note that the arrangement of the interior spaces corresponded to those of the Adelya School, which was considered to be its architectural twin. The entrances were opposite facing each other; moreover, the two domes of the mausoleums were also opposite each other. On the one hand, we can see how the architect was able to adjust the new exterior architectural mass to the older one. On the other hand, the mass of the mausoleum took up a large part of the total space of the building because of the owner's importance. In spite of the Al-Zaheria School's displacement towards the south, the façade looked like a complement of the Adelya School. Thus, the displacement helped to emphasise both domes.

Another important example of twin buildings was Al-Ekhnaeya School (Figure 4) and Al-Jaqmaqyah School (Figure 3). These were built in two very close eras. In the attached pictures, we can see the balance and symmetry of the interior spaces and the courtyards, and the mausoleums mass. Here, we should mention that the courtyard ceiling of the Al-Jaqmaqyah School was covered over in the 20th century, but it had previously been open to the sky. The same is also true of the Al-Ekhnaeya School. Also, the iwans of the Al-Ekhnaeya School were open to the courtyard and they have recently been covered. Furthermore, we can see the symmetry of the prayer halls, whose shape looked like an iwan and an arcade at the same time, as they go around the central courtyard. The only difference we notice is the position of the entrance mass, which was shifted in the Al-Jaqmaqyah School from the middle of the main façade to its west side. The reason for this shift was that the main façade overlooked a narrow street, but the west front overlooked a spacious square. This meant that the entrance was more notable in this position than being in the centre of the façade, because of its height compared to the narrow street.

**The complex buildings outside the walls:** The most important examples are the mosques of Senjekdar, Al-Tawrizy (Figure 9), and the schools of Sabounya (Figure 8) and Saybaeya (Figure 10). Al-Tawrizy Mosque was built in a residential street in the west of old Damascus. Its first function was intended to be only as a mausoleum, and then it was decided to continue construction to add a mosque to it. Thus, we can find a distinguished design in which

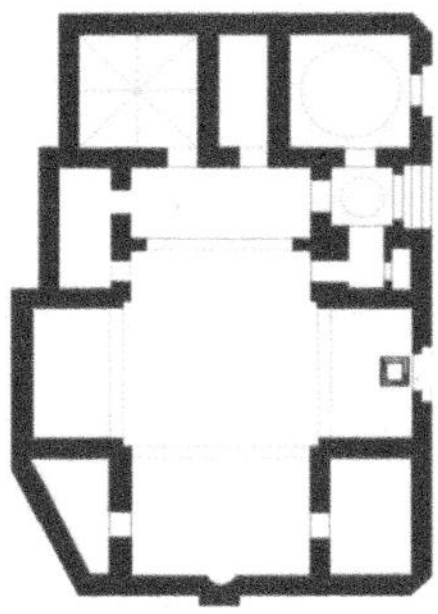

Figure 6. Afridonya School plan.

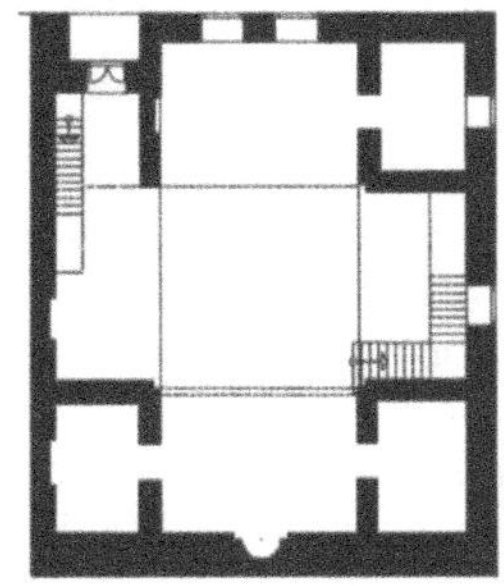

Figure 7. Al-Tankezya plan.

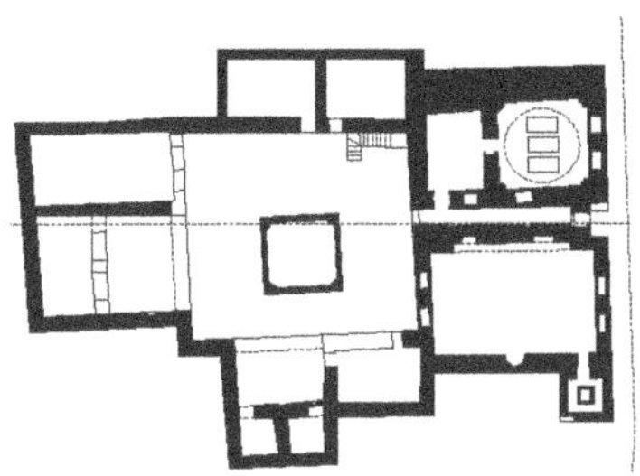

Figure 8. Sabounya plan.

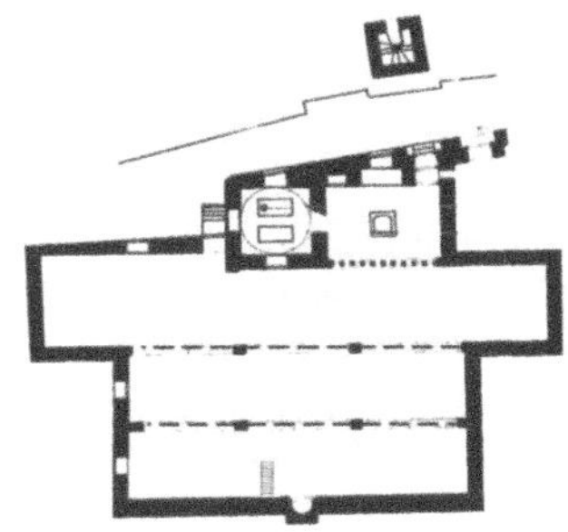

Figure 9. Al-Tawrizy plan.

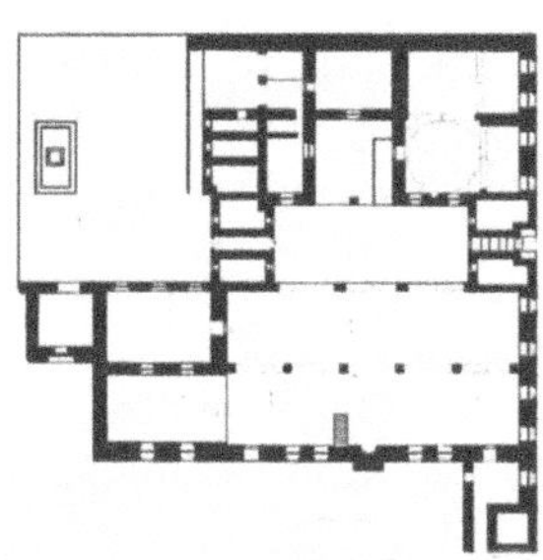

Figure 10. Saybaeya plan.

the mausoleum was located in the middle of the north façade adjoining a small courtyard. It had two entrances, with the main one located in the west of the northern façade and opening directly on to a transverse arcade, which used to be an open courtyard in the original design. But the main impact of the building site with regards to the interior spaces distribution was that there was no place for the minaret. This led to the minaret being put on the opposite side of the street.

We note here that the arcades were transverse due to the availability of space, where the main axes of the building run from east to west, contrary to the previous examples of the iwans plan. The Sabounya School axes goes from east to west, and we also see the same expansion in the Saybaeya School.

In the previous examples there was a transverse passage going from east to west, dividing the plan into the northern part, containing the mausoleum and other rooms, and the southern part, containing the prayer hall and other rooms and iwans. This passage may lead to another interior open courtyard surrounded by four iwans or arcades. This plan was used in the schools that were built outside the walls along the pilgrimage route, where there was not enough space to use the iwans plan, and where the building needed to present the dome to the street, as with the Sabounya School, Saybaeya School and Senjekdar. Also, as these buildings were constructed to be schools, there was no need to put the dome in the main axes of the building, as with the mosques with their arcaded plan.

It became popular in Damascus, at that time, to build the minaret above the main entrance, in order to reflect the continuing escalation of the entrance to the sky. That made it close to the dome of the mausoleum in most cases, when the main façade was not that long. But, when the façade was expanded to be very long, such as in the Saybaeya and Sabounya Schools, the minaret was put on the other side of the dome to achieve a balance.

When we compare the position of the mausoleum to the plan, we see that, in the case of the iwan plan, the mausoleum was put in a place that was separated from the mass of the iwans, not in the corners between them. Whereas, in the complex plans, the mausoleum was part of the entire form of the building plan.

## 4 FINDINGS AND RECOMMENDATIONS

- The urban planning of the old city of Damascus was a result of the political and religious thoughts of the Mamluk rulers, as previously explained.
- At the same time, this distribution of Islamic buildings led to the emergence of new residential regions, such as Saroja.
- The impact of political and religious beliefs on the distribution of the Islamic buildings affected several aspects of the interior architecture of these buildings, which are: choosing an appropriate plan and style for the interior spaces and finding an alternative design for the interior spaces. Thus, the interior architect had the flexibility to change from the original designs of Islamic buildings, according to the construction site.
- The comparative study between the description of the buildings in the historical books and the field study nowadays, showed that the renovation of the ancient buildings in Damascus was not at the required level. A lack of suitable materials and the right construction methods led to the creation of deformed buildings that were unrelated to the old ones and had no aesthetic value.
- The Urban Planning Commission in Damascus must make an effort to preserve the ancient buildings in Damascus by appropriate renovation and reconstruction. It has to create an appropriate urban plan to save the buildings instead of destroying them for the sake of expanding streets, as was undertaken in the middle of the 20th Century by the French architect, Icho Shard.

## THE RESOURCES


- أسعد محمد طلس: ذيل ثمار المقاصد في ذكر المساجد. تأليف: يوسف بن عبد الهادي. المعهد الفرنسي بدمشق، مجموعة النصوص الشرقية، ج3، بيروت، 1943م. Talas, A. (1943). "The mention of the mosques".

- جان سوفاجيه: دمشق الشام، لمحة تاريخية منذ العصور القديمة حتى عهد الإنتداب. ترجمة: فؤاد أفرام البستاني، تحقيق: أكرم العلبي، دمشق، ط1، 1989م. Sauvaget, J. "Damascus, Alsham, the history from the ancient era till the French occupation"
- عبد القادر بن محمد النعيمي الدمشقي(ت978هـ): الدارس في تاريخ المدارس. إعداد: ابراهيم شمس الدين، دار الكتب العلمية، بيروت، لبنان، ط1، 1990م. Al-Nuaimi: "A study in the history of schools".
- دوروتيه، زاك: تطور وبنيان مدينة مشرقية إسلامية. ترجمة: قاسم طوير، المعهد الفرنسي للشرق الأدنى، معهد الآثار الألماني، دمشق، 2005م. Zak, D: "The development and architecture of an Islamic oriental city".
- كارل ولتسينجر وكارل واتسينجر: الآثار الإسلامية في مدينة دمشق. ترجمة: قاسم طوير، تعليق: عبد القادر الريحاوي، معهد الآثار الإسلامية، دمشق، سوريا. Watzinger, C. &Wulzinger, K. "Islamic monuments in the city of Damascus"

## REFERENCES


Al-Harthy, H. (2001). *The concept of space in Mamluk architecture.* In Muqarnas: An annual on the visual culture of the Islamic World, XVIII, 73–93. American University of Bierut, Bierut, Lebanon.

Flood, F.B. (1997). Umayyad survivals and Mamluk revivals: Qalawunid architecture and the Great Mosque of Damascus. In G. Necipoglu (Ed.), *Muqarnas XIV: An annual on the culture of the Islamic world.* Leiden: E.J. Brill.

Rabbat, N. (2010). *Mamluk history through architecture: Monuments, culture and politics in medieval Egypt and Syria.* London: I.B. Tauris.


Figure 1: Diagram by the author.
Figure 2: The map by Zak, D: “The development and architecture of an Islamic oriental city.”
Table 1: The table by the author.
Figure 3: Plan from directorate of the antiquities in Damascus.
Figure 4: From, https://www.arab-ency.com/_/details.artifacts.php?full=1&nid=166851.
Figure 5: https://www.naseemalsham.com.
Figure 6: https://www.facebook.com/pg/مدارس-دمشق-الشام-القديمةMadrasas-in-Damascus-.
Figure 7: https://www.facebook.com/pg/مدارس-دمشق-الشام-القديمةMadrasas-in-Damascus-.
Figure 8: https://www.facebook.com/pg/مدارس-دمشق-الشام-القديمةMadrasas-in-Damascus-.
Figure 9: https://www.facebook.com/pg/مدارس-دمشق-الشام-القديمةMadrasas-in-Damascus-.
Figure 10: https://www.facebook.com/pg/مدارس-دمشق-الشام-القديمةMadrasas-in-Damascus-.

*Cities' Identity Through Architecture and Arts – Catalani et al. (Eds)*
*© 2018 Taylor & Francis Group, London, ISBN 978-1-138-05409-7*

# Usage of Arundo Donax L. as a sustainable material in interior design and architecture

Sarah Noaman
*Faculty of Fine Arts Interior Architecture, Alexandria, Egypt*

ABSTRACT: Using renewable resources has become a considerable solution for most problems in Egypt nowadays. It plays a role in issues such as energy crises, scarcity of natural resources and climate change. This paper focuses on sustainable transformation by using traditional perennial plants, such as bamboo, Arundo Donax and others, as renewable resources in wood manufacture. Egypt is in critical need of alternative raw materials. Thus, this paper focuses on studying the usage of neglected yet affordable materials, such as Arundo Donax L., in buildings and digital fabrication. Arundo Donax has been cultivated throughout southern Europe, Asia, northern Africa and the Middle East for thousands of years. This paper aims to discuss the use of Arundo Donax, both in its original state and after fabrication, in the context of interior design and architecture.

## 1 INTRODUCTION

### 1.1 *Shallow water plants*

They are semi-aquatic plants, similar to the Arundo Donax and bamboo, which grow in humid places, as their growth needs an abundance of water. The amount of water can be determined according to the plant type, facilitating their growth along lakes, streams, drains and wet sites.

### 1.2 *Arundo Donax (Giant reed)*

Scientifically named Arundo Donax L., this tall perennial cane is also known as the giant reed, Mediterranean reed, Spanish reed, Donax cane, Arundo grass and family Poaceae (Gramineae). Arundo Donax is indigenous to areas surrounding the Mediterranean Sea. It was later cultivated and naturalised in other major continents in regions with warmer climates. Egypt, for instance, is one of the Mediterranean countries where Arundo Donax exists.

### 1.3 *Plant structure*

The Arundo Donax is a sturdy upright perennial grass species that grows in many clumps. The stems are 3–5 cm in thickness, 30–60 cm long and 2–6 cm broad with tapered tips and hairy tufts at the base. The giant reed has a widespread network of rhizomes under the soil surface, which are 5–30 cm in depth. The stem is a hollow segmented culm that measures from 1 cm to 4 cm in diameter and is able to branch. The culms' walls range from 2 mm to 7 mm in thickness and the internodes can reach 30 cm in length. Under optimal conditions, stems can grow up to 10 cm per day, as it is one of the fastest growing plants.

### 1.4 *The relationship between Arundo Donax and the surrounding environment*

Arundo Donax is a sustainable plant that causes $CO_2$ sequestering, soil erosion reduction, water regulation, speedy growth, low use of nutrients and primary energy. Furthermore, the Arundo Donax is an economic plant that is produced locally and, therefore, is of a low cost and creates local jobs and additional income for farmers.

Moreover, this plant resists salinity, humidity and wind pressure due to containing a high percentage of lignin, also known as wood fibres, in the plant tissues. Table 1 lists the mechanical properties of the stem.

Table 1. The mechanical properties of Arundo Donax.

| Property | Estimated value |
|---|---|
| Density | 2,295.00 $N/m^3$ |
| Mean tensile strength | $32.17 \times 10^4 N/m^2$ |
| Mean bending strength | $130.00 \times 10^4 N/m^2$ |
| Mean compressive strength | $66.50 \times 10^4 N/m^2$ |
| Mean bearing strength | $26.68 \times 10^4 N/m^2$ |

*Source: Institute of environmental studies and research, Ain Shams University.

## 2 DIFFERENT USAGES OF ARUNDO DONAX

### 2.1 *The use of Arundo Donax in Egypt*

Arundo Donax is used locally to make fences, woven baskets and kites. Ancient Egyptians wrapped their dead in the leaves, and used canes to make fishing rods, walking sticks and writing tools. This plant exists in many places in Egypt, as it is one of the Mediterranean countries, as shown in Table 2 below.

Table 2. The existence of giant reed in Egypt governorates.

| Governorate* | Places |
|---|---|
| Cairo | Qubba Palace—Saffron Palace |
| Giza | Orman Park—Giza Zoo |
| Alexandria | Rural areas and slums |
| Sharqiya | |
| Gharbiyah | 13 shares – 15-karat – 95 acres |
| Dakahlia | Bridges, canals and banks |
| Monoufia | Bridges, canals and banks |
| Behera | Bridges, canals and banks |
| Qalyubiyah | 100,000 Inch/Acre |
| Fayoum | |
| Suez | Large areas cannot be counted |
| Port Said | Al Manzalah Lake |
| Matrouh | Trace amounts at Siwa Oasis |

*Source: The centre of progress and development of small industries, Ain Shams University.

### 2.2 *Usage of Arundo Donax internationally*

- *Energy crops, biofuel and cultivation*

Energy crops are plants that are produced with the sole purpose of using their biomass energy while reducing carbon dioxide emissions. Biofuels, derived from lignocellulose plant material, represent an important renewable energy alternative to transportation fossil fuels. Stem and rhizome have the ability to sprout after removal from the mother plant and are then used for clonal propagation.

- *Musical instruments*

The cane is rendered into reeds, which are used in the production of clarinets, saxophones, oboes, bassoons, bagpipes, flutes and other woodwind instruments. For example, the ancient end-blown flute, ney (nai), is made from the same reeds.

## 3 ARUNDO DONAX IN ARCHITECTURE

### 3.1 *Usage of Arundo Donax as a traditional building material*

Many inherited methods of building rural houses in Egypt using the Arundo Donax exist. The most common building method is the Lattice, or *Chebika* which is named after the method of attaching the rods together. A linkage appears in its construction as an ornamental unit. Since the plant is available locally, it is possible for each resident to build his own house. Similarly, the building process is simple for an unprofessional builder. The building process starts by cutting the reeds, allowing them to dry well, collecting every ten poles in a beam and tying them by simple ropes to form the construction.

### 3.2 *The walls*

The beams are grouped vertically, placed together in batches and attached to proper ropes. Another layer of beams is then added horizontally and attached to the vertical layer. Then, an inclined layer of reed beams is added in order to strengthen the two perpendicular layers. The height of the wall is 6 metres tall, the room's area is 6*4 $m^2$ and its height is about 3 to 4 metres tall. These calculations include the part of the wall that is built below ground level to support the building, as shown in Figure 1. The building is insulated by using plastic and stucco, and then piles of rice straw and papyrus are added. Finally, the walls are covered by mud that is mixed with rice straw and hay and painted with lime.

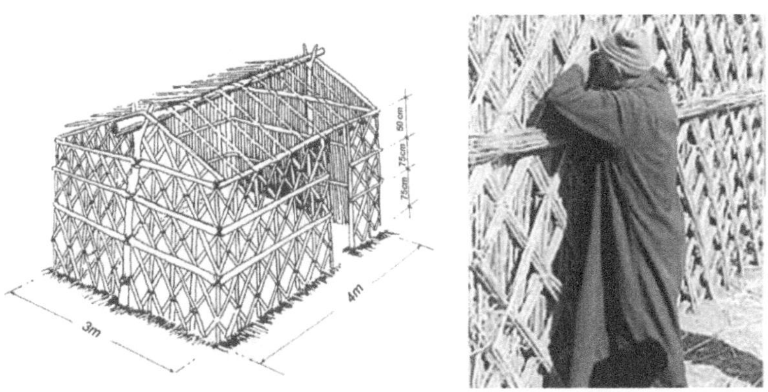


Figure 1. The common method of building in Al Manzalah Lake: Lattice (Chebika).

### 3.3 *The ceiling*

The ceiling may be built by using the same steps as for the building of the walls or by creating a layer of single rods that are attached to each other and hitched tightly by thin robust ropes. This is specifically applicable along the short side of the room. The second layer is then added perpendicularly to the first one and parallel to the long side of the room. Finally, and to carry the ceiling, camphor wooden poles are used.

### 3.4 *Disadvantages and solutions*

The main disadvantage of this plant is its flammable nature. It is also difficult to use reeds in constructions that have wide spans and long heights. However, as a solution to its flammable nature, the use of epoxy, which is a mixed fireproof material that is rich with aluminium, stops it from catching fire and is used to cover the outer surface of the plant. Meanwhile, the latter disadvantage could be dealt with by using wooden trusses for a wider span.

### 3.5 *Advantages*

These houses are suitable for places that have a high temperature, humidity and heavy rains. Economic plants are sold for 35 Egyptian pounds, by the metre squared, in Al Manzalah Lake of Port Said. The plant is suitable for weak clay soil that cannot carry heavy constructions, such as concrete. Soil subsidence causes concrete to collapse, while the lightweight Arundo Donax can be easily carried by the soil. It also facilitates quick construction, so that ten workers can build one hundred housing units in three days. It was used until recently in the making of booths in the markets and for ceiling coverages in the cafes of slums, as shown in Figure 2.

Figure 2. A café whose ceiling is covered by Arundo Donax in the outdoor area, (Middle, Right). A booth in El-Maamora market, (Left). Place: El-Maamora El-Balad, Alexandria, Egypt.

## 4 ARUNDO DONAX IN INTERIOR ARCHITECTURE

### 4.1 *Mixing Arundo Donax with natural wood in interior architecture*

Arundo Donax rods are used in interior architecture, furniture design and ceiling cladding, as shown in Figure 3. The reeds are used as appropriate thermal insulators and sound absorbers as they prevent echoes, especially when they are used as an integrated building unit. This is possible by a distortion in the sound waves caused by their bumpy surfaces.

Figure 3. Usage of Arundo Donax in ceiling design supported by natural wood in two different interior spaces.

### 4.2 *Arundo Donax items in interior design*

The rods are used in the making of decorative interior items and lighting units. The items are produced by forming the proper rods and merging them with other materials, such as fabrics and wood, as in the designs shown in Figure 4.

Figure 4. Arundo Donax decorative and light items, Gift Shop, Alexandria, Egypt.

### 4.3 *Arundo Donax partitions with metal frames*

This is an experiment to examine the ability of Arundo Donax as thermal insulation. The thickness consists of three partitions. Each partition contains Arundo Donax rods and metal frames with lengths of 2.5, 4 or 5 centimetres, a depth that ranges from 100 or 150 to 200 centimetres and a height of 180, 240 or 260 centimetres. The poles are knitted by galvanised metal wire with a diameter of 2 mm. The two sides of the partition, with a length of 5 centimetres, are covered with a layer of mixed cement and sand with a thickness of 2 centimetres, as shown in Figure 5. Thus, this results in a durability against fire that lasts for half an hour.

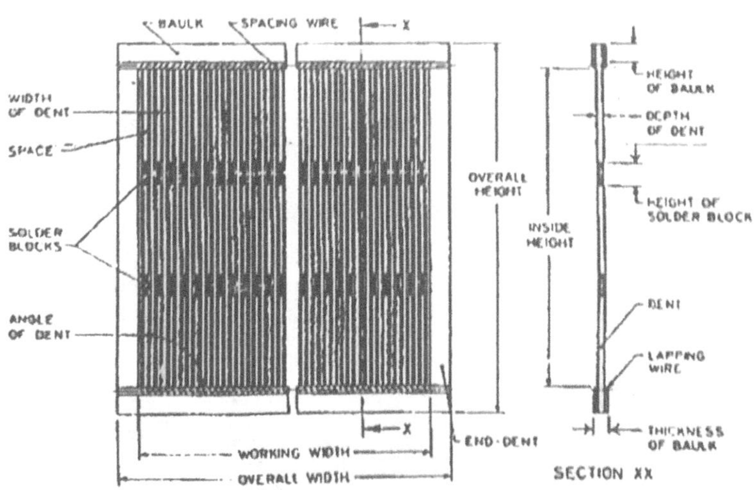


Figure 5. The details of the Arundo Donax partition used to examine combustion and thermal insulation.

## 5 EXAMPLES OF ARUNDO DONAX ORGANIC STRUCTURES

### 5.1 *Water universal exhibition*

This pavilion, known as the water university exhibition, was constructed in Zaragoza, Spain back in 2008. It is a shaded structure that allows access to the Pavilion of Citizen Initiatives. The pavilion was designed by Canayaviva in collaboration with the architect, Ricardo Higueras, as shown in Figure 6.

Figure 6. Pictures for the pavilion constructed in Zaragoza, Spain in 2008.

### 5.2 *Casa de Laila structure*

Another example is the Casa de Laila structure, as shown in Figure 7. It is a multiple use open space that is 9 metres long that was built in the Alhaurín el Grande in Malaga, Spain in 2013. The structure covering is full of canes with mud, lime, hemp and aggregates and was designed by Canayaviva.

Figure 7. Picture of the Casa de Laila structure.

## 6 ARUNDO DONAX A ZERO WASTE MATERIAL

### 6.1 *Zero waste*

The GrassRoots Recycling Network (GRRN) defines the zero waste design principle as one that goes beyond recycling, by taking a whole-system approach to the vast flow of resources and waste through human society.

### 6.2 *Arundo Donax example for zero waste (Radical Ship Pavilion)*

An installation, which resulted from a workshop, was created in September 2016 by the LAN laboratory for natural architecture, within the Festival of Mediterranean Literature, in order to design an intervention of 'Urban Land Art' using natural materials, as shown in Figure 8.

Figure 8. Picture for the Radical Ship.

This structure can be reused by taking the rods and redesigning them in a new way for a new purpose, for example, fences and interior units, such as chairs, tables and lighting units, by using manual or digital construction methods, as shown in Figure 9.

Figure 9. Suggested new forms for the structure.

# 7 ARUNDO DONAX IN DIGITAL FABRICATION

## 7.1 *Digital fabrication*

This is a process that uses Computer-Aided Design (CAD) and additive and subtractive manufacturing machines, such as a CNC router, laser cutting and 3D printing, to allow designers to produce material digitally with great accuracy.

## 7.2 *Applications of Arundo Donax in digital fabrication*

The designs inspired by the pattern are formed when the rods of Arundo Donax are sliced, as shown in Figure 10. The digital method uses Rhino and Grasshopper software and it can be produced by using a CNC mill.

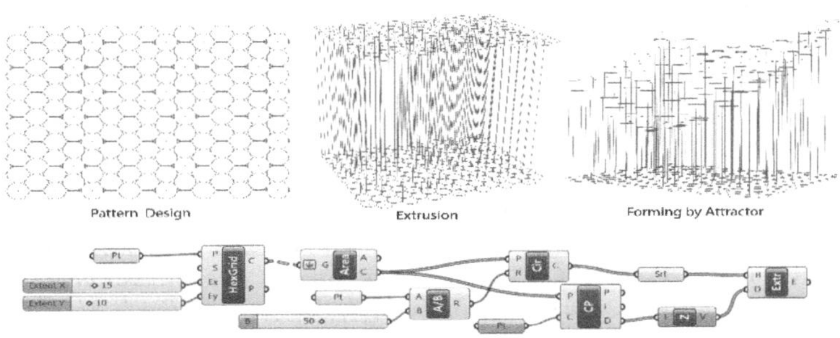


Figure 10. Form developing.

The rods can be linked together with metal connectors or be tangled together using tight ropes. The shape can be used either indoors or outdoors. After developing the form by using Grasshopper, the item can be modified by Rhino, 3Ds Max and other software. To get a suitable design and create a proper environment around it, a rendering of the scene is necessary to observe the overall view, as shown in Figure 11.

Figure 11. The orthogonal projection of the fabricated chair (Left). Final render after editing on 3Ds Max (Right).

## 8 CONCLUSIONS AND RECOMMENDATIONS

### 8.1 *Conclusions*

Arundo Donax can be used with other materials, such as natural wood, bamboo and fabricated materials, as a complementary material in the manufacturing process. This shows the importance of improving visual manifestation by searching for the proper methods of treating forms using local materials. Arundo Donax is one of many sustainable solutions that can be used as a thermal insulator. The importance of reforming shallow water plants, such as the Arundo Donax and bamboo, stems from the possibility of its use in interior design and architecture.

### 8.2 *Recommendations*

Attention must be paid to the digital fabrication methods to find the best ways of implementing modern and parametric designs by using local environmental materials. Proper cultivation of the Arundo Donax by using modern techniques is significant for producing the best types of plants in Egypt. It is essential to implement the proper treatment on the materials and to use them in the manufacture and building processes, especially since such urban fabrication should maintain the identity and nature of these areas.

## REFERENCES


Abissy M. & Mandi L. (1999). The use of rooted aquatic plants for urban wastewater treatment: Case of Arundo Donax. *Revue des Sciences de L'Eau, 12*(2), 285–315.

Alshaal T., Elhawat N., Szabolcsy D. É., Kátai J., Márton, L., Czakó M., Fári M. G. (2014). Giant reed (Arundo donax L.): A green technology for clean environment. *Phytoremediation, Vol I. Springer.* 3–4.

Bailey L. H. (1935). *The standard cyclopedia of horticulture, Vol I. The Macmillan Company, NY.*

Barreca F. (2012). Use of giant reed Arundo Donax L. in rural constructions. *Commission Internationale du Génie Rural (CIGR) Journal, 14*(3), 48–49.

Bell G. P. (1997). Ecology and management of arundo donax and approaches to riparian habitat restoration in southern California. *Plant Invasions. Studies from North America and Europe., Publisher: Backhuys Publishers, Leiden, The Netherlands, Editors: J.H. Brock, M. Wade, P. Pyšek, D. Green, 103–104.*

Christou M., Mardikis M., Alexopoulou E. (2000). Propagation material and plant density effects on the Arundo donax yields. *1st World Conference on Biomass for Energy and Industry, Vol. 1, Sevilla, Spain. 1622–1628.*

Helal, A. A. A. (1989). *An analytical study for Arundo architecture in Al Manzalah Lake* (M.S. thesis). Faculty of Fine Arts, Helwan University.

Hussein, M. M. I. (2000). *Environment & materials in interior design between function and aesthetic values* (PHD. Thesis). Faculty of Fine Arts, Alex. University.

Ibrahim, N. H. (1992). *Study of the appropriate means of utilization of bamboo in structures and environmental industries* (M.S. thesis). Institute of Environmental Studies and Research Engineering Department, Shams University.

Indian Standard Institution. (1965). *Working widths and reed spaces of plain calico looms.*

Lewandowski, I., Scurlock, J.M.O., Lindvall, E. & Christou, M. (2003). The development and current status of perennial rhizomatous grasses as energy crops in the US and Europe. *Biomass and Bioenergy, 25*, 335–361.

Opperman, K. (1956). *Handbook for making and adjusting single reeds*. New York, New York: Chappell & Co.

Sanderson, K. (2006). US biofuels: A field in ferment. *Nature, 444*, 673–676.

Soliman, M. A. H. M. (2009). *Arundo donax L. and its use in thermal insulation in architecture to decrease the environmental pollution* (M.S. thesis). Institute of Environmental Studies and Research Engineering Department, Shams University.
http://www.idrs.org/publications/controlled/DR/DR2.1/arundo.html
http://entheology.com/plants/arundo-donax-giant-reed/
http://keys.lucidcentral.org/keys/v3/eafrinet/weeds/key/weeds/Media/Html/Arundo_donax_(Giant_Reed).html
http://canyaviva.com/English/Proyectos/ExpoZaragoza2008.html
http://canyaviva.blogspot.com.eg/
http://antoineboudin.com/arundo-donax/
http://www.parametriccamp.com/en/what-is-parametric-design/
https://www.rcbc.ca/files/u3/PPI_Zero_Waste_and_Local_Govt.pdf

## *City as a scene*

*Architecture*

*Cities' Identity Through Architecture and Arts – Catalani et al. (Eds)*
*© 2018 Taylor & Francis Group, London, ISBN 978-1-138-05409-7*

# Authenticity of the physical environment that influences a sense of place: A qualitative study at Ampel Street Corridor, Surabaya, Indonesia

Dyah Kusuma Wardhani & Astrid Kusumowidagdo
*Universitas Ciputra, Surabaya, Indonesia*

ABSTRACT: This study aims to explore the physical factors that contribute to the Ampel Street Corridor in Surabaya, Indonesia. Ampel is known as both an historic area and as an area in Surabaya that is inhabited by Arabian communities. The Ampel corridors lead to the centre of activities, the mosque and the tomb of Sunan Ampel, who was the spreader of Islam in Java Island in the 16th century. Qualitative methods of study are used, where the data are collected through focus group discussions. The informants involved are visitors to Ampel. The study reveals that the physical factors that affect the formation of a sense of place for the corridor of the Agung Mosque and and tomb of Raden Rahmat Sunan Ampel are an authentic gate, corridor dimension, commercial space disharmony, the diversity of products and natural lighting. Meanwhile, the physical factors that affect the Ampel Suci Corridor are an authentic gate, ceiling ornament, a connecting road to the settlement, comercial space harmony, artificial lighting and the diversity of product types with religious nuances.

*Keywords*: physical factors; sense of place; street corridor; commercial retail

## 1 INTRODUCTION

A place provides a centre for people's activities and culture (Low & Altman, 1992). Quite often a place also gives people the opportunity to be connected with each other, such as with friends and communities. The value of a place can also be related to its history and culture. However, the aforementioned social factors are not mutually exclusive. The physical factors of a place are also important. Physical factors shape a sense of place (Stokols, 1990; Relph, 1976; Norberg-Schulz, 1985; Kusumowidagdo et al., 2013), so that a place will have the qualities of its physical scope.

A sense of place is associated with physical factors, especially in commercial areas. This has been widely researched, for example, in various commercial locations. Several related studies were conducted in the corridors of shopping centres (Kusumowidagdo et al., 2013; Kusumowidagdo et al., 2015), the corridors of commercial areas in Wilaya underground (Zacharias, 2002), the historic commercial neighbourhood (Shamsuddin, & Ujang, 2008), and also in the coffee shops that functions as social gathering places (Waxman, 2006).

This study focuses not only on the areas with commercial attributes, but also on those with historical and religious attributes. The uniqueness and completeness of these attributes means that the research location, Ampel Street Corridor, is a thought-provoking main topic for this study.

As the earliest streets in the city, the traditional shopping streets preserve their character, both physically and socially, through the street activities and the interaction of the people (Shuhana et al., 2004). They are regarded as important in influencing the city's identity and economic vitality. Research regarding a sense of place has identified the importance of a place's attributes and characteristics in the construction of the sense of place and attachment

(Gieryn, 2000; Stedman, 2003). Therefore, the authenticity of the physical factors that influence the sense of place at the Ampel Street Corridor has become the research focus to be studied by the research team.

## 2 LITERATURE REVIEW

### 2.1 *Physical factors of a sense of place*

Places are composed of three broad interrelated components that give them meaning: the physical setting, the individual's internal psychological and social processes, and the attributes and activities that are done at the place (Canter, 1977; Relph, 1976; Smaldone et al., 2005; Stedman, 2003; Stokols, 1990). People's experience of a place, the physical forms, activities and meanings combine together to form the sense of place and its character (Montgomery, 1998).

Najafi & Shariff (2011) states that the physical characteristics and the environment relate not just to the setting or order layout of a place, but also contribute to the meaning. The elements, which are proportion, size, scale, distance, ornaments, sounds, temperature, visual variety and colour (Steele, 1981), have contributed to the creation of the feelings of happiness, sadness, In the corridor, the form of tenant space, finishing and proportion influence the sense of place. The dimension of a store, for example, also provides convenient visual effects for the product exploration of both visitors and tenants (Wee & Tong, 2005; Kramer, 2008; Kusumowidagdo et al., 2015). Next, the elements that affect the atmosphere of the corridor are materials, dimensions and finishing. Dimensions, such as the width of the corridor and the size of the space, can also affect the perception of space (Zacharias, 2002; Kusumowidagdo et al., 2015). On the corridor, the materials used for floor finishing and ceilings will also create a certain atmosphere (Baker, 1986; D'Astous, 2000; Kramer, 2008; Kusumowidagdo et al., 2015). In addition to the corridor dimensions, there are booths or retail carts that help to create a creative and dynamic atmosphere, through selling attractive, colourful and seasonal products (Wee & Tong, 2005). In addition to colour, lighting also affects perception (Zacharias, 2002; Kusumowidagdo et al., 2015).

As mentioned in the literature, physical factors contribute to the function of a space, by creating meaning for local people or visitors. Therefore, the legibility of a place, as well as people's satisfaction with it, can be assumed to be influential factors, and hence to have a meaning in shaping a connection between people and place (Hashemnezhad et al., 2013).

### 2.2 *The concept of traditional streets*

Streets in an urban context are places of economic and social significance. Great cities and places are, in most cases, identified by their main streets and the character of the streets reflects the image of the cities (Bentley et al., 1992). The streets also represent the people's perception of a city's character and identity (Shuhana et al., 2004) due to the bonding developed by the experience. Traditional streets are strongly characterised by the traditional forms of transaction within the old shophouses and the informal vending activities that take place along the streets, while the modern streets concentrate on modern shopping complexes and formal spots for leisure activities. The form and spatial treatment of the shop frontages influences the manner in which the pedestrians engage in the activities and thus either encourages or discourages attachment (Shuhana et al., 2004).

## 3 PURPOSE

The purpose of this research is to explore the physical factors that shape the sense of place of the Ampel Corridor area as an historic religious site.

Table 1. Question list for focus group discussion participants.

| Topic | Research question |
|---|---|
| Physical factors that shape sense of place | • Why are you interested in this area?<br>• Are you familiar with this area?<br>• What are the physical conditions that characterise this area?<br>• Can you explain that condition? |

## 4 RESEARCH METHOD

This research is qualitative in nature, with observation, documentation and focus group discussion used as the techniques of data collection. Six visitors to the historic Ampel religious site, who share different professional backgrounds, are treated as the informants. The process of holding the focus group was preceded by explaining the aims of the research. In order to streamline the discussion, the participants were firstly asked to write down their personal data as part of the respondents characteristics. Then they proceeded to give their responses to several discussion topics that had been prepared previously. The data obtained from these six respondents are categorised into the following topics:

The researcher analysed the focus group discussion results, at the end of the focus group discussion, the recorded notes were checked and read several times and the important parts were highlighted. The objective was to select phrases containing hidden ideas that would support the research and to compile them.

## 5 RESEARCH OBJECT

The area of the Ampel corridors is located in an Arabian village in Surabaya, Indonesia. The Ampel Corridor is a term for some of the corridors leading to the complex of the Agung Mosque and the tomb of Sunan Ampel. This area is unique, with its specific setting and atmosphere. The number of ancient buildings, the community of Arabian descendants, and the business area dominated by Arabian goods lead to the uniqueness of this area. This research is limited to two corridors, namely corridor A and corridor B, in which both of them have similar characteristics as commercial spaces, or have commercial enclosures on both sides of the corridor.

## 6 FINDINGS AND DISCUSSION

The physical factors that create the sense of place of the Ampel Corridor are as follows:

### 6.1 *Corridor A: Entrance of the Agung Mosque and the tomb of Raden Rahmat Sunan Ampel*

#### 6.1.1 *Authentic gate*

The existence of the gate is a physical marker of the Ampel Corridor, especially at corridor A. The shared understanding about this gate can be found in the discussions, as shown by the following:

> *'Gate is a physical marker for the main entrance of Agung Mosque and the tomb of Raden Rahmat Sunan Ampel religious tour' (F, field supervisor, Sidoarjo)*

#### 6.1.2 *Corridor dimension*

The corridor at Area A has a width of ±3 metres and is used as an entrance and exit for visitors on the religious tour. Its narrow corridor and crowded circulation have marked the

Figure 1. Physical factors of corridor A.

situation of this corridor, and thus it has become a distinctive trait of the Ampel Street Corridor.

*'The street corridor narrow and its circulation is unclear, since it is used as both entrance and exit for all visitors of religious tour' (D, interior designer, Surabaya)*

As stated by Kusumowidagdo et al. (2015), the dimensions of a corridor can affect comfort, and studying the products available in the area surrounding the corridor can help to define its sense of atmosphere. The importance of the corridor's dimensions is similar to the findings regarding the design features, as studied by Baker (1986), Zacharias (2002), Wee and Tong (2005), as well as Kramer (2008).

6.1.3 *Commercial space disharmony*

This area is a main circulation line leading to the mosque and the tomb of Sunan Ampel, which is why traders use the corridor to sell their goods. The activities range from selling goods in the shophouses to selling goods as street vendors. Uniquely, commercial space disharmony is a chaotic situation that marks the identity of this place, and thus creates the sense of place of the corridor.

*'The distance between sellers' booths differ to each other. Many sellers use the street to sell their goods hence the corridor is getting crowded' (M, staff, Surabaya)*

*'This corridor street is main entrance to the Ampel mosque as religious site, and is also close to the bus drop area. That is why many sellers use entrance line to sell their goods' (F, field supervisor, Sidoarjo)*

6.1.4 *Diversity of products*

The goods sold by the sellers in this area are prayer items, date palms and food for the visitors.

*'The products sold are praying stuffs, date palm, and many food sellers for the visitors who do religious tour' (D, university student, Surabaya)*

### 6.1.5 *Natural lighting*

Good lighting provides a boost to the spirit when the visitors to this corridor area are conducting their activities. This is why, in the corridor of Area A, the space is open with maximum natural lighting to support the atmosphere of the religious tour activities.

*'The atmosphere of religious tour activities is strong, since the street corridor leads to the mosque is open and without canopy' (D, interior designer, Surabaya)*

Areni and Kim (1994), Turley and Milliman (2000) and Wee and Tong (2005) state in their studies that appropriate lighting applications are one of the comfort factors of a corridor.

## 6.2 *Corridor B: Ampel Suci*

### 6.2.1 *Authentic gate*

Similarly, there is an authentic gate at corridor B. The impressions of the gate are expressed in the following statements:

*'The gate is designed and shows that we are in the Ampel area' (M, employee, Surabaya)*

*'The gate is vintage yet more organized' (N, employee, Surabaya)*

### 6.2.2 *Ceiling ornament*

This corridor is covered by a ceiling. There are patterned ornaments on the ceiling frame, which helps to avoid a monotonous look. The ornaments on the ceiling of the Ampel Suci are of a flowery pattern and are attached to the canopy frame. Furthermore, the ornaments on the ceiling are also in the form of Middle East-style chandeliers at several points.

*'The area is more organized with flowery ornaments on the ceiling and middle east-style chandeliers' (D, interior designer, Surabaya)*

The ceiling ornaments are important in creating the image and ambiance of the shopping centres (Baker, 1986; Kusumowidagdo et al., 2015).

### 6.2.3 *Connecting road to the settlement*

There are several methods of access from the corridor of the village to the Ampel Corridor. People can walk through these areas and thus become part of the special identity of this place.

*'The atmosphere of Ampel area is stronger, since there are many accesses in this area to the Arab village which is located near the shopping corridor' (S, architect, Surabaya)*

### 6.2.4 *Commercial space harmony*

Ampel Suci is an area specified for shopping. People use the old shophouses and some street vendors use the area near the corridor, yet they do not sell goods in the street, and that is why this area looks neater and appears to be better organised.

*'Ampel Suci is a first area that is specified for shopping area, that is why this area is more neat and organized with lighting, canopy, and distance between booths' (F, field supervisor, Surabaya)*

Store harmony is one of the elements that are considered to shape the sense of place (Kusumowidagdo et al., 2015).

Authentic gate | Ceiling ornament & artificial lighting | Connecting road to settlement

Figure 2. Physical factors of corridor B.

Commercial space harmony | Diversity of products

Figure 3. Physical factors of corridor B.

### 6.2.5 *Diversity of products*

The products or goods sold in Ampel Suci are mostly prayer items and Arabian souvenirs, as expressed in the following statement:

> *'The products sold in Ampel Suci vary from praying stuffs to Arabian souvenirs' (S, architect, Surabaya)*

The key attractions of the traditional shopping streets are the products offered and the shopping activities that are associated with the different ethnic groups in the city, as claimed by Shamsuddin and Ujang (2008).

### 6.2.6 *Artificial lighting*

At the area of Gate B, artificial lighting is very dominant in shaping the sense of place, since:

> *'The atmosphere is more closed because of canopy, which is why each old shop house uses artificial lighting to support the lighting' (M, employee, Surabaya)*

In their research, Kusumowidagdo et al. (2015) also claim that lighting is a factor that influences the corridor of the shopping area. This claim supports the research of Baker (1986).

## 7 CONCLUSION

The traditional shopping streets of Ampel have been important and meaningful for users as places for religious activities, shopping locations and sociocultural diversity, as well as for self and group identity. The physical factors of the area of the Ampel Street Corridor have had an influence on the shaping of the sense of place.

The physical factors that were found in both corridors were the authentic gate and product diversity. At corridor A there are some other distinctive factors, which are natural lighting, corridor dimension and commercial space disharmony. The presence of informal traders on the pavement adds to the diversity of the place and has a significant role in creating a unique atmosphere on the streets. Meanwhile, at Gate B, the distinctive factors are the ceiling ornaments, the connecting roads to the settlement, commercial space harmony and artificial lighting. These physical factors simultaneously shape the sense of place of the Ampel Corridors area.

This research is expected to make a scientific contribution in several areas, including in both academic and practical areas. With regards to the academic aspect, the implementation is expected to become the foundation for further research focusing on a sense of place, commercial street corridors and historical-religious areas. The practical contribution of this research is as a reference for government and urban planners in making and revitalising meaningful urban places.

## ACKNOWLEDGEMENT

The research team expresses their gratitude for the participation of the communities of the Ampel area during the research process, to the seminar reviewers, and to the sponsor. This research is funded by the Grant of Applied Product Research (*Penelitian Produk Terapan*), from the Minister Education and Cultural Affairs, Directorate of Higher Education, 2017.

## REFERENCES

Areni, C. S. and Kim, D. (1994). The influence of in-store lighting on consumers' examination of merchandise in a wine store. *International Journal of Research in Marketing*, *11*(2), 117–125.

Baker, J. (1986). The role of the environment in marketing services: The consumer perspective. *The Services Challenge: Integrating for Competitive Advantage*, *1*(1), 79–84.

Bentley, I., Alcock, A., McGlynn, S., Murrain, P. & Smith, G. (1992). *Responsive environments: A manual for designers.* Oxford: Butterworth Architecture.

Canter, D. (1977). *The psychology of place*. London: The Architectural Press Ltd.

d'Astous, A. (2000). Irritating aspects of the shopping environment. *Journal of Business Research*, *49*(2), 149–156.

Gieryn, T. F. (2000). A space for place in sociology. *Annual review of sociology*, *26*(1), 463–496.

Hashemnezhad, H., Heidari, A. A. & Mohammad Hoseini, P. (2013). "Sense of place" and "Place attachment". *International Journal of Architecture and Urban Development*, *3*(1), 5–12.

Jacobs, A. (1999). *Great streets*. Cambridge, MA: MIT Press.

Kramer, A. (2008). *Retail development*. Washington DC: Urban Land Institute.

Kusumowidagdo, A., Sachari, A. & Widodo, P. (2013). The setting of internal shopping centres' public spaces and their relationship to the visitors. *GSTF Journal of Engineering Technology (JET)*, *2*(1), 211.

Kusumowidagdo, A., Sachari, A. & Widodo, P. (2015). Visitors' perception towards public space in shopping center in the creation sense of place. *Procedia-Social and Behavioral Sciences*, *184*, 266–272.

Low, S. M. & Altman, I. (1992). *Place attachment*. New York USA: Plenum Press.

Montgomery, J. (1998). Making a city: Urbanity, vitality and urban design. *Journal of Urban Design*, *3*(1), 93–116.

Najafi, M. & Shariff, M. K. B. M. (2011). The concept of place and sense of place in architectural studies. *International Journal of Human and Social Sciences*, *6*(3), 187–193.

Norberg-Schulz, C. (1985). The concept of dwelling on the way to figurative architecture. New York: Rizolli.

Relph, E. (1976). *Place and placeness.* London: Pion.
Shamsuddin, S. & Ujang, N. (2008). Making places: The role of attachment in creating the sense of place for traditional streets in Malaysia. *Habitat International*, *32*(3), 399–409.
Shuhana, S. & Ahmad Bashri, S. (2002). The vanishing streets in the Malaysian urbanscape. In P. Miao (Ed.), *Public places in Asia Pacific*. Dordrecht: Kluwers Academic Press.
Shuhana, S., Ahmad Bashri, S., Norhaslina, J. & Maslyana, M. (2004). *Criteria of success for traditional shopping streets in Malaysia: Case study of Kuala Lumpur*. Unpublished Research Report, Universiti Teknologi, Malaysia.
Smaldone, D., Harris, C., & Sanyal, N. (2005). *An exploration of place as a process: The case of Jackson Hole, WY*. Journal of Environmental Psychology, 25, 397–414.
Stedman, R. C. (2003). Is it really just a social construction?: The contribution of the physical environment to sense of place. *Society &Natural Resources*, *16*(8), 671–685.
Steele, F. (1981). *The sense of place*. Cbi Pub Co: Michigan University.
Stokols, D. (1990). Instrumental and spiritual views of people environment relations. *American Psychologist*, *45*(5), 641.
Turley, L. W. & Milliman, R. E. (2000). Atmospheric effects on shopping behavior: A review of the experimental evidence. *Journal of Business Research*, *49*(2), 193–211.
Ujang, N. & Dola, K. (2007). Linking activity and place attachment dimensions in enhancing the sense of place. *ALAM CIPTA, International Journal on Sustainable Tropical Design Research & Practice*, *2*(1), 59–67.tWaxman, L. (2006). The coffee shop: Social and physical factors influencing place attachment. *Journal of Interior Design*, *31*(3), 35–53.
Wee & Tong. (2005). *The 4RS Asian shopping management*. Jakarta: PT.Buana Ilmu Populer.
Zacharias, J. (2002). Choosing a path in the underground: Visual information and preference. In *ACUUS International Conference Urban underground space: a resource for cities* (pp. 14–16). Torino.

*Cities' Identity Through Architecture and Arts – Catalani et al. (Eds)*
*© 2018 Taylor & Francis Group, London, ISBN 978-1-138-05409-7*

# Back into the future: The city improvement board of Hyderabad

Anuradha S. Naik
*Anuradha Naik Associates, Hyderabad, India*

ABSTRACT: The City Improvement Board (CIB) was set up in 1912 by HEH Nawab Sir Mir Osman Ali Khan (1911–1948), the ruler of Hyderabad, the largest princely State in India, with the purpose of developing the city in a holistic manner.

Structured housing schemes, adaptation and re-use of old buildings as well as heritage and water conservation were central to the schemes of the CIB. Specially designated commercial districts were designed; playgrounds and parks were created to bridge economically poorer areas of Hyderabad with newer ones.

The CIB was abolished in 1957, and in the forty-five years of its existence, it changed, along with the skyline of the city, the lives of thousands of its citizens. The positive impact of the work done by the Board is still visible today.

The paper will elucidate some of the key projects of the CIB and also highlight its impact on Hyderabad even today.

*Keywords*: City Improvement Board; CIB; conservation

## 1 INTRODUCTION

Hyderabad State located in the Deccan Plateau was the largest of the five hundred and sixty five princely states of India. Covering an area of roughly 215,000 sq.km it was under the domain of the Asaf Jahi Dynasty from 1724 until 1948. Initially as viceroys of the Moghul Emperor, the Nizams, as the Asaf Jah rulers were known, rose to power, claiming independence in 1857, after the fall of Delhi.

The capital of Hyderabad State was also known as Hyderabad, 'the fourth largest city in South Asia during the colonial period' (Beverley, 2015). Founded in 1591 by Mohammed Quli Qutub Shah (1580–1611CE), it was a well-planned city laid out on a double grid system in the form of a giant cross. Over the course of three hundred years Hyderabad evolved with the famous monument, the Charminar at its centre, but the well-designed city had disappeared under the matrix of a complex, dense historic and multi-layered urban setting.

Located to the south of the River Musi, the city had grown in an organic and haphazard fashion and it was unfortunately a cataclysmic flood and an epidemic of the plague that propelled the need to develop and restructure the city at the turn of the 20th century.

### 1.1 *The floods of 1908*

On 28th of September 1908, the city received 325.12 mm of rain within the first 24 hours and 155 mm in the next (Visvesvaraya, 1909). The discharge rate of water in the swollen river began with an incredible 3115 $m^3$/sec and soon rose to 12035 $m^3$/sec. About 20,000 houses were destroyed, 15,000 people perished and at least 30,000 were rendered homeless (Prasad, 1984). The sixth Nizam, Mir Mahbub Ali Khan (1869–1911) opened up his palaces and set up community kitchens for the citizens of Hyderabad. His most lasting contribution was to request Sir Visvesvaraya (1861–1962) to report on steps to be taken to assess the situation and take appropriate measures for the future. Sir Visvesvaraya, an eminent Indian engineer, scholar and statesman, was then the Chief Engineer of the Public Works Department of Bombay and *Divan* (Minister) of Mysore State. Knighted by King George V for his contributions to public good, he also received the Bharat Ratna, India's highest civilian honour in 1955.

In Hyderabad, Sir Visvesvaraya recommended the construction of two flood catchment reservoirs upstream. He also proposed to strengthen the riverbank along the city and prohibit the construction of houses and dwellings there, instead lining it with boulevards, parks and public buildings. His findings were published in 1909,

> *"The chief aim of the proposals being to provide the most efficient and economical remedy.... By carrying out the works in their entirety a control will be obtained of the river at all times and irrigation will be started on a scale never before attempted in the State. The works will mitigate both floods and famine and will remain an abiding landmark of His Highness' administration" (Visvesvaraya, 1909).*

## 2 THE FORMATION OF THE CITY IMPROVEMENT BOARD (CIB)

The need for urgent action was reiterated when the plague struck Hyderabad in 1911, and the city lost roughly a hundred thousand residents (Beverley, 2015). The State had a new ruler; HEH Nawab Sir Mir Osman Ali Khan (1911–1948) was the last and most dynamic of the Nizams. He set up the CIB in 1912 with a mandate to improve "the social, moral and physical conditions of the citizens" (Beverley, 2015). Aligning with Sir Visvesvaraya's report which stated that "in carrying out improvements on such a large scale, it would not be right to ignore the artistic, economic and sanitary considerations associated with the proposals" (Visvesvaraya, 1909), over the next four decades following its establishment, the CIB reshaped Hyderabad city, with slums giving way to proper housing, parks and gardens, "the boulevards being the lungs of the City" (Visvesvaraya, 1909). The CIB was to develop the city in a holistic manner with the inclusion of aesthetics, as is evident in the Urdu word for the CIB, the *Araish e Baldia*, which roughly translates to 'the Embellishment of the City'.

Constituted of engineers, planners and horticulturists, the CIB sent its members to study the works of municipal bodies and botanical parks in England, Germany, Italy and the United States. Conversely, the CIB's work also drew the attention of planners in Europe, especially in the period between the wars.

The CIB published annual reports of its work, giving detailed accounts of proposals, completed works and financial statements. The dedication and commitment of the CIB is tangible in these reports, and reflects the general enthusiasm of the populace towards the proposals, and as is discussed below, their very effective results.

### 2.1 *Setting the stage: Gardens, roads and infrastructure*

The first task the Board took up was to decongest the city. Sir Visvesvaraya's report stated that streets should be "laid out so as to present a pleasing aspect and open to abundance of light and air, houses...constructed with strict attention to health necessities and public standards of beauty, and parks and playgrounds...within easy reach of the largest number of residents"(Visvesvaraya, 1909).

The CIB was to work within the old walled city of Hyderabad and in the suburbs north of the river as well. Large areas near the Hussain Sagar and Mir Alam tanks and swampy water-bodies like Mir Jumla, Afzal Sagar and Masaheba tanks were drained and converted to parks. The combined area of these new parks was roughly 3175 acres. Smaller parks, playgrounds and *maidans* (arenas) were included even within the closely built walled city.

The CIB laid new roads, opened out crowded areas and laid avenues along the riverbanks. Connections to the south of the river were made broader and axially re-aligned. Roads linked railway stations, bus stations and other public transport routes. A river circuit road connecting the two riverbanks with large tree lined boulevards and newly constructed bridges was completed by 1932 (CIB Report, 1932). The CIB had recommended that an Inner Circular Road of nearly 15 km be constructed to reduce traffic in the centre of the city and an outer concentric Circular Road of nearly 30 km for future expansion. Radial and cross roads were designed to connect centres of business with the new circular roads. Interestingly, the Inner and Outer Ring Road schemes, that were completed as recently as 2012, were envisioned by the CIB, with axial roads, connecting to the concentric circuits. The routes have been modified and enlarged, the Inner Circular or Ring Road, as it is known covering an area of 50 km whereas the eight-lane Outer Ring Road covers an area of approximately 150 km.

The drainage system for Hyderabad was revamped. An underground system was put in place and storm water drains were separated and emptied directly into the River Musi. Feeder channels were constructed to take water away from low-lying areas to prevent stagnation and thereby disease. A new sewerage system and treatment plant was installed and effluents were treated, diluted and used for irrigation.

### 2.2 *Osmansagar and Himayatsagar reservoirs*

There was a boost to irrigation in areas around the city following the construction of the two reservoirs built to dam the Musi and its tributary the Esi. The most well known of the CIB schemes, the reservoirs were named after the seventh Nizam HEH Nawab Sir Mir Osman Ali Khan and the Heir Apparent HH Nawab Walashan Mir Himayat Ali Khan Azam Jah, the Prince of Berar. With capacities of approximately 160 million $m^3$ and 120 million $m^3$ (Bawa, 1984), they were constructed to create a drinking water source for the city in addition to combining irrigation with flood prevention.

### 2.3 *The 'River district' and the business districts*

The river was dammed within the city limits creating a 5 km long lake. The river district as the embankment in the city limits was named, was to be developed as the main civic centre of Hyderabad. Tree lined avenues; gardens and "good buildings with suitable frontages" were designed so that "the whole of the locality ...would ... rise in value and importance.... And give dignity to the riverfront"(Visvesvaraya, 1930).

The schemes of the CIB also included carefully planned commercial districts and purpose built bazaars. The Patthergatti arcade (Fig. 1), still amongst the most popular shopping areas in the old city was designed and developed by the CIB. Moazzam Jahi market, another popular landmark of Hyderabad, and perhaps its first shopping mall, was also a part of the CIB's design.

The work was conceived to benefit business communities and to integrate markets into the new urban fabric that was developing. Markets came up in the proximity of residential

Figure 1. The Patthergatti shopping arcade, circa 1934.

areas. Quiet, lesser-known bazaars like the Begum Bazaar Fish Market, sensitively designed to provide a 'home like' environment, were part of the CIB's urban renewal.

An industrial area called Azamabad on an area of 120 acres, east of the Hussainsagar Lake was plotted and allocated to factories for commercial goods like tobacco, silk and matches. As reported in the 1937 annual report, "economic development would also provide Hyderabad with the infrastructure to compete with both British Indian manufacturers and those of Japan even those of Britain itself" (Beverley, 2015).

Soon, another industrial area Sanatnagar was established in 1941, with factories manufacturing various metal works and tools. Other industrial areas located on urban fringes along railway lines were created.

The urban renewal schemes and the construction of an eclectic mix of buildings, form and function were woven together using a common architectural vocabulary. Starting with majestic public buildings commanding the riverfront to the small functional workshops, the works of the CIB are unmistakably recognisable.

### 2.4 *CIB Architecture: a new identity and skyline for the city*

The High Court of Hyderabad (1916) (Fig. 2) and the City College (1917) were constructed on the south bank of the Musi and the Osmania General Hospital (1925) and the Asafia Library (1932) on the north bank. With the exception of the library, these buildings were designed by the English architect Vincent Esch (1876–1950). Initially categorised as the Indo-Saracenic style, the CIB under the Nizam of Hyderabad was working with a definitive mandate: to draw upon elements from both Hindu and Islamic architecture, creating a unique palate, where aesthetics though important was not the objective. A powerful statement of secularism was the intent. This was a first conscious effort at creating a distinct architectural style. Developed and honed by subsequent designers of the CIB it came to be described as the Osmanian or CIB style. Esch's buildings and those that followed, including the Patthergatti shopping arcade (1930) and the Moazzam Jahi Market (1935), made a conscious effort at incorporating new technology, materials and functions, with architectural influences of over six centuries, from the Kakatiyan through the Qutub Shahi, Moghul and Asaf Jahi forms, and therefore resulting in the eventual emergence of a distinct and recognisable style. Hyderabad's noted historian Dr. Ghulam Yazdani described the style as "modern in conception and planning, yet retaining traditional details based on the old architecture of Hyderabad". (Yazdani, 1944).

Figure 2. The majestic High Court on the South Bank. Gardens and stone embankment on the north bank visible.

## 2.5 *Conservation and re-use*

Known landmarks, buildings, other old structures and a significant portion of the City Wall were repaired by the CIB. The original historic gateways like the Delhi *Darwaza*, the Purana Pul *Darwaza* and the Afzal *Darwaza* were restored. Where the wall was damaged beyond repair (along the river front) a road was constructed along its footprint and the wall extended to the riverbank (CIB Report, 1932).

A dilapidated building in the Bashirbagh locality was purchased by the CIB and converted into a guesthouse. " By doing the work economically and taking advantage of the old structure wherever it was good, a decent building has been made which is now worth about double of what has been spent over it" (CIB Report, 1932). (Fig. 3(a) and Fig. 3(b)) Interestingly, adaptive re-use is a term put forth in more recent times in terms of sustainable building.

## 2.6 *Slum clearance and re-housing*

The largest and most significant of the CIB's work was the clearance of slums and re-housing of people across the city. To adjust urban densities, the CIB acquired any adjoining open lands including reclaimed land from tanks, agricultural tracts in the city and part of the *Sarf-e-Khas* (Crown Lands), where possible. These were built upon first with all infrastructure and amenities and then gradually the slums demolished and residents re-housed. The reason for this is clearly stated.

Figure 3(a). The Bashirbagh guest house before intervention.

Figure 3(b). The Bashirbagh guest house after intervention.

*"It was observed that the slum dwellers that were dis-housed, formed fresh slums in open localities. With a view to check this tendency and to safeguard their health and indirectly the health of the whole population of the city, the Government decided to construct sanitary houses on low rents to house such people. Open lands available round about the city were acquired and construction of buildings was started."(CIB, 1937).*

The policy of the CIB was to retain the *pucca* (permanent or strong) houses and then overlay a scheme to build over the remaining land after the demolition of the *kuccha* (huts or temporary) houses. The overlay implemented by the CIB used only 60% of the area for building, with a clear mandate to build only on two-thirds of each individual plot, leaving a third open "for promoting healthy conditions of the locality" (CIB Report, 1932). The remaining 40% was taken up by roads, playgrounds and *maidans* (Fig. 4).

The localities designed by the CIB were functional and robust, designed in traditional Hyderabadi vocabulary and finished with a flourish of the Art Deco, then in vogue. Houses of various grades were designed to accommodate citizens of all strata of society, so as not to create a class-specific locality. The need for the outdoors, fresh-air, playgrounds and parks was recognised and every planned locality had considerable space for children. Public health care, particularly of children was paramount to the CIB's role and infant welfare centres were set up in the old city (Fig. 5).

True to the spirit of the CIB, the focus was not on individual homes but on the community, from the '*chabuttaras*' (outdoor seating platforms) outside the homes to the playgrounds, parks and reading rooms around which the houses were ribboned. The individual units themselves were constructed in the traditional Indian style, complete with a courtyard and fruit trees, separate *mardana* and *zenana* entrances, for male and female family members and visitors.

Figure 4. The Khairatabad housing scheme, soon after completion. Circa 1933.

Figure 5. The Azampura playgrounds.

Four types of low rent houses were constructed. Type A had an area of 160 $m^2$ with four rooms including two bedrooms, and an enclosed courtyard. Type B had an area of 90 $m^2$ with three rooms in an enclosed courtyard. Type C of an area of 55 $m^2$ consisting of two rooms and a kitchen in an enclosed courtyard and type D with an area of 22 $m^2$ and one room and a kitchen, with detached toilets provided by the CIB free of rent (CIB Reports, 1933–34 & 1939–40).

The CIB began its work in Nampally, where the epidemic of the plague had struck. Slums had sprung up in the small area between the modern development of the Hyderabad Railway Station and the old medieval quarter of shrines, mosques and the tombs of saints. Land was of prime value and open areas like parks could not be provided due to the density of the urban fabric. Instead, public facilities like a hospital and school were designed and constructed.

The Nampally scheme was extended to the north by the Red Hills locality and on the south to Aghapura and they were connected by constructing a network of roads. To the west of Nampally, the open grounds, swampy marshland around the Afzal Sagar tank and some cultivable lands were taken over to create and construct Mallepally, linking the various localities to become a part of the whole development of the city. This scheme was built around playgrounds. The concept of the traditional Hyderabadi home was taken a step further. Only, at Mallepally, the 'courtyard' was the football ground. Mothers could send their children out to play while keeping a watch from their homes. Many sports figures of India coming from Hyderabad are from the Mallepally area. The probable reason behind this is that the area has eight playgrounds, which have surprisingly survived a hundred years of change. The grounds were used mainly for volleyball and football, although cricket later came to be played in a few of them.. Football thrived in Mallepally, and the neighbourhood produced five international players, four Olympians and eleven national players (Survey by author, 2013).

In the old city, an open tract of land known as Musallam Jung gardens was partly taken over to design a new locality known as Azampura. Another locality Malakpet was constructed across the railway line from Azampura. Playgrounds, tennis and badminton courts and a children's park were provided as part of the scheme. The reports include aesthetic value on a par with economic and technical details. The 1932 Report describes the Musallam Jung Garden scheme: 'it gives a very pleasing impression to travellers entering Hyderabad by the Meter Gauge line" (CIB Report, 1932).

In the densely populated walled city, localities like Moghalpura (60 houses) and Bazaar e Nurul Omara (101 houses) a relatively low number of houses were constructed, even though the demand was high. This was because of the high land value and lack of open and available space since the nobility of Hyderabad and well to do tradesmen still resided in the old city at the turn of the 20th century.

The social cause and the moral obligation that the CIB took upon itself is evident in the work it did in settling up a colony known as the Dabirpura Temperance Colony. This was established for those who had pledged to abstain from drinking and was supported by donations from local wealthy traders who funded the community halls and spaces (CIB Report, 1934).

While the houses were originally to be rented or leased from the CIB, schemes were set up for residents to purchase the houses from the government by paying interest free installments (CIB Report, 1939). The CIB continued to work until 1957. The Housing Division was merged with the Andhra Pradesh Housing Board in 1960. In its forty-five years of existence it had built 3319 houses in thoughtfully designed localities in Hyderabad (Bawa, 1984).

## 3 CONCLUSION

For nearly half a century after it was set up, the CIB worked zealously in accomplishing the works set out in the plans proposed by Sir Visvesvaraya.

The impact of these works is still visible in Hyderabad a hundred years later. The Osmansagar and Himayatsagar reservoirs continue to provide drinking water to Hyderabad's ten million residents and the part of Hyderabad dealt with by the CIB has not faced inundation by floods in the past century.

Sadly, in Hyderabad, the recognition and understanding of these efforts have been lost under decades of new urban planning and the city's rapid growth into a metropolis. One of the reasons could be the abrupt disbanding of the CIB in 1957, soon after Indian Independence.

The riverfront buildings still dominate the skyline along the river. The potential of the riverfront however, lies forgotten. Re-visiting Visvesvaraya's original layout, restoring as much as is possible and intervening where necessary would bring back to the city, its original centre and re-connect the south to the 'modern' northern bank.

As demonstrated by the CIB example, integrated housing will always be relevant, especially in the urban context. It is worth mentioning here that the individual units designed by the CIB were made of local building material and also reflected and accommodated for an Indian lifestyle, making them popular and successful. Water conservation and harvesting were concepts the CIB had addressed at a time when it was an option, and not a necessity.

The majority of the work the CIB undertook was in the period between the two wars, a time when countries especially in Europe underwent drastic re-building, slum clearance and infrastructure management as they grappled with internal issues of unemployment and poverty. Its work attracted the attention of planners and designers globally. One such planer was BS Townroe from England who published an article in the Asiatic Review Journal in July 1934.

In his article, "Town Planning: An Indian Example," Townroe compares the CIB schemes to planning schemes in Britain, which "remained as 'castles in the clouds'", whereas Hyderabad "is to be congratulated on possessing both the means and the will to translate ideas into action." (Townroe, 1934).

To conclude by quoting Townroe

> *"In spite of difficulties peculiar to India, the beauties of this Indian city must stimulate the imagination.... It is..." a well laid out capital adorned with treasures of architecture, well provided with open spaces and planned so as to secure increased health and contentment for coming generations"(Townroe, 1934).*

## REFERENCES


Alam Shah Manzoor, *Hyderabad-Secunderabad (Twin Cities): A study in urban geography,* Allied Publishers, New Delhi, 1965.

Beverley Eric Lewis, *Hyderabad, British India, and the World, Muslim networks and Minor Sovereignty c. 1850–1950,* Cambridge University Press, New Delhi, 2015.

Housing Schemes in Hyderabad City—A review by the City Improvement Board, Government of Hyderabad, 1942.

Law John, *Modern Hyderabad, Deccan*, Thacker, Spink, Calcutta, 1914.

Reports of the City Improvement Board, Government of Hyderabad, 1919–1942.

Shorey SP, "An Urban Conservation Study for Hyderabad City" in *Design*, April-June 1985.

Tillotson GHR, "Vincent J. Esch and the Architecture of Hyderabad" in *South Asian Studies*, No.9, 1993.

Townroe, B.S., "The Modernisation of Hyderabad" in *the Asiatic Review*, No. 33, July 1937.

Visvesvaraya M., *The flood of 1908 at Hyderabad: An account of the flood, its causes and preventive measures,* Venoogopaul Pillai & Sons, Hyderabad Deccan, 1909.

Visvesvaraya M., *City Improvement Schemes: Hyderabad, Deccan:* Times of India Press, Bombay, 1930.

Yazdani G., *Hyderabad: A souvenir*, Government Central Press, Hyderabad, 1944.

*Cities' Identity Through Architecture and Arts – Catalani et al. (Eds)*
 *ISBN 978-1-138-05409-7*

# Drama as a conservation tool for architectural heritage

Sara Sabahy
*Faculty of Arts and Design, October University for Modern Science and Arts, Giza, Egypt*

ABSTRACT: Heritage buildings often face many problems, such as omission and the threat of being erased from the memory and concern of the community. In this paper, the researcher introduces drama as a conservation tool, which helps to document heritage buildings and heritage sites by showing them to a large number of people. If these buildings are used in movies then they will be seen for generations to come. This would also help to increase the number of people who know and care about their heritage, increase the number of visitors to these buildings and, therefore, increase the income produced, which can then be used for conserving these buildings.

*Keywords*: Architecture; Conservation; Drama; Heritage

## 1 INTRODUCTION

Our heritage is diverse and includes buildings, monuments, gardens, cemeteries, landscapes and archaeological sites. Each of these places contain elements that help to tell its own individual story. The conservation of built heritage is generally perceived to be in the long-term interest of society. This can be better understood if it is categorised under 'economic', 'cultural' and 'environmental' interest, although they are not mutually exclusive and, indeed, they are often interlocked (CPWD, 2013).

Cinema is basically an exercise in communication and architecture is one of the tools employed by film makers in order to communicate.While architecture contributes to cinema, cinema also contributes to architecture. Carl Dreyer argued that the cinema's closest relative is architecture (Thiruvananthapuram, 2011). In the cinema, architecture appears as a background to the scene and as a background to the action (Mateo, 2012). Architecture, whether in the foreground or background, is an intrinsic part of any film, and cinema holds a position as a transformative reference in contemporary architecture (Kacmazerk, 2009).

## 2 IMPORTANCE OF PLACE IN CINEMATIC SCENES

The first truth in cinema is the place, which is indispensable to the level of the film, scene or cinematic snapshot and which cannot be discharged from spatial content (Yahia, 2007). The place in cinema has a powerful influence on a dramatic scene, one that is as great as the action and the character (Abdel-Aziz, 1994). The place is a distinct personality that has physical, social and psychological dimensions. The physical dimensions are in the form of the architectural place, geographical location and architectural qualities, which have a unique effect. The social dimension is the social level of the place and the relationships that are created between the palce and the film charecters. The psychological dimension is represented in the interaction between the place and the film charecters.

The director should choose the place very carefully in order to achieve the required dramatic effect, whether this choice of place already naturally exists or is built specifically for the purpose of the film. For example, in the Egyptian film 'Life or Death', the director Kamal Sheikh took advantage of the place (the streets of Cairo in the 1950s) to create a dramatic

Figure 1. Two scenes from the Arabic movie 'Life or Death' showing Cairo streets in 1950s (Lotus Film & El Sheikh, 1954).

Figure 2. Two scenes from 'The Mummy' Movie (Cairo cinema production company & Abdel-Salam, 1969).

atmosphere full of suspense, and also to make this film a real record of the Cairo streets at that time that can still be referenced (Yahia, 2007).

In Egyptian cinema, one of the most famous examples of movies that could be considered as a model on the forming level is 'The Mummy', directed by Shadi Abdel Salam, a movie in which every shot is a great composition.

In order to study the relationship between drama and the conservation of heritage buildings, the research will start by defining drama, conservation, historical and heritage sites. A case study will then be used, by taking screen shots from different movies that show heritage sites.

Drama is the specific mode of fiction represented in performance (Elam, 1980). The term 'drama' comes from a Greek word meaning 'action'.

Conservation means all the processes of looking after a place so as to retain its historical, architectural, aesthetic and/or cultural significance and includes maintenance, preservation, restoration, reconstruction and adoption, or a combination of these things (CPWD, 2013).

A historical or heritage site is an official location where pieces of political, military, cultural or social history have been preserved due to their cultural heritage value. Historical sites are usually protected by law, and many have been recognised with official national historic site status. A historical site might be any building, landscape, site or structure that is of local, regional or national significance (Alderson, 1985).

## 3 CASE STUDY

In the case study, the researcher focused on movies that have shown world heritage sites in Egypt in their scenes or on their posters. Screen shots were taken from these movies to show the appearance of these heritage sites.

According to the UNESCO website, there are seven Egyptian sites that are inscribed on the World Heritage List, which are: Abu Mena (1979), Ancient Thebes with its Necropolis (1979), Historic Cairo (1979), Memphis and its Necropolis—the Pyramid Fields from Giza to Dahshur (1979), Nubian Monuments from Abu Simbel to Philae (1979), Saint Catherine Area (2002) and Wadi Al-Hitan (Whale Valley) (2005). All of them are considered to be cultural world heritage sites, except for Wadi Al-Hitan, which is considered to be a natural world heritage site.

First, examples from movies that showed parts of the world heritage site Pyramid Fields from Giza to Dahshur, as shown in Table 1, the researcher chose to view samples from movies that had shown the pyramids of Giza. The chosen samples indicated the appearance of this site in the poster for the international movie 'Jumper', the animated international movie 'Despicable Me' and local Egyptian movie 'Miss Sugar'.

Second, examples from movies that showed parts of the world heritage site Historic Cairo, as shown in Table 2, the researcher chose to view samples from movies that had shown parts of historic Cairo. The chosen samples indicated the appearance of this site in the international movie 'The Spy Who Loved Me' and local Egyptian movie 'Assal Eswed'.

Third, examples from movies that showed parts of the world heritage site Ancient Thebes with its Necropolis, as shown in Table 3.

Table 1. Movies that showed Pyramids.

| No. | Movie information | Movie scene |
|---|---|---|
| 1 | Movie name: Jumper<br>Production year: 2008<br>Director: Doug Liman<br>Movie type: Action, Adventure, Sci-Fi<br>Language: English | <br>Figure 3. Poster of 'Jumper' showing the pyramids of Giza and the Sphinx (IMP, 2008). |
| 2 | Movie name: Despicable Me<br>Production year: 2010<br>Director: Pierre Coffin and Chris Renaud<br>Movie type: 3D computer-animated comedy film.<br>Language: English | <br>Figure 4. Scene from 'Despicable Me' showing the pyramids of Giza (Meledandri, Cohen, Healy, Coffin, & Renaud, 2010). |
| 3 | Movie name: Miss Sugar<br>Production year: 1960<br>Director: El Saied Bedir<br>Movie type: Comedy<br>Language: Arabic | <br>Figure 5. Scene from 'Miss Sugar' showing the pyramids of Giza and the Sphinx (AlMansora Movies & Bedir, 1960). |

Table 2. Movies that showed historic Cairo.

| No. | Movie information | Movie scene |
|---|---|---|
| 1 | Movie name: The Spy Who Loved Me<br>Production year: 1977<br>Director: Lewis Gilbert<br>Movie type: Action<br>Language: English | <br>Figure 6. Scene from 'The Spy Who Loved Me' showing part of Islamic Cairo (Broccoli & Gilbert, 1977). |
| 2 | Movie name: Assal Eswed<br>Production year: 2010<br>Director: Khaled Marei<br>Movie type: Comedy<br>Language: Arabic | <br>Figure 7. Scene from 'Assal Eswed' showing Al-Hakim bi-Amr Allah Mosque (Albatrous film production & Marei, 2010). |

Table 3. Movies that showed Ancient Thebes with its Necropolis.

| No. | Movie information | Movie scene |
|---|---|---|
| 1 | Movie name: The Spy Who Loved Me<br>Production year: 1977<br>Director: Lewis Gilbert<br>Movie type: Action<br>Language: English | <br>Figure 8. Scene from 'The Spy Who Loved Me' showing part of Karnak temple (Broccoli & Gilbert, 1977). |
| 2 | Movie name: Nile Bride<br>Production year: 1963<br>Director: Fateen Abdel-Wahab<br>Movie type: Comedy<br>Language: Arabic | <br>Figure 9. Scene from 'Nile Bride' showing part of Karnak temple (Nagib, & Abdel wahab, 1963). |

As shown in Table 3, the researcher chose to view samples from movies that had shown parts of Ancient Thebes, specifically Karnak temple. The chosen samples indicated the appearance of this site in the international movie 'The Spy Who Loved Me' and local Egyptian movie 'Nile Bride'.

Through the previous examples, the researcher showed how a movie scene could capture and preserve the state of a heritage site.

## 4 CONCLUSION

The researcher started by showing the importance of place in cinematic shots through the literature review. A case study sample of movies that had captured Egyptian world heritage sites in their scenes was also given, in order to prove that movie makers already consider using these sites as background scenery for their shots.

Through these examples the researcher wanted to study the effect of choosing these heritage sites on both the movie and on the site itself.

With regards to the benefits for the movie, the directors chose these sites because they suited the movie's story. On the other hand, there were also useful benefits for the heritage sites that were captured in these movies, as it is considered as a way of documentaion for these sites. Also, by using the same building as background scenery in different movies and in different eras, this gives us the opportunity to study and document the changes that happened to these heritage sites over time.

However, this is not the only benefit. We should also consider that, by showing these sites to the movie's audience, these sites will remain in their memory and they may feel curious enough to visit and explore them. So we can consider these movies to be a good marketing method that can be used to attract visitors from all over the world and to achieve economic benefits that could be used in the conservation process.

## 5 RECOMMENDATIONS

The film makers should be more aware when making their choice of place, by taking into consideration the effects of using the heritage building or site in their scenes, not only on the movie's story but also on the building or site itself.

There should be regulations that protect the heritage sites from any damage during the moviemaking process.

## ACKNOWLEDGEMENT

I would like to express my sincere gratitude to Prof Dr Soheir Hawas (Professor of Architecture and Urban Design Faculty of Engineering Cairo University) for her continuous support, her patience, motivation, immense knowledge and her guidance, which helped me in writing this research.

## REFERENCES

Abdel-Aziz, A. (1994). *The creative recruitment of place in feature film* (Master's thesis). The Higher Institute of Cinema, Arts Academy, Cairo, Egypt.

Albatrous film production (Producer), & Marei, K. (Director). (2010). *Assal Eswed* [DVD]. Giza, Egypt: United Brothers for Cinema.

Alderson, W. (1985). *Interpretation of historical sites*. Walnut Creek, Lanham, New York, USA. Toronto, Canada. Oxford, Uk. Rowaman Altamira & Littlefield Publishers.

AlMansora Movies (Producer), Bedir, A. (Director). (1960). *Miss Sugar* [Motion picture]. Egypt: East Company for distribution.

Broccoli, A. (Producer), & Gilbert, L. (Director). (1977). *The Spy Who Loved Me* [DVD]. Los Angeles, United States: United Artists.

Cairo cinema production company (Producer), Abdel-Salam, S. (Director). (1969). *The Mummy*. [Motion picture]. Egypt: General Egyptian Cinema Organisation.

CPWD, (2013). *Handbook of conservation of heritage buildings*. New Delhi: Directorate general, Central Public Work Department.

Elam. (1980). Retrieved April 8, 2017, from wikipedia: https://en.wikipedia.org/wiki/Drama.

IMP. (2008). IMP Awards / Jumper poster [Image]. Retrieved from http://www.impawards.com/2008/jumper.html

Kacmazerk, G. (2009, August 16). *Architecture in cinema: A relation of representation based on space*. Istanbul: LAP Lambert Academic Publishing.
Lotus Film (Producer), El Sheikh, K. (Director). (1954). *Life or Death* [Motion picture]. Egypt.
Mateo, J. (2012, July 4). Architecture and cinema. dAP. Issue 4. Retrieved from http://www.architecturalpapers.ch/index.php?ID = 35
Meledandri, C., Cohen, J., & Healy, J. (Producers), & Coffin, P., & Renaud, C. (Director). (2010). *Despicable me* [DVD]. California, United States: Universal Pictures.
Nagib, R. (Producer), Abdel wahab, F. (Director). (1963). *Nile Bride* [DVD]. Cairo, Egypt: The United for Cinema.
Properties inscribed on the World Heritage List from the UNESCO official website (NA, ND): http://whc.unesco.org/en/statesparties/eg (accessed on 25th February 2017).
Thiruvananthapuram. (2011). (NA) *Film and architecture* (Master's thesis). Department of Architecture, College of Engineering, India.
Yahia, A. (2007). *Director and place [Power point slides]*. Unpublished manuscript, The Higher Institute of Cinema, Arts Academy, Egypt.

*Cities' Identity Through Architecture and Arts – Catalani et al. (Eds)*
*© 2018 Taylor & Francis Group, London, ISBN 978-1-138-05409-7*

# Human response and the complex city scene

Tanisha Dutta & V.S. Adane
*Visvesvaraya National Institute of Technology, Nagpur, India*

ABSTRACT: The city is a scene, representing life's various facets in all its glory, comprising its physical and subtle elements. This paper draws attention to the similarities and differences in the scenes of historic monuments and cities. It is a comparison between the qualities and characteristics of an architectural masterpiece—Kandariya Mahadev Temple, Khajuraho—and a city as a scene. Such a comparison shall help to understand the essential qualities of interactive cities as compared to communicative architecture.

It is concluded that, just as a painting or any other piece of art, demands a response, so do these higher-order arts i.e. architecture, which generate higher-order complex responses. These responses then become part of the scene and interact to make the scenes more intricate, providing us with a complex manifestation of our behaviour and thoughts.

*Keywords*: historic monuments; historic cities; Kandariya Mahadev Temple; Khajuraho; city as a scene

## 1 INTRODUCTION

### 1.1 *Interactive cities and communicative architecture*

Our cities are an agglomeration of various scenes, as perceived by the residents, heuristically creating a complex 'whole' in the mind of the experiencer. Similarly, highly complex pieces of architecture generate comparable cognitive processes in the mind of the visitor. The vast dissimilarity between these scales impact differently, but nonetheless have similarities which are elaborated upon later. This paper will therefore, focus on the similarities of the cognitive heuristics of these physical environments.

Interactive cities talk about the digitisation of various basic necessities of a city or neighbourhood and the interaction within communities to help build a social whole. But the term 'interactive' etymologically signifies the reaction of people residing in or visiting a city, and the city's reaction towards them. The way a city or a cohesive region redefines itself with changing population, changing needs of the people, and their complex reactions that reflect the intricate social fabric, culture and thoughts of the people, is through interaction.

In an essay by Sam Jacob (2017), architect and director of a London-based architectural company, he beautifully describes communicative architecture as another form of cultural practice and a form of media. Its literal media may be the material it specifies: brick, stone, glass steel, concrete or any other. But it becomes a medium in and of itself through the way it arranges these materials. The arrangement of these materials into form and space turn them, like the arrangement of words on a page, pixels on a screen, or paint onto canvas, into information. We can think of architecture then, as a concentration of information assembled into built form. This elementary introduction to communicative architecture helps one to recognise the impact of Hindu temple architecture upon its visitors, as it packs immense symbolic and metaphorical information on its façades.

It is acknowledged that architecture forms a part of the city scene, but due to its sheer scale, its impact is magnified. Just as social media documents cater to the physical needs of the population by documenting reactions of the public, the interactive dialogue between a piece of architecture and its visitor satisfies an intellectual and psychological need. In a temple, architecture becomes the vehicle of content, and therefore meaning, in all its presentations. As Juan Pablo Bonta and Ludwig Wittgenstein (2006) rightly stated, architecture should not be studied for its meaning, but for its meanings. Having said that, it becomes imperative to mention the areas which describe the foundations of Hindu temple architecture.

As Khajuraho is a small town, sustaining itself through the influx of tourists, it does not exhibit the complexities of a city (by definition). like streets' network, public spaces, landscapes and buildings. The city scene compared to Khajuraho Temple is a conceptual one which historically and theoretically is in sync with the temple concept. This paper, therefore, shows a balance and coordination between traditional ideas governing architecture and conglomeration design alike.

### 1.2 *Methodology of study*

Table 1. Methodology of study.

| Methodology | |
|---|---|
| Study the outline of the Hindu temple concept. | |
| Study the outline of the traditional conglomeration or city design theory in India. | |
| Study the detail of Kandariya Mahadev Temple as an example. | |
| Compare the theories of temple architecture and city design conceptually. | |
| Discuss | How temples are a reflection of society. |
| | The relationship of temple architecture to a city scene. |
| | The human response and thought process. |
| Conclusions. | |

### 1.3 *Highlights of hindu temple architecture*

It is significant that nowhere in the extensive vocabulary of the Indian languages is there a term that corresponds to the term 'religion' – in fact, religious and non-religious matters are never distinguished in Hinduism as it is unimaginable that any activity, impulse or process, can be without some divine potential. The *Rig Veda* used a subtle system of symbolism so that in later Indian thought its meaning was always open to interpretation at several levels (Michell, 1988).

The role of the Hindu artist was to give visible form to the values of their society, rather than to communicate a personal interpretation of these values. In order that certain theological ideas should be translated into art, particularly in the fashioning of sacred images, the priests set out elaborate prescriptions which governed all the details (Michell, 1988). The survey of the artistic activities of about 1,500 years uncovers a consistency in the depiction of these manifold personalities, and this visual coherence is seen in the written canon followed by sculptors and painters. The image making norms in the *Puranas* and *Agamas*, as well as the *Vishnudharmottara Purana*, guided artists to work according to strictly defined rules.

The rituals and ceremonies that lie at the very core of the religious life of Hinduism, as well as the more elusive ideas and beliefs that accompany divine patronages, have fundamentally influenced temple architecture (Michell, 1988). But the rituals are a much later addition to the religious activity and the higher spirit of temple construction remains metaphysical.

### 1.4 *Highlights of symbolic inferences related to hinduism*

The philosophy of Hinduism generally lies beyond the visual realm of art and architecture. In temple sculptures, divinity appears in inexhaustible range of aspects and emanations testifying more to the imaginative potentiality of literary iconographic sources than to the liturgical requirements of worship (Michell, 2000).

Sentences and pictures are concrete representations, but they do not represent inherently. They have power to represent only derivatively, deriving of the states of mind of thinkers who use them. To understand why these things are representations, we need to appeal to the thoughts, intensions, plans and desires of thinkers—their intentional states. Object does not mean thing or entity. Rather, to say that something is an intentional object is just to say that it is an object of thought for a subject (Crane, 2001). An important consequence of the cosmological valence of symbolism is the creation of a person who understands symbols, not only opens himself to the objective world, but at the same time succeeds in leaving his unique condition, acceding to a comprehension of the universal (Crane, 2001). Consequently, 'to live' a symbol and to decipher the messages correctly is equivalent to gaining access to the universal (Eliade, 1985).

It is important to understand that a symbol makes a concrete object 'explode' by disclosing dimensions which are not given in immediate experience. Also, a symbol explodes a particular condition by revealing it as exemplary, that is, i.e. it is indefinitely repeated in multiple and varied contexts (Eliade, 1985). This concept is of immense importance in this context as the symbolism and metaphors in temples pave the way for further imagination; well beyond the physical manifestation; to understand the non-physical aspects of spiritualism. To further enhance the imaginative capacity of visitors, 'The Indian artist imitated the gesture of nature and created on his own account, using a different space and different forms from natural ones' (Eliade, 1985). Therefore, it is clear that Indian classical art does not create works of art, but spiritual models, images to be interiorised through meditation; its action upon man does not conduce to aesthetic feeling, but to a sentiment of reconciliation and perfection, the point of departure for a spiritual ascent which far transcends profane art (Eliade, 1985).

## 2 CASE STUDY—KANDARIYA MAHADEV TEMPLE AND THE CITY SCENE

### 2.1 *Kandariya Mahadev temple: Concept and theory*

The name of Khajuraho is derived from the ancient Sanskrit word of *Kharjuravahaka* (*Kharjur,* meaning date palm) and is situated in the state of Madhya Pradesh in the district of Chattarpur. Arguably one of the most popular and frequented tourist destinations in India, Khajuraho has the largest congregation of temples both from Hindu and Jain religious beliefs. The Khajuraho group of monuments was declared a United Nations Educational, Scientific and Cultural Organization (UNESCO) World Heritage Site in 1986. The temples of Khajuraho in Madhya Pradesh exemplify the technical and stylistic accomplishments of the era – 10th to 13th century – Madhya Pradesh under Chandela and Parmar dynasties.

Figure 1. Mountain view of Kandariya Mahadev temple with cave-like entrance visible.

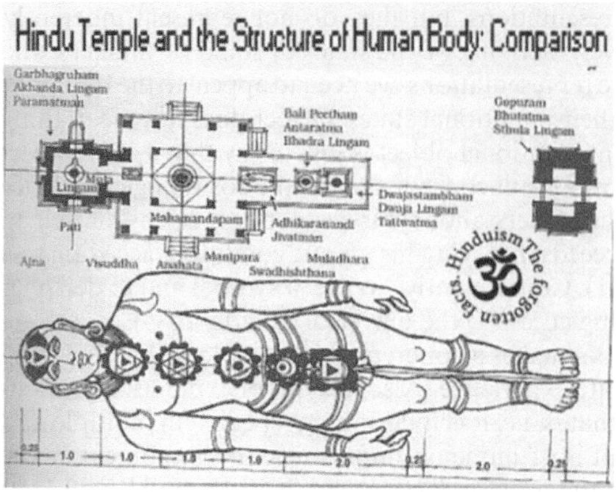


Figure 2. The temple plan depicted as a human body (Mariappan, 2016).

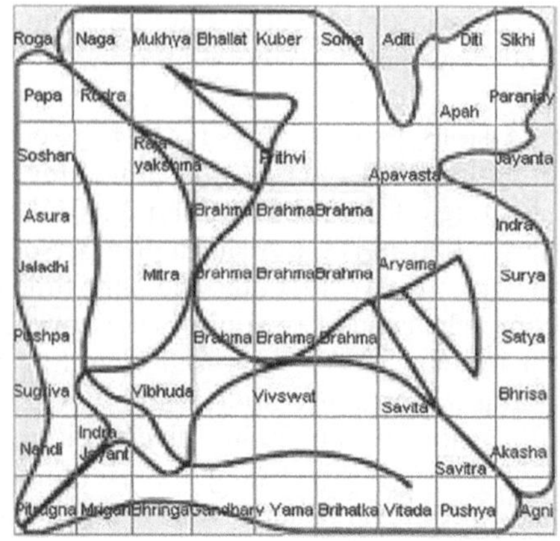


Figure 3. The *Vastu Purusha Mandala* – the basis of ground plan for temple architecture as well as city design (Astrojyoti, 2017).

The Kandariya Mahadev Temple was erected during the reign of King Vidyadhar (1004–1035 AD). This cave-like temple gets its name from the word *kandara* (meaning caves) and it is the tallest of all the temples at Khajuraho, with a height of 30.5 m and a width of 20 m. It was built circa 1030 and is the largest and most magnificent temple in Khajuraho. The elegant proportions of this building and its sculptural detailing are the most refined examples of artistic heritage in central India, towering to a height of 35.3 meters.

These temples are made mostly of yellow and pink sandstone, though in places, pale buff and brown sandstone has also been used together with granite. It is well known that sandstone is a soft stone and, therefore, is more amenable to carving. The temples of Chitragupta, Jagadamba and Laxmana are known for their grandeur; however, the sculpture attains its zenith in Kandariya Mahadev Temple. Distinctive physiognomy is evident in the display of human figures here, which adds grace to the figures of *apsaras*.

Visually, its rhythm and pattern is like a dream in soft-toned sandstone. It consists of a *Garbhagriha* surrounded by a passageway (the *Pradakshina* path) with porches projecting outward on three sides, a columned *mandapa* also with side porches, and an extended frontal porch is reached by a steep flight of steps. The outer walls are raised high on a sequence of basement mouldings and then divided into multiple projections enlivened with superimposed tiers of sculptured figures. These richly textured surfaces are interrupted by projecting porches where inclined slabs form the back of the balcony seating, and are sheltered by eaves above (Michell, 2000). A series of steps on the east leads one to a high platform then through an exquisite *Makaratorana* on the temple's interior. Walking through the porch one finds intricately decorated halls; which leads to a vestibule (*Antarala*) where the visitor can experience a *darshan* of the marble *linga*, placed at the centre of the cella (*Garbhagriha*). On the exterior of the temples are three bands of sculptures representing *apsara* griffins, images of Shiva, *Dikpalas* and snake goddesses in corners where channels of rainwater flow. Alexander Cunningham counted 646 figures on the exterior of this temple and 226 figures in its interior. The famous erotic groups are placed at the junction of the big hall and the sanctum, which corresponds to the walled portion between two balconies (Michell, 2000).

The spire of the sanctum has a series of graded replicas of itself, which cluster around the central peak and create the effect of a mountain range. In the soft evening light, one can experience the rhythm of the ascent and descent of its miniature spires leading the eye upward to the summit. Exactly below the highest point of the spire is positioned the Shiva *linga* in the womb of the sanctum (Desai, 2000). The darker recesses of the wall surface show protective creatures, mythical beings fighting the forces of evil, and serpent queens folding their hands in prayers and blessings. The play of sunlight over the wall surface creates a rippling effect,

Figure 4. Various views of Kandariya Mahadev Temple illustrating, for example, the ornamentation, complexity, texture, human figurines and scenes.

casting deep shadows in some areas and bathing others with pure light. The design of the temple is, therefore, entirely symbolic of spiritual progress, from the mundane to the sacred, and from ignorance to the eternal light of wisdom.

Khajuraho sculpture is mainly divided into five categories. They are:

- cult images;
- family, attendant and enclosing divinities;
- heavenly nymphs;
- animals and other species;
- miscellaneous themes such as teachers, and disciples, dancers and musicians.

Animals and birds were also closely associated with different cults prevalent during the Chandela period. They were shown as *vahanas* for gods and goddesses and became part of religious imagery, mostly assigned according to the nature of their masters. The erotic sculptures, usually considered as a degrading act of the human soul by religious schools, is different from what *tantra* says. As per *tantric* philosophy, sexual energy, if channelized according to methods of *tantric* canons, can unite the *tantra* practitioner with the Supreme Being Shiva. It is looked upon as just another human activity like any other daily chore. But when one moves inside the temples, one notices a great change. The human figurines do not seem even remotely attracted to sex, but are standing apart in deep contemplation.

Even the inner room of the porch and *mandapa* are created to mimic great upside-down pools with flower and leaf motifs. The doorway of the sanctum is profusely decorated with narrow panels of images. The central lintel carries the seated figure of Shiva holding a trident and snake, Vishnu is seated to his left and Brahma to the right. Within the dark unadorned *Garbhagriha*, stands the creative symbol of Shiva the *linga*, which is the final stage. The walls of the sanctum sanctorum do not have any sculptures, there are bare walls and they have a single *pratima* of the deity.

> 'This temple has been highly praised by art historians and connoisseurs for the superb harmony of the graded proportion for its various component unit along with their superstructures' (Michell, 2000).

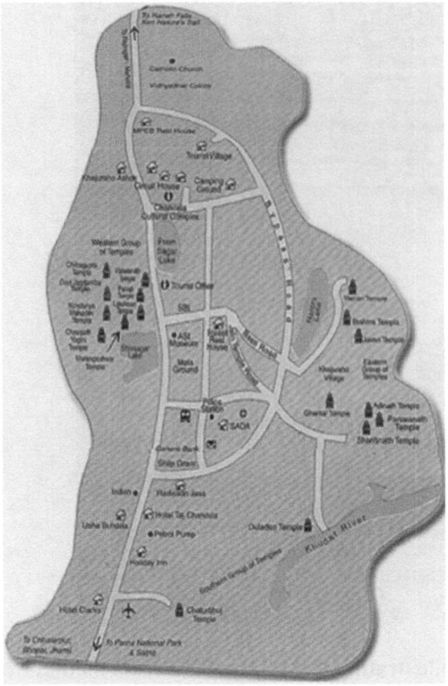

Figure 5. Khajuraho city plan (Narasimhaiah, 2014).

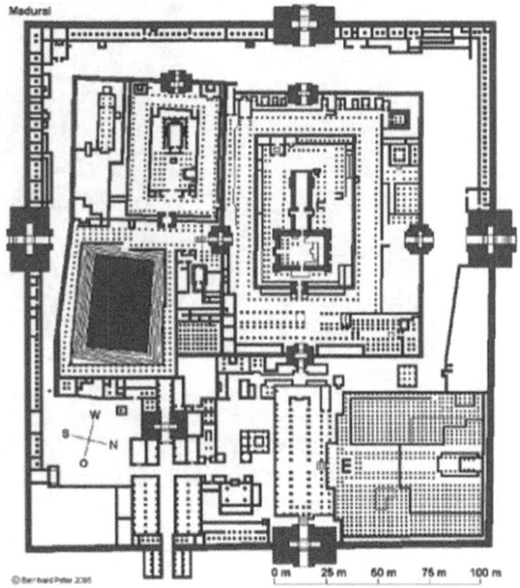


Figure 6. Srirangam city plan (Mariappan, 2016).

## 2.2 *The historic concept of the city form in theory*

Lynch (1984) rightly states that 'impersonal forces do not transform human settlements'. The normative theory guiding the religious centres throughout history has followed the cosmic theory for the layout of cities or conglomerations. The Indian theorists made explicit connections between gods, men, rites and city plans. The series of texts on city planning – the *Shilpa Shastras* – indicated how the earth could be parcelled out and the evil forces of chaos enclosed and controlled (Lynch, 1984). The typical form of *mandala* was used: concentric rings and subdivided squares overlapping to form sub-parts, each significant and appropriate for certain specific activities. This pattern concentrates most power to the centre.

Lynch (1984) points out that with the help of myths, it was explained how a city came to be, demonstrated why a city worked as it did and what could go wrong. If these tenets were followed, they enhanced earthly power and gave people feelings of security, awe and pride. They were complete and operative theories of the city, both functional and normative. The radial perfection of the cities was meant as a symbol of the orderly, mathematical universe. Some common concepts, according to Lynch (1984), are the axial line of procession and approach, the encircling enclosure and its protected gates, the dominance of up versus down or big versus small, the sacred centre, the diverse meanings of the cardinal directions, the regular grid for establishing a pervasive order, the device of organisation by hierarchy, bilateral symmetry as an expression of polarity and dualism, landmarks at strategic points as a way of visibly controlling a large territory, and the sacred nature of mountains, caves and water. These similar features of form were reinforced by similar institutional features, such as regularly recurring religious rights, the organisation of government, disposition of the social ranks, and the dress and behaviour of city people. Space and rite are establishers of behaviour and serve to bind human beings together, just as they do for many other animals. Behind these concepts lie certain primary values: order, stability, dominance, a close and enduring fit between action and form, but above all the negation of time, decay, death and fearful chaos (Lynch, 1984).

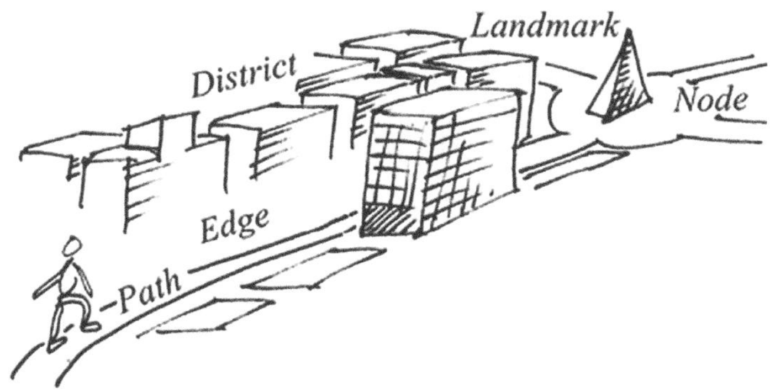


Figure 7. Elements of the city Kevin Lynch—Image of the City (Lynch, The Image of the City, 1960).

It can be seen that these symbolic forms are attractive because they speak to deep emotions of anxiety in people. They do indeed give a sense of security, of stability and continuity, of awe and pride, and while the magical rational of the theory may be discredited, the psychological power of these devices cannot be so easily dismissed. The cosmic model upholds the ideal of a crystalline city: stable and hierarchical—a magical microcosm in which each part is fused into a perfectly ordered whole. If it changes at all, the microcosm should do so only in some rhythmical, ordered and completely unchanging cycle (Lynch, 1984).

Districts have a strong visual identity and an endowment of visible boundary, active centres of special characters, visible and audible landmarks at strategic points and time; and natural features or conservable urban character.

Orienting buildings to the apparent movement of the sun makes the compass direction legible and increases a sense of structure of time. While historic conservation is used to make connection with the past, less thought is given to the connection with the future. It is also possible to increase sensibility by improving the human ability to perceive the environment, and this is less often thought of by designers trained to focus on objects.

The plurality of users in any large settlement will always pose technical problems. Thus sensibility will be easier to attain in more stable and homogenous societies. It is likely to be important both in rich and poor settlements, since human perception is a constant, but the means of achievement will differ.

The city is a 'field of force', but more than anything else, the city is a communication network. Thus it seems plausible to import the concept of a field of force, which is such a powerful metaphor in the physicist's universe and deals with multiple influences acting at a distance. The study of a city has no powerful basic language of its own, but expresses itself by borrowing the devices of geographical architecture.

## 3 DISCUSSION

### 3.1 *Temples are the reflection of the society*

Temples are the reflection of the society we live in, and depict the times along with the social and cultural fabric of the place. As discussed previously, Khajuraho city plan does not exhibit patterns in plan which may be similar to or compared to the plan of the Khandariya Mahadev temple. But in the example of Srirangam city Madurai Figure 6 it strictly follows a particular repetitive pattern in the plan of the temple as well as the city. Srirangam still sticks to the original *mandala* plan, historically laid out for it and the streets and bus routes follow the same pattern and appears in its purest form. The city itself is composed of seven concentric wall enclosures of which the inner four are part of the temple, while the outer three are residential quarters (Kostof, 1991). Similarly, in Banares or Varanasi, the first city to appear after the great cosmic dissolution (*mahapralaya*), Hindu cosmography is charted through holy markers and the pilgrims' routes. The ideal diagram is of a series of concentric circles increasing in sanctity as one nears the centre. But citing these examples does not make the Khajuraho city plan any less interesting and any less bonded with temple architecture. It is bonded through the theory of temple architecture concept, and the historic city theory laid down by various experts.

### 3.2 *Relationship of temple construction to the city scene*

Even though the *Shilpa Shastras* laid down strict rules/canons for structuring the ideal city, each city should be looked upon as a unique story of flowering, growth and evolution; as cities are unique historical processes. The metamorphosis a particular place goes through is pertinent to only that place and does not replicate itself anywhere else. The various structural aspects of a city, network, sense of place, time orientation, transparency and legibility only aid in the organisation of thoughts when describing a city. Moreover, the application of the cosmic theory and the organic theory shows very high levels of organisation and discrete planning on the part of our ancestors. These attributes, most importantly, are congruent with the attributes considered while describing a temple monument. The Kandariya Mahadev, in this case, is the epitome of perfection where the cosmic principles are concerned. The concept of the cosmos, the generic pattern, the rhythm and the wisdom, have been beautifully replicated in this example.

The aspects of temple form to be analysed are broadly borrowed from 'The Theory of Architecture' by Nikos Salingaros (2006) with certain inclusions by the author. These parameters are namely: structural order, scale (coherence), inspiration from nature, ornament, temperature, hierarchical cooperation, concept of metaphor, harmony, memes (architectural) and form language. Similarly, the parameters to be considered for the analysis of city form have been adapted from 'Good City Form' by Kevin Lynch (1984). These parameters are: sense, transparency, time orientation, legibility, semiotics, congruence and organisation. Some of these parameters are directly comparable as their idea remains the same through these scales of temple architecture and city planning, whereas some others relate distant and pass through many layers before forming the mental connecting bridge between these scales. If extended growth is necessary, it should occur by the budding of new colonies (Lynch, 1984).

While both these scales have a sharp external boundary, it is not so easy to divide an organism internally. It does not have differentiated parts, but these parts are in close contact with each other and become a homogenous whole when viewed. They work together and influence each other in subtle ways. Form and function are indissolubly linked. They need to be understood as a dynamic whole alongside recognising the parts and sub-parts thereof. As the rhythmic concept of the cosmos is embodied by its deep relationship between "the whole and the part", emotional feelings of wonder and affection accompany our observation of these entities. One is reminded of and given experience of the natural rhythm in the universe, which is consistent at all imaginable scales from the atom, through humans, temples, cities and, ultimately, to the cosmos.

### 3.3 *The human response, its contribution and intricate relationship with thought processes*

The comparison of these two experiences of the temple structure and the historic city form gives an insight that, although their scales are vastly different, the experiences they suggest create similar heuristics in the mind of the visitor. This is because the parameters of analysis of architectural form target the same senses in our cognition as do the parameters of analysis of city form, therefore giving a similar experience.

Lynch (1984) supports this argument by describing the aspect of 'sense' that is the clarity with which environmental form can be perceived and identified, the ease with which its elements can be linked with other events and places in a coherent mental representation of time and space, and that representation can be connected with non-spatial concepts and values. This is the join between the form of the environment, and the human process of perception and cognition. This quality lies at the root of personal feelings about cities. It cannot be analysed except as an interaction between person and place. Perception is a creative act, not a passive reception.

The experiences give rise to the responses created by the mind and body of the visitor. These responses, in turn, form part of the impact and experience. This event is less dramatic in the temple context and more magnified in the city context as the number of people experiencing the city is multiple times greater than in case of architecture. Higher the number of people around, more are the responses generated. This has a compounded impact on the city scene. The number of responses generated towards a place, activity or event, in turn, contributes to the sense of place. The scene becomes more intricate as the responses generated by the people are all different, owing to their cultural inclinations, background, moods, and perception of the environment. Thus the sense of a particular place will vary for different observers, just as the ability of a particular person to perceive form varies for different places.

Sense, therefore, not only depends upon the spatial form and quality, but also on the culture, temperament, status, experience and current purpose of the observer. Nevertheless, there are some significant and fundamental consistencies in the experience of different people in the same place. These constancies arise from the common biological basis of human perception and cognition, certain common experiences of the real world (like gravity, inertia, shelter, fire and sharpness in the case of cities, and structure, harmony, ornament and form language in the case of temple architecture), together with the common cultural norms that may be found among those who habitually use any particular place. Places have a greater or lesser sense, and so do events. Activities and celebrations associated with a location, support its perception to the extent that they are themselves perceived as vivid and coherent (Lynch, 1984). Human cognition has its limits and the process of cognition is of greater value than the resulting mental structure. There is pleasure to be found in puzzles, ambiguities and mysteries. Lynch (1984) confirms this general proposition in that an interactive place is one which is, in some way, appropriate to the person and his culture, makes him aware of his community, his past, the web of life, and the universe of time and space in which those are contained.

An important aspect here is the difference between the culmination of the two experiences. In the context of temple architecture, the finale is the peace found through enlightenment after walking through the linear path of the various areas of a temple to reach the cella—*Garbhagriha*. Whereas, in the cityscape context, the cognition culminates into mental pictures of small sections and then their integration, which then encourages an understanding of the spirit of the city and people. This is when all the people in the city become a collective whole and are perceived as a population with one essence.

This larger, more complex organism, the city, has many scales and layers to it, which reveal themselves to the visitor only gradually. First the physical environment is perceived and, progressively, the subtle elements of the cityscape present themselves. In the case of temple architecture, the physical environment casts its spell first, increasingly letting the symbolic metaphors touch our minds and take us on the path of enlightenment.

## 4 CONCLUSIONS

It can therefore be concluded that balance and coordination between traditional ideas governing architecture and conglomeration design have historically been the guiding light in constructing our built environments. The sharp and clear connections between architectural expressions and responsive cities' forms show that an observer perceives them in the same manner heuristically. The perception of the built environment is one that is impacted immensely by classical conditioning of a person of that generation and thought. This also points towards the fact that a kind of people will perceive a particular surrounding in a similar manner, to generate similar heuristics and therefore similar responses.

These feelings generated, become multifold in case of a city response, which is magnified due to the sheer presence of numerous people, their experiences, and therefore their responses. These responses generate further responses, making the process and the built envelope more complex. This complexity contributes to the sense of the place. Therefore, any type of space will be observed and responded to with similar intent and result.

An interesting point to be noted here is that modern contemporary architecture is very much a part of the cityscape experience and contributes significantly to the response generation process. Though this aspect is part of ongoing research that does not form part of this paper, it needs to be mentioned that most historic cities have evolved into a beautifully knitted fabric of tradition and technology, presenting both as if amicably sitting beside each other, and speaking for themselves as well as each other. This texture is seen in the physical built form, along with peoples' attitudes towards everything around them.

This paper therefore concludes on the similarities in the cognitive heuristics of these physical environments of historical temple architecture and cityscapes. This beautiful symbiotic relationship enhances each other's cognitive qualities, as they can never exist independently of each other. They are both intertwined in each other, as are the people who are creating them, experiencing them, and responding towards them, thereby creating a new whole and continuing the cycle.

## REFERENCES

Astrojyoti. (2017). Vastu Purusha & Vastu Mandala. *Astrojyoti.com*. Retrieved from http://www.astrojyoti.com/vastumandala.htm.

Bonta, J. (1979). *Architecture and its interpretation: A study of expressive systems in architecture.* London, UK: Lund Humphries.

Crane, T. (2001). Intentional objects. *Ratio*, *14*(4), 336–349.

Desai, D. (2000). *Khajuraho—Monumental legacy.* New Delhi, India: Oxford University Press.

Eliade, M. (1985). *Symbolism, the sacred, and the arts.* New York, NY: Crossroad Publishing Company.

Jacob, S. (2017). The Communicative Mode of Architecture. *MAS Context*. Retrieved from http://www.mascontext.com/issues/14-communication-summer-12/the-communicative-mode-of-architecture/

Kostof, S. (1991). *The city shaped.* London, UK: Thames & Hudson.

Lynch, K. (1960). *The image of the city* (Vol. 11). Cambridge, MA: MIT Press.

Lynch, K. (1984). *Good city form.* Cambridge, MA: Massachusetts Institute of Technology.

Mariappan, S. (2016). Unit 3 – Evolution of Hindu temple architecture. *LinkedIn*. Retrieved from https://es.slideshare.net/SumathiMariappan/unit-iii-evolution-of-hindu-temple-architecture.

Michell, G. (1988). *The Hindu temple: An introduction to its meanings and forms.* Chicago, IL: University of Chicago Press.

Michell, G. (2000). *Hindu art and architecture.* London, UK: Thames & Hudson.

Narasimhaiah, S. (2014, May 25). Khajuraho Group of Monuments [Web log message]. Retrieved from http://sateesh-mane.blogspot.in/2014/05/khajuraho-group-of-monuments.html.

Salingaros, N. (2006). *A theory of architecture.* Solingen, Germany: Umbau-Verlag.

Whyte, W. (2006). How do buildings mean? Some issues of interpretation in the history of architecture. *History and Theory*, *45*(2), 153–177.

*Cities' Identity Through Architecture and Arts – Catalani et al. (Eds)*
 *ISBN 978-1-138-05409-7*

# Interactions between urban dynamics and new spatial patterns: The case of Istanbul

Hulya Turgut
*Department of Architecture, Faculty of Architecture and Design, Ozyegin University, Istanbul, Turkey*

Ozgur Ozten
*Architect, M. Arch, ITU*

ABSTRACT: Cities are increasingly concerned with fluidity and mobility, where social, cultural and economic activities can rapidly be transferred from any one locality to another. Yet, powerful effects of globalisation on economy, society, and urban environment create fragmentation as well as interesting transitions in each system. Whilst urban transformation in response to globalisation creates sharp changes in former urban textures and typologies, new spaces and identities have been produced with the formation of recent networks and encounters. This paper, therefore, examines the implications of urban and housing transformations in the city of Istanbul within the context of recent economic, cultural and political conditions. The authors aim to establish a critical discussion of the city's texture, where separate and overlapping urban functions are easily captured through a site section.

*Keywords*: urban environment, urban transformation, globalisation, housing transformations, city texture

## 1 INTRODUCTION

The rate of change in urban housing environments is continually increasing as the effects of globalisation impact in multiple ways on the contemporary city. Such dynamic processes create what has sometimes been described as a layered construction created over the course of time. Many of the changes relate to the movements and energies of low-income groups whose activities are becoming increasingly dominant in rapidly growing cities throughout the world. There are numerous interrelated factors in the growth of cities. Migrants from rural areas and small-size cities create pressure on existing urban housing stock and frequently on the development of new informal settlements. Natural growth of urbanised groups leads to large populations seeking affordable accommodation. In some cities, historic or older central areas deteriorate through excessively high densities of tenant populations and, in others, high-density multi-storey constructions replace older settlements, changing the social and economic relations of the area. In recent years, especially in metropolitan cities under the umbrella of urban regeneration, informal settlements or deteriorated housing stock have been replaced with high-rise housing and gated communities geared towards the high-income population.

In the case of Istanbul, one of the key factors in the creation of this transformation is the replacement of existing stock (such as informal settlements) with new developments. The new housing stock creates a new chaotic and mixed development which forces different demographic groups to live side by side. Istanbul is a city of layers with a history dating back almost 2,700 years and has served as the capital city for the Roman, Byzantine, Latin and Ottoman empires. Today, with more than 15 million inhabitants, it still preserves the surviving fragments of these civilisations. The city carries significant architectural entities, such as mosques, synagogues, churches, palaces, towers and castles, which represent the various chronological layers, serving as the 'modern' face of Turkey's cultural and financial capital. Istanbul has performed as a 'global city' that connects civilisations and continents for several thousand years. In addition to this, it has been facing a new situation caused by new social and spatial urban dynamics. Its urban texture has been changing like any metropolis that is undergoing the trauma of warp speed urbanisation (Figure 1). According to the international investigation of cities carried out by the Urban Age Project, Istanbul may not be growing at the dizzying pace of Mumbai and Shanghai, nor suffer from the widening social inequality and violence of São Paulo, Mexico City or Johannesburg, but it does face many of the same challenges confronted by all 'Urban Age' cities, including London, Berlin and New York: economic stability, social cohesion and climate change (Burdett & Nowak, 2009).

Turkey's urbanisation started to accelerate in the 1950s. Due to the lack of housing policies and inefficiency in providing housing to the newcomers, the city experienced the *gecekondus*, which are basic shelters for low-income inhabitants. Later, this phenomenon shifted into *apartkondus*, where the *gecekondus* were transformed into apartment blocks. The authorities applied little restriction over this process of newly built illegal settlements on public land, both in terms of quantity and quality. Therefore Istanbul, along with many other big cities, faced urban margins with no infrastructure, public space and legal status. In the 1980s, along with continuing economic development and reforms, these formerly prestigious inner-city districts gained new popularity among higher-income families, attracted by their location close to the financial district (Ergun, 2004). Most of the physical transformation associated with globalisation has taken place with the development of gated communities and five-star hotels. Istanbul has been packaged as a consumption artefact for tourists and new office towers, with the expulsion of small businesses from the central districts, gentrification of the old neighbourhoods, and global images on billboards and shop windows (Oncu, 1997). In the last four decades, Istanbul's sociocultural and urban identities have been undergoing radical transformation. Although Istanbul has always been a city of duality, fragmentation and polarity, it has never before displayed such intense qualities of heterogeneity as it does today (Keyder, 1999). Economic policies seem to have always had a strong effect on urban growth and change in Turkey. In each period, the urban space has been shaped by the economic policies of the state. As in other countries, social and cultural change in Turkey has followed economic cycles. While today's cities are being shaped within the effect of global restructuring processes, urban housing has been evolving by itself, interacting with these changes. The development of housing areas and the creation of the environment are therefore being formed under the influence of a confused interaction between globalisation and the city's own history (Turgut, 2010).

In this context, this paper is based on ongoing research on 'new urban housing concepts' and the Master's thesis of Ozgur Ozten, supervised by Hulya Turgut (Ozten, 2010). The ongoing research seeks to examine various examples of new housing developments in Istanbul by investigating the social and spatial dynamics of its new situation. This provides and creates a platform to discuss the issues of emerging residential patterns and dynamics of the city of Istanbul.

Figure 1. New faces of the city profile of Istanbul.

## 2 URBAN DYNAMICS AND NEW RESIDENTIAL PATTERNS EMERGING IN ISTANBUL

Today's urban dynamics follow changes in global restructuring processes and cities are becoming the stages for transformation. The urban dynamics that arise from global platforms also have an effect on the formation of new urban housing trends. Between the city and its role in the process of globalisation, there is a transactional relationship which arises from economic, political, sociocultural and technological changes. Urban dynamics that affect the transformation can be categorised into six main areas: global economy; being a 'world city'; diversification; new joint ventures and participation; new industries; social exclusion and societal fragmentation (Cakır, 2007; Turgut, 2010; Turgut & Inalhan, 2007). The concept of the global economy brings prestige, competition, economic revival, governance and foreign capital. Thus, by the expansion of the global economy, urban transformations are targeting the 'marketing of place and prestige'. The representational need of the city in this matter gives rise to the concept of a world city in terms of urban identity, image, environmental quality, tourism, consumption and the industry of culture. The focus on urban identity, image and quality also has a direct relationship with urban memory, sustainability, local culture and conservation as components of diversification. The generation of knowledge, advanced technologies and productivity as the dynamics of new industry in the city has an impact on urban transformation. All these interlinked urban dynamics also highlight issues of social exclusion and fragmentation, where the need to bring equality, liveability and accessibility is the aim of an urban transformation. All of these concepts have found their reflections in the formation of a holistic urban vision (Turgut, 2010).

Istanbul's rapid urban transformation during the economic restructuring process of the last four decades has occurred through the interaction of politics, culture and economy, and been directed and legitimised by the global city discourse. During this period, many global city discourses have been developed for Istanbul. Central government and local authorities have proposed various projects to position Istanbul as a global city. These projects have been radically transforming Istanbul's urban identity and the city has been growing with intense heterogeneity, especially in its urban housing (Turgut, 2010). Since the beginning of the 1980s, change in the city has been the driven by the effects of globalisation, liberalisation of the economy, rapid urbanisation and technological advances. This change, while being more prominent in the 1980s, gained momentum in the 1990s. Economic policy, which produced radical changes in the social structure, played a key role in the change and transformation process. It transformed the urban space and created new urban forms.

In the mid-1980s, the peripheries of Istanbul became more popular for residential settlements for the urban middle-high and high-income groups. In the 2000s, inner-city residential areas became affordable and desirable for middle-income groups. According to Kurtuluş (2005), the reason for the security tendency in the case of Istanbul is the urban elites trying to integrate the global consumption culture in their residential areas. Prestigious neighbourhoods guarantee real estate values, security and a sense of privilege and there are, therefore, often gated settlements in Istanbul. Developments caused by changing economic structure and global influences have created a new metropolitan lifestyle for the middle- and upper-income groups, which has resulted in a demand for luxurious new houses. The development of new housing patterns over the last 30 years in Istanbul can be divided into four categories: garden cities—suburbia (beginning in the 1980s): luxurious housing—villa towns and settlements (beginning in the 1980s); multi-storey residences (beginning in the 1990s); mixed city housing (beginning in the 2000s); and mixed inner-city and outer-city housing (Turgut, 2010; Turgut et al., 2010).

In this context and over the last four decades in Istanbul, the urban texture represents different elements of choices, time periods, styles, functions and users with an intense degree of heterogeneity. At this point, the urban section of the city is capable of forming a new type of urban layout: a mixed form of functions and transformations. In order to show this new type of mixed urban layout, this paper will analyse a cross section of the Levent-Maslak neighbourhood in Istanbul, with a focus on housing developments.

## 3 ISTANBUL'S CHANGING URBAN TEXTURE: THE EXAMPLE OF THE LEVENT-MASLAK AXIS

As already described, the new forms of housing area resulting from the effects of the interactions between globalisation and the city's own history give Istanbul several important urban areas to exemplify this structure. The Levent-Maslak axis in Istanbul is an important example in exposing the changing urban texture and housing preferences. It is more of a critical debate about the area of focus where the social and physical fragmentation are clearly exposed. By analysing this part of the city, this paper aims to cover the accessibility and the readability of the different dynamics of the urban environment and their reflections on the process of transformation. Therefore, through the changes in the Levent-Maslak area in economic, political, social, ecological and psychological terms, contradictions, transformations and reproductions are combined within the same location and can be exposed.

The Levent-Maslak axis has an important and critical function as the node point between the Asian and European routes of the city's main roads. Therefore, it is the most important north–south transportation axis within the city.

The Levent-Maslak axis named *Büyükdere Street* started a change in its development path during the 1990s with the rapid construction of high-rise buildings, increasing rates of high-density population, and a shift in profile. The axis is shown in Figure 2: on the west side the street covers the illegally developed neighbourhoods of Ortabayir, Celiktepe and Emniyet Evleri, and on the east side Levent and Konaklar neighbourhoods exist with a shift in their previous functions of being social housing units into mostly business and commercial use (Ozten, 2010). This differentiated development of the axis, therefore, has brought an emerging sociocultural fragmentation, duality, polarity, social exclusion, especially since the 1990s. For a resident of the city, the social exclusion process is becoming more evident via spatial diversification. While holding a spatial sense of belonging towards a location, an exclusion

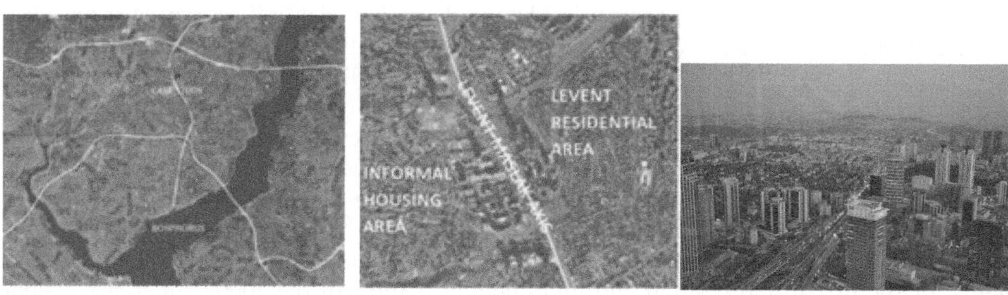


Figure 2. Location of the Levent-Maslak axis and its close surroundings.

Figure 3. Levent high-rise blocks and the Levent's former social housing units on the left, and illegal housing areas on the right.

from all the other elements brings social exclusion and cultural and spatial fragmentation. These developments result in a mix of identities or, in a general context, the loss of the identity of the city itself (Kurtuluş, 2005).

In this framework, the focus on the Levent-Maslak development axis and its cross section aims to expose the loss of identity of the city where social and spatial fragmentation is sharp and sense of belonging is relative. The aim is to create a critical framework for the process of rapid urbanisation of the city of Istanbul and its impacts on the city's different housing areas and preferences. Therefore, urban development phases through the cross section of the Levent-Maslak axis will be described within the time frame in order to pin down key points of spatial and social transformation and fragmentation.

### 3.1 *Formal part of the axis: Levent social housing area*

The area was designated as the northern development path of Beyoglu district and proposed to become a housing settlement. It started to develop in 1947 with the social housing project of Emlak Kredi Bank, completed in 1960 as one of the few exemplary social housing projects of the Turkish Republic. In 1975, single-storey housing units were given permission for a plus one storey and this permission brought the settlement a spatial change. With the 1980s master plan for the metropolitan area, Levent was assigned as a housing development site to promote Istanbul's importance and image within the global arena. According to the 1994 master plan, due to the increased efficiency of highway and transportation axes, the commercial axis of the 1980s master plan was expanded towards the Maslak area in the west. After the construction of the second bridge over the Bosphorus, the joint roads cut through the Levent area and, together with this effect of transportation, the neighbourhood started to be impacted by a second transformation phase. This transformation mainly triggered functional changes as well as an increase in market values. This sharp differentiation in values brought social housing into a luxurious housing development and this change was reflected spatially on site, especially with the Besiktas Municipality's permission in 1991 for preservation of heights of the units, transformation of buildings to tourism-oriented commercial units (Ozten, 2010).

### 3.2 *Informal settlement areas: Ortabayir, Celiktepe and Emniyet Evleri*

While the Levent social housing development started its implementation, the west side of this location was being used for agriculture and animal farming in the 1950s. Because of the rapid

Figure 4. Former and current social housing of Levent on east side of the Levent-Maslak axis (Aru & Gorbon, 1952).

Figure 5. Informal settlement areas: Celiktepe neighbourhood on west side of the Levent-Maslak axis.

migration process from rural to urban areas, at the end of the 1950s these locations—known today as the neighbourhoods of Ortabayir, Celiktepe and Emniyet Evleri—were given to immigrants by the municipality in order to cover the increased need for housing. This was followed by the development of unplanned and illegally built housing units in the area. With the first master plan implementation in 1975 and the permissions given to illegal developments between 1975 and 1985 the spatial texture of this area started changing. With 1985's rehabilitation master plan, the transformation process of low-storey housing units brought medium-sized apartment blocks to the site and the characteristics of the illegally developed settlement were continuously changed (Ozten, 2010). From this period onwards, it is possible to locate spatial formations belonging to different development processes of the area.

### 3.3 *Büyükdere Street as the Levent-Maslak axis*

According to the Prost master plan (1937) for Istanbul, all roads towards the north were supposed to be enlarged in order to create a main artery. With the additional planning of 1955 and the development of a new industrial area in the Levent neighbourhood, low-cost lands in the surrounding area started to develop illegally on the west side, whilst along the road industrial buildings and factories started to get built. From the 1970s, the Levent-Maslak axis was developed along with new urban developments in the city. It is important due to its northerly direction towards water reserves and forests. In the 1980s, office buildings started to get constructed along the road. Besides its orientation towards the north, Büyükdere Street is today used as a prestigious business centre that is different from neighbouring areas. Since the 1990s, the functions along the street have shifted greatly towards business, residences, shopping, commerce and offices (Gulen, 2006). The street is now a mixed-use, high-rise axis of the city of Istanbul (Ozten, 2010).

### 3.4 *Evaluation*

Istanbul as a multifaceted metropolis has gone through a rapid transformation process as a result of local and global pressures. During the 1980s, effects and transformations on the spatial environment became sharp and dynamic. In this context, two main approaches were exposed in the city's development. The first one was, as Ohmae (1995) states, that the world moved towards a singular identity and lost regional constitutions, and the other approach was, as Taylor (2000) underlines, that the global economy created new spatial forms as well as controlling the surroundings. In this response to the new millennium, spatial developments started to be more competitive in order to attract large-scale projects and foreign capital. In this manner, cities such as Istanbul sought to boost their urban environments and their popularity by adapting themselves to the new global urban dynamics.

In this context, this paper aims through the case study to visualise a dynamic spatial development during the last 60 years of the city's development history. Thus, the impacts of all the periodical urban and social dynamics are exposed within a cross section of the

Levent-Maslak axis and its surroundings (see Figure 6). Together with a clear distinction between the fragmented social and spatial characteristics of the neighbourhoods, Büyükdere Street has developed as a catalyst of the global dynamics for the whole area in terms of construction and spatial transformation. Therefore, because of the force exerted by the new type of urban development that fits into the global dynamics, the close surroundings of the street were affected by this influence much more radically. Thus, through a time frame bulleting the differentiation of urban dynamics according to the current needs and expectations, an urban development path is exposed. With the transformation of planning visions and political assents, a sharp change in the social and spatial developments brought fragmented societies and spaces into the cityscapes in return. In this manner among the city of Istanbul's sharp transformation process through local policies since 1940s, sharp changes in the urban identity and the image, social and spatial texture are being exposed for the case study area of the Levent-Maslak axis in this paper. In the 1940s, along with the start of rapid migration and the process of industrialisation, Istanbul, just like many big cities, also faced rapid urban development and new formations in its spatial texture. New urban development plans and implementations also catalysed the formation of the new avenues of Vatan and Millet, new factories and industries in the Levent area, and neighbourhoods like the Levent social housing settlement. As a result of these new developments, the lack of housing policies, insufficient housing for newcomers, and the destruction of housing areas to open up new avenues, the Istanbul municipality gave the Ortabayir, Emniyet Evleri and Celiktepe areas, which were former agricultural lands of Kagithane village, to the immigrants to settle. On the other hand, with the modernisation period of the country, Levent social houses were also built between 1947 and 1958 in order to exemplify a modern neighbourhood development. In this context, the rapid housing developments of the city varied from horizontally growing illegal settlements to new neighbourhood designs of social houses according to modern theories. From the 1960s onwards, urbanisation within the city speeded up and enforced the Levent-Maslak axis as a new urban route, together with its modern housing and industry. Emphasis on motorways and highway transportation also sped up the development of the Levent-Maslak axis. With the opening of the first bridge over the Bosphorus in 1973 and the second one in 1988, the Levent-Maslak axis became a joint venture in the area. Increase in the commercial and industrial developments along the axis catalysed the extension of the illegal settlements around Emniyet Evleri, Ortabayir and Celiktepe neighbourhoods and increased the horizontal and vertical growth in these areas.

With 1980s neo-liberal influences, Istanbul started to become a metropolis and created a vision to establish a modern image within the global arena. One of the peak points in this period is the start of permissions for illegally developed areas, which led to a situation in which horizontal and low-rise housing grew towards medium-sized apartment blocks. The process started the transformation of housing settlements physically as well as functionally. With the new visionary developments of the city, the Levent-Maslak axis started to reshape

Figure 6. Interaction zones within the focal section of the case study area.

its first structure and moved towards commercial and housing in mixed use. From the 1980s until the millennium, fluid and rapid global influences developed the Levent area as an image of the city of Istanbul with its modern housing, and recently implemented commercial uses. By the 1990s, the Levent-Maslak axis had become the modern representative of business and commerce with the first high-rise constructions in the city of Istanbul.

In the last two decades, through the significant effects of the global economy and the political pushes for Istanbul to become a world city, a construction peak has taken place along the Levent-Maslak axis. This construction boom affected the former spatial texture and demonstrates a dynamic transformation process in the case study area. In this respect, along with the axis as the main centre of gravity for the zone, the adjacent, illegally developed neighbourhoods of Emniyet Evleri, Ortabayir and Celiktepe, as well as the Levent social housing area, have adapted their physical environments and sociocultural inputs to the recent global dynamics.

In this manner, the urban transformation of the area exposes a social and spatial fragmentation, with a societal exclusion along the main axis. Thus, although a sharp distinction along the axis exists in physical, social, cultural and economic terms, fragmentation among the society within the area creates a contested heterogeneity. Therefore, across the area as a whole it is possible to expose a variety of physical examples of different periods that developed due to changing urban dynamics. To sum up, from 1940s until today, development of the Levent-Maslak axis has created an urban contest due to global dynamics reflected in the site. This urban contest eventually affected the housing settlements in the area and resulted in socially and spatially fragmented urban development.

## 4 CONCLUSION

The paper concludes its urban development statements for the city of Istanbul, where recent urban dynamics are restructuring the cityscape and the society in physical, social and cultural terms. Based on this framework, recent housing projects, new development trends, former urban patterns and new forms of spatial and sociocultural interactions are reflected in the city's development attitudes and its future development. However, it is also important to underline that the interaction of spatial dynamics, sociocultural values, and their transitions give rise to a rapidly changing urban settlement. In addition to this dynamic movement, a loss of cultural, social and spatial identity takes place through the effects of globalisation. Particularly since the neo-liberal urban dynamics of the 1980s, globalisation has created spatial and social tensions, emphasis on real estate values, and representational struggles for creation of a holistic image.

As a multi-layered or palimpsest city, Istanbul, with its historical and cultural values, undergoes an important and rapid transformation process. The importance of the transformation process of the city is its intense speed. In fact, the speed of the process shifted the urban texture of the city from a small-scale contractor, small-sized plot combination of little architectural input to a high-level social class's market development.

Cities do face continuous changes; however, within this continuous change some manage to transform spaces through development of their unique characteristics. In the case of Istanbul, rapid change of the cityscape has brought little quality into the spatial texture but rather a mixture of interaction and a loss of identity. Today, Istanbul might be characterised as a city based on a multilayered physical structure of a variety of influences and contradictions from Eastern and Western civilisations. However, as a city that faces migration, and constant reshaping over the past 2,700 years, this historical, spatial and sociocultural palimpsest within the urban environment should be analysed with a multi-dimensional and comprehensive approach and accepted as a heterogeneous compound.

To develop a visionary approach for the heterogeneous structure of the city of Istanbul, in which new dialogues and alternative development strategies can facilitate the establishment of social and spatial links, a holistic approach is needed. Thus the formation of spatial environments, housing settlements, representative spaces whilst developing qualified environments for inhabitants of the city, the potential of palimpsests and heterogeneity of spaces should develop inclusive urban environments as well as social interaction.

## REFERENCES


Aru, K.A. & Gorbon, R. (1952). Levent mahallesi. *Arkitekt, 253–254*, pp: 167–173.

Burdett, R. & Nowak, W. (2009). *Istanbul: City of intersections*. London, UK: Urban Age.

Çakir, N. (2007). *Impacts of current urban dynamics on urban transformation* (Master's thesis, Istanbul Technical University, Institute of Science, İstanbul, Turkey).

Ergun, N. (2004). Gentrification in Istanbul. *Cities*, *21*(5), 391–405.

Gulen, M. (2006). *"Stratejik planlama yaklasimi cercevesinde kentsel projeler—Kamusal alan iliskisi: Buyukdere aksi levent plazalar alani ornegi"/Urban projects in the context of strategic planning approach—Public space relation: the example of Buyukdere-levent axis* (Master's thesis, Mimar Sinan Fine Arts University, Istanbul, Turkey).

Keyder, C. (1999). *Istanbul: Between the global and the local.* Lanham, MD: Rowman & Littlefield.

Kurtuluş, H. (2005). Garden City as an Utopia/Bir utopya olarak bahçeşehir. In H. Kurtuluş (Ed.), *İstanbulda kentsel ayrışma/Urban Fragmentation in Istanbul* Istanbul, Turkey: Baglam yayıncılık.

Ohmae, K. (1995). *The end of the nation state: The rise of regional economies.* London, UK: Simon & Schuster.

Oncu, A. (1997). The myth of the 'ideal home' travels across cultural borders to Istanbul. In A. Oncu & P. Weyland (Eds.), *Space, culture and power: The new identities in globalizing cities* (pp. 56–72). London, UK: Zed Books.

Ozten, O. (2010). Evolution of everyday life by place and in place: the structure of the processes of change of cities: Levent—celiktepe example *Gündelik hayatın mekânla vee mekânda evrilmesi, kentlerin degisim süreçlerinin yapılandırması: Levent—celiktepe örneği* (Master's thesis, Istanbul Technical University, Institute of Science, Istanbul, Turkey).

Taylor, P. (2000). World cities and territorial states under conditions of contemporary globalization. *Political Geography*, *5*, 19–32.

Turgut, H. (2010). Urban dynamics and transformations and their impact on urban housing: The case of Istanbul. *Open-House International*, *35*(4), 76–84.

Turgut, H. & Inalhan, G. (2007). *Cultural and spatial dynamics of Istanbul: New housing trends.* Paper presented at ENHR 2007, International Conference-Sustainable Urban Areas, Rotterdam, The Netherlands.

Turgut, H., Yönet, A.N. & Torus, B. (2010). Global dynamics and new residential developments in Istanbul: Trends and expectations. In L. Villegas, O. Ural, V. Abrantes, I. Lombillo & C. Liaño (Eds.), *Design, technology, refurbishment and management of buildings, 37th IAHS World Congress on Housing Science, Santander, Spain.*

*Cities' Identity Through Architecture and Arts – Catalani et al. (Eds)*
*© 2018 Taylor & Francis Group, London, ISBN 978-1-138-05409-7*

# Investigating a sense of place at a historic commercial street corridor: Visitor perception of social aspects

Astrid Kusumowidagdo & Dyah Kusuma Wardhani
*Universitas Ciputra, Surabaya, Indonesia*

ABSTRACT: A strong social relationship between people and a place can trigger the emergence of activities, which can also strengthen the bond between them. This study will examine the Ampel Street Corridor, which is a historic religious area in Surabaya, Indonesia. It is an area with a strong identity and has a bonding relationship with its visitors. This study aims to reveal the types of relationships between the visitor and the place, confirm the scale of 'sense of place' that uld be achieved in this location, and to identify social aspects that are essential for experiencing this area and creating a sense of place. The nature of this research is qualitative. The data collection techniques were by interviews, focus group discussions, visual observations and documentation. The results of this study revealed that the relationship types found between a person and a place are biographical, spiritual, ideological, narrative, commodified and dependent. Another revelation is that the sense of place that the visitors could experience includes the knowledge of being located and involved in a place, belonging to and being attached to a place, identifying the self and the place's goal, and making sacrifices for a place. A third discovery is that the social aspects that create the sense of place are history, memory, crowdedness, religious atmosphere, lifestyle, interaction, activity, and the presence of Arabian merchants.

*Keywords*: social factors; sense of place; street corridor; religion; culture; people–place interaction

## 1 INTRODUCTION

Fischer found that urban-scaled places are generally becoming the centre of diversity, tolerance, socialisation, means of public transportation, cosmopolitanism, and information exchange (as cited in Gieryn, 2000). This can, therefore, trigger the occurrence of spontaneous interactions, freedom and creativity. In many cases, these community processes can be held in community centres through regular interactions, which depend on various factors including social class, ethnicity, taste, lifestyle and culture (Wellman, 1979). Besides physical appearance, people and social conditions are also elements that create a sense of place, and thus these elements are highlighted in this research.

The 'sense of place' has long been studied in the parameters of geography, architecture and urban design. Many studies on the sense of place focus on the relationship between social factors and place ownership, (Bondi 1993; Waxman, 2006), childhood memories (Chawla, 1992; Marcus, 2006) social interaction, opportunity to linger, social being, and the familiar stranger phenomenon, among others (Waxman, 2006). Various locations have also been studied, such as the corridor of a shopping centre (Kusumowidagdo et al., 2015), corridors of underground channels (Zacharia, 2002), corridors of street trading (Shamsuddin & Ujang, 2008), and camping areas (Kyle & Chick, 2007). This research aims to explain the social factors that are linked to the trade corridor of the religious historical area in Ampel, Surabaya, Indonesia.

## 2 RESEARCH PURPOSES

The purposes of this research are as follows:

- to find the types of relationships that occur as a result of the people–place relationship;
- to confirm the sense of place that could be experienced; and
- to explore the kinds of social aspects that trigger the creation of a sense of place.

Theoretically, this research is a preliminary study of the preservation of sense of place in the street corridor of Ampel, Surabaya, Indonesia, as a historic and religious area.

## 3 LITERATURE REVIEW

### 3.1 *Defining the sense of place and relationship type towards a place*

Sense of place can occur as a result of the complete sensing of the environment, either physical or social conditions, which make people feel attached to a given place. Cross (2001) explains that the process of developing a sense of place can occur as a result of six types of relationships, namely biographical, spiritual, ideological, narrative, commodified and dependent. Biographical relationships relate to the chronology of life's journey, and the forms of this relationship include historical and family ties. Spiritual means everything associated with feelings of self-creation, and the forms of this relationship are emotionally tied and invisible. Ideological means that the process is created and runs according to both ethical and religious aspects in society, while it takes the form of morality and ethics. Narrative means the process is formed through a variety of stories, myths, family history, political, and fictional stories. Commodified means the relationship is formed through a place that reflects self-actualisation, lifestyle, and other places that reflect an ideal; the form of this relationship is cognitive—based on choices and desires. Dependent means the emergence of a sense of place is usually because there is no other choice or because of economic factors, for example; the form of this relationship is material. The emergence of different senses of place is due to the different levels of satisfaction, identification and attachment to the community that each person has (Hummon, 1992).

### 3.2 *The scale of sense of place*

When relationships between people and places emerge, there will be self-intentionality of people towards that relationship. According to Shamai (1991), intentionality of the sense of place is categorised into different scales ranging from not having a sense of place (Tuan, 1977), knowledge of being located in the place, belonging to a place, being attached to a place, identifying with the place's goals, being involved in a place, to making sacrifices for a place.

### 3.3 *Social factors of sense of place*

The indicators of social factors discussed in this research are crowd and density (Bell et al., 1997). Density can cause stress and distress (Milgram, 1974, Saegert, 1978). However, crowds at an event in a shopping centre and cafes can be perceived as something positive (Bell et al., 1997) and can encourage other visitors to join in and make them passionate about the place. Meanwhile, the density can be perceived differently from each person's point of view (Baker, 1986; Bell et al., 1997 d'Astous, 2000). Other visitors with an appropriate lifestyle (Astuti & Hanan, 2011; Baker, 1986; d'Astous, 2000) and crowd size and behaviour (Baker, 1986; d'Astous, 2000) can provide a feeling of comfort and security, while ethnicity also affects the sense of place (Kusumowidagdo et al., 2015). Moreover, ethnicity, besides race (Zakaria & Ujang, 2015), is a cultural characteristic that also influences a sense of place. Other social factors include cultural beliefs and past experiences (Hashemnezhad et al., 2013; Low & Altman, 1992; Ujang, 2008), as well as interaction and activity features (Hashemnezhad et al., 2013), and memories and experience (Hashemnezhad et al., 2013).

## 4 RESEARCH METHOD

This research paper is an explorative preliminary study that aims to get one step closer to the overall research objective. This research was qualitative in nature, using interviews, visual observation and documentation, and focus group discussions to obtain opinions related to the objectives of this research. First, interviews were carried out with three people, which included experts and community leaders in the study area. Second, visual observation and documentation was performed for about one month in this area. Third, a focus group discussion was organised and carried out. Ten visitors with various backgrounds were treated as members of the focus group discussion that revolved around three aspects: types of relationships, social factors that trigger a sense of place, and the scale of the sense of place. The number of informants was based on Sari (2012), who suggested six to twelve informants were needed in order to have a mini-focus group discussion. The questions mentioned in focus group discussion are on Table 1.

## 5 RESEARCH PLACE CONTEXT

The research setting, the city of Surabaya, is the second biggest city in Indonesia after Jakarta (the capital city). Surabaya is a melting pot for various cultures in Indonesia, which have developed alongside modern urban life. One of the most interesting places in Surabaya is the 'Arabian village' complex, where many Arabian descendants live. This Arabian village has become a historic religious tourism site, particularly because Agung Ampel mosque and the tomb of Sunan Ampel, the propagator of Islam in Java, are located there. These nodes function as a centre for activities with corridors leading between them. The corridors themselves have become crowded business areas and lead to the centre of the Ampel. The two

Table 1. Questions to guide the discussion activities in focus group discussion.

| No | Topic | Questions |
|---|---|---|
| 1 | Types of relationships in the Ampel Street Corridor | What motivates you to come here?<br>How can you describe your attachment to this place?<br>Do these types of relationships describe all the possible types of relationships? |
| 2 | Scale of sense of place in the Ampel Street Corridor | Can you describe that intensity on a scale?<br>Is this scale of sense of place representative of your attachment to this place? |
| 3 | Social aspects of sense of place in the Ampel Street Corridor | Why are you interested in being in this place?<br>Are you familiar with this place?<br>What are the social conditions that become the characteristics of this corridor? |

Figure 1. The location of Area A in Ampel Street Corridor.

Figure 2. The location of Area B in Ampel Street Corridor.

most crowded corridors were chosen to form the focus of this research. The chosen corridors were crowded during the study with visitors on a religious tour, purchasing things for their prayer needs, as well as Arabian souvenirs. The chosen corridors were divided into Areas A (Figure 1) and B (Figure 2). Area A was the entrance of the Masjid Agung and the tomb of Raden Rahmat Sunan Ampel. Area B was the area called Ampel Suci Corridor.

## 6 FINDINGS AND DISCUSSION

### 6.1 *Relationship types at the Ampel Street Corridor*

The types of relationships between people and place in Ampel Street Corridor included biographical, spiritual, ideological, narrative, commodified and dependent relationships.

> *When you come for the first time, the process of recognising the identity of a place begins. For example, its gate, signage, or any other characteristic can make anyone recognise the identity of the corridor as an area of activity. This identity differentiates between Ampel and other corridors.* (M, Employee).

> *Besides identity, the sense of place can be the feeling of owning the place and feeling closer to the area. The attachment to the place is also possible, for example, it could occur as a result of the willingness to be involved in the development of the area. Other activities such as charity work are also possible since this area is identified as a spiritual area.* (F, Architect, 34)

> *Biographically, someone who is involved in an area must share an emotional bond with that area. Spiritually, religious similarity can trigger the connection between humans and the Ampel corridors area. The similar Muslim fashion, style and religious activities make the place seem familiar.* (D, Architect)

> *It is automatically commodified since the corridors are used as an area for vendors, commerce, and is full of small business activities.* (R, Employee, 30)

> *Narrative bonds have emerged from the history of Sunan Ampel as the propagator of Islam in Java.* (M, Employee, 30)

### 6.2 *The scale of a sense of place*

The sense of place, according to Shamai (1991), can range from not having a sense of place (Tuan, 1977), knowledge of being located in the place, belonging to a place, being attached to a place, identifying with the place's goals, being involved in a place, to making sacrifices for a place. All of these could be cases within this study.

### 6.3 *Social aspects that create sense of place*

The social aspects that influence a sense of place are history and memory, crowdedness, religious atmosphere, lifestyle attributes, interaction and activity, and also the presence of Arabian merchants. History, memory, crowdedness, religious similarity, lifestyle attributes, interaction and activities are social factors were found in both Area A and Area B. The presence of Arabian merchants is what differentiates Areas A and B, as they were found only in Area B. The similar social aspects that are found in both Area A and Area B are set out in the following subsections.

#### 6.3.1 *History and memory*

The memory of Sunan Ampel, as a propagator of Islam in Java, is one of the social factors. This legendary figure was buried in the Ampel area, close to the Ampel mosque. This area is therefore a site for religious tours as well as a tourist spot that preserves the memory and history of Sunan Ampel.

> *Sunan Ampel was one of Walisongo, that is why many people come here for religious tours; they pray here and wish for blessings.* (D, University student).

Memory and history are part of the determining factors, as described in Najafi and Shariff (2011) and Hashemnezhad et al. (2013).

#### 6.3.2 *Crowdedness*

The crowds often make the Ampel Street Corridor seem too crowded; however, this identity is already formed and hence contributes to the sense of place of this area, which is supported by the following statements:

> *This small corridor is full of visitors who thrust into each other, and because of its crowdedness, some visitors even push each other and drop their goods.* (S, Architect)

> *This street corridor is very full. Most of the street is used for selling goods, including rickshaws that gather at the exit and offer lifts.* (N, Employee)

Crowdedness is a determining factor of the identity of a corridor, which was revealed in a related study by Kusumowidagdo et al. (2015).

#### 6.3.3 *Religious atmosphere*

Religious similarity is the presence of the same religious purpose in the different people who visit a place. Visitors to Sunan Ampel are predominantly Muslim. This religious similarity is one of the social factors that shapes its sense of place.

> *I'm familiar with this place because of the similarity of religion and the presence of people who attend religious tours.* (S, Architect)

Religion is a factor that can be considered similar to the cultural factor. A similarity of belief can provide the feeling of familiarity. The cultural factor is one that contributes to the sense of place (Hashemnezhad et al., 2013).

#### 6.3.4 *Lifestyle attributes*

Lifestyle attributes in this study are fashion and appearance, which can encourage a sense of place being felt towards an area. Most visitors share a similar sense of Muslim fashion and style, considering this area as a religious space:

> *I feel I recognise this place because of the similarity of identity and physical appearance or style of fashion.* (N, Employee)

This Ampel religious area serves as a place that builds the identity of Muslim tourists. Davenport and Anderson (2005) state that places play a vital role in developing and maintaining the identities of the people.

### 6.3.5 *Interaction and activity features*

Activities and interactions are distinctive factors in both these corridors:

> *Many people who attend religious tours on their way to the mosque purchase religious ornaments, food, and souvenirs.* (M, Employee)

> *The corridors of the street towards the mosque and tomb of Sunan Ampel are crowded with people attending religious tours, while the Ampel Suci are most travelled by people who want to purchase Arabian souvenirs.* (D, University student)

This is consistent with previous studies on shopping streets, where active engagement is considered very influential. Active engagement can be seen in the direct communication and interaction between buyers and sellers, and the movement of pedestrians from one spot to another (Shamsuddin & Ujang, 2008).

Both street corridors studied are shopping areas that are suitable for all economic classes, ranging from low to medium income, because the price of goods are affordable and bargaining is welcomed. However, the Ampel Suci Corridor better accommodates shopping and bargaining activities between sellers and potential buyers because the corridor is comfortably wide with light circulation. Meanwhile, on the street corridor leading to the tomb and mosque of Sunan Ampel, visitors tend to directly purchase goods without spending much time in the seller's booths because of its heavy circulation:

> *Shopping at Ampel Suci is more fun because the corridor length is more comfortable and the circulation is not too dense, so it is possible to bargain for the goods.* (S, Architect)

The difference in the social factors that shape the sense of place for these two areas is found in the ethnicity factor. In Area A, most sellers are Maduranese, while the visitors share various ethnic backgrounds (dominated by Maduranese and Javanese) given that this area is the entrance of the Agung mosque and the tomb of Sunan Ampel. In Area B, most sellers are Arabian descendants, and the visitors also share various ethnic backgrounds (mostly Maduranese and Javanese):

> *I feel a new atmosphere in the Ampel Suci because most of the traders are Arabian descendants, while in the area of the street corridor to the mosque and tomb of Sunan Ampel, the traders and visitors there vary between Maduranese and Javanese.* (S, Architect)

Activity and interaction are also factors that highly shape the people–place relationship, as revealed in Low and Altman (1992) and Hashemnezhad et al. (2013). The findings of this study support these related studies.

## 7 CONCLUSION

The relationship types found between people and the Ampel Street Corridor area are biographical, spiritual, ideological, narrative, commodified and dependent.

The scales of sense of place that might be found in the (historic religious) Ampel area in the perception of visitors range from not having a sense of place, through knowledge of being located in the place, belonging to the place, being attached to the place, identifying with the goals of the place, and being involved in the place, to making sacrifices for a place.

The social aspects found in the Ampel area (in both areas studied) that developed the bond between visitors and these places are history and memory, crowdedness, religious atmosphere, lifestyle attributes, interaction and activity features. The unique social aspect in Area B, which becomes the differentiator, is the presence of Arabian merchants.

## ACKNOWLEDGEMENTS

This research was facilitated with the aid of Penelitian Produk Terapan, from the Ministry of Education and Cultural Affairs, Directorate of Higher Education, 2017. The research team expresses gratitude for the financial aid and to the paper reviewer for their acceptance.

## REFERENCES


Astuti, S. & Hanan, H. (2011). The behaviour of consumer society in consuming food restaurants and cafes. *Journal of Asian Behavioural Studies*, *1*(1), 72–83.

Baker, J. (1986). The role of the environment in marketing services: The consumer perspective. *The Services Challenge: Integrating for Competitive Advantage*, *1*(1), 79–84.

Bell, P.A., Green, T., Fisher, J.D. & Baum, A. (1997). *Environmental psychology.* New Jersey, NJ: Harcourt-Brace.

Bondi, L. (1993). Locating identity politics. In M. Keith & S. Pile (Eds.), *Place and the politics of identity* (pp. 84–101). London, UK: Routledge.

Chawla, L. (1992). Childhood place attachments. In I. Altman & S.M. Low (Eds.), *Place attachment* (pp. 63–86). Boston, MA: Springer.

Cross, J. (2001). What is sense of place? Paper presented at the 12th Headwaters Conference, Western State College, 2–4 November 2001. Retrieved from http://western.edu/sites/default/files/documents/cross_headwatersXII.pdf.

d'Astous, A. (2000). Irritating aspects of the shopping environment. *Journal of Business Research*, *49*(2), 149–156.

Davenport, M.A. & Anderson, D.H. (2005). Getting from sense of place to place-based management: An interpretive investigation of place meanings and perceptions of landscape change. *Society and Natural Resources*, *18*(7), 625–641.

Gieryn, T.F. (2000). A space for place in sociology. *Annual Review of Sociology*, *26*(1), 463–496.

Hashemnezhad, H., Heidari, A.A. & Hoseini, P.M. (2013). Sense of place and place attachment. *International Journal of Architecture and Urban Development*, *3*(1), 5–12.

Hummon, D.M. (1992). Community attachment. In I. Altman & S.M. Low (Eds.), *Place attachment* (pp. 253–278). Boston, MA: Springer.

Kusumowidagdo, A., Sachari, A. & Widodo, P. (2015). The important factors of corridors settings in shopping center design: A study of Indonesian shopping centers. *Architecture Science*, *11*, 29–43.

Kyle, G. & Chick, G. (2007). The social construction of a sense of place. *Leisure Sciences*, *29*(3), 209–225.

Low, S.M. & Altman, I. (1992). Place attachment. In I. Altman & S.M. Low (Eds.), *Place attachment* (pp. 1–12). Boston, MA: Springer.

Marcus, C.C. (2006). *House as a mirror of self: Exploring the deeper meaning of home.* Berwick, ME: Nicolas-Hays.

Milgram, S. (1974). The experience of living in cities. *Crowding and Behavior*, *167*(41), 1461–1468.

Najafi, M., & Sharif, M.K.M. (2011). The concept of place and sense of place in architectural studies. *International Journal of Human and Social Sciences*, *6*(3), 187–193.

Saegert, S. (1978). High density environments: Their personal and social consequences. In A. Baum & Y. Epstein (Eds.), *Human response to crowding* (pp. 257–281). Hillsdale, NJ: Lawrence Erlbaum Associates.

Sari, W. (2012). *Qualitative research method, theory and practice*. Jakarta, Indonesia: Penerbit Salemba Empat.

Shamai, S. (1991). Sense of place: An empirical measurement. *Geoforum*, *22*(3), 347–358.

Shamsuddin, S. & Ujang, N. (2008). Making places: The role of attachment in creating the sense of place for traditional streets in Malaysia. *Habitat International*, *32*(3), 399–409.

Tuan, Y.F. (1977). *Space and place: The perspective of experience*. Minneapolis, MN: University of Minnesota Press.

Ujang, N. (2008). *Place attachment towards shopping districts in Kuala Lumpur city centre, Malaysia* (Doctoral dissertation, Universiti Putra Malaysia, Serdang, Selangor, Malaysia).

Waxman, L. (2006). The coffee shop: Social and physical factors influencing place attachment. *Journal of Interior Design*, *31*(3), 35–53.

Wellman, B. (1979). The community question: The intimate networks of East Yorkers. *American Journal of Sociology, 84*(1), 182–196.

Zacharias, J. (2002). Choosing a path in the underground: Visual information and preference. In *Proceedings ACUUS International Conference on Urban underground space: A resource for cities, 14–16 November 2002, Torino, Italy* (pp. 14–16). Turin, Italy: Associazione Georisorse e Ambiente.

Zakaria, J. & Ujang, N. (2015). Comfort of walking in the city center of Kuala Lumpur. *Procedia-Social and Behavioral Sciences*, *170*, 642–652.

*Cities' Identity Through Architecture and Arts – Catalani et al. (Eds)*
 *ISBN 978-1-138-05409-7*

# Jammu—the city of temples

Chander M. Seth
*INTACH, Jammu, Jammu and Kashmir, India*

ABSTRACT: The city of Jammu is located on the banks of the river Tawi in a hilly mountainscape of the Himalayas. Jammu is an ancient city which was established by Raja (King) Jambulochan (1320–1290 BC) when he saw a wild goat and a lion drinking water together in one of the waterholes in the forest. He was so impressed with the coexistence of two rival species that he decided to set up a city at this site, where people of all faiths and religions could coexist peacefully and tolerate each other's customs and traditions. Over a period of many centuries, several kings who ruled this part of the country constructed temples on the bank of the river and inside the city. More than 100 such temples adorn the city, which has given it a unique name and identity as the "City of Temples in India". Jammu is also famous for its Bahu Fort, built by Raja Bahulochan and the royal Mubarakh Mandi Palace of the Dogra kings. These sites are designated as heritage buildings.

Jammu is the winter capital in the Jammu and Kashmir State of India. Its present population is about one million. Inhabitants of this city are descendants of an ancient martial tribe of Dogras and their spoken language is Dogri. Dogri is one of the Pahari (hilly) languages that has influence of Persian, Pushto and Devanagri languages because Jammu is located at the cross-roads of the ancient Silk Route, trading with central and west Asia.

All the temples of Jammu have unique architectural styles, describing the traditions and culture of its historical past. These temples are part and parcel of the daily life of its citizens and preserve the cultural heritage of the city. In the past, these temples were centres of social and religious activities. As Jammu is located in the driest part of the country, it has faced shortages of drinking water and so large ponds were constructed to collect rainwater. These ponds are unique sources of traditional knowledge for rain water harvesting and water conservation and are part of the city's heritage.

Jammu attracts between eight and ten million tourists every year. In this paper, the history of the city, its heritage and tourist value, including a description of the architecture of its temples, is discussed.

*Keywords*: Jammu; Raja Jambulochan; City of Temples, Bahu Fort; Palace of Dogra Kings; heritage buildings; Vastu_Shastra; INTACH; Nagara style

## 1 INTRODUCTION

The history of Jammu city can be traced from prehistoric times to the present day. Earlier literature and archaeological excavations have revealed that the city and its region have a history of about 5,600 years. Prehistoric sites at Manda, Akhnoor and other adjoining places provide evidence that Jammu city was a part of the Harappan civilisation (Sharma, 2007). Redware pottery, double spiral headed pins, bone arrow-heads, terracotta bangles and triangular terracotta cakes excavated from these sites, indicate the same culture and traditions in the Harappan cities. There is a well-documented historical record of Jammu rulers from 1600 BC to AD 1947, but with some missing links in-between (Sharma, 2007). After 1947, the State of Jammu and Kashmir became part of the Indian union, like all other princely states of India. At present, Jammu is part of Jammu and Kashmir (J&K) State, located in the Western Himalaya.

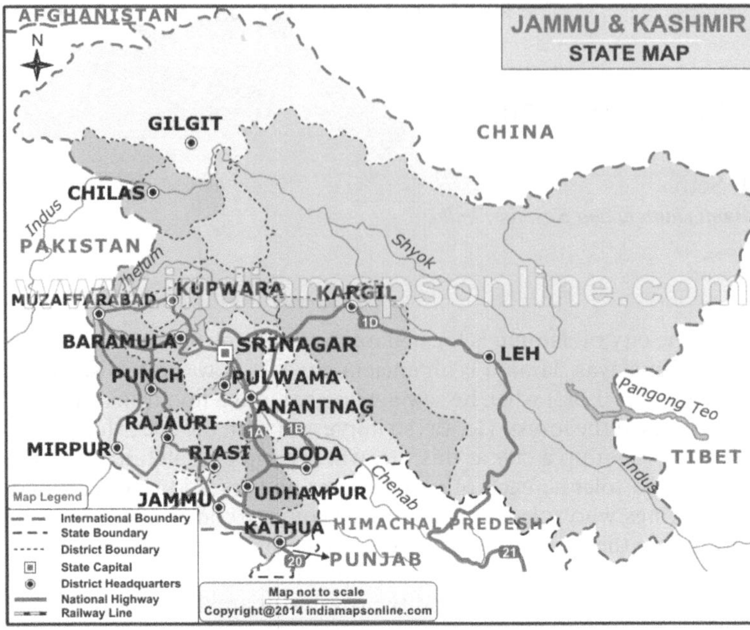


Figure 1. Map of Jammu and Kashmir State—India.

## 2 FOUNDER OF THE CITY

Raja Maldev, fourth King of the Dev dynasty, ruled Jammu from AD 1361 to 1400 (Goswami, 2015) and is believed to be the founder of present day Jammu city. He was a very tall and powerful king of his times. Several legends and ballads in the Dogri language exist in local folk art and depict his bravery, administration and governance. Raja Maldev established his headquarters at Purani Mnadi in the centre of the city. From 1400 to 1733, ten descendants of Maldev ruled the territory of Jammu. The eleventh ruler of the Maldev dynasty was Raja Ranjit Dev, who ruled from 1733 to 1782. He was known as an apostle of justice, chivalry and administration. He was the most secular and religiously tolerant king. His period is remembered as an era of prosperity, peace and coexistence. In his time, Jammu was the largest state in northern India.

## 3 THE PERIOD OF THE DOGRA KINGS

The Dogra Jamwal Kings ruled the state for a period of about 100 years from AD 1847 to 1947 (Charak, 1985). Maharaja Gulab Singh was the founder of Dogra rule and extended his boundaries to enclose the whole of Jammu, Kashmir, Ladakh and Gilgit. It was during the Dogra period that several forts, palaces, temples, educational institutions, hospitals, museums and library buildings were constructed with unique architecture and art. Although this had influence of Mughal, Rajasthani, Kashmiri and British Baroque architecture, it retained the flavour of local Dogra art and architecture. Ganhar (1973) described the Dogras as, "deeply devoted to their land and especially their places of worship, around which they have woven a variegated tapestry of myth and legend." Their rich folklores and achievements in artistic expression were inspired by their devotion to religion and have become part of the vocabulary of Dogra art and architecture.

All the earlier rulers of Jammu from 1600 BC to AD 1947 contributed to the art and architecture of the region, giving it a distinctive place in the architectural landscape of the country.

The town planning of Jammu city started at the time of Raja Jambulochan (1320–1290 BC) and continued throughout the reigns of Raja Maldev (1361–1400), Maharaja Ranjit Dev (1733–1782) and the Dogra kings (1847–1947). From the outset, all rajas, maharajas and kings contributed to the city's planning, layout and architecture in several layers and phases. In studying the planning of the city, it appears that the city's art and architecture has evolved and developed over a period of time. It has been influenced by the art and architecture of mainland India, blended with local materials, traditions, culture and traditional knowledge of times, and also based on the ancient Indian science of architecture, "*Vastu Shastra*" (Vastu Shastra, Google, 2017).

The city of Jammu was established by Raja Jambulochan, after whom the city derives its name, on a hilly slope of the Shiwalik mountain range (Figure 2a). The site was ideal from a defence point of view; on the south-eastern side flows the river Tawi, also called Suryaputri, and on the north side is the naturally fortified Ramnagar Shiwalik mountain range (Drew, 1875). The area was rich in forests, wildlife and sufficient water available from the river Tawi. The site had a typical tropical climate with hot summers and cool winters. On the southern side of city are the plains of the Punjab and on the northern side the undulating Shiwalik hills create a natural amphitheatre with a high-rise backdrop of the Trikuta hills.

The layout of Jammu indicates that while planning the city, the natural drainage system of the hilly slopes, access to the river Tawi, creation of water bodies (locally called *talabs*), parks, temples/- shrines and the security of the city, was kept in view. The city was basically developed as a pedestrian city with small streets, locally called *galis* or *kuchas* connected at *chaougans* with *mohallas* and *ahatas*. The city had several *kuchas* named after prominent persons from those streets (Mangotra, 2013). Zonation of the city into *mohallas* was well planned and there are 11 main *nallahs* (city drains) which drain into the river Tawi on the eastern and southern slopes. Alongwith these *nallahs,* there were 11 *dhakis* (hilly stone/brick pedestrian paths) for the movement of the population from city to river and back. Dhaki of Peer Mitha, Saranja Dhaki and Naina de Dakhi are some that still exist today. On the eastern and northern sides, the city is enclosed by two wildlife sanctuaries, Ramnagar and Bahu. These protected areas help in water conservation and recharging of the river water, as well as contributing to the environment of the city. Palaces and forts were located at the commanding heights of the city at the Manda hills and in earlier times were located in the centre of the city at Purani Mandi and later shifted to Mubarakh Mandi. A *mandi* is a place where Rajputs used to reside. There is no record available to show whether the city was developed as a planned city. However, one copy of the city map is available in the Dogra Art Museum in Jammu and the physical evidence of the city indicates that it was well planned architecturally

a b

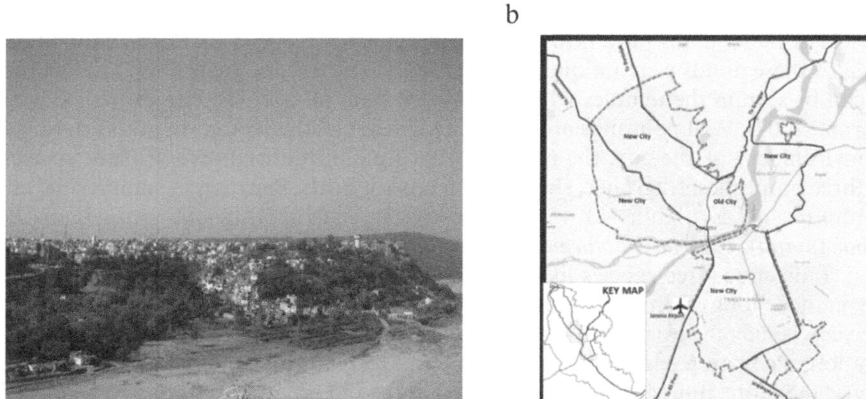


Figure 2. (a) Old city of Jammu, (b) Jammu city master plan.

Figure 3. Old city map 1880–1890 AD.

and developed on the principles of *Vastu Shastra*. The original map printed on cloth in AD 1880, can be seen in the National Museum in Delhi (see Figure 3).

Study of this map indicates that Jammu was a walled city, fortified on three sides, starting from the western side to the southern side and terminating on the eastern side (Chaudhary, 2007). On the northern side, the Manda hills provided natural fortification. The city was accessed from four main gates (locally called *deodis*). These *deodis* were named as Jogi Gate, Mahespura Gate, Gumat Gate and Dennis Gate. Mubarakh Mandi, the seat of governance, was the iconic building of the City, giving it a distinctive identity and characteristic. Mubarakh Mandi and the palaces were surrounded by the buildings of the ministers (*wazirs*) and other high officials of the kings. This was followed by *mohallas* (zones) comprising homogeneous populations of traders, businessmen, skilled workers and other service providing classes at the outer periphery of the city. These *mohallas* were named according to their inhabitants; thus, Afghan Mohalla where Afghans settled, Jullaka Mohalla where the *jullayas* (weavers) settled, Sarajan Mohalla where leather tanners resided, Khadika Da Mohalla, Jain Mohalla (goldsmiths) and bazaars (shopping streets) where the Jains settled, Kachi Chawani where the Army was based, Peer Mitha Mohalla after the Sufi saint Peer Mitha who probably came from Iran and settled here, and *mohallas* of other communities. Prominent bazaars of Jammu city included Jain Bazar where only goldsmiths had their shops, Lakhdatta Bazar and Moti Bazar which had utensil makers (*thatheray*), Kanak Mandi, which had only grain merchants, and Urdu Bazar, which had only entertaining hubs or *kothas*. Many *talabs* were constructed in the city for rain water harvesting to meet the city's requirement. These *talabs* (ponds) were constructed adjacent to temples and shrines. Some of the important *talabs* were Rani Da Talab, Raghunath Mandir Tank, Khadika da Talab, and the *talab* near Gada Dhar temple. All these *talabs* have now been filled in except for that of Raghunath Mandir. The design of these ponds were unique architecturally. The devotional art and architecture of the city can be seen in the temples which have been built all over the city over a period of time (Sharma, 2007). Wall paintings are typical Dogra art of the region. In addition to improving the environment of the city, the rulers of that time introduced several plant species, which had three properties: fragrance, shade and fruits for birds. Because, in summer the city's temperature reaches as high as 40–45°C, plant species such as *Magnolia grandiflora, Jasminum sambac* (*motia*), *Cestrum nocturnum* (*rat ki rani*) and *Mimusops elangi* (*maulseri*) were introduced. Indigenous tree species like *Ficus benghalensis* (*banyan*) and *Ficus religiosa* (*peepal*) were planted alongside roads, in temples and around ponds. To provide ice—cold water from the river Chenab, the rajas of that time constructed the 30 km Ranbir Canal up-to the city for irrigation, navigation, electricity generation and recreation purposes. The detailed layout of the city, therefore, quite clearly establishes that the old city was very much planned, keeping in view all the architectural and town planning requirements of the city and its people, based on the five principles of the science of *Vastu Shastra*.

Modern town planning of Jammu started in 1957. The third Master Plan (2021–2032) covers an area of 650 km$^2$ (see Figure 2b) (Jammu Development Authority, 2016). The city is developing very fast and is becoming cosmopolitan with the new digital culture, modern western fashions and life style, and high living standards. The city's old ethos, values, architecture and art is changing with the times and the westernisation of society. The city is also still evolving, with a new composite culture and architectural trends. Its growth is inclusive, equitable and organic. Evolution is the law of nature and the city continues to evolve with the new ideas, scientific and technical knowledge, innovations and ethos of a multicultural cosmopolitan society (Singh, 2014).

## 5 ART AND ARCHITECTURE OF FORTS, PALACES AND TEMPLES

The Indian National Trust for Art and Cultural Heritage (INTACH) has listed 12 major temples and 158 heritage structures in Jammu city (cited in Agrawal & Gupta, 2009). It is not possible to discuss the art and architecture of each and every building in this paper, but representative structures of a few iconic heritage buildings such as the Mubarakh Mandi Complex, Amar Singh Palace, Bahu Fort, Raghunath Temple and Ranbireshwar Temple have been selected to represent the typical Dogra art and architecture of the city. These heritage buildings are located in old Jammu city (see, Figure 4).

### 5.1 *Palaces and forts*

#### 5.1.1 *Royal Mubarakh Mandi Complex*

The historic Mubarakh Mandi Complex in Jammu city comprises the royal palaces and the secretariat of the Dogra Kings of J & K State (Chaudhary, 2008). In 2005, due to its deteriorating condition, the Archives, Archaeology and Museums Department declared the complex a protected monument, under the provisions of the J&K Ancient Monuments Preservation Act 1977. In 2006, J&K State set up an autonomous body called the - "Mubarakh Mandi Heritage Society" - to ensure its conservation, protection and restoration work by the Archaeological Survey of India is underway. Draft Project Report (DPR) for restoration was prepared by INTACH in 2007 (Agrawal & Gupta, 2009).

There are 25 components of the Mubarakh Mandi Complex. These are: Gol Ghar, Rani Charak's Mahal, Rani Kathar's Mahal, Rani Bandral's Mahal, Mahal of the Ranis of Raja Amar Singh, Rani Guleri's Mahal, Royal Courts building, Raja Ram Singh's Palace, Heritage

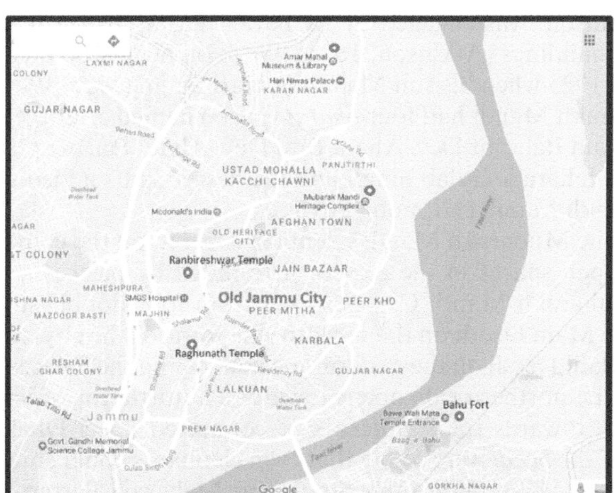


Figure 4. Location map of heritage sites in the old city (Google, 2017).

structures of Niki Deodi, Mahal of the Ranis of Raja Ram Singh, Marble Hall, Smadhi of Baba Ragho, Pink Hall, Baddi Deodi, Darbar Hall, Army Headquarters and Foreign Office, Main Deodi, Old Magistrate Complex, Raja Amar Singh's Palace, Toshakhana Complex, Gada Dhar Temple, Old Pond, Gateway towards Panjtirthi, Darbar-e-Aam and Reserve Treasury Building. Mubarakh Mandi—covers an area of 50,000 $m^2$ or ~12.5 acres, and the built-up area is 4,800 $m^2$. According to one estimate, the total area is 120 *kanal* and 16 *marlas*. The princely state of Jammu and Kashmir enjoyed a unique political position at the national and international level and Mubarakh Mandi signified the power centre of this princely state, a political and administrative headquarters that saw a blending of native and international influences on a regular basis in almost all domains. Similarly, the architecture which emerged at Mubarakh Mandi was a unique provincial style that amalgamated colonial influences with native building traditions, manifested in the erection of buildings such as the Darbar Hall. Developed in stages over a century, several layers of construction, attributable to different Dogra rulers are identifiable in the complex and provide its present shape. There is an extensive vocabulary of semi-circular and Gothic arches with domes inspired by British colonial Baroque architecture co-existing with Mughal—inspired multi-cusp arches and *jharokhas* (balconies). There is also an influence from the Rajasthan style of architecture. Use of river pebbles (*gittian*) in decoration on walls and pillars is a local innovation.

In the present state of Jammu and Kashmir, the Jammu region is culturally very distinct from those of Kashmir and Ladakh because of their separate historical and socio-cultural identities. Mubarakh Mandi, despite the melting pot of several influences, is the highest embodiment of Dogra art and architecture in the Jammu region and a source of pride for the local community of Jammu city.

According to historians, construction of the Mubarakh Mandi was started in the period of Raja Dhruv Dev in 1710 when he shifted his office and residence from Purani Mandi to Darbar Gad, later named as Mubarakh Mandi. This building was constructed on a high plateau overlooking the river Tawi on the eastern edge of Jammu city. The architecture was in the Dogra Pahari style. Its outer walls towards the river Tawi were very strongly supported with *diwalgirs,* some walls are as thick as 15 ft. During his reign, these buildings were called Darbar Gad. His son Ranjit Dev (1753–1781) added more structures to this building. In 1822, Gulab Singh was made Raja of Jammu and once again, after a gap of six years, Jammu city came under the Dogra rulers, when he and his family, came to Darbar Gad. Further construction and development of Mubarakh Mandi took place during 100 years of Dogra rule. The largest contribution to Mubarakh Mandi was in the period of Maharaja Ranbir Singh (1857–1885). Maharaja Partap Singh (1885–1925) constructed Rani Charak Palace and some other official buildings. There was a great fire in 1898 in which Dewan Hall, the Governor's office, the Foreign office and Darbar Hall were burnt and destroyed. British Superintendent Engineer, Alex Atkinson, was engaged by Maharaja Partap Singh to redesign and reconstruct the damaged buildings (Atkinson, 1898). Work on new structures was started in 1913 and continued until 1925 when his son Maharaja Hari Singh (1925–1947) became king.

Originally, Mubarakh Mandi had four *ahatas* (parts) named after the sons of Raja Dhruv Dev. These were Ahata Balwant Dev, Ahata Braj Dev, Ahata Darbar Gad and Ahata Dhansar Dev. Later on, Maharaja Gulab Singh and his descendants added to and expanded the buildings of Mubarakh Mandi (Nirmohi, 2016).

Construction of the Mubarakh Mandi Complex is based on the courtyard style of Indian architecture, with open spaces in the centre surrounded by palaces and official buildings. Entrance to the Mubarakh Mandi Complex was from four double-storey gates or *deodis*. These gates were the Main Deodi on the western side towards the city, connecting Jain Bazar and Pucca Danga, Badi Deodi on the eastern side connecting the Palaces of Rajas and Ranis. The *deodi* towards the northern side opens towards Panjtirthi for officials (*darbaris*) of the Maharaja and *deodi* towards the southern side connected Pacci Dhaki Mohalla with the palace. The doors of the *deodi* were so big that even elephants could enter the palace. Shobha *yatras* (ceremonies) of Maharaja Ranbir Singh and Maharaja Partap Singh would emerge through this *deodi*. The door of this *deodi* had small window doors, which were used in emergency. This *deodi* had Rajasthani—style chhajjas (eaves) and *chattris* (cupolas). In front of the *deodi* an image of the sun had been engraved. The second storey of the *deodi* was occupied by the sentries who would keep watch on the public entering the palace. Important guests were brought through this gate. On the southern side of central courtyard was the Foreign Office and Army Headquarters, one above the other. Grey Hall building adjoins this building towards the southern side of the courtyard and was used for official meetings. On the eastern side, small gates give entry into Gol Ghar. This was a four-storeyed structure with three domes on the top. These domes were visible from a distance and were a great attraction for the Maharajas. In 1985, the eastern portion of the complex was burnt and great damage occurred to the Gol Ghar building and a great heritage was lost. Of the three domes, only two are still intact. The rest of the buildings from inside was totally burnt and damaged. On the eastern side, towards the river Tawi, the Mahals of Rani Kathar, Rani Bandrali Mahal, Sheesh Mahal, Toshakhana, Rani Charak's Mahal, Mahals of Ranis of Raja Amar Singh, Mahal of Rani's of Raja Ram Singh and Rani Guleri's Mahal were located. This area was called as *zanana* (harem) and no male was allowed. Entrance was only through Niki Deodi. Facing towards the courtyard were Marble Hall and Pink Hall. Adjoining this building was Gadvai Khanna (the pharmacy) and the Royal Courts building. On the northern side of the complex was Raja Ram Singh's Palace. On the western side of the courtyard were the buildings of Raja Amar Singh's Palace, Toshakhana and the Reserve Treasury. In the centre of the complex was the Raj Thada (the royal platform), made of marble, which has now been damaged. This platform was surrounded by a beautiful park with fountains.

The construction material used was locally made bricks and tiles, stone bricks from a place near Ramnagar, slate, wood and red-stone from Rajasthan. Local *gittian* (pebbles) from river bed, *surkhi* (crushed bricks) and *chuna* (lime) was also used (Chaudhary & Katoch, 2008). The *jarokhas* and *chattris* are in a typical Rajasthani style. In some places, terracotta *jallis* (trellis) are used on the roofs. The Army headquarters and Foreign Office (see Figure 5a) is of the British Baroque style, with iron grilles and other material from England. Three domes with a big clock in the central dome are, again, a Baroque style of architecture. Many pillars and front portions of the veranda have engraved and sculpted floral images. In these buildings, Kashmiri art such as *khatambandi* (marquetry) and papier-mache are in the ceilings. All papier-mache ceilings are painted with decorative work. The walls and pillars inside are decorated with miniature floral wall paintings and themes drawn from epic Ramayana and Mahabharata stories (Figure 5b). In the old buildings towards the eastern side, the flooring was made of wooden planks with a layer of earthen pitchers and again covered with wooden planks and slates. This unique style is a local architectural innovation for keeping rooms cool during the summer months. Such flooring could act like a thermostat both in summer and winter. In Sheesh Mahal only coloured cut glass was used. Perhaps this was a later addition on the pattern of several Rajasthani and Mughul palaces in India. Another local architectural innovation was the design of stairs. These stairs were of different sizes for people of different ages and for different purposes, and were mainly made of wood and stone slabs.

a b

Figure 5. (a) Foreign Office, and Army Headquarters, (b) Wall paintings.

Small pebbles (*gittian*) decorate pillars, walls and stone plates in the foundation again, a local architectural innovation.

5.1.2 *Amar Mahal*

Amar Mahal was constructed by Raja Amar Singh in 1892. The architect of this beautiful palace was French. He could not complete this palace, perhaps, because of the sudden death of Raja Amar Singh in 1908, and from the outside it appears that only a half portion has been constructed. It was built on the topmost northern part of the city, adjoining the Manda hills and facing the river Tawi, occupying a commanding position from all sides. The architecture of the building is a replica of the French forts (chateaux) and farm—houses of the time. Katoch (2012) observes that this palace is greatly influenced by the Queen Anne style of architecture. The palace is a three-storeyed building with basement. The plan is rectangular, and 150 ft. in length, 100 ft. in width and 50 ft. in elevation. The plinth of the building is 1 m high. The building is open on all sides with lawns in front and to the rear to allow sufficient light and air. The walls are thick to maintain the temperature of the rooms from inside in both winter and summer. Burnt bricks were used in the construction. Mouldings of sandstone adorn the building in different places. The front façade is marked by a series of semi-circular arches running all along the corridor on the ground floor, and rectangular/wheel/bay windows on the first and second floors (see Figure 6a). The rear façade of the building is marked by simple arches running all along the veranda on the ground floor, and bay and French windows with a small balcony on the first floor. On the second floor, windows are bay windows with triangular pediments. An outer veranda is covered with sloping ridged—tin sheets resting on a wooden framework. An arcade of simple arches running all along the veranda is the most attractive structural feature, which enhances the aesthetics of the façade of the building. The pillars of these arches are decorated with various floral designs and stone inlay work (Figure 6b). The roof was made of slate, which in 1955 was changed to steel sheets to protect it from rain and high-speed winds. The double column supporting the wooden framework stands on a simplified brick pedestal emulating the Greco-Roman style. The columns used in the building are of a composite order. The art work is local Dogra style with paintings of lotus flowers, birds and animals. Unfortunately, Raja Ram Singh did not use this palace as his residence because of his sudden death in 1908. At present the palace, has been converted into a museum that thousands of tourists visit every day, and the library is used by scholars from India and abroad.

5.1.3 *Bahu Fort and palace*

As discussed earlier, Raja Bahulochan ruled Jammu and its adjoining areas from 1350 to 1320 BC. He constructed Bahu Fort on the right bank of the river Tawi. After his death, his brother Raja Jambolochan, established the city of Jammu opposite Bahu Fort and town.

a b

Figure 6. (a) Front façade of Amar Mahal, (b) Use of local material and designs.

a

b

Figure 7. (a) Bahu Fort and palace, (b) Aerial view of Bahu Fort.

Bahu Fort was constructed in the era of Raja Bahulochan about 3000 years ago (Nirmohi, 2016). The fort has one entrance gate or *deodi* which is two storeys high and is covered from inside. The *deodi* opens into a large central courtyard with a central passage which leads to the Temple of Mata Kali. The fort is divided into two parts. On the eastern side are confinement cells for prisoners, rooms and lawns, and on the western side is a pond in the centre, which is surrounded by two-storey buildings and rooms. This fort and the palace inside were constructed by several kings over a period of time (see Figure 7b).

The fort was damaged by Sikh rulers of Punjab (Goswami & Gupta, 2015) and was later repaired and reconstructed by Maharaja Gulab Singh when he became ruler of Jammu. Bahu Fort and Mubarkh Mandi are opposite one another on the right and the left banks of the river Tawi. One part of the palace is called Old Mahal and another part New Mahal (palace). All rooms of the palace open onto an internal courtyard. Arches of veranda are of Mughal style, perhaps constructed in the later Mughal period. The palace had 12 rooms. Its roof was made of local wooden sleepers as beams, covered with wooden planks. The construction materials used were local stone bricks and *chuna surkhi* mortar. In some parts, the bricks covered with *chuna surekhi* plaster. All rooms had *bukharis* (heaters) and small lamp ledges (*taukdas*) where oil lamps were burnt for lighting purposes. The walls of these palaces were adorned with beautiful wall paintings on local themes. These paintings were made directly on the plaster of the walls using natural colours. From the outside, the palace is surrounded by fortified high-rise walls (see Figure 7a). The roof of the rooms was used by soldiers for watch, ward and security of the palace. This fort was very strong and was used for defensive purposes.

The fort and palace have been notified as a protected monument by the J&K government and conservation and restoration works were undertaken by the Archives and Jammu Development Authority under the supervision of Mr. K.A. Qadri, then Director General of Archives and Archaeology department in 2007.

### 5.2 *Temples and shrines*

Janmu city has a number of temples and shrines constructed at different points in history (Manohar, 1971). Some of the main temples are Raghunath Temple, Shri Ranbireshwar Temple, Panchbakhtar Temple (also known as Rupaywalla Temple), Diwan Mandir, Mahamaya Temple, Maha Lakshmi Temple (Pacca Danga), Radha Krishna Temple (Rani Park), Hanuman Mandir (Moti Bazar), Mandir Mata Chint—Purni (Dhounthali) and Rani Mandir (Gumat), (Shakar, 2012). Because people from different faiths made Jammu city their home, shrines of Sufi saints have also been constructed. Gurdwara of the Sikh saints have also been constructed, as have Christian churches. There are cave temples too. Almost all 11 *dhakis* from the river Tawi to the city had temples. The art and architecture of some of these temples have been discussed in this paper and gives Jammu its unique name of - “City of Temples” -.

5.2.1 *Raghunath temple*

Raghunath Temple Complex is one of the biggest complexes of several temples in one place (Ganhar, 1973). Raghunath Temple was constructed in the time of Maharaja Gulab Singh in 1835 and was completed 25 years later in 1860 in the time of his son Maharaja Ranbir Singh. At the time of construction, the temple complex was located on the periphery of the city at the southern end below Raghunath Bazar. It is now in the centre of the city. The architecture of the temple (see Figure 8a) is typical of the northern Indian or Nagara style (Manohar, 1971). The manual of the Nagara style of temple architecture, *Brihat Samhita,* written by Varahamihira in the 6th century and describes the design and construction of the Nagara style of Hindu temples (Brown, 2014). This style has two distinct features. In plan, the temple is square with a number of graduated projections in the middle of each side, giving a cruciform shape with a number of re-entrant angles on each side. In elevation, a *shikhara* (steeple) gradually inclines inwards in a convex curve (like a beehive) using a concentric rotating square and circle principle. The temple is dedicated to Lord Rama, who is also called Raghunath. This temple has a quadrangular platform on which the main temple is constructed with an entrance gate, *mandapa* (prayer hall) and *garbhagriha* (inner most sanctum), where *murtis* (stautes) of Lord Ram, Laksmana and Sita are installed. Inside the walls of *garbhagriha* are covered with gold-plated sheets on three sides. The *garbhagriha* is surrounded by parikrama (also called pradakshina or devotional walkways). The *shikhara* of all temples are golden in colour and can be seen from any part of the city. On top of the *shikhara* are *kalasha*. The parikrima is surrounded by many galleries, which houses *saligrams* (ammonites) brought from the Narbada river in Madhya Pradesh. The outer side of the temple contains small temples which depict the various avatars of Lord Vishnu. On the northern side, there is a large water tank to meet the needs of the Temple and for the public (in those early days). The temple complex also has a big library and a school for Vedic studies. On the eastern side of the complex is Hanuman ji Mandir, which is on the left side of the entrance gate. On the right side is the temple of Shiva with crystal *Shivling* (symbol of Shiva). All temples in the complex are constructed in Nagara style of architecture. There are 17 big and small temples in the complex, dedicated to Hindu gods and goddesses (Shakar, 2012). There was another pond outside the temple on southern side in Kaleeth Mohalla to meet the needs of devotees and the students in the hostels. This pond is now filled in and a shopping market has been constructed on its site.

5.2.2 *Shri Ranbireshwar temple*

This temple is located near the central part of city and faces west. The temple is also constructed in the Nagara style of architecture with a two-storeyed raised platform (see Figure 8b). There is a flight of stairs to reach the main temple above the platform. The platform is constructed on the rooms that were used for accommodation of pilgrims (*yatris*), *pujaris* (temple priests) and other staff. It also had a kitchen and toilets. This temple has one

a b

Figure 8. (a) Raghunath temple, (b) Shri Ranbireshwar temple.

of the biggest *lingams* of Lord Shiva in black marble (7 ft high). On either sides of this *lingam* are five small *lingams* each two foot in height. Outside the main entrance of the temple is a sculpture of a Nandi (a sacred bull) facing Lord Shiva. Nandi is companion of Shiva. This temple was constructed by Maharaja Ranbir Singh in 1873 (Ganhar, 1973). On the inner walls of the temple, several paintings of Lord Shiva and stories from Shiv Puran have been written. On top of the *shikhara* there are four big golden-plated *kalasha*.

#### 5.2.3 *Mahamaya temple*

This ancient temple (see Figure 9) was built on top of the Mahamaya Hills and dedicated to the goddess Mahamaya. It overlooks the city of Jammu and the river Tawi at a short distance from Bahu Fort. Historians believe that this temple was perhaps built in the ancient city of Dhara Nagri in the 15th century (Goswami & Gupta, 2015). Today, no city exists and the temple is surrounded by thick forest. It is thought that the city was devastated due to some unknown natural calamity (Sharma, 2007) but the temple has survived to this day. At this site, some old fossils, terracotta utensils and images have been recovered from the ruins around the temple. Initially, the temple was a very small structure with a Mahamaya deity inside. The temple was repaired at the time of the second Dogra Maharaja, Ranbir Singh, and new additions are still being made by local *pujaris* and the Mahamaya Temple trust.

### 5.3 *Cultural centres, archives, library and museums*

At present, there are several cultural centres in Jammu city that preserve and promote the art, architecture and cultural identity of the city and the region. Four government organisations including the Academy of Art, Culture and Languages, Kala Kendra and the Institute of Music and Fine Arts are helping in the education and promotion of preservation of art, culture and architecture of the region and state. Department of Museology University of Jammu and Dogri Sanstha Jammu are also promoting cultural heritage of Jammu. The Department of Archives, Archaeology and Museums is responsible for the identification, conservation and restoration of old heritage structures, manuscripts, historical books and records. In addition, the Archaeological Survey of India is also engaged in excavations, preservation and restoration of sites of archaeological importance. In the modern architectural landscape of Jammu, it is expected that such institutions will find space in the upcoming Jammu Master Plan and the Smart City Plan for the future, (Jammu Municipal Corporation, 2016).

Figure 9. Mahamaya temple.

## 6 CONCLUSION

From the long history of Jammu, it appears that the layout plan of old Jammu was based on the Indian *Vastu Shastra*, an ancient science of architecture and town Planning. Its palaces and forts indicate influences of local Dogra, Rajasthani, Mughal, Kashmiri, French and British Baroque architecture. Wall paintings are in the Basholi miniature painting style. Due to the number of temples and shrines, Jammu is rightly named as the - "City of Temples" - from ancient times. Temple architecture is in the Indian Nagara style. The city is expanding at a very fast pace and new Master Plans are being framed by the government. New colonies are being planned with good road networks, parks and drainage systems. High rise buildings are being constructed. The architectural landscape is changing rapidly from the old style to modern designs and styles. The state government is in the process of upgrading the city to a Smart City, keeping in view the modern concepts of architecture and also preserving its existing old Dogra art and architectural heritage.

## REFRENCES


Agrawal, R.C. & Gupta S., (2009), *Restoration and adoptive reuse of Mubarakh Mandi Complex Jammu,* New Delhi-India: INTACH.

Atkinson, A., (1898), *Report on the outbreak of fire in Mubarakh Mandi Jammu and its restoration,* Retrieved from archives repository, Jammu.

Brown, P., (2014), *Indian Architecture*, (Buddhist and Hindu Periods), Delhi, India: K.R.J. Book International.

Charak, S.S., (1985), *A short history of Jammu,* Pathankot, India: Ajaya Prakashan Publishers.

Chaudhary, P., (2007), *The walled city of Jammu: Changing landscape*, Jammu, India: Saksham Books International.

Chaudhary, P., & Katoch, J.S., (2008), *Mubarak Mandi Palace: Inheritance of an ailing heritage*, Jammu, India: Saksham Books International.

Drew, F., (1875), *The Jummoo and Kashmir Territories: A geographical account.* Jammu, India: JK Book House Publishers.

Ganhar, J.N., (1973), *Jammu shrines & pilgrimages*. New Delhi, India. Ganhar Publications.

Google (2017), Map of Jammu. Retrieved from www.google.co.in/maps/place/jammu.

Google (2017), Vastu_shastra-wikipedia. Retrieved from https://en.wikipedia.org/wiki/Vastu_shastra.

Goswami, O., & Gupta, A., (2015), *Cultural history of Jammu (duggar da sanskritik ithas)*, Jammu, India. J&K Academy of Art, Culture and Languages.

Jammu Development Authority, (2016), Master Plan of Jammu.

Jammu Municipal Corporation, (2016), *1st draft report of Jammu Smart City and city wide concept plan.* Jammu, India. Housing and Urban Development Department J&K Government.

Katoch, J.S., (2012), Amar Singh Palace: A treasure of Dogra architectural heritage. In Poonam Chaudhary (Ed.), *Heritage and Cultural Routes*: An anthology: (pp. 545–552). Gurgaon Haryana, India: Shubhi Publications.

Mangotra, L., (2013), *Chete diana galian*, Jammu, India: Divya Publications.

Manohar, K., (1971), *Kashmir Hindu, Buddhist and Muslim Architecture*, New Delhi, India: Sagar Publications.

Nirmohi, S., (2016), *Duggar ke Raj Mahal*, Jammu, India: Sahitya Sangam Publishers.

Shakar, K.K., (2012), *Jammu ke prachin mandir*, Jammu, India; Sahitya Sangam Publishers.

Sharma, S.C., (2007), *Antiquities, history, culture and shrines of Jammu*, Jammu, Inda: Vinod Publishers and Distributors.

Singh, D., (2014), *City's fate and region:* A historical exploration of Jammu City. In R.C. Chandna, K.D. Sharma & K.S. Singh (Eds.), Immortal cities: (pp. 63–73). Haryana, India: Institute for Spatial Planning and Environment Research.

*Cities' Identity Through Architecture and Arts – Catalani et al. (Eds)*
 *ISBN 978-1-138-05409-7*

# Nature and physical configuration: A study of topographical influences on the physical configuration of mountain settlements in the Iraqi Kurdistan region

Hoshyar Qadir Rasul
*Engineering Technical College, Sulaimani Polytechnic University (SPU), Kurdistan Region, Iraq*

A.I. Ahmed
*Architectural Engineer, Al-Isra'a Engineering Co., Kurdistan Region, Iraq*

ABSTRACT: The art of architecture is a result of a smooth linkage and relationship between building and nature. Several parameters influencing architecture, formed by topographical factors, contribute in determining the form of architecture, through influences upon buildings and their locations as well. This study will explain the influences of topographical factors that are defined as basic variables in the physical configuration of residential buildings. In this sense, the research problem consists of the lack of knowledge regarding the mechanisms of topographical influences on the physical configuration of human settlements, specifically in mountainous areas. The goals of this research were addressed by exploring the nature of topographical influences on physical configuration, identifying mechanisms of configuration according to the topography of the site, and following a descriptive analytical methodology. The study assumed that topography operates effectively and directly on the physical configuration of human settlements, especially in mountainous areas.

*Keywords*: architecture; building and nature; residential buildings; human settlements; topographical influences

## 1 THE NATURAL AND BUILT ENVIRONMENTS

The natural environment, with all its elements and its components, has created a persistent challenge to human presence and its activities, growth and evolution. This challenge presents as a factor on the formation of physical and economic environments for human communities around the world. The natural environment consists of climate, with all its elements, and topography, inclusive of soil, lithosphere and hydrosphere, with all its surface water and groundwater contents (al-Shami, 1999). The elements of the natural environment do not act independently because they constitute an integrated system. However, the built environment is an expression of culture in material form and the land upon which cities are built is a dynamic surface manipulated to enrich urban culture with varying degrees of success (Davids, 2016). Therefore, these two environments (natural and built environment) have been complicated, integrated and matched with each other. This fact has been consistently demonstrated, both historically and in recent times too.

## 2 HUMAN SETTLEMENTS: SHAPES AND LOCATIONS

Human settlement locations have been considered to be distinct decisions in the process of planning, especially in mountainous areas where strategic location become as a limited factor for development of any settlement in the future. Shapes and locations of settlements in mountainous areas have been determined as follows:

1. Settlements inside mountains: settlements that take a medium location between mountain chains, in valleys or in plains;
2. Settlements in the foothills of mountains: settlements in the foothills or at the bottom of mountains; and
3. Settlements in front of mountains: settlements on flat plain areas facing mountains (Al-Janabi, 1987).

The natural factors impacting spatial or locational indicators due to specific factors are described in subsequent sections.

### 2.1 *Topography impacts*

The factor of topography has two main influences. The first is focused, indirectly, on climate elements and the second has a main and direct impact on topography itself (Golany, 1978). Topography represents one of the most important factors influencing the direction of expansion in mountainous regions. Amedi, in the Kurdistan Region of Iraq, is among the most unique examples that portray this impact, as shown in Figure 1. The city is located on the semi-flat top of a mountain and is surrounded by sharp drops in all directions which leads to the impossibility of expansion in any direction (Regional Planning Board, 1989).

### 2.2 *Impacts of climatic factor*

The great variations in the characteristics of local climate, resulting from variations in topography, restrict possibilities for any site. To select the most appropriate choice, the impact of each element in this process must be identified.

Contour lines, for instance, facilitate the determination of circulation paths so that the form of the physical fabric and the type of settlement are determined as well. When the slope percentage range is between 4 and 10, a parallel or perpendicular form of settlement should be adopted. Alternatively, when the slope percentage range is between 10 and 20, the form should follow the contour lines. When the slop percentage exceeds 20%, emerging the certain type will be necessary (Rasul, 1996).

Figure 1. Amedi settlement—views from two distances and locations.

## 3 CONFIGURATION IN ARCHITECTURE

In addition, there exists a process, initiated by designers, that uses visual vocabulary as an essential element and a principle of design to create masses and spaces in a certain system. Physical or architectural configuration begins with exploring the sensual characteristics of various shapes and relationships, either at the horizontal level or in volumetric composition. This may include some configured values that govern the relationships between masses and architectural spaces (Ibrahim, 1987). Thus, configuration is defined as formulating the shape in a way that produces new features and relationships.

### 3.1 *Principles of architectural configuration*

A principle of configuration is the process of organising various elements in forms and images. On a separate note, it is a process of collecting visual elements within a certain order to achieve coherent work and serves the purpose of a particular and dynamic goal. This plan of regulation and organisation determines which way is better in gathering and combining elements to create a certain effect and distinguish one artistic work from another (Riham, 2006). Proper configuration requires an integrated and coherent unit of production for different elements. In other words, contradiction becomes linked and creates harmonious conditions with a defined method.

## 4 CONFIGURATION IN URBAN DESIGN

Configuration in 'urban design' can be described as the art of creating visual unity between elements of the city. Urban design depends on the organisation of material components of the physical environment. Therefore, it is considered to be a formative art related to the appearance of things, aesthetics, and expressive and symbolic values, as opposed to a science concerned with the functionality and efficiency of work performance in equal measure (Al-Hachim, 1993).

As in architectural configuration, a number of concepts govern urban design configuration in order to deliver perceptible values. According to Zevi, configuration is applicable through a special vocabulary, such as identification, contradiction, symmetry, privacy, size, light, shade and shadow, and rhythms of space and mass (Moughtin, 1992). Alongside materials, methods of construction and technology, changes in design patterns owing to various factors including climate and topography play a significant role in the configuration of a settlement and its organisation.

Apart from topographical and environmental factors, socio-economic issues, such as the lifestyle of the locals, safety concerns, privacy, and construction and material limitations, play a major role in affecting the morphology of rural settlements (Philokyprou et al., 2015). Rapaport (1969) believed that socio cultural and natural factors were the main basis for shaping the built environment; other factors were secondary or adjustable. Rapaport (1969) saw that traditional housing is one of the most successful solutions to applying, maintaining and supporting values and social relations; he deemed it possible through the organisation and configuration of modalities. Rapaport (1969) also found that expression includes all social, cultural and symbolic manifestations that have an impact on form and constitute a given symbolic nature.

## 5 THE RELATIONSHIP BETWEEN TOPOGRAPHY AND PHYSICAL CONFIGURATION

Multiple instances exist where the topography of a site is a basis for comparison to establish the degree of harmonisation and appropriateness (Dieter, 1980). The belief in a humane settlement, especially in rural areas, prepares the ground for beneficial gain from nature and

(a) (b)

Figure 2. (a) Architecture as a man-made creation (bringing nature to perfection); (b) Architecture as a part of nature—Palangan village in Kurdistan Province, Iran.

defines a clear relationship between nature and architecture on the basis of two main aspects: architecture as a part of nature and architecture as a man-made creation in nature.

Architecture requires compatibility and harmony with nature (Figure 2a), especially because architecture, as a man-made creation within nature, expects perfection (Figure 2b). In other words, harmony with nature, and perfection within nature, are two entirely reasonable architectural approaches. In the traditional view, sustainability is achieved through a harmonious association with nature and perfect alignment with topography (Adeli & Abbasi, 2015).

These approaches have been adopted in practice through various mechanisms, such as correspondence, similarity and discord, which are also representative of strategies followed by architects in their daily activities.

### 5.1 *Configuration of traditional settlements*

Traditional settlements have always been related to specific localities where meaningful correspondences between climatic conditions take place, and topography and settlement morphology coexist. Different surface reliefs generate different physical forms and growth patterns in rural settlements. For instance, settlements in a valley usually take linear forms parallel to the direction of the land contours. Settlements on a plain may take on the form of a dense cluster or an enclosure, while settlements on a hill often have the form of concentric or longitudinal clusters forming a series of semi-circular terraces perpendicular to the slope (Philokyprou et al., 2015).

## 6 CASE STUDY

The Iraqi Kurdistan region is famous for its distinct location and topography and this is due to geological phases that have occurred through sequential historical periods. The complexity of topography has increased and is oriented towards the north-east from the south-west. As a result, two distinct regions exist and are referred to as the mountainous and semi-mountainous regions.

The Tawella settlement area has been selected as a case study for this paper. It is classified as lying within the mountainous region for the following reasons:

- its distinct topography;
- its distinct climate, categorised as the climate of a mountainous region; and
- its deeply historical nature and classification as a long-standing and ancient human settlement.

Tawella, an ancient human settlement and boundary gateway, is located in the east of the Iraqi Kurdistan Region (Figure 3). Its history as an ancient human settlement dates back more than 4,000 years. It benefits from beautiful scenery within an area known as Hawraman. Moreover, its geographic location, bordering Iran, and its intensive forests (prolific with hickory trees) have promoted Tawella as an agricultural town.

Tawella is composed of three residential districts divided over three downhill areas of the valley. These three districts are composed of residential groups with buildings aligned in the form of overlapping masses, and penetrated by natural rocks and ridges (Figure 4).

Tawella is famous for its distinct climate. It can be covered with snow during the winter months with temperatures dropping below 0 °C. With a moderate temperature during the spring and summer seasons, it makes an appealing tourist destination (Figure 5).

The mean heights of mountains in Tawella are about 2,000 m above sea level. The highest point of the Hawraman mountain chain is 2,548 m above sea level (Figure 6).

The field study in this paper included coordinate readings and topographic surveys using a total station, X-Y-Z determination, transformation of these coordinates in Microsoft Excel, and then processing of all data using a land development software program in order to construct a topographic map. Total station coordinate readings were also supported by the use of GPS to link the coordinates in reference to sea level readings.

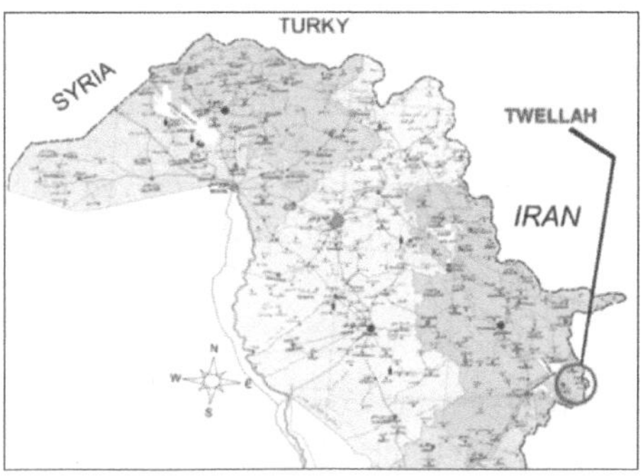


Figure 3. Map showing the location of Tawella.

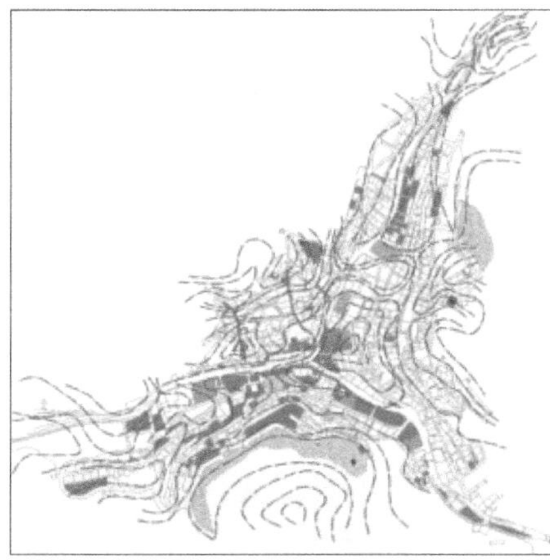

Figure 4. Plan of Tawella with topographical lines.

Figure 5. Tawella settlement.

Figure 6. Topography of Tawella.

Table 1. Basic information for Tawella.

| | |
|---|---|
| Overall area of the site | 262 ha |
| Number of residential sector | 3 |
| Population number | 3,260 |
| Number of dwellings | 430 |

## 6.1 *Tawella physical configurations*

The physical configuration of the study area varies according to differences in the surrounding environment. These variations are reflected through differences in colour, texture, materials and overlap with nature (Figure 7).

In order to discover more about the mechanisms and characteristics of the study area, the formative structure was studied through the analysis of configuration characteristics, the rate of influence for each element, and an analysis of the results. During the field study and the comprehensive site survey, a representative residential district within the study area was selected as a sample.

## 6.2 *Residential units with land formation relationships*

The existing residential units within the study area have two main types. The first type is a perfect correspondence with the land formation and the second is at odds with the land formation. Because of the steep regression of land form the residential units will be free from the ground line, it means that the relationship between the ground line and residential units has been lost.

Figures 8a and 8b illustrate the gradient level, mechanism of configuration and the relationship of residential units with perfect correlation or semi-perfect relationships with the land formation. All of these comparisons show land formation on one side and residential units located on the other. In addition, there exists a dissonant relationship with land formation, as done with the dwelling units set out above the other blocks which is free from any relation to the formation. Huge and rigid bodies of mountains have been carved into, creating small cramped beds with beautiful, lofty and sturdily built structures. The stones

Figure 7. Variation in texture, colours and materials.

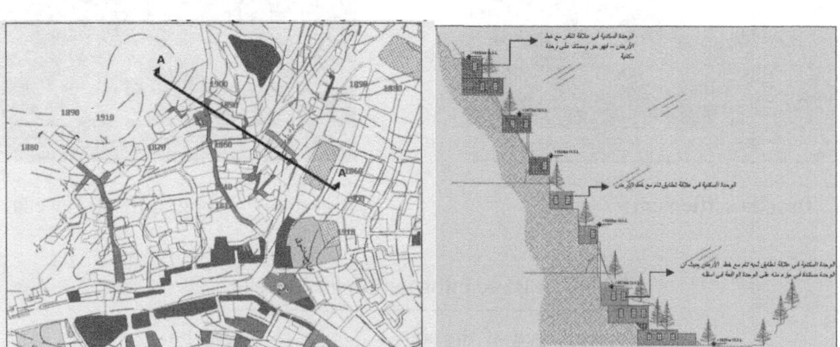

Figure 8. (a) Study area sector plan with topography lines; (b) Section A–A: The topography of the study area and the relationship patterns with land formation.

taken from that area are composed of completely vertical and plain surfaces (Rahmatabady & Amjadian, 2015).

### 6.3 *The relationship between physical configuration and site topography*

To determine the impact values of the existing relationship between physical configuration and site topography, it is necessary to study the relationship between configuration elements and the surrounding environment using descriptive and formal measurements.

#### 6.3.1 *Configuration with site topography*

The relationships between configuration elements were conformed and interconnected. Thus, when the configuration lines were compared with topographical lines, an embodiment of topographical lines within the physical configuration was noticed. The extracted data was supported by a detailed survey, as shown in Table 2.

The analysis process used was as follows:

- a descriptive scale (junk scale) was used, which deals with determining values of (X, Y) where (Y) represents the value of configuration lines in the sector and (X) is the configuration of lines that are directly in contact with the topographical line;
- the data was converted to maps using the Autodesk Auto CAD, Land Development software program; and
- the layers system was used to match lines from existing situations with topographic lines (Figure 9).

The common lines were determined by the ratio X/Y where X = 546.5 ml (the configuration lines were in touch with topographic line) and Y = 723.2 ml (the general configuration lines). In other words, the ratio is approximately ¾ (546.5/723.2 = 0.756). Therefore, the relationship type is 'partially overlapping'.

Table 2. The length of contour lines within the sector and general configuration.

| | Contour lines within the sector | Length of contour line within the sector (m.l.) | Length of general configuration (Y) (m.l.) | Length of general configuration (X) (m.l.) |
|---|---|---|---|---|
| 1 | 1st contour line (1810) | 138.0 | 142.0 | 102.1 |
| 2 | 2nd contour line (1840) | 190.1 | 163.2 | 123.2 |
| 3 | 3rd contour line (1860) | 180.4 | 152.3 | 111.3 |
| 4 | 4th contour line (1870) | 168.9 | 140.0 | 98.5 |
| 5 | 5th contour line (1890) | 120.0 | 83.0 | 68.5 |
| 6 | 6th contour line (1900) | 62.0 | 42.7 | 41.9 |
| | The sum of lines | 859.5 | 723.2 | 546.5 |

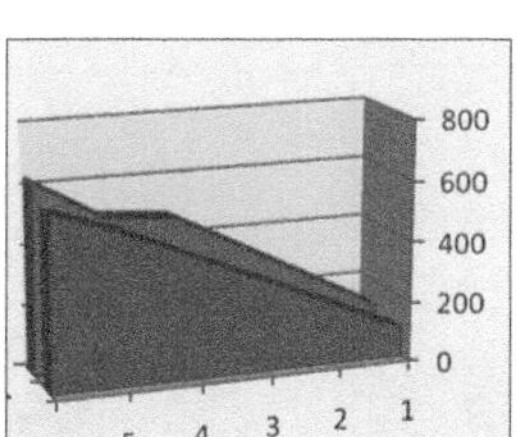


- Blue colour: General configuration lines which are attached with topographic line.
- Red colour; General configuration lines.

Figure 9. The relationship graph for configuration lines with topography.

6.3.2 *The relationship between shapes and site topography*

It was observed that shapes which were generated within the physical fabric, represent the natural environment parameters. These shapes located in a hierarchy situation within the blocks and overlapped with other shapes composing the whole configuration.

To determine the extent to which they overlapped, the formal scale was used for comparisons between the overlapping of shapes, their gradual configuration, and the nature of the overlap with the topographic gradient. The steps followed were:

- samples with six residential units within the general configuration were taken;
- the existing situations were converted to 3D shapes with landform realisation;
- to identify values that were to be analysed in order to determine the overlapping rate among modal blocks, the blocks were analysed and their areas were found using the Autodesk 3D-MAX program;
- the overall areas of shapes—within blocks – were 2,538 $m^2$, which is denoted by (X);
- the area of overlapping shapes within the configuration was 279 $m^2$ (Figure 10a); and
- the area of shapes that overlapped with each other, in the absence of an effective topography, was 414 $m^2$ (Figure 10b).

After studying the ratios, the results were as follows:

- in the case of the presence of topographic impacts, the ratio X/Y = 279/2538 = 0.11;
- in the case of the absence of topographic impacts, the ratio X/Y = 414/2538 = 0.163;

- after comparing these two values, the impact of topographic factors was noticed when identifying the types of shapes overlapping. In the case of a gradient with an angle of 71°, the ratio of shapes that overlapped was decreased by 32.5 per cent; and
- after comparing the block configuration (Figure 11a), the area of blocks overlapping over the land was calculated and compared with the area of configuration overlapping with land (without the topography lines i.e. flat plain area), as shown in Figure 11b.

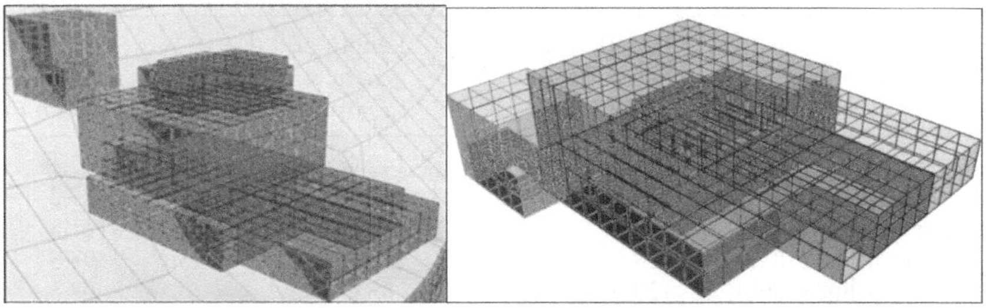

Figure 10. (a) Shapes overlapping together with topography within the configuration; (b) Shapes overlapping together without topography within the configuration.

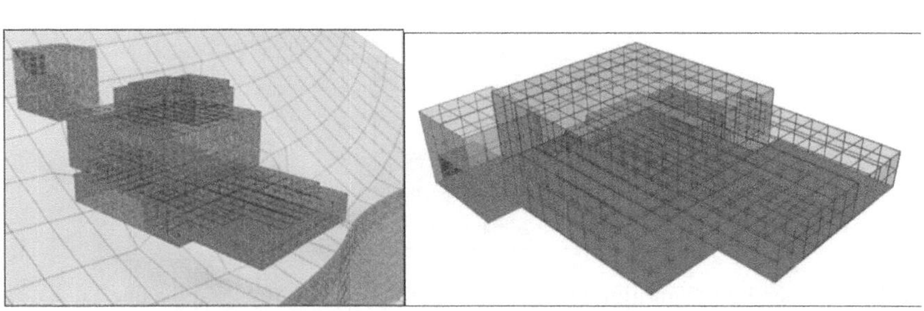

Figure 11. a) Area-shape overlapping with topography within the configuration; (b) Area-shape overlapping without topography within the configuration.

Table 3. he results of the cases studied within the practical part of the study.

| | Main issue studied | Related issues | Results | |
|---|---|---|---|---|
| 1 | Dwelling units with ground line relationships | | Perfect correspondence for units below the road level<br>Discord for units above the road level | |
| 2 | Configuration lines with site topography | | Ratio of relationship = 0.756<br>Relationship type is partially overlapping | |
| 3 | **Configuration—Site topography relationships** | | | |
| 3.1 | The shape with site topography relationships | A: shapes overlapping for masses with each other inside the fabric | Topography considered | Ratio = 0.11 |
| | | | Topography not considered | Ratio = 0.163 |
| | | | After comparison and analysis | Decreased by 32% |
| | | B: shapes overlapping for masses with ground lines | Topography considered | Ratio = 0.11 |
| | | | Topography not considered | Ratio = 0.874 |
| | | | After comparison and analysis | Increased by 37.5% |

The ratio results were as follows:

- In the case where the topographical impacts had been considered:
  - he area of overlap = 409.5 $m^2$;
  - he overall area of masses overlapping = 819 $m^2$; and
  - he X/Y ratio = 409.5/819 = 0.5.

- In the case where the topographical impacts were not considered:
  - he area of overlap = 716 $m^2$;
  - he overall area of masses overlapping = 819 $m^2$; and
  - he X/Y ratio = 716/819 = 0.874.

Therefore, masses with land form relationships is (overlapped relatively), when the topographic factor impacts were considered. The ratio of overlapping decreased by 37.5% rather than the situation if the topographic factor impacts has been neglected..

### 6.3.3 *Relationships between site topography with inside-outside spaces*

The physical configuration of the study area is featured in two types of spaces: interior spaces, which are represented in the dwelling units, and external spaces, which are represented by the axial meeting of zigzag alleys which are linked to one another. These spaces are formed as (private-level) nodes that involved in facilitating movement and circulation toward parts that take concurrent tracks and parallel to contour lines due to site topography.

There are not any types of median spaces due to a steep inclination that sometimes exceeds 65°. Thus, there emerges a need to build retaining walls in order to exploit those areas rather than rely on stairs for movement. In terms of space linkages, the principle of stacking (compactness) for a group of masses within the same fabric was noticed. This is because of mountain canyons that limit continuous stacking and shifting from one level to another (with a noticeable difference leading to a gap between blocks within the fabric, see Figure 12a).

Moreover, if a lump within the fabric is taken into consideration, the ratio of mass overlapping will be too large and will lead to a high partial overlapping ratio such as ¾, as shown in Figure 12b. The lower surfaces of masses used as corridors between lower and upper levels. On the other hand, some of the dwellings (with two floors), which are located in the rough edges, their entrances will be from the upper floor. This is mainly due to insufficient space in front the unit due to topography. Figures 13a and 13b).

Figure 12. a) Dwellings aligned within a block and spacing between existing blocks; (b) An example of dwelling alignment and masses overlapping.

Figure 13. a) Moving from one unit through the use of stairs; (b) A dwelling located on a hillside.

## 7 CONCLUSIONS

This study led to a set of conclusions that encompass several important points. The main aspect of architecture, in terms of relationship to nature, may be summarised in two main approaches. The first approach is architecture as a part of nature (harmony with nature), and the second is architecture as a man-made creation in nature (nature perfection). The models studied in this paper were on longitudinal strips, and consisted of compacted and balanced residential units which represent the outcomes of topographical influences on configuration.

Moreover, the configuration lines of residential units overlapped with land form within the fabric in two patterns. The first created a perfect correlation while the second gave a partial correlation, represented by a configuration of residential units perpendicular to topographical lines. The shapes generated within the physical configuration in the mountainous region were determined by gradient and featured overlaps between each other in ratios less than those in flat areas. Additionally, the overlap ratio between masses and shapes, which were composed from the configuration with land form in the mountainous region, was changed. The ratio, depended on the (rate of changing) in topographical factor. This proportion ratio was also inverted when horizontal overlapping occurred. However, overlapping on a vertical basis increased proportionality.

Finally, spaces of physical configuration in the mountainous region have special features in terms of overlapping with masses. These spaces interfered horizontally, were parallel to each other and were accessed through open front aisles or corridors. Corridors within the physical configuration were in harmony with the topography. In residential configurations with rugged topography (represented by a slope angle above 50°), corridors (movement paths) vanished and access was limited to stairs and the upper parts of the buildings.

## REFERENCES


Adeli, S. & Abbasi, M. (2015). Approaches to nature in Iranian houses in terms of environmental sustainability. In C. Mileto, F. Vegas, L. García & V. Cristini (Eds.), *Vernacular architecture: Towards a sustainable future* (pp. 33–38). Leiden, The Netherlands: CRC Press/Balkema.

Al-Hachim, M.A. (1993). *The impact of urban form on the feeling of place* (Master's thesis, Baghdad University, Baghdad, Iraq).

Al-Janabi, S.H. (1987). *The urban geography.* Beirut, Lebanon: Dar al-kutub Foundation.

Al-Shami, H.A. (1999). *The impact of natural factors on planning and designing of human settlements in the mountainous region* (Master's thesis, Higher Institute of Urban and Regional Planning, Baghdad University, Baghdad, Iraq).

Davids, R. (2016). Working with the lay of the land. *Berkeley Review of Latin American Studies*, Spring–Fall 2016.

Dieter, Prinz (1998). *Stadtbaullches, entwerfem*, (Urban Design) (p. 95). Stuttgart, Germany: Band 13.

Dober, R. (1969). *Environmental design.* New York, NY: Van Nostrand Reinhold.

Golany, G. (Ed.). (1978). *Urban planning for arid zones* (p. 11). New York, NY: John Wiley and Sons.

Ibrahim, M. Abdul-aal. (1987). *Environment and architecture, Arabic architecture chain 5* (p. 115). Beirut, Lebanon: Dar Al-Ratib.

Moughtin, C. (1992). *Urban design: Street & square* (p. 33). Oxford, UK: Butterworth Architecture.

Philokyprou, M., Savvides, A., Michael, A. & Malaktou, E. (2015). Examination and assessment of the environmental characteristics of vernacular rural settlements: Three case studies in Cyprus. In *Proceedings of the International Conference on Vernacular Heritage, Sustainability and Earthen Architecture* (pp. 613–618).

Rahmatabady, S.S. & Amjadian, K. (2015). The impact of mountainous geography of west of Iran on architecture and form of Hawraman dwelling. *ISJ Journal of Architecture and Engineering, 1*(1), 91–100. Retrieved from http://architecture.scientific-journal.com/abstracts/1/14.pdf.

Rapaport, A. (1969). *House form and culture* (p. 69). Englewood Cliffs, NJ: Prentice-Hall.

Rasul, H.Q. (1996). *The optimum approach for planning of residential areas in the mountainous region* (Master's thesis, Higher Institute of Urban and Regional Planning, Baghdad University, Baghdad, Iraq).

Regional Planning Board. (1989). *Problems of Iraqi cities expansion and their impacts on regional settlements distribution* (No. 759, p. 31). Baghdad, Iraq: Ministry of Planning.

Riham, F. (2006). *Proposal for historical fabric rehabilitation in the mountainous areas in Syria* (Master's thesis, College of Architectural Engineering, Damascus University, Syria).

*Cities' Identity Through Architecture and Arts – Catalani et al. (Eds)*
*© 2018 Taylor & Francis Group, London, ISBN 978-1-138-05409-7*

# The effect of social and environmental factors on the urban form of a place: A fishing village in Hurghada, Egypt

Eman Faiez Maher Bassily
*Department of Architecture, Faculty of Fine Arts, Helwan University, Giza, Egypt*

ABSTRACT: This research considers how to improve the urban quality of the city fabric to restore its identity using the hypothesis that environmental and social characteristics of people in certain communities have a great impact on the urban form of a place. This research is based on a number of facts, including that the design process of housing and construction materials used in a fishing village are related to the residents' main activities and social life. The main problem is represented by the unplanned actions implemented in these communities that ignore the social and environmental factors that should be studied before implementing any changes. The research methods used included surveys, site visits to both the existing village and the new one built by the government, analysis and observations. The main findings of the study revealed that unorganised and haphazard changes to these communities cause them to lose their characteristic local socio-environmental urban form and architecture. It is concluded that sustaining the place would preserve the environment and the social life of people and, consequently, conserve the urban form and spatial characteristics of the place, which in turn gives it its heritage, activity and identity.

*Keywords*: place; identity; environment

## 1 INTRODUCTION

The urban form and architectural characteristics of a place are affected by the social life of the people living in it and the general environment of the place itself. In addition, the socio-economic characteristics of a place and the environment affect each other to different degrees; this, therefore, results in the urban form of a place that also reflects its environment. This effect is seen in three main urban components: urban planning (urban form of buildings, streets and services), urban open spaces and the architectural design of residential buildings.

Hence, the research is based on the hypothesis that the environment and the social characteristics of people in any community have an effect on the urban form of a place. Any change in the environment and/or the social characteristics of the people affects the final result, that is, the urban and architectural product, its form, shape and style. It is therefore important for a researcher to recognise the aforementioned concept and apply it to any given community in order to identify its urban environment (Almansory, 2001).

This research hypothesis can be applied to villages that have unique characteristics, such as fishing villages. These villages are characterised by their primitive and traditional communities and ways of life. They have clear social, cultural, and economic characteristics, which are considered to be the most important factors that have an impact on the urban form.

## 2 THE RESEARCH PROBLEM

The initial main problems identified in this study were unplanned actions, lack of studies when implementing changes in the urban form of these communities, and ignoring social and environmental factors. In terms of modern technology, changes that take place in communities are considered necessary, but the social and cultural factors of these communities change alongside such changes, which also alters urbanism. However, the change that occurs should not negatively affect the characteristic features of any community, especially when they relate to heritage and social life. In some cases, the government is responsible for changing the urban identity of certain places through haphazard alterations that are implemented regardless of the culture, social life, heritage of the place, and the natural environment. This in turn causes them to lose their characteristic urban form, which should be preserved as a cultural heritage that brings life to the place.

The problem has four dimensions and is discussed in the following subsections.

### 2.1 *The national dimension*

There is no strategy to upgrade villages of a certain value. Instead, the government applies certain rules which negatively affect the social and cultural character of a place, and neglects its role in the development process.

### 2.2 *The planning dimension*

When renovating and restructuring a community the social factor is ignored, which repels the original residents and results in a new community without people.

When the Egyptian General Authority of Tourism Development planned the development of the coastal districts, they ignored the existence of the local fishing villages and concentrated only on tourism projects. As a result, many fishermen that lived in the new tourist districts along the eastern coast from Safaga to El Qoseir started working in the tourism sector.

### 2.3 *The social dimension*

The original residents are sometimes unaware of their surrounding cultures.

### 2.4 *The economic dimension*

The increasing interest in the fishing village is primarily due to the importance of fishing and the fish crop within Egypt. This is a result of a lack of animal protein sources, and the existence of large areas of water and coasts in Egypt.

## 3 THE AIM OF THE RESEARCH

This research aims to improve the urban quality of the city fabric, to restore its identity. This could be achieved by identifying social, environmental and economic principles that should be considered by planners and designers when starting to plan new communities for people. In addition to the main purpose of the research, there are a number of secondary aims as follows:

- understanding and realising the nature of change in primitive communities and the factors affecting that change as well;
- restoring the architectural and urban character of these communities;
- increasing the effectiveness of such communities in order to increase the national income;
- attaining a degree of balance between modernism and valued heritage. The study of the social life of the local people will help in acquiring the knowledge required.

## 4 THE COMMUNITY

People generally prefer to live in groups because this makes them feel safe. These groups are organised by a number of frameworks that govern the relationships between different individuals within the group. These frameworks are made up of a number of rules, habits, and customs (Mohamed, 2004). The idea of the community originated from the aforementioned concept of groups. A community is defined as a group of people ruled by a group of relationships that have the same framework, characteristics and benefits (Zaki, 1986).

## 5 THE SOCIAL ENVIRONMENT

The social environment is a part of an inclusive environment that contains individuals and groups and their interactions. It also includes social relationships and other types of social order. The social environment therefore, consists of a community with all its social relationships and types, and it is the space that is made up of people with organised social relationships and bonds. In addition, the social environment is considered to be all the social and cultural factors affecting individuals or groups.

Social action is a form of human behaviour and is always variable, differing from one individual to another because it is based on psychological and cultural values, and the surrounding community (Wasfi, 2003).

### 5.1 *The elements shaping the social environment*

The following elements define the differences and similarities between social environments and they are necessary to understand a specific community's social environment. One or two of these elements could work together, the most important of which include:

- demographic characteristics;
- social layers and standard of living scale;
- women's role in the society;
- education.

## 6 THE URBAN ENVIRONMENT

The urban environment is a part of its surrounding environment. It is the physical product resulting from the interaction of people with the environment in order to fulfil their requirements given their cultural and social background.

### 6.1 *Types of urban environments*

The urban environment can be classified as a primitive traditional urban environment or a planned urban environment.

The differences between the traditional and planned urban environments arise because of the dissimilarities between people and the surrounding environment of each. The traditional urban environment is one that has a direct relationship between the people and their environment. On the other hand, the planned environment has no direct relationship between the people and the environment. This is because of the influence of a number of factors,

Figure 1. User participation in the construction phase creates social links between users. (Source: Researcher).

resulting in the final product being considered a planned one according to the policies and attitudes of the government and not fulfilling people's needs of their environment.

The differences between traditional and planned environments are illustrated in the following subsections (Ibrahim, 1999).

### 6.1.1 *The stage factor*

The traditional environment is characterised by its staged growth, since the construction is related to people's needs. Whenever their needs increase, the construction process continues. However, the planned environment is characterised by its completed stages of construction according to previously calculated needs. Consequently, the urban form of the planned environment appears in its final form from the beginning.

### 6.1.2 *The designer factor*

A designer does not exist in the traditional environment, because the user of the place is the one who designs and forms the urban environment according to the needs and requirements of the people. This takes place with the help of local builders. The final product therefore depends on the character of the users and their culture (Figure 1).

### 6.1.3 *The construction factor*

In the traditional environment, the customs, habits and traditions are the main factors in the construction process. This defines the distance from neighbours, openings, heights, direction, use of building materials, and certain building solutions for entrances or roofs. In the planned environment, the rules and regulations of buildings are the main factors in the construction process that define the heights and distances from other neighbours.

## 7 VILLAGE DEFINITION

A village is a social unit with a defined area with strong social characteristics; people's activities are related to the surrounding environment. The village could be a 'country-community' if its activities are related to agriculture, and is not a country-community if it is related to any other sector of activity. However, it will always be defined as an area that has the same characteristics as the countryside (Abd Elhamid, 2000). The village community is classified according to a number of factors as follows:

- area classification according to statistical factors;
- economic classification;
- form classification;
- administrative classification;
- classification based on relationship to the city.

On the basis of these factors, an ascending classification of local communities, as depicted in Figure 2, can be created as follows.

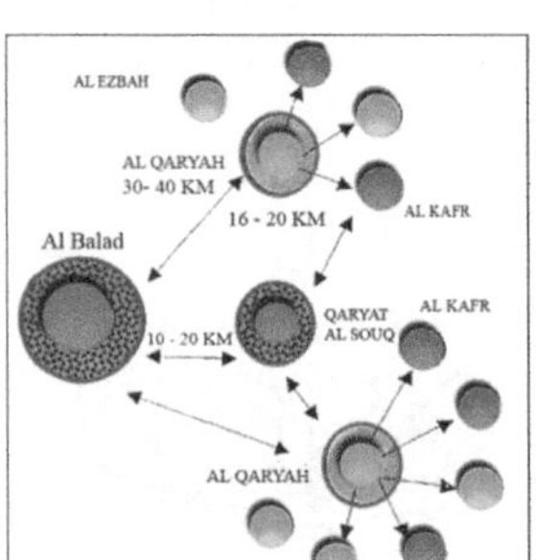


Figure 2. Urban communities hierarchy.

### 7.1 *Al Ezbah*

This is the smallest regional unit spread in the countryside on separate distances from the villages. It depends on surrounding villages for its services and its population is between 30 and 100 people.

### 7.2 *Al Kafr*

This settlement is bigger than Al Ezbah, but it resembles it in function. Its population is between 50 and 100 people.

### 7.3 *Al Qaryah (the village)*

This village serves the surrounding settlements of Al Ezbah and Al Kafr. Many of its residents work in handicrafts, carpentry and blacksmithing, as well as agriculture. There are a number of services, such as businesses, a mosque and a church. Its population is between 300 and 800 people.

### 7.4 *Qaryat al Souq (the market village)*

This village hosts the main market that takes place once or twice a week. It is located between Al Ezbah and Al Kafr and other villages benefit from its services over a distance of 16–20 km. It is a centre for commerce and product exchange. Its population is between 800 and 3,000 people.

### 7.5 *Al Balad*

This settlement is a developed unit between the villages and the city and provides a number of services, but not to the level provided by cities. Its population is between 3,000 and 8,000 people.

## 8 THE FISHING VILLAGES (GENERAL CONTEXT)

These are the villages where fishing is the major economic activity. The main components of the urban form of the fishing villages are as follows:

- building mass – most of them are a group of adjacent buildings with a view of the sea or lake, since fishing is the main activity;
- residential housing for the workers – consist of a reception and living area, bathroom, kitchen, bedrooms, a storehouse for fishing tools, and an open court with fencing. The building materials used are stone and palm fronds (Figure 3);
- main open space – a square that holds the most important elements, such as the market, mosque and shops (Figure 4);
- secondary residential open spaces – found inside the residential clusters or in areas surrounded by houses in order to provide a form of social interaction between people (Figure 5); and
- fishing activity related services – found in the newly constructed villages and concentrated next to the harbour. Refrigeration units are provided for the preservation of fish. Small ice factories, sales centres and small markets can also be found here (Figure 6).

### 8.1 *The factors affecting the architectural and urban form of the fishing villages*

In the history of art and architecture, it is the case that there are a number of factors that affect any architectural or structural style appearing in any place at any time in history. These factors define the character of this style in an indirect way. Among these factors are social and environmental ones. The social factors consists of sub-factors such as kinship, religion,

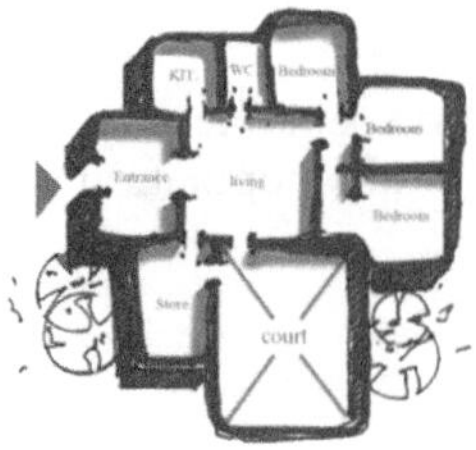

Figure 3. Plan of a typical fisherman's house.

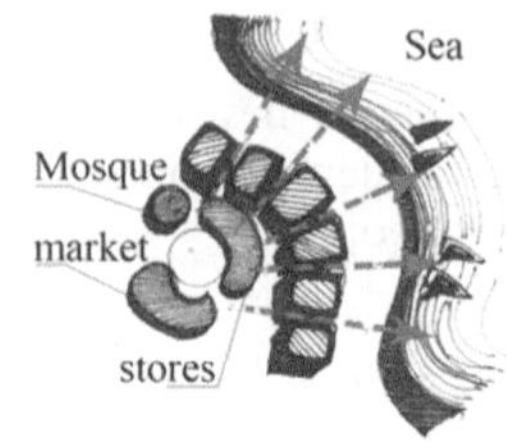


Figure 4. Plan of main open spaces.

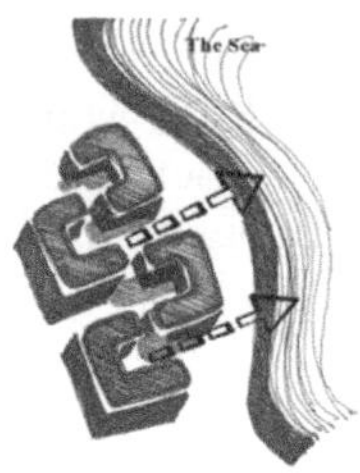

Figure 5. Plan of residential open spaces.

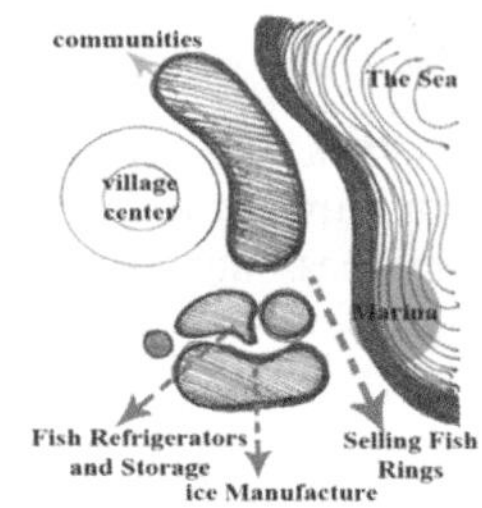


Figure 6. Plan of fishing activity related services.

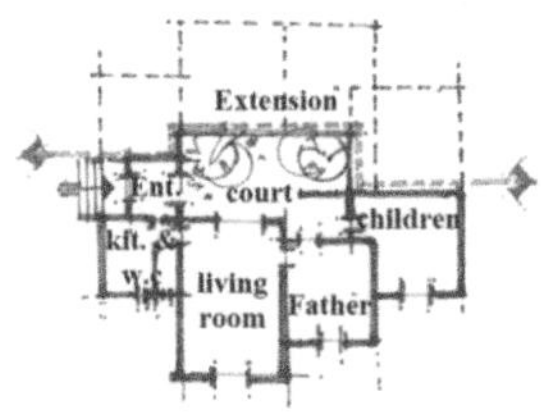


Figure 7. A plan of a typical *diwany*.

culture, customs, traditions, popular beliefs and socio-economic effects. The environmental factors, such as climate, contribute to shaping the place and the materials used in buildings.

### 8.2 *The influence of the kinship factor on architectural and urban form*

The kinship factor plays an important role in village life, because the people within a group of their own kin perform all their activities together. The group has an economical, religious and political role.

This factor plays an important role since it defines the individual's membership in the group, and his or her responsibilities and duties towards others. It has an influence on the architectural and urban form and it takes into consideration the next generation that will later live in the place. For example, the design of the house has a large court for the future generation. The *diwany* is a small house inside another larger future house, consisting of an interior open space such as a living room, a dining area, a kitchen, a bathroom, and a bedroom. The *diwany* (Figure 7) is a future house designed for the owner's daughter for when she gets married (Suzan, 1999).

### 8.3 *The influence of the religious factor on the architectural and urban form*

The religious factor is reflected in the architectural and urban product of a place. Privacy is reflected in the design of entrances to houses, in addition to the existence of another secondary open space to serve the residential buildings (Figure 8). The mosque forms the most obvious and characterful location in a village; its form is very simple and it controls the visual form of the place with its minaret.

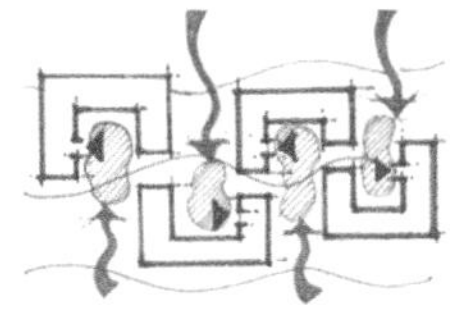

Figure 8. Incorporating privacy in house design.

### 8.4 *The influence of the cultural factor on the architectural and urban form*

It is important to study the culture of any community because it reflects that of its people. It is also important to study the relationship between culture and the architectural and urban form of a place to understand the influence of culture on a community. For example, the indirect entrance of a house prevents people outside the house from seeing others inside; this ensures the privacy of residents and fulfils their characteristic needs in their daily life. Additionally, the drawings on doors and walls convey certain meanings related to the community's culture. For example, the drawing of a lion supposedly prevents evil forces from getting near the house, the alligator represents privacy and protects residents from envy, and triangular shapes protect people from the surrounding natural environment of mountains (Tavakolian, 1993).

### 8.5 *The influence of socio-economic factors on architectural and urban form*

The economic factor is one of the social factors that affects the community; it is made up of secondary factors such as forms of ownership, means of production, consumption, and exchange and work distribution. It is a group of ideas used to fulfil the primary requirements of people according to the surrounding natural environment. The social-environmental factor is reflected upon the architectural and urban form. For example, the planning of a number of open spaces in the urban form of a village is classified according to the activities of people. These villages could have markets, fish products and services to fix fishing nets, and there are also pathways that connect people who work within the same activity. The design of a house has a space for fixing fishing tools and nets, and also has a storehouse for fishing objects. The economic situation is reflected in the building material used for the house, along with the way it is constructed, and the fact that women and the whole family contribute to the construction process.

## 9 THE FISHING VILLAGE OF HURGHADA

### 9.1 *Location*

The village is located in Hurghada, Egypt, and the local residents originally worked in fishing activities due to its location on the Red Sea. However, after Hurghada's main activity was changed to a touristic one, the percentage of people working in the fishing industry decreased and the number of people working in the tourism sector increased (Figure 9).

### 9.2 *Pedestrian pathways in the fishing village*

The pedestrian network has fulfilled the gradient of pathways network, the strength of network and its main function.

### 9.3 *Analysis of general services*

Based on a survey conducted by the authors, the different services related to activities that take place in the village were analysed (Table 1).

## 9.4 *Social and economic characteristics*

The survey also analysed the socio-economic data (Table 2).

## 9.5 *Influence of social factors on the urban form of the fishing village of Hurghada*

Table 3 illustrates the influence of social factors on urban form, as detailed in the following subsections.

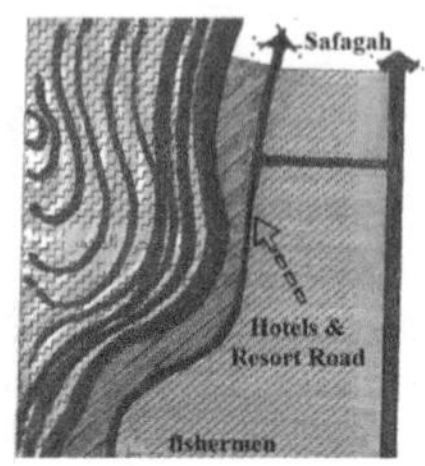


Figure 9. Location of the fishing village of Hurghada.

Table 1. Fishing village activities.

| Percentage | Area | Service type |
|---|---|---|
| 42.5 | 1.7 | Education service |
| 16.5 | 0.65 | General daily activities |
| 12.5 | 0.50 | Fishing-related services |
| 11.5 | 0.45 | Religious service |
| 10 | 0.4 | Health service |
| 7 | 0.35 | Administration service |
| 100 | 4.00 | Total |

Table 2. Socio-economic characteristics (Environmental Characterization for Red Sea Governorate, 2004).

| Socio-economic characteristics | | | |
|---|---|---|---|
| Village area: | 14 acres | Agriculture employment: | 10% |
| Population: | 2,500 | Fishing type: | Collective |
| Density: | 178 p/m | Fishing tools: | Single |
| Average size of family: | 7 persons | Main sales: | For tourist villages |
| Unemployment level: | 0 | Fishing and admin.: | Collective |
| Main economic activity: | Fishing | Income average: | EGP 400 |
| Fishing employment: | 50% | Women in education: | 10% |
| Tourism employment: | 35% | Illiteracy rate: | 50% |
| Commerce employment: | 5% | | |

Table 3. The influence of social factors.

| Village form | Open spaces hierarchy | Architectural design | Residential clusters |
|---|---|---|---|
| 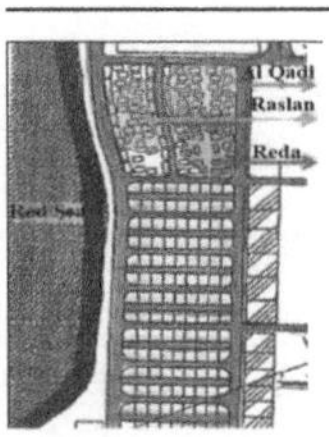 | 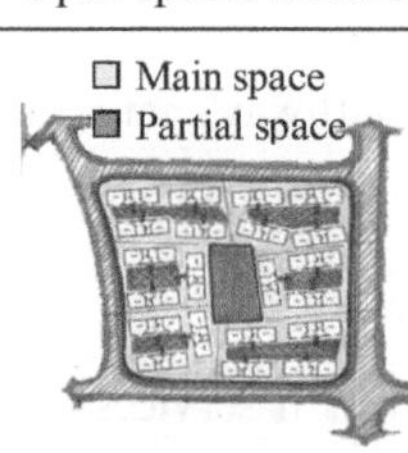 | 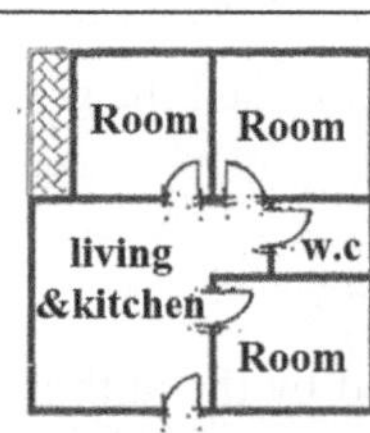 |  |

9.5.1 *Village form*

The kinship factor influenced the form of the village where there are three major families in the old core of the village (the Raslan family, the Al Qadi family and the Reda family).

9.5.2 *Open spaces hierarchy*

The residential buildings are grouped together, where each group is made up of a secondary open space where the fishermen perform their social activities.

9.5.3 *Architectural design*

The extended family influenced the design of the houses in the village.

9.5.4 *Residential clusters*

The locations of entrances to houses were designed in an alternative way and direction to ensure privacy.

## 9.6 *Influence of environmental factors on the urban form of the fishing village of Hurghada*

9.6.1 *Land use*

The old village had different land uses due to its location, and the mosque is located directly on the coast due to its importance to people there.

9.6.2 *Architectural design*

The non-existence of a 'reception' area in the house, which is related to religion, given that fishermen spend months away from home at sea.

9.6.3 *Building materials*

Stone, and palm fronds and trunks were used as the building materials. This was due to the low income of the residents. These materials were brought from the surrounding environment.

Table 4. Comparison of the old and new village.

| | The new village | | The old village | |
|---|---|---|---|---|
| Building material | The building material is mainly concrete. |  | Walls are made from stone and roofs are made from palm fronds. |  |
| Location and urban form | Its location on the sea made fishing the main activity. | 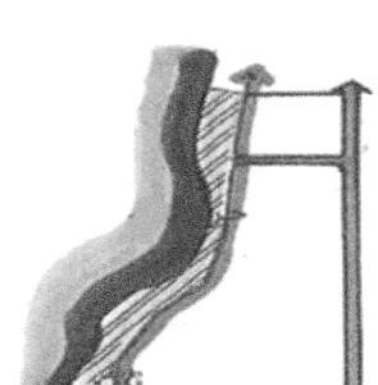 | Random urban form due to ownership system and the surrounding environment. | 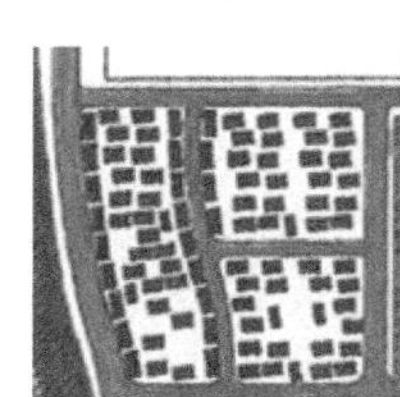 |
| Land use and open spaces | Workshops to fix boats. | 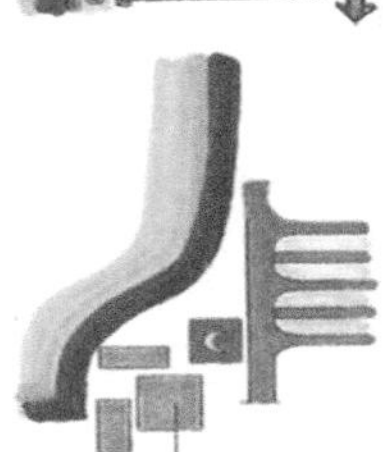 | Secondary open spaces used to fix boats. |  |

### 9.6.4 *Village form*

Due to the ownership system, the urban form and the buildings appeared in an irregular plan; however, this protects the village from different changes in the weather across the year and due to its location on the coast.

### 9.6.5 *Open spaces*

The existence of secondary open spaces, used to fix boats and to create good air circulation, plays an important role in lowering the summer temperatures.

Table 4 illustrates, in brief, the comparison between the old village and the new one built by the local government for the people.

## 10 RESULTS AND CONCLUSIONS

The main findings of the study revealed that the unplanned changes implemented on these communities caused them to lose their unique local socio-environmental urban form and architecture. The results of the relationship between the social factors and the environment showed that proximity to the city and tourism resorts affected urban form. Other factors have also affected the urban form, such as cultural factors, which are represented in traditions and customs. This has especially affected houses, building materials, spaces and their hierarchy. The kinship factor also affects the form of the house and, consequently, the village. However, the religious factor was not found to have a great effect on the overall urban form, but it affects the design of individual houses.

It is concluded that sustaining a place would preserve the environment and the social life of the people and, consequently, conserve the urban form and spatial characteristics of the place. These, in turn, give it its heritage, activity and identity.

## REFERENCES


Abd Elhamid, E.M. (2000). *Adjacented rural communities for modern urban in Egypt* (Master's thesis, Engineering and Technology College, Helwan University, Helwan, Egypt).

Almansory, A. (2001). Environmental dimension in traditional urban communities at south Egypt.

Environmental Characterization for Red Sea Governorate. (2004).

Fouda, A.K. (1991). *Architecture and environment, with an application on countryside architecture* (Master's thesis, Cairo University, Cairo, Egypt).

Ibrahim, A.E. (1999). *Arab architects*. Cairo, Egypt: Center of Planning and Architecture Studies.

Lewis, G.J. (1979). *Rural communities*. London, UK: David & Charles.

Mohamed, G. (2004). *The village community* (p. 45). Alexandria, Egypt: Dar El Maaref.

Sablet, M.D. (1969). *Des espaces urbains agréables à vivre: Places, rues, squares et jardins* (2nd ed.). Paris, France.

Suzan, A. (1999). *Human and environment and community*. Alexandria, Egypt: Dar El Maaref.

Tavakolian, A. (1993). *Spaces for human communications* (Doctoral dissertation, University of Pennsylvania, PA).

Wasfi, A. (2003). *Culture and personality* (p. 36). Cairo, Egypt: Dar El Maaref.

Zaki, B.A. (1986). *Glossary of terms for social science.*

*Cities' Identity Through Architecture and Arts – Catalani et al. (Eds)*
*© 2018 Taylor & Francis Group, London, ISBN 978-1-138-05409-7*

# The nature of cities

Domenico Chizzoniti
*Architecture, Built Environment and Construction Engineering Department, Politecnico di Milano, Italy*

ABSTRACT: As we already know, urban conditions are moving towards an "urban age", where urbanity in its varied forms is going to characterize the human habitat and become our dominant social condition in the "recent future". Despite their differences, many cities around the world face some common challenges as they rapidly urbanize. Social, economic and increasingly spatial inequality is a repeated theme, with cities such as Karachi, Mumbai, Lagos, Jakarta Johannesburg or Mexico City, seriously struggling to find common ground for rich and poor. In both these cities, as in many others worldwide, slums and informal settlements continue to sprawl between gated and secured luxury suburbs. A city may have redeveloped its system, outgrown its capacity or lost it altogether. A growing reliance on shrinking natural resources is one of the most alarming challenges to be faced worldwide.

*Keywords*: urban age; spatial inequality; slums; secured luxury suburbs

## 1 INTRODUCTION

### 1.1 *Nature and cities*

What do we mean, today, by the words nature and cities? Are they now empty and crystallized notions or, on the contrary, do they still have a way to search, meet and explain? In which measure can a person, as historical being, understand and test the nature as a precondition for his/her existence? How can the relationship between nature and cities, nature and technology, nature and art be read? We try to answer these questions through different approaches and styles, starting from the condition of the new city and considering its reason and its artificial structure.

Although the theoretical context has always been improved by several contributions, a number of essential issues should be further highlighted. The first concerns the need for a unitary and simultaneously plural conception of the city as a physical and natural fact. This implies the attempt at rethinking of the impossibility of considering their relationship in dualistic terms. The recognition of the naturalness of the human settlement as the original dimension of the human beings, both as a biological and bodily organism, goes hand in hand with the discovery of their ability to transcend nature, and after all it can be said that the creator of the city, i.e. the human being, is not a nature but has a nature (Schindler 2015).

In the same way, in which neither a pure ego nor a consciousness external to the world and unrelated to nature exists, there is for us no pure nature, conceivable outside the historical condition underlying our understanding of it.

In other words, the schizophrenia characterizing the contemporary works of visual art, that convey the paradoxical conception of a nature as an available and changeable mass and, simultaneously, as a lost origin, primitive and idyllic dimension nostalgically recalled: at urban level, for example, this "pre-technical reactionary longing" (Duque 2007) appears in the creation of spaces delimited inside the city, the new enclaves of postmodernism such as gardens and parks, and even the snow-capped mountains of Dubai, the so-called new Disneyland of emirs, which artificially reproduce an uncontaminated nature.

### 1.2 *Culture and nature*

The general idea is that we are no longer able to understand the essence of the semantic structure of nature without considering the technical, urban or architectural intervention. Nature and artifice are arranged on two different levels in our artistic or technical work, therefore nature is always the "exiled" and sublimated element, which is transcended into and from our work. This also means that the natural link with the land is demonstrated within the historical time and the multiple forms of culture: paradoxically, the creative transformation of nature is not possible without prior understanding of the natural data in a horizon of meaning, thus in a historical and cultural horizon (Schindler 2015).

Spaemann states that the concept of nature is essentially dialectical, not only because it usually implies and contains an against-concept (technique, culture, rule, reason, grace etc.), but also because it represents a home and at the same time an obstacle for the human being: hence, the *conditio humana* is characterized by the tension between the inclination to the supremacy (whose instrument is science as a mere knowledge of the functional links, without interest in the origin and in the *télos*) and that to the understanding and to the relationship (typical, on the contrary, of philosophy).

Although human beings are linked to their terrestrial origin (the word *homo* comes from *humus*), nature always indicates an extraneous dimension from which we are already separated *ab initio* as reasoning beings. But the word "culture" comes from the Latin *colere*, i.e. to cultivate the land, to be meant as humanizing and not deleting nature—Heidegger will say that living means building, "a building nursing and nurturing the things that grow".

Through (technical) action and thought, human beings overstep their natural and physical link to earth, creating a meaning horizon that allows to meet the entity: as a matter of fact, beyond any demonization, the technique is the way in which human beings interact with the natural order and, starting from this, build the human order (Duque 2007).

### 1.3 *Nature and artifice*

The construction as a form of living and the creation of a social context of people are the link between natural and artificial; the risk in this construction of human ensembles, however, is that human beings turn the movable boundary interposed between themselves and nature into a "supremacy frame": this is what happens today in the shift from technique to technology, which is self-oriented and self-directing. It is a conception of science and technology in which they are conceived as equal or even substitutes to nature: a form of techno-nature, i.e. a phase in which the technique plays the same strangeness of nature, its inhuman and artificial character (Duque 2007).

The loss of the sense of *télos* in the conception both of nature and of life implies the transformation of science and technology from products or ways of being human into ultimate horizons of meaning, within which human beings are reduced to an ephemeral presence.

## 2 THE STATUTE OF THE CITY AND THAT OF NATURE

### 2.1 *Embellissement vs chaos*

We have a choice in front of us. Either we continue demonizing the cities or we accept that cities and nature should be harmonized for a sustainable urbanization. A "sustainable" city: a far too much widespread slogan. The idea of a contemporary globalized city aims to express a tension involving social, economic and territorial policies, but it sometimes lacks an integrated vision, i.e. its urban structure that is its essence.

The world-scale imbalances result from reckless support to global unsustainability, generated by discontinuous visions and actions of single independent additions. After all, the individual fragmentary actions are always at risk: Edward Lorenz proved it with the 'chaos theory' according to which the beat of a butterfly's wing in Brazil can cause a tornado in Texas.

In the twenty-first century, the urban growth will experience the most rapid expansion in the history of mankind. In Asian and African cities, in which urbanization is in full swing, the population will increase by several billion. For this reason it will also be necessary to plan a development of urban infrastructures able to support this phenomenon.

While facing this challenge, ecologists, planners, economists and landscapers always tend to consider a particular simplistic relationship between city and nature. In particular, there is always a partial view about the relationship between nature and city, which is most often limited to the role of natural infrastructure, i.e. habitats or natural spaces recreated by human beings, which are rather trivially deemed able to provide users with alleged benefits. Cities offer many economies of scale, reducing the per capita use of some resources.

Several recent conservationists look at the current urbanization of our species with sadness because they claim that this phenomenon results in "the end of nature", to use the words of Bill McKibben.

This incomplete view of the phenomenon looks more at the results than at the causes. Stating that we have reached a point where every square meter of land and every ecosystem have been altered by human beings and therefore the nature itself no longer exists, is a way to ratify the end of cities and the death of nature, because urban spaces are spaces totally created and designed by humans for themselves.

### 2.2 *Rural and urban resources*

However, considering the city from the point of view of its relationship with natural resources means basing the analysis on a conception of the city as a complex system, which is however primarily artificial. In this way, the urbanized city would end up representing a system having a plurality of relations with the rural outside. These relations between the artificial city size and the natural size of the countryside have always showed to be a long-lasting relationship in terms of input (food, resources, water, energy) and output (goods, services, technology, knowledge, etc.).

As a system that drains, metabolizes, works huge quantities of natural and energy resources, the city has shown the historically entropic character of its artificial structure, which since the beginning has stood at the edge of arable land and has built a relationship of dependency on the surrounding areas. But, it was in the 19th century that the city was transformed from a low entropic dissipative system into a high entropic dissipative system.

This course was part of the second great transition defined by Clive Ponting, characterized by the large-scale consumption of fossil fuels, and linked to the great economic, social and demographic upheavals of the contemporary age. If until 1800 the percentage of the urban population in the world was only 2.5%, at the end of the twentieth century it increased to about half of the total population (Ponting 2007).

This increase led to structural changes of extraordinary importance. The index of Rees (urban ecological footprint) has shown the increased need for resources, during the twentieth century, by the inhabitants of the urban realities of the industrialized areas (from 1 to 4–6 hectares) and at the same time the reduced available production area (from 5 to 1.7 hectares).

In addition, it is estimated that in just 35 years, six billion people, equivalent to the entire world population in 2000, will live in cities. With three-quarters of humanity settled in urban areas—a phenomenon, which can be considered the largest migration in the history of mankind—several critical points will arise. In this timeframe, it will be our task to review the role played by cities in an increasingly populated world.

## 3 URBAN DEVELOPMENT AND ENCLAVES. THE METAMORPHOSIS OF THE CONTEMPORARY CITY

### 3.1 *Enclaves and ghettos*

Can we still talk about continuous city when it is taking the shape of residential or commercial enclaves along the great infrastructure, but closed and extremely supervised?

The phenomenon is present—but it is weak in this case in comparison with Los Angeles or some other realities not only in California—and it is well defined in the 'Geographies of fear' of Mike Davis. The enclaves, obsessed with security, are a logical evolution of the pilings of monads which characterize contemporary construction. The undifferentiated urban pattern resulting from the sum of basic elementary structures is not the continuous city, and it has nothing to do with 'Madrid-ciudad lineal' of Soria y Mata or the Linear City, Sosgorod of Miljutin.

This new dimension of urban settlement also seems quite unrelated to the well-founded principle underlying the idea of the city as a community, as successfully demonstrated by those archaeologists for which the city began to be defined as such when the space between the buildings has taken on meaning, or better when the significance of the relationship spaces began to prevail over the individual buildings. After all, when it is clear that not only monuments but also the open space connected with the mobility infrastructure were able to establish relations of reciprocity, a change took place in the historical dimension of the city, with the establishment of production facilities in the urban context.

### 3.2 *Permanence and transience*

This condition, in the city construction process, evolves through a process of transformation determined by the sequence of additions, differentiations, oppositions, polarities and architectural supplements. However, if we thought of the "theories of permanence" as developed by Marcel Poète (1929) and Pierre Lavedan (1959) and subsequently taken over by Aldo Rossi (1984), the knowledge of the city would become essential for the purpose of understanding and critically decoding the urban form and the aggregative substance of primary and secondary elements of architecture. The disorder and the loss of its formal structure is one of the most striking aspects of the urban settlement crisis and its discontinuous fragmentation. Already 40 years ago, the spreading of buildings on the territory for Konrad Lorenz was among the 'deadly sins of our civilization'. The constitutive elements of the urban structure as the block, the square, the streets, the orientation and the shape of the structure, the building types, their combinations and ways for land use, are already part of nature and of the characteristics of a place, a city or a part of it. Theory and practice of architectural design are measured with this awareness in the wake of Marcel Poéte's lesson, establishing the urban science underlying the city as the foundation for the study of architecture.

If the word 'city' connotes a reality which has changed, should it be set aside? Is it still allowed to define that constitutive nature of the city in the contemporary metropolis?

Why should they be called continuous-city or spread-city? A new lexicon would probably be needed, capable of grasping discontinuity and diversity in the urbanized continuum where the built magma is opposed to new infrastructures, hampers the continuity of green corridors and wide-ranging networks useful to the quality of environment and of living conditions, and neglects the necessity of shaping new landscapes.

### 3.3 *Vision and reality*

However, retrieving the nature of the city means not only giving a morally sustainable share of equipment and urban green areas or an ecologically sustainable share of duplicate elements denoted as natural but completely different from the physical environment of the city. The recent cases of some principles of metropolitan forestation are visionary and remain rather naively linked to a stereotyped concept of nature. It is also true that the pathology of the recent urban condition is very clear, and the diagnosis is equally clear. The therapy and the methods to counter a deterioration, which appears inevitable are not that clear. Understanding the contingent causes is not enough: the abandonment of building continuity for the 'blocks' and the 'fences'; anachronistic standards; sector regulations; the oscillation of priorities between structures and infrastructure, no longer in symbiosis; the rate at which obstacles pile up on the territory, far from explaining potentials and landscapes; the size of interventions, the splitting of initiatives, and so on. Excellence and sporadic high-quality measures are not enough. The contemporary habitats show wastage and redundancies due to organizational and business models that have a clear innovation potential. Moreover, it is overall not economical.

There is no utopia that is not based on large scales of intervention, or at least common principles, nor is there a reality without utopian vision.

## 4 THE URBAN-PHOBIA OF THE CONTEMPORARY

### 4.1 *The reproduction of nature*

In the present condition it is now recognized that it is no longer the necessary order of nature (logos) that imposes the laws of the *polis*, but it is the *polis* laws which must take responsibility for the fate of nature. Today, the humans' city that was once an enclosed space in the natural world, has replaced nature, which is now reduced to a mere enclosed space in the artificial world of the city (Galimberti 2000).

After all there, is an irreducible duality, a constant conflict, an opposition between the two settlement ways of the human species, i.e. between city and countryside. It is also an ideological fact that has produced several rural and urban settlement patterns throughout history.

But this opposition, in its pure form, has had a definite edge, a certain boundary and a clear border, the one of the walls, which were a necessary element of the historic city. A boundary, the one of the walls, which influenced the shape of the city and has also influenced its density. But the city has crossed the border of its walls and *faubourgs* arose near the walls. In order to survive, it accepted agriculture inside its walls, without giving up the benefits of the countryside: every day, or rather every night, a continuous flow of products was poured into mainstream markets, "bellies" of the city from the surrounding countryside.

The general markets such as *Les Halles* in Paris, protagonist of the novel *The belly of Paris* (1873) by Emile Zola, are the meeting point between city and countryside. The duality city/countryside, rural exterior and urban interior, brings a different dimension of living and production: however not the one between natural and artificial. It should be emphasized that, as a result of the fact that countryside and city are two artificial forms of the territory organization that have undermined the spontaneous and natural harmony of human beings with nature. These two dimensions are both an expression of supremacy capabilities of our species and its ability to imagine and build the future. Two entities forced to live together, even if the countryside can exist without the city, while the latter, until a few decades ago, could almost never disregard the countryside.

### 4.2 *The ideology of nature*

The relationship human-nature has been governed, for us in the West, by two visions of the world: Greek and Judeo-Christian which, in spite of being extremely different from each other, agreed on ruling that nature fell within the sphere of ethics, whose aim was limited to the regulation of relations between people with no extension to the entities of nature. As a matter of fact, the Greeks conceived nature as an immutable order, a horizon not be crossed, an invincible limit that no human action could infringe. When the Greek culture meets the Judeo-Christian culture, the scenery changes because the biblical religion that views nature as a creature of God, conceives nature as a result of a will: the will of God who created it and the will of human beings to whom nature has been given over to their supremacy. Since then, the meaning of nature is no longer "cosmological" but "anthropological" (Galimberti 2000).

Therefore, when we have to discuss the natural aspects of the city we have to keep in mind that the urbanization scope in recent decades is not the growing expansion of the city (which for the first time exceeds the countryside in terms of population). The most important phenomenon is indeed the progressive disappearance of the countryside: the ground floor is divided between cemented and abandoned soil, while agriculture is reduced; one of the two poles of the dialectics of the humanized spaces disappears, brutally placing the city in front of a nature with no human beings. It is a form of "urban-phobia" (or hatred for the city), which is a complex feeling tracing its roots back in the 18th century, especially in Rousseau. The growth of the modern urban society enhances this feeling. The contemporary city in some of its aspects is perceived as a place of anti-civilization and anti-human (Cavin 2009).

## 5 THE NATURAL ORDER OF THE CITY

### 5.1 *The natural aspect of urban space*

"Buildings are appropriated in a twofold manner: by use and by perception—or rather, by touch and sigh. Architecture has always represented the prototype of a work of art the reception of which is consummated by a collectivity in a state of distraction. The laws of its reception are most instructive." (Benjamin 2008)

Starting from the condition of citizens as users, Kevin Lynch investigated the criteria for the urban image to be assimilated and stored. The very important question established by Lynch is the need to create a systemic language to define, insightfully, the urban form. However, this is discernible exclusively if its reverse shot, the rural environment, is also perceivable.

Currently, this duality is already damaged. In the extensive continuum of a city area it is impossible to find a clear boundary that can define its shape. Therefore, the traditional urban-rural dual structure is not recognizable in its traditional limit. During the past, the danger came from the nature, while nowadays it stems from the human power who tries to prevail on it, crossing all limits, not only by using it, but also exhausting it.

So, this duality must be searched in the relationship between the man-made environment and the abandoned and derelict environment. Large abandoned extensions are as dynamic and necessary, as the destiny of decommissioned micro and macro urban structures. The second duality is in the relationship between two main methods of man-made spaces, the rural and the urban one; it is better that they maintain permeable borders and mutual interference, thus re-establishing dynamics which could contaminate mutually.

The urban explosion, which often manifests in a hostile way, is now identified with the phenomenon of sprawl, which is the formless intrusion of the compact city into the countryside and in replacement of the latter. On the contrary, the rural environment pollution in the city occurs by discontinuous and accidental events, sometimes for a form of satisfaction with the recent environmental recriminations and ecological aspiration of the community. It is a simple aspiration that repeats in a rather sophisticated way a "hyper-naturalized" environment for the anthropic environment, which actually de-naturalizes it instead, because it realizes an "in vitro landscape" which is very inhospitable and little expansive. In other words, from the environmental and urban point of view, the incursion of nature into the city (that is called urban farming), without nostalgic compromises, seems like an evanescent answer to the need for increased quality of life. The clearness and the coherence of the city image become the founding characters to know the urban space. In this sense, there is no difference between the ancient man, who had to re-learn his land through the hunting or war, and the contemporary man trapped into metropolitan traffic; as the ancients, even contemporary men arrange things and landscape names according to functional situations of their location.

### 5.2 *An alternative urban paradigm*

The glorious survival of the city, which is the ecological niche of human species, cannot be realized in case of countryside annihilation: the *reductio ad unum* of forms in human spaces is an assumption (and then consequence) of the city crisis, connected to the other *reductio ad unum*, that is the main reason, i.e. the acceptance of a unique thought as a parameter to rule cities. The identification of parts, design and structure of the city is a necessary precondition to define the elements of identity (physical aspect of object), structure (relational aspect) and meaning (practical and emotional). In the recent condition of the city, these identities are not the same ordering elements of the medieval city (cathedral), renaissance (municipal building), nineteenth-century (factory), or modern era (industry).

Figure 1. Cape Town, South Africa, 2010.

Figure 2. Mumbai, India, 2013.

Figure 3. Nairobi, Kenya, 2010.

Figure 4. Sao Paulo, Brazil, 2013.

Figure 5. Rio\de Janeiro, Brazil, 2016.

Figure 6. Mexico City, Mexico, 2013.

Figure 7. New York, 2015.

Figure 8. Hong Kong, 2012.

The physical identities that we can find and recognize within the contemporary city and characterizing its natural order are various and they all refer to the need to organize the contemporary city. These identities can be related to the infrastructure system that physically organizes the local engineering and the layout of the city, through a mutual incursion of the nature into the city and of the city into the countryside.

The paths (all the public or private connective network of an area: pedestrian and automobile streets, railways, canals and so on) represent the connection able to determine the natural structure of the city. The same applies to the limits, no longer those geometrically defined by the physical structure of a wall, but rather those borders that have certain characters of discontinuity (boundaries, shores, building's lines of continuity, railway mobility areas etc.); they are potentially capable to adopt and duly describe the new urban landscapes of metropolitan cities. It is the new Landmarks that, as a physical structure, are able to receive the reference role in the order of urban and rural landscape (punctual signals composed by a physical object: a building, a signboard etc.). Certainly, the urban hubs are new condensers during the rituals of users behavior (a widening, a square, but also a crossroads, a crossover and so on). Finally, it is also the neighborhoods, as defined urban structures (similar parts—two dimensional—of a landscape with morphological and building characters), not only for their functional status, but also for some settlements and formal aspects.

Re-conceptualizing these elements of the contemporary city means regaining current and fundamental elements of architecture, first of all restoring a new idea for the behaviors regarding the use of space.

## 6 A POSSIBLE WAY OUT: THE DESIGN OF THE CITY BY PARTS

### 6.1 *An integrated architectural and urban approach*

The basic question, which we are called to answer today, is the general role of the incisiveness of the architectural action in the present condition. For too long, the architectural practice has been developing a visual aesthetic, on a purely formal conception of the problems. Visual strength tends to exclude consideration of the structural aspects around urban and rural environment, where visual is not considered on a par with some historic, social and economic criteria for a complete outlook on the state of the city (Aymonino 1993).

On one side we still covet the single building as a chance for self-expression. On the other side, when a project has a wide range of issues, these are broken out into statistical entities with little grasp of the overall framework, with little understanding of its implications and its limitlessness, resulting from the intense cohesion of well structured parts. We need to look for new strategies that allow us to face new challenges. Greater than ever before, our challenge is to solve even increasing needs of our complex society for which no previous model exists.

The design of the city by parts can be considered a fairly clear assumption of any problem arising from the complex structure of the recent urban condition that involves, at different scales, both the knowhow of architecture and town planning as disciplines.

It is however evident that the problem of planning in the city is rather vague even if one accepts the method of the construction by parts and even if these parts are well-defined by their being physical parts. At any rate, to make the question clearer it would be better to redefine the sense of the parts forming the city and, above all, the quality and the structure of the new parts that make up the new town. If one rejects this kind of redefinition, or analysis of the quality of structure or dimension of such parts, one must accept all physical facts as elements of the city, without any qualitative discrimination.

It is possible to agree with this thesis on condition that it may introduce a judgement system, able to include both one physical part of the city and the city as a set of physical parts.

It might not be feasible to broaden this kind of judgement on the physical part of the urban settlement in order to include the whole city as a single structure, although it happens to consider a single building as a defined whole. This kind of paradox has an opposite one. The position of those who conceive and plan the city as a building is equivalent to the position of those who consider every building as a micro-city.

Anyway both of these tendencies are inclined to reject a functionalist approach. In fact, if we look at a construction as a neutral "container", as a space set to define the largest possible number of uses, we generally see that the building is an answer to problems arisen by the request for those uses which are still internal, in particular: these uses, although existing in the city, find in the building a condition for autonomy, in order to define an undifferentiated receptacle of a specific activity which may even not be permanent.

On the other side, if we considered the building as a multi-functional system, we would be able to admit a feverish construction of a presence of the building in the city by means of an overlapping of uses, recognizable inside the urban body by their own formal structure

It is possible to better define the relation city as a building or, as well, the relation city as an artefact, and consequently a part of the city as an independent artefact. It is, in short, a matter of considering a building as an individual and the problem of the role of a part in relation with the whole of parts forming the city (Polesello 1968).

### 6.2 *The order of a multi-scalar problem*

The design of the city is regarded both as an urban and architectural problem. It has been known that town planning is usually addressed to the general structure of the city, while architecture design, through a correct composition of the structures forming the city, is usually intended for stabilizing the relation between soil as use and physical body (a building, a square, a street, and so on). These parts of cities built around a public building or public space, are easily recognizable since their urban structure is clearly affected by this presence. If it is possible to take the category of use as a fixed and unchangeable principle, it could be possible to grant both the institution of a metric common to both scales of planning, the town planning scale and the architectural one, and the consistency with the double value system they involve.

Of course, these criteria are not adequate enough to develop this theoretical assumption. In other words the link between scalar use and physical structure does not develop in a linear way, and, if necessary, confirms Hegel's principle that continuous modifications in the quantity of homogeneous elements involve their quality (Polesello 1968).

But it is also true that from the theoretical point of view, therefore, it does not matter which scale is used thanks to the two-way relation: the urban sense of a building, and the spatial reason of the city (or part of it). But rather than two value systems, this is a problem of metrics which must grant the knowledge on any value concerning the city and its design. If we can interpret the part of the urban structure as a whole, we can operate and act according to one scale only.

### 6.3 *Two aspects of designing the city by parts.*

There might be two aspects that can help us to explain the problem concerning the design of the city by parts.

The first aspect concerns the relation between the object of architecture as an aggregation of elementary units and the existing city. It is true that this problem arises only for some architects that include in the context of the real city a module as a component of a "machine à habiter" (Le Corbusier 2007) or, conversely, the criterion of "building technology" (Archigram 1999). However we can find today in the recent production of the urban design some of these aspects, such as grouping pattern of the building cells, which are real numeral elements of the whole system. It is also true that since these patterns of building organization are based on a functioning complexity very close to urban complexity, they are proposed as a comprehensive alternative to the cities, or as a partial alternative concerning only the class of residence-service equipments.

All of these kinds of examples can neither exhaust nor show the meaning of the idea of possible part of the cities. These structures in the city when considered as a building or as a part of the city are foreign to the city itself, or coexist with it at the most. Their being foreign to the city regards not only the status of independence that consists of a high level of autonomy from the other parts of the city, as an enclave (Davis 1992), but priority from the unwanted rejected formal relation with the existing city, with its artificial nature of human settlement. It is really strange how today that experiences, even more advanced than those proposed some years ago by the utopian vision of Archigram and Metabolism, do not, or cannot, give up the problems of formal perceptions, connected with the use of these technologies, and do not look, on the contrary, for more sophisticated technological uses not involving such a problem. These kinematic devices by the modern and recent technological statements are involved in coping with the nature by reproducing it in vitro, in the guise of comfort, sustainability, ecology, and so on.

The second aspect concerns the relation between (a part of) the city and the nature itself. This can no longer be considered an aesthetical relation, let alone the idea to transfer mechanically a part of the natural environment. Said otherwise, this relation should be found in a manner that attempts to address the ways in which architecture and the construction of space can be aesthetically environmental, that is to say not merely the one containing the notion belonging to the ordinary conception of nature, such as the ecological attitude on managing the eco-friendly structures, or maybe emphasizing the green approach to the urban planning. On the contrary, working around the authentic role of Nature means decoding the artificial configuration of the urban context by its "natural arrangements".

In this way, we must face a very interesting problem: if, in a well-balanced development of towns and territories, together with the targets of planned social and economic balance, we considered as extremely interesting also a suitable relation between natural and artificial landscapes, and then introduced the idea of site, existing town structure, typology and morphology of the different urban elements as material foe and of the planning process. All these materials belong to the artificial structure of the city, and here we can find the nature of human settlement, such as the built space as a physic structure of built architecture or the free space as the natural landscape. In the state of current city environment, it could be useful today to go backwards through the process leading from the city to the architecture and from the architecture to the architectures, trying to find out the deep roots and the connections linking together the architecture in the cities and by means of these connections to define, although partially, the urban problem. The relation between the essence of a territory and the essence of its building becomes a symbolic expression of all opportunities and implicit relations of human settling, the definition of its potential being a project in the future (Canella 1974).

## 7 AN OPEN CONCLUSION

Without massive scale investment in development, the city will continue to stumble along in its state of mediocre urbanism and public architecture whilst blindly eroding its setting.

The city spreads and extends all the way to the point where, while it tends to cover the entire orb of the planet, it loses its properties as a city, and, of course with them, those properties that would allow it to be distinguished as a structure to settle a community.

That should not be related only to a common sense of the ecological aspects of the urban environment, if it is possible to re-conceptualize it in an age of urban sprawl, multiple usage of public space and proliferation of the sites of political, social and cultural expression. To sum up, this essay tries to outline the paradoxical concept of "Nature" in the "artificial" context of the city. Urban activists continue to believe that the ideal of "city beautiful" and "garden cities" and most recently, the project of "urban renaissance" and "new urbanism", implies a return to the conservative conception of urban life, although history shows that building sociality through civic engagement between public and private space is one of the attempts at managing public space. From the classical Greek philosophers, theorists of urban modernity such as Benjamin, Simmel, Mumford, Lefebvre and Jacobs, and contemporary urban visionaries such as Sennett, Sandercock and Zukin, all suggest a strong link between urban public space and urban civic virtue as a way to reconsider the "Nature" of the urban environment. The reconstruction of modern town according to the morphologic and geographic parameters does not constitute a science fiction's vision. From this point of view, assembly quotation and reduction of problems to their paradoxical limits are not ingredients of a strange immoral prophetic flight. On the contrary, the presence of reality and the necessity to propitiate future by provoking it too, justify in this case the recovery of some elements of the vision by reading the city as an entity made by parts.

Working in the city by its constituting parts means recognising firstly the nature of the city, then the importance of the natural settlement as a human artefact in the artificial environment of the city, and finally the building up of a single thought that articulates a sense of nature in the meaning of space, valid both for architecture and town planning.

## REFERENCES


Amin, A. 1994. *Post-Fordism: A Reader*. Maiden MA: Blackwell.

Aymonino, C. 1975. *Il significato delle città*. Bari: Laterza.

Aymonino, C. 1993. *Origini e sviluppo della città moderna*. Venezia: Marsilio.

Benjamin, W. 2008. *The work of art in the age of the mechanical reproduction*. New York: Penguin Books.

Bond, P. 2002. *Unsustainable South Africa*. London: The Merlin Press.

Braudel, F. 1984. *The Perspective of the World*. New York: Harper Collins.

Burdett, R. & Sudjic, D. 2007. *The Endless City: the urban age project by the London School of Economics and Deutsche Bank's Alfred Herrhausen Society*. London: Phaidon.

Canella, G. 1968. Passé et avenir de l'anti-ville universitarie. *L'architecture d'aujourd'hui*, (4–5) 137: 16–19.

Canella, G. 1974. Per una drammaturgia dell'Hinterland. *Casabella*, (11) 395: 20–35.

Canella, G. 1982. Guardare il centro dalla periferie e la citta dalla campagna. *Quaderni di urbanistica della città di Monza*, 67–69, 83–84.

Cavin, S. 2009. *Urbaphobie. La détestation de la ville aux XIXe et XXe siècle*. Pompignac: ÉditionsBière.

Cook, P. 1999. *Archigram*. Princeton Architectural Press.

Davis, M. 1990. *City of Quarts: Excavating the Future in Los Angeles*. London, New York: Verso.

Davis, M. 1992. Beyond Blade Runner: Urban Control, The Ecology of Fear. *Open Magazine Pamphlet Series*.

Davis, M. 1996. How Eden lost its garden: a political history of the Los Angeles landscape. In A.J.Scott and E.W. Soja, *The City—Los Angeles and Urban Theory at the End of the Twentieth Century. Berkeley*, CA: University of California Press.

Scott, J. (1985) Weapons of the Weak. Everyday Forms of Peasant Resistance. New Haven, CT: Yale University Press.

Davis, M. 2002, *Dead Cities*. New York: The New Press.

Davis, M. 2006. *Planet of Slums: Urban Involution and the Informal Working Class*. London, New York: Verso.

Duque, F. 2007. *Abitare la terra. Ambiente, umanismo, città*. Bergamo: Moretti & Vitali.

Engels, F. 1987 [1845], *The Condition of the Working Class in England*. Edited by V.G. Kiernan. Harmondsworth: Penguin.

Engels, F. 1940, *Dialectics of Nature*. New York: International Publishers.

Galimberti, U. 2000. *Pshiche e techne. L'uomo nell'età della tecnica*. Milano: Feltrinelli.

Harvey, D. 1973, *Social Justice and the City*. Cambridge, MA: Blackwell Publishers.

Harvey, D. 1996, *Justice, Nature and the Geography of Difference*. Oxford: Blackwell Publishers.

Jacobs, J. 1962. *The Death and Life of Great American Cities*. London: Jonathan Cape.

Lavedan, P. 1959. *Géographie des villes*. Paris: Gallimard.

Le Corbusier 2007. *Toward on Architecture*. Los Angeles: Getty Research Institute.

Lefebvre, H. 1968. *Le Droit à la ville*. Paris: Editions Anthropos.

Lefebvre, H. 1976. *The Survival of Capitalism: Reproduction of the Relations of Production*. London: Allison and Busby.

Lefebvre, H. 1991. *The Production of Space*. Oxford: Blackwell Publishers.

Mumford, L. 1938. *The Culture of Cities*. New York: Harcourt Brace.

Mumford, L. 1945. *City development. Studies in disintegration and renewal*. New York: Harcourt, Brace and C.

Peet, R. and Watts, M. 1993. Introduction: development theory and environment in an age of market triumphalism, *Economic Geography*: 221–253.

Peet, R. and Watts, M. 2004. *Liberation Ecologies: Environment, Development, and Social Movements*. New York: Routledge.

Poëte, M. 1929. *Introduction à l'urbanisme: l'évolution des villes, la leçon de l'Antiquité*. Paris: Boivin.

Polesello, G. 1978, La progettazione della città per parti, *Hinterland* (4) 18: 30–33.
Ponting, C. 2007. *A New Green History of the World*. Penguin Books: New York.
Rossi, A. 1984. *The Architecture of the City*. Cambridge MA: MIT Press.
Sandercock, L. 1997. *Towards cosmopolis: planning for multicultural cities.* New York: Wiley, Chichester.
Sassen, S. 1998. *Globalization and its Discontents*. New York: The New Press.
Sassen, S. 2001. *The Global City: New York, London, Tokyo*. Princeton, NJ: Princeton University Press.
Schindler, D.C & Schindler, J.H. 2015. *A Robert Spaemann Reader: Philosophical Essays on Nature, God, and the Human Person.* Oxford: Oxford University Press.
Scott, J. 1985. *Weapons of the Weak. Everyday Forms of Peasant Resistance.* New Haven, CT: Yale University Press.
Sennett, R. 1905. *Garden cities in theory and practice: being an amplification of a paper on the potentialities of applied science in a garden city.* London: Bemrose.
Simmel, G. 1971.*On individuality and social forms: selected writings.* Chicago: University of Chicago press.
Smith, J., Chatfield, C. and Pagnucco, R. 1997. *Transnational Social Movements and Global Politics: Solidarity Beyond the State*. Syracuse, NY: Syracuse University Press.
Smith, N. 2002. New globalism, new urbanism: gentrification as global urban strategy. *Antipode*, 427–450.
Smith, L. and Hanson, S. 2003. Access to water for the urban poor in Cape Town: where equity meets cost recovery, *Urban Studies*, 40(8): 1,517–1,548.
Solecki, W.D. and Welch, J.M. 1995. Urban parks: green magnets or green walls?. *Landscape and Urban Planning*, 32:93–106.
Sprugel, D.G. 1991. Disturbance, equilibrium, and environmental variability: what is "natural".
vegetation in a changing environment?. *Biological Conservation*, 58: 1–18.
Vidal, J. 2004. They call this recycling, but it's really dumping by another name. *The Guardian*, 21 September.
Zukin, S. 1995. *The Cultures of Cities, Maiden*, MA: Blackwell.
Zukin, S. 1991. *Landscapes of power: from Detroit to Disney world.* Berkeley: University of California press.
Zukin, S. 2009. *Naked city: the death and life of authentic urban places.* Oxford: Oxford University Press.
Zola, E. 1873. *Le Ventre de Paris*. Paris: Charpentier et Cie.

*Cities' Identity Through Architecture and Arts – Catalani et al. (Eds)*
*© 2018 Taylor & Francis Group, London, ISBN 978-1-138-05409-7*

# The Potential of Pavements in the Identity Conflict of a City

Dalia Abdel Moneim Osman
*Architecture Department, Faculty of Fine Arts, Helwan University, Egypt*
*College of Arts and Design, Princess Nourah bint Abdulrahman University, KSA*

ABSTRACT: Cairo has components that present its distinctiveness; the changes occured in a city are due to the alteration in the urban form of them. Its recognition is achieved when movement occurs through a street space; Walking is a remarkable mode; an interaction that exists between people and a place. Hence, it is essential to design and plan pavements to ensure potential for walking and promote the presence of people in the street space.

This paper focuses upon the significance of pavements in hosting pedestrians and discusses how it contributes to the discovery of a city. It explains the role of pavements in encouraging the presence of people in a street space, which acts as a lens for identifying a city through the recognition of its physical fabric. Finally, the paper recommends tools to address absent or deteriorated pavements, and based on successful designs, demonstrates how good pavement planning and design can reveal a city's identity.

*Keywords*: street space; pavements; deteriorated paths; city identity

## 1 INTRODUCTION

A city is a construction in space. Looking at a city produces an image, and on different occasions and for different people, its identity is explored. The identity of the city is mainly related to its image, in addition to other factors such as its history or function; these reflect on the visual image of a place as well. But its identity changes because cities change over time and only partial control can be exerted over a city's growth and form or function. There is no final result, only a continuous succession of phases. Moving elements in a city, and particularly people and their activities, are as important as other stationary physical elements. Many cities are remembered by their streets and people bring activities to their streets, but there should be space for them. The most common space is the pavement, which is the line between the street space for vehicles and the building line of different land uses. However, due to the increased traffic in the city of Cairo, for instance, people do not use pavements. They are inadequate for walking and in many cases they are absent. Based on a hypothesis that there is a relationship between a city's identity and its streets, the main objective of this paper is to confirm the role of the pavement in the shaping and the recognition of a city's identity, through introducing issues that affect pedestrians along their journey in a street space, and recommending five principal demands to be accomplished to achieve the main objective.

## 2 IMAGE AND IDENTITY

The image of the surrounding space is the result of a two-way process between the observer and the environment. The image of the surrounding environment may be analysed into three components: identity, structure and meaning (Lynch, 1960). An image requires first the identification of an object, which implies its distinction from other things and its recognition as a separable entity. This is called 'identity'. The identity of a place is achieved when the image of a place is figured out, and this takes place by using other components as well the structure and the meaning for the observer (whether this is practical or emotional). Therefore, the identity of a place is related to its image and sometimes this is an overlap of many individual images of a place. There are some common elements that appear in the image types of cities (Lynch, 1960): paths, edges, districts, nodes and landmarks. However, there are other influences on a place, such as its history, function, value or social significance. Hence, there are a number of elements that contribute to the recognition of a city's image and consequently, the combination of these elements all point to the physical form of a city and its identity. The paths or streets are one of them, especially the memorable or valued ones.

## 3 STREETS AND RECOGNITION OF A CITY

Streets are the core elements of a city's structure and should possess unique identity. They experience a wide range of developmental changes through time. They are not just the corridor space between the two sides of buildings, but provide opportunities for other uses and activities, and direct access to buildings and spaces at their edges (Desai, 2014). This whole enclosure and the activities generated through these spaces create a unique experience and perception for users. They also give an opportunity to watch a city, which is an important activity for users. In addition, each street has its own role, unique pattern and historic values. There are various distinguished functions of a street and these functions also give a unique definition to a street. A city's districts can be remembered by the main function of their streets: some serve as a basis for social interaction, others provide business or commercial needs, historical knowledge, and, sometimes, just a beautiful vista of a city. They are memorable places within the space where our day to day life occurs in various ways and celebrate a region's unique qualities.

## 4 PEDESTRIAN PRIORITISATION

A city's streets are better for staying in and not just for passing through (Alexander et al., 1977). Unfortunately, today, streets have become only a path to move through, and not a place to stay or to enjoy passing through. Most streets have become 'centrifugal' not 'centripetal' for people, and although they should attract people to walk through them, they in fact drive people out. Walking is an enjoyable journey or experience, and there are many factors that affect both the possibility of walking and the presence and experience of people in any street. The experience gained by walking is much more than that gained through vehicles; an interaction occurs between pedestrians and their surrounding environment. This is a significant issue and plays a role in the recognition of a city. Historically, all of a street's activities were carried out with each other on one platform without any physical separation; however, now it is essential to arrange room for pedestrians in a street space.

Pavements are the most familiar place for people in a street. Their main function now is to separate pedestrians' movements from motorised ones, providing safety for people and decreasing the risk of accidents. In addition, they create a public space for people in a street space, providing a place for them to interact and socialise (Osman, 2016). They have another role in the recognition of diverse values in a city's streets. Encouraging walking in a suitable space down any street encourages culture identification, and historical and customary values, and the city is recognised by people (Movahed et al., 2012).

## 5 THE CONFLICT IN A PAVEMENT'S FUNCTION

Cairo has experienced a number of morphological changes during its history that have resulted in the formation of a discontinuous experience of the physical fabric. The streets of the newer part of Cairo are generally wide and straight. Conversely, the old city is characterised by its

Figure 1. Colonnaded arcade of a pavement on Ibrahim al-Laqqani Street in the Heliopolis neighbourhood of Cairo: Left, picture taken in 1930; Right, recently taken picture of the same street in 2012.

narrow, winding and intimate streets, where they were exceptionally well suited to movement and outdoor activities. Changes have occurred over time (in other distinguished neighbourhoods of modern Cairo too). Some have lost their distinctiveness and others still have part of their character, mainly reflected in the physical form. However the activities within them have changed (Figure 1). This occurred mainly due to the rapid growth of motorisation; streets were designed to give priority to the requirements of motor vehicles. Other activities were moved to other spaces away from the street space; these corresponded to people's daily activities. Pedestrians' movement has influenced the need for space and the space created was the pavement. Hence, from this point the conflict in the pavement's function has emerged. There were problems concerning its presence and status in streets, the obvious one is that affecting the safety of pedestrians. In Cairo, streets lack the presence of a suitable space for pavements, as a result there is not enough space to accommodate pedestrian flow, and other streets lack the presence of pavements. In some cases they are present, but occupied by other uses which block and hinder pedestrians. For these reasons, a rise in the percentage of accidents has been seen. In many streets, insufficient illumination is present and pavements suffer from a deterioration in their materials.

Also, it is essential to avoid the interference between pedestrians and bicycles on the same space, especially in pavements of short width. Pavements should be well illuminated and free of obstacles that may cause people to fall or be struck. Ramps between the pavement and street space provide better conditions for pedestrians with physical disabilities, the elderly and those with wheelchairs (Osman, 2016). These simple procedures can help to decrease the number of accidents, make it a more relaxing experience for pedestrians, help expand the number of shops, encourage walking, improve the ability to see objects and details of buildings, and provide recognition of a place's identity.

## 6 PERCEPTION OF A PLACE: FINDINGS

The findings of this paper were based upon the hypothesis that the physical fabric of a street has a critical impact on enabling or disabling the recognition of a city's identity, and that pavements are a common space for people in integrated streets. The following subsections cover the main issues identified during the course of this research and are thought to have an impact on a city's identity and its recognition; they are the major aspects of interaction for people during their journey along a pavement in a street space.

### 6.1 *Building fronts*

Streets are shaped by buildings' facades, not only by the buildings' shape or style but also by their angles. There is an interaction between building facades and the pavement space, creating active vibrant edges. The treatment and modulation of the adjacent facades relative to the pavement space creates visual interest along the path. The existing uses at ground level identify the street space; one can watch the different uses while walking. Large openings create visual interest and identification of a place, and may expose activity within a building to pedestrians, helping them to integrate visually with the outdoor space. The colour and texture of the frontages have an impact on people and a place's identity through their contribution to the interest of people passing them.

### 6.2 *Building entrances*

The connection between a building and its adjacent outdoor space is the building's entrance. These should appear convenient and welcoming because they impact on the image and identity of a place, especially if they are featured, face the street and provide access to and from the pavement. Exterior transitions between the pavement and the entrance, such as stairs and ramps, also affect image, but should stay within the frontage. Finishing materials used in the pavement or textures on the front of entrances help characterise a place. Even names, numbers or historical information about entrances help identify a place.

### 6.3 *Street furniture*

Furniture provided on pavements affects the identity of a street; even its absence has an impact. Any element of street furniture should aim to provide comfort and convenience to the path; however, it can also be an obstacle if not well designed or located. Many streets lack the presence of street furniture elements; the most common example is the provision of a place to sit, which is a basic necessity. If comfortable and inviting places are offered in the pavement space, place identity will be affected, the pavement will be changed to a gathering place and its role as a liveable public realm will be enhanced. There are, obviously, different forms and locations of seats but, as a whole, just the idea of seating will allow a space to be identified and characterised.

### 6.4 *Finishing materials*

Pavements share in the reinforcement of a sense of place and they establish a city's identity. The materials used, colours and patterns of finishing have an impact on the identity of a place. New pavements should match those which previously existed, since this affects the image of a space. However, obstacles should be avoided when designing new ones, such as street furniture, retail displays or badly placed trees. In order to present to pedestrians a chance to watch and recognise the identity of a place, pavements should be accessible to people of all ages and abilities. Materials should be selected to minimise gaps and rough surfaces, especially for pedestrians using wheelchairs or prams (City of Boston, 2013).

### 6.5 *Streetscape*

The green space in a street helps to define a place, the greenery providing shade, reducing energy consumption and absorbing greenhouse gases. Besides these environmental benefits, it provides social and psychological ones. Trees present comfort, beauty and attractiveness. They create a focal point along a path and are a symbolic connection to nature. People are attracted to places where greenery is present and they are remembered for their image. In addition to changing light and colour, they create character that helps to reduce stress and restore a sense of calm. In the case of identifying a place, trees are used to fulfil a number of functions. Based on their location and arrangement, they can reinforce the rhythm of the building fronts along a pavement, focus linearity, define spaces, create a sense of enclosure, and add a ceiling to a space, as well as texture.

## 6.6 *Street illumination*

People remember a place through illumination. Appropriate pavement lighting provides a well-lit space to facilitate movement, creates a safe place for pedestrians and gives a street character. Features can be highlighted, as well as land uses and activity spots along the path; this can all be achieved through the use of appropriate street lighting. The lighting reveals the public space and any special area to encourage night-time use; many street pavements are characterised by the activities that take place in them late at night. In addition, street illumination used on pavements enhances the character of a streetscape by using various fixtures that reflect the image of a place and distinguish the unique look of characteristic districts in a city, especially the historic ones.

## 6.7 *People and activities*

Everyone is attracted to various activities that exist in a street space. The pavement supports different activities and not just movement. By contrast, a street's main space mostly only hosts motorised movements. The presence and types of activities depend on a number of factors, for instance, there are daily activities that relate to the presence of people in a place: walking, standing, sitting, watching, listening and talking. From a historical perspective, streets have played a role as meeting places for people. People meet each other, communicate and even sell goods. Every activity is carried out in full public view (Yang, 2012). City space has always served three vital functions: providing a meeting place, a market place and a connection space (Gehl, 2010). People gather and move about with others, seeking to place themselves near other people (Gehl, 1987). They will choose to walk in a lively street rather than an empty or low-density one; they react towards the presence of others in space. Social life and its quality are important for planning and reflect the image of a city. Safe and lively streets are also beneficial in the creation of a lively, safe and sustainable city (Yang, 2012).

## 6.8 *Cafés*

One of the images that pedestrians on pavements remember is the existence of cafés along a street. Most commonly, they are in an area in front of a street wall that is called the frontage zone. However, sometimes they occupy part of the pavement. This is reflected in the pedestrians' perception of the identity of a place. The extension of cafés or restaurants into the public way brings activity to the public realm (City of Boston, 2013). At the same time, attention should be given to the design and layout of street cafés to maintain pavement functionality. Regulations should be in place to organise and monitor the impact of pavement cafés on pedestrian movement and environment, so that a clear accessible path for people can be maintained.

## 6.9 *Street soundscape*

The soundscape of a place is its sonic or acoustic environment, with the receiver or listener at the centre of the sonic landscape (Porteous & Mastin, 1985). There exists all types of sounds; ones that distract or annoy people, based on acceptable risk, and others that are regarded as positive, preferred or desirable (Brown, 2004). The fact is that whether the soundscape of the outdoor space is desirable or not, it has an important effect on the identity of a place. There are various sounds that are heard, such as those made by people talking and walking, nature, musical instruments, vehicles or activities. However, the sounds that convey the identity of a place are the dominant sounds that can be heard. The sound environment of a street plays a significant role. Each city has a unique acoustic profile, the composition of which derives from various sound sources that can be divided into preferable and undesirable sounds.

### 6.10 *The modal split*

The mode of movement within a city is split into a number of categories: vehicle, pedestrian and cycling. The movement along a street is an important factor that has an impact on a place negatively or positively and, at the same time, in other cases can identify it. The identities of many cities around the world have been transformed due to changes in the type of movement within them. For instance, all streets in the medieval city centre of Copenhagen were filled with cars, and as car traffic increased, conditions for pedestrians rapidly deteriorated. However, in 1962, Copenhagen's main street, Strøget, was converted into a pedestrian-only area (Gehl & Gemzøe, 1996). Many other streets and squares followed and were converted into pedestrian areas as well, allowing various activities to take place.

The gradual transformation of the city centre from mainly vehicular traffic to pedestrian has led to a change in Copenhagen's culture and character. Outdoor activities and life were developed, even though the temperate Danish climate makes it more difficult to develop public life. On the other hand, the historical district of the old city of Cairo has suffered up until now from the deterioration of its characteristic urban form. This has been due to an increase in unplanned car traffic and the appearance of new uses. Both have impacted on the city's character, threatening the urban form as well as the historical buildings, thus changing its identity.

## 7 DISCUSSION

The significant findings of this paper revealed that street status has an impact on the conflict in a city's identity. This was determined through the interpretation of the answers to two major questions: what is being watched during the journey along a street, and what is its influence upon the perception of a place? The findings are presented through major interacting aspects that are reached in the paper, and these aspects are illustrated during the people's journey along the street space. The conflict in a city's identity arises because of the absence of dealing with a street as a place. Streets are being managed as a means of access or links only; however, they should be considered as spaces for other activities as well, because this has an impact on people's perception of a city.

So it is essential to consider a street as a place, in order to provide people-oriented activities and flexible spaces. These additions will contribute to creating the identity of a place, be a memorable experience for a city's users, residents or visitors, and will help in maintaining the existing identity.

Pavements are major components of a street and they represent the people's space within it. Poor pavements and their disappearance often create weak environments for walking and hence affect the recognition of a place. It is recommended that consideration should be given to designing and planning pavements with regard to their major function as links and places for pedestrians; a link where movement exists from one point to another and a place where activities are performed. If these two major functions are fulfilled in the pavement's design and planning, observations of a city will occur more readily, and the character of a place will be revealed.

To assure the implementation of the major functions of pavements and to achieve the main goal of creating pavements with positive impacts upon the recognition of a city's identity, it is recommended that five principles be incorporated into their design and planning. The first is to provide accessibility to encourage people to use, walk around or reach their destinations regardless of physical disabilities or age. The second is to address the comfort aspects and these include consideration of climatic conditions, as well as location, morphology, proportion, scale, public infrastructure and services. The third principle is to assess the safety aspects, which refer to the extent to which pavements encourage people to use them and walk without the fear of tripping, falling or being attacked. This includes establishment of well-lit pavements, short walking distances, flat surfaces and maintenance of surfaces. The fourth is the legibility aspect, in terms of the signs, visible features and various paving materials that are used. This reflects the character of a place and a distinctive pavement gives people a clear image of where they are. And finally, the participation aspect, fulfilled when good pavement conditions encourage people to walk, participate in street life and communicate with other people.

## 8 CONCLUSION

Pavements are a major component in a street space; they are the spaces which people use mainly for walking. This paper has revealed that there is a relationship between the presence of pavements in a street and the recognition of the identity of a city; they contribute to the definition of the character and life of a city. It is concluded that this relationship takes place through a number of elements and attributes that are revealed while walking along a street, such as the building frontages and entrances, streetscape, people, activities, modes and soundscape. And this relationship can be easily achieved when pedestrians are encouraged to participate in a street space through the promotion of walking and various related activities. This, in turn, requires a space to be provided for pedestrians, which in the case of integrated streets will be pavements. Thus, although pavements typically exist to fulfil the promotion of walking and providing safety, this paper has shown that they have another function. They have a major role in facilitating the recognition of a city's identity because they promote the observation and recognition of the city. Therefore, this function of pavements must be given consideration in their design and planning.

## REFERENCES


Alexander, C., Ishikawa, S., Silverstein, M., Jacobson, M., King, I. & Angel, S. (1977). *A pattern language*. New York, NY: Oxford University Press.

Brown, A.L. (2004). An approach to soundscape planning. In *Proceedings of ACOUSTICS 2004, 3–5 November 2004, Gold Coast, Australia*. Brisbane, Australia: Faculty of Environmental Sciences, Griffith University.

City of Boston. (2013). *Boston complete street design guidelines*. Boston, MA: City of Boston.

Desai, M. (2014). *Reforming complete streets: Considering the street as place* (Master's thesis, School of Community Planning, University of Cincinnati, OH). Retrieved from http://rave.ohiolink.edu/etdc/view?acc_num=ucin1428070222.

Ezzat, T. (2013). *Neoplastic intervention: Re-imagining Cairo's public realm* (Master's thesis, University of Maryland, College Park, MD). Retrieved from https://drum.lib.umd.edu/handle/1903/14239.

Gehl, J. (1987). *Life between buildings: Using public space*. New York, NY: Van Nostrand Reinhold.

Gehl, J. (2010). *Cities for people*. Washington, DC: Island Press.

Gehl, J. & Gemzøe, L. (1996). *Public spaces. Public life.* Copenhagen, Copenhagen Denmark: Danish Architectural Press.

Lynch, K. (1960). *The image of the city*. Cambridge, MA: Massachusetts Institute of Technology Press.

Movahed, S., Payami, S. & Zakeri, H. (2012). A safe pedestrian walkway: Creation a safe public space based on pedestrian safety. *Procedia, Journal of Social and Behavioral Sciences*, *35*, 572–585.

Osman, D. (2016). A tale of a sidewalk: The conflict of its presence and usage in the streets of Cairo. *European Journal of Sustainable Development*, *5*(2), 1–12.

Porteous, J.D. & Mastin, J.F. (1985). Soundscape. *Journal of Architectural and Planning Research*, *2*, 169–186.

Yang. Q. (2012). *Lively streets and better social life. A case study of three comparable streets in Kristianstad* (Master's thesis, Blekinge Institute of Technology, Blekinge, Sweden).

*Cities' Identity Through Architecture and Arts – Catalani et al. (Eds)*
*© 2018 Taylor & Francis Group, London, ISBN 978-1-138-05409-7*

# Towards sustainable slum development: A performance evaluation approach for slum upgrading plans in Egypt

Shaimaa A. Magdi
*Faculty of Engineering, Fayoum University, Egypt*
*Faculty of Engineering, Cairo University, Egypt*

ABSTRACT: Slums are the worst outcome of urbanisation in Egypt. 'Slum upgrading' became a necessary strategy for improving living conditions, in order to achieve the Millennium Development Goal on Environmental Sustainability. This paper aims to establish a deeper understanding of how 'informal areas' (or 'slums' as they are more commonly known) emerged and were activated in the Greater Cairo region. Moreover, it aims to highlight the major strengths and weaknesses of slum upgrade plans, and may provide a useful approach for performance evaluation of these plans. Measuring the performance of slum upgrades is a complex task because of the diverse processes involved. The research in this paper develops a set of parameters for performance evaluation of slum development plans that were either in-site developments or new reallocation plans. This new approach evaluates the quality of the upgrade through a full review. Residents' questionnaires, aimed at measuring user satisfaction with performance quality aspects, were used across four case studies in the Greater Cairo region.

*Keywords*: Slums; slum upgrade; Millennium Development Goal; environmental sustainability

## 1 INTRODUCTION

Slum development has been and continues to be the dominant mode of urbanisation in many developing countries. It is now well understood that slums, and their related informal settlements, are a spontaneous form of urbanisation consisting of a series of survival strategies by the urban poor, most borne out of exclusion (HABITAT III, 2015). The most serious problem of informal settlements in Egypt is one of economic and social security, which influences the safety and stability of Egyptian society as a whole (Hegazy, 2015). For many reasons, the Greater Cairo region is surrounded by 'informal areas' (or 'slums' as they are more commonly known) and, in 2014, they housed approximately 10.5 million people (CAPMAS, 2016); Cairo has been suffering from its surrounding slums for some 30 years (Kipper, 2014). A multidisciplinary integrated approach to evaluat and upgrading slums in Egypt is a key factor essential to sustaining the city's image and identity.

## 2 CONTEXTUAL FRAMEWORK

### 2.1 *Problem and research context*

Informality is a multidimensional socio-economic problem and has appeared as a disease in the urban fabric. Informality comes in many forms, but in most cases it emerges as a result of the inability of cities to absorb population growth within a formal and planned urban framework because of the absence of affordable dwelling options, inadequate building, inadequate planning regulations and lack of suitable housing finance. Many recent studies of slums have pointed to the fact that the majority of the world's population are concentrated in urban areas, reaching 6.3 billion inhabitants in 2012 (Potter, 2012). About 15.5 million Egyptians live in more than 350 slums and almost 40 per cent of these slums are in Cairo. They emerged against the background of rising house prices and the urban crush of the last few decades (HABITAT III, 2015). Informality is a multidimensional problem emerging as a result of inadequate planning strategies and lack of regulations. Given that the Egyptian government has developed many slum upgrade plans, there is a need to focus on plans which promote sustainable development in the coming years. Many slum upgrade plans have focused only on infrastructure development and rehabilitation strategies, and have neglected sustainability objectives as well as socio-economic initiatives to improve quality of life, and to decrease the gap between community needs and government policies.

### 2.2 *Research hypothesis*

The main hypothesis of this paper is based on a belief that multidisciplinary evaluation approach can play a vital role in slum upgrade plans because it seeks to develop a deep understanding of slum living conditions, and to learn from their socio-economic possibilities in order to respond to their future needs.

### 2.3 *Research objectives*

The main objective of this research was to propose a comprehensive approach to evaluating the performance of slum upgrade plans. Providing a proactive and sustainable development will manage the complexity of slum upgrade plans in the future.

### 2.4 *Methods and methodology*

The research methodology consisted of three main parts, as shown in Figure 1 and described in the following sections. For the third part, this paper developed three levels of investigation. The first was to plan a review (using an observation method) which produced a quick snapshot of a slum upgrade plan. The second one was the use of resident questionnaires within interviews designed to measure the performance of in-site upgrade plans. The third one was a detailed survey of reallocation plans not yet implemented.

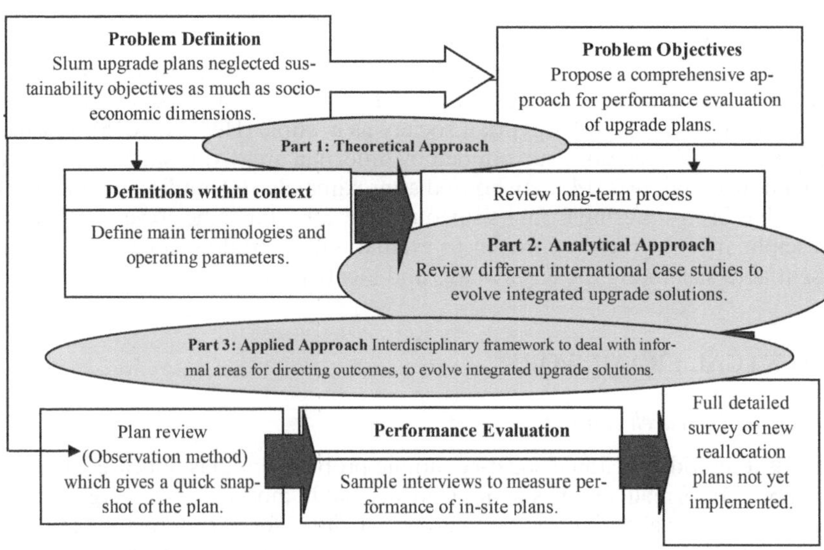


Figure 1. Research methodology.

## 3 THEORETICAL APPROACH

### 3.1 *Definitions within context*

Urbanisation processes have been a significant engine for development in countries, which, in turn, has had important effects on global sustainable development. However, rapid urbanisation has introduced various problems, such as air pollution, traffic congestion, habitat destruction, and loss of available land, which represent threats to the sustainable development of urbanisation in these countries (Shen et al., 2017). In considering the success or otherwise of upgrade plans, it is important to define a number of key terms employed in this context:

- sustainability – a continuous process for the assessment of the existing built situation. The community development of new needs should outline the long-term vision for sustainability. Many cities around the world have developed sustainable urban development plans for leading their urbanisation process towards the desired status of urban sustainability (Fancello et al., 2014). Urban sustainability indicators have been selected as the main elements for determining how successful slum upgrade strategies are (Shen et al., 2013).
- performance evaluation – determining a value for what something is worth. It is important to establish a value for all or part of the built environment because performance addresses the extent to which an upgrade plan has supported goals and objectives and satisfied user needs. Informal areas (or slums) have been products of insufficient housing policies in Egypt since 1950, and a new approach to solutions is needed in order to improve slum quality of life. Informal settlement dwellers are disproportionately affected by ill health, violence and many other socio-economic challenges, which are largely connected to the unhealthy and unsafe physical conditions within which they live (Luthango et al., 2017).
- slum definition – the term 'slum' has generally been associated with any urban society formed outside a legislative framework (El Kholy, 2014).

### 3.2 *Slums – reviewing the long-term process*

The recent definition of informal areas in Egypt has been cited as 'All that is self-built, in the absence of law and urban regulations'. This classification includes areas built on agricultural land and deserts (GOPP, 2007). Urbanisation on agricultural land was the result of a horizontal extension of villages surrounding the capital, combined with a form of urbanisation extending from Cairo city itself (Séjourné, 2012). Informal areas in Egypt were classified into four grades, as shown in Table 1.

Table 1. Classification of slums and approaches to dealing with them.

| Informal Area type | Description | Management strategy |
|---|---|---|
| Grade 1: areas that are threats to life | Under or above sliding geological formations. In floodplain areas. Under threat from railway accidents. | Clearance. Demolition and resettlement. |
| Grade 2: areas with unsuitable housing conditions | Buildings made of makeshift materials, for example, shacks. Sites unsuitable for building, for example, solid waste dumping sites. Derelict buildings. | Reallocation. Demolition and resettlement. |
| Grade 3: areas with health risks | Lacking accessibility to clean drinking water or improved sanitation. Located in the vicinity of industrial pollution or located under electrical power lines. | In-site upgrade. Regulation and legalisation. Development. |
| Grade 4: areas of instability | Areas located on the territory of state-owned land. Areas located on the territory of sovereign quarters. | Development and formalisation. |

In general, slums were built up around three main ideas: construction on someone else's land, construction without following technical rules, and construction without respecting planning regulation (Silva & Farrall, 2014; Sims, 2003).

### 3.3 *Slum perception in Egypt*

In Egypt, the terms 'informal area' and 'slum' have been negatively perceived by the majority of people as places of illegality, problems and crime. However, they act as a housing solution for a major sector of the Egyptian population (Halim, 2014).

Cairo owes most of its physical growth to the development of informal settlements as they inhabit more than 65 per cent of the city's population (Sabry, 2009). According to Khalifa (2011), 'unsafe areas' are territories in which 50 per cent of the housing structures satisfy one or more of the conditions listed in Table 1, ordered according to the degree of risk and thus the urgency for intervention. Approaches to deal with slums have ranged from demolition and resettlement to formalisation.

## 4 ANALYTICAL APPROACH

Sustainability parameters are crucial in spatial analysis and planning in urban development policy as they affect the governmental and community actions in and at local/regional levels. The analytical approach highlights how slum upgrade plans are responsive to the needs of society, and investigates dynamic demands and impacts within international plans.

### 4.1 *Reviewing international slum upgrade plans*

Many slum upgrading plans were reviewed as part of this study in order to identify the most important strengths and weaknesses of informal settlements. Slum upgrades in South Africa had several main objectives. First, to promote social cohesion and improve quality of life for the poor; second, to support the functioning of the entire single residential property market to reduce duality (Magalhães & Villarosa, 2012); finally, to utilise housing as a tool for the development of sustainable human settlements (Napier, 2013). However, it was found that there was an apparent gap between the policy and the reality of implementation, characterised by many problems: the lack of community involvement and choice in the decisions of slum upgrades, lack of access to well located land, limited funding for land acquisitions, lack of capacity and material resource leading to delays in project implementation, poor quality products and settlements, a slowdown in housing delivery, limited or decreasing public sector participation, and the continued growth of informal settlements in the cities (Ziblim, 2013). Kibera in Nairobi has one of the largest slums, with approximately 8,000 residents

Figure 2. Living conditions for slum residents in Kibera before and after upgrade (Ferreira, 2007).

over an area of 2.5 $km^2$. Its upgrade programme offered an affordable opportunity to many residents, which included rehabilitation, ownership/structured owners, and long-term rather than short-term sustainability (Mutisia & Yarime, 2011) (see Figure 2).

From the previous review, it can be concluded that the selected plans were integrated into the city by means of road and infrastructure improvements, combined with social facilities. The solutions proposed for urban and housing problems are compatible with the objective of improving the quality of life and the environment in the cities. Upgrade plans should emphasise society partnerships, involve sustainable development objectives, including economic and social measures (as seen in Brazil).

## 5 APPLIED APPROACH: PERFORMANCE EVALUATION

The past 25 years have witnessed a range of slum upgrade plans for informal areas, which have concentrated on supplying infrastructure and repainting building elevations while neglecting environmental, social and service issues. Involving sustainability parameters is a must in the upgrade actions framework in order to enhance urban context, social and economic cohesion, and changing city identity. Sustainability aspects and specific parameters were selected from the theoretical and analytical approaches for testing in the subsequent part of the methodology.

### 5.1 *Evaluation process and methodology*

As previously indicated, three levels of investigation were developed. The first was to plan a review (using an observation method) which produced a quick snapshot of a slum upgrade plan. The second one was the use of resident questionnaires within interviews designed to measure the performance of in-site upgrade plans. The third one was a detailed survey of reallocation plans not yet implemented. Resident questionnaires aimed to measure user satisfaction with performance quality aspects of two in-site plans (Ezbet Khairallah and Masken Zenhom) in the Greater Cairo region. Statistical techniques were used to assign values to different parameters in a suggested performance matrix.

### 5.2 *Performance evaluation scope and measuring parameters*

A multidimensional evaluation matrix was constructed using parameter values and dimensions ranked according to expert opinions. To assign weights for each selected parameter, a simple questionnaire was designed for local urban and architecture design experts. The following steps were taken:

Table 2. Ranking aspect results.

| Ranking aspects in descending order | 6<br>Urban morphology & housing efficiency | 5<br>Services & facilities | 4<br>Accessibility | 3<br>Environmental & health wellbeing | 2<br>Amenity—safety & privacy | 1<br>Socio-economic facilities |
|---|---|---|---|---|---|---|
| Assignment of proportional weights due to ranking | 1 | 0.8 | 0.6 | 0.5 | 0.3 | 0.2 |

1. a simple online questionnaire method was used among 27 experts in the field of urban planning and upgrade to rank evaluation aspects selected from the theoretical and analytical approach stages; weights for these aspects were then assigned;
2. evaluation aspects were ranked on the basis of the expert questionnaire (see Table 2);
3. a performance evaluation matrix utilised the residents questionnaires to measure the upgrade performance on a two-point evaluation scale (negative or positive influence) then calculating the percentage of positive influenced residnets
4. a field survey was undertaken in order to create a snapshot evaluation of the parameters in reallocation cases (e.g. Imbaba Airport), which have not yet been occupied;
5. the overall performance for each parameter was calculated by multiplying the assigned weights by the percentage of positive resident reporting.

### 5.3 *Case studies within performance evaluation*

Four informal area case studies in the Greater Cairo region were selected (Ezbet Khairallah and Masaken Zenhom as in-site upgrade plans, and Imbaba Airport and Asmarat districts as reallocation plans) according to the following selection criteria:

- areas are located in two different governorates;
- areas are within a similar distance of the capital;
- physical characteristics are representative of real population and living conditions.

These case studies are briefly described below. A total of 65 persons in each selected area were interviewed and investigated. The questionnaire was designed with evaluation parameters seleccted from the analytical stage.

#### 5.3.1 *Ezbet Khairallah*

(Ezbet Khairallah represents one of Cairo's largest informal settlements (see Figures 3 and 4). The population is about 650,000 residents in 480 feddan (~460 acres) on a rocky plateau south of Dar el-Salam. The projects have been funded by different sources: the European Union, the World Bank and the German Agency for International Cooperation (GIZ). In the last 10 years, the government started to install water, sewage and electricity infrastructure in this area.

#### 5.3.2 *Masaken Zenhom*

Masaken Zenhom has 4,000 housing units for 20,000 inhabitants. It was an in-site development on 50 feddan (~48 acres) (see Figure 5). Before upgrading, housing units were wooden units made by the Cairo government for those who lost their units in the Aboelrish area. They stayed in these units for 30 years until the project started in 1998. It was completed in 2007 and was divided into three phases (see Figure 6). Services provided were schools, youth centres, culture centres and a charity hospital. Since 2009, the project has lacked sustainability provisions and rapidly deteriorated in performance.

Figure 3. Open spaces in (Ezbet Khairallah).

Figure 4. Conditions of buildings and streets in (Ezbet Khairallah).

Figure 5. Buildings and streets in Masaken Zenhom after upgrading.

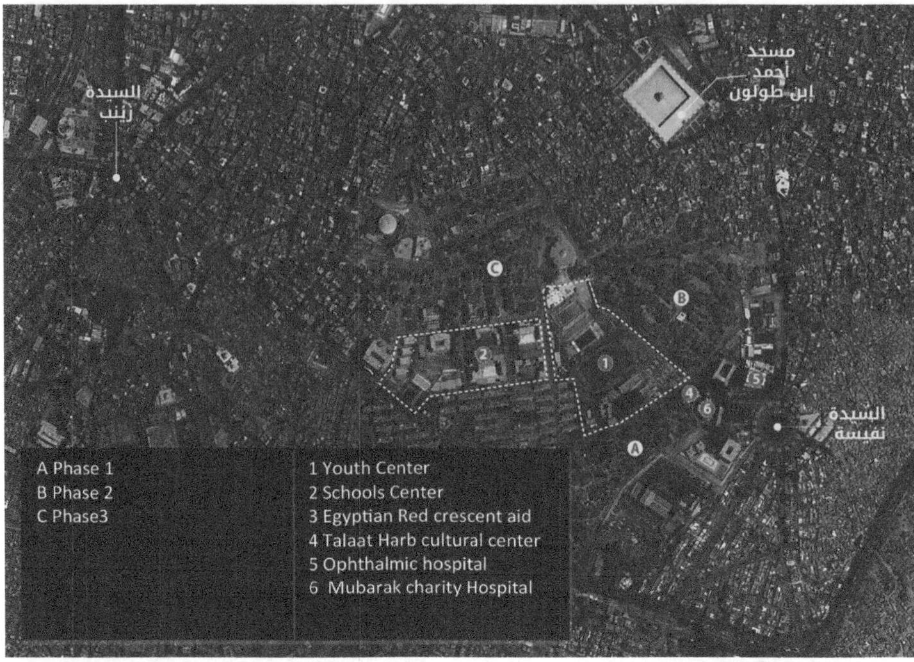


Figure 6. Masaken Zenhom upgrade plan phases.

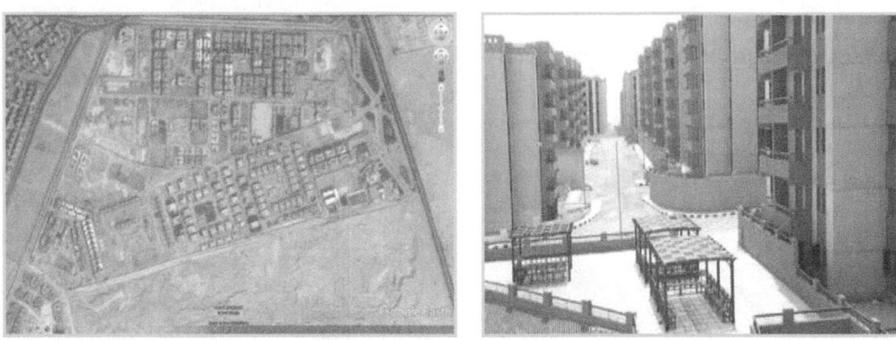

Figure 7. Buildings and streets in the Asmarat district project.

Figure 8. Buildings and streets in Imbaba Airport.

### 5.3.3 *Asmarat district*

In the first phase, a total of 6,258 housing units were constructed on 65 feddans (~63 acres) (see Figure 7). Meanwhile, the second phase included 4,722 housing units, plus hospitals, schools and other public utilities on 61 feddans (~59 acres) at a cost of EGP700 million (financed by the Long Live Egypt Fund). Three schools were also built in the Al Asmarat district that aimed to service the new residents moving to the area. Al Asmarat first and second districts were previously delivered in Moqattam, comprising a total of 11,000 housing units.

### 5.3.4 *Imbaba airport*

Imbaba Airport in the Imbaba area, Giza Governorate, is considered one of the most populated and unplanned urban areas of Egypt, with 700,000 inhabitants. This plan was developed as a cornerstone of the North Giza regional development plan concerning slum upgrades at a local level. It was located in North Giza district within an area of 204 feddans (~196 acres). The upgrade plan was developed with three main phasing levels: first was the airport land development as an action plan; second was West Mounira as a local urban plan; third was the regional level for North Giza as a whole at the action planning level. The development utilised a policy of gradual replacement. Components of the land use plan were: residential area of 52 feddans (~50 acres) for 5,000 families (see Figure 8); public gardens of 38 feddans (~36 acres); 69 feddans (~66 acres) for economic investments; service centres of 26.6 feddans (~26 acres). The proposed upgrade project provided Imbaba with the basic, necessary infrastructure and services. Playgrounds, youth centres, a student sports city, public parks and green areas were also planned, but the target population has not been achieved yet.

## 5.4 *Evaluating slum upgrade plans*

An indicative review utilised an observation method checklist (see Table 3) which gave a quick snapshot of each case study and highlighted major strengths and weaknesses.

### 5.4.1 *Performance evaluation for case studies*

Strengths and weaknesses of sustainability performance evaluation in six dimensions were identified and linked to parameters in their contexts. Table 4 represents the findings of questionnaire

Table 3. Case studies general aspects checklist.

| Parameters | Ezbet Khairallah | Masaken Zenhom | Asmarat district | Imbaba Airport |
|---|---|---|---|---|
| Mixed land use | ✓ | ✓ | – | – |
| Vitality | ✓ | ✓ | – | – |
| Open public green space | – | – | ✓ | ✓ |
| Market availability | ✓ | ✓ | – | – |
| Space organisation | ✓ | ✓ | | ✓ |
| Good smell and noise reduced | – | – | ✓ | ✓ |
| Accessibility managed and secured | – | – | ✓ | ✓ |
| Flexible street shape network | – | – | ✓ | ✓ |
| Crowded vehicle flow | ✓ | ✓ | – | – |
| Availability of transportation nodes | ✓ | ✓ | ✓ | – |
| High densities | ✓ | ✓ | – | – |
| Social mix and backgrounds | ✓ | ✓ | ✓ | – |
| Low maintenance | ✓ | ✓ | – | ✓ |
| Distribution of activities | – | – | ✓ | – |
| Social communication and participation | ✓ | – | ✓ | – |
| Local security and amenity | – | – | – | – |
| Poor income | – | – | – | – |

Table 4. Sustainability performance evaluation results for case studies.

| Aspect | Weighting according to Ranks | Weight per parameter | Upgrade performance parameters | (Ezbet Khairallah % | (Ezbet Khairallah Total | Masaken Zenhom % | Masaken Zenhom Total | Asmarat district % | Asmarat district Total | Imbaba airport % | Imbaba airport Total |
|---|---|---|---|---|---|---|---|---|---|---|---|
| Urban morphology and housing efficiency | 1 | 0.1 | Legible visible structure | 10 | 1 | 87 | 8.7 | 97 | 9.7 | 1 | 0.1 |
| | | 0.1 | Legible visual image | 70 | 7 | 95 | 9.5 | 95 | 9.5 | 1 | 0.1 |
| | | 0.1 | Site quietness | 12 | 1.2 | 17 | 1.7 | 100 | 10 | 0 | 0 |
| | | 0.1 | Playground availability | 93 | 9.3 | 97 | 9.7 | 100 | 10 | 1 | 0.1 |
| | | 0.1 | Public space availability | 10 | 1 | 12 | 1.2 | 100 | 10 | 1 | 0.1 |
| | | 0.1 | Children's play area | 0 | 0 | 0 | 0 | 100 | 10 | 1 | 0.1 |
| | | Aspect total performance | | 19.5 | | 30.8 | | 59.2 | | 0.5 | |
| | | 0.1 | Good housing conditions | 10 | 1 | 27 | 2.7 | 100 | 10 | 1 | 0.1 |
| | | 0.1 | House area > 50 m$^2$ | 10 | 1 | 15 | 1.5 | 100 | 10 | 1 | 0.1 |
| | | 0.1 | No. of rooms > 2 | 12 | 1.2 | 0 | 0 | 0 | 0 | 1 | 0.1 |
| | | 0.1 | No. of family members 4 | 6 | 0.6 | 12 | 1.2 | 20 | 2 | 0 | 0 |
| | | Aspect total performance | | 2.8 | | 2.7 | | 12 | | 0.3 | |
| Services and facilities | 0.8 | 0.26 | Walkability to services. | 37 | 9.62 | 83 | 21.58 | 90 | 23.4 | 1 | 0.26 |
| | | 0.26 | Availability of education service | 0 | 0 | 0 | 0 | 100 | 26 | 1 | 0.26 |
| | | 0.26 | Availability of health service | 0 | 0 | 70 | 18.2 | 70 | 18.2 | 1 | 0.26 |
| | | Aspect total performance | | 9.62 | | 39.78 | | 67.6 | | 0.78 | |
| Accessibility | 0.6 | 0.3 | Accessibility to public transport | 97 | 29.1 | 75 | 22.5 | 95 | 28.5 | 1 | 0.3 |
| | | 0.3 | Accessibility to work | 56 | 16.8 | 34 | 10.2 | 12 | 3.6 | 0 | 0 |
| | | Aspect total performance | | 59.32 | | 77.88 | | 121.7 | | 0.3 | |

| | | | | | | | | | | | |
|---|---|---|---|---|---|---|---|---|---|---|---|
| Environmental and health wellbeing | 0.5 | 0.125 | Cleanliness | 17 | 2.125 | 30 | 3.75 | 100 | 12.5 | 1 | 0.125 |
| | | 0.125 | Air ventilation in units | 14 | 1.75 | 80 | 10 | 100 | 12.5 | 1 | 0.125 |
| | | 0.125 | Pollution minimised | 12 | 1.5 | 40 | 5 | 90 | 11.25 | 0 | 0 |
| | | 0.125 | Waste separation units inside | 83 | 10.375 | 0 | 0 | 0 | 0 | 0 | 0 |
| | | Aspect total performance | | 15.75 | | 18.75 | | 36.25 | | 0.25 | |
| Amenity—safety and privacy | 0.3 | 0.075 | Feeling safe and secure | 32 | 2.4 | 70 | 5.25 | 78 | 5.85 | 0 | 0 |
| | | 0.075 | Increasing resident's control | 78 | 5.85 | 56 | 4.2 | 58 | 4.35 | 0 | 0 |
| | | 0.075 | Local amenity availability | 5 | 0.375 | 70 | 5.25 | 76 | 5.7 | 0 | 0 |
| | | 0.075 | Children play outside units | 30 | 2.25 | 45 | 3.375 | 86 | 6.45 | 0 | 0 |
| | | Aspect total performance | | 10.875 | | 18.075 | | 22.35 | | 0 | |
| Socio-economic facilities | 0.2 | 0.04 | Surrounded by relatives | 84 | 3.36 | 12 | 0.48 | 37 | 1.48 | 0 | 0 |
| | | 0.04 | Liveability and vitality | 94 | 3.76 | 70 | 2.8 | 15 | 0.6 | 0 | 0 |
| | | 0.04 | Supportive neighbours | 96 | 3.84 | 80 | 3.2 | 17 | 0.68 | 0 | 0 |
| | | 0.04 | Community participation | 97 | 3.88 | 45 | 1.8 | 0 | 0 | 0 | 0 |
| | | 0.04 | Social communication | 86 | 3.44 | 65 | 2.6 | 63 | 12.6 | 0 | 0 |
| | | Aspect total performance | | 14.92 | | 10.4 | | 13.88 | | 0 | |
| Overall resident satisfaction | | Satisfied with upgraded housing units | | 10 | | 84 | | 100 | | 0 | |
| | | Satisfied with services | | 0 | | 35 | | 100 | | 0 | |
| | | Satisfied with accessibility | | 98 | | 46 | | 78 | | 0 | |
| | | Empowering district participation | | 13 | | 18 | | 0 | | 0 | |
| | | Overall satisfaction | | 123.984 | | 183 | | 278 | | 0 | |
| | | Total performance | | 132.785 | | 198.385 | | 332.98 | | 2.13 | |

method and graphs were created to evaluate the outcome measurements and compare the four case studies (Figures 9 and 10).

Parameters Numbered as shown in matrix on × axis

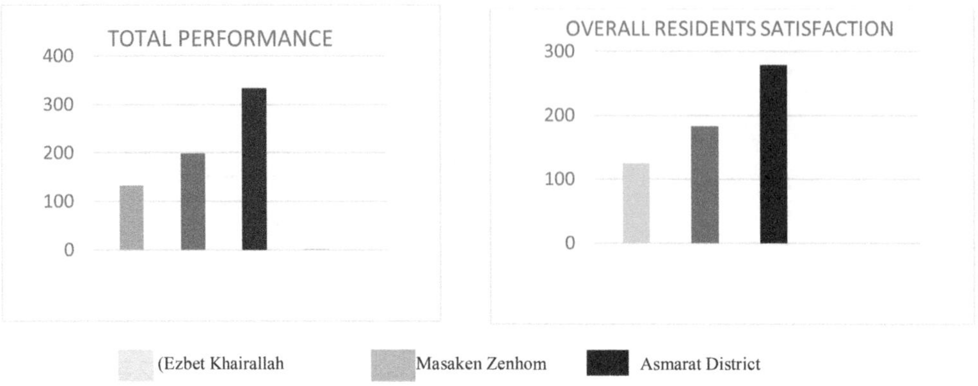


Figure 9. Total performance and overall satisfaction for case studies.

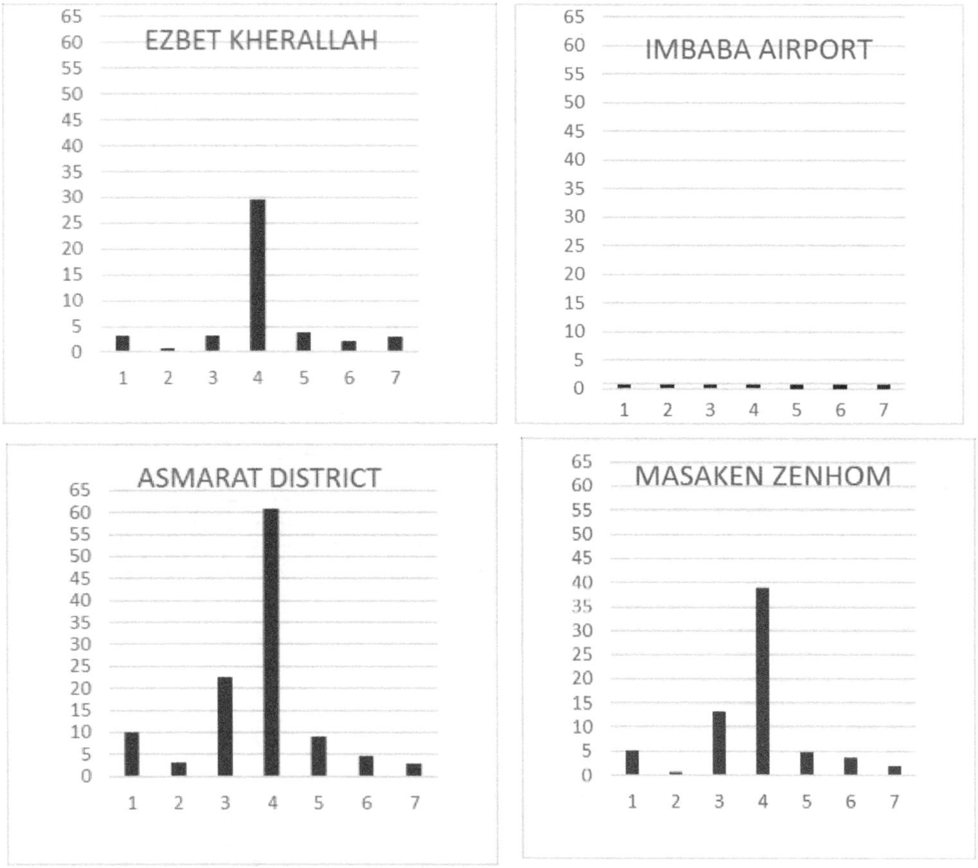


Figure 10. Parameters of performance for each upgrade plan studied.

The Asmarat district achieved the highest value in total performance (333) and resident satisfaction (278), Masaken Zenhom came second (total performance score of 198) while EzbetKhairallah had the lowest performance and resident satisfaction.

## 6 MAIN FINDINGS

1. The applied evaluation approach highlighted the aspects neglected in the upgrading plans: socio-economic facilities and environmental health and wellbeing.
2. In-site upgrade plans were not able to deliver expected outputs. The plans did not move residents' lifestyles towards healthier living.
3. Reading the matrix vertically shows the most neglected parameters that achieved negative influence (shown in grey colour in each studied area). The matrix showed the numerical assessment of the upgrade plan performance evaluation based on community perspective and sustainable performance parameters. It shows the percentage of positive influence of the upgrading parameter. Accessibility, services and facilities achieved a highly positive influence.
4. Reading the matrix horizontally gives a comparative analysis between the four case studies (in-site development and reallocation cases) within each individual parameter.
5. An average of total value for parameters shows the overall performance of each aspect for each case study. Asmarat district, a reallocation plan, achieved the highest total performance with 332.98 points. In terms of in-site upgrade plans, Masaken Zenhom achieved a higher score than Ezbet Khairallah.
6. Asmarat district achieved the highest resident satisfaction and, in terms of in-site upgrade plans, Masaken Zenhom scored more highly than Ezbet Khairallah.

### 6.1 *Analysis of findings*

With regard to in-site development plans in terms of the performance aspects of morphology and accessibility, Ezbet Khairallah and Masaken Zenhom achieved the highest values in availability of playgrounds, accessibility to public transport, safety and security facilities, and walkability to services. Both achieved low values in the aspects of housing efficiency, environment and socio-economics. Over 90 per cent of residents investigated in Ezbet Khairallah were not satisfied with the upgrading plan.

In terms of the performance of the reallocation plans, Imbaba lacked the development dimension as it focused on the urban morphology parameters while neglecting socio-economic aspects. This resulted in no life achievements for the past six years. It is obvious that there was no social participation from residents or Non-Governmental Organisations (NGOs) in the upgrade plan because it was carried out by the government in order to achieve the President's programme for slum development. Asmarat, which was recently finished, performed better in urban aspects, transportation, accessibility and availability of services, such as public open spaces.

### 6.2 *Conclusion*

The above findings reflect the conceptualisation of slum problems not as housing problems, but as the product of an underlying socio-economic problem that needs to be redefined and addressed in Egypt. After analysing the upgrade plan dimensions and evaluating performance with the test matrix, it was found that the case studies were concerned with morphological dimensions and neglected social, environmental and economic dimensions. This showed that in-site upgrade plans did not improve the quality of life for slum residents and did not enhance environmental or social aspects. This supports the hypothesis of this research.

On the basis of the Imbaba Airport case findings, it is concluded that some slum upgrade plans were a form of political propaganda and did not enhance the quality of life for residents nor did they support local community needs. This particular plan was developed from political decisions and was a central plan developed by the Giza Governorate without any coordination or participation from the local community. Reallocation upgrade plans show the most strengths and the fewest weaknesses; otherwise, gradual replacement could be an effective upgrade solution in the case of there is available land for replacement in the same site The results of this research can provide valuable guidance to planners and decision makers of sustainability indicator programmes and planners of new programmes. This evaluation approach for upgrading plans provide an interdisciplinary framework involving social and economic aspects. It developed a sustainable performance evaluation framework by identifying the main objectives and outcomes of upgrade plans. Using the suggested matrix for a sustainable slum development could help manage the complexity of future upgrading plans in Egypt.

### 6.3 *Key recommendations*

- Redefine the parameters of informal areas within the delivered matrix for comprehensive balanced slum upgrade plans.
- In-site upgrading is always preferable to reallocation, in order to maintain social and economic networks. If there is going to be relocation, it should be to well located land.
- Informal settlement upgrades always need to be part of an integrated housing strategy that includes a range of delivery options to meet differing housing needs.
- , It is important to have reliable and up-to-date information about community to work on for an integrated development Real community participation is essential at all levels from strategy to project implementation; participation in allocation processes, layout design and unit design is particularly important.
- Residents are the experts in their own area; they have the best knowledge of their problems, causes and possible solutions.
- Resident's real needs should be investigated through a user satisfaction questionnaire and survey, especially for young people, women and the disabled.

## REFERENCES


CAPMAS. (2016). *Central Agency for Public Mobilization And Statistics statistical year book*. Cairo, Egypt: CAPMAS.

El Kholy, A. (2014). *The response of architectural education to informal areas: Community based learning approaches, responsive urbanism in informal areas*. Paper presented at the 6th International conference—ArchCairo2014, Cairo, Egypt.

Fancello, G.A., Carta, M.A. & Fadda, P.A. (2014). A modeling tool for measuring the performance of urban road networks. *Procedia—Social and Behavioral Sciences*, *111*, 559–566.

Ferreira, J. (2007). *Upgrading in South Africa: A study commissioned by Habitat for Humanity.* Bratislava, Slovakia: Habitat for Humanity International/EMEA Office.

General Organization for Physical Planning (GOPP). (2007). *The strategic urban development master plan study of the greater Cairo region, draft report.* Cairo, Egypt: Nepon.

Habitat III. (2015). *Issue Papers 22 – Informal Settlements.* New York, NY: UN-Habitat. Retrieved from https://unhabitat.org/wp-content/uploads/2015/04/Habitat-III-Issue-Paper-22_Informal-Settlements.pdf.

Halim, K.M.A. (2014). PDP's methodology for participatory urban upgrading. In R. Kipper & M. Fischer (Eds.), *Cairo's informal areas: Between urban challenges and hidden potentials. Facts. Voices. Visions* (pp. 125–129). Cairo, Egypt: GTZ. Retrieved from http://www.citiesalliance.org/sites/citiesalliance.org/files/CA_Docs/resources/Cairo%27s%20Informal%20Areas%20Between%20Urban%20Challenges%20and%20Hidden%20Potentials/CairosInformalAreas_fulltext.pdf.

Hegazy, I. (2015). Informal settlement upgrading policies in Egypt: Towards improvement in the upgrading process. *Journal of Urbanism: International Research on Place Making and Urban Sustainability*, *9*(3), 254–275.

Khalifa, M. (2011). Redefining slums in Egypt: Unplanned versus unsafe areas. *Habitat International*, *35*, 40–49.

Kipper, R. (2014). Cairo: A broader view. In R. Kipper & M. Fischer (Eds.), *Cairo's informal areas: Between urban challenges and hidden potentials. Facts. Voices. Visions* (pp. 13–17). Cairo, Egypt: GTZ. Retrieved from http://www.citiesalliance.org/sites/citiesalliance.org/files/CA_Docs/resources/Cairo%27s%20Infor-mal%20Areas%20Between%20Urban%20Challenges%20and%20Hidden%20Potentials/CairosInformalAreas_fulltext.pdf.

Luthango, B., Reyes, E. & Mntungwa, G. (2017). Informal settlement upgrading and safety: Experiences from Cape Town, South Africa. *Journal of Housing and the Built Environment*, *32*(3), 471–493.

Magalhães, F. & Villarosa, F. (Eds.). (2012). *Slum upgrading: Lessons to be learned from Brazil*. Washington, DC: Inter-American Development Bank. Retrieved from http://www.citiesalliance.org/sites/citiesalliance.org/files/Slum-Upgrading-Lessons-from-Brazil.pdf.

Mutisia, E. & Yarime, M. (2011). Understanding the grassroots dynamics of slums in Nairobi: The dilemma of Kibera informal settlements. *International Transaction Journal of Engineering, Management, & Applied Sciences & Technologies*, *2*, 197–213.

Napier, M. (2013). *Informal settlement integration, the environment and sustainable livelihoods in South Africa*. Pretoria, South Africa: Program for Sustainable Human Settlements, Council for Scientific and Industrial Research.

Potter, G. (2012). *Urbanizing the developing world*. In Worldwatch Institute (Ed.), *Vital signs* (Vol. 20). Washington, DC: Island Press.

Sabry, S. (2009). Egypt's informal areas: Inaccurate and contradictory data. In R. Kipper & M. Fischer (Eds.), *Cairo's informal areas: Between urban challenges and hidden potentials. Facts. Voices. Visions* (pp. 29–33). Cairo, Egypt: GTZ. Retrieved from http://www.citiesalliance.org/sites/citiesalliance.org/files/CA_Docs/resources/Cairo%27 s%20Infor-mal%20 Areas%20Between%20Urban%20Challenges%20and%20Hidden%20Potentials/CairosInformalAreas_fulltext.pdf.

Séjourné, M. (2012). The history of informal settlements. In R. Kipper & M. Fischer (Eds.), *Cairo's informal areas: Between urban challenges and hidden potentials. Facts. Voices. Visions* (pp. 17–24). Cairo, Egypt: GTZ. Retrieved from http://www.citiesalliance.org/sites/citiesalliance.org/files/CA_Docs/resources/Cairo%27 s%20Infor-mal%20 Areas%20Between%20Urban%20Challenges%20and%20Hidden%20Potentials/CairosInformalAreas_fulltext.pdf.

Shen, L., Ochoa. J., Shah, M. & Zhang, M. (2013). The application of urban sustainability indicators—A comparison between various practices. *Habitat International*, *35*, 17–29.

Shen, L., Shuai, C., Jiao, L., Tan, Y. & Song, X. (2017). Dynamic sustainability performance during urbanization process between BRICS countries. *Habitat International*, *60*, 19–33.

Silva, P. & Farrall, H. (2014). Lessons from informal settlements: A 'peripheral' problem with self-organising solutions. *Town Planning Review*, *87*(3), 297–319.

Sims, D. (2003). Urban slums reports: The case of Cairo, Egypt. In *Understanding slums: Case studies for the global report on human settlements 2003*. Cairo, Egypt: GTZ. Retrieved from http://www.ucl.ac.uk/dpu-projects/Global_Report/pdfs/Cairo.pdf.

Ziblim, A. (2013). *The dynamics of informal settlements upgrading in South Africa: Legislative and policy context, problems, tensions, and contradictions*. Bratislava, Slovakia: Habitat for Humanity International/EMEA Office. Retrieved from http://globalhousingindicators.org/sites/globalhousingindicators.org/files/The%20Dynamics%20of%20Informal%20Settlements%20Upgrading%20in%20South%20 Africa.pdf.

*Cities' Identity Through Architecture and Arts – Catalani et al. (Eds)*
*© 2018 Taylor & Francis Group, London, ISBN 978-1-138-05409-7*

# Urban identity and lifestyles of gated communities in Egypt

Mohamed M. Abdelaziz Farid
*Department of Architecture, Faculty of Fine Arts, Helwan University, Cairo, Egypt*

Alaa Mohamed Samy Ahmed
*Canadian International College, El-sheikh Zayed Campus, 6th of October City, Egypt*

ABSTRACT: At the end of the twentieth and the beginning of the twenty-first century, the urban landscape faced a rapid and dramatic increase in Gated Communities (GCs) in most of the cities around the world. As part of the trend in suburbanisation, such changes in the urban environment are often due to change in lifestyles, which is an important factor in urban identity. This paper represents the ways GCs are identified and defined in Egypt, and focuses on identifying the main lifestyle factors influencing their urban identity. These factors should be taken into account by planners, designers and policymakers to minimise negative impacts and maximise positive consequences. The findings confirm that GCs of Egypt have special lifestyle characteristics. The paper makes some suggestions for strengthening and sustaining the remaining urban identity of gated community features that are likely to be part of the urban landscape for a long time.

*Keywords*: urban landscape; urban identity; residential gated community; lifestyle

## 1 INTRODUCTION

The notion of identity is a set of meanings that reflect local traditions, culture and aspirations. It reflects their needs, their successes, their failures and their future. At the urban level, identity can be defined as a process where people interact with places and describe themselves in terms of belonging to a specific place (Lynch, 1981). Its concept has gradually become a major issue in contemporary urban planning due to changing lifestyles in most cities worldwide. Urban identity has an influence on cities and their users, and its continuity is an essential tool in creating meaningful and liveable cities.

The issue of urban identity in Egypt is not a new topic, but reviewing and evaluating urban identity in new gated communities (GCs), which are located in the extensions of Cairo city towards new cities, as their numbers increase makes the challenge of maintaining urban identity appear to be one of the biggest tasks of this century.

### 1.1 *Object of the research*

The main question concerns local identity and the differentiation between GCs, as many have a kind of international identity (Italian, Greek, etc.). Therefore, the research question is 'How to manage the change and sustain the remaining urban identity of GCs?' This entails a review of the ways GCs are identified and defined in Egypt, identification of the main factors influencing their urban identity, and reviewing the various lifestyles offered by specific GCs in Egypt.

### 1.2 *Research methodology*

The research methodology was based on an inductive approach through a theoretical study of the phenomena of GC and the concept behind their spread, by defining the various categories of GCs and their main features. The history of GCs in Egypt and their associated lifestyles, and deducing the relationship between their lifestyle and mass media was also studied. A deductive and analytical approach was used on a case study of various categories of GCs in Egypt. Analysis of lifestyle factors that influenced urban identity was then carried out in order to identify which GCs offered a unique identity and lifestyle.

## 2 THEORETICAL FRAMEWORK FOR THIS STUDY

### 2.1 *The phenomenon of gated communities*

In the late twentieth century, an ancient urban form began to reappear in modern settlements (Judd, 1995). After several centuries, there were many definitions used to conceptualise the phenomenon of GC. This paper defines GCs as self-contained separate communities with carefully constructed identities (Baycan-Levent & Gülümser, 2004) where entry is controlled and public open spaces are privatised (Low, 2003). They have emerged as a new trend in the housing market, appearing comparable but likely differing in their history, reasons and physical features (Low, 2001). According to Blakely and Snyder (1997), there are three main categories with different degrees of facilities, exclusivity and security. The first type of lifestyle communities often highlights leisure activities, with recreational amenities, facilities and shared services at their centre. The second type is a prestige community that often highlights the attractive richness of its environment and the quality of security, concentrates on exclusivity and privacy over community, and does not often include common facilities. The third type is secure communities with controlled traffic and maintenance of property values, which reflected people's fear of crime. In the next section, a review of the types of GCs that were identified and defined in Egypt will be given.

### 2.2 *Gated communities in Egypt*

GCs demonstrate a new phase in the urbanisation of Egypt. They first appeared as summer and 'rest houses' in the coastal zones but, by the mid-nineties, the Egyptian government began to search for new urban developments where the housing communities were the main spine. The government sold large portions of public land to the private sector that offered new options for standards and styles of living with unique urban identities. There are a number of variables which played an important role in reshaping the visual perception of the urban identity through the intensive use of globalised lifestyles; these attracted people by focusing on visions and modernity through media and marketing advertisements. In addition, they appear to guarantee an idealised vision of an appealing lifestyle. The field study will analyse the urban identity and lifestyle of GCs in Egypt.

#### 2.2.1 *Lifestyle of gated communities*

Development companies represent GCs as a complete lifestyle rather than a house in a residential environment. This given lifestyle is pre-designed, planned and constructed to fulfil the demands and needs of every social class. According to Bali (2009), the identity of residents that belong to a city has been replaced by belonging to a lifestyle offered by a GC; it is called 'town citizenship' (Aydın Yönet & Yirmibesoğlu, 2015). Nevertheless, there still appears to be a sense of belonging to or identification with a GC. Mutual relations associated with a specific identity in a sociocultural setting establish a sense of community in combination with physical measures; this can lead to an 'us and them' community. The gates have a double function of social control, as they include look-alikes and exclude those who are different; residents identify themselves with a social form, which is influenced by internal and external forces (Aalbers, 2003).

#### 2.2.2 *Lifestyle and mass media*

In the contemporary world, the impact of mass media or mass advertising on showing individuals' status or the formation of their status cannot be ignored. As a matter of fact, these instruments inspire people to mass consumption. GC advertising commercials often tell of a lifestyle. They promise that residents will belong to a specific social environment thanks to the leisure activities provided within the community, and that children can safely run in the garden and participate in various activities. Generally, the declaration of the architectural style, or a design characteristic of a famous architectural style, and the lifestyle would be enough. On the other hand, GCs have recently started to include middle-income groups in the promotion of these projects, emphasising the price range and the payment schedule as well as the social activities and constructional attributes, while marketing material that targets the upper classes gives information about specialised sports facilities and shopping centres specialised for high-income groups. Thus, the promise of community lifestyle became a good marketing tool.

## 3 APPLIED RESEARCH

As of 2010, there were more than 450 GCs and their numbers are still on the increase (Soliman Muawwad et al., 2011). Most are located in new urban areas found east and west of Cairo, where they form an extension of Cairo city in the direction of the new cities of New Cairo, 6th of October and Sheikh Zayed. Therefore, the applied research within this paper analysed many patterns of urban identity in one of the most important and fast-developing regions on the edge of Cairo city.

### 3.1 *Types of gated communities in Egypt*

Through observations during site visits, analysis of features and promotional data, four categories of GCs were identified. These were primarily based on physical features, characteristics that form each type, residents' lifestyle and identification of housing types.

#### 3.1.1 *Luxurious gated communities*

Luxurious GCs are similar to prestige GCs; however, they differ in the diverse social status of their residents, and segregation from the rest of the community remains the main reason to live in them (Aalbers, 2003). These types of communities are completely gated. The developer targets the upper-high economic class by presenting a new concept in the Egyptian urbanisation experience (Charmes, 2012). The resident lifestyle in this kind of GC offers groups of features; for example, security, facilities, amenities, richness of landscape views, privacy and the best property values. An example of such GCs are Allegria, Westown City, PalmHills.

#### 3.1.2 *Moderate gated communities*

Moderate GCs are similar to the lifestyle-type GCs but with some different characteristic elements. The developers are highly concerned about the quantity of units in the community in order to achieve the best profitability, so apartment building is the predominant type. They also have an exclusive medium-cost residential subdivision which has moderate features and amenities, some of which are not completely gated. This means that they are partially open to the public and depend on the potential business of non-residents, as well as residents, in the use of shared recreational facilities (Almatarnah, 2013). This sharing of lifestyle choice with others from outside the gated community does bring with it security issues (which may be a minor secondary factor). Examples of this type of GC are El Rehab city located in New Cairo city.

#### 3.1.3 *Lower-high gated communities*

The form of housing in this type of GC appears to be different. Apartment buildings are clearly the affordable answer for such GCs. Residents with lower-high economic status can afford to live there; however, the price of properties and the presence of amenities and features also differ from one gated community to another. While lower-high-income GCs remain a relatively expensive choice compared to city properties, they provide residents with security and a better lifestyle. Examples of this type of GC are El Masraweya, Al Karma.

#### 3.1.4 *Affordable gated communities*

These are an exclusive economic housing development with essential amenities and small lot sizes intended for normal-income families. They vary from other GCs primarily in terms of their selling value, the minimum lot size, the developed area, the amenities and street networks provided. Many residents of these subdivisions earn too little to purchase a home in GCs or in surrounding areas, making the demand for them low. However, the number of affordable GCs is increasing today because many employees want to live in affordable homes that are close to their jobs. Examples of this type of GC are Ashgar Heights, Continental Gardens.

### 3.2 *Reasons for case study selection*

The four categories of GCs described are based on their physical features and the characteristics of lifestyle associated with each type. The following GCs were chosen as case studies: Allegria, El Rehab, Al Karma and Continental Gardens. The selected case studies are located mainly in New Cairo, Sheikh Zayed and 6th of October. The analysis was carried out by the use of observation gathering methods based on the difference in lifestyle factors that were most prominent across the GCs. These factors can be grouped under the following subcategories:

- security (gates, fences, traffic limits and social control);
- exclusivity and privacy;
- social (sense of community, social activities and relationships);
- environmental (landscape elements); and green areas
- architectural and urban character.

### 3.3 *Observations of the case studies*

The following analysis was completed by undertaking site visits, perception, a survey of GC designs and an examination of the lifestyle factors.

#### 3.3.1 *Security*

Security was seen as an important factor in choosing a residential environment. It was used to promote a lifestyle in advertising material for most of the GCs.

Figure 1 shows the security used in different GCs based on controlled gates supported by 24-hour private guards and controlled traffic. Fences surrounding the GCs also provided social control. Figure 1 indicates that developers used the different types of security to instil a sense of security. Luxurious GCs appear to have the highest level of security.

#### 3.3.2 *Exclusivity and privacy factors*

Exclusivity and privacy factors are promoted to address the needs of residents in terms of their desired lifestyle. The use of gates and walls fulfil a dual function of offering social control as well as exclusive activities, such as special sports, swimming pools, water features and landscaping. Developers create an exclusive lifestyle and offer immediate membership of a community where residents identify themselves with a social type. An exclusive lifestyle and private environment are highlighted in the slogans and images of the GCs' marketing material, so that it instils a sense of exclusiveness and privacy as a ready-made lifestyle.

Figure 2 indicates that privacy and exclusivity are more important factors in luxurious GCs than in any other types because they have the most valuable amenities. For example, Allegria has a world-class golf course.

#### 3.3.3 *Social factors*

GCs use social factors in their promotional marketing materials. They often display images of a happy family with elegant style to respond to a potential resident's emotional need for social activities, a social relationship between residents, and a sense of community.

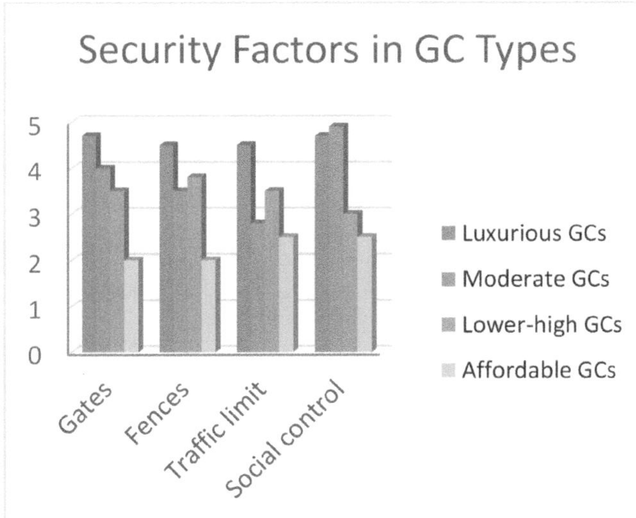


Figure 1. Percentage of security factor in each GC type.

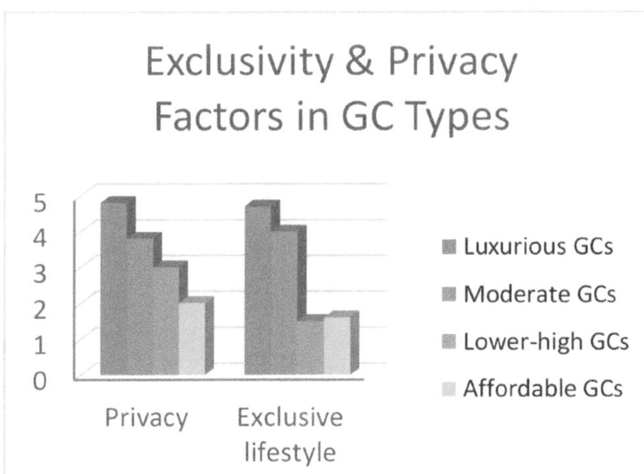


Figure 2. Percentage of exclusivity and privacy in each GC type.

Their hierarchy of open spaces that create social networks gives their residents a sense of community. As in moderate GCs (Al Rehab), building apartments are arranged in clusters and surrounded by courtyards. They are interconnected by a pedestrian network bordered by green strips which increases the social interaction between residents.

Figure 3 indicates that residents of luxurious GCs highlighted social activities as the most important social factor.

#### 3.3.4 *Environmental factors*

The environmental factors are often displayed as green areas, richness of views, good climate and ease of maintenance in each type of GC.

Figure 4 shows the various environmental elements assessed. Green areas were present in about 70 per cent of luxurious GCs and 40 per cent of moderate GCs. The amount of green space per person decreased in affordable GCs.

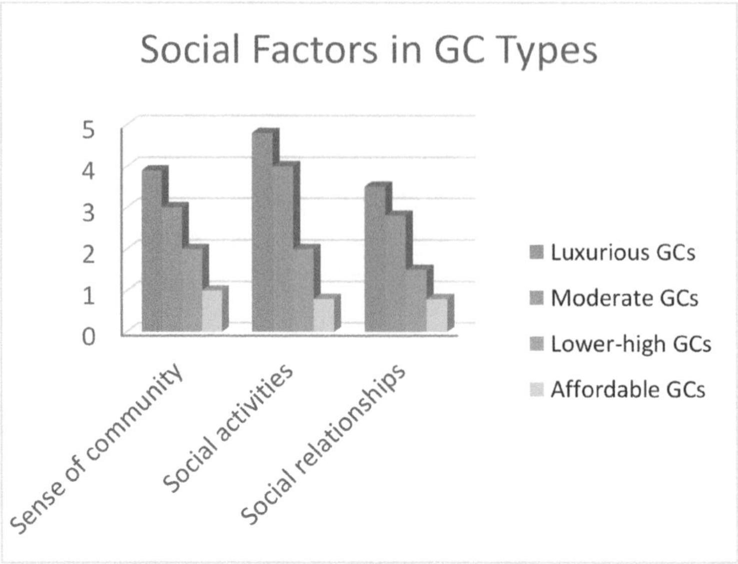


Figure 3. Percentage of social factors in each GC type.

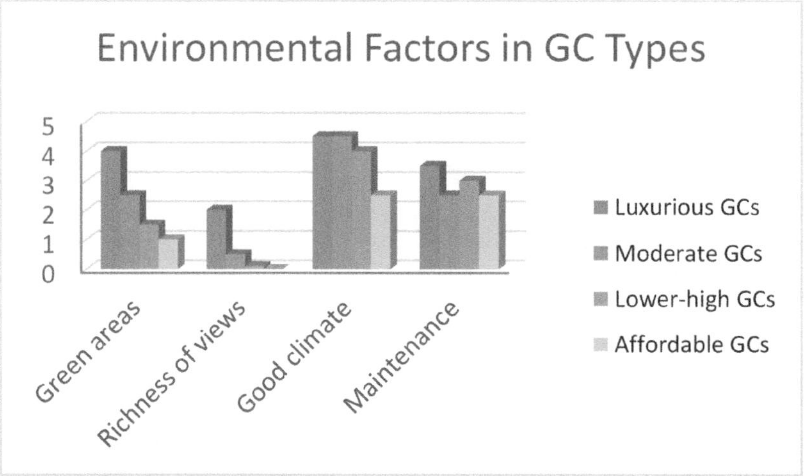


Figure 4. Percentage of environmental factors in each GC type.

### 3.3.5 *Architecture and urban character factors*

Urban character can be grouped under the subcategories of land use facilities, urban fabric, road network, and squares and open spaces. Figure 5 indicates that planned land use involves a variety of facilities, especially on luxurious and moderate GCs. Also, GCs Urban fabric which determins the features of the urban structure, various between linear, semi-pointed. The hierarchical road network used to define each zone depends on the design concept employed.

Architectural character elements can be grouped under the subcategories that affects the visual perception of building character, as the housing type, modernity of the architectural style, the balance of the skyline, opening shape, relation between building scale with human scale, level per apartments, quality of finishing materials, exterior paint colour, exterior texture, and porosity which indicate the ration between the opening to solid walls.

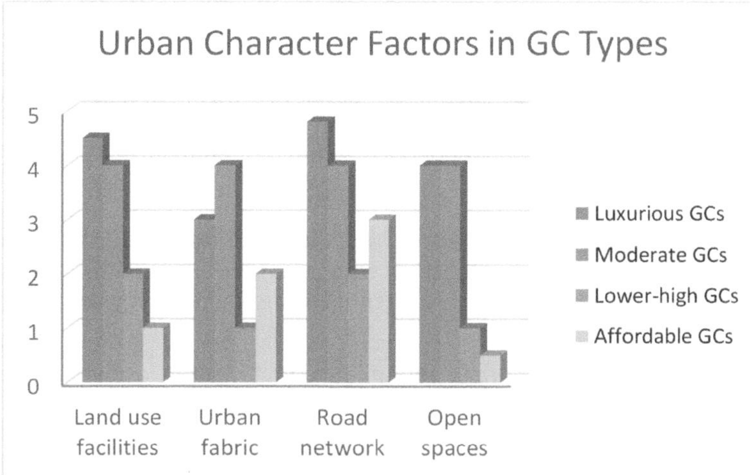


Figure 5. Percentage of urban character factors in each GC type.

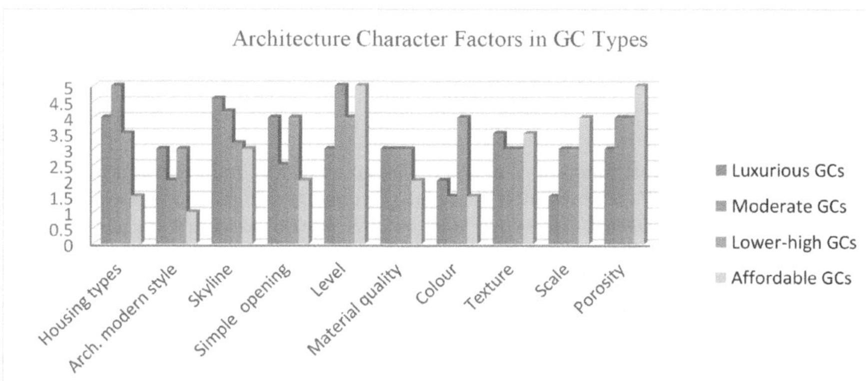


Figure 6. Percentage of architectural character factors in each GC type.

Figure 6 indicates that moderate GCs have various housing types and consists of a group of neighbourhoods each with a different design, character, and urban housing pattern. Furthermore, most of the GCs used a Western architectural style. The building ratio decreased in luxurious and moderate GCs so that they had a smooth-flowing skyline, while affordable GCs have dynamic and balanced movement in their skyline. Luxurious and lower-high GCs have the simplest of opening design shape than other GCs. The number of levels in affordable and moderate GCs ranged from four to five. The exterior material finish ranged from stone to paint, with affordable GCs using only paint; a light colour was used except for the lower-high GCs, which used darker colours. Human Intimate scale used in most of the GCs, the affordable GC have the most range of porosity than other types.

## 4 CONCLUSION

The study concluded that developers of GCs highlighted their housing as special compared to other property. They purport to offer their residents a unique lifestyle by using the various lifestyle factors to convey a sense of community. Finally, the identity of a GC resident's has social belonging to the lifestyle offered.

Developers used lifestyle factors to promote a unique lifestyle by offering a GC property as a package for a balanced life. This encourages the audience to desire the lifestyle and sells the concept of GCs.by instil a sense of community and thus increase their investment potential.

In their promotional marketing material, GC developers highlight that their developments offer a more secure community than non-GCs. They represent GCs as a specific social environment thanks to the leisure activities provided within the community. Many of the GCs are expanding and are becoming large-scale, which actually reduces the sense of community and isolates them from the main city.

Finally, the study concluded that, theoretically, GCs have the potential to create a strong local identity as they bring citizens together with common interests. In order to establish architectural and urban identity, there are essential terms for planners, designers and policy makers:

- a comprehensive definition of urban and architectural identity, explaining the theory of identity for modern society;
- reinforcing the factors of identity and recognition;
- creating factors which reflect the urban identity and ignore meaningless elements; designing symbolic components, elements and prominent architectural signs in the built environment.
- reducing turbulence of urban features and preserving similar spaces;
- constituting close relationships between people and urban society through direct and routine contacts.

## REFERENCES


Aalbers, M. (2003). The double function of the gate. Social inclusion and exclusion in gated communities and security zones. Paper presented at *Gated Communities: Building Social Division of Safer Communities, 18–19 September, University of Glasgow, UK*. Retrieved from http://www.bristol.ac.uk/sps/cnrpapersword/gated/aalbers.doc.

Almatarnah, R. (2013). Choices and changes in the housing market and community preferences: Reason for the emergence of gated communities in Egypt – Case study of Great Center Cairo region. *Ain Shams Engineering Journal*, *4*, 563–583.

Aydın Yönet, N. & Yirmibesoğlu, F. (2015). The new mode of housing production: Gated communities in Istanbul. *Intercultural Understanding*, 5, 7–14. Retrieved from http://www.mukogawa-u.ac.jp/~itcs/publications/IU_vol5/pdf/IU_vol.5_1-1_article.pdf.

Bali, R.N. (2009). *Way of life to life style: New elites, new places, new lives.* Istanbul, Turkey: *8th edition. iletişim yayinlari.*

Baycan-Levent, T. & Gülümser, A.A. (2004). Production and marketing of gated communities in Istanbul. Paper presented at the *44th European Congress of the European Regional Science Association, Regions and Fiscal Federalism, Porto, Portugal, 25–29 August 2004* [CD-ROM].

Blakely, E.J. & Snyder, M.G. (1997). *Fortress America: Gated communities in the United States*. Washington, DC & Cambridge, MA: Brookings Institution Press and Lincoln Institute of Land Policy.

Charmes, E. (2012, October 1). Gated communities: Ghettos for the rich? *Books and Ideas*. Retrieved from http://www.booksandideas.net/Gated-Communities-Ghettos-for-the.html.

Judd, D.R. (1995). The rise of the new walled cities. In H. Liggett & D.C. Perry (Eds.), *Spatial practices: Critical explorations in social/spatial theory* (pp. 144–166). Thousand Oaks, CA: Sage.

Low, S.M. (2001). The edge and the center: Gated communities and the discourse of urban fear. *American Anthropologist, 103*(1), 45–58.

Low, S.M. (2003). *Behind the gates: Life, security, and the pursuit of happiness in Fortress America.* New York, NY: Routledge.

Lynch, K. (1981). *A theory of good city form.* Cambridge, MA: MIT Press.

Soliman Muawwad, M.A., Zamle, H.E.L., Amin, K. & Ghonimi, I. (2011). The contribution of gated communities to urban development in Greater Cairo Region new towns. *Journal of Al Azhar University Engineering Sector*, *6*(21), 1349–1364.

*Graphic*

*Cities' Identity Through Architecture and Arts – Catalani et al. (Eds)*
*© 2018 Taylor & Francis Group, London, ISBN 978-1-138-05409-7*

# From national disgrace to cultural heritage and international film set. The case of Matera (Italy)

Ina Macaione, Antonio Ippolito, Anello Enrico & Roberto La Gioia
*Nature_city lab, DiCEM, University of Basilicata, Matera, Italy*
*Are_lab, Massafra, Italy*

ABSTRACT: The change in Matera was born when it was declared a 'national disgrace' by De Gasperi and the 'the shame of Italy' by Togliatti[1]. While Italy was recovering and developing from the Second World War, Matera had an agricultural identity. This case was brought to light in the book 'Christ Stopped at Eboli' edited by Carlo Levi and others sociologists, anthropologists and began to be seriously interested in the city until the laws on '*Risanamento dei Sassi*' started a process of renewal, bringing Matera to UNESCO Heritage. A cultural and architectural change supported the ability to be natural film set. An intuition of Carlo Lizzani emerged with Pier Paolo Pasolini and its 'Il Vangelo secondo Matteo', Mel Gibson with its 'The Passion' until 'Ben-Hur' in 2014. Now Matera is an open-air movie set, a city full of art, history and architecture. It is unique and part of the world's heritage.

*Keywords*: Matera; agricultural identity; architectural change; cultural change; world's heritage

## 1 INTRODUCTION

To regenerate a city is to satisfy urban, socio-economic and environmental needs expressed by the population. Citizens, through actions and reversible processes, are able to influence the development of the city. In this perspective, besides having an active role in shaping the surrounding environment (Bandura, 1999). The city can't be compared to an 'immutable institution' but to a set of interactions between citizens that create a constantly changing system. 'Places, temporalities, and processes of change are the product of social interactions; they are the result of different processes, and the contextual character of change found a plurality of paths and diverse narratives constitutively heterogeneous, and they are socially constructed, an achievement never definitively accomplished, in a tangle of stories simultaneously becoming' (Cremaschi, 2008).

'The city is something more than a congeries of individual men and of social conveniences-streets, buildings, electric lights, tramways, and telephones, etc.; something more, also, than a mere constellation of institutions [...]. The city is not, in other words, merely a physical mechanism and an artificial construction. It is involved in the vital process of the people who compose it; is a product of nature, and particularly of human nature' (Park et al., 1925).

It is now known that the urban or environmental landscape is a system where the processes and elements that compose it are in continuous interaction between themselves, and Man's ability is, in fact, to make changes in order to make a place were living Consequently, man can be considered an important agent: a modifier of the landscape.

Therefore, cities need to be understood as process of flows that with physical actions are able to valorise the existing environmental, geographic and logistic conditions of the territory. In fact, actions of different stakeholders are constantly affected by the political processes. In this

1. This definition does not have a precise date, because it was born following a visit to Matera.

context, the definition of resilience is important because it 'is the ability of a social system to respond and recover from disasters and includes those inherent conditions that allow the system to absorb impacts and cope with an event, as well as post-event, adaptive processes that facilitate the ability of the social system to reorganize, change, and learn in response to a threat' (Cutter et al., 2008). This concept appears in ecology where it is defined as the capacity of an ecosystem to respond to anomalies without changing the processes of self-organisation and basic structures. It is also considered to be the capacity to regain a steady state after a disturbance.

This is the case of Matera, a town that, throughout its history, has experienced many changes. It is a real example of a resilience change that has enabled the city of 'Sassi' to achieve, like few others, the regeneration that has made it famous worldwide.

«to where among the mindful traces of past stories appears the resilience. In its elastic fluidity, one discovers the bio-diverse population, in the crisis, in the regeneration of the city. So, in the various times of living and in spacious types, the great enters the small: humanity, landscape, nature and the world, penetrate as icons, well within the matter. In living the limit, the 'mind' of architecture expands. In the heterotopia of the Nature-City the almost nothing nature is more where you expect it (Sichenze, 2014)».

Matera is a city in which is present a «'strange form' of existence is nothing other than the 'time to live'. A time which is always different. Like life. Materialised in Matera in its being a hybrid, from the Gravina to the streets of the Sassi, to the Corso, to the 'alternative' street of the Beccherie, to the pathways in the modern neighbourhoods of the Renaissance (Sichenze, 2014)».

In our research, Matera, is a city that meets the requirements of a Nature-City. A city that possesses within itself the germ of a 'new beginning' and hence has a vital character that, beyond the difficulties of surviving the moment, allows it to resist the devastating impacts that are putting a strain on the world's urban economy.

### 1.1 *Background. About the research*

To measure the degree of resilience of a city is complicated and passes through different indicators. Linking again to a definition of a biological nature, you might say that architecture, especially in a city, is like an complex 'autopoietic' organisation. This definition merges with that of resilience: it is, once again, the identification of a system that has an affinity with biological nature and looking to the city as an organism that can grow and eventually collapse after a state of shock.

We define resilience not only as the property of recovering from an unpredictable event but also an innate capability to find new resources to react against a negative situation. It has been several years that, due to the general economic conditions, we have been in a situation of uncertainty. As depicted by the legal and social scientist, Charles Sabel, who was one of the first to address to expose the concept from the industrial production theory to the government politics, the rigidly vertical organisational model typical of modern western democracies, and their institutions, was to be replaced by an experimental continuous learning one, just like it had been done by the innovative enterprises (Sabel & Zeitlin, 2010).

From the management point of view, it is necessary to choose an organisational system that favours the independence and responsibility of the people where the results and not the processes are evaluated. From the planning tools point of view, we should use the most flexible tool possible (e.g. strategic territorial development plans) and define the principles of planning through a set of integrated tools. (This shift has been particularly relevant for urban planning in Italy, a country that used to have a very strict 'architecture' of interdependent yet different types of plans).

Identity and new technologies are both important to build resilience.

In his independent report Fabrizio Barca (former Minister of Cohesion Policy of the Italian Republic) describes a 'model for the territorial development (economic and social) that contains the definition of a resilient community' (Barca, 2010).

This approach sees, in the conflict between endogenous (local society) and exogenous (know-how, technologies) elements, the possibility to produce innovation, starting by social equality. This is crucial for our discussion because enhancing the human and social capital is one of the actions that can create a resilient community. It is, once again, the quality of the people that determines the result. So, while the local people care about the city they are living in more than outsiders do, their capability of being smart and having a set of light tools, instead of a pedantic series of procedures, makes them able to solve a situation of uncertainty, even producing innovation.

What conditions must be met in order for a city to be defined as a Nature-City?

We use, once again, the research and study of Nature-City theory[2] to find out that there are 10 key of reading, that also allow us to assess the degree of resilience of a city.

They are Naturalness (linked to nature), Landscapeness (linked to the earth), Representativeness (linked to the city), Domesticity (linked to the house), Insularity (linked to the boundary), Co-Existentiality (linked to ecology), Topicity (linked to archaeology topicity), Time Depth (linked to time), Centrality (linked to the world), Initiality (linked to phenomena) (Macaione, 2007; Macaione, 2016; Sichenze, 2000; Sichenze, 2006).

The key feature of sustainability is resource efficiency, for which a solution could be the principle of using without possessing applied to each type of resource (buildings, roads, vehicles, offices and people); the important thing is knowing how and where to find it.

Sustainability also means creating structures to live in a beautiful place and not to keep them under control with technology.

Sustainability is also declined to pay attention:

- to think in advance about the consequences of our design actions, with great attention to natural systems, industry, culture, etc.;
- to carefully consider the flows of matter and energy in the different systems that we develop;
- to give priority to human activities and not consider living beings as simply 'factors' within a process;
- to design by providing 'real values' to users;
- to treat the 'content' as something to be created, to be transformed, not as a thing to sell;
- to treat the environment, time, cultural differences, as positive values;
- to focus on services, not on things, so as not to invade the planet with unnecessary items.

"Therefore, we need a culture that is based on a sense of community and connectivity and is, at the same time, fun, challenging and responsible, following four fundamental principles:

- to eliminate the concept of waste (any waste from a cycle must become raw material for another cycle, just as happens in ecosystems);
- to reduce the transportation and distribution of goods;
- to involve as many people as possible while consuming less raw materials;

to use the natural energy flows (Thackara et al., 2000)."

To all this we must add a strong desire to make a 'cultural'. Thus emerges the relationship with resilience: resilient is a city that respects its identity and complexity (an almost oxymoronic combination), making each involved process is respected.

2. This research is part of a series of design experiences in which the relationship between architecture and the city, which featured the most significant lines of research of the Italian School of the project, are currently evolving into a new School of the project, in which it is decisive the theme of the emptiness, that has a major importance also in Eastern cultures. According to this new approach of the disciplines of the project, nature takes on a new centrality. This means that the thought of the *limit in architecture*, as a result of the analysis that goes down from the city to the building, has an opposite correspondence (from the particular to the general) in the *translimitation* that occurs through a sensible *void-making* in the built. The *Italian School of Nature-City*, proposed and established twenty years ago by Armando Sichenze and Ina Macaione, has been developed along two parallel lines of research. The first concerns the comparative reading of about 150 cities in the Mediterranean Europe, especially in Basilicata, among whom Matera, where over the millennia has been taking place a close comparison between western and eastern cultures. This reading is useful to discover the conditions of existence of the urban qualities through which a settlement recognizes itself as a city. The second study concerns the reinterpretation of the thinking about the city of the Italian architects in the late twentieth century.

## 2 MATERA: THE NATURE-CITY

### 2.1 *Regeneration strategy*

The most interesting example of this regeneration strategy is the environmental and architectural rescue of the Sassi of Matera. Matera is, in fact, being regenerated into a new form of housing, through the creation of new cultural places as well as widespread hospitality. These ways of regenerating, combined with the environment, the archaeology and the city structure, can be the beginning of tourism in the nature-city world.

Matera could be considered to be one of the most representative examples of a city where tourism is the way of changing. Matera is now a symbol. How many cities in the world have been regenerated in their nature as well as in their civilisation? A lot of cities have been rebuilt, for example those that were bombed or were theatres of natural disasters. But which city has been able to recover from the shame of its history, and how? Only Berlin has done so much! The regeneration processes of Matera and Berlin were obviously based on different circumstances. The subhuman living conditions present in the historical Centre of Matera in the Sassi, in contrast to the extraordinary human and environmental landscape. From the early fifties, Matera has demonstrated that it has an ecosystem of its own. A self-reliant structure, with regenerative functions and the ability to make exchanges with the outside world.

This paper cannot describe how Matera and the Sassi became part of the UNESCO world heritage list or European Capital of Culture in 2019, but can only summarise the the adopted process. The houses, which are the basic entities of a city, have been regenerated through contact with both the depth of the earth and the openness of the landscape, so that the value of the whole overcomes the sum of its parts.

The accommodation facilities are turned into residential buildings and represents a model of a different way of living the past experience.

The resulting is a complex system of varied accommodation facilities (bed & breakfasts, five-star hotels, beautiful mansions, etc.), and of places of knowledge. This condition is the characteristic of the process of Matera. Where a rich ethnic, eco-cultural and gastronomic condition, is able to connect the different insular habitats of which a city is made up of. In this way the domesticity thus becomes the engine for co-existentiality, increasing possibilities and lifestyles, uncovering inner resources and encouraging external (foreign) exchanges, all typical features of the Nature_City theory.

In Matera, the variety of tourist offerings and housing is matched with the historical urban centre, as bio-diversity is to the gorge, which is the natural historical centre of the city. Among these two centres the "operators of exchanges" in the ecosystem of the nature-city can be found: the birds (hawks, buzzards, swallows, doves, etc.) because are able to make the environment alive.

Eyesight can be used as a coup d'oeil, a tool, to describe the landscape. Through this coup d'oeil the city discloses its time depth, showing its archaeological sites, ruins, museums, parks. The narrative culture, therefore, regenerates itself and becomes representativeness.

The nature-city of Matera has an inner ability to sustain tourism due to the existence of a natural archaeological park (the Murge), as well as its historical architecture and urban landscape. These basic parts are so closely tied together that they are inclined towards a deeper mutual connection, at times completing each other, while at others limiting each other.

This would explain, at least partially, why today even famous personalities from the world of culture remain superficial when visiting Matera. With the exception of those equipped with a cultural understanding capable of thinking, as Carlo Levi or Pier Paolo Pasolini.

### 2.2 *Our discovery: Matera's DNA*

Matera has been designated European Capital of Culture 2019 not just for what it is or for what it has done, but also for what it is expected to do.

In this context, for example, the space set aside for contemporary cultural forms would be the mark of new features to promote the territory and its international personalities. The program should be supported through a cultural dynamism, which enhances the identity of the places that are developing innovative cultural products.

But there are some problems!

Now Matera is very famous in the world, but few people study and understand the Matera phenomenon, its strategy and its true identity.

It happens to be the case that the spokesmen of today's post-industrial society but the 'potable water', as we know, has a different history in Matera. To which the following written pieces refer.

From the 1980s onwards, within Matera, during and after the restoration of the Sassi, new vital energies were born, and the way of planning itself was changed, transforming the old poverty into the richness of a unique biodiversity of living, where citizens learn to 'seek oneself' in the city.

This vision of the future, deeply rooted in 'Matera's DNA' (Sichenze, 2014), has guided our research. Also, despite its extraordinary visibility, this living capital has unbelievably vanished from the scene in the last years, melting into a sea of events more related to consumption than to cultural production. Maybe to anesthetize and domesticate the forces which, at least in the last fifteen years, have produced the regenerative value of Matera, is necessary implement is necessary to implement the resilience capability. This too will be an interesting phenomenon to study. And Matera will become, once again, an advanced observatory. This time it is going to be an observatory of the phenomenon the 'European Capital', in addition to a laboratory of urban regeneration. Here the richness of the life rooted in nine thousand years of history must inevitably be faced, even in Italy today, with the 'new poor' in power.

Thinking about local inhabitants and to those who come to Matera to learn life strategies because it is a city that can regenerate itself because here, life is changing once again.

The new change in Matera is related to its ability to be a an heritage place. A natural movie set that, since the 1950s, has inspired some of the most famous national and international directors; in this place, in the past called a 'national shame', major Hollywood producers have decided to set their own screenplays.

Initially the Sassi were used as a symbol to document the backwardness and poverty of the South. Then, thanks to anachronistic aspect that characterises them, the old district of tufo (stone) have been able to host the most diverse film productions. Among the most important productions should certainly be remembered, in 1964, that of 'Il Vangelo secondo Matteo' by Pier Paolo Pasolini. The director from Bologna was the first to notice the similarities that unite the Sassi of Matera with the Middle East locations at the time of Jesus. This film began a long series of religious-themed productions related to the birth and death of Christ (APT, 2007; Bencivenga et al., 2013). In the following decades, the city of Matera began to be better known, becoming an ideal set for many other Italian and foreign productions, even Hollywood productions. Some of the most famous actors and directors on the world stage have come to Matera. It is impossible not to mention the famous Hollywood star Richard Gere, Mel Gibson, Catherine Hardwicke, Morgan Freeman and so forth.

That films are good for the tourism in the city that represents a constantly evolving phenomenon for the territory. And the places saw on the big screen are important in promoting cultural tourism and for the economic growth of Matera.

For the crisis of Italy, film tourism is a rapidly growing phenomenon and is an important safety valve for economic growth.

## 3 CONCLUSION

The most important questions should be: what can Matera and its territory tell to other cities in Europe? And what can the local cultural realities share with European ones. For example, the development of creative projects in response to the challenges that Europe faces: smart growth (integrated urban planning of a city of culture and knowledge which is able to combine economic development, culture, creativity and digital technologies), sustainable growth, or (the consideration of scarce environmental resources) and inclusive growth (promotion of intercultural dialogue—considering the changes taking place in the Mediterranean basin—and social inclusion).

We need to involve the city and the citizens. A European Capital of Culture should be attractive (able to converge the local and national population, but also foreign tourists) and active (involving the entire city, region and more distant places). The city should be encouraged to develop a program with lasting effects that can fit into the long-term growth of the city itself. It should not be just an ephemeral outburst of cultural events. For example, cities are asked to build projects and long-term partnerships, characterised by economic and organisational sustainability. The event will be an opportunity to change or to consolidate and develop cultural activities in the city.

Within this context, characterised by a series of initiatives and events sponsored by the city of Matera, we want to create an international working group for the establishment of a new kind of School. The desired result is to create, through this new centre of cultural aggregation, a collaborative network of people and cities that want to share experiences related to urban regeneration projects.

The project is being developed in collaboration with other institutions.

The Nature-City_LAB Dicem, University of Basilicata, co-ordinates a section of the project that deals with 'Architecture and City.' This section is part of a broader framework consisting of parallel units—divided by themes. Each unit still complies with the general ideas of cultural development, but it is autonomous in terms of training courses offered, time and organisational rules.

The 'Architecture and City' section intends the school as the centerpiece of cities and a multiplicity of architectural cultures that meet in Matera: a great place to research new possibilities for urban regeneration, on which different communities (local, urban, scientific) can compare themselves.

## REFERENCES


APT. (2007). *Basilicata in scena.* Potenza: Azienda di promozione turistica di Basilicata.

APT. (2016). *Ciak si viaggia. Location cinematografiche in Basilicata.* Matera: APT.

Bandura, A. (1999). Self-efficacy: Toward a unifying theory of behavioral change. In Roy F. Baumister (Ed.), *The self in social psychology* (pp. 285–298). Philadelphia: Taylor & Francis.

Barca, F. (2010). *An agenda for a reformed cohesion policy.* European Commission http://www.europarl.europa.eu/meetdocs/2009_2014/documents/regi/dv/barca_report_/barca_report_en.pdf.

Bauman, Z., (2003). *Liquid Modernity (Italian version).* Bari: Laterza editore.

Bencivenga, A., Chiarullo, L., Colangelo, D. & Percoco, A. (2013). *Cinema e initerari in Basilicata.* Matera: Proprietà editorial riservata Editore.

Cremaschi, M. (2008). *Tracce di Quartieri: il Legame Sociale nella Città che Cambia.* Milano: FrancoAngeli.

Cutter et al. (2008). *Resilience is the ability of a social system to respond and recover from disasters and includes those inherent in Cities and disasters.* New York: Taylor & Francis.

John, T., McDonough, W. (2000). *In the bubble, The Hannover principles: Design for sustainability, prepared for EXPO 2000,* the World's Fair, Hannover, Germany.

Macaione, I. (2007). *Architettura e Management della Città-Natura.* Milano: FrancoAngeli.

Macaione, I. (2016). *Nature City. Visions of Nature-City in Italian architecture.* Trento: ListLAB.

Park, R.E., Burgess, E.W. & Mckenzie, R.D. (1925). The City: Suggestions for the investigation of human behavior in the urban environment. In Morris, J. (Ed.), *The heritage of sociology.* Chicago, Il: University of Chicago Press.

Parsons, T. trans. (1971). *Societies: Evolutionary and comparative perspectives.* Bologna: Il Mulino.

Sabel, C.F. & Zeitlin, J. (2010). *Experimentalist governance in the European Union, towards a new architecture.* New York: Oxford University Press.

Sichenze, A. (2000). *Città-Natura/nature-city in Basilicata.* Novara: DeAgostini.

Sichenze, A. (2006). From cultural heritage to sustainability: architecture and the nature-city. In *7th European Commission Conference 'Sauveur' Safeguarded Cultural Heritage Understanding & Viability for the Enlarged Europe.* Prague.

Sichenze, A. (2014). *Into Matera.* Matera: Edizioni Giannatelli.

*Cities' Identity Through Architecture and Arts – Catalani et al. (Eds)*
*© 2018 Taylor & Francis Group, London, ISBN 978-1-138-05409-7*

# Strategies of creative thinking for solving design problems in the field of graphic design

Maha Mohammed Khalil Mohammed Elboksomaty
*Design and Communication Graphics, Faculty of Fine Arts, Alexandria University, Egypt*

ABSTRACT: The research discusses strategies of creative thinking for a graphic designer to use in his or her designs. It presents and explains the different strategies that can be used in the creative thinking process such as the brainstorming strategy, strategies to solve problems and more. The strategies discussed are mainly presented for the designer to identify design problems and come up with solutions through his or her designs as deemed important by the researcher. The research will cover information about creative thinking, as well as its types, such as critical thinking, innovative thinking and creative thinking skills which encompass originality, flexibility, fluency, and ability to think outside the box. Additionally, it will focus on the concept of creativity, thinking, and the scientific divisions of creativity levels such as expressive creativity, productive creativity, inventive creativity, innovative creativity, and imagination creativity. The research will explain, as well, creative thinking through the theory of gestalt and the gestalt principles in the field of graphic design. It will introduce the theory of Guilford for creative thinking as well. The aim of this research is to study the strategies of creative thinking and how designers may use them in their designs.

*Keywords:* graphic design, creative thinking, brainstorming strategy, critical thinking, innovative thinking

## 1 INTRODUCTION

There are strategies of creative thinking for the graphic designer to use it while starting in his or her designs, which could be obtained by learning the creative thinking strategies such as the brainstorming strategy, strategies of solving problems, and more. The researcher aims to manoeuvre through two kinds of strategies of creative thinking known as the brainstorming strategy and the strategy to solve problems. These strategies are especially and significantly beneficial for the designer and are being defended in this research through results of questionnaires. Details for such measurements shall be thoroughly explained in this research. Similarly, creative thinking skills and types such as critical thinking, creative thinking, innovative thinking; skills that encompass originality, flexibility, fluency, and the ability to think abstractly will all be covered by the researcher. The research will focus on the concept of creativity, the concept of thinking, and different scientific levels of creativity. The research will explain, as well, creative thinking through the theory of gestalt, the gestalt principles in the field of graphic design as well as the Guilford theory for creative thinking.

## 2 RESEARCH PROBLEM

The problem and results presented in this research is centred around the concept of creative thinking and its usefulness in solving design problems in the field of graphic design, and discovering the ability, or lack thereof, of the designers utilising it.

## 3 RESEARCH IMPORTANCE

Firstly, the graphic designer has to be aware of the concept of creative thinking and the strategy behind creative thinking. The graphic designer must follow a strategic method of creative thinking that is suitable for him or her in the creation of a certain design.

## 4 MEASURING RESEARCH TOOLS

A questionnaire for graphic designers, specifically senior students in the graphic design department, will be distributed to measure the awareness and utilisation of creative thinking strategies in a designer's work.

**The concept of creativity**

Gardner, a scientist and a professor in Educational Graduate Studies College at the University of Harvard, stated that a creative individual who can regularly solve problems and develop new findings, or ask questions in a particular area, is characterised by seriousness and gains the acceptance of the community. *(Grown, Fathi.(2002). Creativity. Jordan: Dar Al-Fekr. Page. 26)*

**The different divisions and levels of creativity according to Taylor**

- Expressive creativity: was defined as free expression that the originality and the efficiency is not important for it, it does not need skill, and it is important for the appearance of the other levels of creativity.
- Productive creativity: A type of creativity that is linked to the development of machines, a product or a service.
- Inventive creativity: invents new styles that includes the work of explorers and inventors who show their genius by using innovative materials and styles.
- Innovative creativity: Refers to the continuous development of ideas, and results in the acquisition of new skills. It is concerned with the development and improvement of previously existing methods through the use of appropriate skills.
- Imagination creativity: Is also known as the imaginary and the rare. It encompasses what it takes to develop new ideas and assumptions, where the highest levels of creativity and achievement facilitate the emergence of a new theory. (Abdelhamid, Shaker. (1987). Creativity process in painting art. Volume 109. Kuwait: National Council of Culture, Arts, Fine Arts.)

**The concept of thinking**

The concept of thinking is the realisation and identification of the problem of accessibility as a first step to a solution.

**Types of thinking**

1. Critical thinking: This type is defined as accurate examination to a work or an idea that aims to development.
2. Creative thinking: The ability to create ideas characterised by originality, flexibility and fluency that are applied to some or all of the following points:
   - To develop previously unidentified facts and data
   - Finding solutions to difficult problems that cannot be solved instantly
3. Innovative thinking: Is an authentic thinking that produces ideas, works, and solutions that was never previously thought about with considering the age, the time and the environment.

**The concept of creative thinking**

According to Guilford, back in 1959*, thinking in an open format was characterised by the diversity of the produced answers that do not specify given information.

*(Al-Titi, M. (2001). *Development of thinking capacities and creativity*. Amman: Dar Al-Masirah for Publishing, Distribution and Printing).

**Creative thinking skills**

Creative thinking skills encompass originality, flexibility, fluency, and the ability to think abstractly.

**Creative thinking from the perspective of the gestalt theory the whole and the parts**

The gestalt school of psychology, which began in Germany in around 1912, investigated how we see and organise visual information into a meaningful whole. The conviction developed was that the whole is more than the sum of its parts. This whole cannot be perceived by a simple addition of isolated parts. Each part is influenced by those around it. This theory consists of:

Similarity—Proximity—Continuation—Closure.

**Gestalt principles in the field of graphic design**

A designer works not simply with lines on paper, but with perceptual structure. By learning the gestalt perceptual principles, one can take advantage of the way objects, the eye, and graphic creation interweave.

**Similarity**

When we see things that are similar, we naturally group them. Grouping by similarity occurs when we see similar shapes, sizes, colours, spatial locations (proximity), angles, or values. All things are similar in some respects and different in others. In a group of similar shapes and angles, we notice a dissimilar shape or angle. In the Figure 1 the grey square is drawn to our attention because it is different from the squares surrounding it. The three letter 'i's in 'similarity' are shown to be similar by being treated in a similar fashion to each other but differently from the other letter forms.

Grouping by similarity is true for realistic subject matter as well as nonfigurative design forms. The symbol and logotype created for Alcoa by Saul Bass (1920–1966), a renowned American designer, relies on similarity of shape. Count the triangles in Figure 2.

**Proximity**

Grouping by similarity in a spatial location is called proximity, or nearness. The closer two visual elements are, the more likely they will be seen as a group. In Figure 3, the four squares on top seem to form a group whereas the eight squares on the bottom appear to belong to a different group. Figure 4 uses a close and careful placement of all the elements to create a

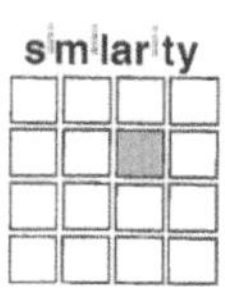


Figure 1

Figure 2*

*(*Arntson, A.E. (2012). Graphic design basics. Wadsworth Cengage learning.*)

Figure 1. The use of similarity also draws attention to differences.

Figure 2. Saul Bass. Trademark for Alcoa. Courtesy, Aluminum company of America.

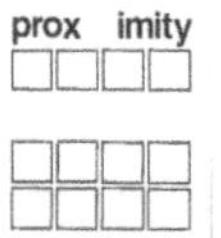

Figure 3*   Figure 4*

*(Arntson, A.E. (2012). Graphic design basics. Wadsworth Cengage learning.)*

Figure 3. **(left)** Proximity grouping is grouping by similarity in spatial location.

Figure 4. **(right) A3 Design** created this logo for Valley Winery. Similarity, continuation, and reversible figure/ground all unite this strong design. Courtesy of A3-Design.

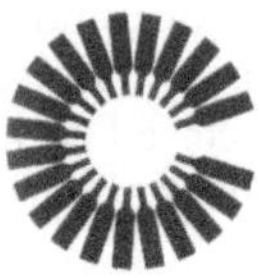

Figure 5*   Figure 6*

*(Arntson, A.E. (2012). Graphic design basics. Wadsworth Cengage learning.)*

Figure 5. **(left)** Nationally renowned designer and educator Michael Vanderbyl created this logo for a wine distributor in Toronto. It incorporates several gestalt principles.

Figure 6. **(right)** Stefan Kantscheff, Bulgarian designer, created this beautiful example of rhythm and repetition in symbol design. Courtesy of the artist.

winery logo. Similarity is important, but so is the close proximity and careful placement of the elements, as shown in Figures 5 and 6. These figures place shapes in close proximity to each other, they never touch but form a dynamic whole through proximity and the use of more gestalt unit-forming principles as listed next. Proximity of lines or edges makes it easier for the eye to group them to form a figure. Having finished this chapter, come back to these marks and analyse the combination of unit-forming principles at work.

**Continuation**

The viewer's eye will follow along a line or curve. Continuation occurs when the eye is carried smoothly into a line or curve that links adjoining objects. The diagram in Figure 7 shows how the eye follows the interruption of the black outline, seeing a continued, implied shape (in this case an X). This principle is used extensively in layout design to unite various elements, often by placing them along invisible grid lines.

Shapes that are not interrupted but form a harmonious relationship with adjoining shapes please the eye. The symbol of the U.S. energy extension service, as shown in Figure 8, uses continuation to emphasise the moving, dynamic nature of energy. In this example, the ends of the 'e' line up with the ends of the arrowhead, forming a continued line that harmoniously unites the shapes. The Family Circle logo by Herb Lubalin and Alan Peckolick, as shown in Figure 9, creates continuation by lining up the verticals of the two 'i' letterforms and also lining up the 'l' and 'r' forms. The eye draws a line down these vertical shapes that makes a new whole out of two different words. American graphic designer and photographer, Herb Lubalin (1918–1981), was the editorial design director for several distinguished publications. His work is shown several times throughout this text.

**Closure**

Familiar shapes are more readily seen as complete and as incomplete. When the eye completes, or rather closes over, a line or curve in order to form a familiar shape, closure occurs. The diagram in Figure 10 shows white circles appearing as the eye and brain close the open

Figure 7* Figure 8 Figure 9

**(Arntson, A.E. (2012). Graphic design basics. Wadsworth Cengage learning.)*

Figure 7. **(left)** Continuation occurs when the eye is carried smoothly along a suggested line or curve.

Figure 8. **(middle) George Jadowski**, designer, **Danny C. Jones**, art director. Symbol for the U.S. Energy Extension Service. This symbol illustrates energy with its use of continuation.

Figure 9. **(right) Herb Lubalin** (art director) and **Alan Peckolick** (designer). Family Circle. 1967. This magazine logo makes quiet but elegant use of placement and continuation. Courtesy of the Herb Lubalin study Center of Design and Typography at the Cooper Union.

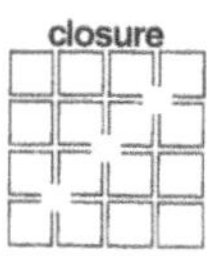

Figure 10* Figure 11* Figure 12* Figure 13* Figure 14*

**(Arntson, A.E. (2012). Graphic design basics. Wadsworth Cengage learning.)*

Figure 10. The eye and brain close the white areas into circles. This demonstrates a highly active partnership between the eye, the brain and the graphic image that is at the heart of visual gestalt.

Figure 11. **Pat Hughes** and **Steve Quinn**. This elegantly constructed symbol for 1+1 Design uses a rich combination of similarity, reversible figure/ground, and closure.

Figure 12. **Stefan Kantscheff**. symbol for the Staatliches Operettentheater in Sofia, Bulgaria.

Figure 13. **Herb Lubalin**. 1965. This creation by an important 20th-century designer relies on an anthropomorphic identification with the shape of the letterforms to bring closure. Courtesy of the Herb Lubalin Study Center of Design and Typography at the Copper Union.

Figure 14. The wide ranging Memphis Design Firm **Tactical Magic**, created this design for The Eyewear Gallery. Samples of this design are shown on the accompanying web site. Applied to billboard, web and signage applications.

areas into a familiar circular shape. Figure 11 is a symbol created by the 1 plus 1 design form. Do you see the white plus sign created by the figure/ground relationship? This is a visual closure as our eyes finish the form. Part of the closure in this example includes a sudden conceptual connection and understanding of the name of the form. This sort of connection is especially useful in trademark design. Closure is sometimes accompanied by a reaction filled with realisation and awe on the observer's part. An elegant editorial statement is made in Figure 12 in this opera symbol when the link between a musical note and a heart is recognised.

Figure 13, by Herb Lubalin, calls for active conceptual participation by the viewer to achieve an intellectual closure with the O shape and a womb. The playful shape in Figure 14, for an eyewear company on closer inspection forms spectacles.

**Creative thinking from the perspective of the theory of Guilford**

Guilford's perception theory about the phenomenon of creativity through his theory of mental training, known as the structure of intellect, was where he selected the three dimensions of mental activity for a person, known as operations, content and product. *(Alsultani, A.M. (1984). The relationship of creative abilities to some of the personal characteristics of middle school students. PHD. Bagdad University.)*

**Guilford division of the three dimensions into different mental processes**
1-Cognition: is defined as an individual who is an expert in his/her own field. 2-Memory: is intended to keep and retrieve a person's experiences when needed. 3-Evaluation: is the issuing of decisions through the previous experiences of the person. 4-Production: is intended to give a solution to the problem faced by a person, and is divided into two types known as convergent production and divergent production.

Convergent production includes the production of incorrect or specific information in advance and specifically agreed upon. Divergent production includes the production of a variety of information without prior agreement and is associated with right and wrong.

**Creative thinking strategies**
This paper presents 2 types of strategies that can be employed in the teaching and learning of creative thinking, brainstorming strategy, problem solving strategy.

**The brainstorming strategy**
Alex Osborn, a journalist back in the 1950s, has contributed to the development of several creative thinking strategies. The brainstorming strategy is one of the most powerful strategies in the development of creative thinking. The brainstorming strategy aspired to break the usual thinking of the individual and to contribute to the production of a list of diverse ideas. The main idea for this strategy depends on the separation of the production of ideas from the evaluation of ideas. The reason for this separation is to prevent mental control over the flow of ideas, which prevents a lot of ideas from escaping, especially considering that some of these ideas may seem unacceptable.

**The concept of the brainstorming strategy**
It is an individual yet collective thinking process that produces the largest amount of ideas and solutions to a problem.

**Levels of brainstorming strategy**
Individual brainstorming is a strategy used by graphic designers in different forms such as dual brainstorming, small group brainstorming and large group brainstorming.

**The strategy of solving problems**
Strategies of problem solving include intended operations performed by the individual using the information and knowledge already learned and taught and skills acquired in overcoming new and untraditional situations. The designer is then allowed to control, access, and implement a solution.

**Steps to resolve a problem**
1. Sensing the problem: This step can be summarised as identifying an obstacle to achieve a specified target. 2. Identification of the problem: Can be summarised as defining and accurately describing an issue, allowing the designer to define the outlines of the problem. 3. Analysis of the problem: Realising the basic elements of the problem and excluding all that is irrelevant. 4. Collection of the relevant data: Determining the best available sources from which to gather information and data in relation to the problem which could help in reaching an effective solution. 5. Proposing of solutions: Represents the person's ability to master and determine a number of ideas proposed to solve a problem. 6. Studying the proposed solutions as a critical study: choose the best one between the proposed ideas (the solutions). However, preference is usually given to the idea with the most positive impact and least amount of negatives according to the data collected.

### Measuring research tools

### Question No. 1

1. Do you know that there are strategies of thinking that the graphic designer must follow at the start of his/her design? Yes, I know. No, I don't know. If your answer is no for the previous question, please answer the next question.

### Answer for Q.1

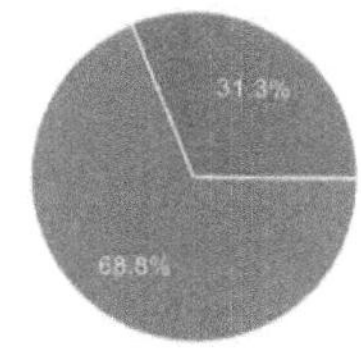


| | | |
|---|---|---|
| Yes I know | 22 | 68.8% |
| No I don't know | 10 | 31.3% |

### Question No. 2

2. If you were aware of strategies of thinking for the graphic designer that can be followed to find a solution for a design, would you follow one of them? Yes, I would. No, I would not.

### Answer for Q.2

| | | |
|---|---|---|
| Yes, follow | 10 | 100% |
| No, Don't follow | 0 | 0% |

### Question No. 3

3. Do you follow a strategy of thinking when you start your design? For example, when you design a logo for a company, do you follow a strategy of thinking for the solution design? Yes. No. If your answer was yes for the previous question, please answer the next question.

### Answer for Q.3

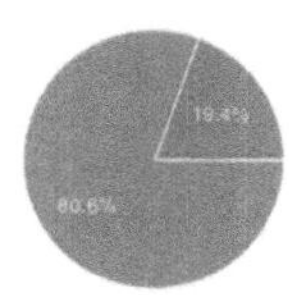


| | | |
|---|---|---|
| Yes | 25 | 80.6% |
| No | 6 | 19.4% |

### Question No. 4

4. Select the strategy that you followed (strategy of brainstorming—strategy of solving problems—other (............................)

### Answer for Q.4

32 responses (in Google) the sample senior students in the Graphic Design department in Princess Nora Bint Abdurrahman University.

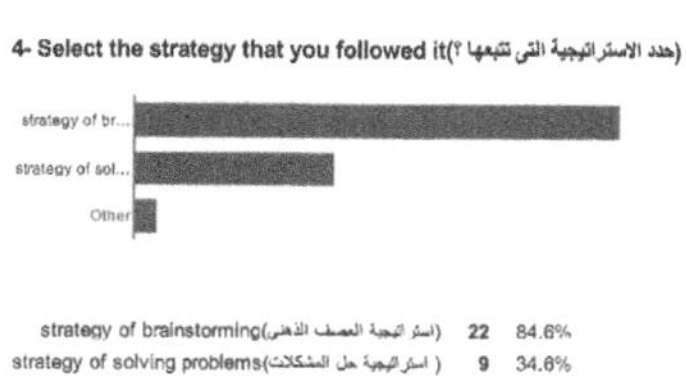


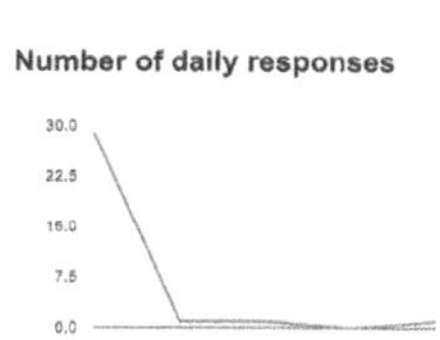

## 5 RESULTS OF MEASURING RESEARCH TOOLS

1. Some graphic designers did not know that there are strategies of creative thinking that the graphic designer must follow when initiating a design.
2. All graphic designers lacked knowledge with regards to the range of strategies of creative thinking but agreed with regards to the need to follow one to find a solution for a design.
3. Most graphic designers that follow a strategy of creative thinking prefer the strategy of brainstorming over the strategy of solving problems, while only a few frequently used other strategies.

## 6 RESULTS OF THE PRESENTED RESEARCH

1. That the graphic designer has to be aware of all strategies of creative thinking and use at least one of them when designing.
2. The research presents, as well, a conclusion that all innovative ideas in the field of graphic design, with a positive impact on society, rely on the use of creative thinking and its strategies.

## REFERENCES


Abdelhamid, Shaker. (1987). *Creativity process in painting art*. Volume 109. Kuwait: National Council of Culture, Arts, Fine Arts.

Al-Titi, M.H. (2001). *Development of thinking capacities and creativity*. Amman: Dar Al-Masirah for Publishing, Distribution and Printing.

Arntson, A.E. (2012). *Graphic design basics*. USA: Wadsworth Cengage Learning.

Costa, A. (1985). *A glossary of thinking skills developing minds: A resource book for teaching thinking.* New York: Mc Graw.

Grown, Fathi. (2002). *Creativity.* Jordan: Dar Al-Fekr.

Issa H.A. (1993). *The psychology of creativity between theory and practice*. Tanta, Egypt: The Cultural Center in the Middle East.

Solso, R.L. (1988). *Cognitive psychology* (1st ed.). USA: University of Nevada.

*History and arts*

*Cities' Identity Through Architecture and Arts – Catalani et al. (Eds)*
*© 2018 Taylor & Francis Group, London, ISBN 978-1-138-05409-7*

# A process of urban regeneration from below. The case of Taranto (Italy)

Ina Macaione, Antonio Ippolito, Anello Enrico & Roberto La Gioia
*Nature_city lab, DiCEM, University of Basilicata, Matera, Italy*
*Are_lab, Massafra, Italy*

ABSTRACT: This research focuses on the crisis of Taranto, and starts by analysing the current state of shock caused by the presence of the steel factory, ILVA. The factory has caused disastrous effects on environmental, social, economic and urban conditions. To deal with this problem local citizens, joined by voluntary associations, initiated processes of urban regeneration from the bottom, reactivating small portions of urban land (abandoned structures, disused areas, etc.). The citizens of Taranto proved themselves to be able to live in life-threatening conditions, awkward conditions they would have never thought they would have to cope with. The purpose of the activities carried out by the associations was to educate citizens to become smart and sustainable.

*Keywords:* Taranto, ILVA, urban regeneration, sustainability

## 1 INTRODUCTION

To regenerate a city is to satisfy urban, socio-economic and environmental needs expressed by the population. Citizens, through actions and reversible processes, are able to influence the development of the city. In this perspective, besides having an active role in shaping the surrounding environment (Bandura, 1999). The city is not, in other words, merely a physical mechanism and an artificial construction. It is involved in the vital processes of the people who compose it; is a product of nature, and particularly of human nature (Park et al., 1925).

It is now known that the urban or environmental landscape is a system where the processes and elements that compose it are in continuous interaction between themselves, and Man's ability is, in fact, to make changes in order to make it liveable. Consequently, man can be considered an important agent: a modifier of the landscape.

Many of the realised district, especially those of the industrial city, appear to condense serious problems relating to social marginalisation and degradation of urban planning and construction. Similar conditions assign, these days, the 'public city' to the role of a fertile laboratory for the testing of integrated social and spatial redevelopment.

Thus the city becomes a public design laboratory to gather knowledge, ideas, and proposals, in order to contribute creating urban regeneration actions from the bottom aimed at ensuring the improvement in living conditions of residents in the neighbourhoods affected by the interventions.

Therefore, cities should be understood as a process of modelling and development that determine the physical design of the territory. In fact, actions of different stakeholders are constantly conditioned by the processes of development and change in environmental, urban and socio-economic settings. If these balances can't persist, they face systemic crises 'that, by analogy, we could assimilate to the crisis of a society to the persistence of a period of social change, or of settling between new forces or new stakeholders' (Parsons, 1971).

The city must therefore be considered to be a set of interactions between citizens, aiming at a sustainable environment through each concrete action to change the entire system. 'Places, temporalities and processes of change are the product of social interactions; they are the result of different processes, and the contextual character of change found a plurality of paths and diverse narratives constitutively heterogeneous; and they are socially constructed, an achievement never definitively accomplished, in a tangle of stories simultaneously becoming' (Cremaschi, 2008).

Urban regeneration is therefore understood as an organic process, as the paradigm of the city-nature, rich in projects aimed at creating urban environments that are respectful of diversity, preserving the built cultural and historical heritage and transferring it to future generations.

In this context is important the definition of resilience. It is defined as 'the ability of a social system to respond and recover from disasters and includes those inherent conditions that allow the system to absorb impacts and cope with an event, as well as post-event, adaptive processes that facilitate the ability of the social system to reorganise, change, and learn in response to a threat' (Cutter et al., 2008).

In ecology, the capacity of an ecosystem to respond to the anomalies, also to regain a steady state after a disturbance.

'Where among the mindful traces of past stories appears the resilience. In its elastic fluidity one discovers the bio-diverse population, in the crisis, in the regeneration of the city. So in the various times of living and in spacious types, the great enters the small: humanity, landscape, nature and the world, penetrate as icons, well within the matter. In living the limit, the 'mind' of architecture expands. In the heterotopia of the Nature-City the almost nothing nature is more where you expect it (Sichenze, 2014)'. A sustainable city is also a resilient city. A resilient city is an urban system that not only adapts to climate change (especially global warming) which, in recent decades, has made cities increasingly vulnerable with ever more dramatic consequences and rocketing costs. A resilient city is a city that changes itself, building new social, economic and environmental responses that enable it to withstand the stresses placed by the environment and history.

This is the case of Taranto, a city in which is present a 'strange form' of living, linked to the presence of a steel factory. On the one hand we find a purely industrial reality in which many citizens are also workers and on the other hand the citizens want the closure of the factory.

In this paper we are less interested in assessing 'community' participation within urban regeneration policies in Taranto; nor do we wish to deconstruct the rhetoric of participation that has recently become so overwhelming (Jones, 2003).

This study allows us to know and understand how citizens are responding to industrial presence which marks the transition from a town devoted to agriculture and fishing to a city which relies on the industrial sector. This allows us to have knowledge of the factors that caused the socio-economic and urban impact and how citizens relate with them. So, the aim of this paper is to widen the debate about urban regeneration with reference to the initiatives promoted by local, self-organised civic networks (Morandi & Rabbiosi, 2012). We focus on urban regeneration 'from the bottom-up' (Morandi, 2008) as a tool for empirical research about urban resilience and the restructuring of urban governance. 'In so doing we also connect, and test, a new concern in urban policies as well as urban studies which consists of matching the effects of the economic recession with a renewed interest in what is often called the 'hidden potential' of local areas within cities (Unsworth et al., 2011)'.

In this paper we are interested in better understanding «the mobilisation of actors who do not have direct policy commitments of their own» as process of «productive outcomes on the organisation of space. It is argued that these actors not only make claims in the public sphere, but also actively contribute to the dynamics of space production that trigger the processes of spatial change at the urban level» (Rossi, 2004).

This paper tries to understand the role of social associations and bottom-up initiatives. In particular, the study conducted by Are_lab and Nature-City Lab refers to locally based initiatives that consider the territory not only as a frame, but also as a resource and an object of intervention. Adopting a multi-level approach, the role of bottom-up experiences in urban regeneration process and their connection with broader dynamics will then be discussed through the analysis of the specific case-study of Taranto.

### 1.1 *'Bottom-up' processes*

The last few decades have seen renewed attention to spatial issues, both in public debate and within social sciences. A revived awareness that society does not take shape in an abstract space but in specific times and places has risen and cities, being one of the particular settings where flows become concrete, have come back to the centre of the analysis. In this sense social issues become urban issues and broader changes are reflected in the transformations of cities.

Bottom-up urban regeneration is a term that identifies a variety of experiences which may differ in inspirational principles from so-called 'integrated' public policies to business-led economic development strategies or popular grassroots and neighbourhood-based efforts to capture the benefits of urban restructuring for local residents (Pacione, 2005). In any case, urban regeneration from the bottom-up presupposes a certain degree of cooperation, if not of participation, among the actors involved in the policy process (Healey, 1997). The generally restricted meaning of participation in urban policy, as well as the ideology and implications behind it, was the subject of enormous debate at the beginning of the 2000s (Raco, 2000).

Bottom-up initiatives and associations are therefore gaining a new centrality in the urban regeneration process. However, this new role may also involve some critical questions and sensitive issues. Firstly, bottom-up initiatives, grass-roots mobilisations and civil society associations express specific needs, desires, or visions of what their city or their neighbourhood should be. What often happens is that, as time goes by, these actors find themselves torn between their original vision, the needs and requests that they express and the urgency of attracting public or private funds for their actions. In order to do so, which is especially significant in a period of scarce resources, they need to be able to 'intercept larger trends and build their own image to fit within the strategic vision of the city' (Bolzoni, 2012).

Secondly, the presence of different bottom-up initiatives and movements asks for the consideration of the question of legitimacy and representativeness. This pluralism remarks the high level of diversity characterising urban contexts, but if 'the freedom to make and remake our cities and ourselves is [...] one of the most precious yet most neglected of our human right' (Harvey, 2008).

Urban regeneration has been associated with various approaches throughout history, sometimes involving massive physical reconstruction of neighbourhoods with many negative side-effects. A common denominator of such traditional urban regeneration approaches is that they have been centrally run. On the contrary, bottom-up regeneration stands for practices that also contribute to better living conditions in cities, in particular in relation to open public space, but are bottom-up, initiated by the citizens and local entrepreneurs, who use the public space daily, and are implemented through events and interventions in the public space.

These projects can also be a bottom-up process (initiated by cultural non-profit organisations, artists, and other cultural operators) that can draw local government's attention, and attract other actors from the community to the project (e.g. universities, schools, community members), and push for policy measures that would accommodate their interests. The bottom-up, culture-led urban regeneration process has great potential to contribute both to the development of the cultural field and to urban regeneration, and challenges the traditional way of policy-making and implementation, in which the local government has the central role. At the same time, its bottom-up character makes it more vulnerable and less sustainable in the long run (Stoica, 2010).

The bottom-up approach is an alternative to the top-down strategies of inner-city regeneration and encompasses any individual or collective revitalisation actions undertaken by members of local communities in a given neighbourhood. Its key distinguishing feature is the genuine engagement of inner-city residents which follows on from willingness to improve their quality of life. Even if activities directed towards this goal are assisted by public authorities, it is the initial grassroots initiative which makes all the difference.

According to P. Clay (1979), the author of the term, 'incumbent upgrading' stands in opposition to gentrification as it describes spatial improvements in degraded blue-collar and working-class inner-city neighbourhoods which are initiated by the sitting residents and not newcomers. Hence, it is claimed that the influx of a higher-status population may not need to be the necessary key to positive changes in run-down urban areas. Clay's way of reasoning was subsequently resumed by several authors (Downs, 1981; Baldassare, 1984; Varady, 1986) who enriched the theory of incumbent upgrading with further clarifications.

### 1.2 *The case study*

The role of social participation (citizens and all social actors in the local community) in the achievement of community well-being is one of the main aspects through which you can activate regeneration processes and the regeneration of places and spaces. The present and active involvement of all parties in the community—citizens, families, institutions and volunteers—can respond to the situation of great economic uncertainty. The urban reality becomes the place where actors seek to answer emergencies. This through shares and reversible processes that can influence the development of the city.

The research approach will be followed by an analysis on Taranto and further, two bottom-up initiatives will be discussed. Even if they are quite different, they all share a grassroots origin and a relevant, and sometimes unexpected, role in the transformation of the city. This paper is based on the current PhD research project conducted by Antonio Ippolito, that in collaboration with Nature-City Lab and Are_lab, has adopted an approach with a combined use of participant and naturalistic observation, in-depth interviews and documentary analysis. The fieldwork started in October 2014 and is still in progress. As part of the research we moved into the city, taking part in events, to see every day how cultural and social associations worked.

Taranto is a city with 200,000 inhabitants and is the sixth largest city, by population, of all of southern Italy. It has a strategic geographical position, between the *Mar Grande* (Great Sea) and the *Mar Piccolo* (Small Sea) and the *Mar Mediterraneo*. In the early 1960s it was decided to build a steel plant. Its choice depended by logistical reasons and by economic crisis of Taranto and on the 9th July 1960 the first stone of the factory was laid. The first products of Italsider were sent to the Soviet Union in exchange for oil and this economic exchange has contributed to an enlargement of the establishment and the operation of other systems. The production capacity in that period was about two million tons per year of crude steel. Also, in 1967 the Eni refinery was built, located in areas adjacent to the factory on an area of 270 hectares. In the early 1970s the production capacity of the steel plant reached 4.5 million tons per year and it was decided to expand it to make it up to 10.5 million tons per year.

This led to an expansion of the plant that changed the urban plan of the city. In the mid-1980s the European steel crisis hit Taranto which was forced to make many redundancies. In 1987 the company was changed in Ilva. In the early 1990s, the group was still in crisis, therefore, the European Union mandated that the factory had to be purchased by a private investor and in 1994 the sale process started. In 1995 the Riva family bought Ilva for about 1,750 million euro. In subsequent years, the Riva family started an internal restructuring process with a reduction of work. Following investigations by the magistrate on the high pollution in Genoa, all steel production was transferred to Taranto which become the only steel producer of the company, and also the biggest steel factory in Europe. In 2006 production capacity reached an historical record of 14 million tons per year, while in 2007 the group made a profit of 900 million euro. The first investigations by the magistrate of Taranto has shown that the company generated profits at the expense of safety, worker health and environmental protection.

In 2005 the first conviction was issued for pollution against executives of Ilva and Emilio Riva. In 2009, by order of the administration of the Puglia Region, nearly 2,000 head of cattle were killed because of dioxin-contamination and one year later an order was issued prohibiting grazing within a radius of 20 kilometres from the steel plant. In July of 2012 the investigation *Ambiente Svenduto* lead to the seizure of the hot area. This lead to several arrests, including that of the owner of Ilva, Emilio Riva, his son and former CEO, Fabio Riva, the management team including the Head of External Relations, Jerome Archinà, and politicians. In 2014 the conviction for manslaughter was made against 27 defendants, many of them leaders of Ilva, including Fabio Riva, for the deaths among the workers caused by exposure to asbestos. But the first condemnation against Riva came in 2005 for air pollution and dust emissions.

In 2012 the Judge Simone Orazio, in his judgment of 23 May 2014, declared 'Ilva's workers died because of asbestos present in the factory and they could have been saved if only the company had acted promptly'. He also declared that if there had been a medical appropriate, they would have 'diagnosed the disease for the workers'. According to the magistrate, 'The policy of the factory had always been set to achieve maximum profit, even at the cost of compromising the health of the workers'. But to save the health of the employees, the leaders of the plant could at least provide adequate equipment but, instead, testimony in court made it clear that the workers were provided only 'disposable' respiratory masks which experts have called inadequate.

### 1.3 *Two bottom-up initiatives*

#### 1.3.1 *'Ammazza che piazza' project*

The analysed project is located on the eastern outskirts of Taranto, an area characterised by a low presence of green. The *Pinetina Antonio De Curtis* is set among purely residential lots. The 'hidden area' by neighbouring buildings is not readily identifiable and is reached from the main road. An area able to contain a greater number of users. The area had been neglected for years and was always seen as a place of little relevance, a remnant of old housing development plans. In the early 2000s, the administration approved a reorganisation plan for the site, allocating the area as a public green space, providing it with a lighting system and making a series of paths and parking areas. The reorganisation of the actions still cannot be possible and hasn't generated attractive effects for new users. In the last months of 2012 around the area was born a spontaneous movement of citizens that comes together under the name '*Ammazza che piazza*'. The goals of this unofficial organisation were to revitalise the area and regain possession of the abandoned spaces of the city, returning them to citizen ownership. The movement '*Ammazza che piazza*', made up of volunteers, is not organised by any form of association, and it is self-funded. The project to reorganise the '*Pinetina Antonio De Curtis*', proposed by the committee, was a first step in this direction, and became the pilot project of the movement. In fact, the objectives of '*Ammazza che piazza*' are to recover all of the degraded and abandoned areas of the city, involving a great number of people. The meetings held by the committee concerning the activities are public and often carried out in open spaces in order to increase engagement. For the purposes of active participation and sharing of information about these initiatives, they have organised open dissemination of activities to all citizens, so as to showcase progress and to be able to get advice and ideas to contribute to the continuous improvement of the work. The measures put in place by the committee are directed towards the reorganisation of the areas through the enhancement of spontaneous uses and the inclusion of a plurality of functions and activities. The arrangement of the green space with street furniture, created especially with lightweight materials, is also used to enhance and 'formalise the spontaneous paths that have gradually been created over time. The functional mix is a key element: an area for dogs, a children's play area, spaces for sports, and the offer of recreational activities that are an attempt to attract people from outside the neighbourhood. The diverse functions and uses promotes enjoyment of the area for different kinds of users, belonging to different age groups and that frequent the area at different times of day. These areas become in this way liveable and secure places. The majority of materials used are wood, pallets, tyres, and metal from dried paint tins and recycled plates. With these materials the group has made benches, tables, rubbish bins and more. In this case the only actor involved in the process of redevelopment and revitalisation of these areas is '*Ammazza che piazza*'. The reorganisation of the areas is free of the implications associated with the acquisition of the land because it belongs to the public domain. The project was entirely managed and implemented by the movement while the government has limited itself to let the members speak in the normal management of business.

#### 1.3.2 *'A Tamburi battenti' project*

Another bottom-up action, it was achieved through the interception of the funds made available by *Fondazione con il Sud* under the name *Ambiente e Sviluppo*. The winning project was *A tamburi battenti* and it was fielded by associations, ecclesiastical institutions and educational institutions operating in the area (Airone onlus; Associazione Marco Motolese; Associazione Solirunners; Associazione culturale teatrale Sant'Antonio; Confraternita Maria SS.ma della Scala; Learning Cities; nessuno escluso onlus; onlus caritas christi; XI Circolo 'G.B. Vico'). The goal is to provide the restructuring of the *Teatro San Francesco De Geronimo*, which belongs to the church, to give rise to a common 'house' collectively designed through innovative and sustainable practices, enabling the testing of functional and decorative solutions through eco-design and the reuse of waste materials from local companies. The work will involve disadvantaged people with manual skills and expertise related to carpentry and upholstery.

Then there will be a job training program for citizens to deepen the acquisition of skills related to the redesign and reuse of furniture and decorative elements. '*Tamburi battenti*' will be a lasting network for the establishment of a multidisciplinary group bringing together architects, curators, artists, designers, theatre companies, choreographers, and ecologists, who can interact to start actions able to transform the district Tamburi in Taranto, starting with artistic and cultural practices. Through the programming of 'residences' of artists, production of new works, organisation of workshops and seminars, archiving and dissemination of works with publications and audio-visual documentation, the collective will work to strengthen the relationship between the various forms of art and life of the community, helping to spread the participatory and collaborative approach to creative processes and sustainable practices. The goal is to create an experimental process which will define a kind of 'hub'. A shared and open hub used to rethink the relationship between public space, citizenship and artistic and cultural communities.

The network will be structured through actions that will take place in relation to the presence of the theatre. The actions will be coordinated by the multidisciplinary team. In this way, the project wants to implement the social relations between city and province, public and private sector, that are linked to consolidated cultural products. In relation to networking, it is expected the activation of participatory practices (Social and participatory practices) and procedural (context-based) to reflect on the most urgent issues related to the relationship between the residents and the environment, to identify solutions and methods interaction in relation to the 'theatre' environment.

The theatre will be able to count on a series of 'satellite' spaces belonging to the project leader, which will form functional areas to implement the measures. Among them, the John Paul II multipurpose centre with congress and conference facilities; the Emporio space on a nearby street; laboratories in the *Exhibition Nasisi*. The theatre will be a social incubator in which to bring together an exchange of services and expertise useful to the territory and it can act as a driving force for employment development interventions in the neighbourhood, also working through the active involvement of the schools. It will include the production of 'Made in Tamburi' with the creation of clothing, accessories and curtains, which will involve ten women of the neighbourhood, selected according to the degree of need and personal inclinations, which will subsequently be integrated in a cooperative work. The musical activities developed within the theatre provides for the establishment of a percussion orchestra called '*Tamburi di Taranto*' that focuses on its composition and originality using handmade instruments.

These actions are aimed at encouraging the promotion of the project at a national and international level. A permanent service orientation to work and start-up development will be activated in the urban area, a place dedicated to disadvantaged people. The crowdfunding work will support the deployment of strategies for the sustainability of individual projects, workshops, performances, exhibitions.

## 2 CONCLUSIONS

Identity and community are both important to build resiliency. This is crucial to enhance the human and social capital to create a resilient community.

This paper has tried to show the role that bottom-up initiatives and social associations may have in a process of urban change and regeneration. They all identify particular needs in shaping the urban space. This work can help to find a new way to conceive a new idea of sustainable development.

Before these actions, strategies for mending physical urban areas were missing in town, as were paths towards participative forms of urban regeneration framed by at least a few criteria of social, environmental, economic, and institutional sustainability. Indeed, «recognis(ing) untapped areas of potential by challenging and going beyond the business-as-usual urban policy orthodoxy, and how to enable communities to realise this potential to build their own resilience strategies and improve well-being» (Unsworth et al., 2011) is not only of interest to self-organised civic networks but also, and more often, to institutional actors, which might be very sincere in their desire to take the path of listening to bottom-up proposals, promoting 'active citizenship' and legitimate forms of urban regeneration interventions coming from the bottom-up. The effective presence of many different stakeholders proposing urban regeneration projects 'from the bottom-up' represent only a form of urban resilience.

## REFERENCES


Baldassare, M. (1984). Evidence for neighborhood revitalization: Manhattan. In J. Palen and B. London (Eds.), *Gentrification, displacement, and neighborhood revitalization* (pp. 90–102). Albany: State University of New York Press.

Bandura, A. (1999). Self-efficacy: Toward a unifying theory of behavioral change. In R.F. Baumister (Ed.), *The self in social psychology* (pp. 285–298). Philadelphia: Taylor & Francis.

Barca, F. (2010). *An agenda for a reformed cohesion policy*. European Commission.

Bolzoni, M. (2012). Bottom-up initiatives and urban regeneration. An exploration of the case of San Salvario's neighbourhood, Turin. In *Rethinking urban inclusion: Spaces, mobilisations, interventions*. Coimbra: Press University.

Clay, P.L. (1979). *Neighborhood renewal: middle-class resettlement and incumbent upgrading in American neighborhoods*. Lexington: Lexington Books.

Cremaschi, M. (2008). *Tracce di Quartieri: il Legame Sociale nella Città che Cambia | Neighborhood Tracks: The Social Bonding in the Changing City*. Milano: FrancoAngeli.

Cutter et al. (2008). *Resilience is the ability of a social system to respond and recover from disasters and includes those inherent in cities and disasters*. New York: Taylor & Francis.

Downs, A. (1981). *Neighborhoods and urban development*. Brookings: Brookings Institution Press.

Harvey, D. (2008). The right to the city. *New Left Review*, *53*, 23–40.

Healey, P. (1997). *Collaborative planning: Shaping places in fragmented societies*. Vancouver: UBC Press.

Jones, P. (2003, March). Urban regeneration's poisoned chalice: Is there an impasse in (community) participation based policy? *Urban Studies*, *40*(3), 581–601.

Morandi, C. (2008). Politiche Pubbliche e Networking Pubblico-Privato per la Valorizzazione del Commercio Urbano | Public Policy and Public-Private Networking for the Enhancement of Urban Commerce. *Territorio*, *46*, 152–164.

Morandi, C., Rabbiosi, C. (2012). Rigenerazione Urbana dal Basso Nel Sud-Ovest Milanese: Criticità e Proposte | Urban Regeneration from the Low in South West Milanese: Critics and Proposals. *Territorio*, (123–132).

Pacione, M. (2005). *Urban geography: a global perspective*. London: Routledge.

Park, R.E., Burgess, E.W. & Mckenzie, R.D. (1925). The city: Suggestions for the investigation of human behavior in the urban environment. In Morris, J. (Ed.), *The heritage of sociology*. Chicago: University of Chicago Press.

Parson, T. (1971). *Societies: evolutionary and comparative perspectives* (T. Parson, Trans.). Bologna: Il Mulino.

Rabbiosi, C. (2013). S16. Resilience cities and the crisis: Local responses, governance and citizens actions. In *RC21 Conference 2013: Resourceful cities*. Humboldt: University of Berlin.

Raco, M. (2000). Assessing community participation in local economic development—Lessons for the new urban policy. *Political Geography* (573–599).

Rossi, U. (2004, April). The multiplex city: The process of urban change in the historic centre of Naples. *European Urban and Regional Studies* (156–169).

Sabel, C. & Zeitlin, J. (2010). *Experimentalist governance in the European Union, Towards a new architecture.* New York: Oxford University Press.

Sichenze, A. (2014). *Dentro Matera* [Into Matera]. Matera: Edizioni Giannatelli.

Stoica, A.C., (2010). *Towards a culture-led urban regeneration strategy for Romania from a cultural policy perspective. The case of the paintbrush factory.* Budapest: Central European University.

Unsworth, R., Ball, S., Bauman, I., Chatterton, P., Goldring, A., Hill, K. & Julier, G. (2011, April). Building resilience and well-being in the margins within the city: Changing perceptions, making connections, realising potential, plugging resources leaks. *City*, (181–203).

Varady, D. (1986). *Neighborhood upgrading: A realistic assessment*. Albany: Suny Press.

# *City as an organism*

*Architecture*

*Cities' Identity Through Architecture and Arts – Catalani et al. (Eds)*
 *ISBN 978-1-138-05409-7*

# Developing sustainable urban growth in Egypt towards the location of renewable energy resources

Wael Ahmed Aboneama
*Department of Architecture, Faculty of Fine Arts, Helwan University in Cairo, Egypt*
*Faculty of Engineering, King Khalid University, Abha, KSA*

ABSTRACT: Sustainable urban growth nowadays is not an option for big cities all over the world. It is mandatory to protect the environment and our planet for the future of generations to come. We should work through a strategy that will eventually satisfy future cities' needs for energy. This strategy depends on generating energy from different resources, in ways that go beyond energy generation and utilisation. In addition, the strategy should also aim for environmental conservation and improving local manufacturing of energy equipment. We should study the potential locations in Egypt for generating renewable energy from wind, solar energy, waves, and biomass to ensure sustainable urban growth. Such strategies typically involve three major technological changes: to save energy consumption on the demand side, efficiency in the energy production, and replacement of fossil fuels by various renewable resources. This paper discusses the perspective of the renewable energy potential in Egypt and its impact on sustainable urban development for the 2050 Egyptian vision.

*Keywords*: sustainable; urban development; renewable energy; energy resources

## 1 INTRODUCTION

The Egyptian future plan must include an ambitious commitment to renewable energy as a means to diversify its energy supply options. This will reduce the environmental impact of fossil fuel based power generation. It is expected to enlarge the renewable energy target to 20 percent, according to the aims of the supreme energy council, by increasingly using wind, hydro and solar energy as substitutes for non-renewable energy. The council also has to approve key policy steps, which are:

- Increasing the size of renewable energy resources to satisfy the demands of the people.
- Encouraging the use of sustainable transportation, instead of normal fuel busses.
- Encouraging people to generate clean energy by reducing customs and taxes on such projects and buildings.
- Supporting investment in manufacturing equipment for clean energy production to reduce cost.

On the other hand, new sustainable urban development should be supported by renewable energy resources and sustainable transportation. The standard of living of the people in a country, and in most countries, is determined by the energy consumption per capita. It reflects the need for energy for industrial purposes. Unfortunately, as illustrated in Figure 1, energy production helps to create new economic potentials and lead the whole country to be able to compete in the global market. Even agricultural activities need more energy for mass production since the production industry supports these agricultural activities.

Figure 1. World energy consumption per capita (2013) (upload.wikimedia—World_Map Energy_Use_2013.png, 2013).

## 2 LITERATURE REVIEW

### 2.1 *Lack of energy power types in Egypt*

No one can deny that Egypt is pressured by a lack of energy sources, including fuel for cars, electricity for domestic and industrial use, and gas for domestic use.

By the year 2022, it is expected that other natural resources will become available. This plan is based on the following forecasts:

- The energy production by fossil-based fuels will remain at the same level
- The usage of wind energy will increase by 20 per cent to contribute to electricity generation (Figure 2) (El.Sobki, Future of renewable energy in Egypt, 2015)
- Nuclear energy would contribute up to 6 per cent of energy use.
- The energy supply mix after five years is expected to be as follows: (Figure 2 illustrates energy consumption now)
  - 40 percent from non-renewable sources (20 per cent fossil fuel and 20 percent natural gas (NG))
  - 8 per cent from energy efficiency implementations
  - 9 per cent from renewable energy (2 per cent from hydro and 7 per cent from wind)
  - 6 per cent from nuclear energy
  - 37 per cent additional energy is needed

### 2.2 *Lack of use of renewable energy resources and energy consumption in Egypt*

Egypt is considered to be a country that is rich in natural resources. According to the US Energy Information Administration (EIA), it is the largest non-OPEC oil producer in Africa. Moreover, it is the second largest dry natural gas producer. However, energy production in Egypt has been steadily declining since 2009. This is due to the fact that the natural gas supply and oil production are unable to keep up with the demand for energy. The shortages have led to decreasing exports (Figure 2) because main Egyptian industry based on high energy demand such as cement, ceramic, and fertilizers production. Egypt's primary energy supply was 78,214 ktoe in 2012.

### 2.3 *Egypt's problems as historical country*

Several problems are causing cities in Egypt to struggle, these problems include, but are not limited to, the following:

- Cairo is an expanding city that is already huge in size which naturally leads to countless environmental issues. Air pollution in Cairo is among the serious environmental problems

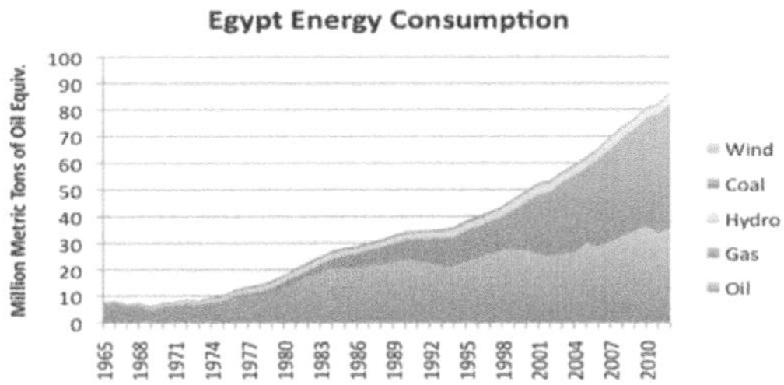


| Total Final Energy Consumption | | | |
|---|---|---|---|
| | | ktoe | % |
| Industry sector | | 13.357 | 23.90 |
| Transport sector | | 16.980 | 30.39 |
| Other sectors | | 18.479 | 33.07 |
| of which | Residential | 11.910 | 21.31 |
| | Commercial & public Services | 2.913 | 5.21 |
| | Agriculture/Forstry | 2.841 | 5.08 |
| | Non-Specified | 814 | 1.46 |
| Non-Energy use | | 7.062 | 12.64 |
| Total | | 55.879 | 100 |

Figure 2. (a) on the left, illustrates energy consumption in Egypt (source: climateobserver, 2016), (b) on the right, total energy consumption in Egypt, 2012.

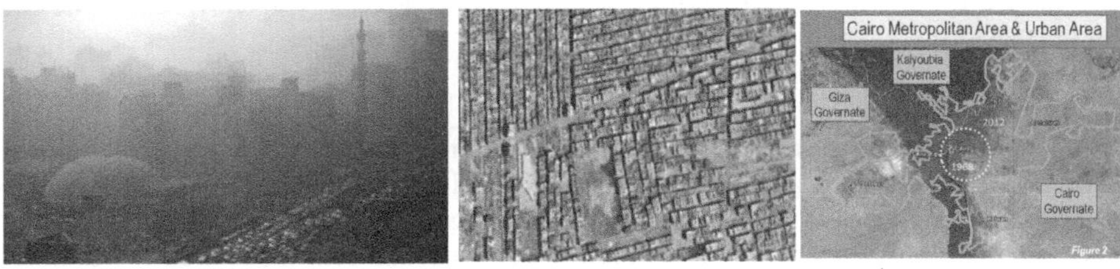


Figure 3. (a) on left, pollution in Egypt (http://media.philly.com/, 2014) – (b) middle, slums in Cairo (researcher by google earth) – (c) on right, Cairo expansion. (source: Cairofrombelow, 2011).

and is caused by the higher-than-normal levels of lead, carbon dioxide, sulphur dioxide, and suspended particulate matter concentrations from vehicle emissions, urban industrial operations, and the burning of rubbish. The streets of Cairo have over four and a half million kilometres. (Figures 3 and 11). (United Nations, 2014)

- The total prime agricultural land lost due to urbanisation between 1952 and 2002 amounts to 1,260 square kilometres, (Figure 3). (United Nations, 2014)
- The population increase per year is approximately 1.35 million.
- There is a shortage of affordable housing for the poor despite the presence of 5 million vacant units.
- About five hundred thousand new housing units are needed annually between now and 2020.
- Optical pollution.
- The underground metro is the only tool for sustainable transportation.
- Economic problems create an obstacle to achieving sustainable growth and environmental goals.
- There is a high residential density of more than double the allowable international rates.
- Traffic problems are increasing year by year.
- Natural resources are being destroyed.
- Mediocre living conditions in some areas.
- Limited green spaces (0.3 $m^2$/person within the ring road and 1.5 $m^2$/person in the region as a whole).
- The challenge is not the size, but the population distribution over the whole region.

## 3 METHODOLOGY STEPS: APPLY NEW SUSTAINABLE URBAN DEVELOPMENT

### 3.1 *Available renewable energy resources in Egypt*

There are many potential sources of renewable energy (solar power, wind, rain, tides, and waves). Renewable energy often supplies energy in four important areas: electricity generation, air and water heating or cooling, transportation, and rural (off-grid) energy services (Figures 4 and 5).

### 3.1.1 *Solar energy*

Egypt, given its location, is one of the regions in the world that has high levels of solar power (Figure 4). It was the first country in Africa to build a solar power plant. All studies have claimed that, with the present consumption rate, all fossil fuel resources will be depleted within the 21st century. Solar energy is the only resource that competes with wind power as a cheaper source of renewable energy.

As illustrated in the map in Figure 5, many areas in Egypt could be powered using solar energy. Sinai is one of the most suitable regions in Egypt that could be powered using this technique. All the cities situated on the Red Sea coast can be powered using solar energy. Toshka also has this potential. (El.Sobki, New & renewable energy agency, renewable energy in Egypt, Feb. 2015)

### 3.1.2 *Wind power*

Egypt's location gives it some of the best wind resources in the world, especially along the Gulf of Suez. Wind speeds and power densities of 7–10.5 m/s and 350–900 W/m$^2$ have been estimated at a height of 50 m above ground level (Figure 5). The Gulf of Suez has a small number of inhabitants, this region can therefore easily host several thousand MW of installed wind capacity which means that it can supply several cities. However, Egypt still needs huge investments to create wind power plants in different regions, according to wind speed studies. (El.Sobki, Wind Energy in Egypt, 2016)

### 3.1.3 *Hydroelectric power in Egypt*

Wind power produces five per cent of clean renewable energy compared to hydroelectric energy, this is clearly depicted when looking at Egypt's High Dam. However, Egypt has more

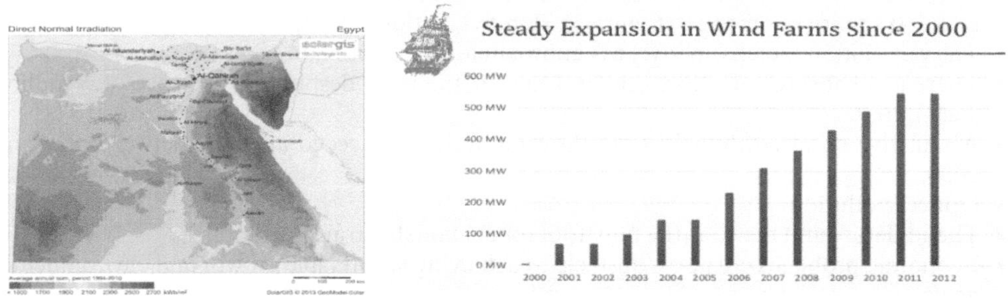


Figure 4. (a) on left, illustrates potential solar power plant locations in Egypt (source: helioscsp, 2016) – (b) on right, illustrates the increasing number of wind farms over 12 years. (source: company, 2011/2012).

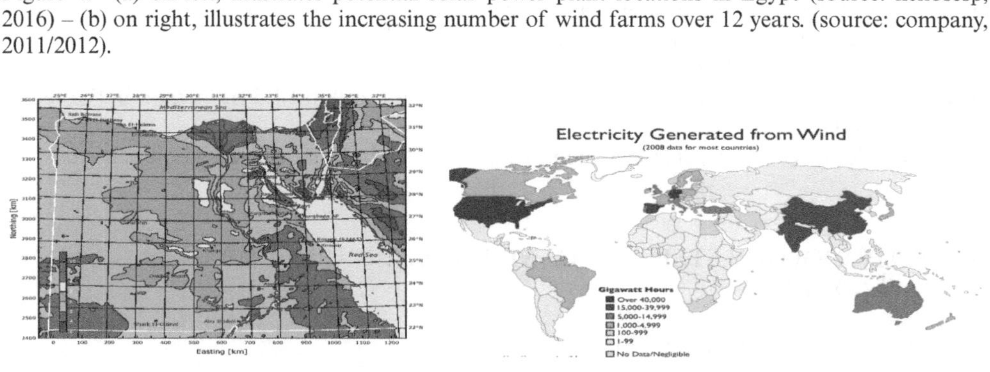


Figure 5. (a) on left, illustrates wind strength in Egypt (source: ars.els-cdn, 2009) – (b) on right, illustrates the distribution of countries generate power from wind. (source: geocurrents, 2012).

than 3,000 kilometres of sea shore (Survey of energy resources, 2007). There is more than one method of generating hydroelectric power (Figure 6). (Al-Youm, 2015) The energy of the sea tides come from the gravitational pull of the moon and the sun upon the Earth. In fact, sea energy comes from a number of sources. In addition to tidal energy, there's the energy of the sea's waves, which are driven by both the tides and the winds. (Survey of energy resources, 2007)

#### 3.1.4 *Biomass energy in Egypt*

Biomass is made up of biological materials that could be derived from living or dead bodies of living organisms. The criterion of biomass as a resource for making energy is often based on plants or plant-based materials. They are not used for food, and are specifically called lignocellulose biomass (Figure 6). Biomass can either be used directly through combustion to produce heat, or indirectly after converting it to various forms of biofuel. It can be converted to other usable forms of fuel such as methane gas or transportation fuels such as ethanol and biodiesel. (Sriram & Shahidehpour, 2005)

### 3.2 *Cairo vision 2050 (Cairofrombelow, 2011)*

#### 3.2.1 *Main goal*

Sustainable development proposed in 'Cairo vision 2050' (Figure 7) is a good example of sustainable urban growth. It includes social, cultural and economic fields integrated with the urban field.

#### 3.2.2 *Road and transportation networks connection*

- Egypt should have access to a road network linking the Cairo suburbs to regional roads.
- Egypt should have an excellent and sustainable transportation plan and network that rises up to the level of European countries.

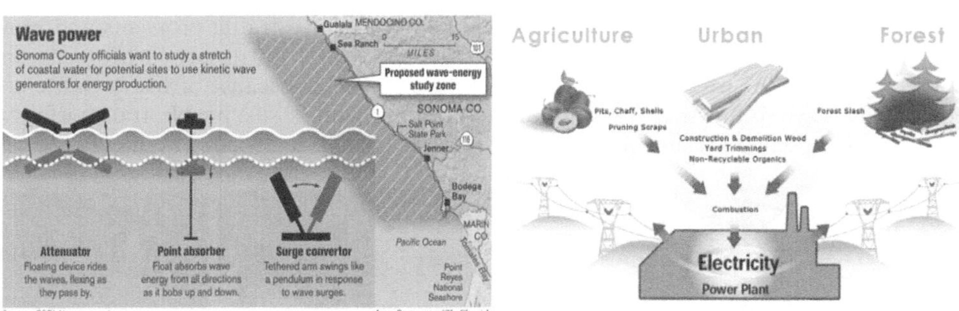


Figure 6. (a) on left, hydro alternative resources (source: ESRI, 2014) – (b) on right, biomass resources (source: Calbiomass, 2013).

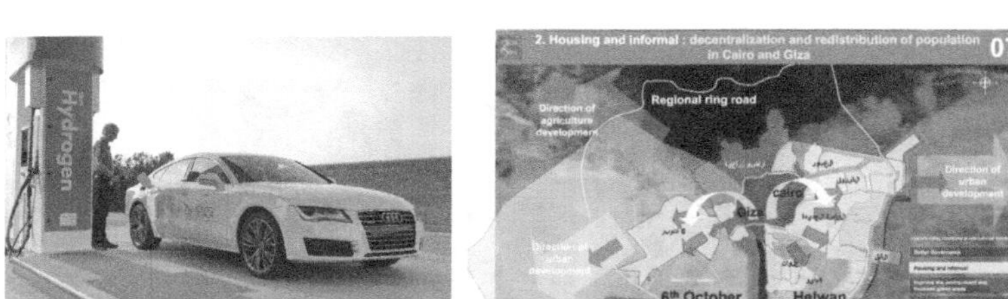


Figure 7. (a) on left, zero emission car (source: Insideevs, 2016) – (b) middle, Cairo vision 2050 (source: Cairofrombelow, 2011).

– Egypt should have a pedestrian and bicycle network linked with public transportation.

## 3.3 *Sustainable transportation*

### 3.3.1 *Zero emission vehicles*

Green vehicles are powered by alternative fuels. They include hybrid electric vehicles, plug-in hybrid electric vehicles (Figure 7), battery-powered electric vehicles, compressed-air vehicles, hydrogen and fuel-cell vehicles, neat ethanol vehicles, flexible-fuel vehicles, and natural gas vehicles. The Egyptian government can encourage people to buy these kinds of cars by removing any taxes or customs imposed on them. (Mui, 2010)

### 3.3.2 *Zero emission buses 'Electric Bus'*

These are very similar to zero emission cars. A zero emission bus is designed to be a fully-clean solution in transportation. It is highly used in Europe to create a competitive and sustainable transport system. (Rami, 2015)

### 3.3.3 *Installing trolleybuses as sustainable transportation*

Trolleybuses have previously existed in Egypt and were installed in districts that were highly populated. Unfortunately, like tram lines, trolleybus lanes are no longer an option ever since the implementation of a plan that ended most of Egypt's sustainable transportation (Figure 8). (Ehab, 2015)

### 3.3.4 *Developing bike lanes for district connection*

A bikeway is a lane, or path which is specifically designed for bicycles (Figure 8). Bike lanes should be visible and marked using paint, these markings are quite common in many cities. In some European countries cycle tracks, bollards or boulevards are quite common. (Ehab, 2015)

### 3.3.5 *Installing new tram lines as sustainable transportation and stop demolishing old tram lines*

Old city solutions should be taken into consideration to solve traffic problems and move towards sustainable urban growth. The 2004 development of Rabat city in Morocco and Athens in Greece are examples. New cities that face problems because of heavy traffic and crowdedness have found that the solution to their problem is sustainable transportation, meanwhile Egyptian government demolished most of tramlines in Cairo (Figure 9).

Figure 8. (a) on left, zero emission bus "source:", (b) second left, trolley bus (source: Hildalsolis, 2017), (c) second right, bicycle rack (source: Pinimg, 2014), (d) on right, pedestrian pavement (source: sportifycities, 2016).

Figure 9. (a) on left & next, Mustafa el-Nahas St. in Cairo after demolishing tram lines (source: Elwatannews, 2014) – (Tadamun, 2015) – (b) next, monorail (source: inhabitat, 2014) – (c) river taxi (source: Offtolondon, 2011).

3.3.6 *Installing monorail for heavy traffic places*
A monorail is a railway in which the track consists of a single rail, typically elevated. The term is also used to describe the beam of the system, or the vehicles traveling on such a beam or track. Its vehicles often appear similar to light rail vehicles. Monorails can be driven by linear induction motors like conventional railways and other advanced rapid transit systems (Figure 9).

3.3.7 *Developing Nile river bus (taxi) and stations and installing zero carbon engines*
A water taxi or water bus, also known as a sightseeing boat, is a watercraft used to provide public or private transport, usually, but not always, in an urban environment. The service may be scheduled with multiple stops, operating in a similar manner to a bus (Figure 9).

## 4 RESULTS AND FINDINGS

We can conclude from this study that we are facing five challenges that we should work on, including the absence of long term planning since we cannot solve the existing situation without long term planning. We do not have a vision for the next twenty or fifty years based on analysis of all facts (Figure 10). The lack of coordination between authorities is also a challenge given that it is very clear that some authorities in Egypt are working separately. An example of this in Egypt is all too known; after the road asphalt is finished and ready to use, another governmental department concerned with infrastructure would dig up the road to install the necessary pipelines. Another challenge is the lack of experience amongst decision makers. Without applying the philosophy of reward and punishment, decisions will be taken without consultation. The final challenge is the insufficient available grid capacity and coverage; the electricity production and structure plan should be restudied and planned.

More than 80 per cent of the population of Egypt live in slums. Inhabitants are forced to live in inhumane settlements, owing to a severe shortage of affordable housing in the cities. These inhabitants suffer from lack of electricity and sewerage services, and are subjected to mistreatment by the state. Thousands of underprivileged Egyptians who survive in slum areas are left on their own to deal with extreme heat in the summer or treacherous rainfall in the winter, such as a recent storm that drenched shanty towns. The ever-growing number of slum dwellers highlights the huge disparity in the distribution of wealth, residential units, and unequal access to housing options (Figure 10). Since the 1970s, policies have always been biased towards big capital and profit accumulation rather than the needs of society's lower tranches. Governments literally ignored informal housing. It is very clear that all urban growth in Egypt is focused on only one item which is agriculture around the Nile River

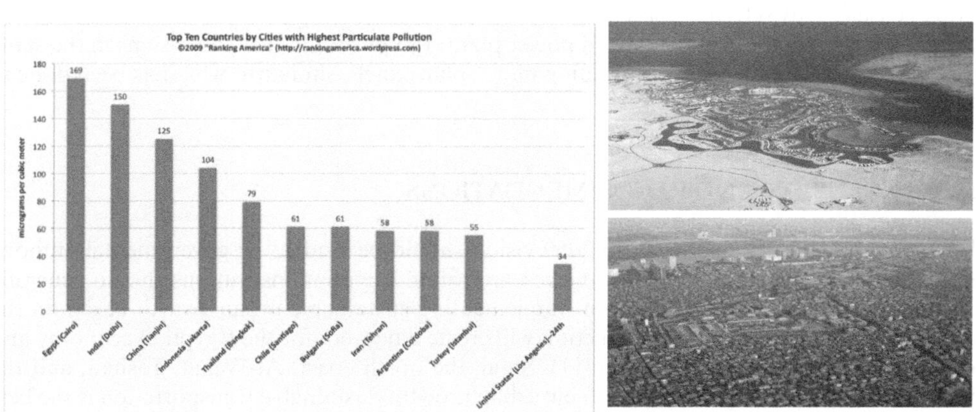


Figure 10. (a) on left, highest polluting countries (source: Wordpress, 2010) – (b) top right, El-Gouna village – (c) bottom right, Slums in Egypt.

Figure 11. (a) top left, solar plant, (b) middle left, wind plant, (c) bottom left, world installed capacity – (d) on right, Urban growth locations "source: By researcher".

which was the central focus of historical urban growth all over Egypt. (Reuters, 2015) The possibility of generating renewable energy should be the main objective for the new urban growth plan. The vision of creating an industrial country and starting new approaches and development directions should also be part of this growth. (Thorpe, 2015) Authorities should also be aware of the main pillars of sustainability which are the environment, people and the economy. In other words we have to restudy the location of new cities based on the location of renewable energy and water resources. New cities will be the basis of a strong economy based on agriculture and industry. From all previous studies we can conclude that the north coast, Al-Wahat, Toshka, the Red Sea coast and Sinai are empty areas with excellent potential for renewable energy and water (Figure 11).

## 5 DISCUSSION AND ANALYSIS

Sustainable urban development requires long term planning and vision that could be divided into short, five-year stages to be able to confront our serious challenges. Sustainable urban growth and development is the only way to ensure a bright future for the coming generations. (Pascale, 2015) Renewable energy production is increasing all over the world and Egypt should not stop at the Zaafrana wind power plant. Egypt should take more steps in the same direction and in the direction of creating more solar panels, similar to what has been done in Morocco (Independent, 2015).

## 6 CONCLUSION AND RECOMMENDATIONS

Long-term plans such as the Cairo 2050 vision should be studied by governmental authorities in co-operation with academic and research staff. Egyptian institutions should generally work together in the form of a team. Implementing sustainable urban growth based on the potential of renewable energy production will create a new era for the Egyptian economy and will elevate living standards (Figure 11). Sinai, the north coast, Al-Wahat, Toshka, and the Red Sea coast are the main areas for new urban growth. Sustainable transportation is the best way to connect and develop new urban areas such as Heliopolis and Nasr city. We should use all means of sustainable transportation which are; tram, trolleybus, monorail, underground metro, zero emission buses, environmentally friendly cars, river taxi, and bikes. The demolition of tram lines was a disaster given its effects on Cairo's transportation and against Cairo vision 2050. (Bank, 2015) Publishing information about long term planning will help researchers to provide new directions and develop ideas (Aboneama, 2012). Authorities should also apply the sustainable rating system and push all designers and developers to build upon its recommendations; this will push Egyptian projects to become environmentally friendly (Aboneama, Enhancing LEED as a sustainable rating system by applying its regional priority for all environmental regions, 2015). Finally, the authorities should eliminate taxes and customs on environmentally friendly buses and cars and provide refill stations for them.

## REFERENCES


Aboneama, W. (2012). Protect the planet through sustainability rating systems with local environmental criteria—LEED in the Middle East. *ASIA Pacific International Conference on Environment-Behaviour Studies.* Cairo: Future Communities: Socio-Cultural & Environmental Challenges.

Aboneama, W. (2015). Enhancing LEED as a sustainable rating system by applying its regional priority for all environmental regions. *International conference and exhibition on Satellite* (pp. Wael. (2015, August). ). Houston: OMICS group.

Al-Youm, A.-M. (2015). *Egypt Plans New Hydroelectric Power Stations in Cooperation with China.* Cairo: Al-Masry Al-Youm.

ars.els-cdn. (2009). *Wind strength in Egypt.* Retrieved from https://ars.els-cdn.com/: https://ars.els-cdn.com/content/image/1-s2.0-S2090123210000068-gr1.jpg.

Bank, T. W. (2015). *Who Gains and Who Loses from Plunging Oil Prices in the Middle East and North Africa Region.* Washington: The World Bank Group.

Cairofrombelow. (2011, https://cairofrombelow.files.wordpress.com/2011/08/cairo-2050-vision-v-2009-gopp-12-mb.pdf). *Cairo vision 2050.* Cairo: https://cairofrombelow.org/. Retrieved from https://cairofrombelow.files.wordpress.com.

Calbiomass. (2013). *biomass fuel sources.* Retrieved from Biomass fuel: http://www.calbiomass.org/wp-content/uploads/2013/02/biomass-fuel-sources.jpg.

climateobserver. (2016). *egypt-energy-consumption.* Retrieved from Egyptian energy consumption: http://climateobserver.org/.

company, E. E. (2011/2012). *Annual report.* Cairo: Egyptian Electricity Holding company.

Ehab, F. (2015). *Egypt Aims to Halt Gas Imports in 2020 – Minister.* London: Thomson Reuters.

El.Sobki, M. (2015). *Future of renewable energy in Egypt.* Cario: Minsry of ecetricity & renewable energy.

El.Sobki, M. (2016). *Wind Energy in Egypt.* Cairo: Minsry of ecetricity & renewable energy.

El.Sobki, M. (Feb. 2015). *New & renewable energy agency, renewable energy in Egypt.* Cairo: Minstry of Housing.

Elwatannews. (2014). *Elnahas street.* Retrieved from img.adminlive: http://img.adminlive.elwatannews.com/old_news_images/large/266563_Large_20140909015505_11.jpg.

ESRI. (2014). Environmental Systems Research Institute. *Nature magazine.*

geocurrents. (2012). *Distribution of countries generate power from wind.* Retrieved from http://geocurrents.info/: http://geocurrents.info/wp-content/uploads/2012/07/electricity-from-wind-map.jpg.

helioscsp. (2016). *Potential of solar power.* Retrieved from helioscsp.com: http://helioscsp.com/wp-content/uploads/2016/11/Egypt-dni.png.

Hildalsolis. (2017). *Zero emission bus.* Retrieved from zeroemmissh: http://hildalsolis.org/wp-content/uploads/2017/07/zeroemmissh.jpg.

http://media.philly.com/. (2014). *Cairo air pollution.* Retrieved from media philly: http://media.philly.com/images/airpollutionciaro.jpg.

Independent, E. (2015, Mar 4). Egypt Plans New Hydroelectric Power Stations in Cooperation with China. *Egypt Independent.*

inhabitat. (2014). *Paulo-Monorail.* Retrieved from inhabitat.com: https://inhabitat.com/wp-content/blogs.dir/1/files/2014/01/Sao-Paulo-Monorail-2-lead.jpg.

Insideevs. (2016). *Zero emission cars.* Retrieved from insideevs.com: https://insideevs.com/wp-content/uploads/2014/12/A7h140026.jpg.

Mui, S. (2010, May). *The Zero Emission Vehicle Program.* NRDC, Clean Vehicles and Fuels.

Newgeography. (2014). Retrieved from http://www.newgeography.com/files/cox-cairo-f2.jpg.

Offtolondon. (2011). *London river taxi.* Retrieved from www.offtolondon.com: http://www.offtolondon.com/images/thames-clipper.jpg.

Pascale, E.-K. (2015). *Egypt Aims to Produce 20% of Power from Renewables by 2020*. Aman.
Pinimg. (2014). *Bike bicycle*. Retrieved from Bicycle: https://i.pinimg.com/736x/8e/99/9b/8e999b8723c22931c5c6bb2c47b8bade-mtb-bike-bicycle.jpg.
Rami, G. (2015, Jan 6). Falling Oil Prices Both Blessing, Curse for Egypt. *Al-Monitor*.
Reuters, T. (2015). *Egypt to Build 4,300 MW Solar, Wind Plants in 3 Years*. London: Reuters.
sportifycities. (2016). *Sport cities*. Retrieved from Pedistrian: https://sportifycities.files.wordpress.com/2016/07/bicyclelanepavement_report_2.jpg?w=730&h=541.
Sriram, N., & Shahidehpour, M. (2005). *Renewable Biomass energy.* New York: IEEE.
*Survey of energy resources.* (2007). London: World Energy Council. Retrieved from http://www.Wikipedia.com.
Tadamun. (2015). *Elnahas street*. Retrieved from www.tadamun.co: http://www.tadamun.co/wp-content/uploads/2015/04/Capture.jpg.
Thorpe, D. (2015). *How Solar Energy Can Solve Egypt's Electricity Crisis.* Sustainable Cities Collective, Social Media Today LLC.
United Nations, E. c. (2014). *Promoting energy efficiency investment for climate change mitigation and sustainable development.* Geneva, Switzerland: UNECE for a stronger and greener economy.
*upload.wikimedia—World_Map Energy_Use_2013.png*. (2013). Retrieved from wikipedia Web site: https://upload.wikimedia.org.
Wordpress. (2010). *Highest polluting countries*. Retrieved from Rankingamerica: http://rankingamerica.wordpress.com.

*Cities' Identity Through Architecture and Arts – Catalani et al. (Eds)*
*© 2018 Taylor & Francis Group, London, ISBN 978-1-138-05409-7*

# Enhancing the housing industry in Egypt through the application of new design and construction techniques aimed towards sustainability and actual market demand

Wael Ahmed Aboneama
*Department of Architecture, Faculty of Fine Arts, Helwan University, Cairo, Egypt*

ABSTRACT: The housing industry and housing production in Egypt are facing a catastrophic situation. Nobody can deny the housing problems in Egypt. More than 80% of Egyptians are living in slums. Around 20% of all Egyptians are living in non-healthy homes ((ECES), 2010). Billions of US dollars are spent every year in building new, high-style compounds in Egypt which leads to a recession of selling these units. The whole world is now focusing on sustainability and has started renovation and replacement programmes for housing. On the other hand, in Egypt we are still looking to meet the housing demand. Housing problems in Egypt are not related to low investment in this sector, but to the wrong direction for this housing investment. In other words, the production is focused on only one housing sector, which is only around 10% and the supply is more than the demand, however it costs more and more than all other types of housing. In addition to the wrong policies, all house construction is focused on one type of construction which is reinforced concrete (post and lintel), casting on-site (Stephen Everhart, June 2006). This research will present different types of sustainable construction for all housing types as one of the key solutions, and also to meet market demand and to solve this problem in the right way.

*Keywords*: Housing industry; Sustainability; Design and construction techniques; Market demand

## 1 INTRODUCTION

The housing industry in Egypt has been damaged since the middle of the last century. It was affected by several factors. Political decisions such as the reduction of rental value several times and the elimination of time framed rental contracts were the major factors that destroyed the housing industry in Egypt in a few years after those laws due to first and second world war. In addition, economic failure after the nationalisation of the private sector was the second knockout for the housing industry in Egypt. Wars in Sinai and internal immigration increased the demand for private houses. The public sector was unable to meet all market requirements following the dramatic changes of social classes in this very specific period of Egyptian history ((ECES), 2010). The result was a disaster and the slums in Egypt were born (Figure 1). A second era for housing in Egypt, with some positive features, started after the 1973 war, including positive changes for enhancing the economic situation and creating new cities to solve housing problems (Figure 2). The housing market is now saturated for certain levels of housing and low income housing demand is increasing dramatically. The only way to find a real solution is to apply environmental designs with new construction techniques under a real vision for all housing levels.

Figure 1. (a) on left, illustrates El-Rehab city – (b) on right, illustrates slums area where 80% of Egyptians live. (Source: Google earth by researcher).

Figure 2. (a) on left, illustrates slums in Egypt (https://inhabitat.com/, n.d.) – (b) in middle & (c) on right, illustrate luxury compounds design style in Egypt. (http://top10cairo.com, 2016) – (Al-Ahram, 2015).

## 2 LITERATURE REVIEW

Solving the housing problems in Egypt starts by identifying the problems that we are confronting and the relationship between housing and economic situation and changes in Egypt.

### 2.1 *History of housing in Egypt (Aboneama, 2002)*

#### 2.1.1 *Housing in Egypt before 1952*

No one can deny that before 1952 Egypt did not have a housing problem (Figure 3). The architecture and urban design of Cairo before 1952 could compete with all European capitals. All illustrated photos (Figure 3) present sustainable transportation and the middle class society housing in Egypt in the first decades of twentieth century. Unfortunately, we could not even protect those buildings from damage. (Aboneama, 2002).

#### 2.1.2 *Housing in Egypt from 1952 until the war of 1973*

After 1952 many changes happened to the society, economy and laws in Egypt. The housing industry in this period was affected by some catastrophic decisions, starting with the regulation of the relationship between owners and tenants, followed by a series of actions which were focused on the nationalisation of most of the private sector icons in Egypt. Finally, a series of wars dramatically affected the economic situation in Egypt. The result was eliminating private sector investment in housing. It was the responsibility of the government to produce houses for people under pressure of the weak economy and internal immigration due to wars. The housing construction of this period is illustrated in Figure 4. (Aboneama, 2002).

#### 2.1.3 *Housing in Egypt after the war of 1973 until the economic growth of 1992*

The 6th of October war was the final war between Egypt and its enemy which helped to start the recovery of the Egyptian economy. We can divided this period into two decades (1970s and 1980s). The opening of the Suez Canal, petroleum exploration and currency transfer from Egyptians working abroad all helped to start a new era in modern Egyptian history. The housing problem has been started and governmental solutions could not present actual solutions to stop it. Interest groups has their input to build for their members. For example, syndicates, co-operative housing, and the military started building on a huge plot a number of residences such as El-Obour, Masaken Sheraton, and Agha Khan (Figure 5) (Aboneama, 2002). On the

Figure 3. (a) on left and middle illustrate the heart of Cairo in Egypt. (https://i.pinimg.com/, 2006) (https://i.pinimg.com/, Old Cairo, 2006) – (b) on right illustrates Heliopolis design in the beginning of 20th century (http://www.egy.com/, 2008).

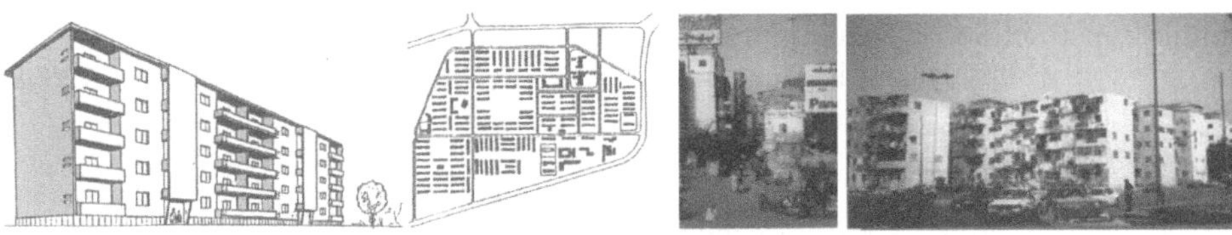

Figure 4. (a) on left & right, illustrates housing prototypes after 1952. (b) in middle, illustrates integration between slums and housing prototypes. (Aboneama, 2002) (Photos captured by researcher).

Figure 5. All illustrate big housing projects that have been constructed by interest groups in the 1970s and 1980s in Egypt. (Panoramio, 2012), (media.linkonlineworld, 2014), (Panoramio, 2012), (Aqarmap, 2016).

Figure 6. (a) on left, illustrates types of high-class compound – (b) middle, First residence housing units & hotel – (c) on right, tourist houses. (top10Cairo, 2016), (B-static, 2010), (images.sunshine, 2013).

other hand, developing new cities all over Egypt (satellite cities such as 6 October, Sadat … etc.) took a serious actions after retrieving Sinai. (Rania Nasr Eldin, December 2012).

### 2.1.4 *Housing in Egypt since 1992*

This period started from the beginning of the last decade in the 20th century until now. The beginning of the 1990s had positive economic signs which affected housing industry (Aboneama, 2002). The housing industry became an attractive field for investors and developers which seemed positive. Unfortunately, there was no national plan for housing production with numbers based on actual demand divided according to social classes. All developers invested in high-class houses and import invasive planting environment into Egyptian ecology such as golf courses (Figure 6). It ended with a recession in high-class housing and expensive empty cities. The northern and eastern coasts were covered with tourist villages (Rania Nasr Eldin, December 2012).

### 2.2 *Housing problems in Egypt*

This paper will focus on the housing types and problems in Egypt. The low-income class solves its own problem through constructing on farm lands their own homes in one of the worst slums in the world. (EIPR, July 2014).

#### 2.2.1 *Informal housing in Egypt*

According to USAID report of housing study for urban Egypt, "The changes to the informal housing production processes, and the shift to a profitable mode of illegal construction, affected the production housing typologies. They have evolved from self-built, low-rise structures to semi-professionally built, fifteen-story towers. Initially, housing structures followed a simple construction of concrete frames and brick infill, with street patterns registering agrarian subdivisions, generating (urban canyons)" (Figure 7). (USAID, December 2008).

#### 2.2.2 *Living in tombs*

According to Hamza, Hani (*The Northern Cemetery of Cairo*) "In the south-eastern part of Cairo, a city that is home to nearly 18 million people, lies El-Arafa cemetery. It is estimated that around 1 million Egyptians live in the 6.4-kilometre-long cemetery, which has led to it being dubbed the City of the Dead (Figure 8). In the cemetery, families, power lines, multi-story buildings, and a post office mingle with the graves, tombs, and mausoleums. There are often three generations of Egyptians living in the necropolis, many of whom have been there since the 1950s. The settlement is far from legal, but the Egyptian government has long since given up on evicting residents. Doing so would require the unattractive proposition of moving the inhabitants to state-built apartments or forcing the creation of more slums". (Hamza, 2001).

#### 2.2.3 *Living in tinplate nests and on the streets without any shelter*

The lowest income class live in the worst housing type. The aim is to create any form of shelter and partition using garbage sheets of metal, wood or plastic (Figure 9). This unit cannot protect people from bad weather conditions such as rain or sand storms. Usually, the location for such housing is out in the countryside without any kind of services. Social and physical diseases are endemic, which too many to listed and not applicable to this paper. (Edgar Goell, 2008).

Figure 7. (a) on left, middle and right, illustrate slums and informal housing in Egypt even with high rise buildings. (i.pinimg, 2009), (cairofrombelow, 2011), (MEI, 2011).

Figure 8. (a) on left and (b) middle, children are playing and a woman is hanging cloths between tombs stones – (c) on right, a living room above a tomb (static5.businessinsider, 2010).

2.2.4 *Some housing compounds against Egyptian ecology*

Environmental conservation is based on many items. One of the most important is potable water conservation, especially after the construction of the Grand Ethiopian Renaissance Dam. On the other hand, the increase in the building of luxury houses was combined with the construction of golf courses (Figure 10). So, Egypt suffers from a lack of potable water resources and a lack of services in all slums, and yet they use enormous amounts of water to irrigate golf courses, which is the favourite game of millionaires (El-Shafei, 2001). There is no excuse, even if the water used is treated grey water, because it could be used in many other areas that would be helpful for the economy and the people.

2.2.5 *Losing agricultural land for the construction of informal housing*

Egyptian agricultural land surrounds the Nile river, in an area of approximately 25 thousand square kilometres (6 million acres). It is about 2.5% of the total area of Egypt. When, for decades, the economic and governmental situation could not meet their need for homes, Egyptians started to build for themselves.

In doing this, they override laws, planning regulations, engineering standards, and everything. They are building on agricultural land without government supervision. The result is the loss of more than 20 thousand acres every year from the best farmland. It is one of the worst blows to the Egyptian economy, society, and future (Shokry, 2009). Corruption and big profits for land brokers are increasing this operation accumulatively. The result is shown in Figure 11 which looks like Figure 10 but the two are totally opposite. In Figure 10 rich people use natural resources (water) for a game, whereas in Figure 11 poor people also use natural resources (farmland) to build slums and informal houses without following any regulation or engineering standards (Malterre, 2016).

2.3 *Egypt's problems as historical country*

- Rapid urban expansion resulted in encroachments on agricultural land. Total prime agricultural land lost to urbanization during the period 1952–2002 amounts to 300,000 acres and pollution all over Egypt (Figure 12).
- Shortage of affordable housing supply for the poor despite the presence of 5 million vacant units.
- About 440,000–600,000 new housing units are needed annually between now and 2020.

Figure 9. (a) on left and right, illustrate the tinplate nests housing type. (nmisr, 2015), (El-badil, 2014).

Figure 10. (a) on left and right, illustrate golf courses in new satellite cities. (Ladyegypt, 2014), (realestateegypt, 2013), (businessandnews, 2015).

Figure 11. Illustrates slums and an informal attack on farm land. (metropolitiques, 2013).

Figure 12. (a) on left, pollution in Egypt – (b) middle, slums in Cairo – (c) on right, pollution in Cairo. (rackcdn, 2007), Google earth by researcher, (scidev, 2009).

- Optical pollution.
- High residential density in main existing agglomeration.
- Limited green spaces (0.3 $m^2$/person within the ring road and 1.5 $m^2$/person in the region as total).
- The challenge is not the size, but the population distribution over the whole region area. (Aboneama, 2002).

## 3 METHODOLOGY STEPS TO ENHANCE HOUSING INDUSTRY IN EGYPT TOWARDS SUSTAINABILITY AND ACTUAL MARKET NEEDS

### 3.1 *Prefabricated housing types*

#### 3.1.1 *Polystyrene construction: Eco-friendly light weight structure*

Polystyrene or EPS is a better and cheaper building material for the provision of housing in Egypt (Wael, November 2012). According to Shawn O'Donnell "Modern, eco-friendly building materials are currently emerging to satisfy the requirements of sustainable developments and mitigate environmental challenges" (O'Donnell, 2009). While it was discovered that there is little or no availability of the material in most Egyptian building markets due to lack of awareness, it x-rayed the need for a paradigm shift from the conventional system of building characterised by waste, cradle to grave (demolition) and global warming to a waste-free, recyclable, eco-friendly and cheaper alternative way of building (Figure 13).

#### 3.1.2 *Precast housing construction*

According to Zuhairi bin Abdel Hamid "Precast concrete is a green material because it contributes to green building practices in significant ways. The thermal mass of concrete allows shifting of heating and cooling loads in a structure to help reduce mechanical-system requirements (Stephen Everhart, June 2006). It reduces construction and manufacturing waste and debris on site, reducing construction IAQ (Indoor Air Quality) concerns. The load-carrying capacities, optimised cross sections, and long spans possible with precast concrete help eliminate redundant members, and concrete readily accommodate recycled content. It contains recycled content. Insulation in precast concrete saves up to 25% on heating and cooling costs (Figure 14). (Hamid, 2014).

Figure 13. Left, illustrates the shoot Crete and right, illustrates slab construction (designenterprises, 2010), (technopol.co.za, 2017).

Figure 14. On left & right, illustrate housing compounds constructed with precast concrete. (Zuhairi bin Abdel Hamid).

Figure 15. (a) on left, environmental design concept for hot weather – (b) middle, construction methods – (c) on right, environmental housing forms. (i.pinimg, Adobe construction, 2011), (pbs.twimg, 2007).

### 3.2 *Sustainable designs*

#### 3.2.1 *Passive designs (Hassan Fathy housing prototype)*

Hassan Fathy's housing design is one example. His style focused on the environmental issues of hot weather and he believed that the answer must lie in the recreation of forms true to the region as well as to local methods of construction (Figure 15). According to Abdelmoniem El-Shaorbagy "He valued architecture that is rooted in the location and the culture of a region, as opposed to an imported internationalism, rooted in a common technology rather than a common humanism" (El-Shorbagy, 2010). The architecture should be shaped by a conceptual framework which develops an understanding of contemporary responses to modern environmental, urban and social conditions of existence. (Mohamed Soliman, 2005).

#### 3.2.2 *Adobe architecture for rural agricultural villages*

Adobe bricks are most often made into units that are small enough to quickly air dry individually without cracking and can be subsequently assembled (Figure 16). According to this type definition in Wikipedia "Modern methods of construction allow the pouring of whole adobe walls that are reinforced with steel. In dry climates, adobe structures are extremely durable, and account for some of the oldest existing buildings in the world. Adobe buildings offer significant advantages due to their greater thermal mass but they are known to be particularly susceptible to earthquake damage if they are not somehow reinforced". The only disadvantage of this type is dredging fertile soil, as it is only recommended to take needed clay from underneath the footprint of the building not for mass production. (Rania Nasr Eldin, December 2012).

Figure 16. Left and right, illustrate adobe blocks & architecture. (claysandstraw, 2005), (d2wpnc0 srowh1f.cloudfront.net, 2014).

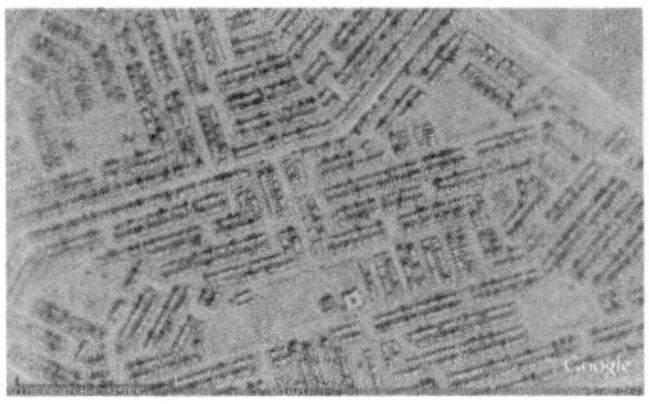

Figure 17. Left, Ebny Betak Google Earth view—right, photo from the project. (youm7, 2014).

### 3.3 *Build your home to protect green fields* 'مشروع ابني بيتك'

The informal sector proved that people have the ability to build their own homes, however it has had a negative impact on the national economy because they are building on agricultural land without following any regulations or sense of architecture and urban design (Aboneama, 2002). The idea of the 'Ebny Betak' project was focused on giving the informal sector the land for free, with free architectural designs and urban design but with restricted building regulations. The people built the given designs exactly (Figure 17). The poor people succeeded again, but the problem remains that the authorities have let people build without providing any facilities or services.

## 4 RESULTS AND FINDINGS

- More than 80% of the population live in Egypt's slums.
- Inhabitants are forced to live in inhumane settlements, owing to a severe shortage of affordable housing in the cities.
- The informal sector proved that it has the ability to build their own homes, however it has had a negative impact on the national economy because they build on agricultural land without following any regulations or sense of architecture and urban design.
- We are using only one method of construction which is cast in situ using reinforced concrete.
- Housing policies and economic failure after 1952 were the reasons for the housing problem.

## 5 DISCUSSION AND ANALYSIS

- Informal housing in slums proved the power of the informal sector to build their own homes but the absence of building, planning & engineering regulations produced harmful units for people, economy, society and the whole country.
- The private sector has many targets, the first priority is to gain profit. This sector depends on loans from national banks to build high-class housing through which some of them destroy natural resources.
- Slums could destroy all farmland in Egypt.

- The Ebny Betak 'مشروع ابني بيتك' project proved, once again, the power of low-income people to build their own homes.
- Sustainable construction techniques and urban growth are the only way to create our grandsons' future. (Mohamed Soliman, 2005).

## 6 CONCLUSION AND RECOMMENDATIONS

- Polystyrene construction, pre-cast concrete and other sustainable techniques should be the most prevalent methods of construction not the least.
- Architects and designers should learn and borrow from their own history, not from opposite environments and cultures.
- Complete and extend the Ebny Betak 'مشروع ابني بيتك' project and supply it with services. Authorities should solve its problems instead of announcing falsehoods to kill its success.
- Applying the sustainable rating system and pushing all designers and developers to build upon its recommendations will push Egyptian projects towards supporting a better environment.
- Sustainable techniques and concepts, quick construction methods, passive design and the use of recycled materials are the only ways to build new homes and recover the damaged areas.
- Create laws to prevent the waste of natural resources such as water, energy, and farmland.
- Eliminate taxes and customs on the private sector when it invests in low-income housing.

## REFERENCES


(ECES). (2010). *Dead capital and the poor in Egypt (appendices)*. Cairo, Egypt: The Egyptian center for economic study.

Aboneama, W. (2002). *The imapct of economic changes on housing in Egypt between 1973–1998*. Cairo: Helwan University.

Al-Ahram. (2015). *Madinaty*. Retrieved from Al-Ahram English: http://english.ahram.org.eg/Media/News/2015/3/17/2015-635622014954485688-448.jpg.

Aqarmap. (2016). *Neighborhood*. Retrieved from Cairo: https://img-1.aqarmap.net/wompoo/aqarmap/neighborhood_slider/1604/57023853c63d0.jpg.

B-static. (2010). *Hotels in Cairo*. Retrieved from https://q.bstatic.com/: https://q.bstatic.com/images/hotel/max1024x768/212/21246355.jpg.

businessandnews. (2015). *Golf compounds*. Retrieved from http://businessandnews.com/: http://businessandnews.com/uploads/thumbs/f1c/f1c113481be1220a20ad322d9ece9004.jpg.

cairofrombelow. (2011). *Cairo-photographed*. Retrieved from Al-Masri Al-youm: https://cairofrombelow.files.wordpress.com/2011/08/cairo-photographed-by-john-harris-al-masri-al-youm.png.

claysandstraw. (2005). *Adobe construction*. Retrieved from Clay construction: http://www.claysandstraw.com/projects/mtncttge/timeline/2005/uptoOct10/fromGeri/cuttingadobes.jpg.

d2wpnc0srowh1f.cloudfront.net. (2014). *adobe-bricks-baking*. Retrieved from Adobe construction: https://d2wpnc0srowh1f.cloudfront.net/Mainblog/wp-content/uploads/2015/08/adobe-bricks-baking-in-the-sun.jpg.

designenterprises. (2010). *Polystyrene*. Retrieved from sustainable construction techniques: http://www.designenterprises.co/home/shotcrete-walls-for-homes.html.

Edgar Goell, N. E.-L. (2008). Sustainable Cities in Egypt, Learning from Experience: Potentials. In C. f. Studies. Cairo: The Egyptian Cabinet—Information and Decision Support Center.

EIPR. (July 2014). *Drafting a Fair Housing Policy for Egypt: EIPR Housing Policies Paper II*. Egypt: Egyptian intiative for personal rights.

El-badil. (2014). *Egypt Slums*. Retrieved from Slums: https://elbadil.com//.

El-Shafei, Z. (2001). *Comparison of core housing designes proposed in Egypt*. Cairo—Egypt: Cairo University.

El-Shorbagy, A.-m. (2010). Hassan Fathy: The unacknowledged conscience of twentieth century architecture. *International Journal of Basic & Applied Sciences IJBAS-IJENS Vol: 10 No: 02*, 29–35.

Hamid, Z. b. (2014). *Green materials—What are the standards and specifications for green materials certifications?* Malaysia: "Precast concrete".

Hamza, H. (2001). The Northern Cemetery of Cairo Cairo. *The American University in Cairo Press*, ISBN 977-424-618-7. Retrieved from https://en.wikipedia.org: https://en.wikipedia.org/wiki/City_of_the_Dead_(Cairo).

http://top10cairo.com. (2016). *Allegria*. Retrieved from Housing compound in Egypt: http://top10cairo.com/wp-content/uploads/2016/02/allegria.jpg.

http://www.egy.com/. (2008). *Heliopolis*. Retrieved from http://www.egy.com: http://www.egy.com/P/heliopolis/razzouk-6.JPG.

https://i.pinimg.com/. (2006). *Old Cairo*. Retrieved from https://i.pinimg.com/: https://i.pinimg.com/originals/fc/53/ff/fc53ff00074f580877a2bd6682a8cfdf.jpg.

https://i.pinimg.com/. (2006). *Old Cairo*. Retrieved from https://i.pinimg.com/: https://i.pinimg.com/736x/3e/07/4d/3e074df452b3a0bf212ab62b36bfa915-s.jpg.

https://inhabitat.com/. (n.d.). *Sulms in Egypt*. Retrieved from Inhabitat: https://inhabitat.com/files/garbagecity-ed04.jpg.

i.pinimg. (2009). *Cairo*. Retrieved from Slums: https://i.pinimg.com/originals/d9/85/04/d98504aff6c3b485528b00c777be8bf0.jpg.

i.pinimg. (2011). *Adobe construction*. Retrieved from Hassan Fathy: https://i.pinimg.com/736x/69/d7/ab/69d7ab85405c761f8c4c3eac36ae1560.jpg.

images.sunshine. (2013). *Tourism in Egypt*. Retrieved from https://images.sunshine.co.uk: https://images.sunshine.co.uk/ss/htl/0/2/3/2311-12-l.jpg.

Ladyegypt. (2014). *Golf compounds*. Retrieved from High class housing in Egypt: https://www.ladyegypt.com/files/large/dbb516eb573fe3d/720/478/fit.

Malterre, C. (2016). *Housing Cairo—From small-scale informal housing construction to semi-professional speculative urban schemes.* Zürich: Architecture Dep., Chair of Marc Angélil, Neunbrunnenstrasse.

media.linkonlineworld. (2014). *Cairo*. Retrieved from Images: http://media.linkonlineworld.com/img/large/2017/7/22/2017_7_22_22_42_58_87.jpg.

MEI. (2011). *Alexandria*. Retrieved from Middle East Institute: https://www.mei.edu/sites/default/files/publications/Alexandria%20housing.jpg.

metropolitiques. (2013). *Slums in Egypt*. Retrieved from Destroying planting farms in Egypt: http://www.metropolitiques.eu/IMG/jpg/illu-sims-2.jpg.

Mohamed Soliman, I. H. (2005). *New urban communities in Egypt (Policies & useful lessons)*. Cairo.

nmisr. (2015). *Slums images*. Retrieved from Egypt-slums: https://www.nmisr.com/wp-content/uploads/2017/04/2015-09-10T155244Z_2102912587_GF10000200351_RTRMADP_3_EGYPT-SLUM.jpg

O'Donnell, S. (2009). *Informal Housing in Cairo: Are Ashwa'iyyat Really the Problem?* Portugal.

Panoramio. (2012). *Cairo*. Retrieved from Photos: http://static.panoramio.com/photos/large/17003604.jpg.

pbs.twimg. (2007). *Hassan Fathy*. Retrieved from Hassan Fathy: https://pbs.twimg.com/media/CLIj5ntUsAAi35g.jpg.

rackcdn. (2007). *Pollution in Cairo*. Retrieved from Polluted cities: https://9c998969b63acdb676d137595348221e1b716e1a6cfee3ed7891.ssl.cf1.rackcdn.com/almpics/2015/10/RTR2TSBT.jpg/RTR2TSBT-870.jpg.

Rania Nasr Eldin, H. A. (December 2012). *Residential mobility in Egypt…a must or a myth.* Egypt: OPIC.

realestateegypt. (2013). *Allegria*. Retrieved from http://www.realestateegypt.com/: http://www.realestateegypt.com/images/new_property/Allegria/ProjectHome/1.jpg.

scidev. (2009). *pollution*. Retrieved from Cairo: http://www.scidev.net/objects_store/thumbnail/18F364A8AD44D9E85A94BC409CC89C08.jpg.

Shokry, H. (2009). *Is a Sustainable Urbanism Possible In 21 Century Egypt? A Future Direction for Walkable Settlements.* Cairo.

static5.businessinsider. (2010). *Housing in tombs*. Retrieved from Housing in Egypt: http://static5.businessinsider.com/image/5466245d69bedde269bdaeee-1200-800/1-1811.jpg.

Stephen Everhart, B. H. (June 2006). Overview of the Housing Sector. *OPIC*.

technopol.co.za. (2017). *Building Construction*. Retrieved from Sustainable construction: https://www.technopol.co.za/wp-content/uploads/2017/02/BuildingConstruction2.png.

top10Cairo. (2016). *Cairo compounds*. Retrieved from http://top10cairo.com: http://top10cairo.com/wp-content/uploads/2016/02/allegria.jpg.

USAID. (December 2008). *Housing study for urban Egypt.* Egypt: USAID.

Wael, A. (August 2015). Enhancing LEED as a sustainable rating system by applying its Regional priority for all environmental regions. *International conference and exhibition on Satellite at Houston.* Houston: OMICS group.

Wael, A. (November 2012). Protect the Planet through Sustainability Rating Systems with Local Environmental Criteria—LEED in the Middle East. *Procedia—Social and Behavioral Sciences,* 752–766.

youm7. (2014). *Ebny betak project*. Retrieved from low income houses: http://img.youm7.com/large/1720132718828.jpg.

*Cities' Identity Through Architecture and Arts – Catalani et al. (Eds)*
*© 2018 Taylor & Francis Group, London, ISBN 978-1-138-05409-7*

# Reclaiming the city: Guerrilla Gardening in Nairobi

Syeda R. Hussain
*United States International University in Africa, Nairobi, Kenya*

ABSTRACT: The identity of the city is constructed through its people's interaction with the physical design and aesthetic of their surroundings. A crucial part of both the aesthetic and design is the environment; where its strategic use affects the politics, culture and everyday life of the people in the city. This paper shall highlight the use of the environment as both an aesthetic and method to resist the politics of poverty in the city. It shall focus on guerrilla gardening in Nairobi, Kenya; as an environmental aesthetic and movement that shapes the identity of the city. The paper shall therefore explore the nexus between poverty and political ecology with emphasis on the garden city movement and its influence on Nairobi's 1948 Master Plan as well as Nairobi Vision 2030.

*Keywords*: environment; guerrilla gardening; environmental aesthetic; Nairobi Vision 2030

## 1 INTRODUCTION

The idea of space is not new to academic literature, as both urban space and the public space have both been widely theorized and contested (Massey, 2005; Butler, 2009; Gehl, 2011). These scholars focus on the theory of spaces. Common to these scholars is the realization that space, place and the environment are in constant transformation as a result of certain foundational processes.

Lefebvre (1991) in his book *The Production of Space* presents a triad of spatial definitions that demonstrate the complexity of the creation and maintenance of space. Central to space according to Lefebvre is the different modes of its production from natural or absolute space to complex spaces such as social space. For Lefebvre space is primarily a social product based on value systems and meanings which affect practices and perceptions of that space. Therefore, he introduces the perspective that space is produced, allowing research to analyze the conflicts, politics and character of the process of spatial production.

Harvey (2009) highlights in his book *Cosmopolitanism and The Geographies of Freedom* the idea that space is about multiplicity. He asserts that space is of particular interest because there is something ungovernable about it, in that it is constantly escaping from very tight regulation in spite of increased surveillance on the streets. In his book, he asserts that one cannot talk about space in isolation from concepts of place and environment because three concepts interrelate to each other. That is, if there is a process of place and space making it means you are shaping an environment. This shaping of the environment he further asserts has political implications for how people in the world think. He cites the example of place making in the United States after World War II where the policy moved towards the creation of suburbs. The suburb he argues perpetuates capitalist ideology and contributes to political attitudes in the United States.

Further, Harvey (2009) asserts the importance of spatial relations through his observation that political movements often use spatial strategies as part of their politics. These strategies often involve occupying spaces and places in certain ways, which act as a means to counter the dominant forces.

Since the popularity of critical studies and radical theories, there is a large amount of literature on the relationship between state and space (its creation and function in society).

These bodies of work highlight in specific the idea of public space as both a material and virtual space (Sithigh, 2012). Additionally, in the study of International Relations, space is a central concept as it is through studying different spaces and relationships within or between spaces, that we form our perceptions of world order and politics.

Lefebvre (1991) describes 'State space' (which I will argue in this paper is what we refer to as public space); as a space that is created, constrained and controlled by the State through exchange and production. What happens then in the public must be approved and regulated by the State. The effect is that the public is then free only to the extent that State institutions allow it to be free-the public hence becomes more restricted, less public and a less democratic space.

Thus, while scholars use Lefebvre as a point of departure for their research by acknowledging the existence of public/State space. They often continue to attach another description to public space and one which has become popularized through the increased use of liberal discourses-that public space is free, a place to study the citizen and a space central to any democratic State (Marcuse, 2013). Looking deeper into these assertions, one however finds the contrary.

According to Sithigh (2012), it is becoming ever more evident that there is growing tendency for the public to become more restricted, less public and hence less democratic. Susan A. Phillips (Pitzer College author of Wallbangin) in *Bomb It*; notes that through these restrictions the State in effect is excluding people. This realization prompts wider questions concerning, what kind of space we are resisting; what kind of space do we want and what kind of people do we want to be in that space?

From the above, it is clear that the production of space (in specific public space) is not divorced from politics. However there additionally exists a symbiotic relationship between environment and the politics behind the production and control of public space. The garden city movement in particular highlights this relationship and emphasizes the need to consider an environmental dimension to city planning.

This paper highlights the use of environmental movements as a method to resist public space and policies that use it to maintain poverty. It begins by providing an introduction to political ecology and an outline of the relationship between politics, the environment and space. It then briefly discusses the nexus between poverty and political ecology. Last, it shall look at the case study of guerrilla gardening as an environmental phenomenon and movement that contests a public space that is environmentally unfriendly and maintains poverty.

### 1.1 *Political ecology and space*

Today with increasing environmental degradation and issues like climate change, there is a growing need for politics that moves towards grounding in the understanding of natural resources, the environment and sustainability. According to O'Neil (2009), at the core of political science is the concept of power. Political science is therefore constantly engaged in answering the following questions: who has power? How did they get it? How do they use it and what are the ramifications of the existing and potential future distributions of power on influence? Environmental problems under the concept of power are therefore a symptom of the distribution of power—where it is always an 'other' group that has to deal with it (environmental problem) while other groups enjoy the benefits of the activity (leading to the environmental problem). In some sense, therefore, environmental problems are an exercise of power—where environmental degradation is a result of an oppressive relationship.

Zanetti (2007:3) discusses the history of political ecology as "the study of the ways nature could be both party to, and an outcome, of political process" analyzed exclusively as a "Third World phenomenon". She identifies two main reasons for this. She states that the first is due to the centrality of colonial legacy on power relations in these places. She notes that political ecology is strongly influenced by Marxist conceptions of economic relations. In this perspective, therefore political ecology on space makes reference to economic and class divisions either as the cause or result of production of spaces. Harvey (2009) makes reference to capitalism as a foundational principle that perpetuates the creation of suburbs and urban spaces which maintains a class divide.

The second reason highlighted by Zanetti (2007:3) was the "understanding of nature [...] based on a fetishisation of expansive wilderness areas apparently not tainted by human interference" which were "generally signified by economic exploitation" (Cronon, 1995 cited in Zanetti, 2007:3). Zanetti asserts that these spaces were seen as to be located in places of low economic development.

In the Third World, therefore, what one sees is the proliferation of cities with urban spaces for the middle class and suburbs for the elite and slums for the proletariat. All of these spaces are in proximity with each other and take on different environmental aesthetics. With the slum being characterized as a waste land with rampant pollution; the urban area as a crowded cement jungle and the suburbs a green utopia. To quote Fanon and to emphasis the centrality of colonial legacy on the production of spaces in cities, Fanon (2005: 3–5) in his book The Wretched of the Earth describes the colonial world as divided into two. That of the colonizer and the colonized; he describes the colonist's sectors as "a sector built to last, all stone and steel. It's a sector of lights and paved roads, [...] where the streets are clean and smooth". While the colonized sector, the shanty towns is "a world with no space, people are piled one on top of the other, the shacks squeezed tightly together."

Zanetti (2007) further notes that Political Ecology has broadened from focusing on rural areas to including urban spaces. Citing David Harvey's book Justice Nature and Ecology of Difference, Zanetti stresses the acknowledgement of Political Ecology that nothing is unnatural about cities and urban spaces. This broadening of Political Ecology she states; allows researchers to focus on seemingly "unnatural, or human-made, ecosystems like parks or gardens though the same epistemological lenses we might use for natural areas such as forests, prairies and so on" (Zanetti, 2007:4).

Reverting back to a Marxist perspective of ecology, the relationship between space, the environment and politics means that; the production of spaces either as public spaces or green spaces is a reflection of the system (for example a capitalist system). This relationship gives rise to specific spaces with specific characteristics and specific meanings. Zanetti (2007) explains that in capitalist societies nature is used to maintain social and economic class distinctions. In that, marginalized urban areas tend to be devoid of natural landscapes, and those which do exist are often poorly maintained and perceived as unsafe (Hymen, 2003 and Brownlow, 2006 cited in Zanetti, 2007:4).

### 1.2 *Poverty as 'planned misery'*

The United Nations (1995) defines poverty as; "lack of income and productive resources to ensure sustainable livelihoods" including "hunger [...] unsafe environments and social discrimination and exclusion." The United Nations further states that "it occurs in all countries [...] as mass poverty in many developing countries [and] pockets of poverty amid wealth in developed countries" (United Nations, 1995). Therefore, at the United Nations Millennium Summit (New York, Sept. 2000) the world adopted a global initiative to end poverty by the year 2015-the Millennium Development Goals.

The global fight against poverty is an initiative based on the international community's commitment to fight inequality and promote human rights everywhere. Since the adoption of the Millennium Development Goals action plan, United Nations reports' have emphasized its achievements. Yet, severe poverty still exists alongside rising affluence.

The 'development world' comprised of Non-Governmental Organizations, Civil Society and Governmental Organizations is plagued with the practice of determining 'root causes'. Using neo-liberal discourses, it sets out to stamp out hunger, eradicate poverty and bring democracy to all corners of the world. Marks (2011:5) notes that root causes are currently a prominent feature of the discourse of international human rights, where the rationale behind 'unearthing' them is that; by identifying them you can "mark the level at which an intervention would be effective" and hence "bring about significant and lasting change". From a legal perspective, she asserts that the practice of determining root causes is key in establishing "state responsibility-for failure to comply with obligations" (Marks, 2011:62). Marks however argues that; while the discourse of 'root causes' is valuable by attempting to answer why violations occur and what it will take to bring about change. It stops its analysis too soon by treating effects of violations of issues as if they were causes. The result is, rhetoric that concentrates on 'bad' procedures or rules therefore suggesting that, by replacing them with 'good' ones, the problem will be solved.

Further, according to Marks (2011:3), the over-reliance on institutions and 'root causes' by the 'development world' does away with the role of 'grassroots movements' in the discourse. It does this by maintaining a system where change must be "channeled through the state". The minimized role of grassroots movements has a significant effect on the way human rights functions and the way they are understood in international law. Throughout the history of human rights, it has been the ordinary person that defies all odd to assert his rights. It is the ordinary person that is affected by human rights violations and therefore it is the ordinary person that human rights apply to. Therefore, by downplaying the role of these movements we create a gap in our overall understanding and hence our solutions end up being unsustainable and impractical.

Marks (2011) describes as alternative approach- 'planned misery'. The premise behind this approach is that for every person that suffers from a violation, there is another that benefit from it. This approach is valuable for understanding the nexus between poverty and the environment as it builds upon Heynen (2003) and Brownlow (2006) (cited in Zanetti, 2007:4) observation that marginalized urban areas are void of natural landscapes. As well as Marxist understanding of Political Ecology that asserts 'class divide' is maintained and perpetuated by the production of spaces that use the environment to create and maintain poverty.

Planned misery as an approach proposes several advances to 'root causes'. Firstly, it alters questions from "what are governments not doing" to "why governments and others are doing what they are doing" (Marks, 2011:29). Secondly, it draws attention to the link between deprivation and privilege by associating violations to direct actions. It further recognizes the importance of critically analyzing violations as ideas; hence scrutinizing the conditions as stated by Marks (2011:39) "within which those ideas were able to develop and gain influence". Finally, it encourages a deeper analysis of the conditions under which 'remedies' are proposed; therefore, allowing us to formulate realistic solutions.

What needs to happen is; awareness of the fact that as noted by Pogge (citied in Marks, 2009:12) "the poor are systematically impoverished by the present institutional arrangements and have been so impoverished for a long time during which our advantage and their disadvantage have been compounded". Therefore, poverty is not simply an occurrence with root cause but instead it is a construction of conscious policy options and projects by certain groups of people to maintain their advantage. In terms of this paper, policy options and projects are directed towards restricting and controlling environmental spaces such as parks, fields, verges and roundabouts.

## 2 GUERRILLA GARDENING AS RESISTANCE

This section presents Guerrilla Gardening as a movement that adopts an environmental strategy to counter the polluting influence of urbanity and class divide. Zanetti (2007:2) defines Guerrilla Gardening "as any voluntary, and potentially illicit, gardening in space which can in some way be deemed public, over which the gardeners hold no direct or explicit ownership". Adams, Hardman and Larkham (2014:2) state that the prefix "guerrilla" has a military connotation in which rebels are in conflict with an oppressive dominant power. This point raised by Adams, Hardman and Larkham in the beginning of their discussion about Guerrilla Gardening, is significant as it positions the movement as resistance rather than contestation. This positions the movement as a force greater than reformist movements. Guerrilla Gardening therefore through its tactics demand a re-imagination of spatiality in urban spaces and the right to cultivate land. Guerrilla Gardening as noted by Mikadze (2014) represents a tactical spatial practice that contributes to 'in-between' spaces. In-between spaces are spaces that lie between the public and private dimensions. They are spaces of uncertainty such as verges, roundabouts abandoned plots and slums.

The majority of literature on Guerrilla Gardening and environmental tactics to resist poverty and global inequality relies heavily on a historical narrative that begins in the West. Guerrilla Gardening as an illicit act of cultivating someone else's land often makes reference to acts started in New York (Zanetti, 2007 and Mikadze, 2014). However, there is a long history of what is today termed Guerrilla Gardening in the global south. The specific term 'Guerrilla Gardening' was however first used by Liz Cristy in 1973 (Zanetti, 2007). Yet for century's people in the global south cultivated land for the good of the community before the imposition of contemporary property rights and divisions of public and private by colonial regimes. There is therefore a direct link between Guerrilla Gardening and groups in the global South that fight for the right to reclaim their land.

Guerrilla Gardening is a response to the unequal distribution of land, and capitalist agricultural practices (such as mono-culture) aimed at profiting the rich which have led to rampant food insecurity forcing people into poverty. As agricultural land and space become increasingly limited peoples basic right to grow their own food is being taken away through the justification of development and privatization. Mono-culture in particular has led to soil exhaustion and is a key player in climate change. Additionally, the transformation of the farm into large scale private factories has made agriculture a major polluter (Allen, 2009). Richard Reynolds (2008) in his book On Guerrilla Gardening asserts that the unequal distribution of land and resources is at the root of poverty, hunger and war. An assertion that is common in political ecological perspectives.

Increasingly people are reclaiming private and public spaces for the greater good. Author and self-styled de-growth activist Charles Eisenstein (2011) in his book *Sacred Economics*; argues that private property, when viewed in its historical context, largely arises from the theft of the commons, which is land or resources that once belonged to a whole community. He calls for land to be utilized in ways that work best for people and the environment. It is not hard to see the logic of this argument—while fertile land is left barren, chronic food poverty is a serious issue around the globe. So-called food deserts, where fresh and healthy foods are hard to access, exist in many poor urban neighborhoods, with the associated health costs borne by local residents. In Nairobi, Kenya the term 'Guerrilla Gardening' is spreading, but it has been a common practice in Nairobi's urban centers and slums since British colonial rule.

The city of Nairobi started with the construction of the Mombasa-Uganda railway line in 1896. It was intended to be a railway town housing the railway headquarters and eventually the colonial administrative center. As the population of the town grew, the function of Nairobi developed to being the center of corporate power.

Underlying the development of Nairobi as an industrial and colonial administrative hub was the growing sentiments of segregating the city either in terms of race or class. Torres (2010) notes that the 1926 Plan for a Settler Capital was clear about its intention to create and maintain racial segregation. Nairobi was a city only for Whites. It was a White city, geographically designed to create and maintain racial segregation. There was no assigned residential area for Africans as they were treated as an inconvenient necessity to 'run' the city. Africans therefore settled in informal settlements with perpetual threat of being forced out.

The 1948 Master Plan for a Colonial Capital on the other hand was presented as a scientifically and technically neutral design. However, it contributed to racial segregation through its ideology of separated development felt throughout the design (Medard, 2010: 27). That is, while the 1948 Plan did not overtly advocate for racial segregation it replaced it with social segregation in that, residential areas were designed and assigned according to class. The 1948 Master Plan was used as a guide to the development of the 1973 Nairobi Metropolitan Growth Strategy and later Nairobi Vision 2030. Both plans maintain and perpetuate class segregation. That is, after independence in 1963 the colonial design of the city was maintained but the new African elite replaced the European colonial suburbians while poor continue to occupy the informal settlements (Torres, 1998). Bigot (2010: x) best describes Nairobi today as a "fragmented and highly paradoxical city since it is upon the city's empty space and peripheries that urban dynamics are structured".

Central to the aesthetics of these plans is the aesthetic principles of the garden city movement based on the idea of functionalism. Thorton (1948: 45) acknowledge that Nairobi's

1948 design was influenced by the garden city movement described as "town planning as landscape gardening". As such, the garden city movement has become used as an aesthetic method to design the urban environment. Emphasis in this regard is on the functionality of the city and the abundance of 'open green spaces'. The idea of an urban green environment presented in the background of a society that is facing threats from land grabbing, deforestation and climate change is a welcome initiative. However, who is designing the city and for whom remains a central concern.

The 1926, 1948 and 1973 plans for Nairobi have consistently marginalized the poor. Kibera, one of Kenya's largest slums and the largest slum in urban Africa is emblematic of the influence of design on poverty. Kibera's first population were Sudanese soldiers of the British East African army formally known as the King's African Riffles. In 1904 the British colonial government allocated a forest area in Nairobi as a barracks to house the Sudanese soldiers. They called it 'Kibra' meaning forest. The location of the barracks was central to the design of British colonial rule as it was located close to the city center for ease of deployment in case of unrest or resistance. However, as the soldiers began to get old and retire and settle in Kibera, the government began to view this unplanned residential area as a hindrance because the land was too valuable and too near to European settlements (Parson, 1997). Since colonial rule, successive governments have continued to attempt to relocate, reclaim and demolish Kibera with the descendants of the Sudanese soldier's being excluded from citizenship. Kibera is therefore made up of a large population on stateless people who have no access to citizenship rights and therefore amenities such as education and healthcare.

As a response to a history of oppression, neglect and forced poverty the residents of Kibera have responded with several methods of resistance from art to civil disobedience. Guerrilla gardening therefore is an additional tactic and expression of resistance. With little open space to farm on and no legal right to residence; the people of Kibera are growing vegetables in soil filled sacks. These gardens are a response to increased food insecurity, growing encroachment on open spaces and pollution in the city. The gardens are situated on former dumpsites and counter hunger and poverty as well as sustainably addressing environmental problems (Zuckerman, 2011).

It then follows that; community gardens cultivated illicitly, acts as an avenue for survival and resistance against controlled land use and food production. However, Guerrilla Gardens as a means to resist are not restricted to the slums. Throughout Nairobi one is able to see illegal roadside nurseries and gardens that challenge constructed norms of boundaries (between public and private), legality and the environment. All of which not only reclaim the city from those that design it; but additionally, redefine what should be meant as a 'garden city'.

## 3 CONCLUSION

The garden city movement is emblematic of a movement in urban planning to aesthetically 'green' the city and provide open spaces for the public. However, it additionally demonstrates the extent to which class and politics has the capacity to infiltrate the right to the city and in particular the right to open green spaces. To the extent that the idea of public becomes more restricted in meaning and use.

Nairobi's history of urban design demonstrates that urban aesthetics is designed by a few for a few. This in turn leads to the creation of peripheries and peripheralization of people. Therefore, perpetuating and maintaining poverty in the city. Guerrilla Gardening thus proves to be a form of resistance that highlights and addresses the historic links between city planning, environmental degradation and poverty as 'planned misery'. Hence reclaiming the right to the city and the right to the aesthetic of green spaces.

## REFERENCES


Adams, D., Hardman, M. & Larkham, P. 2015. Exploring Guerrilla Gardening: Gauging Public Views on The Grassroots Activity, Local Environment. *The International Journal of Justice and Sustainability* 20(10): 1231–1246.

Allen, R. 2009. *The British Industrial Revolution in Global Perspective*. New York: Cambridge University Press.

Bigot, C.H. 2010. Preface. In Torres, D.R. (ed.) *Nairobi Today: The Paradoxes of a Fragmented City*. Dar es Salam: Mkuki na Nyota Publishers.

Eisenstein, C. 2011. *Sacred Economics: Money, Gift, and Society in the Age of Transition*. California: Evolver Editions.

Fanon, F. 2005. *The Wretched of the Earth*. New York: Grove Press.

Gehl, J. 2011. *Life Between Buildings: Using Public Space*. Washington: Island Press.

Harvey, D. 2012. *Rebel Cities. From the Right to the City to the Urban Revolution*. Verso.

Lefebvre, H. 1991. *The Production of Space*. Donald Nicholson-Smith trans., Oxford: Basil Blackwell.

Marcuse, P. 2013. The paradoxes of public space. *Journal of Architecture and Urbanism* 38(1): 102–106.

Marks, S. 2009. Human Rights and The Bottom Billion. *European Human Rights Law Review* 1:1–18.

Marks, S. 2011. Human Rights and Root Causes. *The Modern Law Review* 74(1): 1–33.

Medard, C. 2010. City planning in Nairobi: the stakes, the people, the sidetracking. In Torres, D.R. (ed.) *Nairobi Today: The Paradoxes of a Fragmented City*. Dar es Salam: Mkuki na Nyota Publishers.

Mikadze, V. 2015. Ephemeral Urban Landscapes of Guerrilla Gardeners: A phenomenological Approach. *Landscape Research* 40(5): 519–529.

O'neil, K. 2009. *The Environment and International Relations*. Usa, Cambridge University Press.

Parsons, T. 1997. "Kibra is our blood": The Sudanese military legacy in Nairobi's Kibera location, 1902–1968. *The International Journal of African Historical Studies* 30(1): 87–122.

Reynolds, R. 2008. *On Guerrilla Gardening: A Handbook for Gardening Without Permission*. London: Bloomsbury.

Síthigh, D.M. 2012. Virtual walls? The law of pseudo-public spaces. *International Journal of Law in Context* 8(3).

Torres, D.R. (ed.) 2010. *Nairobi Today: The Paradoxes of a Fragmented City*. Dar es Salam: Mkuki na Nyota Publishers.

Torres, D.R. 1998. Nairobi, between Muthaiga and Mathare Valley. In Grignon, F. & Prunier, G. (ed.) *Le Kenya contemporain*. Paris: Karthala-IFRA.

United Nations. 1995. The Copenhagen Declaration. *World Summit for Social Development Copenhagen*.

Zanetti, O. 2007. Guerrilla Gardening: Geographers and Gardeners, Actors and Networks: Reconsidering Urban Public Space (Master's Thesis). *Queen Mary University of London*. London, United Kingdom. http://www.guerrillagardening.org/books/ZanettiGG.pdf.

Zuckerman, J.C. Nov, 29 2011. The Constant Gardener's. *Mother Nature Network*. http://www.mnn.com/your-home/organic-farming-gardening/stories/the-constant-gardeners.

*Cities' Identity Through Architecture and Arts – Catalani et al. (Eds)*


# Recording the intangible heritage of the city in the Metropolitan Area of Porto

Renata Barbosa & Fernando Paulino
*ISMAI—University Institute of Maia, Portugal*

ABSTRACT: How collaborative platforms can contribute simultaneously to the recording of the intangible heritage of the city and the production of an intangible heritage that collaboratively is being built and transformed. To research how the various agents of recording/production influence and shape the way we produce and record the intangible heritage, either as external or as participating agents, in that the image of this heritage as reflected by others and reflected by the self, not being necessarily one of opposition, is non coincidental. The research of this issue brings about, in what strategies are concerned, some contributions on the inclusion of Intangible Cultural Heritage in e-museums and the use of collaborative platforms as a means of production and registration in this context.

*Keywords*: intangible heritage; cultural heritage; e-museums

## 1 INTRODUCTION

In Portugal there are two Metropolitan Areas, one around the city of Lisboa, another around the city of Porto. They group the municipalities around these major cities. Metropolitan Areas embody cooperation between municipalities and seek to catalyze broader institutional networks with strategic importance for the promotion of economic, cultural, social and environmental development and for regional affirmation.

The Metropolitan Area of Porto has been developing a strategic planning process of an operational nature, with a view to preparing the participation of this sub-regional space in the different sectors of society. This metropolitan area is located in the North Coast of Portugal, and embraces a geographical area nowadays composed of 17 contiguous municipalities, with an area of approximately 1,900 $Km^2$ with a resident population of around 1,800,000 inhabitants. All these councils assume their particularities but converge to a complementarity for diversity, in which the metropolitan area is undoubtedly a bearer and promoter of this cohesion.

The municipalities of the Metropolitan Area of Porto did not have a tradition of working together in projects of common interest. This situation started to change in 2007 with the launch of the iPorto project (Henriques et al. 2008) that for the first time provided a common online platform that all of these municipalities could use to provide information about cultural events occurring in their territory.

Six projects were worked upon that covered different areas, namely:

- Cultural events: iPorto, dissemination of information about cultural events (10 themes) in the Metropolitan Area of Porto. A common online platform that supports the feed of the iPorto web site, the iPorto newsletter, and the nEventos database. Its nEventos database supports other projects, for instance it is used by some components of the Official Tourism Portal of the city of Porto, helping promoters of cultural and/or tourist events. It can be used to build derived contents, such us newsletters or custom magazines and exported to other information systems.

- Heritage points of interest: survey of the museums and architectural and archaeological heritage of the Metropolitan Area of Porto. The information collected covers 79 Museums and 95 points of interest of the architectural and archaeological heritage. The site is dynamic, new items can be easily added through the backoffice. It allows cross referencing the information in this web site with that in the iPorto web site (many of the cultural events take place in the museums or heritage points of interest collected here). Besides presenting the information from each point of interest this web site allows the items to be added to a personalized itinerary made with the help of the Google Maps API. The itinerary can then be printed or sent to a given email.
- Tourism: dynamics of tourism supply and demand in the Metropolitan Area of Porto. This metropolitan project for the time being is only targeting the technical staff of the different municipalities, that is all the data is accessed through the backoffice and no data is directly made available to the general public. Still there is an API that allows the different municipalities to extract the relevant information and use it to power tourism web sites, or the tourism section of their institutional web site, or to create mobile Apps. This is the most strategic project being developed in the context of the Metropolitan Area of Porto due to its possible impact on the tourism industry and its potential to help develop an ecosystem of cooperation between the different stakeholders, as already happened in the case of the municipality of Porto.
- Nature and environment: information about the network of national parks and nature sites of the Metropolitan Area of Porto. It also included information about a selected set of natural sites and studied and described the 14 rivers that constitute the hydrographic network that crosses the area. Another output of this project was a report with a set of good practices for designing, building and maintaining green spaces targeted at the technical staff of the municipalities.
- Education: information about the educational spaces of the Metropolitan Area of Porto.
- Intangible heritage: information about a selected set of themes from the Intangible Heritage of the Metropolitan Area of Porto.

The goal of all of these projects was to achieve the following:

- That whenever possible the information be directly filled-in in the online platforms by the users from the different municipalities. If that was not possible or if a common set of editorial guidelines was wanted, then the project team took care of collecting the initial set of information that was then made available to the users of the different municipalities for maintenance and updating. It was considered critical that the responsibility for the maintenance of the information rested in the municipalities.
- That the information provided by a given municipality should be useful for that municipality but at the same time that the aggregated information from the 17 municipalities could be useful at a metropolitan scale.
- That the information could be reused in other contexts. For that purpose a set of Application Programming Interfaces (API's) were developed to allow programmatic access to the databases, and the platforms were also configured so that the information could be exported in structured formats, such as JSON (JavaScript Object Notation), CSV (Comma Separated Values), or spreadsheet format (Excel).

With the exception of the intangible heritage project that was developed in the context of—University Institute of Maia—ISMAI the remaining five projects were developed in the context of the Porto Digital Association with support from staff and students of ISMAI.

The evaluation of the degree to which it is possible for municipalities to produce cooperative work at a metropolitan scale supported by online platforms is of major importance to determine how subsequent collaborative projects should be adapted to the reality of the collaborative working environment. In the context of the, at the time, sixteen municipalities of the Metropolitan Area of Porto, between September 2010 and December 2013, six projects were developed, covering the areas of cultural events dissemination (iPorto), information to support school visits to educational spaces (EDU), information about national and natural parks (SITIOS), tourism dynamics of supply and demand (DINTUR), information about museums and heritage sites (PIN), and documenting the intangible heritage (PIAMP). The common feature of all of these projects was that they were launched by a central authority (Junta Metropolitana do Porto) but then were executed, supplied with information and maintained at a local (municipality) level, therefore their success or failure depended on the interest that they could originate at the municipality level. One of us participated in the development of five of those projects (providing support for the users of the different municipalities, and for those projects for which there is already enough data, assessing the degree of success of each initiative), and both of us participated on the remaining one, PIAMP.

## 2 COLLECTING THE INTANGIBLE HERITAGE OF THE METROPOLITAN AREA OF PORTO

The PIAMP project consisted in each municipality identifying a topic of their intangible heritage that they would like to see recorded, and then to address that topic by creating a 15-minute video that, as best as possible, represented that theme, placing more emphasis on the anthropologic point of view in detriment of the component of tourism marketing.

The project of *Collecting the Intangible Heritage of the Metropolitan Area of Porto* had as its main goal collecting information and registering audiovisual information about initially sixteen (16), now seventeen (17) intangible heritage themes. Besides this general goal, the objective was to choose a theme that represented a practice or collection of practices, of cultural, religious/pagan, human-activity or of a community identity nature, that constituted a certain tradition, paying special attention not so much to its birth but to its development, current practices, and means of transmission to future generations. In January 2013 each of the sixteen municipalities was invited to choose one theme, the project started being developed in March 2013 with a deadline set for December 2013. Because the municipality of Paredes joined the Metropolitan Area of Porto in September 2013, in 2015 the project PIAMP was extended to also cover the municipality of Paredes.

### 2.1 *Cultural Diversity of the Metropolitan Area of Porto*

Due to the vast and diverse contexts particular to each municipality, their links to the local community, operating autonomy and regionally-based production systems, the Metropolitan Area of Porto although it pretends to be inclusive and cohesive, still is constituted by a multitude of regionally distinct territorial units grouped around three major clusters: coastal/urban, hinterland/agrarian, and hinterland/industrial; with their own set of priority interests and rules of action to achieve them. These diverse contexts are reflected upon the theme that was chosen by each municipality.

Additionally, the fieldwork, led us to conclude on some important characteristics of these types of projects:

- Some themes are recurrently explored by the municipalities, in some cases involving the same set of key informants (that are suggested again and again);
- Some themes are chosen without prior field investigation and do not represent the current state of affairs, but that of fifteen or more years ago;
- Some themes, of second level of relevance, are an effort to deviate from themes of first order of relevance already been explored by other projects, but, in some cases, fail to be representative of the community as a whole (the municipality), representing instead a very small group of people;
- Some themes reduce the possible sources of information to one, making it difficult to undertake the operational development of the project since it is based on one main informant that may have absolute control over the context and over the access to subsidiary

informants, therefore the project is too dependent on the degree of availability and cooperation provided by that informant;

- Although it was agreed from the start, that the municipalities would contribute resources to assist the development team with the fieldwork, notably, making the initial contacts with the informants and subsequent explanation of the project objectives and what was required from them, in the end that rarely happened;
- The municipal (council) elections that occurred at the end of September 2013, were a stressful event for the project operation, due to the unavailability of key personnel before—because they were preparing for the elections or trying to finish ongoing projects –, during—in some cases complete unavailability –, and after—due to changes in leadership, influencing the interaction or lack of it in a non positive way. This lasted from mid July to the end of October 2013.

The themes initially chosen were developed, transformed, adapted, or eventually discarded and replaced by other themes in some way connected with the initial themes.

As referred above, the project had as its main goal collecting information and registering audiovisual information about the intangible heritage themes that were chosen by the different municipalities. This information is the main source of construction of narratives in the format of documentaries, with the approximate duration of fifteen (15) minutes that in turn, are the main source of the teasers (short version of the documentaries) with the approximate duration of five (5) minutes, of the seventeen themes chosen. An eighteenth documentary was also constructed, from the aforementioned seventeen, by selecting one (1) minute of footage from each one, and thus creating a video that summarized the richness of the intangible heritage of all of the Metropolitan Area of Porto.

All information collected, namely the audiovisual information in raw format, was delivered to the Metropolitan Area of Porto, that in turn, delivered it to its municipalities. The objective here is to grant access to source material so that, an audiovisual database can be constructed, and augmented in the future, allowing the creation of new narratives, based on unused images/video images or the creation of derivative products based on the ones constructed in the context of this project. For this reason all the material was collected and is delivered according to the Creative Commons License that allows the creation of derivative products including for commercial use, that is, the license "*Atribuição-CompartilhaIgual 3.0 Portugal (CC BY-SA 3.0 PT)*". This facilitates access to audiovisual information that can be used on other contexts by the municipalities, mainly since, with every passing day, it is more difficult due to budget restrictions caused by the current economical situation of Portugal, to accommodate human resources for this kind of work.

The relative lack of homogeneity on the information collected was also strongly influenced by the aforementioned six reasons. Although we tentatively tried to persuade some of the informants to fully contribute, it was not always possible and, in some cases, the contribution was restricted to the participation in the documentary, not to the full usage of raw material. In the following sections is a description of each of the seventeen subprojects and respective themes.

### 2.1.1 *Arouca*

The theme chosen to be registered by the municipality of Arouca was the communal goat herd of Regoufe. Regoufe is a hamlet located at the hillsides of Serra da Arada at approximately 615 m of altitude. Serra da Arada is part of the Gralheira Massif, along with Serra da Freita and Serra do Arestal. It lies in the transition from Beira Litoral to Beira Alta with its higher elevation at 1071 m of altitude (Alto da Cabria). It serves, in part, to divide the basins of the rivers Paiva and Vouga. It is approximately 20 km long and 15 km wide.

Thirty families inhabit Regoufe, although the hamlet has more housing units, a consequence of the population aging due to the migration of the younger people to the coastal cities to study and work, although occasionally, we can see the inverse movement due to the current economical situation but, those young people that come back, come to reintegrate their respective nuclear families, and not to create new ones. The communal goat herd, nowadays, is owned by five families, where the animal grazing is ensured by the women (five woman shepherds). It has been like this since the existence of the communal goat herd more than a hundred years ago, according to Regoufe's inhabitants.

### 2.1.2 *Espinho*

Espinho is a small city located on the north coast, which is limited to the west by the sea. The theme that the Espinho municipality chose to be registered was the art of trawl fishing (Xávega) practiced in Espinho. In fact, the area were Xávega fishing is done stretches from Espinho to Vieira de Leiria. The documentary Arte da Xávega em Espinho, explores and registers this reality as it is practiced nowadays, using as its main informants the crews of the three last '*companhas*' (fish societies) in activity at Praia de Espinho, namely the Companha '*Nelson e Sérgio*', the Companha '*Vamos Andando*', and the Companha '*Vicking*'. Since this activity, trawl fishing (Art of Xávega) is still practiced in a traditional way, we found only a few major differences, to know, the use of boat motors for propelling the boat forward, the use of tractors instead of ox teams to push the boats into the sea, to pull the boats from the sea, and to pull the nets. Some complementary activities were also modernized.

### 2.1.3 *Gondomar*

The theme chosen to be registered by the municipality of Gondomar was the filigree craft industry present in the Gondomar region, and practiced in the small-scale workshops, where filigree-crafters, produce articles of goldsmiths' or silversmiths' wares using the filigree techniques, passed on from each generation to the next. These small-scale workshops are mostly family based. This traditional industry is still present in this area, but the continuous need for modernization has been transforming the work processes and what is achieved by it.

In order to preserve the knowledge and expertise accumulated by humankind over the centuries, nowadays, it is of uttermost importance that the procedures for the transmission of this knowledge should be remembered and registered. That was one of the main objectives of this project.

### 2.1.4 *Maia*

The theme chosen to be registered by the Maia municipality was the traditional flower baskets (flower hampers) street parade of the Nogueira da Maia village. Nogueira da Maia is a village east of the Maia city where it is still noticeable its agricultural background.

In the early days of these festivities, it was tradition for the population to offer flowers to the Virgin Mary, which were carried in flowering baskets denominated '*canastras*'. These baskets were about a meter tall, were made in *Cana-da-India* (indian cane) and were covered with myrtle and fresh flowers picked in the gardens. This was a tradition that was lost over time, and that was recovered in the nineties, and is currently one of the high points of these festivities, lending to the pilgrimage a different colored and showy ambiance (Maia 2011). Currently, the street parade of the flower baskets takes place between the main church of Nogueira and the Chapel of Our Lady of the Hour on Mount Calvary, where they are exposed for a week, then been integrated in the procession in honor of Our Lady of the Hour. The documentary *Maia em Festa, Canastras Florais*, explores and registers this reality, using as its main informants the inhabitants, mostly from Nogueira da Maia, belonging to private or institutional groups that directly participated in this activity. These groups, nine, correspond to the nine parishes represented in the 2013 festivities.

### 2.1.5 *Matosinhos*

The Matosinhos municipality chose the theme of its traditional canning industry, in this case exemplified by the factory *Pinhais*. Matosinhos is located on the northern coast of Portugal and the county is organized spatially along the Atlantic coast, never very far from the sea. Since always the fishing activity and all other derived were crucial from the county's economic point of view. The sea and its resources, with special emphasis on the fishing of sardine (Sardina pilchardus), the most important in Matosinhos, are uncertain. This relationship between accessibility to the sea and its products and canning existed at least since Roman times, who built the tanks, chiselled into the rock outcrops of Praia de Angeiras, commonly called, salting tanks or cetarias. "*We have travelled, from fishing to the canning industry in Matosinhos, retaining a wealth of knowledge that changed slightly over time. Lets us now take over the past, reconstructing, step by step, the major phases of the development of the canning industry*" (Nunes 2003). In the documentary, the manufacture processes of traditional canning industry were recorded from the moment the fish baskets enter the factory. The women fish handlers formerly named "*empreiteiras*" begin by pouring the baskets of sardines in the marble tables, spreading it. These fish handlers are joined by others in this *escorcha* section in which in one stroke, fish head and guts are plucked out. The byproducts (*escasso*) of this phase are used in the production of fishmeal at their own factories and when they were abundant, gave a characteristic smell to Matosinhos. The inhabitants of Matosinhos, depending on the proximity to the factories and the relative location of their residence, identified the wind direction by the presence or absence of that smell.

#### 2.1.6 *Oliveira de Azeméis*

The theme chosen to be registered by the municipality of Oliveira de Azeméis was the Park of La Salette, the origins of its religious context and its present organization and uses. The documentary *Parque de La Salette*, explores and registers what is the Park of La Salette, how it was born and its association to an alleged religious miracle, how it is used nowadays and what kind of community equipment it contains.

#### 2.1.7 *Paredes*

The municipality of Paredes chose the manual art of working the wood by the master carpenters and carvers of ancestral tradition in this county that has still a few remnants in some family owned carpentries.

#### 2.1.8 *Porto*

The theme chosen to be registered by the Porto municipality was the poet Eugénio de Andrade remembered by his friends. Those friends were initially chosen due to the close relationships developed amongst them, and that came into existence from differentiated contexts. It was not always possible to keep the initial group together due to the non-availability of their friend. For this reason the initial objective of trying to represent a different context with each friend was impossible to obtain.

#### 2.1.9 *Póvoa de Varzim*

The theme chosen to be registered by the municipality of Póvoa de Varzim was the traditional fishing boat denominated *Lancha Poveira*, namely the present usage of an exact replica for learning purposes, tourism usage and community identity motives.

The wealth of a land of tourist attractiveness is based not on the repetition of the same global offer, but on the distinctive mark, as a carrier of a cultural identity, immersed in the deepest traditions of their community.

The context of Póvoa de Varzim as an essentially fishing community has been transformed, particularly in the last decades, and the one that was the most important economic activity ceased to be, giving way to others, in particular those linked to the tourism sector integrating Póvoa as an important seaside resort. And yet, according to (Saldanha 2008) a "*[...] social cohesion element inseparable of the change process, the maritime heritage is revealed in the unique cultural diversity of our fishing (and riverside) communities, that consolidate autonomous projects which embody the uniqueness and plurality of their identities [...]*". One such project is the Lancha Poveira do Alto 'Faith in God', initiated by Manuel Lopes, that claims on several occasions the importance of seeing and studying the sea and the activities related to it, not from the land but from the sea itself, as participants or observers. The vessel lancha poveira do alto was chosen for the fact that the identity of the maritime community has in this "*[...] one of its most significant and bright symbols [...] embodying and signalling the memory of the journey and of the naval construction techniques diaspora, of the fishing activities and of the arts of sailing of the poveiro's fishermen*" op. cit. in (Costa 2011). The *lancha poveira* was built in about seven months and was launched on September 15, 1991.

### 2.1.10 *Santa Maria da Feira*

The theme chosen to be registered by the municipality of Santa Maria da Feira was the tradition of the '*fogaceiras*', a traditional activity, that has its origins in a legend, that at a subsequent time developed into a more complex format, always with a religious motivation.

It can be attested that one very relevant aspect of this project is how the municipality administration took into their hands the task of guaranteeing the continuity of this tradition, promoting activities that help spread the tradition, as much as guaranteeing that the children get to know its origins and present actions. The documentary *As Fogaceiras*, explores and registers the activities developed by the community, in order to help spread and transmit to the future generations how the tradition was born, and how the municipality administration takes into their care these tasks, using as its main informant an old '*fogaceira*', the administrative representatives, and the students and teacher of a group being taught about this relevant tradition.

### 2.1.11 *Santo Tirso*

The theme chosen to be registered by the municipality of Santo Tirso was the textile industry, its past, present and future. Motivated by several vicissitudes it was not possible to treat all facets of this theme, therefore the work was mostly concentrated on the past, where the factory model associated with cotton ruled, together with its decadence and the exploration of new alternatives for its present and future where a new role is being created and developed within the textile context. The former Spinning and Fabrics Factory of Santo Thyrso, since it was deactivated in 1990, was acquired by the Municipality and rehabilitated, integrated into a revitalization and urban development strategy. Rehabilitation maintained the architectural traces of the original factory typical of the industrial architecture of the late nineteenth century. It was then spatially reorganized to function as a "*workspace, business, experimentation and innovation space for culture, enjoyment and pleasure*", and prepared to receive incubating companies and to create a modern unit for the promotion of textiles, this multi-functional space is organized into sub-areas with different features designed for each of them.

### 2.1.12 *São João da Madeira*

The theme chosen to be registered by the municipality of São João da Madeira was the history of the Hat Industry. São João da Madeira is a city located on the extreme north of Aveiro's district, belonging to the Beira Litoral region, centrally positioned on the sub-region of Entre Douro e Vouga (between the rivers). The city itself is crossed by the Ul river. The city of São João da Madeira is known in Portugal for its tradition in the industrial area, particularly in relation to the manufacture of hats and footwear, and it is recognized in the country as the Shoe Capital. Following this line, São João da Madeira developed new strategies for the tourism industry, namely those concerning the industrial tourism.

> *"Inside these walls we keep machinery, tools, raw material and hats. We also keep stories that were preserved by common memories. [...] Inside this building, that was once the Empresa Industrial de Chapelaria, one of the most important enterprises in town, the Hat Industry Museum was born, in this building where heavy machinery was first introduced, in this city that once was one of the main and more important centers of hat production in the country" (Câmara Municipal de S. João da Madeira. 2013).*

The stories of the hat industry are preserved in a set of common memories. The documentary *Indústria da Chapelaria*, explores and registers these memories, using as its main informants the former workers of Empresa Industrial de Chapelaria that nowadays houses the Hat Industry Museum.

### 2.1.13 *Trofa*

The theme chosen to be registered by the municipality of Trofa was the production of sacred art on the Vale do Coronado area. Vale do Coronado is an area along a valley that includes the councils of São Mamede do Coronado and São Romão do Coronado.

"The workshops of the saint-makers [...] have become an important source of work for the region (mainly agricultural), since the people there had no industries capable of giving them a dignified and remunerative work. The poor abounded everywhere; [...] Around this tough situation, someone stood out and triumphed in life: they were the saint-makers masters (*mestres santeiros*). They were called the 'landlords'. The profession of saint-maker offered guarantees. Furthermore, it was a less harsh profession because their workers were not exposed to the weather conditions and it was not required of them an effort as great as it was required of the farmers and other crafts" (Tedim 1978). As the model is to be reproduced countless times, it is critical that it satisfies conditions of strength and durability and hence the preponderance of wood for their production, although clay or plaster models are common in the workshops. Sometimes, the models were finished and sold as a final piece, and a copy in plaster or clay was made thus becoming the new model. This was also the technique used to copy models of competing workshops. The documentary *"A produção de arte sacra no Vale do Coronado"*, explores and registers the activity of the wood sculptors of sacred art (saint-makers), using as its main informants those saint-makers masters (*mestres santeiros*) in activity in the area of the Vale de Coronado.

2.1.14 *Vale de Cambra*

The theme chosen to be registered by the municipality of Vale de Cambra was the production of corn bread and wine and the traditional agricultural practices associated to attaining these end products. The cultural knowledge and practices to traditionally obtain them were thus to be registered.

> *"In the corn plant everything is useful: breaks up the flag and plucks up the leaves for livestock, before catching the ear; [...] the ears are harvested in late summer, they dry in the floor and are stored in granaries, the husking of the corn is a festive occasion among neighbors; grain, reduced to flour, makes the corn bread [...]" (Ribeiro 1991).*

Similarly, the annual cycle of the vine, follows the rhythm of the seasons. The regions richness in water allows agriculture species requiring a lot of water, often supplemented with irrigation, such as corn. In the village of Paraduça belonging to the parish of Arıes, between the Sierras Arestal and Arada in full highland, where every piece of land is worked by man and used for agriculture, it is cultivated, beyond what is necessary for the daily feeding of the community that dwells there-in, corn and vines. With minor changes the agricultural rituals are the same as in the past, but all fear that by being increasingly abandoned, the land and its cultivation, will disappear at the same time as the generation that today still works the land. In an effort to preserve and transmit these practices and traditional knowledge, the village of Paraduça has been organizing itself, to acquaint those inside and those outside, on how it produces the corn bread, the main type of bread produced and consumed in this region of the highlands. For that, a few years ago began the restoration of the community oven, although most people have their own wood oven for baking bread, and the restoration of the five community mills, all fed by the same taken, diverted from the Ribeira Paraduça.

In what concerns the making of cornbread, three of our informants, in order to convey this knowledge to the new generations, show the public how bread was made with corn and rye flour, water and yeast, stating that, as of today, no longer the younger can do it without help.

In Macieira de Cambra it was recorded the theme of the wine, describing the harvesting and treading of the grapes. After the grapes harvest is finished the berries of the grapes are separated from their clusters in the *desengançadora* and the press mills, and then one proceeds to treading the grapes by the traditional method, i.e., with the feet. Groups of men, in a convivial and carousal environment, tread the grapes for hours. After a fermentation period, that can last in the case of red wine up to two and a half days, it is passed through the press and is placed in barrels or vats to ferment, thus concluding the cycle of the wine. The need to adapt the methods, to facilitate the activity and to have a better performance, is a recurring theme. States our main informant of Macieira de Cambra that only that way is it possible to "*save the heritage and see how was the old culture*". Being necessary to follow the new developments, to modernize and move forward

> *"with some regret, it was recently said that [...] in elementary school and in middle school, [one comes to the] children today [...] and ask them if they know where do the grapes come from, or where does the corn come from, or where does an egg come from, and they say—it comes from the supermarket -, they are no longer connected to the earth, we are losing a lot in terms of understanding the value of the earth".*

It is this concern with the transmission of knowledge about the man, the earth and the interactions between them, that leads to the fact that Vale de Cambra, seat of a county still largely rural, has organized the feast of the removal of the ear of the corn in the city centre. It thus tries to convey to the younger generations, how it was one of the activities performed in the end of the corn agricultural cycle, in which everyone can participate, removing the ear of the corn and joining the banter and songs associated with this activity. The removal of the ear of the corn was always a communal activity, a job and simultaneously a social event and the associated games were the few times when physical proximity was not only permitted but also promoted.

#### 2.1.15 *Valongo*

The theme chosen to be registered by the municipality of Valongo was the flowering mat of the festivities in honor of Nossa Senhora do Amparo in Alfena. The religious activities begin one week before the main festivity, with the Candle (candlelight) Procession between Capela de Nossa Senhora do Amparo and the Main Church, and end with the all night flowering mat construction by the community that is used to cover the streets where the Nossa Senhora do Amparo Procession passes by, this later procession uses the inverse path of the first. Thus, the construction of the flowering mat has a strong religious motivation. The inhabitants of Alfena participating in the planning, preparation and construction activities of the flowering mat, usually live by the streets receiving it. They are organized by street, and in some cases the same street is divided by more than one group.

The exploration and registration of the flowering mat construction was divided into five phases. The planning of the activity (one group), the collecting of raw materials (two groups), the preparation (almost all groups), the construction of the flowering mat (all groups), the epilog or destruction by the rain and the use of the mat under the feet of the passing procession. The documentary *Tapetes Florais de Alfena*, explores and registers the mat planning, preparation and construction activities, using as its main informants the inhabitants of Alfena council, mostly the ones living by the streets where the procession passes by and where the flowering mat is constructed. They belong to the construction groups (private or institutional) and have directly participated in this activity. These groups, in a total of ten, correspond roughly to the streets receiving the flowering mat in the 2013 festivities.

#### 2.1.16 *Vila do Conde*

The theme chosen to be registered by the Vila do Conde municipality was the experiences and memories of men and women from the Caxinas community in Vila do Conde that were deeply connected with the fishery activity and the sea, whether that activity was land based or done at the sea. Initially, the theme proposed by the municipality was the Caxinas's parlance but, when the team fieldwork began, it was concluded that this particular parlance was not spoken any more. Although a particular pronunciation between inhabitants of the area could be identified, it can be said that the Caxinas's parlance was not in use any more. For this reason, and given the informant contacts that were provided, it was decided that what united our work were the memories connected with the fishery activities and that were practiced by men and women.

> *"I want to bring to mind those who have already departed or aged, like the old vessels they manned, with whom I often talked. I want to contribute for mankind never forgetting that it passes transiently, but it should always tie the past to the future. I feel*

*that it is my duty to report here the hard life of this humble people, as a lesson to the descendants and those to come, so that they can keep as a document, as a memory, as example and advice, the legacy of these men, these people of my homeland—Caxinas" (Cova 2008).*

A strong sense of identity between the inhabitants of Caxinas, that originated a long time ago, and that is still true in the present times, is always connected with the sea, or with the activities developed around it. A large part of the population was engaged and still engages in activities related to fishery. Those people, of both genders, are the informants that within this documentary share their memories from different points of view, thus the documentary "*O Homem e o Mar*" follows the life of the fishermen, men and women, their hopes and fears.

#### 2.1.17 *Vila Nova de Gaia*

The theme chosen to be registered by the municipality of Vila Nova de Gaia was the associative movement, experiences and memories of the community. People belonging to theatre associations from some councils of the Vila Nova de Gaia municipality were invited to describe a life story, triggered within this context, that changed, in some particular way, their life. Based on their stories, they were selected and their memories were registered. Some stories have a very personal involvement; others represent an institutional involvement. "T*he experiences of individuals within the communities and the positive impact on their lives of these aspects are often overlooked. Habits and ways of life of hundreds of people who feed the associative sector do not manifest themselves in the representations of a play or a recital. Let's say these are just the tip of the iceberg*" [Vítor Silva Pinto, personal statement, C. M. Vila Nova de Gaia].

### 2.2 *Online platform*

The PIAMP (Património Imaterial da Área Metropolitana do Porto) project results are available in the web site http://piamp.amp.pt. This site contains links for the videos (15-minute full version and 5-minute teasers) of each of the 17 municipalities, and the video of the Metropolitan Area of Porto that was created combining clips from the videos of each municipality. The videos are stored in a Vimeo Plus account (www.vimeo.org), include subtitles in four idioms—Portuguese including Portuguese hard of hearing, English, Spanish, and French –, a short text in these four idioms describing the projects of each municipality, and a set of photographs of each project.

## 3 DISCUSSION

*"Intangible cultural heritage can only be heritage when it is recognized as such by the communities, groups or individuals that create, maintain and transmit it—without their recognition, nobody else can decide for them that a given expression or practice is their heritage." (UNESCO 2011).*

The project Collecting the Intangible Heritage of the Metropolitan Area of Porto (AMP) highlights the recovery of the collective imaginary of AMP, through their social practices, rituals and festive events, as well as its role in the identity construction of the territory, evoking three determining pillars throughout the project: the collective imaginary, memory and identity.

The Intangible Cultural Heritage, as defined by UNESCO, manifests itself among other areas in the oral traditions and expressions, social practices, rituals and festive events, and in knowledge and practices related either with nature or with the universe. It is presented as well as a wide range of manifestations and expressions of intangible character that have the memory as a means of preserving and orality as a transmission medium. Then arises, encompassed within this immateriality, legends, myths, folktales, but also the rituals and festivals, as well as the entire universe of knowledge and experiences of popular cosmogony.

Contrary to the designated material cultural heritage, the intangible differs from the first with respect to its supports of an extremely fragile nature, and therefore, easily perishable. In this sense, there is an urgent need to recover, collect and preserve them as part of projects such as this one from the Metropolitan Area of Porto. These turn out to be decisive, with its objectives of making the inventory, and treatment—we refer specifically to interpretive studies—as well as its disclosure, and are determinant for the collective memory and identity of a group or society.

### 3.1 *Intangible cultural heritage: collective imaginary, memory and identity*

The notion of intangible heritage appears as an expression of reaction, as an expression of concerns within an area not covered by those who worked the material heritage. Thus, the distinction between material and immaterial is a purely political distinction, first debated and worked within UNESCO, which is supra-national in scope, and becomes a part of the list of States' concerns, in order finally to reach the academy. Far from the debate remain the societies, the groups and the individuals holding the patrimony.

We understand that notions of material and immaterial are culturally determined notions, and that it is up to each culture, society or group to establish a boundary between what is material and what is immaterial. It seems appropriate to underline the lack of involvement of local communities in taking the initiative to preserve and safeguard their local heritage.

There are, therefore, two assumptions that must be present when working in the realm of intangible heritage. Since the definition of intangible cultural heritage is almost an anthropological definition of culture, it must therefore be recognized as something dynamic, something that is permanently evolving. A feast, a ritual, an agricultural practice, should be recognized as processes, reflections of the way of life of communities, of groups, but which will only exist as long as they make sense to individuals or communities. This feast, this ritual, or this agricultural practice, can never be fixed, crystallized, in the sense of an imposition. They will have to be regarded as dynamic processes, subject to re-significations, subject to changes due to socio-cultural context changes—and that the socio-semiotics explores and deconstructs scientifically—in a process very similar to the myths and their characteristics (evolutionary and non universal).

The second assumption is the need for an integrated approach to tangible and intangible heritage. In fact, separating tangible from intangible heritage is something that shakes Marcel Mauss's (1988) theory by which each cultural manifestation is considered as a "total social fact". If, on the one hand, the intangible heritage arises dependent on a material space—a territory—that gives it meaning, or a landscape that evokes it, intangible heritage always shows a subject/space relationship, even if both are implied. In other cases, this dependence results from the connection of an object, a tool, with a traditional practice. Facing intangible heritage as an element per se will result in de-contextualization and, consequently, loss of meaning. It will therefore be impossible to work an activity/knowledge, a memory, a myth or a legend decontextualizing it geographically, it is the territory itself, the space that also confers sense to it (see Edensor, 1998).

### 3.2 *Cultural identities identification, construction and transformation*

Likewise, this second question, translated into a material/immaterial separation, raises yet another problem. Intangible heritage by implying a collection (carried out by means of a written, audio, video or photographic record) leads to a process of materialization of the immaterial, that is, the record sanctifies intangible heritage, granting it a status that will allow it to exist as a tangible object. This process of sacralization is in line with the way how things end up operating, the objectives include a logic of action in the economic sphere, mainly for tourism purposes. This need for preservation arises when the tradition begins to disappear, and it is only from that moment on that it is an object of concern. In some cases, even when tradition does not exist as such, the process of preservation ends up becoming a process of patrimonium-alization, of museum-alization, becoming almost an object of "authenticity" (staged authenticity) with objectives of tourism exploitation.

This whole set of problems leads us to the methodological question in the collection of intangible heritage. It is difficult to define a methodology of its own, however, as already mentioned, the identification and collection process will have to constitute itself as an activity. A process of stimulation in which the researcher participates with who holds the patrimony—the informant or social actor. In this process that passes through the evocation of memory, it is the responsibility of the researcher to bring the social actor into the present, in some cases, it would seem, to rescue him from the past. Likewise, it implies that the researcher is aware that the collective memory is always conditioned by belonging to a community or group, as well as by the physical transformations of the territory. When it transforms, there are no more spatial references in which memory is supported and dissipated.

Thus, it seems appropriate, in the context of collecting the so-called intangible heritage, to apply the methodology used by visual anthropology, because collecting a knowledge / activity, a memory, a myth, a legend or a short story does not only mean collection of a content, but also in a way, the way in which it is verbalized. The participation of gesture, the emphasis placed on words, facial expressions, among others, constitute a performance that is also part of the narrative. In this way, we have access not only to the text, but to a wide set of elements that participate in the story and the interpretation of it, bringing the social actor to the scene, giving it an active role.

The use of audiovisual media thus assumes a predominant role in the collection to be made. The relationship established on the ground is essentially a sharing relationship. A relationship that allows access to local knowledge, to the point of view of the subjects, a privileged time of constant relations and interactions. A time of active participation in the construction or collection of a knowledge. The use of the image in the research processes, is the basis of this construction or re-collection. The images made, as well as representing the starting point of the research process, are also the result of this same process. If the use of the image is fundamental at the beginning of the investigation—as a process to start the research, and later as a process for the development of the same research—its use also leads to the way in which the researcher uses the images in the production of knowledge, bearing in mind an axis polarized by two visual cultures: local visual culture and academic visual culture (Pink, 2001).

### 3.3 *Communication strategies of the cultural diversity of the Metropolitan Area of Porto*

This project was not only the collection of intangible assets but also the dissemination of the cultural and immaterial heritage of a region.

The (re) discovery of the local heritage by the populations that inhabit this space is of decisive importance. It will be important for local inhabitants to become aware of the existence of an identity within their region, hitherto unknown to the subjects. This implies to enjoy one's own identity in a direct way, to become aware of the unnoticed because it was daily experienced and as such considered pedestrian. Thus, local development policies should have repercussions in increasing the collective consciousness of the population as a member of a territory and its integration within it.

Dissemination can play an important role in this perspective, integrating the local populations, the identity, the values of a region, in the respective advertising messages with a tourist scope. By increasing the self-esteem of the populations and projecting it abroad, as members of a territory and the role played by them in the representation and valorization of the same, they become aware of the potentialities of the local heritage and the role they hold in their respective knowledge of the region. In this perspective, the local populations as well as the local heritage present themselves as central and inseparable elements.

Pedro Hellín Ortuño states that "local heritage represents a key element, an instrument of development capable of being integrated into the policy of territorial planning, which generates well-being, is an engine of economic growth and creation of jobs, besides favoring the creation of a differentiated brand image" (Hellín Ortuño, 2009: 219).

## 4 CONCLUSION

As stated in the introduction this was the first time that these municipalities tried to work together in a systematic way to build common information systems and record cultural assets.

Of this set of projects, the projects that have the potential to have higher impact are PIAMP (due to the strong involvement of different user communities) and DINTUR (due to the size and scope of the information being collected about the tourism dynamics of supply and demand). Now it is too early to tell, still both look promising. For instance, the decision in PIAMP to produce all materials according to a Creative Commons License that allows the creation of derived content can have a profound impact on the added-value of that project. In practice the raw materials can even become more important than the finished product.

As of now all six projects remain in active development or are being actively maintained, and this, comparing to the panorama of just five years ago, shows a completely different attitude and a completely different mindset. Now it can certainly be said that the 17 municipalities of the Metropolitan Area of Porto learned how to work together for the common good.

## REFERENCES


Câmara Municipal de S. João da Madeira. 2013. Leaflet "Museu Chapelaria S. João da Madeira".

Costa, M. (Ed.) 2011. Notícias da Lancha 1991–1992 facsimile. Município da Póvoa de Varzim.

Cova, J. V. 2008. Caxinas: A minha terra e a minha gente. Câmara Municipal de Vila do Conde.

Edensor, T. 1998. Tourists at the Taj. Performance and Meaning at a Symbolic Site, London: Routledge.

Hellín Ortuño, P. A. 2009, "Estrategias publicitarias en la promoción del turismo de interior", in Enoturismo e Turismo em Espaço Rural, Maia: GEHVID / Edições ISMAI.

Henriques, E. C. & Rocha M. & Sousa, A. 2008. iporto 2.0: a collaborative software platform for low-cost dissemination of cultural and tourist information. I Jornadas Internacionais sobre Enoturismo e Turismo em Espaço Rural, Porto / Maia/S. João da Pesqueira

Maia, V. M. d. S. 2011. Santa Maria de Nogueira, Monografia, Volume 1. Nogueira, Maia: Author's edition.

Nunes, S. 2003. As Pescas e a Indústria Conserveira, Volume 7 of Matosinhos: Monografia do Concelho. Câmara Municipal de Matosinhos.

Pink, S. 2001. Doing Visual Ethnography. London: Sage Publications.

Ribeiro, O. 1991. O Mundo Rural. In Opúsculos Geográficos, Volume 4. Fundação Calouste Gulbenkian.

Saldanha, J. B. 2008. A causa do mar em Manuel Lopes. In M. da Conceição Nogueira (Ed.), Póvoa do Varzim Boletim Cultural, Volume 42, pp. 139–148. Câmara Municipal da Póvoa de Varzim.

Tedim, J. M. A. 1978. Os Santeiros da Maia. In F. de Oliveira, F. Ferreira, and E. Guimarães (Eds.), Bracara Augusta—Revista Cultural da Câmara Municipal de Braga, Volume XXXII (73–74), pp. 337–356. Câmara Municipal de Braga.

*Cities' Identity Through Architecture and Arts – Catalani et al. (Eds)*
 *ISBN 978-1-138-05409-7*

# The urban regeneration of the peripheral areas. The case study of Tor Vergata (Rome, Italy)

Luciano De Bonis
*Dipartimento di Bioscienze e Territorio, Università del Molise, Italy*

Giuseppe Di Benedetto, Maria Luisa Germanà & Ferdinando Trapani
*Dipartimento di Architettura, Università degli Studi di Palermo, Palermo, Italy*

Maurizio Petrangeli
*Dipartimento di Architettura e Progetto, Università di Roma Sapienza, Roma, Italy*

Chiara Tonelli
*Dipartimento di Architettura, Università di Roma Tre, Roma, Italy*

ABSTRACT: The 'Urban Regeneration of Peripheral Areas' workshop experimented with an innovative design model for the urban regeneration of Rome. The working method of the workshop is based on diachronic analysis of the evolution of the combined natural history and culture of the area, the architectural, technological, environmental and landscape situation and relation-ship with the surrounding territory. The workshop dealt with the mutual relationships, which run through and characterise the dimensional and relational scales of the environments of metropolitan ecosystems as structures and landscapes. The strategy for urban and territorial integration is left to the procedures and policies of the community activism organisations and the administration because only they play central roles in the implementation of the design proposal. Even in the most critical cases it is possible to regenerate the outskirts, preserve their value by trans forming them with the integration/recombination of their components and oppose the shared project of an urban landscape with an increasing number of buildings outside the proper context appropriate to their socio-cultural identity.

*Keywords*: Urban Regeneration; diachronic analysis; landscape; architecture; metropolitan ecosystems

## 1 INTRODUCTION

On Friday 14th October 2016 the 'Urban Regeneration of Peripheral Areas' workshop at the "Roma Tre" University came to an end. This workshop experimented with a new sustainable urban regeneration model for the case study of Tor Vergata in Rome. The area has been recommended by the Olympic Committee for the construction of the Olympic village as part of the candidature of Rome for the 2024 Olympic Games where the great works for the 2009 World Aquatics Championships had been started and then abandoned. The workshop was directed by Ferdinando Trapani and started on the basis of the scientific contributions of Maria Luisa Germanà and Giuseppe Di Benedetto. Maurizio Petrangeli led the work group with the help of Chiara Tonelli and Nabil Mohareb. Luciano de Bonis produced a study of the historic, environmental and regional situation together with the tutors Barbara Cardone, Nicola Moschena, Francesca Giangrande, Michele Porsia and Stefano Simoncini. Mieke Oostra, Adolf Sotoca and Ibrahim Maarouf took part in the discussion and final assessment.

## 2 THE ARCHITECTURE DESIGN AND URBAN PLANNING INTEGRATION DEBATE

The Tor Vergata district is a symbolic and exemplary case on account of the various topical questions arising in the debate on the contemporary city undergoing transformation. It is an area in the outskirts of Rome, definable and recognisable as a fragmentary, dispersed and weak area but at the same time also dynamic and open to a new morphological and urban identity. The meaning of "the territory of architecture" (Gregotti, 1966) must be extended as in this case study there are several vantage points because the approach to the subject of urban regeneration of extensive metropolitan outskirts must necessarily be interdisciplinary. In these multiple ambits, the architecture modifies the relationships of meanings of the physical space which once again becomes a place opposed to the idea of terrain vagues: a space which is unbounded, of a hereditary disposition and morphologically vague (Augé, 2004; Gregotti, 2013). The combined architecture-town planning concept, (Samonà, 1975) which interprets the complex construction of physical space in the transformation process of urban settings by radical regenerative actions, comes again to the fore.

Architecture reconsiders the themes connected with the "form-content" relationship and reassesses the interweaving of relations as "connective tissue of the content and form of the results of the same process" (Samonà, undated). The combined architecture-town planning concept makes it possible to include the cultural, social and political interactions that control the physical configuration of every transformation. The renewed unification of urban architecture and building architecture has more sense, since it assimilates the very idea of the city with the place, with the environment and with the manmade landscape. Architecture affects the entire setting on which it aims to confer aesthetic value and can be understood only considering all the dimensional scales. It is preferable to speak of reclamation and redevelopment rather than a new process since these activities are related to the existence of values produced by a possible regeneration in the urban dimension (Gregotti, 2011; Ricci, 2011). Architectural-town planning design operates continuously through various scales and reconnects the fractures in the periurban and infrastructural landscapes of the contemporary city.

Urban outskirts are sometimes transformed through horizontal development with dissemination of buildings over vast areas (OMA—Rem Koolhaas; Villette Park; Melùn Senart; Bernard Tschumi and the "office park" covering more than 200 hectares at Chartres; Jean Nouvel with the "Active city" of Nîmes). In other cases it is vertical condensation of buildings that prevails in order to free as much ground as possible and strengthen the open structure character as in the Atlanpole of Nantes (Di Benedetto et al., 2017). Gregotti, Battisti, Gabetti, Isola and others base the architectural strategy on the enclosure archetype or principles of initial anthropisation (Busquets in the outskirts of Grenoble or the headquarters of Fiat by Gabetti and Isola in Candiolo), in an attempt to give character to the places by delimiting them and giving the design a founding role.

This peripheral city can no longer be interpreted in terms of finished parts. The "unfinished" condition results in the need for an empirical approach based on reconstruction and regeneration of the incoherent and morphologically weak building forms. Reflection on the architecture of this city, which started in the 1990s, adopted geographical and socio-political approaches (Lanzani, 2003) without viewing the project as an instrument for examining and becoming acquainted with spatial transformations (Torricelli, 2012).

## 3 URBAN IDENTITY BY THE BUILT ENVIRONMENT: THE INCOMPLETENESS OF THE PERIPHERAL AREAS

'Urban identity' represents a decisive and extremely concise factor in regeneration, both as an analytical focus and a design objective for works in historic town centres, outskirts, disused areas or residential districts (Lo Piccolo, 1995). Even if the regeneration processes are driven by an organic view of cities in continual evolution and typically have resilience in dealing with the many critical aspects of the question, there is a need for clear and shared guidelines. This does not imply the need to differentiate between approaches and strategies for the parts of the contemporary city that were developed in different periods, but rather to be able to adapt them to the specific circumstances of regeneration (Germanà, 2013/a; Trapani, 2016). Numberless portrayals, of an artistic, scientific, essayist and literary nature, comment on the ambiguous character of the outskirts, anonymous places perennially waiting for a meaning, no longer countryside and not yet city, where "a kind of environment resignation„ a daily familiarity with the unfinished, which makes even thoughts incomplete" (Fois, 2015, 194) seems to have concentrated. A similar characteristic which can be found in any outskirt takes on a special meaning in the case of Tor Vergata, characterised by the City of Sport (also known as the "Sails") designed by Santiago Calatrava and never completed. For this reason some considerations on the incompleteness in the field of the Architectural Heritage (Germanà 2015/a; 2015/b) can be reviewed.

Incompleteness may refer to material or immaterial aspects which often interweave or interact. Incompleteness in the process may refer to gaps attributable to single phases (planning, design, execution and management) or operators (clients, designers, contractors and users) or—in the more serious cases—the entire process, when due account has not been taken of the continuity of the long term vision, of the integration of the cultural, social, economic and environmental spheres or of the interactions between experts and end-users. The incompleteness of the product is found in the physical state of the buildings which were once intact but are no longer (such as the thermae building covering more than 500 sqm, probably connected to a luxury residence of the first century A. C. recently excavated at Tor Vergata by the University) or buildings never completed (such as the buildings for the 2009 World Aquatics Championships, just a few yards from the example mentioned above). Material incompleteness is measured on various scales, in the building component and/or the urban and landscape context, the consequences of which may lead to three main types: the intrinsic vulnerability of buildings to natural and anthropic agents; impossible or limited utilisation, starting with accessibility; the difficulty in attributing coherent understandable meanings.

A second consideration concerns the extent of the incompleteness, which manifests itself in relation to the various circumstances in which it is found. A mainly recently built environment prevails in the outskirts, having typological, morphological and material characteristics disconnected from the specific location. Although this, archaeological remains are found in peripheral areas very often, especially in the Mediterranean areas (Germanà, 2013).

For these traces of the past ages, as for many other much less patchy seattlement„ the state of incompleteness may reach various levels of which varying degrees are more or less identifiable: from the evident ruin to the less perceptible effect of the interruption of contextual connections and irreversible transformation of the production processes that originated them. In any case, incompleteness in the Architectural Heritage must be considered inevitable and not necessarily something negative but rather an acceptable identity condition to be dealt with each time by searching for a temporary unity of the knowledge, meanings, relative physical conditions, appropriate uses, participation of the users and of the community.

In the case of the 'Sails' of Tor Vergata designed by Calatrava, an example which for its dimensions and wishful thinking may be considered symbolic of the widespread plague of unfinished public works in Italy, any proposal of urban regeneration must include some form of completeness for this relic of contemporary times. Some theories provocatively attribute an aesthetic value to the unfinished buildings and infrastructures of recent decades, almost wishing ironically to put right the shameful legacy of our corruption and inefficiency (Arboleda, 2017). In effect, the imposing metal structure of the closed Calatrava site may produce an aesthetic impact but the main effect of this type of unfinished project, in Tor Vergata as elsewhere, is another: a deep resentment for the enormity of the wasted natural, financial and human resources which overall make the acceptance of these abandoned works unsustainable.

## 4 WHAT KIND OF CITY WITH WHAT ENERGY

With the prospective of Rome's candidature for the 2024 Olympic Games and Paralympics an opportunity has arisen in the city to provide an answer also for the housing emergency.

The foreseeable accommodation for the athletes, like those built for the event in 1960, could be allocated later to council housing. This would help to meet the demand for housing which, for intrinsic reasons due to the economic crisis, impoverishment of the middle class and pressure exerted by migratory flow, is becoming a pressing matter in the capital city.

In the aftermath of these profound transformations, which make it necessary to formulate new residential and urban models, the Olympics could have started a concrete experiment in: a) high energy efficiency of the housing and management models for living in them, in accordance with NZEB standards, constructing highly efficient buildings, raising awareness in the inhabitants regarding their consumption, with the aid of sophisticated monitoring systems but conceived to increase awareness of how to manage energy; b) innovation in dimensional standards and the use of spaces in the home, since current building standards are by now obsolete, evolution of the archetypes of home design developed in the past century in order to satisfy a housing demand that is different from what they were conceived for; c) new production and construction systems, with advanced prefabrication, meeting the needs of economy, certainty of completion dates and rapidity, correspondence to the quality levels of the designs; d) a cultural conscience for environments which are healthy and highly performing, for the most part using natural and renewable materials.

The answers should have come through the international Solar Decathlon international competition (Bellingeri, Tonelli, 2016), conceived by the US Department of Energy in 1999 which selects twenty universities from all over the world to compete every two years in the conception, construction and management of housing prototypes of the future, powered by solar energy. The prototypes are assembled together in the same competition field where a kind of smart village is created for the duration of the competition which, through a single information network, monitors the performance of the houses during the ten contests of the Decathlon. The best solutions would then be the housing models to be used in the entire Olympic Village in the area of Tor Vergata and near the "Sails" site thus providing urban regeneration of an area that is extremely incomplete.

The structure was conceived for accommodating about 18,000 people during the games and then for conversion into housing for 7,000 with a part being destined for residences to complete the campus project of the Tor Vergata University and another part to extend the buildings of the University Teaching Hospital for the relatives of hospitalised patients.

In line with the recent European trend which encourages the saving of ground and recovery of empty urban spaces, the proposals should have identified in the urban regeneration of these rundown areas of the capital the way forward for the coexistence of archaeological remains and unauthorised building, with the construction of multi-story buildings in the spaces present to complete the already consolidated developments.

## 5 THE CHARACTERISTICS OF THE AREA AND ITS URBAN INTERPRETATION

The expression "urban regeneration" immediately raises questions of definition, not only in the overall sense but also in reference to the meaning of its component terms, namely "regeneration" and "urban". Without getting into the now widespread technical and scientific discussion on the topic here, it is however useful to give account for the approach to the subject used in the workshop considered in these notes.

First of all the idea of tackling any kind of regenerative project in the chosen area (Tor Vergata) was not considered possible, especially but not only if giving it a "green" connotation, unless by placing it in a temporal dimension including not only its cultural history but also its natural history which should be handled in an integrated way and not separately and in a markedly cross-scale spatial dimension and therefore completely conforming to the above "inclusive" temporal dimension.

The evolutionary process of the Tor Vergata area, in accordance with the approach adopted, is thus attributable "at its origin" (Faccenna et al., 1995) to the initial phase (about 600,000 years ago) of the activity of two volcanic districts in the area of Rome, the Monti

Sabatini to the northwest and the Alban Hills to the southeast, the products of which extend progressively across the pre-existing plain, altering the morphology considerably and causing variations in the paleo-river bed of the Tiber. It is precisely to the hydrogeological unit of the Alban Hills (Cafiero, 2003) that the area under study belongs. From the biotic viewpoint to that unit corresponds the vegetation series of Carpino orientalis-Querceto cerridis sigmetum of the plains and pyroclastic slopes, as can be seen in the map of the vegetation series of the new 2008 Local Plan of Rome.

The structure of the environmental system of the Rome area overall is completed after the last glacial period (Wurm, about 18,000 years ago), when the watercourse of the Tiber before deepens and is then filled with alluvium in the following interglacial period (Faccenna et al., 1995). It is with this abiotic and biotic environmental structure that historically human interaction commenced in the area of Rome. Although this concerns mainly the alluvial deposits at the core of the city, in practice it also concerns all the rest of the system including the hydrogeological unit and vegetation to which the area under study belongs, an area which in fact still conserves a heritage which is often of inestimable value (see par. 3). But this heritage certainly does not represent the only sign that the historic evolution, especially recent, has left on the environmental sub-system of the Tor Vergata area. Or to put it in a better way, the historic and environmental sub-system of the area under study overall (Calzolari, 1999[1]) certainly does not now manifest only the features to which we are inclined to attribute value (because, take note, it is always a question of our attribution of values though it may be widely supported and justifiable). Faced with the by now "typical" (at Rome and elsewhere) forms of urban sprawl recently involving also this area the abovementioned new Local Plan of Rome, the D2 "Structures of the Plan and metropolitan strategies" map in particular, formulates an approach based substantially on: i) metropolitan outlook; ii) decentralisation and polycentrism; iii) protection of the environment and heritage; iv) more services and urban functions in the peripheries; v) privileging the railway network for mobility (the so called "iron therapy").

Tor Vergata represents one of the centralities at the metropolitan level on which the polycentric strategy is hinged. It also is the site chosen by the 2024 Rome Olympics Committee for the Olympic Village (see par. 4), in contrast however with the residential dimensioning of the Plan and with the orientation of the Marino Administration (prior to the current administration and to the subsequent compulsory one) which places it, together with a river park and without residential "legacies" (but directional instead), in an area gravitating around the northern section of the Tiber within the Grande Raccordo Anulare (orbital motorway encircling Rome). The new and current Administration of Rome, on the other hand having declared itself contrary also to the approach of the 2008 Plan, definitively resolved the question, to put it one way, with an announcement on 21 September 2016 by the Mayor Raggi to the effect that the candidature of Rome for the 2024 Olympics would be withdrawn.

The activities of the workshop, with the conceptual preconditions explained in the preceding paragraphs and with the methods and visions described further below, has been done constantly in a state of complete and free immersion in the complex situation described so far, dating back from our time to 600,000 years ago (at least) and extending spatially throughout the historic and environmental system of Rome and its environs well beyond the limits of the study area but also the boundaries of the municipal territory. Acknowledging that it is more convenient today to define the term "urban", going beyond any traditional definition and residual city/countryside dualism, as a kind of physical and mental reference system consisting of material and immaterial networks as well as technical objects, manipulation of which implies introducing a stock of images and information circulating in loops regarding the relationships our society has with space, time and men, and that urbanity basically consists of a dynamic of mutual adjustment between a form of urban fabric and a form of conviviality (Choay, 1994).

---

1. See also Calzolari M.V. (ed), Roma: Permanenza e fragilità del sistema storico ambientale (https://goo.gl/q54zH099).

## 6 URBAN ANALYSIS WITHIN THE LOGIC OF PLAN

The didactic method used in the workshop is generally based on the prior planning experience for social innovation with the support of ICT (Marsh et al., 2013; Concilio & Rizzo, 2016; Eskelinen et al., 2014), on an analysis of the socio-economical context in terms of urban value (De Bonis & Trapani, 2016). In particular the analysis took account of the multidisciplinary studies on the relationship between the centre and outskirts (Di Benedetto et al., 2017) and urban regeneration in the historic evolution over a long period (Prescia & Trapani, 2016). The key used to exploit the legacy of cultural, infrastructural and environmental resources in the ambit of sustainable tourism was the theoretical-design model of integrated interactive tourism (Trapani & Ruggieri, 2010). Lastly, consideration was given to the aspects of possible resistance to the collapse of this great urban sub-system also considering the aspects of urban resilience (Minozzi & Trapani, 2016). The urban context was studied very carefully with special attention being given to recent planning and planning forecasts as well as public debates at the municipal and local levels concerning the decisions of the administration and the advocates of Tor Vergata as a sports hub. The work group explored in particular the possibility of distinguishing the various components in the features of the urban and semi-rural landscape as follows:

a. Points/hubs: Centres/large attractors (University, Hospital, Shopping area, unfinished sports area); these points represent the hubs and metropolitan turbines of the research and study immersed in the semi-urban landscape of Tor Vergata
b. Residential zones: planned housing/redeveloped areas of urban disorganisation; the project provides for reuse of the houses built in accordance with the above technical norms and mainly those built without a plan (later redeveloped) which require maintenance, urbanisation, energy consumption improvements and revision of consumption cycles (water and refuse).
c. Rural zones: Large abandoned or underused rural areas and large trees in private gardens
d. Lines/corridors: urban and territorial mobility infrastructures (motorways, roads, pedestrian paths)
e. Lines/corridors: Waterways which have undergone soil bioengineering works for connection of the grid of urban vegetable gardens in the area of Calatrava's 'Sails' project. In the final vision the residents who practise zero consumption living styles constitute an urban laboratory, the integrated model of which (centrality/mobility/accessibility/"green" policies and sustainable planning and design regarding water resources) is adaptable to all the outskirts of the great cities of the world with the due precautions being taken.

## 7 FOUR VISIONS FOR THE URBAN REGENERATION

Four planning opportunities in differing scales, methods and instruments can redefine the Tor Vergata area, assigning it a role and influence in the urban panorama of Rome.

The first consists of a possible relationship between architecture and infrastructures. The developed area, which is heterogeneous in morphology, fabric and quality, is in fact all within the existing city and features alternating open and built-up areas deeply imprinted by the Rome—Naples railway line. Redesigning the infrastructures to interact in various ways with the surrounds—underneath, above or beside—would transform the outskirts to produce a city with a futuristic appearance as seen from the motorway when passing through.

The motorway also passes Calatrava's "Sails" project which is an urban landmark of great strength and visibility and the new gateway into the city for those arriving from the south. As something to complete, rethink or leave as an unfinished ruin, the "Sails" project represents an opportunity and a challenge. It could represent a changeover from a traditional single-centred development model to a network system where community facilities—the "Sails" project, the teaching hospital, universities—would become new hubs.

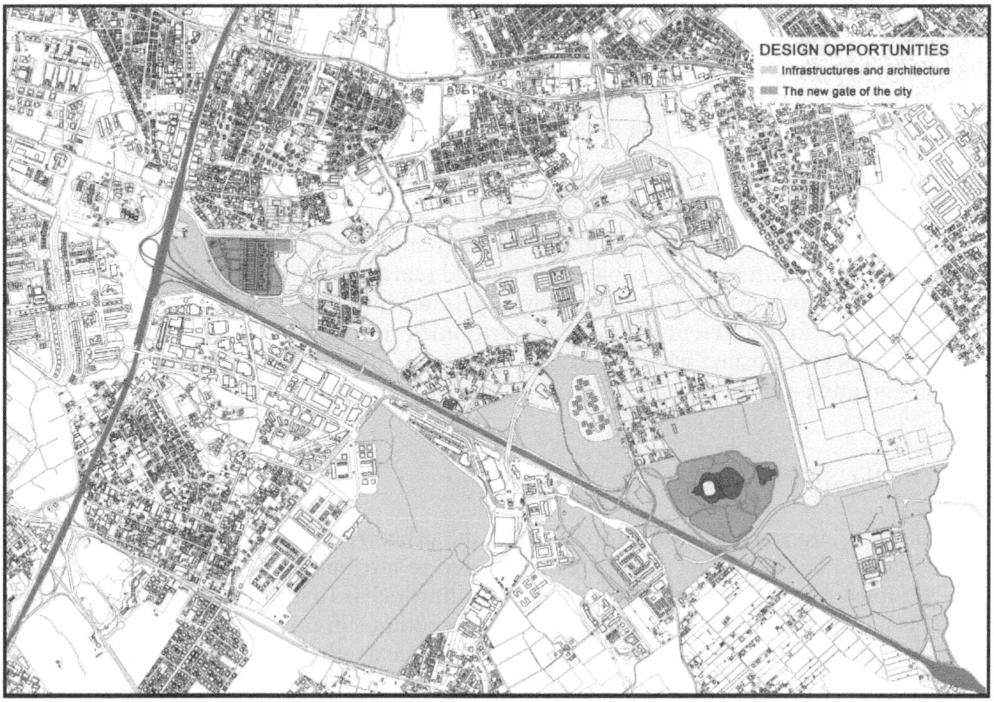


Figure 1. The areas of the possible: the new infrastructure and architecture relation; the new Town Gateway localization.

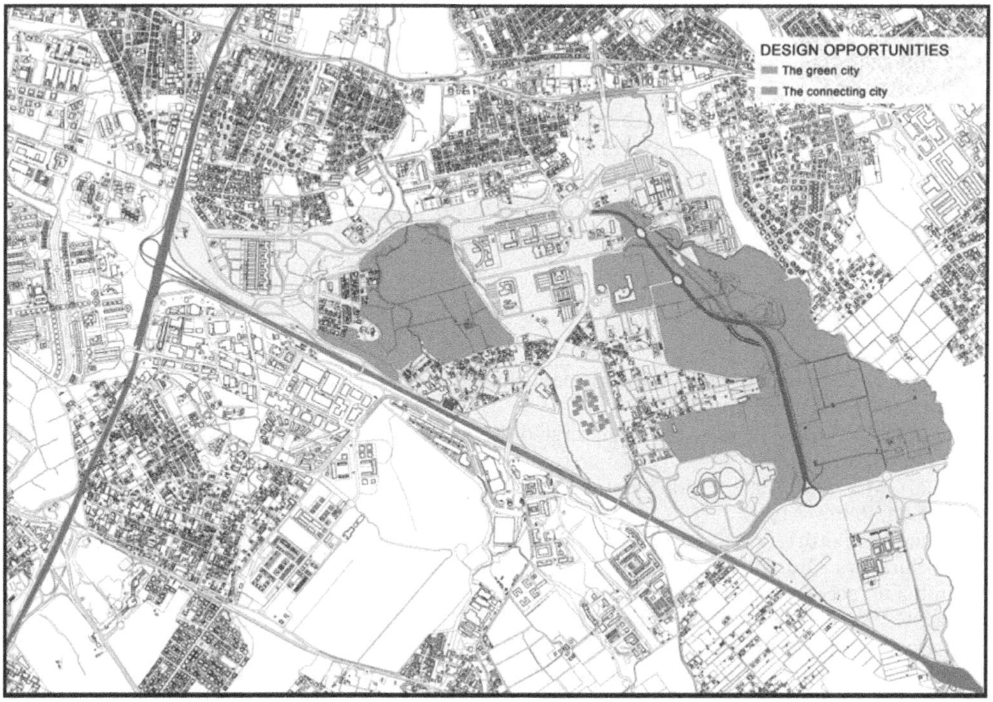


Figure 2. The green & connecting new city fields.

The third opportunity consists of the possibility of transforming Tor Vergata into a green city. The free areas destined by the General Urban Development Plan to be used as a botanical garden, equipped green area and agricultural concerns could become a single large integrated and homogeneous setting, a park with urban traits and attractivity, capable of forming relationships connecting the main features of the area—the sports facilities, university buildings, residences—and of giving shape to a complex multipurpose part of the city.

Lastly, in contrast we have the connecting city: the rejoining of the pre-existing developments by planning the still free areas which wind through the developed areas assumes a style based on the differences, lack of homogeneity, alternation of full and empty, the denser areas and sudden rarefactions. A new urban design could reinvigorate what already exists, rescuing the area from its present situation of anonymity, fragmentation and randomness.

These are the four different planning opportunities, four approaches which provide some possible answers regarding what could be the future of the outskirts redeemed from a state of isolation and anonymity.

## 8 CONCLUSIONS

The short duration of the Workshop in which the urban regeneration plan for Tor Vergata was developed did not allow the first concrete step to attribute completeness to Calatrava's "Sails" project to be taken, at least from a viewpoint which is first conceptual and then physical. This is a step that only the Municipal Administration can take, perhaps with procedures involving all the stakeholders and the community, concentrating on examining the originally intended use and consequent determination of an alternative use. The general planning option developed and briefly outlined here consists of systematising the whole—currently uncoordinated—of what already exists in Tor Vergata'a area: the points/hubs (current and potential centres), the zones (residential and rural) and the lines/corridors (road infrastructures, waterways and ecological itineraries).

Most of the effort went into determining the connective relationships among already existing developments, which are not only physical or visual but also regard meaning. An example of this is the connection, which has lasted for two thousand years, between the thermae building that came to light in the archaeological excavations and the Calatrava's swimming pool, rethought in accordance with the principles of Environmental Design and integrated with developments connected with health and beauty. The urban regeneration of Tor Vergata then can be interpreted as an exercise in attributing completeness of meanings and of uses which are not measured in terms of filling space or land but in the fullness of the cyclical nature of the processes, which is a precondition for every sustainable transformation in the built environment.

## REFERENCES

Bellingeri, G. & Tonelli, C. 2016. Strategie per l'alta efficienza energetica in clima Mediterraneo. Monfalcone, Gorizia: Edicom.

Cafiero, G. 2003 (ed.), Le nove riserve naturali di RomaNaturaUrbanistica. Urbanistica Quaderni, 37. Roma: INU Edizioni.

Calzolari, M.V. (ed.), 1999. Storia e natura come sistema. Un progetto per il territorio libero dell'area romana, Roma: Àrgos.

Calzolari M.V. (s.d.), Roma: Permanenza e fragilità del sistema storico ambientale. In Documenti di Ar-chitettura. Ordine degli Architetti di Roma e provincia. Available on website: http://www.architettiroma.it/ (accessed on 2th February 2017).

Choay, F. 1994. Le règne de l'urbain et la mort de la ville. In J. Dethier, A. Guiheux (eds.), La Ville, art et architecture en Europe, 1870–1993: 26–35. Paris: Centre Georges Pompidou.

Concilio, G., Rizzo, F. 2016. Human Smart Cities. Rethinking the Interplay between Design and Planning. Dordrecht: Springer.

De Bonis, L. & Trapani F. 2017. For a "Living (Lab)" Approach to Smart Cities. In E. Riva Sanseverino, R. Riva Sanseverino & V. Vaccaio (eds.), Smart Cities Atlas. Western and Eastern Intelligent Communi-ties, pp. 143–158. Dordrecht: Springer.

Di Benedetto, G., Germanà, M.L., Trapani, F. (in press). Peripheral centralities: an integrated approach; Proc. intern. conf., Green Urbanism, Rome, 12–14 October 2016. Amsterdam: Procedia Environmental Science-Elsevier.

Fois, M. 2015. Luce perfetta. Torino: Einaudi.

Germanà. M. L. 2013/a. Processi gestionali della città storica tra specificità e necessità di strategie unitar-ie: l'esempio dell'accessibilità. In Castagneto F., Fiore V. (eds.), Recupero, Valorizzazione Ma-nutenzione nei Centri Storici. Un tavolo di confronto interdisciplinare: 22–25. Siracusa: Lettera Ventidue.

Germanà, M.L. 2013/b. Archaeological construction and its relationship with place: the theme of shelters. In M.C. Ruggieri Tricoli & M.L. Germanà (eds.), Urban Archaeology Enhancement: 180–207. Pisa: ETS.

Germanà, M.L. 2015/a. How to face the incompleteness of the Urban Archaeological Buildings. EDA. In-ternational Journal of Architecture and Engineering: 70–81.

Germanà, M.L. 2015/b. L'incompletezza del Patrimonio Architettonico: testimonianze in terra cruda a Solunto. In Sposito, A. & Mangiarotti, A. (eds.). Project Soluntum. Tradition and Innovation in ancient Contexts: 91–94. Roma: Ermes.

Gregotti, V. 2013. Il sublime al tempo del contemporaneo. Torino: Einaudi. Lanzani, A. 2003. I paesaggi italiani, Roma: Meltemi.

Gregotti V. 1966. Il territorio dell'architettura. Milano: Feltrinelli. Gregotti, V. 2011. Architettura e post-metropoli. Torino: Einaudi.

Lo Piccolo, F. 1995 (ed.). Identità urbana. Materiali per un dibattito, Roma, Gangemi.

Marsh, J., Molinari, F., Trapani, F. 2013. Co-creating Urban Development: A Living Lab for Community Regeneration in the Second District of Palermo. In B. Murgante, S. Misra, M. Carlini, C.M. Torre, H-Q. Nguyen, D. Taniar, B.O. Apduhan, O. Gervasi, (eds.). Computational Science and Its Applications. ICCSA 2013; Proc. 13th intern. conf., Ho Chi Minh City, Vietnam, 24–27 June 2013, (III): 294–308. Hei-delberg Dordrecht London NewYork: Springer.

Ricci, M. 2011. Nuovi paradigmi: ridurre, riusare riciclare, città e paesaggi. In P. Ciorra (ed.) Riciclo. Mi-lano: Maxxi Electa.

Samonà, A. (s.d.). Storia, tipologia, morfologia nel processo architettonico e ipotesi di unità ar-chitetturaurbanistica, dattiloscritto, Roma.

Samonà, G. 1975. L'unità architettura-urbanistica. Scritti e progetti 1929–1973, Lovero, P. (ed.). Milano: Franco Angeli.

Tanzilli, M. 2015. Incompiute: Ruderi e Macerie della Contemporaneità. Available on website: http://www.tari-architects.com/portfolio-item/crac/ (accessed on 09/01/2017).

Torricelli, A. 2012. Ritorno alla costruzione. In Architettura civile, (5): 1–3.

Trapani, F. & Minozzi, L. 2015. Policies of Resilience in the New Institutional Process. The Case-Studies of Palermo and Siracusa in the South of Italy. In TeMA, Journal of Land Use, Mobility and Environment, Special Issue ECCA 2015. Smart and Resilient Cities Ideas and Practices from the South of Europe, European Climate Change Adaptaption Conference; Copenhagen, 12–14 may 2015: 47–62.

Trapani, F. & Ruggieri, G. 2009 (eds.). Integrated and Relational Tourism. Territories and Development in the Mediterranean Area; Proc. Third intern. conf. Helwan-Cairo—Egypt. 24–26 October 2009. Conference. Palermo: Gulotta.

Trapani, F. 2016. Rigenerazione urbana e innovazione sociale. In R. Prescia, F. Trapani (eds.) Rigenerazione urbana, innovazione sociale e cultura del progetto: 9–18. Milano: Franco Angeli.

*Cities' Identity Through Architecture and Arts – Catalani et al. (Eds)*
 *ISBN 978-1-138-05409-7*

# New ornaments' influence on the character of modern cities

Amr Elgohary
*Higher Institute of Engineering, Shorouk, Egypt*

ABSTRACT: Across eras and civilisations, architectural vocabulary and mechanisms have always played a great role in connecting architecture with culture, expressing the character of a city in a way that reflects the culture of its society. Consequently, ornaments are a major component of the architectural vocabulary that allows architecture to connect to culture. Nowadays, the new architectural theories deal with the concept of ornament by following a new vision and methodologies that reflect the new cultural turn of the twenty-first century. In this context, the paper discusses the role of ornament in connecting architecture with culture in new cities as well as the great influence of the new digital theories and technologies that form new ornaments in the twenty-first century, aiming to link the architecture and urban typology of modern cities with the new cultural turns.

*Keywords*: ornaments; culture; urban typology; digital; digital era; architectural software

## 1 INTRODUCTION

Ornament is an artistic form that is one of the main interests of architects and artists, because it gives new aesthetic, cultural and symbolic values, not only at the scale of architectural form, but through expansion to include the urban scale, where it can contribute to the formation of a city's character and theme.

By the end of the twentieth century and the beginning of the twenty-first century, ornament had undergone many changes in its form, characteristics and role in forming a city's character. This has occurred as a result of the new architectural design-fabrication-construction technologies that were developed during the new digital breakthroughs and have had consequences for various fields including architecture. In this context, this research aims to determine the meaning and concept of ornament, and the link between ornament and a city's character. It concludes by identifying the concepts and features of these new ornaments and how they contribute to forming the characters of modern cities.

## 2 THE CONCEPT OF ORNAMENT

The term ornament can describe many kinds of objects or forms of art. These can be clarified using the various definitions of the term as follows:

- Oxford Dictionary: *a thing used, or serving to make something look more attractive but usually having no practical purpose* (Ornament, n.d. [a]). Examples are listed in terms of buildings, objects, one's character and music.
- Cambridge Dictionary: *1. an object which is beautiful rather than useful; 2. a formal decoration which is added to increase the beauty of something* (Ornament, n.d. [b]). Examples are listed that include buildings, objects and urban design.

- Merriam-Webster Dictionary: *1. something that lends grace or beauty; 2. a manner or quality that adorns; 3. An embellishing note not belonging to the essential harmony or melody* (Ornament, n.d. [c]).

From these definitions we can conclude that ornaments are supposed to add a positive, decorative or some other aesthetic value to something else.

### 2.1 *Ornament in nature*

There has always been a strong coherence between art, architecture and nature. While nature represents the basic source of inspiration for artistic and architectural theories and forms, which appear in most architectural thoughts and theories, such as 'Art Nouveau', 'organic' 'functionalism', and etc., till the new digital architecture theories.

Therefore, through deep observation of nature, it can be seen that objects, like galaxies or planets, are recognised not only by their own body but also by the levels of visual information surrounding them. therefore that cosmic picture can be considered as a space pattern, which represent a kind of natural growing ornament system with no limiting border condition. The levels of "ornament" help to prevent the eye from being distracted by chaos in the field of vision; the swarm of stars, which can be interpreted as ornament, is enabling the human brain to focus on an important area in space.

Figure 1. A swarm of stars within a spiral galaxy looks like a natural patterned ornament (pinterest, 2016).

Figure 2. Using mimetic ornaments within the Trevi Fountain to add aesthetic value to the building and the urban space (stefano, 2016).

## 2.2 *Ornament and architecture*

Ornament in architecture refers to any element added to an otherwise merely structural form, usually for the purposes of decoration or embellishment. Three basic and fairly distinct categories of ornament in architecture may be recognised: mimetic or imitative ornament, the forms of which have certain definite meanings or symbolic significance; applied ornament, intended to add beauty to a structure but extrinsic to it; and organic ornament, inherent in the building's function or materials (Ornament, n.d. [d]). Figure 2 shows the use of mimetic ornaments within the Trevi Fountain, the largest Baroque fountain in Rome, which adds aesthetic value to the building and the urban space.

# 3 CITY CHARACTER

A city is made up of many functional aspects and components. On the other hand, it can be considered that the character of a city is formed according to optical and symbolic aspects and components, which appear within the urban typology through a strategy of repetition of building elements such as massing, openings and ornaments.

## 3.1 *Buildings' character*

The character of a city's buildings is considered as a major factor in the creation of the city's character. It can be achieved through many aspects, such as their height, the way they are massed, their architectural vocabulary, materials, openings and ornaments (Alliance, 2007).

## 3.2 *Ornaments' character*

The main difference between ornaments and the other elements that form a building's character can be clarified by determining the functional purpose that each item serves. Ornaments, which are constructed mainly for aesthetic and symbolic purposes, have no functional purpose. Their meanings therefore change with changes in the culture and civilisation. Nevertheless, ornaments can be seen as playing a key role not only in the creation of a building's character, but also in creating that of a city as a whole.

## 3.3 *Strategy of repetition to create city character*

It can be considered that, the strategy of repetition can prioritises compatibility and minimises differentiation. This strategy will likely sustain the character of an existing setting so long as the historic elements to be replicated are well understood, 'the technical means to

Figure 3. The Acropolis of Athens: a strategy of repetition of masses, roofs, columns, materials and ornaments creates the Greek city's character (The guardian, 2010).

effect replication are available, and as long as the scale of the replication is modest relative to the original building. Despite frequently expressed disapproval of this strategy by many contemporary preservation theorists and officials, it has the sanction of history. Architects have often chosen to add to existing buildings by reproducing a previous architect's work, sometimes even centuries afterwards, usually for the sake of completing an intended but unrealised symmetry or extending a pattern already established. In such cases, the resource is defined as the design concept as a whole rather than any isolated part of it as it appears at a given time' (Alliance, 2007, p. 4). Figure 3 shows the repetition strategy at The Acropolis of Athens, which is achieved by rhythms of masses, roofs, columns, materials and ornaments to create the character of the Greek city.

## 4 USING ORNAMENTS TO FORM THE CHARACTER OF CITIES

Ornaments can be seen as playing a key role in a city's character because they reflect its culture and its moral values. This can be illustrated through the following effects of ornament:

- Forming the urban character: an ornament is usually associated with the facades or the plasticity of buildings; yet it also establishes relationships between the building and the urban fabric, to form a new ornamental scale that can be observed through an aerial view.
- Adding symbolic meanings: the symbolic aspect of an ornament is widely viewed as representing several messages, which may be functional, aesthetic, political, social or economic. Jencks (2008) argues that an iconic building has to carry a plurality of meanings and mixed metaphors in order to continue its distinctive presence as a landmark.
- Forming the urban fabric character: ornaments at the scale of the city resemble gigantic carpets or textiles rolled across vast tracks of intensely used lands. Figure 4 shows how the urban fabric can be perceived as a growing digital, ornamental carpet. It demonstrates that buildings can be viewed from different scales and distances, as well as from various points of view, through which they reveal different ornamental dimensions.
- Relating regions and buildings: the human mind evolved in part to recognise and analyse hierarchical structures in nature, which is why artificial structures that are not hierarchically organised are often perceived as alien. 'A recognition mechanism based on hierarchical subdivisions is built into the human consciousness. We therefore need hierarchy in complex structures to perceive a structure as a coherent whole. Ornaments can help to relate different-scale regions to each other. It can also embody a level of scale itself in order to allow people to relate to an object or building' (Salingaros, 2000).

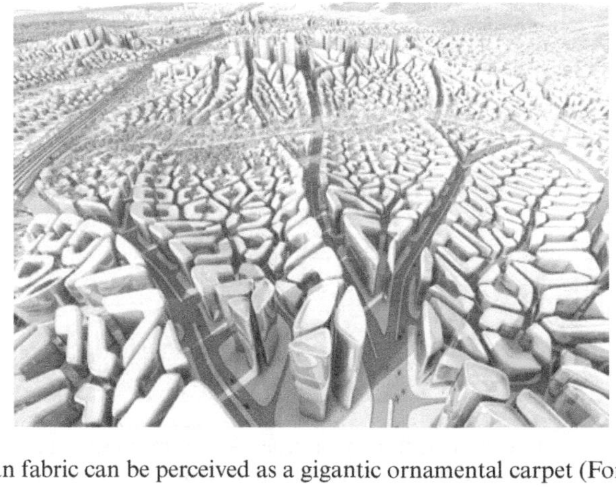

Figure 4. The urban fabric can be perceived as a gigantic ornamental carpet (Formakers, 2012).

- Defining transition zones: an ornament is often found in areas that attract attention, whether used to disguise particular parts or to amplify specific features. Often, an ornament is used in a transformation zone, in between things, at the margins or perimeter; where a door meets a wall, a facade touches the ground, or even when a square meets the street. These parts define the border of a designed shape or object.

## 5 ORNAMENT, CULTURE AND THE CHARACTER OF CITIES OVER TIME

Over the centuries, architects and designers have vied with each other to make their buildings and their cities as ornate as possible, to represent a culture, religion, or any other aspect of human life that gave each civilisation its own style and character.

### 5.1 *Ornament and character of the ancient cities*

Looking at the main ancient civilisations, such as the Egyptians, Greeks and Romans, we see that Egyptians embellished their temples with hieroglyphs and ornamented their buildings with pictures and sculptures of gods, which were inspired from the surrounding nature as represented in humans, animals and plants. Greeks and Romans celebrated the human form, harmony and proportion, which they represented in sculptures, ornamented columns and facades. All of these items formed the character and style of their respective ancient cities, such as Athens and Thebes.

### 5.2 *Ornament and character of the medieval cities*

Islamic civilisation can be considered one of the brightest of the medieval era. However, Islam forbade likenesses of the Prophet and people, but imposed no restrictions on other forms of decorative art. Buildings such as the Alhambra in Granada, in southern Spain, and the Topkapi Palace in Istanbul are among the most beautiful in the world. In addition, the Islamic urban fabric represented a form of organic ornament, which appeared in the unity of courts, building forms and the hierarchy of the street networks. Figure 5 shows a sample of the urban fabric of an Islamic city, suggesting an ornamental organic fabric characterised by harmony and unity of scale.

### 5.3 *Ornament and city character through the Renaissance era*

During the Renaissance era, artists, architects and engineers, especially in Italy, rediscovered a range of ideals and architecture from the classical world, from Greek and Roman to the Gothic style. They tried to reveal it in their buildings' facades, openings and internal

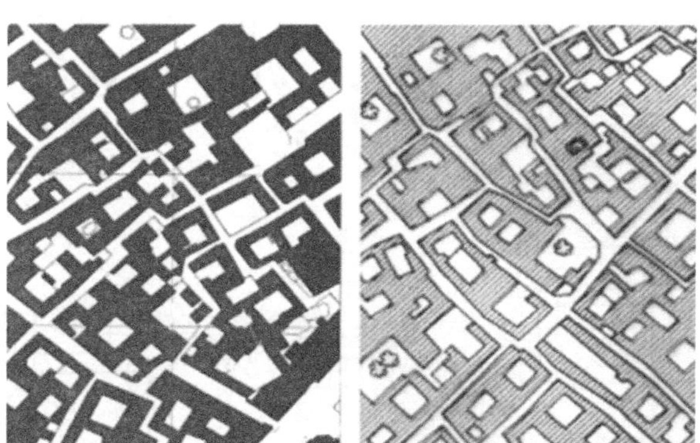

Figure 5. Samples of the urban fabric of an Islamic city (Researchgate, 2013).

decorations. Not only that, but they applied it widely in open spaces such as plazas, approaches, streets and gardens, all of which gave Renaissance cities a unique new classical character. Figure 6 shows many examples for the ideas of the Renaissance cities which resembles a geometrical ornamental fabric.

### 5.4 *Ornament and city character through the twentieth century*

By the turn of the twentieth century, many architectural trends called for a new aesthetics based on a new vision of using ornaments. 'Art Nouveau' was one of these trends, in which building facades formed floral pictures through the use of sculpture and ornament. In the 1930s, Art Deco flourished, characterised by geometric shapes, streamlining and unrestrained ornamentation. New York City's Chrysler Building is the most famous example (Pittsburgh, 2009, p. 14).

### 5.5 *Ornament and city character during the era of modernism*

Following World War I and the loss of almost an entire generation, Europe sought a break from tradition. Architects and designers eschewed decoration, which they saw as a hangover from the old world, and sought purity of form. Practitioners of the German Bauhaus movement were among the foremost exponents of this philosophy. Nevertheless, their buildings are by no means devoid of character. The materials themselves, the careful juxtaposition of forms, and strong simple lines make the best modernist buildings from this era somehow timeless (Pittsburgh, 2009, p. 15).

Architectural ornament, the art of decorative patterning, is commonly perceived as an historical characteristic of architecture that declined in the era of modernism. This modernist emphasis on unadorned form, combined with the upcoming international style and the replacement of craftsmanship by the rise of mass production, yielded a systematic elimination of ornament (Picon, 2011).

The resulting relationship between architects and the public has sometimes become confused. Architects are expected to employ current engineering, materials and technologies in the creation of "beautiful" design; however, the general public most often reveres the ornamented characteristics of historical architecture. New and improved are not always considered 'good architecture' by the common man. A frequent critique of modernist or standardised architecture is that it is 'boring', or lacks character. By contrast, historical, religious and cultural architecture embellished with ornamentation expresses character and

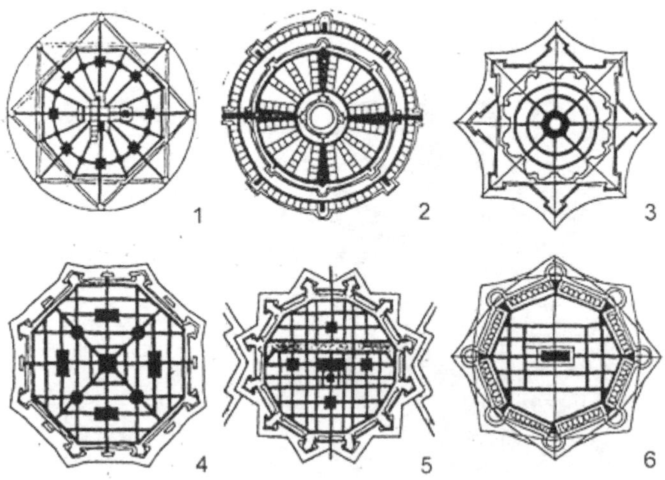


Figure 6. Examples for many ideas of the Renaissance cities (Quadralectic, 2013).

can reveal the relationship between advances in technology and materiality and the resulting evolution of that society. Above all, the ornamental narrative is visible, impressive, and appreciated by the public.

## 6 ORNAMENT AND TECHNOLOGY IN THE TWENTY-FIRST CENTURY

In the twenty-first century, contemporary architecture experienced a great technological breakthrough that forced a radical transformation in the whole architectural field, eventually creating a revival of the use of ornament. Many patterns are generated by software and transformed into, or applied to, construction elements. Complete facades are covered with these patterns, complex forms are generated in order to stand out, and buildings are subjected to experiments with new technologies, concepts or structural challenges.

It is remarkable that ornamental interventions that respond to the shape of a building, as well as the city fabric and character, have been changed radically using these new tools, methods, and techniques. The ideas of seamlessness and fluency have become the current paradigms for the exuberant use of ornament in the digital age. The integration of Computer-Aided Design (CAD) and Computer-Aided Manufacturing (CAM) has allowed the ready introduction of the concepts of scripture, algorithm, morphology, deformation, distortion, evolution, formation, mutation, generation, transformation and variation (Loveridge & Strehlke, 2006, p. 35).

### a. *Role of digital technology in transforming the concept of ornament*

According to Loveridge & Strehlke (2006), it can be argued that the new digital technologies are challenging the traditions of architectural design methodology, the relationship between context and design, and the dependency on skilled workmanship for the fabrication of beautiful and complex architecture. Intellectually, applications of digital technologies are also allowing for the reinvestigation, reinterpretation and redevelopment of historical concepts, theories and skills (Loveridge & Strehlke, 2006).

Consequently, digital technologies can play a key role in the generation of intelligent and responsive patterns that would enhance building designs so that the city fabric can become more complex and interactive. In addition, 3D computer modelling and scripting tools can now define very complex formal language and facilitate mass customisation processes, from Computer Numerically Controlled (CNC) milling and laser cutting to interactive systems and robotics.

### b. *Lighting technology for new virtual ornaments*

Within the twenty-first century, lighting technology has undergone many developments, especially when combined with the new digital technologies, which feature strongly in many architectural applications such as the media facades that basically depend on combined systems of LEDs, projected lights and digital programming systems to give building facades new virtual ornamental effects, changeable according to the programming system. Figure 7 shows how the digital lighting systems (projections, screens or LED media facades) in Times Square, New York, gave the buildings' facades as well as the city's character a new ornamental and changeable digital image.

### c. *Economic constraints of the new technologies*

The character of the contemporary city is more limited by economics than by design or style. Developing more efficient concepts, methods and technologies for the process of design-fabrication-construction will foster a rational and creative approach to ornament and architecture (Loveridge & Strehlke, 2006), combining the new concepts of design and new technologies with historical and cultural aspects.

Figure 7. Digital lighting systems' effects on the facades of buildings (Jeremy, 2011).

## 7 NEW ORNAMENTS FOR A NEW CHARACTER

Nowadays in contemporary architecture we see a revival of the use of ornament. But this has tended to be patterned ornament rather than the traditional sculptured ornament. Thus, pattern in its broadest sense has been included in this research into ornament, as it consists of aspects that may be described as sequential, dynamic, configurative, informative, performative, morphogenetic and parametric. Many new terms have been added to the vocabulary of ornaments, such as 'pixelisation, porosity, fractal, digital, and virtual' (Balik & Allmer, 2016). On this basis, we can classify the new ornamental trends and their impact on city character into three main categories:

- Patterned facades: the new contemporary architectural ornament tends to utilise patterns rather than individual forms in the traditional sense. It is also much more superficial, or rather surface-bound, than traditional ornament, which might take the form of carvings jutting out from the plane of the wall. The new ornament also seems to form the actual skin of the building. It may have a textured effect, but regardless of the form it takes, it essentially stays on the surface of the wall without ever breaking away from it (Picon, 2011). Figure 8 shows how Alibaba's headquarters buildings are linked by an ornate latticed roof that wraps around their exterior and serves as a sun shade.

Consequently, Contemporary ornament can also exhibits a tactile quality, as if its function were to encourage the viewer to go beyond the visual experience and literally caress the surface. This tactile quality is present in many projects, sometimes to the point of being a caricature.

- Fractal ornaments: the term 'fractal' describes any of a variety of extremely irregular curves or shapes, of which any suitably chosen part is similar in shape to a given larger or smaller part when magnified or reduced to the same size (Fractal, n.d.).

Nature is full of fractal ornament; it always involves a great deal of irregular complex shapes that are oscillating and undergoing change, such as leaves, mountains, clouds, or the distribution of galaxies in the universe.

Fractal ornaments are characterised as rough or fragmented geometric shapes that can be split into parts, each of which is (at least approximately) a reduced-size copy of the whole form. Consequently, thanks to the new digital technologies and their great impact on architectural theory and technique, a new cascade of fractal forms as well as fractal ornament has emerged as more dynamic, flexible and nonlinear (Jencks, 2008). Figure 7 shows a digital sketch of the Kartal–Pendik masterplan for Istanbul, which was designed by Zaha Hadid Architects. The designers applied the concept of digital fractals to shape an entire masterplan for the city that gives it a new digital character.

- Virtual ornaments: the concept of virtual ornaments is a concept of virtual reality, which is defined in the online Cambridge Dictionary as 'a set of images and sounds, produced

Figure 8. Ornate latticed roof and facades connect the Alibaba headquarters buildings (Inhabitat, 2011).

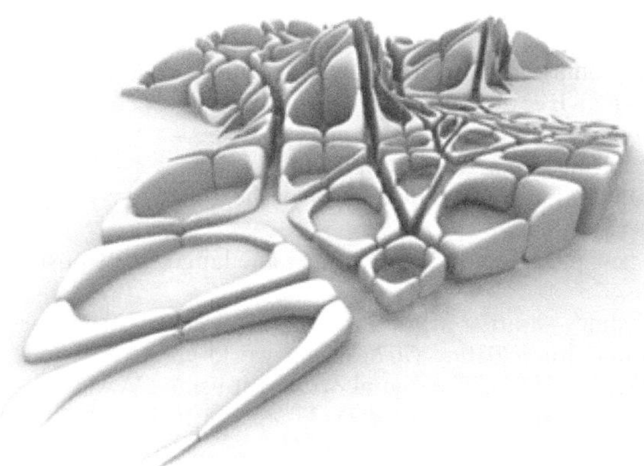

Figure 9. Fractal ornamental fabric in the Kartal–Pendik masterplan (Zaha-hadid, 2016).

by a computer, that seem to represent a place or a situation that a person can take part in' (Virtual reality, n.d.). It can also be argued that virtual ornament is one of the most unique achievements of digital lighting technologies, whereby it can be applied to any part of a building as well as the city, whether it's a new city or an old one; it can change the character of the building facades and thus the entire city.

## 8 CONCLUSION

This research set out to clarify the concept of ornament and how it contributed to the creation of the character of cities through the ancient and medieval eras, the Renaissance and the era of modernism. Based on the above, the following conclusions can be drawn:

- Ornament contributes to forming the character of a city and reflecting its culture and its moral values in many ways and through techniques such as:
  - Forming the urban character
  - Adding symbolic meanings
  - Forming the urban fabric character
  - Connecting regions and buildings
  - Working as transition zones
- The last decade has continuously highlighted that ornament in contemporary architecture has a new definition and aspects that can be described as follows:
  - Sequential
  - Dynamic

    - Configurative
    - Informative
    - Performative
    - Morphogenetic
    - Parametric
- The vocabulary of ornament has been expanded to include many new types of ornamental types and techniques, which can be classified into three main types:
    - Patterned facades
    - Fractal ornaments
    - Virtual ornaments

## REFERENCES


Balik, D. & Allmer, A. (2016). A critical review of ornament in contemporary architectural theory and practice. *ITU A|Z, 13*(1), 157–169.

Formakers. (2012). Urban field adabtive urban fabric. Retrieved from http://www.formakers.eu/project-587-uto-ursula-frick-thomas-grabner-urban-field-adaptive-urban-fabric.

Fractal. (n.d.). In *Merriam-Webster Dictionary online*. Retrieved from http://www.merriam-webster.com/dictionary/fractal.

Inhabitat. (2011). Alibaba headquarters wrapped in a Spider web—like solar shade. Retrieved from http://www.inhabitat.com/hassell-architects-alibaba-headquarters-bring-a-sense-of-community-to-an-office-complex/.

Jencks, C. (2008). The new paradigm in architecture: The language of post-modernism. London, UK: Yale University Press.

Jeremy. (2011). United states—tourist attractions: Times Square—New York. Retrieved from https://www.theguardian.com/culture/2010/nov/07/ancient-world-greece.

Loveridge, R. & Strehlke, K. (2006). The digital ornament using CAAD/CAAM technologies. *International Journal of Architectural Computing, 4*(1), 33–49. doi:10.1260/147807706777009000.

Ornament. (n.d.[a]). In *Oxford Dictionary online*. Retrieved from https://en.oxforddictionaries.com/definition/ornament.

Ornament. (n.d.[b]). In *Cambridge Dictionary online*. Retrieved from http://dictionary.cambridge.org/dictionary/english/ornament.

Ornament. (n.d.[c]). In *Merriam-Webster Dictionary online*. Retrieved from https://www.merriam-webster.com/dictionary/ornament.

Ornament. (n.d.[d]). In *Encyclopaedia Britannica online*. Retrieved from https://www.britannica.com/technology/ornament.

Picon, A. (2011). Ornement et Subjectivité, de la Tradition Vitruvienne à l'âge Numérique. *Le Visiteur, 17*, 65–75, 176–180.

Pinterest. (2016). Las galaxias del universo. Retrieved from https://www.pinterest.co.uk/explore/las-galaxias-del-universo/?lp = true.

Pittsburgh university. (2009). modernism, ornament, reform. University of Pittsburgh press: 14–15.

Preservation Alliance for Greater Philadelphia. (2007). Sense of place: Design guidelines for new construction in historic districts: 4–5 Retrieved from http://www.preservationalliance.com/publications/SenseofPlace_final.pdf.

Quadralectic architecture. (2013). Ideal cities from the Renaissance. Retrieved from https://quadralectics.wordpress.com/4-representation/4-1-form/4-1-4-cities-in-the-mind/4-1-4-1-the-ideal-city/ Research gate. (2013). Old Islamic urban pattern examples. Retrieved from https://www.researchgate.net/figure/259169693_fig6_Old-Islamic-urban-pattern-examples.

Salingaros, N.A. (2000). Hierarchical cooperation in architecture, and the mathematical necessity for ornament. *Journal of Architectural and Planning Research, 17*, 221–235.

Stefano's romecabs. (2016). The Trevi fountain. Retrieved from http://www.romecabs.com/tours/rome-in-a-day/.

The guardian. (2010). Greece: Birthplace of the modern world. Retrieved from https://www.theguardian.com/culture/2010/nov/07/ancient-world-greece.

Virtual Reality. (n.d.). In *Cambridge Dictionary online*. Retrieved from http://dictionary.cambridge.org/dictionary/english/virtual-reality.

Zaha-hadid. (2016). Zaha hadid parametric design. Retrieved from http://www.zaha-hadid.com/wp-content/files_mf/1245_karta_rend_03.jpg.

*History and arts*

*Cities' Identity Through Architecture and Arts – Catalani et al. (Eds)*


# Arabian urban text art: Between cultural identity and artistic identity

Shereen Alharazy
*College of Art and Design, King Abdul-Aziz University, Kingdom of Saudi Arabia*

ABSTRACT: This research is an analytical study of the aesthetics of the city's art in the form of written text, such as calligraphy, typography, graffiti and other types of art that involve the use of Arabic letters. This research also studies the impact of this art on Arabian cities by examining how they are influenced by intellectual and religious thoughts and by studying the impact of art on society.

In the present era, Arabian cities are drifting away from their local identity and have started to take an international path that is influenced by existing ideas, elements, volumes, typographies, ornamentations, colours and materials that coincide with globalisation. Given that Arabs have calligraphy that can be used to protect local identity and can be contemporised, this issue can be ameliorated by portraying the beauty of local identity through calligraphy and using it to enrich Arabian cities and protect their cultural identity.

This study hypothesises that using urban text art in Arabian cities reveals their beauty and the richness of their heritage and helps in maintaining their local identity. It is also considered to be one of the only art styles that uses calligraphy and carries a significant and pure ideology associated with Arabian culture.

Thus, highlighting the importance of urban text art in Arabian cities is a way of ensuring that a city's national identity and specificity can continue to exist. Arab nations are in dire need of this, especially in light of the challenges they face, as it will enable them to adhere to their culture and traditions. This is considered an urgent requirement to help these nations defend and maintain themselves because their cultural heritage is still suffering from neglect, given the lack of awareness among people. It is also important that urban text art be integrated into the educational system, as it constitutes a coherent whole in terms of the knowledge and goals on which it is built.

*Keywords*: Arabic letters; Arabian cities; local identity; typographies; heritage

## 1 INTRODUCTION

The development of urban text art has gone through many different phases that coincide with the emergence of cultural and social evolution. Text art is considered one of the most important types of visual art because it involves rules and characteristics of excellence and uniqueness not seen in other language arts.

Since ancient times, people have documented their history, times and events on the walls of caves and ancient temples, wooden panels, animal bones and skin, and paper in different parts of the Arab and Islamic world. Writings on walls have been found in various languages, including Arabic, Turkish, Farsi, Indian and African.

This original art is linked to the unique Arabian culture and language, which has overthrown other art styles, even though times have changed and the existence of technology has made Arabic calligraphy available all over the world.

Urban text art types differ from other types of art seen in streets and outdoor structures for many reasons, including the cultural experience of the people who see it and regard it as a primary part of the old and modern city. These art forms are also part of an Arab country's identity, and can be found on banners and walls in the form of random love and hate phrases. All these art forms provide each Arab country with its own signature, leading many visitors to tour city streets and enjoy these text art pieces in terms of font, colour and meaning (2).

This study deals with one of the most important characteristics of Arabian cities, urban text art, which began to lose its identity given the interference of modern technology and the effects of modernisation and globalisation that damaged the identity and independence of Arabian cities. Urban text art is no longer seen as part of the heritage and identity of the city but as an aspect of pollution and an indication that the city is underdeveloped (8).

Many examples of neglect indicate the absence of vision, the continued marginalisation of the Arab city, and the disregard of its roots in attempts to keep up with the development of civilisation. This is done without considering the aesthetic and geographical identity of Arabian cities.

## 2 CONCEPTS OF TEXT ART: CHARACTERISTICS AND OBJECTIVES

Every nation is proud of its heritage and considers it a connection to its history. Cultural heritage encompasses aesthetic, historic, scientific, social, economic and spiritual values of the past. The role it plays in Arabian cities shows the continuous need to assess its importance and study it as the present.

As the main topic of this research talks about the Arabic font, a brief listing of other Arabian font styles must be included. The art of writing has characteristics, rules and types that make it stand out among other visual arts. Urban text art is divided into two main types depending on the font used in writing, as described in Sections 2.1 and 2.2.

### 2.1 *Traditional Arabic fonts*

The evolution of Arabic letters has been a long journey, as at first they were simple and basic but later changed to more complex forms with the development of society and the emergence of Islam.

Later on, fonts of Arabic letters were named after Arabian countries, people, or the pens they were written with. The lines and the fonts started to overlap, and some of them were derived from each other, until calligraphers were able to invent new lines, fonts and styles.

Arabic fonts continued to develop over time and new modern fonts were invented, which were free in style and had no rules. Some of the most important types of calligraphy are as follows:

a) Al-Tholth

b) Al-Kufi

c) Al-Naskh

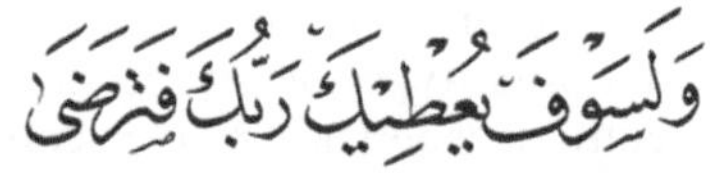

d) Al-Diwani

e) Al-Rukaa

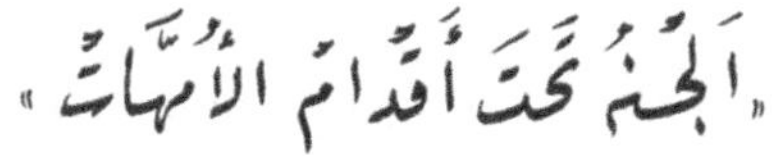


f) Al-Andalusi

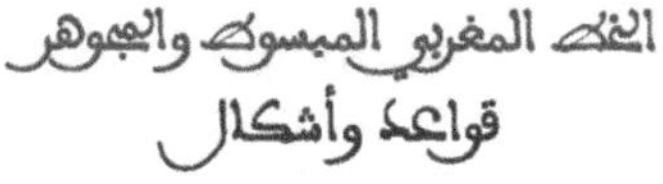


g) Al-Farsi

## 2.2 *Modern Arabic fonts*

The evolution of Arabian calligraphy has continued in the modern era, with calligraphers inventing patterns and new dimensions until the Arabic font became art with its own elements and characteristics. Additionally, the appearance and development of computers as the new writing method led to the emergence of the following fonts:

a) Free modern font

b) Egyptian Reqaa

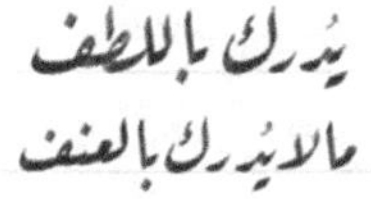


c) Computer font

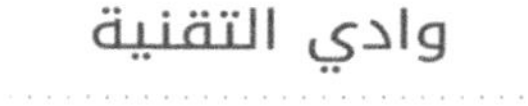


The aim of urban text art is to emphasise expressive writing. While following urban text art in all its forms, shapes and cultures, we find an interesting journey filled with lines and colours that are drawn all around the streets of Arabian cities. Some of them tell a story about heritage and some show a place or product. Some even express sorrow, grief, the hardships of life, or the beauty of love.

The art of writing in all its forms represents creativity in one way or another, which manifests itself on walls, cars, or store signs and is connected to the geographical identity of a place or city.

## 3 FEATURES OF URBAN TEXT ART: PATTERNS AND TECHNIQUES

Text art in all its forms represents creativity and excellence that is shown wherever it appears and is much related to the geographical identity of the place or city, and the artistic identity of the society. Sometimes we find theses writings on cars as a way to confront the unknown or to express the thought of its owners. And other times we find them on public walls as a way of encouragement or guidance. Sometimes, this type of art appears on top of stores in different fronts, all of these writings reflects meanings driven from spatial and geographical identity (5). The thing that makes each text art stand out, depends on the writer, whether he is an artist or a hobbyis. The text arts are divided into three different expression styles according to the method and the type of writing, as discussed in Sections 3.1–3.2–3.3.

### 3.1 *Calligraphy*

Calligraphy is the art of writing and drawing using Arabic letters and decorating them in geometric forms using design elements such as harmony and consistency in lines and curves that have an Eastern characteristic. It is known as the Islamic Arabic script and is considered to be one of the oldest types of handwriting. This type evolved along with Islamic architecture and ornamentation and managed to maintain its identity and beauty. It continues to develop with the evolution of technology, where design programs and software are used to create these types of visual arts. Pens and paintbrushes are the most widely used tools for calligraphy (1).

### 3.2 *Typography*

Typography is the adjustment in font to form a composition of letters depending on the content. It is often designed using software such as Adobe Illustrator.

### 3.3 *Graffiti*

Graffiti is the art of drawing on walls, especially street walls, using special tools in an artistic way, often representing contemporary words and free expressions. This kind of art appeared in the early 1970s when it was associated with hip-hop music in New York, and was known as 'lower-class art'. However, this label did not last long as the middle classes began to express and rebel in the same way. Graffiti is of two types: (a) the written and (b) the signed. One of the most important tools that graffiti artists use is a spray can and/or a graffiti pen (3)..

When digging deep into studies of urban text art and its forms, we find that all its types share some of the characteristics that resonate with the reader and observer. It targets individuals and communities, and it must catch their attention in a few seconds or minutes. It should be catchy enough to attract a person passing by or someone stuck in traffic. Therefore, urban text art must leave an impression that makes people think about what is written: the words written must be clear, consistent in colour, notable and impressive.

## 4 VISUAL CRITICAL STUDY OF URBAN TEXT ART IN ARABIAN CITIES

Urban text art and calligraphy has taken control of visuals in Arabian cities, and they are nearly impossible not to notice. This art defined the visual character of these places and spread across billboards and signs, walls, floors and buildings.

According to sociological and anthropological studies, artistic identity is considered a principle of harmony and conformity. On the social level, artistic identity plays a big role in traditional culture because it is the base of creativity and a method of personal expression.

These countless writings with all their colours and forms have become a tourist attraction in Arabian cities. They have also shaped the identity of these cities (9).

There is no doubt that urban text art, in all its forms, has become a part of the Arab city's landmark that fascinates people and draws in tourists. In order to recognise the geographic identity of calligraphy and its relationship with Arabian cities, it is necessary to study and analyse the social, economic, intellectual and religious bases and factors that have a clear role in the formation of this kind of art. The following section includes examples of one of the most important functional and aesthetic subjects affecting the diversity of urban text art forms. This art became a symbol of the geography of a place throughout different eras (4).

### 4.1 *Folk heritage*

Folk heritage and customs play an important role in urban text art, as they have inspired well-known authors to write popular, well-established phrases that are part of the Arab identity and have been inherited by generations. These widespread ideas and phrases are reflected in the form of comic writings or popular songs, followed by popular expressions and warnings. These are based on religious and social traditions (Figure 1), with other sentences being based on the cultural heritage of the community (Figure 2). These folk thoughts can be reflected in the form of comic writings or folk song writings (Figure 3), followed by folk expressions that are sometimes written on the back of vehicles, for example, expressing the driver's philosophy in life (Figure 4). Some of the sentences written on walls reflect the values and methods of thinking. Popular folk phrases express injustice, cruelty, lost love, anger and envy (Figure 5), and call for patience, surrender, or the need to be near protectors. Some call for mercy, love and cooperation, while others reflect religion, virtue and the acceptance of destiny (Figure 6). All these phrases represent cultural characteristics that are deeply rooted in history.

Figure 1. Egypt.

Pilgrimage is justified and guilt is forgiven

Figure 2. Saudi Arabia.

Eid and winning

Figure 3. Egypt.

Take care of Zuzu

Figure 4. Egypt.

Be with god and god will be with you

Figure 5. Egypt.

Hay evil people enough

Figure 6. Saudi Arabia.

Saying: no god but allah is better than anything in the world

Figure 7. Yemen.

Excuse me I am human

Figure 8. Saudi Arabia.

There is a light never fades

Figure 9. Jordan.

We love life as much as we can

## 4.2 *Expression of social problems*

This type of art is expressed using succinct phrases that are often written in the local dialect and vary according to the nature of each society and the level of suffering (Figure 7). It is represented in written slogans and is used to express a reality full of sorrow and sadness (Figure 8). There are many writings that show deep grief, an unfulfilled dream, or injustice (Figure 9). They also convey other issues in the lives of citizens such as the admission of love (Figure 10), advocating and supporting a particular team, or expressing daily preoccupations such as housing and work (Figure 11). These writings often turn into graffiti paintings that carry a social, cultural and political message (Figure 12).

Figure 10. Jordan.

Going up come down and don't be arrogant

Figure 11. Sudan.

Flavor defying time

Figure 12. Palestine.

I will not go back until I plant in my garden

Figure 13. Egypt.

Groppi

Figure 14. Jordan.

Alshater cooctail

Figure 15. Algeria.

Barber of fashion

Figure 16. Morocco.

Fanta

Figure 17. Dubai.

Don't waste our time and take us quickly to alzaroob

Figure 18. Kuwait.

Fruition cookie

### 4.3 *Publicity and advertising*

Paintings used for advertising are an important source of income for a country and they enhance its general appearance. These paintings are linked to the status of individual cities, urban planning and population concerns. Billboards significantly contribute to the formation of the aesthetic culture of the city because of the bright colours used (Figure 13). Some could say that Arabian cities suffer from a chaos of advertisements and random writings that result in severe visual contamination; however, one could view an Arabian city as an art gallery rather than a city that lacks identity. Looking at different billboards and advertisements, we see pictures of products displayed for sale (Figure 14), whereas others indicate the name of a shop or the services it offers (Figure 15). Some signs have names that will catch the atten-

Figure 19. Egypt.

Do not forget to mention God

Figure 20. Bahrain.

Say: "Are those equal, those who know and those who do not know

Figure 21. Saudi Arabia.

Motherhome in our hearts

Figure 22. Bahrain.

Love you Bahrain

Figure 23. Saudi Arabia.

High morals

Figure 24. Algeria.

We will declare war on drugs

tion of consumers (Figure 16), whereas others choose funny names (Figure 17). We can also see signs that localise foreign names (Figure 18).

### 4.4 *Guidance and advice*

A lot of the art that is found around Arab streets represents guidance and religious advice, such as always remembering God's name, 'Allah' (Figure 19). There are also some of the verses from the Book of Allah, the Qu'ran (Figure 20), or phrases that emphasise patriotism (Figure 21) and the raising of the national flag (Figure 22) accompanied by the King's image. Some of the art aims to spread a culture of awareness, respect and the ethics of national development (Figure 23) or discuss social, behavioural or economic problems and situations (Figure 24). Art also offers guidance in psychological, moral and educational aspects so that an individual might become a better-rounded member of society.

## 5 CULTURAL IDENTITY OF URBAN TEXT ART

Cultural identity indicates uniqueness in everything that is connected to culture, including habits, behavioural patterns, values and visions for life.

The cultural identity of a nation is an essential and common feature that makes the national character a unique one among all other national characters and figures.

Urban text art that involves calligraphy has always been an essential pillar of cultural identity, and it stimulates a sense of pride. Calligraphy has always inspired artists, poets and philosophers and, eventually, it has ended up being a cultural heritage map that connects the nation to its past and reinforces its presence in the global arena (6)

The definition of heritage includes every concept that relates the history of human experiences in the past, present and future. Cultural heritage is the legacy left to us by our ancestors. Its ties could be financial or non-financial bonds with the current people.

There is no doubt that urban text art represents a living memory of the individual and society, and therefore represents the identity of races that can be easily recognised by people across the globe. This cultural heritage that we aim to maintain has a positive relationship with creativity between individuals and communities. As each race has its own heritage that is passed on verbally or through common practices, text art resulted from the interaction between individuals and the community with the surrounding environment through the ages. With the progression of time, it turned into an art form that holds the experiences of individuals and groups and is a source of existence and continuity (7).

The loss of cultural heritage across Arab cities implies a loss of identity. Cultural heritage, as known to researchers and specialists, has two sides:

1. Tangible objects or surfaces produced by former writings and graffiti on buildings and around cities.
2. Non-tangible bonds that include beliefs, customs, rituals and traditions, which are known as traditional heritage that artists and calligraphers use in their work.

Maintaining these two elements means maintaining a nation's identity and history. They also represent the nation's civilisation.

## 6 CONCLUSION

Today, in light of modern scientific principles that highlight the importance of preserving cultural heritage and shed light on the role it plays in the formation of the basic components of any Arab country, cultural heritage in Arab cities must be studied and analysed more closely.

### 6.1 *Results*

Given the importance of studying art and cultural heritage, the following conclusions have been drawn:

1. Planning the cities according to its identity without disregarding its heritage and culture.
2. Maintaining urban text art and studying it will enrich human culture and preserve the cultural diversity of the Arab people.
3. The study of urban text art is an educational, scientific, artistic and social source of information.
4. Urban text art has, in places, become a tourist landmark and a source of attraction for people, and also a source of income in many Arab countries.
5. The removal and loss of this type of art means a loss of identity and a lack of cultural, social and economic development in Arab countries.

## 7 RECOMMENDATIONS

In an attempt to shed light on the positive aspects of urban text art, the following recommendations and observations are made:

1. It is necessary to deal with artistic and cultural heritage as an integral part of the present.
2. Urban text art reflects and also forms the Arab identity.
3. Calligraphy used in urban text art in its various forms is a reflection of the type of visual art associated with the aesthetic vision of a distinct geographical identity.

4. Reconsider the urban planning of modern cities without ignoring the identity of Arabian cities, and develop this identity on all functional, structural, aesthetic and environmental levels in a balanced manner without marginalising the role of aesthetic text art.
5. Avoid planning ideas that are far removed or different from the local environment and cultural traditions; rather use ideas that have distinct historical and cultural patterns.
6. Focus on urban text art as part of the cultural identity.
7. Conduct intense educational campaigns for citizens and encourage them to reconsider the environment and local culture instead of regarding urban text art as environmental pollution.

## REFERENCES


[1] Abdel Salam, A. (2002). *Encyclopedia of Arabic calligraphy*. Amman, Jordan: Dar Osama Publishing.

[2] Al-Aiderous, A.H. (2011, 6 April). Identification place & place of identification. *City Publications*. Retrieved from http://www.al-madina.com/article/78238/

[3] Al-Harbi, A. (2017). *Graphite art*. Riyadh, Saudi Arabia: Obeikan Publishing.

[4] Al-Saqr, I.M. (2009). *Graphic design*. Amman, Jordan: Dar Al Ahlia for Publishing and Distribution.

[5] Bellon, D. & Bellon, K. (2010). *Typography for the people: Hand-painted signs from around the world*. Avon, MA: HOW Books.

[6] Berer, J. (2008). Arabic calligraphic design. Retrieved from http://www.arabiccalligrapher.com/

[7] Reza, A. (1986). *Letter calligraphy*. Cairo, Egypt: Publishing Encyclopedia of Art History and Architecture.

[8] Villagomez, N. (2015). *Culture + typography: How culture affects typography*. Blue Ash, OH: PRINT Books.

[9] *Zaman Al Wasl Newsletter*. (2011, 1 June). Visual pollution threatens the oldest inhabited city in history. Retrieved from https://www.zamanalwsl.net/news/19742.html.

*Cities' Identity Through Architecture and Arts – Catalani et al. (Eds)*


# The remaining houses of the two pearl merchants Ahmed Munawar Rifai and Hussein Bin Yahiya Rifai in the Farasan Islands in the 14th century AH/20th century AD: A study of aesthetic and artistic values

Eman Ahmed Aref
*Art History Department, Faculty of Fine Art, Helwan University, Helwan, Egypt*

ABSTRACT: The houses of the two pearl merchants Ahmed Munawar Rifai and Hussein Bin Yahiya Rifai in the 14th century AH/20th century AD, in the Farasan Islands, south-west of the Kingdom of Saudi Arabia, witnessed a cultural shift and great development in the processing of stucco art decorations. This study discusses the phenomenon of artistic enrichment, its mingling with the local heritage arts, its investment according to common sense, and the rationale behind using some elements of Indian–Buddhist architecture in the Al Najdi Mosque. This study aims to explore the factors that helped in distinguishing the Farasan stucco art style from the contribution of other artistic styles. The Ottoman arts accompanied their actual presence in the region, by developing the units of Farasan stucco art, in spite of the isolation of the Farasan Islands. Through a descriptive, analytical method, this study links historical events to show that the Farasan Islands are considered one of the shipping routes linking the Far East with the Far West. This helped in the mixing of local and imported cultures, as reflected in the appearance of the unique Farasan stucco style.

*Keywords*: Ottoman patterns, engraving stucco, Farasan Islands, Rifa'i house, Al Najdi Mosque in Farasan, heritage of Farasan

## 1 INTRODUCTION

Before the advent of Islam, commercial trips around the world extended from the Far East to the Far West where the arts of China, India, Iran, and Middle Asian civilisations mixed with the arts of the Middle East and Western civilisations in Africa and Europe.

With the centralisation of pearl fishing and trading in the Arabian Gulf and Farasan Islands, south-west of the Kingdom of Saudi Arabia, and the presence of the Asir port,[1] where 'various collections of blue and alkaline green porcelain, which was known in the Abbasid period, were found in their ruins, as well as models of porcelain, ceramic and stone pottery imported from China' (Alzele'i, AH 1423). This helped in the emergence of the Farasan stucco art decoration style. The façade and entrance walls of the houses of the two pearl merchants Ahmed Munawar Al Rifa'i and Hussein Bin Yahiya Al Rifa'i are considered as important heritage monuments, where most of the emerging influences on the art of wall processing with stucco decorations through the Islamic eras converged.

The style of decorations on the façades of religious and residential buildings was not common in all parts of the Kingdom of Saudi Arabia, in accordance with the Islamic teachings of simplicity and abstraction. However, the heritage Najdi style of the region offered a fertile field for the mingling and mixing with the arts of new cultures. Stucco decorations

1. Built in the Umayyad era, it is located 40 km from Jizan City on the coast of the Red Sea, south-west of the Kingdom of Saudi Arabia. Asir port is considered one of the most important active points of communication with the African coast, Asia and China.

developed at the beginning of the 14th century AH/20th century AD. At first, stucco decorations adorned the frames of apertures and windows with white colour, and were limited to the guest reception halls (coffee sitting areas) and their associated doors. Then, attention shifted from privacy to showing off the lavishness of the building; this stage was accompanied by the foundation of the Jizan Idrisid state, which was interested in decorating its palace façades and gates. After the Ottoman sovereign took over ports of the Red Sea and Hejaz countries, Ottoman taste in art leaked in, which carried in its folds Iranian and Chinese art forms, especially in the Safavid and Timurid eras. This was because Tamerlane hired artists of the Ottoman Empire to work in his kingdom where their artistic creations appeared in Samarkand city, the radiation centre of Timurid art with Iranian–Chinese style dyes.

The Farasan Islands contain more than 500 houses that were characterised by their façade decorations with stucco, such as Sayer village Castel, Al Najdi house, Al Najdi Mosque, the two houses of Ahmed Munawar Al Rifa'i and Hussein Bin Yahiya Al Rifa'i, Jarman house, and the Ottoman Castle. Nevertheless, only the two Rifa'i houses and the Al Najdi Mosque had a significant extent of stucco decorations, which helped to show the cultural features and uniqueness of stucco decorations not evident on other buildings.

### 1.1 *Statement of the research problem*

This study discusses the phenomenon of architectural and stucco art enrichment in helping to distinguish the two houses of the most famous pearl merchants and the Al Najdi Mosque in Farasan Islands in the 14th century AH/20th century AD, despite the spread of Yemeni and Najdi styles and their closeness to Sebia city, the centre of the Jizan Idrisid state. However, it is difficult to gather enough information about these houses due to the rarity of resources and of studies about their artistic influences.

### 1.2 *Research objectives*

This study aims to determine the characteristics and features of the components of Farasan stucco art by studying and analysing stucco decorations of some buildings on the Farasan Islands. It highlights the contribution of the Farasan stucco art style in the enrichment of the arts of excavated walls in the south-west of the Kingdom of Saudi Arabia.

### 1.3 *Research limitations*

#### 1.3.1 *Topic limitations*

The contribution of the trade routes and pearl trading to the enrichment and development of stucco decoration art in the Farasan Islands.

#### 1.3.2 *Geographical limitations*

The Farasan Islands are located off the south-west coast of the Kingdom of Saudi Arabia, in the Red Sea.

#### 1.3.3 *Time limitations*

The study covers the period of the 14th century AH/20th century AD.

### 1.4 *Research assumption*

The most famous pearl merchants hired the best artisans from many countries in the 14th century AH/20th century AD to build and decorate their houses in the best artistic and architectural styles.

### 1.5 *Research hypothesis*

The study examines the impact of geographical and geological factors on the culture and the ways of thinking of the wealthy people of the Farasan Islands, which is reflected in the decorative wall art and the conversion of natural raw materials into a precious decorative form.

### 1.6 *Research methodology*

The study follows the analytical, descriptive method that depends on observation and historical links.

## 2 HISTORICAL PREFACE

Islam arrived in India in the Umayyad Caliphate period (1st century AH/7th century AD), and it was deployed effectively by Qutayba Ibn Muslim Al Bahli (AH 49/AD 669 to AH 96/AD 715), the prefect of Khorasan under the reign of Al-Hajjaj Ibn Yusuf Al Thaqafi during AH 86/AD 97. The Umayyad were followed by the Abbasid, who ruled Transoxiana (Uzbekistan—Tajikistan—Turkmenistan—Kyrgyzstan—Kazakhstan). As a result of the weakness of the Abbasid Caliphate, some states, such as the Samanid Empire (AH 261–389), obtained their independence. The most prominent of succeeding rulers/states were the Seljuk Empire in the 5th century AH/11th century AD, 'the Khwarazmian Empire that was pulled down by the Mughals at the time of Genghis Khan, then the Timurid Empire that was founded by Tamerlane, prince Targai's son 771 AH'/1369 AD (Amer, 1998, p. 119), 'and then [they] were pulled down by the Uzbeks in the 10th century AH (16th century AD) led by Shaybani Khan, who founded the Mughal Empire in India' (Mones, 1983, p. 244), and conquered Bukhara, Samarkand and Tashkent cities. Despite these conflicts, the architectural movement was not affected adversely, but it was more enriched. On the other hand, the commercial movement succeeded on the 'silk route that was there before the advent of Islam (4th BC) and which connected the Far East with the Mediterranean Sea' (Leven, 1998) and passed through the Arabian Peninsula until the British invasion in 1857 (13th century AH/19th century AD).

The two centuries, 7th–8th century AH (13th–14th century AD), are considered a stage of flourishing and dominance of the Mughals over the states and centres of Islamic civilisation, such as China, Iran and Mesopotamia, extending to the ends of Asia. Sultan Muhammed Ghouri's era (AH 602/AD 1205–1290), and especially the era of his deputy Qutb Al Din Aibak (AH 603–688/AD 1206–1290) who ruled after his death and was the first Mamluk sultan from Turkish origins with a Persian culture, witnessed a huge architectural breakthrough in Delhi city, where various Indian cultures mingled and unified with the Arab Islamic culture, especially during the reign of Sultan Ghiyassudin Balban (AH 664–684/AD 1266–1287). Several families, such as Khalgi and Tughluq, succeeded after Sultan Ghiyassudin Balban's reign until India joined the Mughal Empire in AH 932/AD 1526 under Babur Shah's reign (AH 888–937/AD 1483–1530) (Al Nadawi, 2012). Throughout history, cultural, architectural and artistic enrichment grew across sprawling India, producing art with special features, including Persian, Sasanian, Turkish, Iranian and local art features. Muslims were devoted and gave their best in all fields to raise the profile of these faraway countries until they became part and parcel of them. Soon, the notion of the Islamic doctrine moved to India and the number of Muslims increased in Tughluq Abad city. South India became the seat of the Bahmani Sultanate (AH 748–826/AD 1347–1423). The trading routes (i.e. the spice and silk routes) linked the Farasan Islands, located in the territorial waters of the Red Sea, with the Far East Asian countries, which facilitated the transmission of different artistic influences and is reflected in some of the religious and residential buildings under the reign of the Ottoman Empire. The closeness of the trading routes to Eritrea and the Dahlak Islands, which are located near the East African coast, helped in enforcing commercial and cultural relationships between the regions, via a sea lane linking all continents from the Far East to the Far West. In addition, both regions united in the course of pearl trading, selling to India, China and countries of Middle Asia, thus bringing an impact on their architecture that was influenced by Indian and Mughal arts.

Figure 1. The trading routes passing through the Farasan Islands.

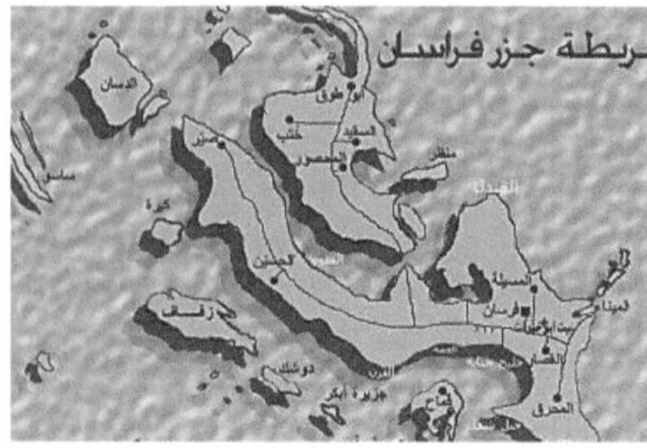


Figure 2. The Farasan Islands.

## 3 GEOGRAPHICAL LOCATION OF THE FARASAN ISLANDS

The Farasan Islands are considered the largest islands in the Red Sea, located off the south-east coast (Figure 2), from where they extended steeply towards the west. They are located to the south-west of Jizan, 30 miles distant, and opposite Eritrea and the Dahlak Archipelago on the East African coast.

The Farasan Islands are characterised by the diversity and breadth of their geography and agricultural lands between hills, valleys and soft sands. Their location has earned them a high strategic status since the early ancient ages, where they were among those islands close to the lanes of the marine fleets that passed through the Bab El Mandeb strait that connects the Arabian Sea, the Indian Ocean and the Red Sea (Figure 1). This revived their economic activity with many countries, such as China, India and countries in the Far East and Middle Asia.

## 4 TRADING AND PILGRIMAGE ROUTES OF THE FARASAN ISLANDS

The Portuguese and Spanish discovery of India and the Far East route in the 10th century AH/16th century AD negatively affected the economy and the trading of countries in the Middle East. The trade routes were also affected by a state of insecurity due to pirate attacks and, thus, the region became a stage for military conflict between the Portuguese and the Ottomans, in order to secure their maritime journeys, during Hadim Suleiman Pasha's time in the 10th century AH/16th century AD. Despite these conflicts, trading continued in the Red Sea, helped by the ease of communication with the countries located in the Atlantic and Indian Oceans. Thus, trading was revived in all parts of Africa through the trade movement among the Nile Basin countries, such as Egypt, Sudan and Abyssinia. The Farasan Islands, with their geographical location on the south-east of the Red Sea and west of the Kingdom of Saudi Arabia, gained commercial, economic and cultural benefits by being located near the two trading routes of the coastal Tihama and the central Sultani avenues, which connected Yemen, Hejaz countries, Mesopotamia, the Levant and the Nile Basin countries. The emerging and local cultures mixed there, producing a unique and distinct type of architectural art and stucco decoration that was not found in the rest of the region. This was particularly due to the abundance of the stucco material formed from the geological composition of the limestone content of the local coral reefs.

## 5 THE OTTOMAN CASTLE IN THE FARASAN ISLANDS (10TH CENTURY AH/16TH CENTURY AD)

Many events took place on the Farasan Islands as a result of the colonial ambitions of Britain, France, Italy and Germany, who made it one of their stores for supplying their ships with coal. However, the Ottoman Empire under the reign of Sultan Abdul Hamid II, which

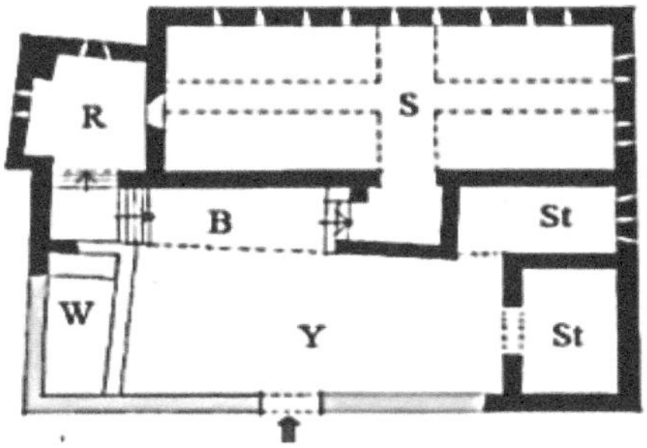

Figure 3. A plan of the Ottoman Castle.

Figure 4. The Ottoman Castle.

extended for more than 30 years 1292–1326 AH/(AD 1876–1909), followed an enlightened policy that enabled the sultan to maintain the islands—rich in natural resources, such as coal, natural gas and pearls—as they were one of the gateways of the trading routes in the Red Sea and the strategic Suez Canal port.

In the 10th century AH/16th century AD, the Ottoman Empire left behind one of its castles in the north region 'between [the] Farasan Islands and the region of Messila and Meshiref' (Saban, 2010), on a hill that gave it a strategic location because it overlooked most parts of the island. The castle was designed in a rectangular form (Figure 3) over an area of 44 square metres, and it was built with the stones and stucco abundant in the Farasan Islands. Its flat roof was covered with palm leaves based on columns made of thick iron bars. The general design was limited to achieving the goal of protection and practical defence of island properties and trade movements (Figures 3 and 4). The castle had a small courtyard on the ground floor, located at the front entrance; to its left was a store believed to be used for keeping supplies and ammunitions. To the left of the courtyard, there was a place for saving water. Next to the courtyard, on the opposite side, there was a stone bench seat and then a small room that was apparently a place for guards. Upstairs, there was a large room allocated for resting soldiers, apertures that revealed parts of the island in many directions, and watchtowers.

## 6 MODELS OF WEALTHY PEOPLE'S BUILDINGS ON THE FARASAN ISLANDS

This research deals with the interpretation of the origins of artistic influences that were considered one of the factors shaping the Farasan stucco decoration art style. Some distinct models were chosen and analysed, wherein their influences could be seen. This is why the Al Najdi Mosque was the starting point despite its later building date of AD 1929 compared to the two houses of Ahmed Munawar Al Rifa'i and Yahiya Al Rifa'i of AD 1922. The origins and sources of the emerging artistic influences clearly met and multiplied in this mosque, which is why it is necessary to start with an analysis of its elements.

### 6.1 *Al Najdi Mosque of Sheikh Ibrahim Ibn Ibrahim Al Tamimi (AD 1929)*

The mosque of Sheikh Ibrahim Ibn Ibrahim Al Tamimi (AH 1336–1386), also known as Al Najdi, is located in the middle of the largest of the Farasan Islands. The mosque was built with mud bricks in AH 1347. Ibrahim Al Tamimi was one of the sons of Hotat Bani Tamim in Najd and one of the pearl merchants who settled on the Farasan Islands to look after their trading. Fishing and diving were among the most honourable professions at that time; thus, after Sheikh Ibrahim Al Najdi obtained the biggest pearl and sold it for a large sum of money, he decided to build a mosque more glorious than any other on the Farasan Islands. He brought manufacturers and raw materials for the mosque's foundation from all over the country, where he travelled for pearl trading. This was reflected in the diversity of the elements of its architectural and decorative structures that were accurately handmade, which gave it a religious dignity and prestigious character.

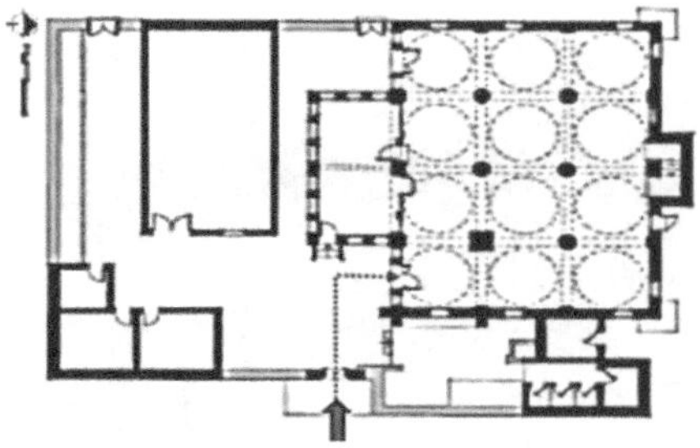

Figure 5. A plan of Al Najdi Mosque (14th century AH/20th century AD) (Mortada, 2014).

Figure 6. Al Najdi Mosque courtyard.

6.1.1 *The mosque design*

The mosque was designed in a rectangular form (Figure 5). In its foreground, there was an exposed *sahn* with a limited exposed area, which almost intertwines with the outer space of the street. It is surrounded by a simple fence painted in white stucco, about 2 metres high (Figure 6). Its bases are built with stones, surmounted by mud bricks. In the fence, there are two entrances, one on the east side and the other on the west, both leading to the *sahn*. There are also remnants of an octagonal minaret base on the south side of the mosque. Next to the east entrance from the inside, there is a place for ablutions (Figure 7). In the middle of the north wall of the *kiblah* from the outside, there is a rectangular salient protruding structure, which is an extension of the *mihrab* and the pulpit with a small window (Figure 8), which is believed to be a louvre used in the hot seasons. Two thick pillars protrude in the north and west corners of the mosque (Figure 9). The mosque is surmounted by 12 shallow domes (Figures 5, 6, 8 and 9), which look like snow-white pearl beads with their smooth texture. There are two gates on the south wall, leading to the inside of the mosque. The Najdi Mosque design was apparently influenced by the most prevalent Indian mosque designs, which are characterised by only one *sahn* and one courtyard, as in the Taj Mahal mausoleum, Jama Masjid and Sunehri Masjid (Golden Mosque, Red Fort). However, this cannot be confirmed because of the small area of the Al Najdi Mosque courtyard (Figure 7). It is more likely to be a mosque without a *sahn*, where it is composed of a totally covered space (one porch) divided into several tiles and domes, opening directly onto the street, like the Moti Masjid (Pearl Mosque) in the Red Fort in Delhi, India, which was built by the Mughal Emperor Aurangzeb in AD 1659–1660.

6.1.2 *The dome*

The mosque is surmounted by 12 shallow domes (Figures 5, 6, 8 and 9) that end in a simple pointed part with an iron column that was believed to be used in fixing the *hilal* ornament. Each dome is based on intersecting Persian broken arches that were used in Persian, Abbasid and Mughal–Indian architecture. The end of the arch is based on thick rectangular abutments. These domes are similar in their appearance to the shallow onion-shaped dome of the Quwwat-al-Islam Masjid (Strength of Islam Mosque) in Delhi (Figure 10), built by Qutb Al Din Aibak, one of the first Mamluk sultans in India (AH 589/AD 1193). This was one of the characteristics of Mughal architecture in India, where features of Islamic and Hindu arts mixed. The shallow dome was previously used in the Ashraf Mosque, the big Jama in Abu 'Arish city in the island (AH 1002/AD 1594), which was built by Sharif Hamoud Bin Muhammed Al Khairati Abu Musmar at the time of the first Saudi state. This emphasises the nature of the ancient mutual influence between Middle Asia and the Farasan Islands, even before the Arab Islamic conquest in India.

The linings of the domes were painted with oil colours and decorated with abstract drawings inspired by nature and its different elements, some of which take the form of a natural scenery of the sky (Figures 11–13), as if the artist was emphasising the symbol of the upper heavens—which the hearts long for, hoping and pleading to Allah—by raising the inner

Figure 7. Al Najdi Mosque courtyard.

Figure 8. The pulpit and the *mihrab*.

Figure 9. The middle of the north wall of the *kiblah*, Al Najdi Mosque; the front entrance.

Figure 10. Quwwat-al-Islam Masjid (Strength of Islam Mosque), at the time of Qutb Al Din Aibak, the first Mamluk sultan in India (AH 589/AD 1193).

Figure 11. The decoration of the inner dome based on four pointed arches, Al Najdi Mosque.

Figure 12. The decoration of the inner dome, Al Najdi Mosque.

Figure 13. The decoration of the inner dome based on four pointed arches, Al Najdi Mosque.

Figure 14. Madhubani *rangoli*, handmade Indian tribal art/folk painting from Mithila, Bihar.

curvature of the dome. These decorations are similar in appearance to the Chinese cloud decoration known as *Chi*, and also the miniature arts in the Timurid era in Iran, which was creative in showing the sky and depicting it with stories from the prophetic Sunnah and the Holy Qur'an. One of the dome linings included drawings (Figures 11–13) containing a floral bunch that appeared to be influenced by Ottoman art, characterised by the drawing of units and elements of three- and four-petal flowers with bending branches in simple lines (see Figures 11 and 12). The delicacy of these floral bunches is similar to those on the walls of the Taj Mahal, drawn with henna on the red sand and known as *parchinkari* (Figure 13). The Hindu *rangoli* influences are seen more clearly in Figure 14, represented in the assortment of decorations in the form of floral plates with a big four-petal flower at its centre, surrounded by floral frameworks, widening towards the base of the dome. It is noticeable that these floral decorative patterns were commonly used on heritage coffee and tea pots in all parts of the Arabian Peninsula. Thus, the origin of these decorations is attributed to the local arts, in spite of their similarity to Asian heritage arts. As artisans were hired from India, it was normal that they carried their heritage with them, which reflected spontaneously in their technical work.

6.1.3 *Windows*

The position of windows and their architectural functions vary. The four upper windows on the *kiblah* wall were ornamented with stucco interspersed with stained glass (Figure 12) to allow light to enter the mosque. Wooden shutters decorated with wooden frameworks perforated with vegetal patterns of an Indian influence were used internally (Figure 11). The lower windows located on the façade of the mosque have hollow stucco mesh, allowing the entrance of light and air, and look like oriels, as in Figure 16 showing the Jama Mosque of Champaner, built by Mahmud Begda (15th century AD), as commonly seen in the Mughal–Indian era.

6.1.4 *The external façade*

a. Overlooking the street: Despite the attention given by the architect to the internal decorations of the mosque, the external shape was simplified for convenience. Therefore, only the top of the narrow entrance is decorated with a blind broken arch containing an ornament of a palm tree, the symbol of the third Saudi state, and a simple wooden door with a horseshoe arch (Figure 6).
b. Overlooking the inner courtyard: The design of the façade of the house of prayer (Figure 18) uses different methods of stucco decorations, both hollow and solid. The façades of the two entrances of the house of prayer was designed so that a semicircular arch is located above the entrance gate (Figure 17) followed by rows of square stucco sunshades. Underneath them, there are big stucco windows that are hollowed-in square figures, containing rhombus figures resembling oriels, with Persian and Mughal–Indian influence, as in Figure 16 in the Jama Mosque of Champaner, built by Mahmmoud Begda in 15th century AD. The stucco artist used incised stucco decorations on both sides of the Mughal–Indian broken arch, and simple excavated wooden gates to minimise the sense of bulkiness of the architectural mass and to maintain its lightness and coherence.

6.1.5 *The house of prayer for women*

It is located directly opposite the *kiblah* and the pulpit (Figures 17 and 18). The stucco artist maintained the privacy of the place of prayer for women while keeping the balance of the design. A hollow stucco barrier, about 2 metres high, was used taking into account audio-visual communication for the prayer (i.e. performing and listening to the Friday sermon or *khotba*); this was also to reduce and limit the size of the big pillars located between the *mihrab* and the pulpit and the house of prayer for women. A wooden door with a frieze excavated with high- and low-relief decorations was added to the stucco barrier. This type of hollow

Figure 15. Al Najdi Mosque, Farasan Islands (14th century AH/20th century AD).

Figure 16. Jama Mosque of Champaner, by Mahmud Begda (15th century AD).

Figure 17. The façade of the pulpit and the *kiblah* wall, Al Najdi Mosque.

Figure 18. The house of prayer for women, Al Najdi Mosque.

stucco barrier was commonly used in most Middle Asian countries and India. However, in most of the Middle East regions and Arab Morocco, bars of lathe wood were used to separate the places of worship for women and men inside the mosque.

6.1.6 *The concave* mihrab

The *mihrab* is located in the centre of the *kiblah* wall (Figures 19 and 20). It was designed deep into the wall. The stained and excavated wooden pulpit was formed of two parts.

a. Façade decorations: These consisted of two types of two pairs of cylindrical columns surmounted by fine decorated crowns, which differed in height and size (Figure 20). The two inner columns were connected by a circular arch, surmounted by a frieze made of a collection of vegetal patterns and fruits and pots known in India in the Buddhist arts (Figure 21), as seen in the Thanjavur Palace in the Thanjavur region of Tamil Nadu, India, from AD 1674 to 1855. The wood of the *mihrab* was also painted with the same colour group used in Indian buildings, such as red, blue, yellow, white and green, which is definitive evidence that it was imported from India.
b. Concavity of the *mihrab*: The concavity of the *mihrab* extended to almost a metre deep, which is a Mughal–Indian influence (Figure 19). It ends with two alcoves having a stained glass shutter in the form of windows, where one is located in the front and the other is located on the left side of the *mihrab*. On the right side, there is an aisle provided with a staircase leading to the balcony of the pulpit.

6.1.7 *The suspended pulpit*

This is an elevated room inside the wall, with a balcony overlooking the place of prayer (Figure 20). Its façade is in line with the *mihrab* wall, connected to the *mihrab* with an unseen side staircase, and decorated with a frieze identical to the one decorating the façade of the *mihrab*. This pulpit design was rarely used in Islamic architecture. However, it appeared in the 20th century AD in some mosques in Middle Asia, and moved to the modern mosques, such as Sidi Muhyuddin Mosque in the town of Thiruppanandal in Tamil Nadu. It also moved to Yemen and is seen in the Ahmed Bin Omar Al Jami Mosque, in Ibb Governorate in the 14th century AH/20th century AD (Figure 22). Soon enough, this design moved to most of

Figure 19. The *kiblah* wall and the pulpit, Al Najdi Mosque, Farasan Islands.

Figure 20. The *kiblah* wall, the pulpit, and the wall pulpit, Al Najdi Mosque, Farasan Islands.

Figure 21. Thanjavur Palace, Thanjavur region, Tamil Nadu, India (AD 1674–1855).

Figure 22. Ahmed Bin Omar Al Jami Mosque, Ibb Governorate, Yemen (14th century AH/20th century AD).

the Arab Gulf countries. The architect was creative in adding a door and a staircase to the pulpit, as sometimes a door was added to the same *kiblah* wall beside the pulpit, and at other times there was an unseen aisle inside the *mihrab*. This was probably due to the lack of space and not wishing to disconnect those praying in the first rows from the preacher (*Khateeb*), by filling the space between them with the pulpit gate and the staircase. On the other hand, it was for achieving the security of the prefect at a period of time when conflict prevailed, as in the time of the Mamluk reign. Hence the pulpit was the symbol of ruling and religion, where rulers, judges, scholars and sheikhs preached to the common people. The tolerance and brotherhood of Islamic teachings was evident in using a *mihrab* and pulpit that were made in the Buddhist–Indian art style used in temples and palaces, proving the success of the Islamic conquest of India in eliminating racism and spreading brotherhood and equality. Islamic art mingled with Hindu and Buddhist art forms, producing at the time of the Khalgi dynasty a genuine Hindu–Islamic art style consisting of *mihrabs*, pulpits, domes and calligraphy.

### 6.2 *The Ahmed Munawar Rifai house (AD 1922)*

The Ahmed Munawar Rifai house is located in the middle of the Farasan Islands. It was built in AH 1341/AD 1922 under the Ottoman reign of the Kingdom of Saudi Arabia, by two accomplished builders, Ali Hassan Badr and Muhammed Mekki Muharam, who carved their names and the building date on the top of the gate of the excavated segmental arch located inside the forefront of the house (Figure 24). Different types of rocks available on the Farasan Islands, such as mountain rocks and calcareous coral rocks, were used in the building. Tamarisk and palm wood were also used in the ceilings, gates and windows.

#### 6.2.1 *The house design*

The house was designed in a rectangular form (Figure 23), surrounded by different courtyards on the north and west sides for healthy ventilation and lighting, as humidity is very high in this region. Window openings were varied; some were directly opened, whereas others were ornamented with stucco decorations. This building had a special architectural design following the Ottoman style, composed of several storeys, where the ground storey was for receiving men, whereas the upper one contained a harem and rooms for women only. Tamarisk, oak wood and palm leaves were used in covering the ceilings.

#### 6.2.2 *Decorations of the gate overlooking the street*

At the forefront of the house, there is a huge main gate with a fence fused to it from both sides in a triangular form. The fence extends to surround the house at a height less than that of the gate. Above the fence, there are windows in the form of a leaf (Figure 24). The fence was built with mud bricks with a stone base painted in white stucco (Figure 25). The gate was formed of a horseshoe arch, which is considered one of the oldest, most widespread and commonly used arches in Islamic architecture because of its ability to withstand heavy weights. The two arch abutments are based on the arcature wall and use two pairs of simple cylindrical columns as motifs, like those found in the arcatures of the Ahmed Bin Tolon Mosque (AH 263–265) in Egypt, which reflects Persian–Abbasid influence. The arch was decorated with a floral frieze and a trefoil design surmounted by the word Allah in the form

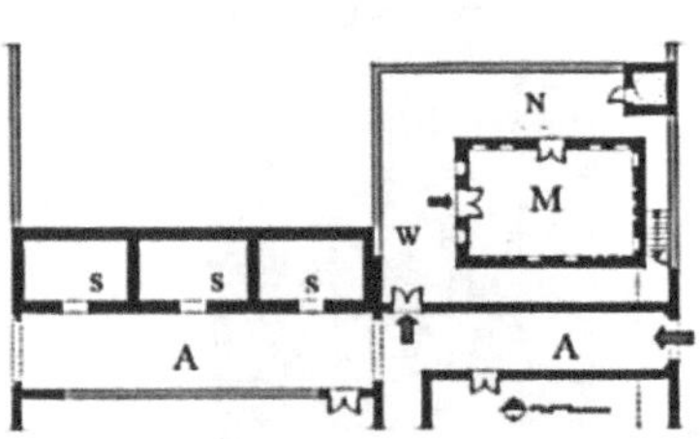


Figure 23. Plan of the Ahmed MunawarRifai house (AD 1922).

Figure 24. Outer gate of the Ahmed Munawar Rifai house (AD 1922).

of a pineapple with a Sasanian influence (Figure 25). The hollow interlaced vegetal branches were used as a background for the two medallions, which contain a motif that was commonly used in different materials in Mughal buildings in India, as in the Tomb of Bahlol Lodi, built by his son Sikandar Lodi in India in 15th century AD (Figure 26), and the mausoleum conspicuously located at Lado Sera, built in the Lodi era (AD 1451–1517) (Figure 27).

It is well known that the decorative medallions were used symmetrically in the corners on the façades of buildings, especially gates, since the Abbasid era. This tradition moved through the times to countries belonging to the Arab conquest, which were preceded by the Greeks and Romans in the triumphal gates where these medallions were.

6.2.3 *Decorations of the façade overlooking the inner courtyard of the house*

The façade of the house was designed in the Ottoman style, which consists of a gate in the middle of the wall, with two smaller windows on both sides (Figure 28). The stucco artist showed lavishness and luxury in the fineness of the stucco decorations, which were like coral reefs in their texture. Motifs were inspired from Far Asia, where the Mughal–Indian and Chinese arts were found, with the spirit of Moroccan–Andalusian art, which was characterised by covering the wall with fine stucco arts, with minimal use of calligraphic elements.

The Farasan style used most of the Indian arch types (Figure 29) in all their stages before and after the spread of Islam, particularly the Mughal–Indian stage, where it succeeded in implementing it in inner and outer surfaces in a balanced way. The style also excelled in repeating them and creating harmony and balance between the high and low reliefs on the stucco decorated wall surfaces, which helped in breaking the monotony due to repetition and overcrowding of decorations. The Ahmed Munawar Rifai house contained windows made of red wood excavated with geometrical and vegetal patterns. The wood and glass were imported from outside as they were not available in the Kingdom of Saudi Arabia. The blind arches were used in a decorative way, especially above the rectangular door openings inside and outside the walls of the sitting hall (Figures 30 and 31). The stucco motifs of the sitting

Figure 25. Front façade of the sitting hall of the Ahmed Munawar Al Rifa'i house (AD 1922).

Figure 26. Tomb of Bahlol Lodi, built by his son Sikandar Lodi, India (15th century AD).

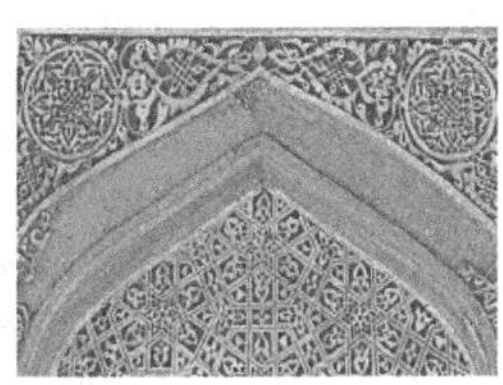

Figure 27. Mausoleum conspicuously located at Lado Sera, Lodi era (AD 1451–1517).

Figure 28. Façade of the Ahmed Munawar Al Rifa'i house (AD 1922).

Figure 29. A collection of the most famous arch types used in Indian buildings.

Figure 30. Stucco decorations, calligraphic streaks, and stucco sunshades with stained glass in the sitting hall of the Ahmed Munawar Rifai house (AD 1922).

Figure 31. Details of the upper wall in the sitting hall of the Ahmed Munawar Rifai house.

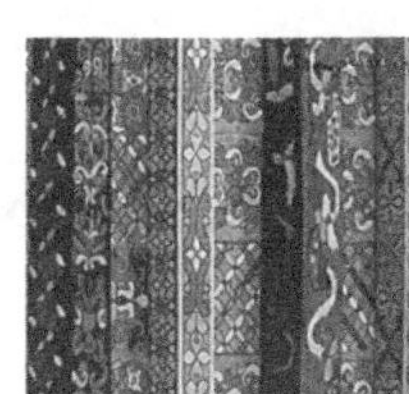

Figure 32. Details of ceiling decorations of the Ahmed Munawar Rifai house.

hall of the Ahmed Munawar Rifai house were greatly influenced by the Sasanian, Seljuk, Indian and Ottoman arts; the wall decoration method depended on multiple levels. Thus, the façades became protruded with various embodied shadows, as in the Seljuk sculptures. They were also characterised by great accuracy despite their minute details. Stained glass was used in the hollow spaces of the upper windows, allowing penetration of light while maintaining privacy. The Ottoman style was evident in the drawings of the parallel wooden column decorations covering the ceiling of the sitting hall (Figure 32). All patterns of Ottoman decorations, with their known colours of turquoise, blue, green, ultramarine and yellow, were used. The Ottoman influence was also reflected in ornamentation of the façade above the guest reception room, with calligraphic streaks of Qur'anic verses (Figure 31). This was seen in the sitting hall of the Ahmed Munawar Rifai house. The top of the door from the inside was ornamented with an arch surmounted by short verses of poetry and calligraphic streaks containing Qur'anic verses, such as the *Al korsi* verse and a verse from the *surat Al Tawba* among others (Figure 31). The walls of the sitting hall were decorated from the inside with a collection of stucco patterns in the form of successive double arches, some of which are segmental arches and the others ogee arches crowned with a flower of Mughal–Indian influence, which were previously used in building decorations in the Abbasid era and in the Ahmed Bin Tolon Mosque.

### 6.3 *The Hussein Bin Yahiya Rifai house (AD 1924)*

The Hussein Bin Yahiya Rifai house was built under the Ottoman reign in the Kingdom of Saudi Arabia. Its building date was probably AH 1343/AD 1924, two years after the establishment of the Ahmed Munawar Rifai house, which is adjacent to it on the north-west side. However, no plan was found for this house; its main form is similar to the plan of the Ahmed Munawar Rifai house (Figure 23).

#### 6.3.1 *The house design*

The house was designed in a rectangular form and its walls were built using a regular-shaped stone course technique. A large main portal (Figure 33) was located on the west side of the house, connected to the fence surrounding the house. The height of the fence was almost 6 metres. The portal was followed by an exposed inner courtyard known as *hosh*. The house consisted of several storeys; a separate one-storey building was established for the sitting of men. The house was designed in the Ottoman style, where its façade was characterised by a door in its centre with two big rectangular windows on both sides, followed by two small rooms located on the right and left sides and used for storage and for keeping drinking water jars, followed by the huge sitting hall. The house ceiling was covered with wooden columns from India and tamarisk wood available locally.

Figure 33. Front façade of the Ahmed Munawar Rifai house, with sunk or recessed arch, overlooking the street (14th century AH/20th century AD).

Figure 34. *Peshtak* of Qila-i-Kuhna Mosque, in the Lodi era.

Figure 35. Marble *mihrab* belonging to the Abbasid era, Islamic Museum in Baghdad.

Figure 36. Fragment of stucco, excavated in Iran, Tepe Madrasa in Nishapur (4th century AH/10th century AD) The Metropolitan Museum of Art.

6.3.2 *Decorations of the gate overlooking the street*

The decorations covered most of the outer gate surface. There was obviously some linear calligraphy on the right side that may have been eroded due to weather factors. The gate was in the form of a rectangular panel (Figure 33) with a pointed arch; its extrados was ornamented with two decorating streaks, one high- and the other low-relief, with Abbasid and Mughal–Indian influences as seen in the *peshtak* of the Qila-I-Kuhna Mosque in the Lodi era (Figure 34). The pointed arch of the gate was surmounted by a blind segmental arch, influenced by the Mughal–Indian art style that has several styles of recessed or sunken arch. The arch was surrounded by a collection of vegetal patterns with low-relief bending and intertwining branches, which was reminiscent of Samaraa decorations in the third stage, which is characterised by the abstraction, bending, and the bevelled corners with the clarity of the ground depth, as seen in the decorations of Bulkuara Palace in Samaraa, Iraq, and a *mihrab* decoration from the Abbasid era (Figure 35). The top of the gate was surmounted by a series of balconies in the form of spruce cones with a Byzantine influence. However, they were closer to a Sasanian figure (Figure 36) in one of the Nishapur schools in both form and lining.

6.3.3 *Decorations of the façade of the sitting hall overlooking the inner courtyard*

The outer main portal led to a courtyard through a narrow corridor, where a sitting area was allocated on the south side for men only. The southern corridor was supplied with facilities used in ablutions and purification. The façade of the sitting hall was a simple one-storey building with a door centred between two low windows, similar to the Ottoman architectural style (Figure 37). Above the door, there was a rectangular area of low-relief decorations that look like Mughal–Indian decorations, and on both its sides were four frameworks containing blind ogee arch patterns.

Several entrances were annexes to the house. On one of the side gates, the creativity of the stucco artist is evident, with decorations that look like the outstanding stone sculptures in buildings in Agra and India and in the Lotus Mahal (see Figures 38–40). The cleverness and creativity of the stucco artist are obvious in the variations between the segmental

Figure 37. Details of the main portal of the Hussein Rifai house.

Figure 38. One of the side entrances of the Hussein Rifai house.

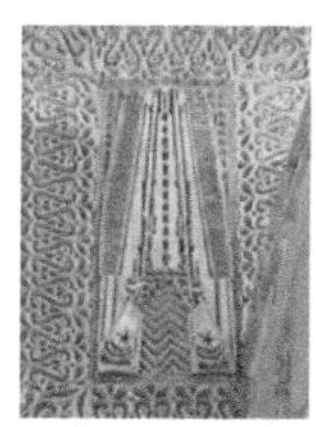
Figure 39. Stucco details of one of the side entrances of the Hussein Rifai house.

Figure 40. Pavilion of the Lotus Mahal (12 century HA/18th century AD).

Figure 41. General view of the reception room in the Hussein Rifai house.

Figure 42. A detailed corner of the wall slots inside the Hussein Rifai house.

Figure 43. Wall slots inside the Hussein Rifai house.

Figure 44. Diwan-i-Khas Mahal, inside the Agra Fort, during the reign of Mughal Emperor Shah Jahan, India (14th century HA/17th century AD).

arches and the ogee arches, with the surfaces becoming finer and more complicated in the multiple levels and differences in the hollow spaces, thus matching the creativity of the Seljuk *muqarnas*.

### 6.3.4 *Decorations inside the sitting hall*

The sitting hall in the Hussein Rifai house (14th century AH/20th century AD) contained different stucco arts that covered two thirds of the upper wall in the form of wooden friezes excavated with Arab calligraphy from Qur'anic verses and prayers, most of which were erased with time. In the front, stucco wall shelves that were aligned in successive rows appeared (Figures 42 and 43). They were decorated with soft lined arches with various levels and shapes in a Mughal–Indian style, similar to the Diwan-i-Khas Mahal (Figure 44) inside Agra Fort, during the reign of Mughal Emperor Shah Jahan, in India in the 14th century HA/17th century AD. These wall slots and shelves were known by the Mughals as *pishtaq*, and were commonly used to keep things, such as books and lanterns, during the time of the Lodi Sultanate in 9th century AH/15th century AD. They were also seen in Mamluk and Ottoman houses in the 9th–13th century AH/15th–19th century AD, and were characteristic of heritage Najdi architecture, especially in the sitting area on the *kamar* wall, a wall in the middle of the sitting hall, which has stucco decorations, shelves and a place for the fire used for preparing coffee. The crowns, inner surfaces and sides of the arches were decorated with plenty of vegetal patterns that were excavated with soft lines in the low-relief style.

## 7 FARASAN PLASTER PATTERN ART STYLE

The houses of the two pearl merchants are characterised by stucco decorations that reflected their taste and showed the craftsmen's wide knowledge of the advantages of plaster material and its processing methods, which led to innovation and creativity. This advanced leap in stucco decoration could be due to many factors, such as close contact with other art forms, acquisition of knowledge, and the influence of cultural openness as a result of bringing skilled workers from other regions and the use of trade and pilgrimage routes. Stucco artists presented patterns in an organised way despite their complexity, which is one of the features of Mughal–Indian and Moroccan–Andalusian art styles.

The appearance of natural elements in stucco designs, where the textures and bending and intertwining lines in an accurate and regular way resemble natural coral reefs, spread in this region. The stucco material gained aesthetic value, transforming it from a simple, crude material into a valuable, high-quality material. This was utilised to maintain the balance and the harmony of the total unity of the wall surfaces internally and externally, through the distribution of the stucco patterns over the walls of the religious and residential buildings. It is worth noticing that the calligraphic streaks were not used on the outer stucco façade of buildings, whereas the sitting halls of the house had calligraphy containing Qur'anic verses, poems and prayers.

Stucco artists used the following two methods in decorating the inner and outer walls of the two houses discussed:

a. The casting method using wooden moulds, which were pre-made in the form of hollow decorated or calligraphic patterns, where the plaster was poured into them and after drying was taken out and fixed on the walls in the form of streaks, windows or balconies. This method goes back to the Abbasid era in the 8th century AH.
b. The direct method that depended on the drawing and hollowing of surfaces using special tools, where the surfaces of the stucco decorations were hollowed out in the form of geometrical frameworks and horizontal and vertical streaks.

## 8 RESULTS

This study reveals the most important factors that contributed to the Farasan plaster pattern art as follows:

a. Pearl trading contributed to the movement of people to the Far East and Far West, saving a lot of money that helped wealthy people in choosing the best methods and materials to establish their buildings.
b. Farasan stucco decorations were free from the abstract elements of the Najdi art style, characterised by geometrical and vegetal patterns inspired by nature, such as palm trees, stars and *hilal*. They were also free from the scratches made by repeated broken lines and arches that gave a rough texture.
c. Farasan artists did not copy the Jizan Idrisid style in stucco excavation, which, sometimes, was limited to streak-form frameworks and, at other times, to decorative abstract intertwined vegetal patterns. They inspired boldness and the freedom of expression on most wall surfaces from that style.
d. Stucco artists found inspiration for their decorative units from most cultures that they knew about through trade movements, using them according to their needs and heritage. They tried to unify different artistic styles to reach a human and environmental balance, which created a sense of internal joy when seeing these innovative stucco decorations.
e. The features of Mughal–Indian and Ottoman arts dominated the buildings that remained from the time of the most famous pearl merchants in the 14th century AH/20th century AD, also carrying Persian art features.
f. The Farasan stucco excavation art was inspired by distinguished cultures, and interest was focused on methods of using these enriched stucco decorations on the surfaces of internal and external walls. In spite of using many architectural elements and stucco decorations on the external walls of the Al Najdi Mosque, it looked plain because of the uniform colour of the stucco, depending on the high- and low-relief texture in the stucco formulation, which was reminiscent of the Najdi architectural spirit and depended on the nature of the material and its formulation according to use. However, stucco artists took good care in showing the artistic and decorative lavishness inside the mosque, using Indian–Buddhist elements that were represented in the pulpit and the *mihrab* and were brought from India as witness to the tolerance of Islamic civilisation and its ability to employ some artistic architectural elements that were contrary to Islamic teachings.
g. Artists attempted to maintain the best decorative product and to manipulate the stucco material. As they reached maximum creativity in the decorations of the houses of Ahmed Rifai and Yahiya Rifai in 14th century AH/20th century AD, their accuracy was equal to that seen in the stone sculptures of the Seljuk era, the Ottoman arts, Arab Moroccan countries and Andalusia, whose stucco arts were characterised by the feature of filling the walls with decorations.

## 9 CONCLUSION

The history of the Farasan Islands since prehistoric eras has been full of many great civilisations, such as the Greeks, Romans and the Yemini Himyarite Kingdom extending to the Ottoman Empire in the 16th century AD, that considered the islands one of the strategic security points for marine fleets. They all had one goal: to secure their trading fleets heading towards Middle Asia and India through the Bab El Mandeb strait and the Red Sea against the Arab pirates of the South. Thus, they left their traces in this region. The profession of diving for pearl fishing contributed to improving the income of the Farasan people. And despite the rough terrain of the islands, their inhabitants were famed for knowing the secrets of the sailing routes towards the East African coast where the Dahlak Islands, which were under Italian control at that time, were. This contributed to the enrichment of their cultural contact. The Farasan people travelled to the Dahlak Islands, which were rich in pearls, for about three diving journeys a year. The Farasan divers also dived in the Arabian Gulf, which eased their cultural contact with Persia; thus, the Farasan artists were exposed to a variety of civilised cultures.

The geological environment, rich with coral and calcareous stones, contributed to producing this creative artistic energy, especially when the artists discovered the properties of the plaster material, which is characterised by resisting and absorbing humidity and thermal insulation. This enabled the Farasan people—through their experiences in stucco decorations since the time of the Aid and Khairat families, and especially in the Idrisid dynasty era—to change the limited use of stucco decorations in the inside of the house in a form of privacy to full coverage of wall surfaces, especially the external walls of buildings. The time of the Jizan Idrisid dynasty witnessed great freedom in stucco decorative art. This was reflected in the styles of the Farasan artists who lived in this period. They innovated and created various methods of stucco excavation and hollowing, covering internal and external architectural surfaces, while maintaining the national identity and harmony of performance. Thus, they transformed the cheap plaster material into a valuable one full of aesthetic value and equal in its magnificence to that of the stucco arts in India, Middle Asian countries, Morocco and Andalusia. This is why the two houses of the two pearl merchants Ahmed and Hussein Al Rifa'i, as well as the Al Najdi Mosque, are considered the best witnesses to the contribution of the different styles and methods of the 14th century AH/20th century AD from the Far East to the Far West, in the emergence of a new distinguished style in the field of stucco decorative art, which came to be known as the Farasan style.

## REFERENCES


Al Nadawi, A.A. (2012). Islamic architecture and the consolidation of Islamic culture in the Sultanate of Delhi, 602–932 AH/1205–1526. In *Yearbook of the Islamic History and Mediator* (No. 2, p. 277). Egypt.

Alzele'i, A.B.O., et al. (Eds.). (AH 1423). Jizan region monuments. *Monuments of the Kingdom of Saudi Arabia series*. Riyadh, Saudi Arabia: Ministry of Education and the Agency of Antiquities and Museums.

Amer, I.I. (1998). Architecture in Samarkand in the Timurid period: A research in the monuments of the East of the Islamic world (p. 119) [Seminar]. Giza, Egypt: Faculty of Archaeology, Cairo University.

Ettinghausen, R., Grabar, O. & Jenkins-Madina, M. (2001). *Islamic Art and Architecture, 650–1250*. New Haven, CT: Yale University Press.

Fuhring, P. (1994). Renaissance ornament prints: The French contribution. In K. Jacobson (Ed.), *The French Renaissance in Prints, 1994*. Los Angeles, CA: Grunwald Center, UCLA.

Leven, B. (1998). *The New about the Ancient East* (Trans. Jabir Abi Jaber, p. 121). Moscow, Russia.

Mones, H. (1983). *Atlas of the history of Islam* (p. 244). Cairo, Egypt: Al-Zahraa for Arab Mass Media.

Mortada, H. (2014). Urban and architectural features of traditional environment of Farasan Islands, Saudi Arabia. In F. Pena & M. Chavez (Eds.), *International conference on Structural Analysis of Historical Constructions, Mexico City, Mexico, 14–17 October*.

Nath, R. (1995). *Colour decoration in Mughal architecture: India and Pakistan*. New Delhi, India: Historical Research Documentation Programme.

Saban, S. (2010). German attempts to exploit Farasan Islands in the south of the Red Sea in the light of the documents of the Ottoman archives: 1318–1320/1900–1902. In *Jordanian Journal of History and Antiquities* (p. 131).

*Planning and approaching the city*

*Architecture*

*Cities' Identity Through Architecture and Arts – Catalani et al. (Eds)*
*© 2018 Taylor & Francis Group, London, ISBN 978-1-138-05409-7*

# Architecture of the 21st century museum as a catalytic phenomenon in the evolving cultural identity of a city

Nina Ugljen-Ademović & Senka Ibrišimbegović
*Department of Architectural Design, Faculty of Architecture, UNSA, AFS, University of Sarajevo, Sarajevo, Bosnia and Herzegovina*

ABSTRACT: Cultural buildings play an important role in enhancing the image of a city as a tourist attraction. It is well known that museums, galleries, art centers, concert halls identified with the city have created a part of cultural revitalisation strategies. Museums also illustrate the evolution of architecture and an accelerated transformation of trends in architecture, and are often planned to be generators of the urban development. Taking care of the development of the city and how city's cultural identity evolves is a sign of respect to the built cultural heritage, so that it presents the distinctive aspects that clarify its uniqueness, distinguishing it from any other city. Creation of such cultural responsible (museum) architecture would contribute to defining architectural processes, if we aim to build or preserve buildings, which respect the past and open new visions of the space for the future. Such processes are very sensitive, since in this age of cultural transformation they may become irreversible and result in losing the cultural identity. In this paper we will focus on the catalytic architectural proposals for the reconstruction and conservation of the History Museum, 'the people's museum', in Sarajevo, Bosnia and Herzegovina.

*Keywords*: cultural buildings; tourist attraction; cultural revitalization; cultural identity; cultural heritage

## 1 INTRODUCTION

Cultural objects, especially museums, as well as architecture by which they are represented and communicated, play an important role in promoting values of a city.

Although it may at first appear that this is a frequently discussed topic of reactivating the space in this way, the truth is that it always and ever anew contributes the planning of important development strategies that promote the importance of cultural and spatial identity. Taking into consideration the inevitable economic importance that goes along with a correct approach to the design and construction of museums, by which it becomes an active participant of the city's developmental processes, this time accent will be placed on the role museum architecture has in raising awareness on the importance of culture and the development of cultural identity of a specific area, a city. To claim that, in accordance with the contemporary architectural discourse, museum architecture exists the sphere of the "closed" relationship the observer – the exhibit, and opens up to other phenomena, is really simple. Although those phenomena, pressured by the global processes, frequently lose their true sense, they can be rescued by the correct strategies applied in valorisation, reconstruction and new construction of these chests of knowledge, history, tradition and memory. One of the approaches would certainly be a culturally aware architecture that does not necessarily entail a sensation in space. It rather becomes a space for contemplation of time, place and traditional values embodied in the cultural identity.

## 2 NEW DIMENSIONS OF SPATIAL FLEXIBILITY

Before we define the principal question and place the issue in a specific context, it is necessary to position the contemporary views on the process of changing the conceptual role of museums. Namely, in the long history of the museum and exhibit space development in general, it is only in the 18th century (1793) when the Louvre, the first public museum opened, that the individual (the visitor) – public (national) domain significantly changed. That moment of "reshaping" the museum into a public space symbolically established a new, more intimate relationship with the daily life, and initiated the processes of change of the past understanding of the museum as an isolated, elite and chosen place. Instead, a dialogue started in which museum exhibitions and exhibits communicate with the visitor, and influence the rise of awareness on the importance of cultural identity, open and dynamically changeable. Yet, it was not before the First World War that modernisation processes encompassed all segments of the society and that the previously disintegrated world started to integrate into a new, modern society. Art and architecture were then more than ever called upon to help those processes and offer a complete liberation of the thought through the liberation of space from the previous obstacles, both physically and spiritually. Mies van der Rohe's pavilion in Barcelona is one of the objects that symbolically reflect new tendencies towards "democratisation of architecture" and "democratisation of art" (Fig. 1). The dynamic and flexible permeation of spaces that flow and erase the borders of the open and the closed are even today considered a matrix in the conceptual sense.

However, since democratisation also carries with it a new and a greater responsibility, the process was not easy and a lot of time passed before it took off and gained its present scope, as well as before the notion of "public" responded to the complex demands. Many authors agree that the departure from the traditional understanding of the museum started in the 1960s, however, the moment that certainly set new standards with regards to the symbolic and functional role of the museum in an urban context was the construction of James Stirling's Neue Staatsgalerie in Stuttgart in the 1980s (Fig. 2). Suggesting the *new dimensions of spatial flexibility*, this object actively participates in the life of the city and remains one of the most important examples of conceptual complexity, which functions as a living organism.

> *"New shapes – forms, apart from being based on pure intuition of the creator, also need to have a respectable base in the author's knowledge of the former solutions applied in the concrete problem. From that arises a creation that can be called the process of adaptation of the forms gained from the past aesthetic ideologies that could serve the needs of the present." (Ugljen-Ademović, p. 37.)*

Although we would later witness the fascinating, globally-recognised architectural structures, we consider this example to be a relevant answer to the complex demands, for, unlike others, it does not create tensions in space, rather, it questions the thesis according to which only the dazzling, sensational and overly dimensional architectural structures can attract

Figures 1 & 2. German Pavilion in Barcelona, Spain, designed by Ludwig Mies van der Rohe for the International Exposition in 1929; Neue Staatsgalerie, Stuttgart, James Stirling, 1984.

attention and thus be generators of progress. On the contrary, by creating an analogy between the object-museum and the city, and by respecting the urban pattern, the Neue Staatsgalerie is permeated in functional, formative and symbolical sense, reflecting timeless values that can be evaluated even decades after the construction, as Michael Levin states:

> *"Museum is an instrument that defines, represents and makes transparent changing cultural trends. The museum, almost by definition, does more than express current social values and tastes; it also makes a cultural statement which goes beyond its own place in history." (Giebelhausen M, 2003)*

Such an approach is seen as a starting point of the research, and the messages it carries are applicable to the case study of the History Museum in Sarajevo (the former Museum of the Revolution), and will serve as an answer to the question – in what way to approach a reactivation of the existing context, keeping in mind the preservation and protection of valuable architectural structures within it, and which demands does the museum need to fulfill in order to become a catalyst of progress and enhancement of a specific cultural identity? Regardless if the context is global or local, polarising attitudes towards the approach to architectural shaping of the museum, keeping in mind its symbolical reflexion, are more than ever prominent. Thus, today, the museum can be a *representative* of time and place, or an *attractor* in a certain point in time, or, finally, a *corrector* and critic of the society, simultaneously offering completely different spatial implications. Finally, should a museum protect the existing (referent) cultural and spatial identity, or promote its spontaneous development and changes?

## 3 DEVELOPMENT OF CULTURAL AND SPATIAL IDENTITY – SPECIFIC CHARACTERISTICS OF A PLACE

Finding an answer to the posed questions is even more important in the present moment, marked by prominent processes of commodification of the daily life, applicable to architecture more than ever, for spiritual, non-material traits of our existence, the creativity, the inventiveness, are becoming a commodity. That is one of the reasons why we make cultural identity in synergy with architecture the focus of the issue. In that way, architecture visualises cultural values, and, as such a powerful medium, it gives rise to new ideas, sets criteria and by doing so it determines its place in history. However, a dynamically changeable cultural and spatial identity that we witness possesses certain implications both at the local and at the global level, and introducing balance between them becomes imperative. Namely, the existing context has its local characteristics that do not suggest a closed concept; rather they are an open process. It is important to enter that process gradually, without any predefined global signifiers that disturb the vital, semiotic relationship between the sign and its meaning. By gradually developing local characteristics, we enable a spontaneous development towards the global goal, enriching it by the new and the special. In that case, a redefinition of the local by other, superordinate terms characteristic of globalisation is unnecessary; rather, one can speak of their synchronic action. If we were to accept the fact that "there is no final state in the life of a city" (Perković, 1969), we, in fact, reveal anew the different meanings in architectural forms, which construct an urban space as "the place of a permanent, repeated activity" (Rossi, L'Architettura della Citta) that came as a result of an endless process of shifting of different aspects – developmental, historical, social, architectural. That also liberates urban memory, as a specific juncture of individual memory and the memory embodied in architecture, while spontaneous, continual development of cultural and spatial identity is encouraged as well.

> *"Space is locally unique, but that local specificity is produced through its socio-spatial relations to other places both local and global. Crucially therefore the positive and rich multiplicity of places is only realised through practice in space which connect, react, and change the political, economic, and cultural relations that are infused in local and global spatial relations." (Bower, p. 181)*

## 4 MUSEUM ARCHITECTURE – A SPACE THAT INTEGRATES, OR A "WHITE BOX" – A NEUTRAL, EMPTY SPACE

In the years of considering this issue, a dual relationship is posed, which, on the one hand, observes the museum as a place that intrigues and provokes by its architectural expression – the place of "new monumentality", or, on the other hand, as a "white box" – a neutral, empty space, showing thus the relationship between the architectural form and context, as well as the relationship between the architectural form and art that it contains. Due to such a simplified treatment of the complex issue, based on opposition and polarisation of the approach, it is necessary to establish valorisation criteria by which architecture should preserve its own vitality, even in the time of ever increasing shifts of the cultural aspect of architecture. The case study will show that the existing spatial flexibility offers *new dimensions of spatial flexibility* through integration with the context at several levels – aesthetic, usable, flexible, symbolic, and, instead of a simplified juxtaposition or superposition of the new architectural structures, it is spatially integrated and becomes a mediator between the past and future meanings and a sub-creator of identity of a place.

## 5 CASE STUDY – THE HISTORY MUSEUM OF BOSNIA AND HERZEGOVINA IN SARAJEVO

The architectural heritage of Modernism in Bosnia and Herzegovina is valuable in many respects. It shows a strong mark of tendencies characteristic of its era, but also of signature styles of renowned regional, European and even globally acknowledged architects. Mostly, these are buildings whose visual identity is often even ahead of its time. (Ugljen-Ademović, Turkušić, Ibrišimbegović, p. 365).

### 5.1 *Architectural heritage of Modernism in Bosnia and Herzegovina*

The City of Sarajevo has for centuries been a meeting point for different cultures. Layers of Ottoman and Austro-Hungarian heritage, all the way to the socialist era and urbicide of the 1990s, are omnipresent in the city culture. Today, 22 years after the war, a question is asked: What is Sarajevo today?

Leaning on urban and architectural experience of the past periods (Ottoman and Austro-Hungarian architecture), architectural Modernism, although autonomous and functioning beyond spatial and temporal realities, is a true bearer of cultural memory and a constructive creator of Bosnian national and cultural identity. Architectural work as an integral unit is defined by its parts, function, construction, materialisation, form; priorities and approaches to emphasising certain parts of these units have changed through history. That is why it appears that architecture survives only when it manages to preserve its own character, by negation of the form that the society expects of it. That is why Tschumi suggests that there never was a reason to doubt the necessity of architecture, for the necessity of architecture is its "non-necessity". (Tschumi in Tschumi 2004: Fireworks, 1974) (Ibrišimbegović, 2015)

### 5.2 *Architecture of the History Museum – new dimensions of spatial flexibility through merging with the context*

The specific combination of transcendentalism of modern architecture, and sensitivity towards history and cultural heritage, is reflected in the building of History Museum of Bosnia and Herzegovina that is located within the historically-sensitive, urban fabric of the Sarajevo city centre, the Marijin Dvor area, a natural link between the historic centre and the modern city zone, between the old and the new. Architectural design of the museum was based on a disciplined adoption of modernist practices not only in the field of architecture and art, but also from the culture in general. Such procedures are prominent in shaping the museum space on the principles of reducing the expressive language of the architectural and of autonomy of the artistic form in respect to the content. White, stone-lined cube, erected on a pedestal, which also highlights the monumentality of the building, is part of the universal formative vocabulary, which will later appear in other modernist centers (Fig. 3). Such elevated main exhibition space is also emphasized by the appearance of floating volume, which is achieved by its separation from the pedestal by a modular grid of glass surfaces, which at the entrance level very subtly connects visitors with the environment (Fig. 4).

The specific combination of transcendentalism of modern architecture, and sensitivity towards history and cultural heritage, is reflected in the building of History Museum of Bosnia and Herzegovina that is located within the historically-sensitive, urban fabric of the Sarajevo city centre, the Marijin Dvor area, a natural link between the historic centre and the modern city zone, between the old and the new. Architectural design of the museum was based on a disciplined adoption of modernist practices not only in the field of architecture and art, but also from the culture in general. Such procedures are prominent in shaping the museum space on the principles of reducing the expressive language of the architectural and of autonomy of the artistic form in respect to the content. White, stone-lined cube, erected on a pedestal, which also highlights the monumentality of the building, is part of the universal formative vocabulary, which will later appear in other modernist centers (Fig. 3). Such elevated main exhibition space is also emphasized by the appearance of floating volume, which is achieved by its separation from the pedestal by a modular grid of glass surfaces, which at the entrance level very subtly connects visitors with the environment (Fig. 4).

The inner atrium of the main corpus of the building on the second floor forms the ideal "white box". (Ugljen-Ademović, Turkušić, Ibrišimbegović, p. 365–370) Bernard Tschumi emphasizes, on the one hand, architecture as a matter of mind, a dematerialised or conceptual discipline with this typological and morphological variations, and, on the other hand, as an empirical event, focusing on the senses, on the experience of space. Architecture is finally a meeting place. It flourishes in its dual position, between cultural autonomy and advantage, between contemplation and dwelling. (Tschumi, 2004, p. 70) Even in Le Corbusier's time, architecture was defined and viewed upon as a phenomenon of emotions, hence, he, as a representative of the very beginnings of Modernism. It is important to understand that in order to understand the role of architecture in the society, be it the material role, artistic-aesthetical, and spiritual. Acknowledging a certain building stems from the relationship established between architecture and the way in which the social community percieves that architecture in cultural, functional or emotional sense. That relationship between architecture and tis perception is conditioned by heterogenous imageries and expectations about the "artistic" values architecture should represent, especially in situations when they are marked by a strong identitary charge. *"A building becomes a part of the social landscape only when it absorbs the imageries of a social community."* (Mrduljaš, 2009, p. 86–88) Architectural theoretician Norberg-Schulz refers to phenomenology in architecture as a method that urges us to *"return to the natural things"*, unlike the abstraction and mental construction.

Figures 3 & 4. History museum: View of exterior garden, View from courtyard towards the street.

As for his claims, he made several phenomenological studies of the environment. The potentional of phenomenology is identified in architecture as an ability of the surounding to gain sense through creation of special places. He again introduces the old Roman idea of **genius loci**, the "spirit of a place." (Ibrišimbegović, 2015, pp. 37–39)

### 5.3 *Architectural proposals for History Museum in Sarajevo*

The "Reactivate Sarajevo" initiative, together with domestic (History Museum and Faculty of Architecture, University of Sarajevo) and international (Urban-Think Tank at ETH Zürich and Baier Bischofberger Architects) experts, presented the project "Sarajevo Now" at the 15th Venice Biennale of Architecture in 2016, entitled "Reporting from the front". The project analyses challenges and potentials of the city of Sarajevo, as well as a reactivation of the existing infrastructure through development of various conceptual designs. The Historic Museum was in the focus, treated by domestic experts as a concept, both architectural and comprehensive, that needs to stay in its original state and should, as the pearl of Modernism in Bosnia and Herzegovina, be an indicator of evolutionary development of the City of Sarajevo's cultural identity through architecture. The intention of Baier Bischofberger Architects, inspired by demonstrations of solidarity and popular ownership, is that the exhibitors redefine the museum not as an elitist institution dedicated to the display of objects, but as a catalytic urban space. A temporary strategy of adaptive reuse can compel a new reading of the site and its surroundings. Their intention was sheathing the museum in transparent vinyl skin leaves, the decay and patina of the original building intact, while juxtaposing the old and the new (Figs. 5, 6). The overall effect is a détournement, with the design projecting an oppositional message. The historic structure itself is enclosed as an uncanny artefact, subverting conventional notions of a museum as a sterile container.

The new proposal of the authors of is considered to be a very ad-hoc approach:

"*The envelope would actually be covered with a special material made to shrink-wrap scaffolding. We would be shrink-wrapping the structure around the building so that there is no further deterioration from its current state. The covering also offers new spaces between the existing building and the scaffolding structure, which can be activated with programs and attractors. We envisage the project as a short-term ten-year plan until the money can be found to restore the building. We are interested in the space in-between, which will be filled with new programs, for using the building in a completely new way.*"

"*What we are doing is trying to create a system that protects the building from further decay. Instead, we are interested in the possibility of creating a functional space that can still be used while repairs take place. The future of this building depends on how the city adopts it, and embraces its expanded function as a cultural centre.* But if you really understand modernity it is a series of layers. Cities are built up over time in layers, and Sarajevo represents those layers. *We are designing a reactivation strategy that crosses into the realm of the arts and intellectual life. Architects have always played an important role in these areas throughout history.* Architects are reinventing themselves through participation in local and national politics that impact the urban context." (Figs. 7, 8).

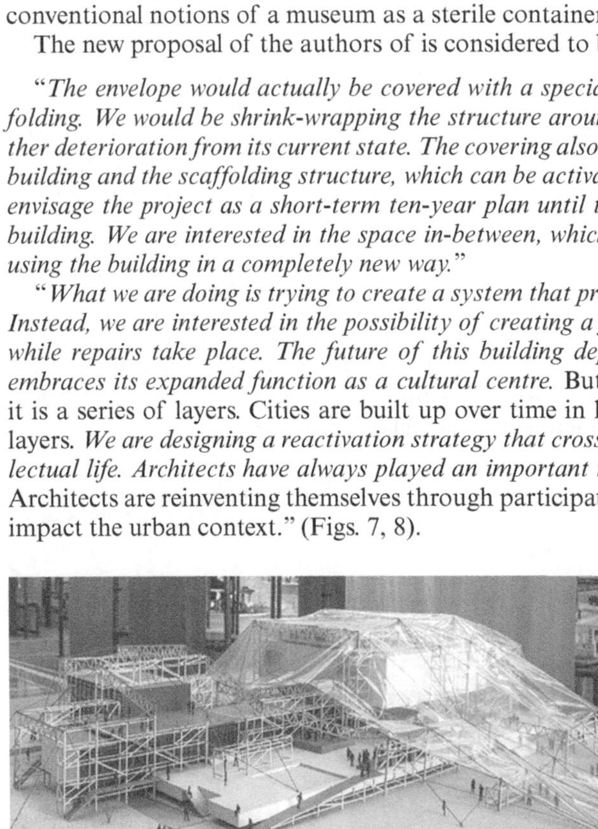
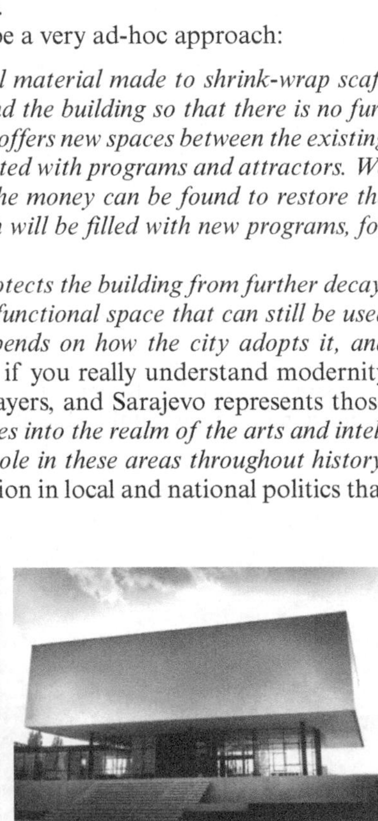

Figures 5 & 6. The existing structure is contained within a transparent vinyl skin; History Museum.

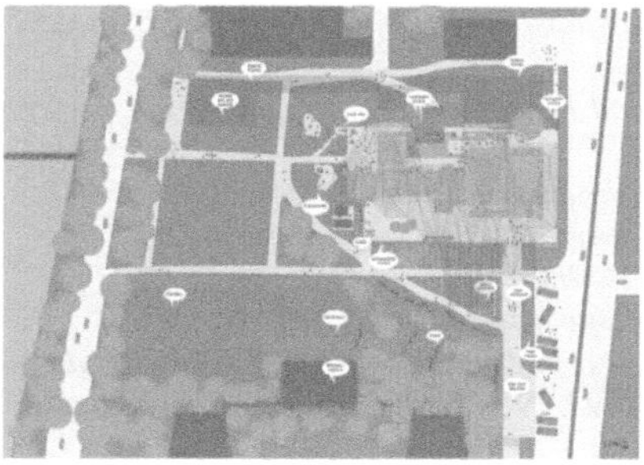

Figures 7 & 8. Site plan illustrating the proposal in more detail; The decay of the original building is preserved image.

Such a solution for preserving the History Museum is "questionable" as it is opposed to the idea of new dimensions of space flexibility. By this, the significance of architecture as a sub-creator of a cultural identity is neglected, since by the fulfilment of strictly functional conditions, it is set as a closed and a separate entity, which performs only its basic function. The whole structure and architecture of Museum hast to be visible, as it is an attractor to the visitors by its Modernist architecture.

## 6 CONCLUSION

Research has shown that the existing approach suggests the contemporary tendencies of integration of the "high" and popular culture, as well as democratisation of the audience, without the danger of integration into the consumer culture. It also reflects upon the thesis that one should not create tensions in space in such a place, and this proposal most certainly does that. When successful, architecture takes into consideration joint activity in a senseful action, enabling participants to understand their place in the world. In other words, it opens space for experiencing purpose of an individual through participation in cultural institutions. In that way, architecture offers a place for existential orientation to the societies, and its meaning is time-related. (Perez-Gomez, 2009, p. 143) Architecture offers understanding of one's place in the universe and in the civic world; it changes the life of an individual and provides grounds for one's own being. Indeed, inspite the seductive abilities of the modern technology and telecommunicational capacities aimed to strengthen democratic processes, it is important to understand how the local artistic and architectural practices resemble the precious endangered species. That means that architecture is neither a clear form, nor is it exclusively determined by socio-economic and functional limitations; rather, the quest for its definition should always be within urban dimensions, that is, within a spatial context as well.

*The surrounding with its material and non-material facts becomes a part of an architectural strategy in which a museum is a catalyst of progress and cultural identity enhancement. Perhaps a new architectural strategy lies there, a strategy that warns us, in the time of the global culture, that only through critical thinking of architecture can we construct, react, and change the expected urban succession, not by the ad hoc, hermetically closed concepts, which turn the existing form into their opposite.*

## REFERENCES


Bower, R., 2016. *Architecture and Space Re-imagined: Learning from the Difference, Multiplicity, and Otherness of Development Practice.* s.l.: Routledge.

Giebelhausen, M., 2003. *The Architecture of the Museum: Symbolic Structures, Urban Contexts.* s.l.: Manchester University Press.

Ibrišimbegović, S., 2015. *Arhitektura muzeja savremene umjetnosti kao kapsula vremena.* Sarajevo: PhDissertation.
Liebeskind, D., 2003. Traces of the unborn. In: *Architecture and Revolution: Contemporary Perspectives on Central and Eastern Europe.* London: Routledge, p. 256.
Mrduljaš, M., 2009. Otvorena arhitektura. *ORIS časopis za arhitekturu i kulturu: broj 55*, pp. 86–97.
Mrduljaš, M., 2010. Vrijeme muzeja. *ORIS časopis za arhitekturu i kulturu; broj 63.*, pp. 8–28.
Mrduljaš, M., 2012. Naoshima. *ORIS časopis za arhitekturu i kulturu: broj 76,* pp. 8–39.
Norberg-Schulz, C., 1979. *Genius Loci, Towards a Phenomenology of Architecture.* s.l.: Rizzoli.
Norberg-Schulz, C., 1996. The Phenomenon of Place. In: K. Nesbitt, ed. *Theorizing a new agenda for architecture: An anthology of architectural Theory 1965–1995.* New York: Princeton Architectural Press.
Norberg-Schulz, C., 2009. *Intencije u arhitekturi.* Zagreb: Naklada Jesenski i Turk.
Perez-Gomez, A., 2009. Izgrašeno na ljubavi: arhitektonska čežnja za etikom i estetikom. *ORIS časopis za arhitekturu i kulturu: br. 55*, pp. 140–147.
Rossi, A., 1984. *The Architecture of the City.* Boston: MIT Press.
Tschumi, B., 2004. *Arhitektura i disjunkcija.* Zagreb: AGM.
Ugljen-Ademović, N., 2004. *Sistematično proučavanje starog da bi se moglo izraditi kreativno novo.* s.l.:Arhitektonski fakultet Univerziteta u Sarajevu.
Ugljen-Ademović, N., 2007. *Dvojnost pirstupa problemu integriranja novog u postojeće u arhitektonskom oblikovanju.* s.l.:Doktorska disertacija, Arhitektonski fakultet Univerziteta u Sarajevu.
Ugljen-Ademović, N., 2012. *Kritika - stimulans arhitektonskoj ideji.* Sarajevo: Dobra Knjiga d.o.o.

## FIGURES

Figures 1 & 2. German Pavilion in Barcelona, Spain, designed by Ludwig Mies van der Rohe for the International Exposition in 1929; Neue Staatsgalerie, Stuttgart, James Stirling, 1984. http://www.jasonmkelly.com/2013/09/30/mies-van-der-rohe-barcelona-pavilion/; http://kultur-online.net/node/17486

Figures 3 & 4. History museum: View of exterior garden, View from courtyard towards the street www.muzej.ba; http://u-tt.com/exhibition/sarajevo-now-at-the-venice-biennale/

Figure 5 & 6. German Pavilion in Barcelona, Spain, designed by Ludwig Mies van der Rohe for the International Exposition in 1929; Neue Staatsgalerie, Stuttgart, James Stirling, 1984. http://archinect.com/features/article/149947453/examining-the-2016-venice-biennale-sarajevo-now; http://muzej.ba/sarajevo-now-peoples-museum/

Figures 7 & 8. site plan illustrating the proposal in more detail; The decay of the original building is preserved image http://u-tt.com/exhibition/sarajevo-now-at-the-venice-biennale/; http://www.designboom.com/architecture/venice-architecture-biennale-urban-think-tank-sarajevo-now-the-peoples-museum-05-31-2016/

*Cities' Identity Through Architecture and Arts – Catalani et al. (Eds)*
*© 2018 Taylor & Francis Group, London, ISBN 978-1-138-05409-7*

# Conserving historical areas through the roles of main cities: Urban identity in the era of globalisation

Ahmed Elewa
*Department of Architecture, Faculty of Fine Arts, Helwan University, Helwan, Egypt*

ABSTRACT: Throughout the last three decades, the process of globalisation has impacted the spatial urban structure of the world's main cities; however, this impact has negatively affected cities that still have historical hubs and are looking to preserve their local urban and architectural heritage. Meanwhile, adhering to the common requirements of urbanisation for the sake of globalisation has been mandatory for economic and political reasons. The cities that managed to adapt to the impacts of globalisation and, in turn, conserved their urban and architectural heritage, are currently the ones that have a significant and unique urban identity. Cities like Rome, London, Paris and Istanbul have generated good examples of dealing with this issue.

The hypothesis behind this study is that urban identity in main cities can balance between the common requirements of globalisation and the conservation of their architectural heritage. This paper will discuss the connection between the urban requirements of globalisation, the conservation of urban local heritage and urban identity.

The methodology used depended on an analytical comparison of selected case studies to evaluate their urban experiments. The results showed that cities with a significant urban identity have created successful urban experiments, showing that they have been able to find a balance between globalisation and the conservation of local urban heritage.

*Keywords*: globalisation, spatial urban structure, historical hubs, heritage

## 1 INTRODUCTION

It is the era of globalisation. Over the past decades the world has witnessed the results of globalisation as a process. Globalisation has become a commonly known term, particularly in the field of contemporary world economics and culture. Meanwhile, there are other parallel processes in addition to globalisation that have affected the formation of the world's main cities (WMCs) of today throughout the recent decades. The most significant process is the rapid growth of the urban population, since it is projected that the urban areas in the world will host 70 per cent of the world's population by 2050. According to the United Nations (2014), this future urban population growth is expected to be concentrated in the WMCs that consist of at least 10 million inhabitants.

Globalisation and the rapid growth of urban population are continuous processes that are strongly linked to the WMCs. A new type of WMC has risen; this is what is now known as the so-called global city. Today, WMCs compete for global resources, capital and talented human resources, positioning themselves as platforms for innovations and ideas to grow. Global cities are sharing a global urban identity, which reflects the new spatial urban plan as well commonly urban features. The transformation process of the WMCs has negative impacts on the local urban identity and poses a threat to historical areas given that they may be neglected and lose their importance. This is why there is growing awareness among scholars and decision makers about the necessity of conserving these areas, not only to preserve buildings but also to protect the historical urban context as a physical component of the local urban identity in the era of globalisation.

While the majority of research that concerns globalisation focused on its economic and cultural understanding, this study illustrates this phenomenon from an urban perspective. The study discusses the possibility of using the roles of urban identity as a tool to conserve the historical areas of today's WMCs that transformed to be global cities. This hypothesis is based on the practical results extracted from actual cases of global cities that managed to conserve their historical areas while still responding to the mandatory urban reforms of globalisation.

## 2 GLOBALISATION AND URBAN IDENTITY: AN OVERVIEW

The rise of global cities has led to a new type of urban identity that can be observed in today's WMCs. This new global urban identity has impacted the WMCs and has imposed urban features, which has led to reforms in the urban spatial structure of those cities to be global cities.

This globalised urban identity may be suitable (or ideal in some cases) for some of the WMCs, such as in the case of Sydney, Rio de Janeiro and Dubai. This particular type of WMC involves new cities in comparison to the world's main historical cities, which include cities such as London, Paris, Vienna, Istanbul and Cairo. These historical WMCs already have their local urban identity; however, policy makers in these cities have to respond to the mandatory urban requirements of globalisation, which may cause negative impacts on the original urban heritage of these cities. There are some good examples of WMCs that managed to balance the need to conserve their local urban heritage and the need to be global cities. This study raises specific questions about the nexus between globalisation, global cities and urban identity:

- What is the meaning of globalisation from an urban perspective?
- Are there new types of WMCs due to the impact of globalisation?
- What is the impact of globalisation on the urban identity of WMCs?

### 2.1 *Globalisation from an urban perspective*

Globalisation is a recent term, the current meaning being established in the 1970s (James & Steger, 2014). Globalisation as a phenomenon has impacted all international anthropogenic activities, including international economy and culture. As a result, daily urban life has changed in the WMCs, having lost a lot of their own local essence. In addition, the term 'globalisation' represents an interdisciplinary concept of uncertain definition due to differences in views and interpretations. Thus, one of the main questions of this research has to do with a clear definition of 'globalisation' from an urban perspective. There is a strong nexus between globalisation as a process and WMCs of today or what are so-called global cities. These cities are threatened by the process of globalisation, in particular, the type of cities in which there is a unique local urban identity; for example, the main historical cities of Europe, such as London, Paris and Rome. These cities are threatened by shifts in the global economy and this process has already started early in the last three decades of the 20th century. The shift in the global economy has impacted the WMCs, leading to physical changes in the urban spatial plan of those cities; however, these cities were and still are in the arena for the processes of globalisation. The definition of globalisation from an urban perspective can be extracted from the understanding of its influences on the urban spatial plan of global cities; globalisation has the same effects over the cities. 'Cities that are getting shaped according to the new world order are tending to show similarities between themselves within globalisation' (Hergül, 2014).

Globalisation, from an urban perspective, can also be understood as the phenomenon by which today's WMCs are being reformed to a new urban spatial plan, which enables them to perform their key role as arenas for the various anthropogenic activities that occurred due to the globalisation process. This definition illustrates the strong nexus between globalisation as a process and WMCs (global cities) as they are the urban spaces in which globalisation takes place.

### 2.1.1 *Rise of global cities (WMCs of today)*

Through the recent decades, and under the impact of globalisation, the WMCs transformed into global cities; the new millennium witnessed the rise of global cities, which can be seen as the building blocks of globalisation (Charnock, 2013).

In this study, it is important to determine a specific definition of global cities from an urban perspective. This will be necessary to understand the influences of globalisation on WMCs and how they transformed into global cites. It will also help scholars to recognise the new urban spatial features that form their globalised urban identity.

### 2.1.2 *Definition of global cities*

The definition of global cities in this study is based on the understanding of globalisation from an urban perspective; in the light of this, global cities represent the physical part of globalisation. They are the urban production of globalisation. However, there is a broad definition of global cities from an economic perspective; for example, the definition by Sassen (1991), '[a] global city is a significant production point of specialised financial and producer services that make the globalised economy run', as well as the definition by Charnock (2013), 'an urban centre that enjoys significant competitive advantages and that serves as a hub within a globalised economic system'.

In actuality, this idea of defining global cities from an economic point of view does not in any way oppose the definition of global cities from an urban perspective. On the contrary, both points of view investigate the integrated relationship between globalisation as a process and global cities as the urban nodes where the operations of the global economic system takes place; this is in line with Keil and Brenner's (2006) definition of global cities: 'they are exactly the interplay between globalisation and urban development'.

## 2.2 *Urban identity of WMCs (global cities)*

Urban identity, place identity or what is known as urban character are terms that refer to the same concept that is concerned with the recognition of urban areas through their significant urban features. These features reflect the socio-economic and cultural functions of urban areas. The term 'place identity' was identified by many authors with a background in urban studies. Lynch (1984) defined place identity as 'the simplest form of sense of place. Identity is the extent to which a person can recognise or recall a place as being distinct from other places-as having a vivid, or unique, or at least a particular, character of its own.' This definition is in line with the views of Proshansky et al. (1983) and Hague and Jenkins, (2005), which identify urban identity as a term that concerns the meaning and significance of urban places for their inhabitancy and users. Urban identity has become a significant urban issue during the last three decades as a result of the awareness of specialists and the local authority of the necessity to protect the local urban heritage. This took place notably under the impact of globalisation and the rising fears about the loss of individuality and distinctiveness between urban areas (see Figure 1). 'There has been a growing concern that local communities, towns, cities and regions are losing their identity in the midst of rapid globalisation and urbanisation' (Kim, 2000).

Global cities are those that have responded to globalisation requirements and, as a result, their urban spatial plans have already changed to a new one, with new urban characteristics that can be summarised as follows:

- An active central business district (CBD) or districts that offer a variety of international financial services, including finance, banking, insurance, real estate and other related activities (logistic services), such as hotels and conference halls.
- An advanced infrastructure system, notably the communication sector.
- An advanced transportation system that offers multiple modes of transportations and serves a large mass transit network.
- One or more major international airport(s) that make the city globally well connected.
- A liveable downtown area that offers multiple socio-economic, recreational and cultural activities attracting global tourism.

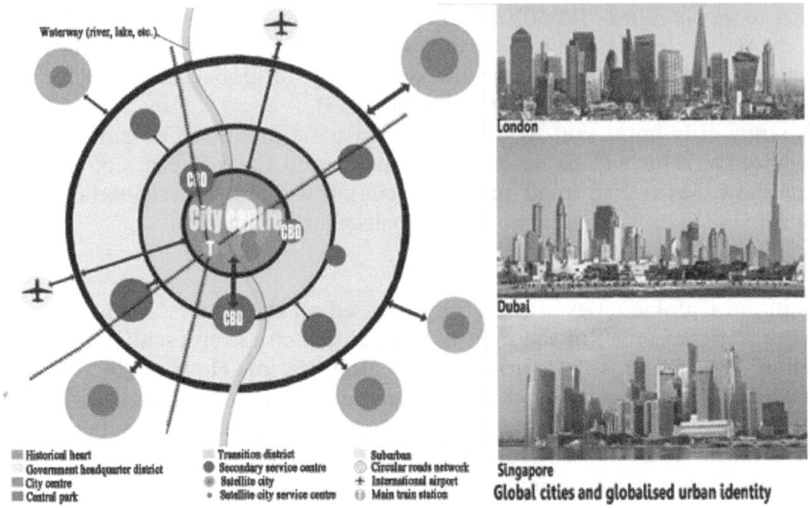


Figure 1. Global city model and the globalised urban identity (Source: Author).

- Having a prominent skyline that characterises its CBD area and summarises its economic power as a global city.

Globalisation has impacted the urban spatial plan of the WMCs to be transformed into global cities; see Figure 1, which shows a model of the urban spatial plan of global cities.

### 2.3 *Urban identity of historical WMCs (historical global cities)*

WMCs have transformed into global cities with a new urban spatial plan that includes some significant urban features forming its global urban identity. This urban identity symbolises the impact of globalisation on today's WMCs through a dominant identity that characterises all global cities. This new globalised urban identity was not a problem in the case of global cities that have no significant local urban identity, such as Dubai (there is a consensus on considering it a global city) (A.T. Kearney, 2016), which has no deep history. However it is a real challenge in the case of WMCs with a significant local urban heritage (due to their historic legacy) to appropriate between globalisation requirements and the conservation of its local urban identity. Obviously, there are two types of WMCs (global cities) in terms of the urban identity issue:

- First type: Contemporary global cities such as Singapore, Dubai and Abu Dhabi;
- Second type: Historical global cities such as London, Paris, Berlin and Rome.

#### 2.3.1 *The intended meaning of conserving historical areas in this study*

In this study, the intended meaning of conserving historical areas is broad. It is not only about the preservation of buildings and the urban context as they represent the physical aspect of the historical areas, but also about how to protect the scenic views of historical areas and individual valuable buildings. This understanding of conservation means to protect the view of a specific place or historic building from another location.

## 3 URBAN IDENTITY AS A CONSERVATION TOOL

Based on the definition that has been reached for urban identity, using urban identity as a tool to conserve historical areas is a reasonable concept; there is a strong nexus between the urban identity of a city and its local urban heritage. Urban identity concerns the meaning and significance of urban places that form the local urban heritage and give each city its own urban identity. In other words, the existence of urban identity requires the existence of significant urban features such as the historical urban areas; thus, any strategy that aims to maintain a distinguished urban identity ought to take into consideration the conservation of historical areas.

### 3.1 *The roles of urban identity in historical WMCs (historical global cities)*

The discussion about urban identity explains the main role that it plays in WMCs, which is to protect the local identity. It also assumes other roles that support the conservation of historical areas in direct and indirect ways as follows:

- A remarkable urban identity relies on the existence of historical areas as a physical criterion, according to Lynch (1984): 'Design scholars also focus on physical aspects of local identity of place in their theoretical studies.' This is why urban identity plays an important role in dealing with the undesired effects of modern urbanisation that are impacted by globalisation. As a result of this, urbanism is almost meaningless and lacks any unique identity (Tavakoli, 2010).
- Urban identity serves as a reference point in terms of conserving a sustainable urban image and in terms of the wishes of the society. It is also important to note that the recognition of the value of an urban area is a fundamental component in urban studies. 'The sustainability of a place depends on a series of factors, which contribute to the quality of life, sense of place and recognition of identity' (Sepe, 2006).
- Urban identity acts as an index for the quality of urban life and the socio-economic conditions of each city; moreover, a unique urban identity can be a source of community pride and satisfaction with their city (Ghavampour et al., 2008).
- Urban identity is the reflection of the historical context of the city as a part of both its physical components and cultural spirit. The urban identity of cities grows from the continuous relationship between the place and its residents by forming the inhabitants' collective memory; at the same time, it is a means of making the community aware of the conservation of the historical context (Oktay, 2002).
- Urban identity can be an honest indicator of the development rate of a city. According to Torabi and Sima (2013), it is a measure of growth for the city's identity as a factor for the development and promotion of environmental quality.
- A significant unique urban identity can also act as an economic tool that can boost city branding among the other competitive global cities. This essentially means more foreign investments and attracting the talents of various people. 'A strong sense of identity can be an attractor, bringing new investment and talent into an area' (Watson & Bentley, 2007).

## 4 CASE STUDIES

Two cases of European cities were selected to represent examples of historical WMCs (historical global cities) that still have historical areas (this includes historical buildings and urban contexts from different ancient ages up to the 19th century). An analytical comparative study was done to clarify how the decision maker managed to use the roles of urban identity in the era of globalisation to conserve historical areas. The criteria of the analytical comparison study included the following aspects:

- A brief note about the case.
- Recognition of the current urban identity through the study of the urban spatial plan of the city and how far it responded to the globalisation impact (through the existence of the urban characteristics of global cities that was highlighted before); also, exploration of the current situation of historical areas and recognition of the current skyline of the city as a reliable indicator of the current urban identity.
- Explanation of how the decision maker, through urban strategies and regulations, dealt with the issue of conserving historical areas in this age of globalisation through the roles of urban identity.

### 4.1 *Case study of London*

London is the capital city of the United Kingdom. It is also the largest metropolitan area and the largest urban zone in the European Union by most measures. The city is an example of the historical capitals of Europe. London still has significant historical areas, notably from the 18th and 19th centuries. Further, the city was ranked as the most important global city in the world in 2016 (A.T. Kearney, 2016).

#### 4.1.1 *Urban spatial plan of London*

London is one of the most important global cities in the world according to most urban scholars, such as Sassen (1991). London's urban spatial plan is a clear example of plans of global cities, as the common urban features of global cities exist in it. London has two active CBDs: London city centre and Canary Wharf (i.e. the transformed docklands area); both these form an important hub for the global economic system. London is one of the most important destinations for international flights; the city is served by six international airports and several smaller airports, and together they make one of the busiest airport systems in the world in terms of passenger numbers. The city has a liveable downtown area, forming London's socio-economic, cultural and historical heart. It is an important global centre for media and innovation. It also has an advanced public transportation system and infrastructure (see Figure 2).

#### 4.1.2 *Historical areas in London*

Although London has a rich urban and architectural history, there is no patchwork of historical areas of ancient ages; however, like other historical European main cities, London has its classic urban context in the downtown area (City of London and Westminster) that reflects Victorian London in the 18th and 19th century. London's scheduled sites and listed buildings are individual structures, in many cases assembled gradually by parts from many different periods (Roumpani & Hudson, 2014).

#### 4.1.3 *Recognition of the current urban identity of London*

London has no prominent skyline that characterises the urban identity of the city as a contemporary global city or as a historical city with a local urban heritage that it relies on. The urban identity of London is derived from its long history and its contemporary role as a major global financial centre. These factors have resulted in a city skyline of great complexity and diversity. It is an embodiment of the conflict between the global urban identity and the local urban identity that reflects its local urban heritage.

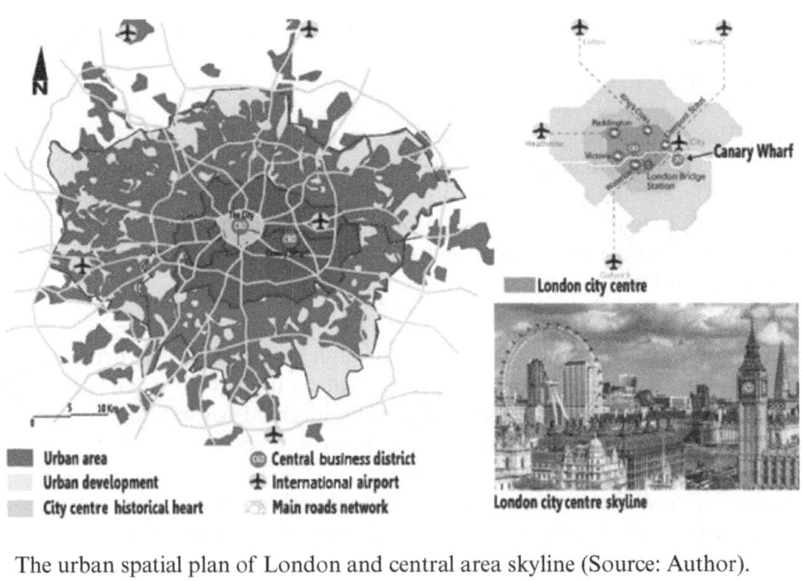


Figure 2. The urban spatial plan of London and central area skyline (Source: Author).

#### 4.1.4 *Urban strategies and regulations that form the current urban identity*

The decision maker's awareness of the urban changes that are related to globalisation started in the 1990s through the successive urban development plans of the city. In the 1994 Unitary Development Plan, the issue of conserving the historical areas in London was the main concern and a priority target. This plan was based on previous plans in 1989 and the supplementary guidance for London on the protection of strategic views in 1991. The plan defined the historical areas as 'areas of special architectural or historic interest, the character of which it is designed to preserve or enhance' (English Heritage, 2016). The plan also regarded the review of the boundaries of conservation areas as the city's duty with a priority to protect London's skyline, especially its landmarks and historical buildings.

In the year 2000, the Greater London Authority (GLA) was established. In the updated plan of 2009, the mayor of London described his vision and objectives to conserve historical areas as the main physical component of the local urban heritage and to ensure that London is a leading global city. This vision demonstrated how to use the roles of urban identity as tools to conserve the historical areas. According to the mayor of London, over the years to 2031 and beyond, London should be ensuring the following aspects:

- An internationally competitive and successful city with a strong and diverse economy and an entrepreneurial spirit; a city which is at the leading edge of innovation and research and which is comfortable with—and makes the most of—its rich heritage and cultural resources.
- A city that delights the senses and takes care of its buildings and streets, having the best of modern architecture while also making the most of London's built heritage.
- Supporting a high-quality urban living space—including protection of London's heritage from air and noise pollution (Blowers & Evans, 1997).

### 4.2 *Case study of Paris*

Paris is the capital city of France and a socio-economic and cultural hub. It has a special place in people's imaginations because it owns a significant urban identity that is derived from its historical downtown (the Paris of Haussmann) (Ehrlich et al., 2016). The city is also one of the first global cities (A.T. Kearney, 2016).

#### 4.2.1 *Urban spatial plan of Paris*

While Paris is ranked as the third among the top 15 global cities (A.T. Kearney, 2016), the city still has areas that reflect its local urban heritage. The urban spatial plan of Paris represents the same common urban features of global cities, with a unique central area known as the Paris of Haussmann (Enright, 2016). It is an active liveable downtown that represents a good example of the socio-economic and cultural activities that should exist in a WMC and, of course, in a global city as well. Paris is a centre of innovation in fashion, arts and other cultural aspects. The city has a main CBD, La Défense, which is the hub for global economic activities. It is home to no fewer than 1,500 corporate head offices (Lang, 2005) and is located to the north-west of the Paris of Haussmann. The city is a major hub for advanced transportation systems that connect the city locally and with Europe; it is also connected globally by two international airports, Charles de Gaulle, which is the second busiest in Europe, and Orly (Ehrlich et al., 2016) (see Figure 3).

#### 4.2.2 *Historical areas in Paris*

Paris has a unique and long history; the city has grown over the centuries from a central point that dates back before Roman times, this point is in Îll de la Cité. The whole central area is the Paris of Haussmann, which has a unique urban context of the 18th and 19th centuries. There are also sites and historical buildings of various eras, notably the medieval times. The central area also contains the historical districts of La Cité, Louvre, Champs-Elysées, Les Grands Boulevards and Le Quartier Latin. The area is almost exactly delimited by the route of the Paris inner ring road, or Boulevard Périphérique, which circles the city centre.

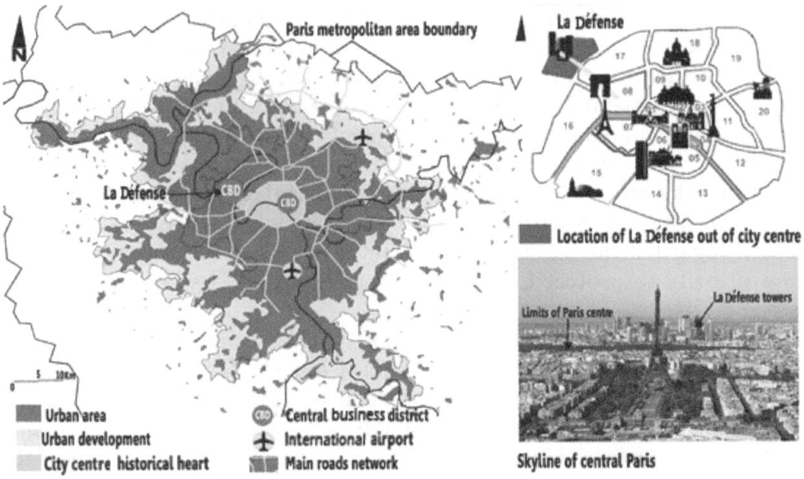


Figure 3. The urban spatial plan of Paris and the central area skyline (Source: Author).

4.2.3 *Recognition of the current urban identity of Paris*

Paris has many faces, in other words the city has clear areas and each one has a dominant urban identity. So far, the city, through its urban spatial plan, has managed to keep its historical heart protected from the invasion of contemporary tall buildings of globalisation. The city centre still has its own classical urban identity, but the city has another face that represents its global urban identity which is clearly observed in the areas that are located out of the city centre, such as La Défense where the main CBD of Paris is located. Thus, this is why the city has diverse skylines, each one with a dominant urban heritage, and that is what marks the city as a unique example of historical global cities.

4.2.4 *Urban strategies and regulations that form the current urban identity*

The current iconic skyline of central Paris is owed to Georges Haussmann because his urban modification (started in 1853 and continued to the end of the century) remodelled the medieval urban spatial plan of the city in keeping with the significant urban context of the 19th century. It is characterised by long straight boulevards, magnificent open spaces and elegant building façades. Haussmann also set urban regulations that were imposed on building façades and heights. He set a decree that buildings should not be more than five storeys high (Girouard, 1985). These buildings allow for a clear line of sight to the Eiffel Tower and other monuments that define Paris architecturally (Karmelek, 2015).

The decision to establish the CBD La Défense in 1958 west of central Paris (now a part of the Paris metropolitan area) was a successful one. La Défense has a towering skyline that represents the global identity of the city. This project protects the historical heart of the city that is mostly free of tower blocks, except the Montparnasse Tower built in 1973. This 59-storey, 209-metre-high eyesore has had few champions since it was built. Following this experience, the City Council in 1976 set a height limit of 37 metres (121 feet) for new buildings within the city limits. This allowed famous monuments, notably the Eiffel Tower, to form the unique skyline of Paris and to give the city its significant urban identity. However, this unique skyline and urban identity is threatened by the urban requirements of globalisation, given that in 2010 the City Council raised the height limit to 50 metres in certain central areas and 180 metres in the city's outer areas (Karmelek, 2015).

4.3 *Evaluating the cases through a comparative study*

For a brief comparison of the two cases, see Table 1.

Table 1. A brief comparison of the cases of London and Paris.

<table>
<tr><th>Aspects of comparison</th><th>Case of London</th><th>Case of Paris</th></tr>
<tr><td>Urban spatial plan</td><td>The city's spatial plan contains the main urban components of global cities.<br>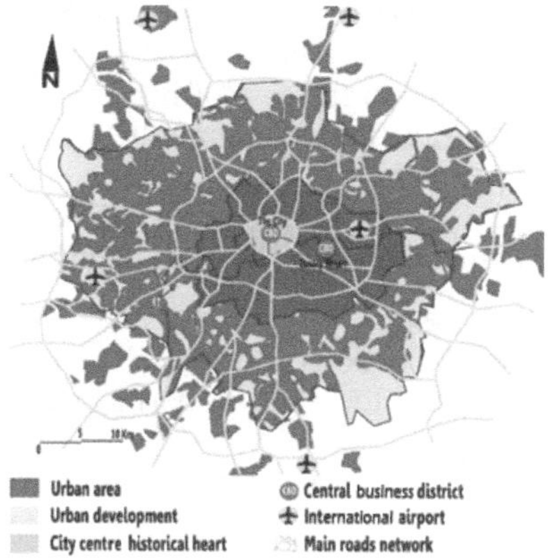<br></td><td>Although the city's spatial plan contains the main urban components of global cities, the city centre (the historical heart) still has a classical urban spatial context.<br>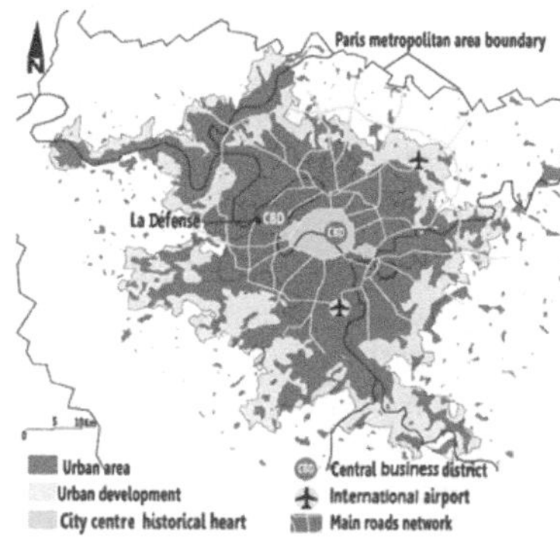<br></td></tr>
<tr><td>Historical areas</td><td>Like other historical European main cities, London has its classical urban context in the downtown area (City of London and Westminster). London's scheduled sites and listed buildings are individual structures that are distributed randomly in the city.</td><td>The whole central area is the Paris of Haussmann, which has a unique urban context of the 18th and 19th centuries.</td></tr>
<tr><td>Recognition of the current urban identity</td><td>The city has no prominent skyline that characterises the urban identity of the city as a contemporary global city or as a historical city with a local urban heritage. The urban identity is derived from London's long history and its contemporary role as a major global financial centre.<br></td><td>The city has clear areas and each one has a dominant urban identity. The city centre still has its own classical urban identity. The global urban identity is clearly observed in the areas that are located outside the city centre, such as La Défense where the main CBD of Paris is located.<br></td></tr>
<tr><td>Urban strategies and regulations</td><td>The supplementary guidance for London set regulations that protected strategic views in 1991. The 1994 Unitary Development Plan defined the historical areas and regarded the review of the boundaries of conservation areas as the city's duty. In 2000, the Greater London Authority was established. In the updated plan of 2009, the mayor of London described his vision and objectives to conserve historical areas and to ensure that London is a leading global city.</td><td>Haussmann set urban regulations in central Paris that were imposed on building façades and heights. He set a decree that buildings should not be more than five storeys high and roofs should have a 45-degree pitch to allow daylight to reach the sidewalks. In 1976, The City Council set a height limit of 37 metres (121 feet) for new buildings within the city limits. In 2010, it raised the height limit to 50 metres (160 feet) in certain central areas and 180 metres (590 feet) in the city's outer areas.</td></tr>
</table>

## 5 CONCLUSION

WMCs have transformed to today's global cities under the impact of the globalisation process. We live in an era of globalisation where there is socio-economic and cultural competition among the world's global cities. The cities that meet the mandatory urban requirements of globalisation are facing a serious challenge about the possibility of achieving a balance between conserving their local urban identity and the new globalised urban identity. This challenge is not a problem in the case of global cities that have no significant historical areas, such as Dubai; however, it is more complicated in the case of historical global cities. Generally, there is a dominant global urban identity that is commonly shared among the global cities.

The analytical comparative study of the cases of London and Paris clarified urban lessons about the issues of conserving historical areas through the roles of urban identity in today's global cities as follows:

- The two cases showed clearly that the conservation of historical areas is essential to obtain a significant urban identity that reflects the local urban heritage, notably in the era of globalisation.
- The conservation process is not limited to its direct concrete meaning, which concentrates on the preservation of historical buildings and the whole urban context. It also includes the protection of these areas to ensure the formation of a unique skyline, as clarified in the case of Paris, through urban strategies that aim to offer an integrated vision of the conservation process. This requires some restrictive policies such as a restriction on the height of new buildings and to offer designated areas for high-rise buildings; these areas have no negative impact on the skyline of the historical area, as was clear in the case of Paris through the establishment of the CBD, La Défense, out of the historical central area.
- In London's case, the GLA managed to conserve historical areas; however, this conservation was only to preserve the physical case of buildings and urban context and not to offer complete protection for the skyline of historical vistas, even under the 1991 supplementary guidance for London on the protection of strategic views. London competes to have a skyline that reflects globalisation (through high-rise buildings). This vision in the historical urban context means a city skyline of great complexity and diversity. It is an embodiment of the conflict between global urban identity and local urban identity.
- The urban identity of cities is derived mainly from their unique urban features. This means that there is a strong and direct nexus between urban identity roles and the conservation of historical areas, as those areas are the main physical components that form a significant urban identity. The urban identity of Paris is well known, even to non-professionals, through its unique skyline that is characterised by iconic historical buildings such as the Eiffel Tower, which is still almost the tallest building in central Paris.

It is obvious that the protection of urban identity is the responsibility of urban policymakers. Cities that have a significant urban identity also have significant conserved historical areas. The cases of London and Paris illustrate how the roles of urban identity of WMCs can be used to conserve these historical areas.

## REFERENCES


Blowers, A. & Evans, B. (1997). *Town planning into the 21st century* (p. 71). London, UK: Routledge.

Charnock, G. (2013). Global city. *Encyclopaedia Britannica*. Retrieved from https://www.britannica.com/topic/global-city.

Ehrlich, B., et al. (2016). Paris. *Encyclopaedia Britannica*. Retrieved from https://www.britannica.com/place/Paris.

English Heritage. (2016). Retrieved from http://www.english-heritage.org.uk/.

Enright, T. (2016). *The making of Grand Paris: Metropolitan urbanism in the twenty-first century*. Cambridge, MA: MIT Press.

Ghavampour, E., Rabiee, Z., Norouzi, H. & Mansoori, S.A. (2008). Identifying effective elements in the landscape of historical districts (Typical sample: Oudlaajaan neighborhood in Tehran). In *2nd International Conference on Built Environment in Developing Countries* (pp. 1693–1705).
Giap, T., Thye, W. & Aw, G. (2014). A new approach to measuring the liveability of cities: The Global Liveable Cities Index. *World Review of Science, Technology and Sustainable Development*, *11*(2), 176–196.
Girouard, M. (1985). *Cities and people: A social and architectural history* (1st ed., p. 285). London, UK: Yale University Press.
Hague, C. & Jenkins, P. (Eds.). (2005). *Place identity, planning and participation*. London, UK: Routledge.
Hergül, O. (2014). Globalization: Effects to urban identity and relation with slow city movement. Retrieved from http://www.inter-disciplinary.net/critical-issues/wp-content/uploads/2014/04/CandanCengiz-up 4dpaper.pdf.
James, P. & Steger, M. (2014). A genealogy of globalization: The career of a concept. *Globalizations*, *11*(4), 417–434. doi:10.1080/14747731.2014.951186.
Karmelek, M. (2015). Will skyscrapers ruin Paris? *Europe Newsweek*. Retrieved from http://europe.newsweek.com/will-skyscrapers-ruin-paris-330683?rm=eu.
Kearney, A.T. (2016). *Global cities 2016*. Retrieved from https://www.atkearney.com/documents/10192/8178456/Global+Cities+2016.pdf.
Keil, R. & Brenner, N. (2006). *The global cities reader*. London, UK: Routledge.
Kim, J. (2000). Understanding elements of local identity of place: Physical vs. personal-social attributes. In Wells-Bowie, L. (Ed.), *88th ACSA Annual Meeting, Heterotopolis* (pp. 451–456). Retrieved from http://apps.acsa-arch.org/resources/proceedings/uploads/streamfile.aspx?path=ACSA.AM.88&name=ACSA.AM.88.79.pdf.
Lang, J. (2005). *Urban design: A typology of procedures and products* (pp. 219–225). London, UK: Architectural Press.
Lean, G. (2015). Does our history have a future in the hands of the English Heritage Trust? *The Telegraph*. Retrieved from http://www.telegraph.co.uk/news/earth/environment/conservation/11441254/Does-our-history-have-a-future-in-the-hands-of-the-English-Heritage-Trust.html.
Lynch, K. (1984). *Good city form*. Cambridge, MA: MIT Press.
Oktay, D. (2002). The quest for urban identity in the changing context of the city: Northern Cyprus. *Cities*, *19*, 261–271.
Proshansky, H.M., Fabian, A.K. & Kaminoff, R. (1983). Place-identity: Physical world socialization of the self. *Journal of Environmental Psychology*, *3*, 57–83.
Roumpani, F. & Hudson, P. (2014). The evolution of London: The city's near-2,000 year history mapped. *The Guardian*. Retrieved from https://www.theguardian.com/cities/2014/may/15/the-evolution-of-london-the-citys-near-2000-year-history-mapped.
Sassen, S. (1991). *The global city: New York, London, Tokyo* (1st ed.). Princeton, NJ: Princeton University Press.
Sassen, S. (2005). The global city: Introducing a concept. *Brown Journal of World Affairs*, *11*(2), 27–43.
Sepe, M. (2006). Sustainable walkability and place identity. *International Journal of Architectural and Environmental Engineering*, *1*(4), 50–55.
Tavakoli, N. (2010). The role of physical identity of city in urban sustainability (the case study: Yazd Iran). 14th International Planning History Society Conference. Retrieved from http://www.iphs2010.com/abs/ID216.pdf.
Torabi, Z. & Sima, Y. (2013). Urban identity in the entrance of city. *International Journal of Architecture and Urban Development*, *3*(4), 47–54.
United Nations. (2014). *World urbanization prospects: The 2014 revision, Highlights*. ST/ESA/SER.A/352. New York: Department of Economic and Social Affairs, Population Division.
Watson, G.B. & Bentley, I. (2007). *Identity by design*. Oxford, UK: Elsevier.

*Cities' Identity Through Architecture and Arts – Catalani et al. (Eds)*
*© 2018 Taylor & Francis Group, London, ISBN 978-1-138-05409-7*

# Entrance gateway of Kerala temples: Assessing the form of a Kerala temple gopuram through material and construction

M. Yamuna Vijayan
*MARG Institute of Design and Architecture, Swarnabhoomi, Tamil Nadu, India*

ABSTRACT: Every style of building construction reflects a clearly distinctive principle that represents a particular culture and era. In India, the temple and its associated structures evolved with time and the most identifiable difference is in their structure. Specifically in Kerala, a southern state of India, temple architecture is different from that in neighbouring states due to the obvious reason of its geography and various other factors. Gopurams (entrance gateways) are the principal structure of any temple complex. This led to the investigation of the difference in the form of a gopuram in terms of its material and construction. The topic demands a literature study of Dravidian and Kerala gopurams, followed by case studies of different types to support the investigation. In this research, the major reason behind the differences was found to lie in the geography of the state, which itself can be categorised into three different regions within the state such as Malabar, Kochi and Travancore. In addition, an analysis is done on the proportions of the gopurams. This is done by taking into account the dimensions of the temple as well as the structural elements of the gopuram. The conclusion summarises the principles of planning, differences in form, and the scope for future studies.

*Keywords*: Dravidian; Kerala; temple architecture; gopurams; proportions

## 1 INTRODUCTION

Symbolically, the gopuram or the entrance to a temple represents the feet of a deity. A devotee bows at the feet of the Lord at the entrance as one steps into the temple and proceeds towards the sanctum. Compared with the gopurams in the neighbouring state of Tamil Nadu the gopurams of Kerala are insignificant in height and dimensions (Sarkar, 1978). The purpose of this research is to investigate the origin of its form for an academic audience from the field of art and architecture either across the world or from the state of Kerala.

Study of the morphological differences in temple gopurams helps in understanding the underlying logic of their materials and construction. The research aims to give an insight into the practices of Dravidian architecture. The four questions that guide this work are as follows:

1. How did the form of the Kerala temple gopuram evolve in terms of its material and construction?
2. What are the types of temple gopurams seen across Kerala?
3. What are the materials and system of construction used in temple gopurams?
4. What are the rules and systems of proportion of temple gopuram construction?

The scope of the research analysis of morphological differences and studies the form of temple gopurams in relation to their materials and construction methods. Given limitations of time, specific case studies were chosen accordingly, avoiding the iconographic study of space and the details of craft in the structure. The study is limited to the typical architectural style and the importance of temples in the region. The investigation is purely based on the observations of the sites and the existing literature. Any renovations to the structures maybe a threat to the validity of the study.

## 2 METHODOLOGY

The research was done in three phases: a literature study, data collection with case studies, and data analysis. The methodology adopted was a combination of desk, field and case studies. Primary data were collected through the study of the literature in the form of published books, mainly classical texts on temple construction and other books relevant to the topic. Secondary data were collected through scheduled interviews with experts in temple construction. Site visits were carried out to various temples as part of the case studies, which helped to reveal the visible peculiarities of various temple gopurams. The study was conducted on the temple gopurams of Travancore, Kochi and Malabar in Kerala.

## 3 BACKGROUND STUDY

The 'gopuram' is a Sanskrit word meaning a town gate or the gateway to a temple. Its etymology is uncertain and prominent scholars such as acharya, harle have suggested that, in early times, it referred only to the superstructure above the gateway. These structures are generally raised upon one or more of the cardinal axes of a temple. Gopurams provide a means of access through the outer walls that delimit successive enclosures. After their evolution in the 12th century, gopurams had larger significance in the Dravidian region. The smallest and simplest form of the gopuram is no more than a doorway pierced through a wall, whereas the largest can have a measurement of 150 feet or more on one side and a height of 300 feet (Brown, 1959).

### 3.1 *Characteristics of Dravidian gopurams*

The characteristic feature of agopuram is the entrance passageway at the centre. This entrance reaches up to its cornice, dividing the portion below into two equal and separate portions.

In most gopurams, the entrance consists of doorways. Two identical rooms on either side open into the central portion of the entrance. Termed the *garbhagriha*, the holy of holies in a shrine, according to Harle (1995), these are vestibules. They are very deep recesses rather than rooms. The vestibules of larger gopurams are divided into two storeys of the same height. The floor of the upper vestibules is frequently supported by columns. A staircase usually permits one to mount one of the upper vestibules and to the upper storeys. The floor of the lower vestibules is commonly raised as much as 4 feet above the level of the entrance.

An interior circumambulatory/semi-circumambulatory corridor runs around each vestibule. In all later South Indian temples, the upper storeys of gopurams are made of brick, plaster and stucco, whereas the lower portions are stone or faced with stone. The upper storeys form a pylon-shaped tower; these are surmounted by oblong-shaped pavilion with barrel-shaped roofs. Every storey has a large opening on either side above the entrance. It has been noted in practice that gopurams usually have an uneven number of storeys (Harle, 1995). The sides of the tower have a decreasing batter as the number of storeys increases.

The number of gopurams in a temple roughly depends on the number of enclosures. The earliest temple in South India had a single enclosing wall (Brown, 1959).

### 3.2 *Keralan gopurams*

Compared with the gopurams in Tamil Nadu, the gopurams of Kerala are insignificant in height and dimensions. Yet, they are endowed with grace and nobility, and the architects of Kerala have never allowed them to outshine the temple. Like the temples, they are also built of laterite and wood. Roofs are made of tile and rise to a maximum height of three storeys.

The evolution of the gopuram in Kerala temple architecture cannot be traced as the evidence is minimal, but a study on its similarity with the architecture of the temple was done. Even though the temple architecture of Kerala is within the mainstream of the Indian temple-building tradition, it has always followed its own indigenous method (Panikkar, n.d.).

### 3.3 *Characteristics of Kerala temple gopurams*

The characteristic feature of a gopuram in a Kerala temple is the form itself, because it is different from the rest of the region's temple gopurams. Like other Dravidian gopurams, this also has a passage that allows one to enter into the temple complex. This passage consists of a doorway that has a height twice its width and reaches up to the ceiling in certain cases. What is different is the location of the doorway: it is not in the middle of the gopuram, but placed towards the outside, in line with the compound wall. The entrance with the doorway reaches its cornice, dividing the area below into two equal and separate portions.

The entrance divides the gopuram into two identical spaces, as there is no room in the case of Kerala gopurams. This space on either side opens into the central portion of the entrance. The vestibules are deeper than the entrance and sometimes wider too. The vestibules are supported by columns over the plinth that normally reach a height of more than 3 feet in the Kerala temple gopuram. A stairway from the ground permits one to mount the plinth.

The upper storeys of Kerala temple gopurams are made of laterite, wood and, sometimes, granite, whereas the lower plinth is granite. For the walls, wood as well as granite is used. The roof, which rests on wall plates, is made of a roof frame consisting of a ridge beam, collar beam rafter, and purlins and tiles laid on top of it. India's second tile factory was started in Calicut (Kozhikode) in the year 1873 (Mani, 1990). Before that the gopurams had thatched roofs. In Kerala, the gopurams are always seen as a two-storeyed building. The side of the upper storeys is always recessed from the ground and the rest of the space is covered with sloping eaves. Consequently, gopurams have survived from earlier than the 7th or 8th century AD. A background study of both Dravidian and Kerala temple gopurams was done to trace the evolution of the structure (Figure 1).

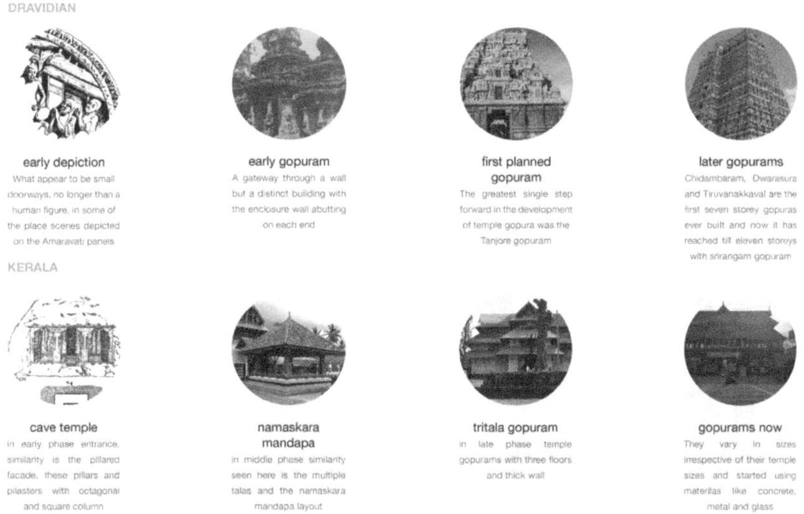


Figure 1. Evolution of Dravidian and Kerala temple gopurams.

## 4 PRINCIPLES OF GOPURAM PLANNING

### 4.1 *Dimension of gateways*

The *panchaprakaras* (five enclosures) are sometimes built as five solid walls. In such cases, entry gate structures (gopurams) are provided in each enclosure and are named *dwarasobha*, *dwarasala*, *dwaraprasada*, *dwaraharmyam*, and *dwaragopura* from the innermost to the outermost walls. The dimensions of these gopurams as well as the number of storeys, increases as one proceeds outwards (Prabhu & Achyuthan, 1996).

The width of the *dwarasobha*, gateway of the first enclosure, is six-sevenths that of the main temple; those of the gateways of the second, third and fourth enclosures are, respectively, seven-eighths, eight-ninths and nine-tenths. The width of the fifth enclosure, the gopuram, is ten-elevenths that of the main temple. This applies when the temple is small or very small (Dagens, 2007). The length of the gopuram is double the width for the fifth enclosure (Namboothiripad, 2013). The projection of the gateway is a quarter and two-fifths that of the outside of the enclosure walls in the third enclosure (Figures 2 and 3).

Heights calculated from their widths should be ten-sevenths, six-quarters, nine-fifths, and double. The height of the gateway is calculated as follows: the height of the base is three-quarters, four-fifths, five-sixths, six-sevenths, seven-eighths, eight-ninths, nine-tenths, ten-elevenths or eleven-thirteenths that of the base of the main temple.

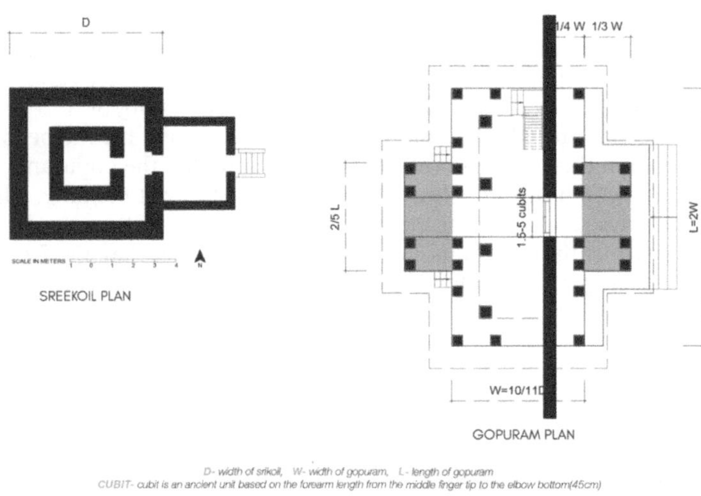


Figure 2. Proportion of the gopuram in relation to the temple plan (Source of base drawing: Kasturba, 2014).

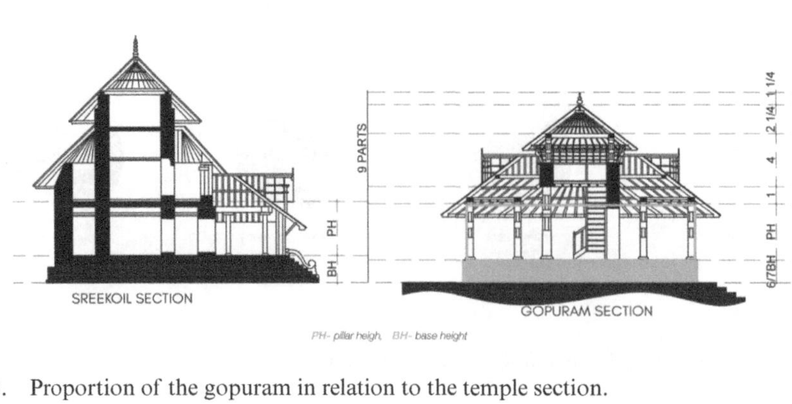


Figure 3. Proportion of the gopuram in relation to the temple section.

The height of the gateway pillars is the same as that of the main building or may be seven-eighths, eight-ninths, or nine-tenths of it; the pillars are sunk into the base at the bottom right down to the regulating course.

## 5 CASE STUDY

### 5.1 *Types of gopuram across Kerala*

There are three types of gopurams in Keralan temples, namely, open-type, closed-type and combination-type gopurams, depending on the form (Figure 4).

With study criteria including the form, material and construction of this region and the availability of adequate documentation, the following three temple gopurams were chosen for the case studies: 1) Tali temple of Calicut; 2) Vadakkumnathan temple of Thrissur; 3) Kaviyoor Mahadeva temple of Pathanamthitta (Figure 5).

### 5.2 *Tali temple gopuram, Calicut*

The Tali temple is a Siva temple situated in the northern part of the state of Kerala. Its gopuram form can be classified as the combination type, as we have an open ground floor and a closed upper floor with brackets and reapers.

The plinth and the pillars of the Tali temple are made of granite. The enclosure wall is made of laterite. The wall plate and the secondary wall plate above the pillars are made of timber. The framework for the ground-floor porch roof forming the gable is also made of timber. On the first floor, the walls are made of laterite and the pillars above it are granite covered with timber brackets and reapers. The first-floor roof structure is also made of timber with Mangalore tiles laid over it.

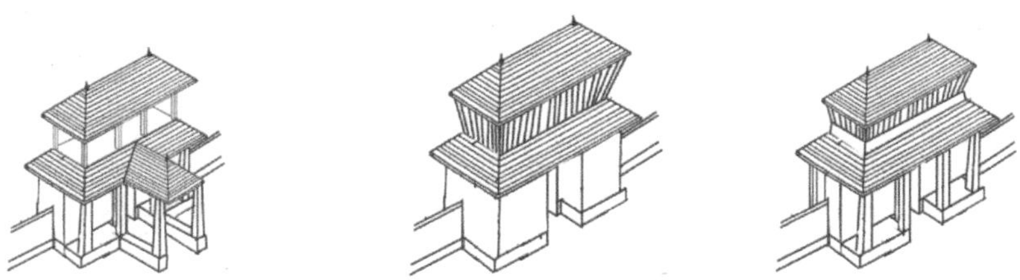

Figure 4. Types of gopuram: (from left) open type, closed type and combination type.

Figure 5. Case studies: (from *left*) Tali temple of Calicut, Vadakkumnathan temple of Thrissur, and Kaviyoor Mahadeva temple of Pathanamthitta.

### 5.3 *Vadakkumnathan temple gopuram, Thrissur*

The Vadakkumnathan temple, situated in central Kerala, is the first Siva temple believed to be created by Lord Parasurama. Its gopuram form can be classified as the closed type, as all three floors are closed with thick walls.

The plinth that forms the base structure is made of granite. The enclosure wall is made of laterite. The wall plate and the secondary wall plate that comes above the wall are made of granite and wood. The eaves of the ground floor are made of timber with Mangalore tiles laid over it. The framework for the ground-floor ceiling is also made of timber. On the first floor, the walls are made of laterite and the pillars on both sides are made of granite. This is the arrangement on the exterior wall. The first-floor eaves have the same timber framework with Mangalore tiles. The first-floor ceiling is also made of timber. The same materials are repeated on the second floor and the main roof frame is made of timber alone.

### 5.4 *Kaviyoor Mahadeva temple gopuram, Pathanamthitta*

The Kaviyoor Mahadeva temple is one of the important Siva temples, located in Kaviyoor, Pathanamthitta District in southern Kerala. Its gopuram form can be classified as the combination type, as there is an open ground floor with pillars and a closed upper floor with brackets and reapers.

The plinth and the pillars, which form the base structure, are made of granite and timber, respectively. The wall plate and the secondary wall plate that comes above are made of timber. This continues to its eaves as well. The framework for the ground-floor porch roof forming the gable, the pillar above the wall with the brackets and reapers, and the first-floor roof structure are all made out of timber.

## 6 ANALYSIS

### 6.1 *Analysis of proportions*

The height of the three temple gopurams varies dramatically from north to south, especially the one from central Kerala. This is because of the temple it belongs to. The gopuram takes its plinth and pillar height from its *garbhagriha* (innermost sanctum).

The height of the gateway pillars is the same as or in proportion to that of the main building; the pillars are sunk into the base at the bottom, right down to the regulating course. The door rises to the architrave and has a width half its height.

The gopuram of the Vadakkumnathan temple has the largest size compared with those of the Tali and the Kaviyoor Mahadeva temples.

The width of the fifth enclosure, the gopuram, is ten-elevenths that of the main temple. This applies when the temple is small or very small. In the three case studies, the gopurams of all three temples have a proportion of ten-elevenths that of the main temple.

Once the width of the gopuram is finalised, the rest of the dimensions are determined from the width of the gopuram; for example, the length is twice the width of the gopuram. The length should be two and two-thirds of the width of the gopuram as prescribed by Mayamatam, and double as prescribed by Tantrasamuchaya.

### 6.2 *Analysis of materials*

The materials used for gopuram construction were similar to those of the main temple. In the case of temple gopurams in northern Kerala, this is evident because of the availability of granite. The temple gopurams in southern Kerala largely used wood for the *garbhagriha*. The roofs are always in wood covered with Mangalore tiles.

The roof is the most important element in any traditional structure and that is what distinguishes it from region to region. Local variations in the arrangement of the rafters and the ridge beam are seen. These should be indigenously developed. In earlier methods, all the rafters were slanting towards the ridge. The measurement and joinery details should be

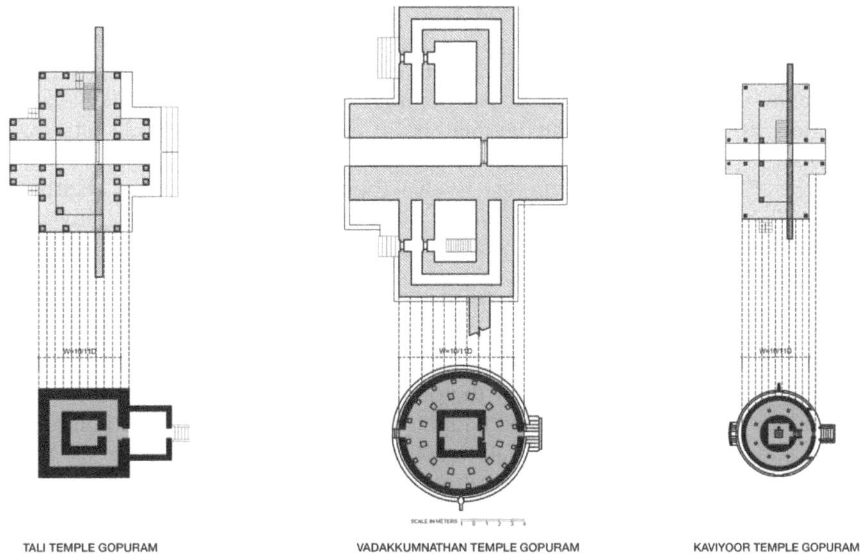


Figure 6. Temple gopurams in relation to their respective *garbagrihas*: (from *left*) Tali, Vadakkumnathan, and Kaviyoor Mahadeva temples.

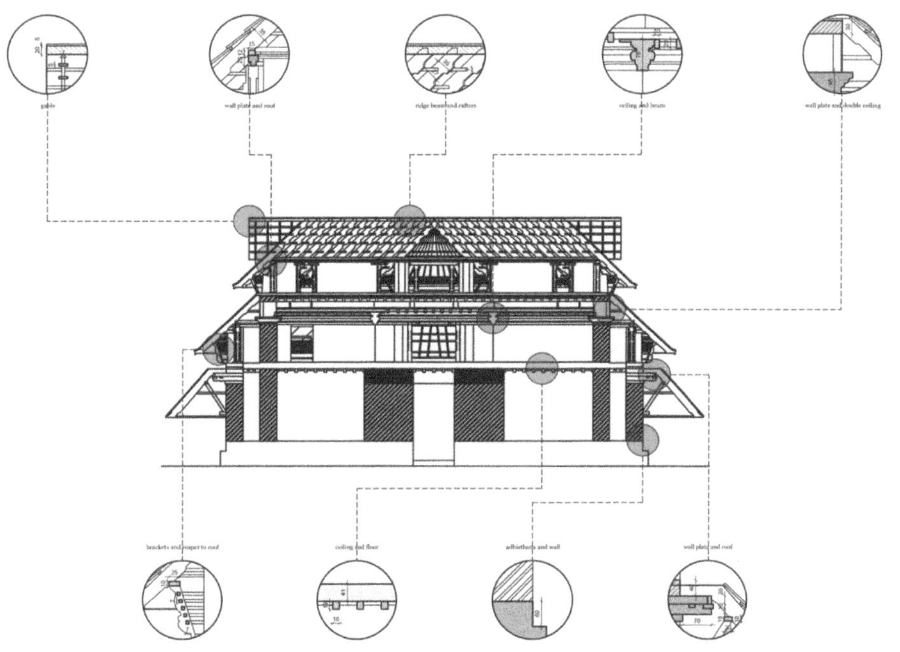

Figure 7. Construction and building material details of the Vadakkumnathan temple gopuram (Source of base drawing: Archeological Survey of India, Thrissur).

worked out independently for each rafter. This is seen in the Tali temple gopuram and the Vadakkumnathan temple gopuram. The last stage of roof development avoided the slanting roof system; here, the rafters were perpendicular to the ridge beam. This is seen in the Kaviyoor Mahadeva temple gopuram. Even though all three had different systems, their appearance is the same except that the Kaviyoor Mahadeva temple does not have the gable (Figure 7).

Generally, temples in Kerala use wood as a building material, mostly in the superstructure; blocks of laterite are used for constructing walls, and granite and wood are used for pillars. The widespread use of laterite as a building material in South Indian architecture is by no means a distinctive trait of Kerala temples. In fact, laterite is available in areas of heavy rainfall, and as an abundant raw material it inevitably finds its way into traditional architecture.

A concentration of wooden temples is also found in the districts of Kottayam and Alleppey in Kerala. In such temples, even the walls are made of timber, bearing beautifully carved panels devoted to themes from various epics. However, plinths are invariably built of granite mouldings. The roofs are made of timber.

### 6.3 *Analysis of construction*

The dimensions of the gopurams follow the principles of proportion, until the width of the wall in the case of the Vadakkumnathan temple and until the width of the column in the case of both the Kaviyoor Mahadeva and Tali temples. These proportions are not followed in the case of the outer columns that form the verandah space. This could be because of the eaves that they support, as in the case of the inside columns that need to support all of the upper storeys.

As far as roof construction is concerned, the members of the roof system take their dimension from the width of the columns. This is not visible in the three different cases of temple gopurams studied here, which have the same roof structure. The dimensions of members are decided by the carpenters building the roof. The dimension given to them is the perimeter, according to which they build the roof frame.

The roof members of any gopurams are made beforehand and assembly is carried out on site. What make the Kerala gopurams different from the Dravidian ones are the portico and the roof. The portico of temple gopurams, as prescribed by the architect, is one-third of the width of the temple gopuram. This could be an influence of residential structures of the region, where importance is given to the sit-out or verandah that greets guests (Figure 8).

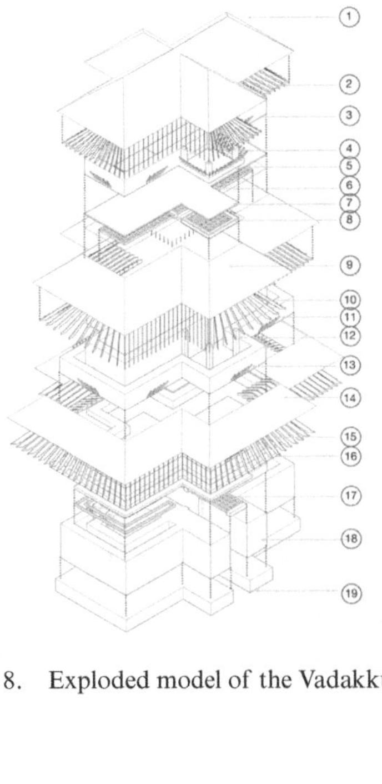


Figure 8. Exploded model of the Vadakkumnathan temple.

## 7 CONCLUSION

Temples were built in Kerala before the Christian era. The built form of the *garbhagriha* of Kerala has undergone evolution through centuries, as have the temple gopurams following their introduction in the 13th century, which is clearly evident from the three cases studied here. Keralan temple gopurams differ from those in other Dravidian regions in several ways.

### 7.1 *Principles of planning*

The principles and canons are strict in the planning, design and construction of temple gopurams, yet have flexibility for adaptation. There is an evident lack of proportion in the roof system or first-floor elements such as the *jalis* (latticed screens) of temple gopurams, which have not been mentioned in any temple architecture literature.

The main differences of Kerala temple gopurams from other regular Dravidian gopurams are the plan shape and interior and exterior spaces. There are differences in aspect ratio, axes of symmetry, and degree of difficulty in the construction.

### 7.2 *Differences in form*

The environmental factors prevalent in a region play a major role in determining the structures of a building, as seen in all the three case studies. The terrain, available materials and microclimate of these regions vary considerably and these variations are seen in the buildings. The choice of material has been gradually rationalised to suit their availability, structural strength, and economy. The upper floor of the Kerala gopuram became the significant space, which was different from the general principle of gopuram planning. This theory excludes the outer dimension of the gopuram being strictly followed as per the rules set in the classical books of Tantrasamuchaya and Mayamatam.

### 7.3 *Scope of future studies*

This study, arguably a pioneering effort, has its own uniqueness, fresh outlook, and innovative ideas. The study should be helpful to those who want to do research in this or similar topics, such as architecture students, teachers, engineers and architects. It contributes to the body of knowledge of structural spaces and forms of temple gopurams. The study may be taken as the beginning of a fresh approach to the study of Keralan temple gopurams with respect to their structural spaces and forms. It is evident that further studies are required on such topics.

## REFERENCES


Acharya, P.K. (1946). *An encyclopaedia of Hindu architecture*. London, UK: Oxford University Press.

Acharya, P.K. (1980). *Architecture of Manasara: Illustrations of architectural and sculptural objects with a synopsis*. New Delhi, India: New Delhi Oriental Books Reprint Corp.

Begley, W.E. (1964). *Book review of Temple gateways in South India. The architecture and iconography of the Chidambaram gopurams* by J.C. Harle. *Journal of the American Oriental Society*, 477–480.

Brown, P. (1959). *Indian architecture: Buddhist and Hindu periods*. Bombay, India: D.B. Taraporevala Sons & Co.

Coomaraswamy, A.K. (1931). *Early Indian architecture. Eastern art*.

Creswell, J.W. (2003). *Research design: Qualitative, quantitative and mixed methods approaches*. Los Angeles, CA: Sage Publications.

Dagens, B.E. (1994). *Mayamatam: Treatise of housing, architecture and iconography* (Vols. 1 and 2, 1st ed.). New Delhi, India: Indira Gandhi National Centre for the Arts.

Dagens, B. (2007). *Mayamatam: Treatise of housing architecture and iconography* (Indira Gandhi National Centre for the Arts) (Vols. 1 and 2, 2nd ed.). New Delhi, India: Motilal Banarsidass

Publishers. Fischer, K. (1955). Some discoveries in Indian art. *Transactions of the Archaeological Society of South India*.
Hardy, A. (2007). *The temple architecture of India*. Chichester, UK: John Wiley & Sons.
Harle, J.C. (1995). *Temple gateways in South India. The architecture and iconography of the Chidambaram gopurams*. New Delhi, India: Munshiram Manoharlal Publishers.
Jouveau-Dubreuil, G. (1914). *Archéologie du sud de l'Inde* [Archaeology of South India]. Paris, France: Paul Geuthner.
Kasturba, A.K. (2014). *Architecture of Kozhikode Tali temple*. Kozhikode, India: Vastuvidya-pratisthanam.
K.P., D.S. (2013). Timber temples of Kerala: Ettumanoor temple. *Journal of the Indian Institute of Architects*, 21–25.
Kramrisch, S. (1976). *The Hindu temple*. New Delhi, India: Motilal Banarsidass Publishers.
Mani, K.P. (1990). *Tile industry in Kerala: Economics, problems and prospects*. Cochin, India: Cochin University of Science and Technology.
Mathew, E. (2014). *Space making elements as reflection of religion and culture in traditional houses of Kerala*. Ahmedabad, India: CEPT University.
Michell, G. (2001). *Encyclopaedia of Indian temple architecture: South India Dravidadesa: Later phase, C.A.D. 1289–1798*. New Delhi: American Institute of Indian Studies.
Namboothiripad, C.N. (2013). *Tantrasamucayamshilpabhagam*. Thrissur, India: Panchangam Press.
Nayar, V.S. (1998). *Vadakkumnathan temple complex*. Mumbai, India: Manan Prakashan.
Panikkar, S. (n.d.). *Temple tradition in Kerala*. Thiruvananthapuram, India: Kerala Council for Historical Research.
Pichard, P. (2002). *Tanjavur Brhadisvara: An architectural study*. New Delhi, India: Aryan Books International.
Prabhu, B.T. (2003). Preface to Manusyalayacandrika. *Sthapati*, *5*(4), 50–52.
Prabhu, B.T.S. & Achyuthan, A. (1996). *A textbook of vastuvidya*. Kozhikode, India: Vastuvidya-pratisthanam.
Rao, P.R. (1956). *The art of Nagarjunikonda*. Madras, India: Rachana.
Sarkar, H. (1978). *Architectural survey of temples of Kerala*. New Delhi, India: Archaeological Survey of India.
Thampuran, A. (2001). *Traditional architecture forms of Malabar Coast*. Kozhikode, India: Vastuvidya-pratisthanam Academic Centre.
Thampuran, A. (2002). Methodology for conservation of ancient temple structures in Kerala. *Sthapati*, *4*(2), 28–29.

*Cities' Identity Through Architecture and Arts – Catalani et al. (Eds)*
*© 2018 Taylor & Francis Group, London, ISBN 978-1-138-05409-7*

# Expression of cultural identity in the contemporary urban built form of Kathmandu

Brinda Shrestha
*Independent Architect/Urban Planner, Nepal*

ABSTRACT: The paper is aimed at studying the contemporary urban built form and the present contemporary scenario of Kathmandu Valley. A descriptive study, through site observations and literature studies, has been used as a research method to understand culture, identity and built forms as well as support its significance. The paper then proceeds with discussing the historical urbanism of Kathmandu and its attempt at showing the collective identity and discussing further the changing contemporary environment in Kathmandu. Moreover, it discusses the symbolic message of identity and all that it resonates. The research then reviews some of the issues and challenges faced by contemporary Kathmandu. However, it attempts to emphasise that traditional institutions, which gave meaning to the city, are not yet obsolete. The paper then concludes that the expression of cultural identity in developing historical and cultural cities like Kathmandu is challenging. It undermines its ancient forms to participate in modernism and questions its legitimisation, especially when contemporary changes are set within the already existing built forms.

*Keywords*: contemporary urban built form; historical urbanism; collective identity; cultural identity

## 1 INTRODUCTION

Kathmandu Valley is the historical and cultural centre of Nepal. Being the capital city, Kathmandu is also the economic and administrative hub of the country. In the case of Kathmandu Valley, urbanisation is the outcome of autonomous growth (Shrestha, 2013) that occurred as a result of the pressures caused by increasing population, given that Kathmandu Valley alone has a population of more than 2 million people with an average population growth of 4 per cent per annum (Shrestha, 2013; World Bank, 2013). Studies indicate that the homogeneity of the city has been much affected due to various pull factors (Thapa et al., 2008; Subba, 2003; Shrestha, 2013), while the challenges of the urban built form of Kathmandu lie in finding a balance between modernism and the essence of the city. Amidst its significant historical urban settings (Tiwari, 1999), Kathmandu today stands as a city struggling to balance its past and present needs.

### 1.1 *Significance*

The issue of cultural identity (Frampton, 1983; Zhang, 2010; Gospodini, 2002; Maghub, 2006) is becoming a universal concern in contemporary urban societies. The clash of traditional notions of urban built forms with modernism has made it essential that social notions be changed in terms of urban space. Thus, the issue of establishing identity in contemporary development becomes critical in dealing with the urban built form in a developing city like Kathmandu, which has witnessed the ancient history of city planning through cultural and religious manifestations.

## 2 THEORETICAL BACKGROUND

Culture is defined as a set of values and beliefs that bond people and associate humans with time; culture also shapes people's relationship with their natural environment and governs their daily lives. In simpler language, culture has been explained as the summation of our everyday practices and the way we live our everyday lives, our behaviours, beliefs, ideologies, or the institutional frame of social interactions. The process of city formation, therefore, revolves around these values and beliefs that help shape the habitat and tailor it according to the needs of its inhabitants. Culture also sets the unique attitude of a city that has been created as a result of history and the people's interaction with space.

Identity, however, is sense of belonging to a place; it is the means by which people establish themselves as members of communities or groups and it influences how they define their place in society. The shared collective values associated with culture affect the forms of buildings and places and this distinguishes one urban form from another. The identity of a city includes the physiological construction of identity that is driven by the common features of its community, which shapes the city form.

### 2.1 *Identity and the built form*

Mumford (1938) states that the city is a product of time with its tangible and intangible forms layering the past times through cultural and social activities, while humans play the role of actors in setting cities as the stage to showcase dramas related to human life. In this regard, Correa (1983) defines identity as a process and not as a fixed object. However, because architecture is a tangible and static concept, it does not change with the intangible moments of cultural change in accordance with time. Hays (2016) believes that the act of creating architecture is an imaginative process that produces memories as well as a possible future and also suggests dynamism of cities. Frampton (1983), with his theory on critical regionalism, argues that architecture is about mediation between modern architecture and values associated with the place and its geography, although the thought has been criticised by visionary architects and those who argue for more high-tech economic advancement. These definitions raise complex questions of identity and the relevance of the city with time dimensions. Identity is relatively an abstract visualisation rather than a tangible concept. The identity of a city is embedded within the layers of sentiments and meanings derived from various symbolic sources within the city, such as the built form, culture, history and the meaningful association of all these. Finding the identity of developing cities is challenging, as on the one hand there has to be a link with roots of the city and its people, and on the other hand, the changes in the city should coincide with newer concepts. Thus, highlighting a city's cultural identity is considered a major defensive tool in re-establishing its lost identity.

The built form (Gehl, 2001; Low & Lawrence, 1990; Lynch, 1960; Zhang, 2010) is an abstract concept that describes the physical building activities carried out by humans while taking into consideration their utilitarian needs and also mediating human relations with the natural environment. Through the built form, humans establish both tangible and intangible relationships with nature allowing them to create spaces for living. Tangible entities mostly include buildings, walls, or roads while the intangible relationship with built forms is the human association with these forms and the application of their cultural beliefs.

Hence, the built form is not only what is visible or can be touched, but it also represents the society, the community and the beliefs of people. Therefore, every built form has two dynamics of approach. First, it is built for the utilitarian purpose of the individual; second, it associates with the collective purpose of the community or group. While socio-cultural needs, defensive needs and ornamental needs were the primary purpose of the collective identity of the built form in the past, the purpose of today's city is continuously discussed between various professional groups or among various stakeholders.

## 3 HISTORICAL URBANISM OF KATHMANDU

Kathmandu witnessed a prolonged history of urban development and city planning, which led to the urban development in the valley that was able to sustain itself over a period of more than a millennium (Tiwari, 1999). The history of urbanisation in Kathmandu dates back to as early as the 6th century with four major civilisations, namely, Kirat, Lichhavi (100 BC–AD 1000), Mallas (AD 1257–1768) and Ranas, that contributed to various aspects of urban planning. Historical records state that settlements existed in Kathmandu in the Kirat dynasty. Settlements later existed in the Licchavi, Malla, Rana and Shah periods, respectively, which witnessed the transformation of the city coinciding with the development of modernism (Thapa et al., 2008).

The historic urban form, especially in the Malla period, is one of the significant urban forms that gave the city its identity. The compact built form, the mixed-use concept and uniformity in design gave the traditional urban form its well-defined identity. Residential buildings established around courtyards or in open spaces aligned sunlight and gave people a common space for interaction, which enhanced the strong feeling of unity within the community. The mixed-use concept, with mostly interactive ground-floor space for shops in prime axial roads, gave the built form a visually interactive space for people to walk and wander with inquisitiveness. Elevations of buildings were inspired by the symmetrical design concept, the ornamental approach through the use of bricks and wood, and the uniformity of design in the neighbourhood. Most building structures were created on a regular human scale for human use; however, other monuments like palace buildings were built on a much larger scale so that they could visually stand out. In the traditional urban form, streets were lively with strategic integration of social, religious and cultural activities within the physical construction of streets and street landscapes. The ratio of the width to height was generally 1:1.5 (Shrestha, 2011a), which made the streets seem livelier with a sense of openness. The legitimacy of the built form was understood in terms of elements of the city like edges, landmarks and nodes. The city expressed a collective identity rather than an individual one and was driven by social and cultural needs rather than individual needs (Lynch, 1960) (Figure 1).

Scholars reveal that one of the many reasons that the traditional urban neighbourhoods have remained and were sustained, is the application of socio-cultural practices in the development and continuity of society. The integration of intangible philosophies of culture and

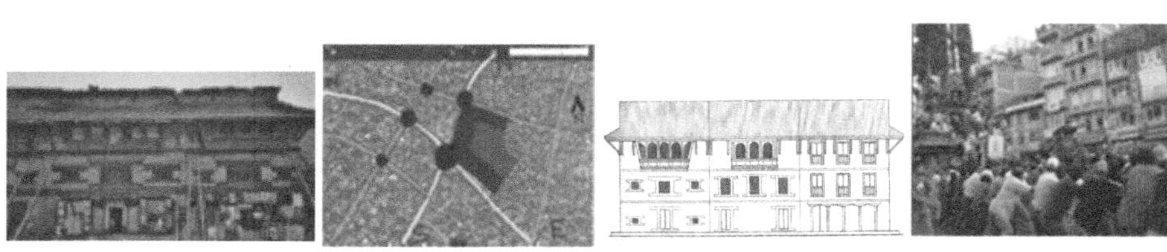

a b c d

Figure 1. (a) Temple-/palace-centric ancient settlement concept; (b), (c) Traditional Malla residential house form; (d) Age-old chariot festival of Malla period, still practised as a communal festival in the urban core of Kathmandu Valley (Sources: (a) Google Maps 2016, modified; (b) Site visit; (c) Korn (1977); (d) Photographer Prashant Shrestha, retrieved from http://bit.ly/2ofDd7o).

Figure 2. Mix of western neoclassical architecture and Malla architecture of the valley, as observed in Durbar Square of Kathmandu (Source: Site visit).

socio-religious beliefs (which are continuously being practised by local inhabitants, even today) with physical settings, through various communal festivals and rituals of life, is what made these urban forms survive until today. The historical traditional cities (Malla cities and towns) had major objectives of forming a collective identity by creating built forms for socio-cultural, defensive and ornamental needs. Their phenomenal town planning norms, which were based on the social and ritual planning system, mostly conformed to the history of urban planning in these eras. However, in the later period of history during the Rana regime, the importation of the Western neoclassical form interacted with the local identity. This period of urban form basically contributed to the grandness of buildings, large Rana palaces, and open-fronted gardens. There were also wide roads axially connected to the palace buildings (Wright, 1990). However, this did not comprehensively affect any of the urban practices; it only added neoclassical architectural vocabulary to medieval buildings (Figure 2).

## 4 CONTEMPORARY KATHMANDU

In the contemporary history of Kathmandu Valley, the years 1950 and 1990 mark the important phases. The process of urbanisation beyond the historical urban form started after 1950 with the end of the autocratic Rana regime and the start of the democratic movement (Thapa et al., 2008). However, with the absence of local architects and professionals, many buildings established during that time were significant contributions from renowned foreign architects, including Carl Pruscha, Tadao Ando, Louis I. Khan and Robert Weise (Shah, 2014) (Figure 3). Nevertheless, during that time, modernism mainly flourished through governmental and international projects financed by foreign aid and technical assistance; this did not greatly affect the local civic context. The decade-long political conflict after the 1990s contributed to the rapid and autonomous urban growth of Kathmandu with significant pressure from the migratory influx (Shrestha, 2011b). Today, the contemporary urban form of Kathmandu, is characterised as the chaotic urban growth in the central core with an increasing urban sprawl in its peripheral land, with no new modern planned form.

For this study, broadly contemporary changes are studied as transformations in residential, commercial and government-owned buildings through site observations and literatures studies.

### 4.1 *Transformation in residential built forms*

Residential neighbourhoods constitute the major urban built form of the city. Broadly, these can be categorised as residential forms in the core of the city, haphazard organic sprawls, and planned residential developments.

In the urban core, the transformation of residential built forms can be observed through interventions made with added floors in brick concrete works, replacement of sloped roofs with flat terraces, and vertical division of houses due to property division. Floor storeys mostly changed from three storeys to five or more, either by replacing sloped roofs with additional floors or by new construction, while floor height changed from the traditional height of 7 feet to a height of 9 feet. This resulted in the change of the skyline of the urban core,

Figure 3. Contribution of foreign architects in construction of modern buildings in Kathmandu: (*left*) Taragaon Hotel, architect Carl Pruscha; (*right*) Ministry of Health, architect Louis I. Khan (Sources: (*left*) www.taragaonmuseum.com ; (*right*) Retrieved from www.landolia.com).

which intervened with traditional physical forms. Other aspects of change can be observed through the adoption of reinforced concrete as a technique for structural construction. This allowed for increased flexibility in height and the use of large glass openings that replaced the traditional wall structure system that used bricks and wood. Significant changes can also be observed in finishing materials where most new constructions have plaster finishes, as opposed to the traditional urban form that involved bricks. The adhoc vertical and horizontal division of houses, among their legal heirs, have further expressed the conflicting nature of built forms. To meet the demand of the migratory population, the functional dynamics of houses have changed to multiple rental households instead of individual family households.

Growing residential developments in the valley's suburban agricultural land, which is mostly self-desired and self-developed by individual residents, follow traditional practices of owner-built developments (Shrestha, 2010). Unplanned sites and infrastructure development are characteristics of residential built forms in Kathmandu Valley. Another form is the planned residential development. According to a research study conducted by Shrestha (2013), these planned residential developments clearly lacked the features of a 'residential neighbourhood' in terms of the poor sense of place, low opportunity for socialisation, poor community coherence and ineffective physical planning, as they fail to address local aspiration. For example, they lack community interaction, functional open space and a cohesive sense of identity influenced by religious and social manifestations.

Thus, residential built forms take on various architectural forms and design vocabularies. As Castells (2010) describes, these residential developments reflect 'individual identity' driven by the choice and decision of the owner.

### 4.2 *Transformation in commercial built forms*

Commercial built forms constitute an important part of the urban city. They are mostly observed in two distinct forms: first, as the continuity of the open-market conceptualisation of traditional market squares; and second, as emerging commercial complexes of modern days, intermingling within the old planning concept of the city.

Even today, traditional market squares are occupied by people who are shopping for daily needs. According to Gehl's (2001) concept of public place, these squares, to some extent, execute the concept of place with interactive ground-floor activities, open roofs, sense of belonging, community coherence and diverse pedestrian activities. These public built forms are still viable today. In the contemporary environment where commercial complexes are emerging to participate in modernisation, the viability of these commercial concepts give different dimensions to the city's identity. (Figure 5 a, b)

a b c d

Figure 4. Changing residential forms in Kathmandu: (a) inside urban core; (b) in government-intervened land development site; (c), (d) different individual architectural forms in new developments (Sources: Site visits).

a b c d

Figure 5. (a), (b) Ancient concepts of commercial markets; (c), (d) Stretch of old commercial buildings transformed to new forms through various modern interventions (Sources: Site visits).

In modern times, there is a growing trend of developing commercial built forms as indoor commercial complexes and multiplexes as per the need and demand of time. Broadly, these can be categorised as commercial built forms in the core and new commercial complexes, synchronising with autonomous development patterns. In the urban core, commercial built forms are set within the available vacant land or are established by transforming old buildings to new ones. The transformation of old buildings to new ones can be observed through various interventions such as changing the functional dynamics of the old mixed-use concept to new commercial use, changing the skyline of the old uniform row housing, and replacing traditional building materials with new ones. As a consequence, these built forms reflect the degree of complexity and variance in visual perceptions (Figure 5 c, d).

In new commercial buildings, one can visually perceive various architectural designs and concepts. Individually, these built forms are driven by the utilitarian need of economic transitions and they are set within the old comprehensive urban settings, which include the existing roads that were developed after the 1950s. These commercial complexes lack the abstract relationship with their urban environment. However, due to the crucial challenge of availability of land, their height is mostly limited to five or six storeys. Functionally, the attempt to transform open commercial activities to indoor economic functions, mediated by modern facilities and services, has made these buildings reflect the change in the lifestyle of new generations in society. Individually, most of these buildings coincide with the perceptions of modernism when it comes to style, form and materials used, but less so with regard to the concept of place, as very few of the recently emerging commercial complexes are attempting to contemplate the intangible relationships with the local urban environment.

### 4.3 *Transformation in government-owned buildings*

According to Spanish sociologist Castells (2010), government buildings assume the role of creating a nation-centred identity through the legitimisation process. In the case of Kathmandu, most government-owned buildings in the contemporary environment function inside the historical Rana and Malla palaces. Using the adaptive reuse principle, these built forms mediate between modern interior functions and historic forms of the past. The reuse of historic palaces for contemporary institutional purposes reflects the recognition of the need to re-establish old built forms to establish the local identity in the contemporary environment. Courtyard planning, passive building technology and open spaces in buildings maintain the interactive urban environment against the conflicting modern identity of the city. Open spaces like courtyards, gardens and water fountains of neoclassical Rana buildings have provided breathing space in government-owned buildings, which otherwise would have lacked the open spaces that are required by people.

On the other hand, most newly constructed government buildings have a blended architectural identity that is usually inspired by traditional elements, shapes and principles. Visually, these buildings seem to contribute to the symbolic representations of significant traditional concepts, trying to reassert local identity in new built forms (Figure 7).

### 4.4 *Ambiguity of contemporary changes in Kathmandu*

In the absence of modern planning implementations, or with plans limited to paper, the contemporary changes in Kathmandu can be visually perceived as self-driven, problem-solving

Figure 6. New commercial built forms in the major commercial hub of Kathmandu (Sources: (*left, middle*) Site visits; (*right*) Retrieved from www.labimmall.blogspot.com).

Figure 7. Government-owned buildings: (*left*) Central administrative office of the prime minister; (*middle*) Nepal Tourism Board; (*right*) Newly constructed presidential office (Sources: (*left*) Retrieved from www.bit.ly/2p589Eh; (*middle*) *The Himalayan Times*, 12 February 2016; (*right*) Roshan Shrestha, Architect/Urban planner, Nepal Government).

activities given the people's desire for change. In the inner core of the city, treating the age-old buildings merely as property has generated ad hoc divisions, structural changes, and a degree of complexity that mostly caused the changes in form. In new constructions, there seems to be a trend of being externally influenced and generating individual built forms, creating a mixture in the vocabulary of both style and design. Intermingling with already existing built forms, these new urban forms produce a mixture of variance and diversity. In the lack of holistic vision, contemporary urban forms express the identity of autonomous changes. The existing patterns of changes suggest that contemporary built forms lack shared collective associations and intangible relationships with the place. However, emerging changes in the skyline, adoption of new material technology and the demand for a new concept of commercial space all indicate the readiness of Kathmandu to implement modern changes and look for new future possibilities.

Even though there has been an attempt to reflect the local identity that has been inspired by the symbolic use of traditional elements and architectural properties, there is a trend of treating buildings as fragmented units and not as part of a holistic urban form. This has further contributed to the ambiguity generated by contemporary built forms in Kathmandu.

## 5 INSTITUTIONAL WEAKNESS

At the national level, Nepal has experienced continuous instability since 1990 (Maoist revolution), with remarkable changes brought about by the conflict and a complete restructuring of the nation from a constitutional monarchy to a federal system. At the state level, urban development is not a prioritised goal (Shrestha, 2011b). The establishment of the Nepal Engineering Council and the Society of Nepalese Architects in the 1990s does prove that engineers have been recognised as professionals for years; however, these engineers are struggling to be socially recognised and accepted by the people. The physical development plan of Kathmandu Valley prepared in 1969 is the only comprehensive planning document that used a rational approach to address urban development in Kathmandu Valley (Nepal Government, 2003); however, the implementation of this plan has not been successful. At the local level, building bylaws and the Nepal National Building Code are the only documents that guide and control the development process, but these documents were not enough to address the contemporary change in the holistic approach. The National Urban Policy of 2007 (Nepal Government, 2007) does try to revive recommendations of the physical development plan of 1969, but it suffers from implementation and coordination challenges among various stakeholders. The establishment of the Ministry of Urban Development in 2013/14 clearly suggests that a rational planning approach through plans and policies for the contemporary building process is new and is just starting to emerge.

## 6 ISSUES AND CHALLENGES

The contemporary urban built form in Kathmandu is observed as the outcome of the pressure of the people's demand as opposed to the outcome of modern planning efforts.

The major challenge of the contemporary urban built form in Kathmandu is that it is implemented either on already existing buildings or on spaces of green-field development that have already been produced, hence creating dissociations in built forms. The recent recognition of modern urban development as a holistic and comprehensive process by the government shows how untimely the established institutions are in intervening in the urbanisation process, even though the city itself continued to grow on its own long before any intervention occurred. Ongoing capital-centric development with no new planned city outside the valley has further accelerated the fragmented build activities rather than the holistic approach.

The sustenance of the ancient urban form, even today, indicates that although Kathmandu is undergoing a process of autonomous modernisation, the social and cultural values embedded within the local inhabitants of the place have remained. Hence, ancient forms cannot be simply neglected because of the influence of external forces that add to the pre-existing challenges of expression of identity by the contemporary built form. There has to be a link between tangible and intangible philosophies of the traditional form with relevant modern changes for its legitimisation process. Although some architects and planners have tried to reassert the local identity in the architectural expression of buildings, they were placed against the backdrop of the chaotic modern environment. This certainly intensifies the challenges faced by contemporary built forms in taking their place in creating a unique identity in this competitive environment.

## 7 CONCLUSION

The fact that contemporary changes are inevitable, in modern urbanisation, does not seem unusual, even in a historic city like Kathmandu. Sweeping modernism, in Kathmandu has brought about transitional changes to the valley's architectural and urban forms, where valley's urban built form varies between two imaginations: one is linked with traditional urban built forms that are interdependent on the social urban sphere, and the other is the modern urban landscape that conceptualises urbanisation with time dimension, representing heterogeneous, ambiguous and transitional identity of change. Thus, the expression of cultural identity in developing historical and cultural cities like Kathmandu is challenging, as undermining its ancient form in order to conform to modernism will disturb its legitimacy, especially when contemporary changes are set within already existing built forms.

## REFERENCES


Bennett, T. & Frow, J. (Eds.). (2008). *The SAGE handbook of cultural analysis, culture and identity* (Chapter 24). London, UK: SAGE Publications.

Castells, M. (2010). *The power of identity, The information age: Economy, society and culture* (Vol. II, 2nd ed.). Chichester, UK: John Wiley & Sons.

Chitrakar, R.M., Baker, D. & Guaralda, M. (2014). Urban growth in the Kathmandu Valley: The transformation of public space. *International Conference on Art, Architecture and Urban Design, Bologna, Italy, 25–27 June.*

Corcoran, P.M. (2002). Place attachment and community sentiment in marginalised neighbourhoods: A European case study. *Canadian Journal of Urban Research, 11*(1), 47–68.

Correa, C. (1983). Quest for Identity. In Architecture and Identity, edited by Robert Powell. Singapore: Concept Media/Aga Khan Award for Architecture.pp-10–13.

Frampton, K. (1983). Towards a critical regionalism: Six points for architecture of resistance. In H. Foster (Ed.), *The anti-aesthetic: Essays on postmodern culture*. Port Townsend, WA: Bay Press.

Gehl, J. (2001). *Life between buildings: Using public space* (4th ed., Trans. J. Koch). Copenhagen, Denmark: The Danish Architectural Press.

Girard, F.L. (2013). Toward a smart sustainable development of port cities/areas: The role of the 'historic urban landscape' approach. *Sustainability*, *5*(10), 4329–4348.

Gospodini, A. (2002). European cities and place-identity. *Discussion Paper Series*, *8*(2), 19–36.

Guardiola-Saenz, L. & Yamada, F.M. (2009). Culture and identity: Culture and Identity: Recasting the Bible for the 21st Century—Workshop at the Book Festival 2009 – STM, February 7, 2009 Chapter 1, pp. 4–5.

Hays, K.M. (2016). Architecture's appearance and the practices of imagination. *Log*, *37*, 205–213.

Korn, W. (1977). *The traditional architecture of Kathmandu Valley*. Kathmandu, Nepal: Ratna Pustak Bhandar.

Mumford, L. (1938). The Culture Of Cities. A Harvest Book/HBJ Book. Harcourt Brace Jovanovich, Inc.

Low, M.S. & Lawrence, L.D. (1990). Built environment and spatial form. *Annual Review of Anthropology*, *19*, 453–505.

Lynch, A.K. (1960). *The image of the city*. Cambridge, MA: MIT Press.

Maghub, Y. (2006). Architecture and expression of cultural identity in Kuwait. People in place in people. *1st International Symposium on Environment, Behavior and Society, University of Sydney*.

Mercer, C. (2006). Cultural planning for urban development and creative cities. Retrieved from www.Kulturplan-Oresund.Dk/Pdf/Shanghai_Cultural_Planning_Paper.Pdf.

Nepal Government. (2003). *Kathmandu Valley long-term development plan*. Kathmandu, Nepal: Department of Urban Development & Building Construction (DUDBC).

Nepal Government. (2007). *National urban policy* (pp. 9–16). Kathmandu, Nepal: Ministry of Physical Planning and Works.

Nepal Government. (2009). *Building byelaws for constructing activities within municipalities and urbanizing villages of Kathmandu Valley* (Chapter 2, p. 21; Chapter 7, maps). Kathmandu, Nepal.

Rai, R. (2008). *Threats to the spirit of the place. Urban space and squares, historic city core, Kathmandu*. Cologne, Germany: Institute for Technology in the Tropics and Subtropics. Retrieved from https://www.icomos.org/quebec2008/cd/toindex/80_pdf/80-FAsv-73.pdf.

Shah, B. (2014). The architecture of foreign architects in Nepal. *Spaces Magazine*, Nepal.Vol.10.No.07.

Shrestha, B.K. (2010). Housing provision in the Kathmandu Valley: Public agency and private sector initiation. *Urbani Izziv*, *21*(2), 85–95.

Shrestha, B.K. (2011a). *Land use changes in urban fringes of Kathmandu: Looking through food security perspective* (Master's Thesis, Tribhuvan University, Nepal).

Shrestha, B.K. (2011b). Street typology in Kathmandu and street transformation. *Urbani Izziv*, *22*(2), 107–121.

Shrestha, B.K. (2013). Residential neighbourhoods in Kathmandu: Key design guidelines. *Urbani Izziv*, *24*(1), 125–143.

Subba, M. (2003). *Urban containment policy: Does it present a hope to manage an impending urban crisis of the Kathmandu Valley?* (Doctoral thesis, Norwegian University of Science and Technology, Trondheim, Norway).

Thapa, R., Murayama, Y. & Ale, S. (2008). City profile: Kathmandu. *Cities*, *25*(1), 45–57.

Tiwari, S.R. (1999). Kathmandu Valley urban capital region and historical urbanism. Paper prepared for presentation to 13th Biennial Conference of Association of Development Research and Training Institutes of Asia and the Pacific (ADIPA), on the theme of 'Managing Asia-Pacific mega cities: Policies to promote sustainable urban development in the 21st century', 29 November–1 December.

Tiwari, S.R. (2016). Learning from traditional towns: Spaces and networks for sustainable multicultural urbanism. Retrieved from http://www.kailashkut.com/wp-content/uploads/2016/05/martinchaptertiwari.pdf.

Tiwari, S.R. (2016). Characters and challenges of public spaces of Kathmandu Valley towns: A sociocultural perspective. Retrieved from http://www.kailashkut.com/wp-content/uploads/2016/05/publicspacesofkathmandu.pdf.

World Bank. (2013). *Urban growth and spatial transition in Nepal: An initial assessment*. Washington, DC: World Bank.

Wright, D. (1990) [1877]. *History of Nepal. With an introductory sketch of the country and people of Nepal* (Reprint). New Delhi, India: Asian Education Services.

Zhang, L. (2010). *Preserving cultural identity through a morphological approach* (Master's thesis, College of Environmental Science and Forestry, State University of New York, Syracuse, NY).

*Cities' Identity Through Architecture and Arts – Catalani et al. (Eds)*
 *ISBN 978-1-138-05409-7*

# Reimagining city identity through safe and sustainable public environments

Deniz Deniz
*Izmir University of Economics, Izmir, Turkey*

ABSTRACT: Cities have their own characteristics and identities, which may be affected by various dynamics or subject to change through the years. However, these changes could be positive or negative. The unique nature of each city is revealed through its social, ecological and environmental form and reflects the cultural and visual characteristics of the city. Safety and security are essential for successful sustainable communities and cities. Thus, improving safety through environmental design and creating sustainable cities have become increasingly important all over the world. Places where people experience crime and fear of crime affect the image of the city in a negative way and diminish people's quality of life. In public areas where opportunities for crime or fear of crime is high, all fields of design have a crucial role to play in reducing opportunities for crime and improving safety in urban spaces, from environmental planning to the design of products. Reimagining cities and public spaces in a positive way, it is crucial to consider that cities as living organisms need to be safe and sustainable to live and work in. Although crime prevention through concepts of design is complex and hard to deal, with it is vital for designing safe and sustainable city images. In this respect, to achieve better solutions for communities as well as successful city images, public and environmental regulations have to be considered carefully. Planners and designers must be aware of obstacles and deficiencies in terms of crime prevention, and they need to understand the basic relationship to develop design and applications in a better way. In this respect, designers, planners, politicians and decision makers must be able to assess which design solutions are better for the city and its image.

*Keywords*: city identity; city image; safety; sustainability; quality of life

## 1 IMPROVING CITY IDENTITY BY SAFETY THROUGH ENVIRONMENTAL DESIGN

Sustainable communities have to be fulfilled by requirements of the environmental, social and economic needs of the future. Since they are defined as well-designed places where people feel secure and fear of crime does not undermine quality of life, environmental safety and security are essential issues for sustainable and liveable communities. In addition, sustainable public environments are considered well-designed and attractive environments to live in, where freedom from crime and fear of crime adds positive value to the city identity.

In urban environments, people experience feelings of insecurity in various formats that reduce their quality of life (Gertsakis, 2001). However, sustainable urban spaces need to improve people's quality of life by providing safe and healthy environments to live and work in; well-designed and well-used public spaces also encourage 'a sense of place'. Sustainable communities embody the principles of sustainable development by integrating social, economic and environmental components of their community and meet the needs of existing and future generations (Colquhoun, 2004). Sustainable communities also need to develop the city image by improving perceived safety, which requires working with the police and wider local community to tackle crimes, such as street crimes (Wekerle & Whitzman, 1995).

Safety and security issues need to be considered for creating sustainable and safe communities because it is obvious that design of the built environment in a safe way has a great effect on improving the city image.

Sustainable cities must include safe and well-planned communities and should offer equality of opportunity for all (ODPM, 2005). These requirements are also essential for better public image and city identity. Such places are not only well-designed and attractive environments to live and work in, but they are also where freedom from crime and the fear of crime improves the quality of life. It is crucial to understand how the city image can be affected through the fear of crime and public environment for creating a better city image (Loewen et al., 1993). To do that, at first, the components of the sustainable communities need to be considered carefully. Crime prevention through the concept of environmental design is an approach combining psychology and behavioural and learning theories, and focuses attention on the physical environment as well as on fear of crime in public places. It also advocates effective use of design for developing a built environment to decrease crime and fear of crime for improving the quality of life (Herzog & Smith, 1988).

Community ownership is fundamental for creating a better city image because places that feel owned and cared for are likely to be used and revisited; thus, a safe public space sends positive signals to the community (Parente, 2015). Community members are more likely to be protective of places with which they feel some connection. Effective design, planning and space management help to stimulate natural improvement (Riger & Gordon, 1981). In addition, people feel safe in public areas when they can see and interact with others, particularly when they connect with that space. Criminals often hesitate from committing crime in places that are well supervised (Clarke & Mayhew, 1980).

Natural surveillance can be achieved by creating effective sightlines between public and private spaces, which include access ways and meeting places. Matching types of lighting with crime risk and using attractive landscaping are also crucial (Nasar, 1982). Space management is linked to the principle of territorial reinforcement. It ensures that space is well used and maintained. Strategies include activity coordination, site cleanliness, rapid repair of vandalism and graffiti, and the refurbishment of decayed physical elements (McCamley, 2001). Thus, sustainable communities should be fair for everyone and cohesive, with a strong local culture as well as other shared community activities. They should offer a sense of community identity and belonging, which can be developed by low levels of crime and anti-social behaviour with visible, effective and community-friendly policing (Warr, 1984).

Sustainable communities also enable inclusive, active and effective participation by individuals and organisations and a sense of civic values. Sustainable communities actively seek to protect the environment in accordance with good applications that make efficient use of natural resources, enable a lifestyle that minimises negative environmental impact, and create a safe and successful city image. Sustainable communities offer a sense of place. They are user-friendly public spaces with appropriate design and layout that complement the distinctive local character of a community; they also promote health and are designed to reduce crime and make people feel safe. These communities recognise an individual's rights and responsibilities to be sustainable and have due regard for the needs of future generations in current decisions and actions. Considering these features, it is possible to say that creating safe and liveable public spaces is crucial for creating sustainable communities. Thus, safety issues should be considered carefully from the planning to the design stage of public spaces.

## 2 FUNDAMENTALS OF SUSTAINABLE COMMUNITIES FOR IMPROVING PUBLIC IMAGE AND CITY IDENTITY

Public spaces play an essential role in the life of cities and residents. Built environments and urban public spaces are often the signature spaces that constitute a city's distinctive identity because they are the settings of everyday life, including streets, parks and transfer stations, where the city's diverse communities interact with each other and where there are spaces for the practice of citizenship. Public spaces and streets, as well as urban furniture and other

Figure 1. Discussions of the relationships between safety, quality of life and city identity (first session).

Figure 2. Presentations on components of cities and fundamentals of city image.

Figure 3. Discussions of the relationships between safety, quality of life and city identity (second session).

components, make up the image of a city. The quality of design and organisation of these components directly impacts the safety, perception and experience of a place because the design of the built realm plays an important role in the way people live in and experience the city.

If cities and public spaces are designed well and also connect people to amenities, residents are more likely to feel safe and secure; this affects the image of the city in a positive way. In urban environments, people experience crime or fear of crime in public places that diminishes their quality of life. Today, there are increasing rates of street crime and violence against persons in every major city. Unfortunately, every person who lives in the city is a potential victim for any crime incident. Therefore, fear of crime can be reduced by better design and maintenance of the built environment. In areas where opportunities for crime and fear of crime are high, environmental design plays a crucial role in reducing opportunities for crime and improving perceived safety.

To develop the dynamics of cities and also to discuss how perceived safety and sense of ownership affect the image of a city, a day-long workshop was conducted with the volunteers who have been educated as environmental designers (Figure 1).

As key actors, designers should be aware of their roles to improve city identity via design decisions, including environments, buildings and products. In this case, it is difficult to be aware of all possibilities that surround design decisions and create negative effects on sustainable improvement of the city image.

Through the first half of the workshop session, designers shared presentations on the components of cities and city image (Figure 2). Because cities have characteristics that may be affected by various dynamics or subject to change through the years, understanding these components was crucial to having a wider perspective (Figure 3).

Although each city has its unique nature according to its social, ecological and environmental form, the impacts of perceived safety and security varies through the cultural and visual characteristics of cities. Thus, improving safety through environmental design and creating sustainable cities have become increasingly important all over the world. Places where people experience fear of crime affects the image of the city in a negative way by diminishing quality of life.

## 3 CONCLUSION

Since the concept of urban transformation through the physical environment and its components became popular, the relationships between safety, sustainable community and urban

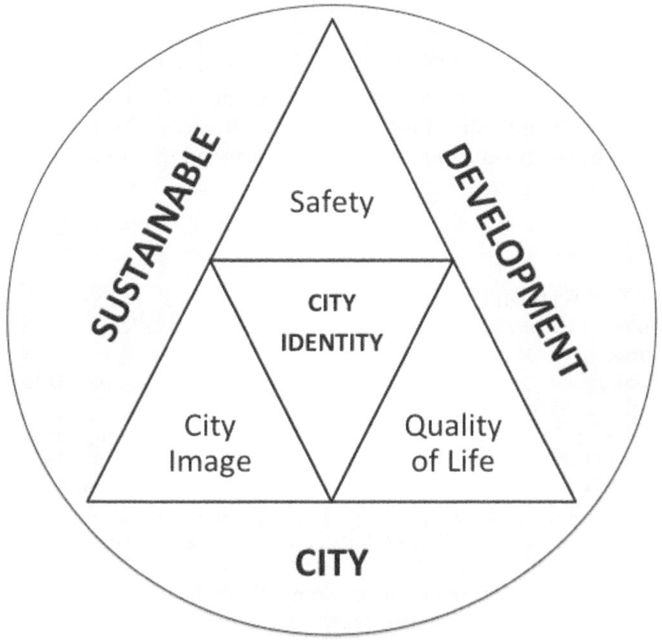


Figure 4. The importance and components of city identity for sustainable city development.

identity have become crucial aspects that need to be discussed. In this particular study, attention has been drawn to the fact that sustainable urban implementation in today's cities is urgent, and decisions made for transforming urban areas could destroy a city's social, economic and cultural values.

Through the study, it is also noted that loss of urban identity brings monotony. To sustain urban identity, characteristics of the city should be preserved and continued in a better way. In this context, identical past, present and future features of a city should be questioned and the features of its identity should be preserved and maintained in more sustainable and safe ways. In public areas where opportunities for feelings of insecurity or fear of crime are high, all fields of design have crucial roles to play, reducing opportunities for crime and improving safety in urban spaces from environmental planning to the design of products.

Reimagining cities and public spaces in a better way, it is crucial to consider that cities as living organisms need to be safe and sustainable to live in. In this respect, although crime prevention through the concept of design is complex and hard to deal with, it is vital for designing safe and sustainable cities to achieve better solutions for communities as well as successful city images. In addition to that, public and environmental regulations also need to be considered carefully in wider perspectives for achieving sustainable solutions (Figure 4).

Environmental designers must be aware of these obstacles in terms of city image to understand the basic relationships for developing safer solutions and applications. In this respect, designers, planners and decision makers must be able to assess which design solutions are better for safe cities and improved city images. Although creating safe and secure cities is a complex issue, it is possible to make a significant difference by considering the key principles crucial for improving quality of life for better and improved city images. In addition, studies should also consider how this concept needs to be combined with all fields of design discipline, given the specific obstacles, national and local conditions, and characteristics of cities.

This study has tried to develop a better understanding of how to deal with this particular issue. The analysis of the local situation and defining the real problems in a better way is crucial for achieving successful solutions for people and cities as their habitat. In order to do that, every aspect of planning and design decisions must be made in full consultation with all partners through powerful collaborations, as working together from the beginning of the process is essential. Furthermore, the whole process from decision-making to the application must be controlled and supported by feedback for a great achievement. Therefore, it is possible to create and improve city images that help to create safe and sustainable communities. As a result, adaptable design solutions, particularly when coordinated with other measures, can make a significant contribution to safety and improved quality-of-life issues.

## REFERENCES


Clarke, R.V. & Mayhew, P. (1980). *Designing out crime*. London, UK: HMSO.

Colquhoun, I. (2004). *Design out crime: Creating safe and sustainable communities*. London, UK: Architectural Press.

Dougherty, D. (2006). *Embodying the city: Identity and use in urban public space* (Master's thesis, Virginia Polytechnic Institute and State University, Blacksburg, VA).

Gertsakis J et all, Linking Innovation, Design and Sustainability—Learning from Real World, EcoDesign Projects, Centre for Design at RMIT University and EcoRecycle Victoria, 2001. p. 11–15.

Herzog, T.R. & Smith, G.A. (1988). Danger, mystery and environmental preference. *Environment and Behavior*, *20*(3), 320–344.

Loewen, L.J., Steel, G.D. & Suedfeld, P. (1993). Perceived safety from crime in the urban environment. *Journal of Environmental Psychology*, *13*, 323–331.

McCamley, P. (2001). Crime, design and urban planning: From theory to practice. *The Royal Australian Planning Institute Magazine*, New Planner, June.

Nasar, J.L. (1982). A model relating visual attributes in the residential environment to fear of crime. *Journal of Environmental Systems*, *11*(3), 247–255.

ODPM. (2005). *Sustainable communities: People, places and prosperity. A Five-year plan from the Office of the Deputy Prime Minister (ODPM)*. London, UK: ODPM.

Parente, M. (2015). Designing the city identity: Strategic and product design for new experiential ways of living, enabling, and interacting with the urban context. *Design Management Journal*, *10*(1), 62–71.

Riger, S. & Gordon, M.T. (1981). The fear of rape: A social control. *Journal of Social Issues*, *37*(4), 71–92.

Warr, M. (1984). Fear of victimization: Why are women and the elderly more afraid? *Social Science Quarterly*, 65, 681–702.

Wekerle, G.R. & Whitzman, C. (1995). *Safe cities: Guidelines for planning, design and management*. New York, NY: Van Nostrand Reinhold.

*Cities' Identity Through Architecture and Arts – Catalani et al. (Eds)*
 *ISBN 978-1-138-05409-7*

# Smart grids, smart cities, interior spaces and human behaviour: An interactive development process

Adnan Mohamed Elhalwagy
*Assistant Lecturer, Interior Architecture Department, Faculty of Fine Arts, Helwan University, Cairo, Egypt*

ABSTRACT: Architecture starts from internal spaces; as the nearest component to the human, the interior architecture space plays the linking role between direct human needs, buildings and cities. Interior architecture space features are changing as a result of the digital revolution, that is, the age of Information and Communication Technologies (ICT) and the expected growth of connected objects as part of the Internet of Things (IoT)

This research aims to shed light on these changes, especially those related to interior space features and the evolution of human behaviour, by reviewing the main benefits of these changes on data, communication, cyber spaces, and energy efficiency. Moreover, it reveals the new control features of these developments in interior space.

*Keywords*: smart city; smart grid; IoT; intelligent buildings; smart homes; interior space; human behaviour

## 1 INTRODUCTION

Cities are large and permanent human settlements where civilisations are incubated; buildings are considered the most substantial components of the city, reflecting its public identity, in addition to containing the functional spaces that enable humans to live effectively and securely.

The smart grids and smart buildings of cities illustrate the quintessential changes that occur through the interactive development process of the successive concepts of interior space, then buildings, and, ultimately, a city.

This progression comprises a developing level of cultural, social and service aspects of society, changing human behaviour and the features of the interior space from many perspectives, starting from functional requirements, dimensions, data and communications tools, and ending in energy and performance efficiency.

Information and Communication Technologies (ICT) and the growth of the Internet of Things (IoT) have changed the way that physical space is created to meet functional purpose in the face of the new data and facilities available, as well as the way humans interact with each other. The anticipated expansion of connected objects as a part of the IoT will have a substantial impact on our global and urban design, and will influence all industries in the next few decades [1].

In addition, this will be the core of a successful smart cities concept, and buildings that will change human behaviour and thinking, because everything will become connected in a network, starting with personal things and ending with equipment and central services areas.

The current research is presented as follows. Following this introduction, Section 2 explains the concept of the smart grid and the link between the smart grid and the buildings of the city. Section 3 illustrates the concept of IoT, data gathering and analysis, monitoring, controlling methods and services. New spaces dimensions and types have been clarified in section 4. The optimization of performance, energy efficiency are presented and discussed in Section 5. Section 6 describes the possible effects on the social and communication aspects of human behaviour, and a set of similar applications are incorporated in each application. Section 7 explores the idea of *Mama Home* as a possible application for a services facility in relation to IoT technology. Finally, the conclusions are summarised in Section 8.

## 2 SMART GRIDS, CITIES AND BUILDINGS

A **smart grid** is an electrical grid which includes an assortment of operational and energy measures. It comprises smart meters and appliances, renewable energy and energy-efficient resources, electronic power conditioning and the production and distribution of electricity control [2], as well as communication and wireless smart sensor networks and controllers, amongst other things [3].

Smart grids use computer-based remote control and automation. These systems rely on two-way digital communication technology between devices in the field and the computer processing of the utility's network operations centre.

All of this provides many advantages to utility providers and consumers, most predominantly in large enhancements to energy efficiency in the electricity grid and in the energy users' homes and office spaces. Each device on the network can be equipped with sensors to gather data (power meters, voltage sensors, fault detectors, etc.). Automation technology represents the key feature of a smart grid because it supports the utility provider in adjusting and controlling each individual device or, indeed, millions of devices from a central location [4].

The generated data in the smart grid is noticeably larger than the ones done in the traditional grid. This could be interpreted due to the continuous two-way communication between the utility provider and the smart meter in the customer's building. It is also a hurdle the expansion of the smart grid, especially where the infrastructure is not in its capacity for such communication. Here, IoT technology plays a remarkable role; it can help streamline the transfer of high volumes of data via the Internet [5]. In addition, a smart grid combines different technologies that support the existence of smart cities, in terms of cyber and physical systems [6].

A smart grid typically consists of three networks: the first one is Home Area Networks (HANs) that connect the devices in the individual home or building and connect smart meters, and distributed renewable energy sources. The second one is Neighborhood Area Networks (NANs) that connect a variety of HANs and convey the collected information to a network gateway. However, the third one is Wide Area Networks (WANs) that function as a communication pillar supporting the process. In fact, the new technologies of communication play a cardinal role in successful smart grid operation (see Figure 1) [7]. They are adoptable in the sense of using it according to the certain features required by the HANs,

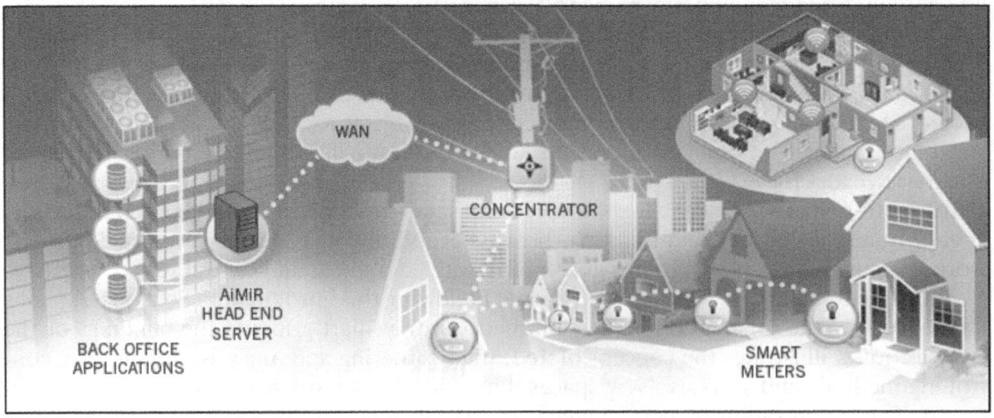


Figure 1. The connection between single devices, buildings, and a city's smart grid.

NANs and WANs. It is worthy to mention that either the wireless or the wired technologies of communication could be easily used in a smart grid. As for controlling the transmission lines or even monitoring it, distributing the facilities or generating the energy plants, along with the video monitoring of consumer premises, could be conducted through the use of wireless sensor networks [8]. The envisioned smart grid provides social, environmental, ethical and economic benefits.

**Smart cities**: The language of urbanisation policy has incorporated the terms "smart" and "intelligent" to refer to the advanced use of IT to enhance the productivity of a city's substantial infrastructure and services, as well as reducing $CO_2$ outputs as well as the inputs of energy in order to reduce the impact on global climate change. In the 21st century, there has been a shift from sustainability assessment to smart city goals. Thus, everything becomes connected and smart, and a smart city is a new model for the city, based on the use of ICT, with the aim of improving its economic, social and environmental sustainability, characterised by smart power grids that will be capable of creating a better balance between electricity demand and supply. This will start with buildings that detect occupants' energy needs, integrate vehicle batteries into their energy forecasts, respond to changing weather conditions, and automatically change their sittings to optimize the performance of the building equipment [9]. In addition, digital city strategies create the infrastructure and services that enable the city to function and also create an environment within which digital society initiatives can flourish [10].

**Smart buildings**: Smart buildings limits were exceeded the typical range of the normal building equipment. They are responsive and connected to the smart power grid and simultaneously interact with the building operators and occupants to empower them with extra levels of visibility and actionable information [11]. In addition, smart buildings provide beneficial services for their occupants, such as space illumination, thermal comfort, air quality, physical security, and sanitation, at the minimum possible cost and lowest environmental impact over the building's lifecycle. Realising this vision necessitates the incorporation of intelligence from the beginning of the design phase through to the end of the building's useful life. Smart buildings depend on the use of information technology during operations to connect an assortment of subsystems, which typically operate independently but can share data to optimise the total performance of the building.

Most new commercial buildings include a growing number of sensors, controls and other devices as standard equipment. In addition, modern buildings have built-in control systems, assigned as Building Management Systems (BMS) or Building Automation Systems (BAS), allowing building engineers and facilities and real estate managers to control their infrastructure. Most buildings have some level of intelligence built in, whether it is Heating, Ventilation and Air-Conditioning (HVAC), lighting, or fire safety, and competitive management and automation services can provide best-in-class climate control, energy savings, and maintenance [12]. So, today, it's possible to obtain more from building data and, ultimately, make better decisions as a result.

## 3 INTERNET OF THINGS (IOT)

IoT is a new technology that provides a network connection between things, that is, devices and vehicles, equipment and smart devices, buildings and other items. These connected things have more advanced electronic pieced embedded in it or sensors. It might also have software's actuators to facilitate network connectivity between them and to enable them to gather and exchange data [13–14]. IoT-GSI[1], proclaim the IoT description as "the infrastructure of the information society" [15]. Where objects can be remotely sensed or controlled through existing network infrastructure, this can provide opportunities for additional direct integration of the tangible physical world into computer-based software systems, which leads

1. Internet of Things Global Standards Initiative.

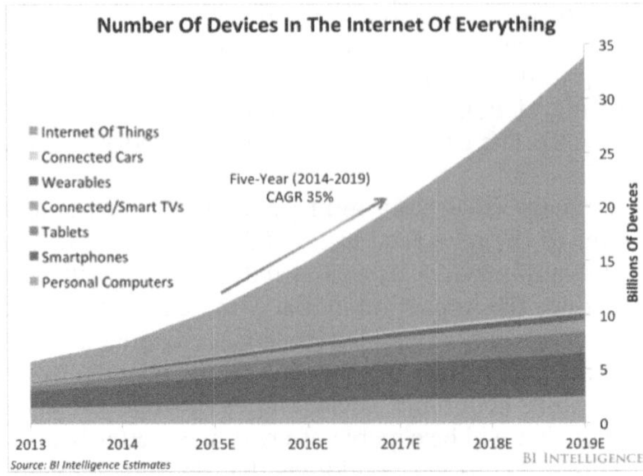


Figure 2. Estimated numbers of objects in the internet of things.

to enhanced efficiency, reliability and economic benefit, in addition to reducing the need for human intervention [16]. When IoT is boosted with sensors and actuators, this technology becomes a basis for wider system integration of cyber and physical systems, which also encompasses technologies such as smart grids, smart homes, intelligent transportation and smart cities. In IoT, each thing is individually distinguishable as a unique object through its embedded computing system but is, at the same time, able to interoperate within the existing Internet infrastructure. Experts estimate that IoT will involve nearly 50 billion items by 2020 (see Figure 2) [17–18].

Radio-frequency identification (RFID) tags are the major way used to connect things and embed sensor and actuator nodes; RFID is considered to have been a vital breakthrough in the realm of merged communications, empowering the design of microchips for wireless data communication.

They help in the automated recognition of anything they are attached to, as well as providing an electronic barcode [15–16]. RFID tags are powered passively as they do not have an internal power source; it is done by extracting the signal of the interrogation from the RFID reader which is utilized again through send their ID to the readers. RFID-enabled applications that can be found embedded in doors to recognise the movement of things, as well as in transportation, like registering the stickers or replacing the tickets or even controlling the accessed applications. Active RFID readers include a battery supply and can initiate the communication of the various applications.

Active objects in IoT are enabled to communicate and cooperate between themselves and the environment in a certain method via exchanging the sensed data revealing the environmental conditions. They can react independently to incidents in the physical world. They can also influence it by embarking on a processes which automatically initiate the action and create the service. IoT makes use of cloud computing as well as artificial intelligence.

## 4 THE NEW SPACES

ICT and IoT considerations have changed the traditional features of interior space; starting with the reduction of functional areas in some spaces, through to the merging or elimination of other spaces. This is because information and digital solutions have either replaced or compensated for some functional requirements, leading to the spread of virtual and cyber spaces, interactive spaces, and hybrid spaces, in addition to the increasing trends for e-commerce, e-tourism, e-health, technology parks and e-government.

Internal spaces have changed from being static to being dynamic, and from having three-dimensional axes to having five-dimensional (X, Y, Z, T, E) coordinates through the addition of time and energy factors. Furthermore, new spaces can offer the flexibility to avoid various construction and material restrictions that opened the mind for more creative designs for spaces.

The new generation of internal space has combined physical material with information technology, changing the entire standard criteria of traditional space, which becomes totally free in terms of dimensions, and adaptation to motion and function.

The integrated and connected technology in the building and the internal space facilitate data gathering and monitoring and adapt or save resources and equipment. In addition, it became possible to analyse and control the whole internal environment to provide the optimum service in relation to the resources available. Thus, spaces become smaller, smarter, more aware and better connected.

## 5 PERFORMANCE OPTIMISATION AND ENERGY EFFICIENCY

Globally, buildings consume 38 to 40% of entire energy usage and CO2 emissions [19]. Reducing the consumption of energy either new on existing buildings is considered as a the major important aspect for all stakeholders, namely, developers of buildings, owners governments, and users or operators to guarantee the protection of environment sustainability and highlight the increasing expense of energy. Thus, The ROI, or return on investment, is very significant in the case of energy-friendly buildings. For instance, the largest element of energy consumption by commercial buildings is expended on lighting (26%), followed by heating and cooling (13% and 14%, respectively). Investment in low-energy-consumption light bulbs and insulation materials can anticipate short payback periods. Furthermore, simple automation and control, such as of window shading, demonstrates obvious reductions in energy demand for cooling and lighting [20].

Wireless sensors can be used to run devices in coordination with periods of reduced grid load. In addition, IoT can assure that the communication among various devices and also the communication between the technological and human dimensions, like residents, are functioning properly. The grid is able to share different types of data with the user—including the real time one—where the user can distantly control the devices. Accordingly, sensors are to be utilized in smart buildings in order to monitor the different parameters of the place, like the solar gain, humidity, carbon dioxide level, etc. It is also to improve them if needed by controlling the HVAC systems [21]. For example, given that the electricity supply can have shortfalls at some times and surpluses at others, depending on electricity demands and peak hours, the personal and home IoT can produce electricity usage data for the house and inform the electricity company, which can, in turn, optimise the supply and demand through the utility's IoT. In more detail, efficient energy consumption can be accomplished by constantly monitoring every electricity point within a house and using this data to modify the way electricity is consumed. At the city scale, this data can be used to maintain the overall load balance within the smart grid, guaranteeing a high quality of service [22]. Thus, the Internet facilitates data sharing between different service providers in a seamless manner, creating multiple business opportunities [23].

## 6 SOCIAL AND COMMUNICATION ASPECTS OF HUMAN BEHAVIOUR

The integration of smart technology in the surrounding environment with humans will lead consumers to become involved in the IoT revolution just as they have been in the Internet revolution [24–25]. Social networking is expected to sustain another shift, with billions of interconnected objects [26–27]; a substantial development will be using a Twitter-like concept where separate 'things' in the house can periodically tweet their readings, which can be simply followed from anywhere by creating a Tweet. Although this provides a virtual framework using the cloud for data access, a new security system will be required for this to be entirely fulfilled [28]. Within a work environment, we refer to the 'Network of Things' as an enterprise-based application. The owners solely utilize the collected data from these networks, and the date is to be selectively released.

Developments in how we communicate have changed the very nature of how we behave – from the emergence of the electric telegraph in the 16th century to the public Internet used at the end of the 20th century, and the advent of Wi-Fi at the start of the 21st century.

Wi-Fi was a tremendous step change in human interconnectedness and behaviour – people are now creating, consuming and sharing data 24 hours a day, 365 days a year, in many languages, covering millions of topics. The rapid adoption and development of wireless technologies has led to 'Internet of Things' as one of the major technology trends of the past ten years [29].

Social and communication aspects of human behaviour have been affected by wireless technology. For example, people have become more eager to catch the available mass of information that rapidly developed and emerges throughout every day, reducing traditional social contact and replacing it with social media websites and visual distant communication. In addition, such contact caused work and education to become more international in scope, at the same time requiring special skills and equipment to address this mass of information and the associated communication opportunities. Such type of communication caused a weakness in direct human contacts that needs a special social activities to compensate this missed habit that balances human social life.

To be compatible with the newly available facilities, smart spaces need smart occupants, able to use these facilities to save resources, time and life stress at the same time as increasing knowledge, productivity, creativity, and communication opportunities. In addition, Smart solutions are minimizing needed function spaces, providing and managing information that supports decision-making and impacts on the efficiency and productivity of human life.

## 7 MAMA HOME

With the advent of IoT technology, *Mama Home* is one suggested application for a services facility. This home is like a mother for its baby. It provides a comfortable internal environment for all aspects, physical (e.g. thermal, acoustic, visual) and psychological, contingent upon the identification of its occupants and data about their comfort repeated word. Thus, it is an interactive home system that aims to provide entire care for each individual separately, or for all interior space users simultaneously.

In addition, *Mama Home* is able to facilitate healthcare services through utilising IoT technology that gives a perfect platform to health follow-ups by using body area sensors and an IoT back-end to transfer the data to remote servers [30]. For instance, a connected smart mirror is able to scan and check facial features, such as eyes and colouring, and send a report to the hospital or family physician about the user's health condition. Similarly, urine analysis tools can be embedded in the equipment of the internal space equipment. Furthermore, an expansion of this personal body area network might be a development of a home monitoring system for the care of patients and the elderly. By using such methods as a smart mattress that monitors weight, movement and heart rate, allowing a physician to monitor them in their homes, and accordingly reducing the expenses of hospitalisation through the early intervention and treatment of the problem. All of these facilities and more can provide continuous medical care and advice according to health status and reports. This is providing health welfare and increasing performance and productivity.

*Mama Home* is also able to contact the marketplace to order necessary goods. For example, a smart refrigerator can contact the grocery store and place an order according to a programmed list, as well as knowing the product expiry dates and other data, and notifying the user if there is any relevant product advice in the meantime. Likewise, a car's program can make contact with the maintenance centre and obtain or arrange a fix through the Internet, connecting without human intervention.

In addition, *Mama Home* is able to save energy and conserve the environment through controlling all home appliances and equipment, like fridges, air conditioners, and washing

Figure 3. IoT-enabled control and monitoring.

machines, providing better management of the home and better energy performance (Figure 3) [31]. Moreover, utility providers can remotely control some equipment in the building through the Internet in order to alleviate grid loads.

Environmental observation is the initial common application that is carried out to adapt the internal environmental conditions and control or manage the inside-building utilities (e.g. HVAC, lighting), contingent upon occupant location.

Sensors have always been an essential part of the programming system for applications such as security and automated climate control. These will ultimately be substituted by a wireless system, allowing the flexibility to make changes to the programmed system whenever required. An entire IoT subnet might be dedicated to factory maintenance.

The entire *Mama Home* system is enabled by the user and they can control it through their smartphone, communicating via protocols such as Bluetooth to interface with sensors measuring physiological parameters. At time of writing, there are already several applications available for smartphone operating systems such as Apple iOS, Google Android and Windows Phone that measure different parameters.

## 8 CONCLUSIONS

The Integration between smart grid, city, building spaces and skilled human and society will lead to the most efficient performance of the process for things to provide better life for human.

Information technology future trend lie beyond traditional desktop devices. In the Internet of Things paradigm [29], the majority of the objects that surround us will be on the network in one form or another. Radio-frequency identification, sensor network technologies and Cloud computing will rise to meet this new challenge, these shift of technologies has affected our social and practical life in addition to city, building, space concept. Which need more ideas and activities to meet these changes and to achieve life balance in addition to be compatible with the new coming lifestyle.

## REFERENCES

[1] Lauren Moores. "The IoT Impact on Human Behavior", article, cioreview may2016, http://media-entertainment.cioreview.com/cxoinsight/the-iot-impact-on-human-behavior-nid-14822-cid-6.html.

[2] Federal Energy Regulatory Commission, "Assessment of Demand Response and Advanced Metering", Staff Report, December 2008, https://www.ferc.gov/legal/staff-reports/12–08-demand-response.pdf.

[3] Subhadip Mahalanabish, Anindya Pradhan, Aniruddha Mukherjee White paper, "Smart Grid and the Internet of Things" https://www.tcs.com/content/dam/tcs/pdf/Industries/energy_resources_and_utilities/Smart%20Grid%20&%20the%20internet%20of%20Things.pdf.

[4] Office of Electricity Delivery & Energy Reliability, "Grid Modernization and the Smart Grid" https://energy.gov/oe/services/technology-development/smart-grid.
[5] Subhadip Mahalanabish, Anindya Pradhan, Aniruddha Mukherjee White paper, "Smart Grid and the Internet of Things".
[6] Rehmani, M.H., Kantarci, M.E., Rachedi, A., Radenkovic, M. & Reisslein, M. (2015). Smart grids: A hub of interdisciplinary research. *IEEE Access*, *3*, 3114–3118.
[7] http://www.nuritelecom.com/solutions/graphics/solutions-AMI-diagram.jpg.
[8] Office of Electricity Delivery & Energy Reliability, "Grid Modernization and the Smart Grid".
[9] Picture of the Future "Heading for Urban Energy Internets" https://www.siemens.com/innovation/en/home/pictures-of-the-future/infrastructure-and-finance/smart-cities-smart-buildings.html.
[10] Steve Hodgkinson, (2011) "Is Your City Smart Enough?", analyst insight, https://pdfs.semanticscholar.org/e5c0/ee5ef760206c64ecdf112b37cbaf5200d337.pdf.
[11] Building Effeciency Initiative, (2011), "What is a Smart Building? "Article, http://www.buildingefficiencyinitiative.org/articles/what-smart-building.
[12] Casey Talon, Noha Goldstein, (2015), "smart offices: How Intelligent Building Solutions Change Occupant Comfort." http://www.intel.com/content/www/us/en/smart-buildings/smart-building-scenarios/overview.html.
[13] Brown, E. (2016, September 20). 21 Open Source Projects for IoT. *Linux.com*. Retrieved from https://www.linux.com/NEWS/21-OPEN-SOURCE-PROJECTS-IOT.
[14] ITU. (2015). Internet of Things Global Standards Initiative. Geneva, Switzerland: International Telecommunication Union. Retrieved from http://www.itu.int/en/ITU-T/gsi/iot/Pages/default.aspx.
[15] Harvard Business Review. (2014). Internet of Things: Science Fiction or Business Fact? Retrieved from https://hbr.org/resources/pdfs/comm/verizon/18980_HBR_Verizon_IoT_Nov_14.pdf.
[16] Vermesan, O. & Friess, P. (2013). *Internet of Things: Converging technologies for smart environments and integrated ecosystems*. Aalborg, Denmark: River Publishers.
[17] Evans, D. (2011). The Internet of Things: How the Next Evolution of the Internet Is Changing Everything (White paper). San Jose, CA: Cisco Systems. Retrieved from https://www.cisco.com/c/dam/en_us/about/ac79/docs/innov/IoT_IBSG_0411FINAL.pdf.
[18] http://www.ironpaper.com/webintel/articles/internet-things-market-statistics-2015/.
[19] USGBC, " building and climate change", report, http://www.eesi.org/files/climate.pdf.
[20] Chen, H., Chou, P., Duri, S., Lei, H. & Reason, J. (2009). The design and implementation of a smart building control system. In Proceedings ICEBE 2009: IEEE International Conference on e-Business Engineering, 21–23 *October 2009, Macau, China* (pp. 255–262). Los Alamitos, CA: IEEE Computer Society.
[21] Subhadip Mahalanabish, Anindya Pradhan, Aniruddha Mukherjee White paper, "Smart Grid and the Internet of Things".
[22] Chen, Han, et al. "The design and implementation of a smart building control system." e-Business Engineering, 2009. ICEBE'09.
[23] Gubbi, J., Buyya, R., Marusic, S. & Palaniswami, M. (2013). Internet of Things (IoT): A vision, architectural elements, and future directions. *Future Generation Computer Systems, 29*(7), 1645–1660.
[24] Alkar, A. & Buhur, U. (2005). An Internet based wireless home automation system for multifunctional devices. *IEEE Transactions on Consumer Electronics, 51*, 1169–1174.
[25] Darianian, M. & Michael, M.P. (2008). Smart home mobile RFID-based Internet-of-Things systems and services. In *Proceedings ICACTE 2008: International Conference on Advanced Computer Theory and Engineering, 20–22 December 2008, Phuket, Thailand* (pp. 116–120). Los Alamitos, CA: IEEE Computer Society.
[26] Ning, H.S. & Wang, Z.O. (2011). Future Internet of Things architecture: Like mankind neural system or social organization framework? *IEEE Communications Letters, 15*, 461–463.
[27] Atzori, L., Iera, A. & Morabito, G. (2011). SIoT: Giving a social structure to the Internet of Things. *IEEE Communications Letters, 15*, 1193–1195.
[28] Li, X., Lu, R.X., Liang, X.H., Shen, X.M., Chen, J.M. & Lin, X.D. (2011). Smart community: An Internet of Things application. *IEEE Communications Magazine, 49*, 68–75.
[29] Guy Strafford, (2014) "The Internet of Things—challenging human behavior", http://www.proximagroup.com/catalytic-thinking/internet-things-challenging-human-behaviour.
[29a] Gubbi, Jayavardhana, et al. "Internet of Things (IoT): A vision, architectural elements, and future directions." Future generation computer systems 29.7 (2013): 1645–1660. http://www.sciencedirect.com/science/article/pii/S0167739X13000241.
[30] Atzori, L., Iera, A. & Morabito, G. (2010). The Internet of Things: A survey. *Computer Networks, 54*, 2787–2805.
[31] http://blog.ourcrowd.com/bigger-better-try-smarter-investing-in-the-internet-of-things-infographic/.

*History and arts*

*Cities' Identity Through Architecture and Arts – Catalani et al. (Eds)*
*© 2018 Taylor & Francis Group, London, ISBN 978-1-138-05409-7*

# Identity in transformation of rural Egyptian villages

Francesca Giangrande
*Sapienza University of Rome, Italy*

Luciano De Bonis
*University of Molise, Italy*

ABSTRACT: This research analyses how spatial planning is changing in the village of Kafr Kela al-Bab (Gharbiyya, Egypt), from which many people have migrated to Rome (Italy). The experiences acquired by transmigrants feed the local imagination about the transformation of the rural environment and, together with the flows of remittances, are leading to an abandonment of traditional building production (the *fellah* houses) in favour of new villas and apartments that are spreading all over the agricultural land, transforming it into a form of "rurban" milieu. If identity is an evolutionary process it is crucial to understand the outcomes arising from the interaction between the vernacular and the new. And it is up to the planner to identify those practices and policies able to produce long-term balance between natural, built and human environments, taking into account the important role that "translocal identities" could play in such a new rurban context.

*Keywords*: spatial planning; Kafr Kela al-Bab; rural environment; fellah houses; translocal identities; rurban

## 1 INTRODUCTION

This paper focuses on identity changes (cultural, social and spatial) that are taking place in rural villages in the Nile Delta region. This reflection is part of recent PhD research in Urban Planning, conducted by one of the authors, under the supervision of the co-author, of this paper. The hypothesis is that the historical contrast between rural and urban settlements is gradually vanishing, to allow new forms of settlement that incorporate the features either of cities or of villages, and that the ongoing transformation is the result of the impact of regional trends and global cultures on traditional communities, who acquire a new "rurban" identity.

Throughout the ages, rural areas have received little attention and few national resources and this has created a huge gap between the development of "backward" villages and "advanced" urban centres. But if a typological rural settlement has survived thousands of years of history without being completely overwhelmed by globalisation, might it make sense to observe and learn from those "spurious outcomes" that occur today in such an environment?

The subject of this research involves the use of a "translocal" (Brickell & Datta, 2011) lens that allows investigation of the interaction and the interconnectivity among the places where migration generates a stream of actors, resources, information, ideas and identities (Levitt, 2001; Pries, 1999; Portes, 1997). Migrants, exercising their agency, are "located and dislocated" subjects, acting as translocal placemakers, subjected to tensions and conflicts arising from their multi-positioning. Robin Cohen (1997) defines them as "transmigrants", who develop an identity that connects them to two or more nation states. They never leave the homeland, but circulate between territories of origin and arrival, as though aiming at a "hybrid" way of life.

If we consider identity in terms of an evolutionary process in which environmental, urban, social and cultural aspects interact, it is crucial to pay attention to the mix of modernisation and preservation that creates tension in rural areas, also making any interpretation questionable but nevertheless necessary to understand this new identity. The research involved six months (in 2015) of fieldwork in Egypt, mainly in Kafr Kela al-Bab (hereinafter 'Kafr'). Specifically, it has seen the development of a specific ethnographical programme and subsequent spatial interpretation. In particular, a multi-site approach has been adopted for the interviews and participant observation, based on "*follow the people; follow the thing; follow the metaphor; follow the plot, story, or allegory; follow the life or biography; and follow the conflict*" (Marcus, 1995).

The ethnographic survey highlighted the capacity of migrants to live and organise themselves between geographically dispersed areas by implementing socio-spatial practices (from housing to places of worship, of socialisation and so on), either in the context of origin or arrival. In Kafr, the migration has created a new way of life, characterised by an emerging "translocal" identity of places, generated by "co-production, hybrid, cosmopolitan experiences" (Clifford, 1991).

## 2 RURAL EXODUS AND URBAN ENCROACHMENT IN THE NILE DELTA

In Egypt, the agricultural sector employs 34% of the workforce and accounts for 12.5% of national GDP (World Bank data[1]). Agriculture had been present in ancient Egypt since the Neolithic Age and farmers contributed to the creation of the first settlements, concentrated mainly in the Nile basin.

Rising sea levels and rapid urbanisation, especially the informal development after 2011, have made the Nile Delta a highly vulnerable region, where pollution, desertification and population growth combine to define a state of emergency for the land (Redeker & Kantoush, 2014). In the Gharbiyya Governorate, where Kafr is located, the arable land area is decreasing by an average of about 3,200 $km^2$ per year (Zaghloul & Elwan, 2011).

Kafr is a village of about 25,000 inhabitants, 80 km north of Cairo and 150 km southeast of Alexandria, at the core of the Nile Delta. According to its strategic plan of 2007, the total area of Kafr is 3,183 ha, of which 2,710 ha is agricultural land, accounting for 85.3% of the total area. Private property prevails over public. The plan complained that urban expansion threatened to overwhelm the agricultural land. This expansion is helped by the fact that the associated legislation is contentious; Agrarian Law criminalises the construction of buildings on agricultural land yet the Constitutional Court does not condemn it. This and other problems have done nothing but encourage deregulation and a random urban sprawl.

The agricultural world in Egypt changed because of market liberalisation policies that had a great impact on the rights of farmers, causing the deterioration of rural livelihoods, increases in rates of poverty, marginalisation and social immobility.

In the 1970s, under Anwar el-Sadat, a policy of opening to the West and economic liberalisation (*Infitah*), aimed primarily at attracting foreign investment, was pursued in Egypt. In 1971, Article 52 of the Egyptian Constitution authorised emigration: the first flows of temporary workers went to the countries of the Arabian Gulf, the Middle East and North Africa (MENA) countries (mainly Libya), and to the West (USA, Canada, Australia and Italy). After the 1973 Yom Kippur War, there was a phase of expansion and immigrant flows that have become a priority on Egyptian governmental agendas.

It is not surprising that, due to the mass of reforms aimed at the modernisation of the country, the four metropolitan areas of Cairo, Alexandria, Suez and Port Said became attractive to a large proportion of migrants, coming mainly from Menufiya, Sohag, Asyut and Sharqia.

1. See the link https://goo.gl/nXFzgN.

During the 1990s, a great exodus from the countryside started, especially towards the northern shores of the Mediterranean, mainly to Italy and France. In the Egyptian context, remittances sent home from abroad are especially important; in fact, they contribute greatly to national development. In 2008, they accounted for 5.3% of national GDP (IOM, 2010) and, in 2009, Egypt placed seventh in the world among countries with high flows of remittances, amounting to 7.8 billion dollars (which could be an underestimation because there is also an unregistered flow of money, sent through informal channels and not through banks).

In Egypt in the late 1980s, various economic and sociological studies of the impact of remittances were conducted. Adams (1989) highlighted a worsening of rural income distribution because remittances flowed mainly to those villagers with high existing incomes. Research in Kafr confirmed similar trends; remittances create dependency upon Italy, causing massive growth in the building sector but few other productive investments. At the same time, they generate higher demand for goods and services and the inflationary pressure on local economies increases. However, the longer the history of migration in a village, the more it lowers costs and opens up new opportunities for the most disadvantaged classes.

In other words, the returning migrants are potentially carriers of both financial and human capital, technology and entrepreneurship, and remittances are the new development mantra, considered as "the development panacea" (Kapur, 2003). All of these factors can contribute to economic development but can also affect social inequality if they are not well regulated. Furthermore, migration, domestic and transnational, combined with population increase and poor governance in rural areas, has led to a decrease in arable land and its conversion to more profitable non-agricultural uses. To date, Egypt has lost about 10% of its agricultural land due to the ongoing urban sprawl (Wahdan, 2013).

## 3 *FELLAHIN*: FROM RURAL TO "RURBAN"

The traditional planning of a village (following the Islamic conquest) started with an hill (*al-koum*) at whose core was positioned a mosque surrounded by a very dense road network, a Koranic school (*kuttab*) located among the housing, and a ring road to the edge of the hill, chosen to avoid seasonal flooding of the Nile. The cemetery, the grain silos and the market were positioned on the outskirts of the residential area (Mahgoub, 2001).

At the beginning of the nineteenth century, Egypt's Governor, Muhamad Ali, carried out some interventions for the control of the hydrological regime of the Nile that had a great impact on the morphology of such villages. In the resulting absence of constraints imposed by flooding, the *al-koum* expanded in the agricultural land below, and new estate villages (*'ezbah*) were built on the plains along the canals and branches of the Nile. The introduction of the car in the early 20th century and its intensive use during World War II by British troops also had a great impact on the development of Egyptian villages (Mahgoub, 2001). Following the 1952 Egyptian revolution, peasants were allowed—for the first time in Egypt's history—to own agricultural land: this stimulated landowners to build new homes on their land, imitating the typical informal construction methods of areas around Cairo and other cities. All these events have contributed over time to the transformation of the village system (Figure 1), which has dramatically changed the quality of life of traditional settlements: giving rise to overcrowding, water pollution from waste, inadequate sanitary and sewer systems, air pollution, inadequate lighting systems and ventilation of houses, and so on.

Kafr experienced all of the transformation processes described above. It seems to be an "urban village" (which is distinct from a "peri-urban" settlement, that is, the meeting between urban and rural, building up gradually through ever denser metropolitan fringes). More specifically, according to unstructured interviews with some inhabitants of the village (especially a pharmacist and a butcher), the Sheikh Mahroos (who is celebrated with a local *Mawlid* or birthday celebration) and other families were the first to buy and build on those lands, then divided them among multiple owners. Traditionally, the social status and position in society

Figure 1. *Fellahin* in Kafr Kela al-Bab, 2015 (Photo: F. Giangrande).

of the *fellah* (farmer or agricultural labourer) were determined by land ownership and the position occupied in the system of his extended family.

With the expansion of the school system in the village and the increase in education levels at the beginning of the 1970s, a different hierarchy of power was established, no longer dependent on the land and the "Sheikhs al-Balad" (community heads), but with the highest role occupied by "doctors", the educated men of the village. In fact, during the interview, the pharmacist stated: "This is called the village of scientists, because we have many educated people at the University of Tanta who are important in their expertise!" Many doctors contributed to the social development of the village, providing money for projects, services and infrastructure. During the 1990s, with the collapse of the education system and rampant social immobility, the peasants and the young graduates have preferred to look beyond the village for the resources to empower themselves. Thus, economic liberalisation and cross-border migration have become the cornerstone of a new individualism, and of private initiative in rural society. We believe that in Kafr we can speak of a "rurban" environment as the most appropriate description of what has been observed:

> *The term 'rurbanisation' was coined in the 1960s, meaning the 'settlement in the rural municipalities of people coming from cities, where they often continue to work and ... thus ... become, in fact, commuters' (JB Charrier, quoting G Bauer and JM Roux, "La Rurbanisation ou la ville éparpillée", Editions du Seuil, 1976).* (Perrella, 2007)

The rurban consist not only of commuters from Cairo, but also of returning transmigrants (re-exodus), bourgeoisified in Italy or elsewhere, who would like to apply a post-modern urban style to the village to which they now have a sense of belonging.

From numerous interviews and observations, it appeared that transmigrants never question their duty, inherent to the migration project, to transfer money to the village, which has significant consequences for the population:

> *One of the most important consequences of the migrants' remittances and their pattern of investment is the great rise in the land prices in the village of Mit Badr Halawa. Another aspect of change that is more readily visible is the presence of a large number of huge multi-storey houses whose architecture is extremely decorative and very lavish-looking [...] On the one hand, Badrawis generally talk of the changes in the village, especially the "new architecture" in positive terms and regard this as a sign of progress. Others who are not from Mit Badr, however, particularly those who come from an older generation of migrants and who were generally more educated are very contemptuous and cynical about such changes [...] and see in their new architecture blind and distorted imitation of what they saw in France.* (Saad, 2005 p.13)

The village environment is changing very fast, both in physical appearance and in its socio-economic conditions. The new houses in Kafr (Figure 2) would seem to be a visible sign of

Figure 2. Villa of a successful transmigrant, Kafr Kela al-Bab, 2015 (Photo: F. Giangrande).

a more abstract, invisible process of creation of translocal class and identity. The lifestyle is increasingly influenced by urban examples, yet the proliferation of new homes does not necessarily imply that the living conditions are improved for everyone in the village. These individual (or family) house builders are increasingly a new class of people, who could be called the "new rurban bourgeoisie" of the villages, who have benefited from remittances invested in real estate and the construction industry or in cash-cropping. Although the urban/rural boundaries are fading, the emergence of this "rurban space" is not necessarily synonymous with dispersed urbanity:

> *In large part, these agglomerations are still dominated by agricultural activities, a feature which defines them, in the words of some experts, as "agro-towns" and, although diversity is growing, it is still limited. Conventional urban services (such as paved roads, piped water, garbage collection and sewer systems) are largely absent and the illiteracy rate, especially among women, is quite high. However, it must be noted that urbanisation of the countryside should be seen not as a uniform spread of urbanity in the hinterland but, rather, as a new trend of polarisation at the level of small cities and large villages, or urban villages.* (Bayat & Denis, 2000, p. 195)

## 4 TOWARDS A NEW IDENTITY: THE NILE DELTA AS A MESOPOLIS

In Kafr, translocality has multiple effects: the village is transformed into a place for consuming imported goods and for the creation of a workforce to be exported, instead of a place of agricultural production. Widening this analysis, this situation is present in other Delta villages too. The owners of land had the power to transform the best agricultural soils into a bigger, residential "rurbanity", where huge villas remind poorer residents of the village of the success obtainable through remittances. Translocal contexts thus represent the places of opportunity and limit that the migrant faces, and the modalities of their connections through migrations are important elements in understanding his agency in improving the spatial environment.

In her book *Villes du delta du Nil: Identités citadines et émergence d'une région urbaine: Tanta, Mahalla et Mansura*, the researcher Delphine Pagès-El Karoui (2002) draws a comparison between the region of the Nile Delta and the region of Emilia Romagna in the centre of Italy, which has been defined by Franco Farinelli (2003) as a "mesopolis".

According to this interpretation, the Delta could be an Egyptian "mesopolis", an urban system where none of the cities (Tanta, Mahalla or Mansoura) is dominant, so that their populations are almost equal, as occurs in the area of the:

Figure 3. Rurban Kafr, 2015: Slow encroachment into the countryside (Photo: F. Giangrande).

> *…Appennino, almost in a straight line from Rimini to Piacenza. Taken together, they form a conurbation, in the sense [of] Patrick Geddes, who invented the term in 1915: not a seamless urbanisation but a "galaxy of cities", a natural "city alliance". In short, a "city-region".* (Farinelli, 2003)

Similarly, the Nile Delta can be understood as an area defined as "mesopolitan", where the rural and urban areas do not have a distinct identity anymore, although their environmental differences are still present in the Egyptian memory (Pagès-El Karoui, 2002).

Thus, there are two entities that are still conceived in public opinion as separate: the agricultural land and its system of irrigation, and the urban pattern with its cities and one thousand villages with their roads. Given the lack of integrated management, these two patterns are in opposition to each other. The urban pattern of the Delta is continuously expanding, consuming fertile land and creating a dangerous conurbation with all the risks that this transition implies:

> *In this widespread mesopolis, two types of transformations are occurring: the completely spontaneous micro-urbanisation of villages and the polarisation of cities that drain all functions. Both these are threatening the rural space, which for Egypt is a subsistence space.* (Maldina & Tonnarelli, 2013, p. 53)

Nevertheless, despite the risks, some potential can also be identified: the Nile Delta region can, in fact, become a place in which to experiment with new forms of urban management that aim at strengthening middle-sized cities, decongesting the large metropolises like Cairo and Alexandria, and implementing a system of connections to establish the cultural value of anthropic and natural landscapes, create a better quality of life ((Maldina & Tonnarelli, 2013), and confront real urban sprawl (Figure 3).

## 5 CONCLUSIONS: HYBRID IDENTITY, TRANSLOCAL SPACE AND SELF-SUSTAINING DEVELOPMENT

In conclusion, due to the pressure of modernisation and globalisation, the transformation of the traditional environment of the Egyptian countryside conveys an important lesson: it is necessary to include the aspirations of the people who live in poverty, and to solve their housing problems. Moreover, it is needed to look at the impact of migrations in these contexts not only for their negative impact on the environment, but also as a result of the will to change the identity of a place. Perhaps it is the lack of networking among that single local identities, due to the incapacity of government, that produces a legal framework that is fitting more urban contexts than rural ones, while not considering forms of mobilisation that aim to provide better living conditions and that do not take into consideration the fragile nature of the Nile Delta region.

In our opinion, the transformative potential of migration activity should be taken into account. Some hope for improving the evolution of the region comes from the fact that there is a form of emulation between the villages in promoting what has been gained elsewhere in terms of better living conditions. This process has produced some effects that may not have been intentional, but have resulted from the ability of the residents to "muddling through" (Lindblom, 1959) in the absence of any governance capable of attracting capital in support of more sustainable local development. Thus, it is necessary to identify practices and territorial policies that are not only rooted in the culture of the residents of that context, but that are capable of strengthening some form of virtuous hybrid development.

Globalisation and "mesopolisation" do not erase everything, including a division which is not spatial but mental; the descendants of the *fellahin* that nowadays live in the cities or in urbanised villages that are no longer engaged in economic activities that are agricultural, remain *fellahin*, and their territories remain rural to the eyes of the urban residents. From the time of the famous novel by al-Hakim, 'Diary of a Country Prosecutor', to 'The Yacoubian Building' by Al-Aswany , the perception of rural areas does not seem to have changed much. Perhaps this is a good thing. And by this we do not mean to reinforce the stereotypical image of the countryside, but that it is interesting to notice how some traditions remain in the form of a hybridisation process that includes the rapid changes of the "translocal countryside".

We need a strategy that leaves room for changes of identity of a place and the empowerment of migrants (more generally, of the entire population) that gives strength to those who live in and create the territory, promoting in them a new culture of "self-sustaining local development", where the term "local" highlights the enhancement of local resources and the identity of a place, and "self-sustaining" refers to the importance of seeking settlement, economic and social policy rules for producing long-term balance between the natural, the built and the human environments, through "*the activation of 'Lilliputian' strategies, weaving non-hierarchical networks (South-South, South-North, between cities and regions), in a dense network that can counter the major networks, highly centralised, of economic globalisation*" (Magnaghi, 2000, p. 91). The dense network to which Magnaghi refers can also be found in the "translocal space" that is produced by migrants and their "globalisation from below" (Ambrosini, 2008; Portes, 1997).

## REFERENCES


Adams, R. (1989). Worker remittances and inequality in rural Egypt. *Economic Development and Cultural Change*, *38*(1), 45–71.

Ambrosini, M. (2008). *Un'altra globalizzazione: Il transnazionalismo economico dei migranti*. Working paper 5/08. Milan, Italy: Department of Social and Political Studies, Faculty of Political Science, University of Milan.

Azzam, O.A. (1960). *The development of urban and rural housing in Egypt* (Doctoral thesis, Swiss Federal Institute of Technology, Zürich, Switzerland). doi:10.3929/ethz-a-000097781

Bayat, A. & Denis, E. (2000). Who is afraid of ashwaiyyat? Urban change and politics in Egypt. *Environment & Urbanization*, *12*(2), 185–199.

Brickell, K. & Datta, A. (2011). *Translocal geographies: Spaces, places, connections.* Burlington, MA: Ashgate.

Clifford, J. (1991). Traveling cultures. In L. Grossberg, C. Nelson & P. Treichler (Eds.), *Cultural studies* (pp. 96–116). New York, NY: Routledge.

Cohen, R. (1997). Global diasporas. An introduction, London; New York: Routledge.

Farinelli, F. (2003). *Geografia: Un'introduzione ai modelli del mondo.* Turin, Italy: Giulio Einaudi.

IOM (2010), A Study on Remittances and Investment Opportunities for Egyptian Migrants, Cairo, IOM.

Kapur, D. (2003). Remittances: The New Development Mantra? *G-24 Discussion Paper Series*.

Levitt, P. (2001). *The transnational villagers*, University of California, Berkeley.
Lindblom, C.E. (1959). The science of "Muddling Through". *Public Administration Review*, *19*(2), 79–88.
Magnaghi, A. (2000). *Il progetto locale.* Turin, Italy: Bollati Boringhieri.
Mahgoub, Y. (2001). The transformation of traditional rural settlements in Egypt, Second International Symposium TRADITIONAL ENVIRONMENTS IN A NEW MILLENNIUM: Defining Principles and Professional Practice Amasya—Turkey, 20–23 June 2001.
Maldina, S. & Tonnarelli, F. (2013). *Frozen Cairo* (Master's thesis, University of Ferrara, Italy).
Marcus, G.E. (1995). Ethnography through Thick and Thin. Princeton: Princeton University Press.
Pagès-El Karoui, D. (2002). *Villes du delta du Nil: Identités citadines et emergence d'une région urbaine: Tanta, Mahalla et Mansura.* Paris, France: CEDEJ-IISMM-Karthala.
Perrella, D. (2007). *Abitare il paesaggio agricolo periurbano. Esperienze a confronto* (Doctoral thesis, Department of Urban Design and Planning, Federico II University of Naples, Italy).
Portes, A. (1997). *Globalization from below: The rise of transnational communities.* ESRC Transnational Communities Programme, Working Paper Series, WPTC-98-01. New York, NY: Princeton University.
Pries, L. (2001). New Transnational Social Spaces. International Migration and Transnational Companies, London; New York: Routledge.
Redeker, C. & Kantoush, S.A. (2014). The Nile Delta: Urbanizing on diminishing resources. *Built Environment*, *40*(2), 201–212.
Saad, R. (2005). Egyptian Workers in Paris: Pilot Ethnography. Development Research Centre on Migration, Globalisation and Poverty.
Wahdan, D. (2013). Planning Egypt's new settlements: The politics of spatial inequities. *Cairo Papers in Social Science*, *32*(1).
Zaghloul, S.S. & Elwan, H. (2011). Water quality deterioration of middle Nile Delta due to urbanizations expansion, Egypt. In *Proceedings Fifteenth International Water Technology Conference, IWTC-15 2011, Alexandria, Egypt.* Retrieved from http://iwtc.info/wp-content/uploads/2011/07/G40.pdf.

*Cities' Identity Through Architecture and Arts – Catalani et al. (Eds)*
*© 2018 Taylor & Francis Group, London, ISBN 978-1-138-05409-7*

# Metaphysics and identity in architecture: Peter Eisenman's Wexner Center for the Arts as case study

Houssameldeen Bahgat El Refeie
*Al Shrouk Academy, Cairo, Egypt*

ABSTRACT: The aim of this paper is to discuss the impact of metaphysical content in the formation of architecture, which subsequently confers identity within the urban context of the city. This metaphysical content includes the visions and perceptions that seek to explain the existence of God (absolute) and the universe, as well as the content of metaphysics and religious beliefs in the formulation of the identity of architecture. Thus, through critical analysis of artefacts of contemporary architecture, the impact and manifestations of metaphysics in the formation of architectural features and notation can be studied. This research selects Peter Eisenman's Wexner Centre for the Arts, Columbus, Ohio, USA, as an example with which to study the relationship between the representation of metaphysical factors and architectural identity in the modern era. The research follows an analytical methodology to study the terminology and build on the knowledge of semiology, which basically depends on the translation of forms and physical symbols. The research also depends on the philosophy of interpretation and explanation to extrapolate the impact of metaphysical reflection in all elements of design such as horizontal projections and three-dimensional design, as well as details and vocabulary that have left their impact on the drafting process and configuration. The main purpose of the research is to analyse the relationship between the metaphysical religious content and the identity of contemporary architecture.

*Keywords*: metaphysics; deconstruction; symbols; icons

## 1 INTRODUCTION

This paper fosters a hypothesis that metaphysics has been the only motive in the formation of ideals and visions throughout history. Religions, beliefs and myths provided answers to the questions of existence, the universe and the absolute. Those metaphysical beliefs were significant in the formation of architecture. These metaphysical beliefs have left their influences on different types of architecture in different civilisations at different times. That is, they occur consciously through the influences on humans being of education and tradition, and unconsciously through collective memory. Thus, metaphysical beliefs are responsible for the continuity and the accumulation of visions and expertise.

### 1.1 *Identification of metaphysics*

Metaphysics is the branch of philosophy that concerns itself with the nature of existence, natural phenomena, types of entities, and the formation of the universe. Alexander Gottlieb Baumgarten identifies metaphysics as "the science which studies the first principles of human knowledge; these principles are ontological (notion of existence), cosmological (notion of cosmos), psychological and theological" (Ghaly, 2003).

Metaphysics leaves its influences on adherents through consecutive generations. It passes regulating sociocultural principles from one generation to the next through collective memory.

### 1.2 *Branches of metaphysics*

The word "metaphysics" comes from the Greek words *meta*, which means 'what follows', and *phusika*, which refers to the physical and natural researches of Aristotle in ancient times. Originally, it referred to the sequence of Aristotle's works in which he discussed those things that transcend the physical and natural, although he did not apply the term 'metaphysics' to his work. Thus, metaphysics does not refer to the text's classification but to the arrangement (Baumer, 1977).

Besides small elements of natural science, Aristotle considered metaphysics to involve three essential branches: the ontological, the cosmological, and the theological. Mankind got all this knowledge from religion, myths and traditions, which interpret natural phenomena and establish the perceptions of God, existence and the cosmos.

## 2 METHODOLOGY

From previous research, we understand that man's perception depends on religion and myths told in different ways and in different languages. In fact, all such mental activities depend on the language that expresses beliefs and feelings (Augros & Stanciu, 1984). We think, communicate and express ourselves through language and words. Language works on a basis of understanding and interpretation through signs, and hence is able to transfer thoughts from one person's mind to another (Kemmer, 2009). Semiology focuses on linguistic analysis, especially the meaning of signs and words.

Not only does semiology focus on language, but also on all forms of communication activities in life, such as habits, social traditions, architecture and poetry. There are two branches of semiology; one of them depends on language interpretation (hermeneutics) and the other depends on the interpretation of form and physical icons. These branches are, in fact, inseparable, but each branch has its own methodology and identity.

### 2.1 *Hermeneutics*

Hermeneutics is a branch of philosophy that concerns itself with the interpretation of texts, originally that of scripture or religious texts. As such, its application depends on the intentions of the critics and authors and the points of view of those reading and interpreting the text.

### 2.2 *Semiology*

Semiology is the study of symbols, which focuses on meanings, signs and icons. It aims to establish a better understanding of text and language. Because signs can convey meanings spontaneously, people can derive meanings and perceptions through signs.

There are different approaches to understanding and analysing semiology. In this paper, we depend on Ferdinand de Saussure's interpretation as a basis for reading and interpreting signs in architecture. Saussure called semiology a "theory of signs". What he meant by the word signs was anything that refers to something indirectly or is a reminder of it. This could be through readings of text, gestures, forms, photos or buildings: a "sign is something, anything, which stands for or reminds us of something else" (Hoffmann, 2001).

Saussure discusses the differences between language and speech in the use of words, grammar and linguistic expression. He discusses how meanings and conversation define man's culture and identity. Saussure goes further and explores how language and words that refer to meanings and linguistic expression depend on syntax and the arrangement of words and phrases to create a specific meaning. For example, the word architect reminds us of architecture, building, construction and structure. Roman architect Vitruvius stated that, "In all matters, but particularly architecture, there are these two points: the thing signified and that which gives it significance" (McEwen, 2003).

## 2.3 *Signifier and Signified*

Saussure argued that a concept has two components: the signifier and the signified. They are connected and entwined together to form the "sign", which acts as a reminder of something else. Saussure claimed that the signifier is the physical thing which transfers the meaning or concept (drawing, picture, building, etc.). The signified is the subjective thing which completes the big picture. Thus, the signifier and the signified are the main elements of any concept. Saussure also claimed that the relationship between them is arbitrary (Glover, 2013). Additionally, he distinguished between synchronic studies, concerned with signs in a specific time and specific location, and diachronic studies, concerned with studying signs and their evolution across time (Glover, 2013). Diachronic studies depend on a collection of descriptive chronicle studies that compare old signs to new ones, meaning that the chronicle is changing with time (Russell, 1984).

# 3 WEXNER CENTER FOR THE ARTS CASE STUDY

The Wexner Center for the Arts in, Columbus, Ohio, was constructed between 1983 and 1989 (Figure 1). It was designed by Peter Eisenman and has become an iconic buildings, which incorporates many signs and a unique vocabulary that gives it a distinctive form, using three-dimensional modules in its design of indoor and outdoor spaces, especially its corridors and paths (Jencks, 1988). The building provides a distinctive features not only to the site but also to the whole urban context. Eisenman embraces his own theories about identity, community, culture and globalisation. He is one of the pioneers of the deconstructivist movement of postmodern architecture, and has his own thoughts about traditional buildings and fragments working together according to the dynamic system of the cosmos (Kolb, 1998).

Eisenman used a multigrid system to express the interaction between domestic and international streams (Figure 2). He interpreted his thoughts in the form of mysterious interlocking grids (Corbo, 2015). Eisenman was inspired by the physical dynamics of planets and used this on multiple levels and multiple layers, where he sought a utopia in architecture through the aesthetic values of forms and their alignment in symphony that matches the movement of the planets and their orbits (Jencks, 1988).

Eisenman was influenced by the deconstructivist philosophy of the French philosopher Jacques Derrida, who frequently argued that western philosophy has uncritically allowed metaphorical depth models to govern its conception of language and consciousness. He saw these unacknowledged assumptions as part of the "metaphysics of presence" to which

Figure 1. Wexner Center for the Arts, Columbus, Ohio designed by, Peter Eisenman (Source: http://fitnessnyc.files.wordpress.com/2009/06/wexner-center05.jpg).

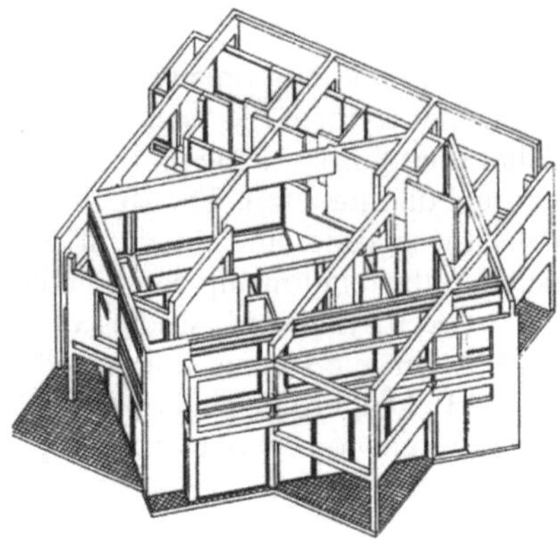

Figure 2. Peter Eisenman using grids in three dimensions (Source: http://www.dkolb.org/ht/ou-dk.html).

philosophy binds itself. This "logo centrism", Derrida argued, creates "marked" or hierarchised binary oppositions that have an effect on everything, from the conception of speech's relationship to writing, to the understanding of racial differences. Deconstruction is an attempt to expose and undermine such metaphysics. This is reflected in Eisenman's handling of forms and architecture using a double code (Jencks, 1988). Eisenman's study of ideologies and beliefs induced him to explore the users and their ideologies, beliefs and background (historical, religious, political and social) before starting the design process (Jencks, 1988).

He interprets these factors on different layers using multiple grids in the design process. From his point of view, history has the ability to "trace memory", dig into the past and search in history. Building upwards represents the future, while the site itself represents the present. All his designs are inspired by this philosophy (Jencks, 1988).

### 3.1 *Metaphysical influences in Peter Eisenman's Wexner Center for the Arts*

In the following subsections, I trace the metaphysical influences on the features and vocabulary of architecture and how they shape the distinctive form of the Wexner Center for the Arts.

#### 3.1.1 *Theological influences*

Eisenman applies his philosophy "Symbolic Digging in Site", where he explores the prevailing beliefs in his community. He finds that Christianity is the major religion; thus, he formulates the essence of Christian beliefs in new symbolic architecture features (Corbo, 2015). This appears through his engagement with the tragic "Crucifixion" scenario. In addition, he announces his passion by the artistic paints addressing crucifying Jesus, especially "Crucifixion" by the painter Matthias Grünewald (see Figure 3).

This image of the torture of Christ inspired Eisenman to present a combination of geometric forms and offset grids to create incomplete representations. The clearest example is the main entrance which symbolises the anthropomorphic three dimensions of the crucified Christ; Eisenman split the entrance into two halves (see Figure 3), with a warm red colour to represent the blood of Christ (Corbo, 2015).

#### 3.1.2 *Cosmological influences*

Eisenman was influenced by Christianity and the creation of the universe. In the light of these beliefs and thoughts, Eisenman was inspired by religious icons and converted them into architecture through a number of deliberately awkward and discordant moments, complicating the intersection of built space with human occupation (Andonian, 2012). In its concept

Figure 3. The main entrance symbolises an anthropomorphic representation of the crucified Christ (Sources: http://www.galinsky.com/buildings/wexner/wexner1.jpg; http://www.i-youniverse.net/wp-content/uploads/2009/04/crucifixion.jpg).

Figure 4. A number of deliberately awkward and discordant moments complicate the intersection of built space with its human occupation (Sources: http://prelectur.stanford.edu/lecturers/eisenman/projects.html; http://www.bluffton.edu/~sullivanm/ohio/columbus/wexner/eisenman.html).

and process, the Wexner Center is an exemplary illustration of Eisenman's unique approach to architecture.

Eisenman assigns divinity to the major grid and gives it dominance over all other architectural vocabulary and features, allowing it to rule all the indoor and outdoor spaces (see Figure 4), and using white colouring to refer to light, transparency and the shining of God (Rogers, 1998).

3.1.3 *Ontological influences*

Eisenman was influenced by Christian belief and the biblical creation narrative from the book of Genesis in which God breathes life into Adam, the first man. Eisenman announces his influencing by the painting by Michelangelo, which forms part of the Sistine Chapel, in which the image of the "near-touching hands" of God and Adam has become iconic to humanity (see Figure 5). This painting is part of a complex iconographic scheme and is chronologically fourth in the series of panels depicting episodes from Genesis (Rizzatti, 1967).

Eisenman magnifies structural elements and creates a strong grid system that dominates the formal language of the building (Figure 5). The urban grids of the city of Columbus and of the university, slightly off-kilter from one another, overlap within this project. The 12.5° of variation between the two results in an axial rotation within the museum, with corresponding

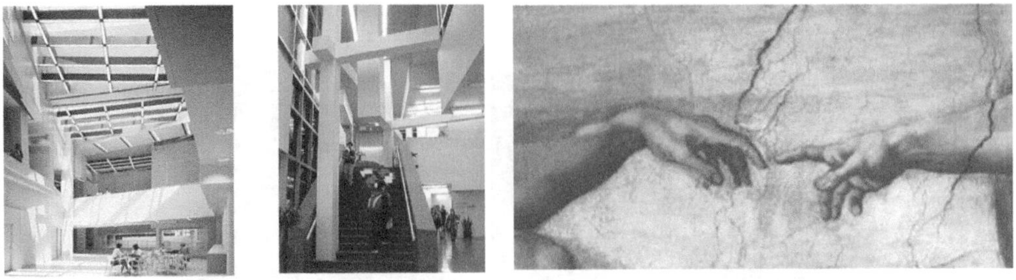

Figure 5. Magnifying structural elements giving the major grid dominance (and representative divinity) over all other vocabulary; the "near touching hands" of God and Adam have become iconic of humanity (Sources: http://www.architetturaamica.it/Biblioteca/recens/Eisenman.html; http://images.worldgallery.co.uk/i/prints/rw/lg/7/1/Michelangelo-The-Creation-of-Adam-7157.jpg).

tectonic elements creating jarring moments of intersection as the two systems compete for primacy (Ansari, 2013).

## 4 CONCLUSIONS

- Metaphysics leave their impact on architecture, and it is one of the major factors in forming and shaping architectural vocabulary.
- Eisenman was inspired by the scenario of the torture of Christ and presented a combination of geometric forms and offset grids to formulate incomplete representations of them.
- Eisenman interpreted the essence of Christianity using different factors on different layers using multiple grids in the design process, which gives the building a distinctive architectural features.
- Eisenman was inspired by metaphysical religious icons and turned it into architecture through a number of deliberately awkward and discordant moments that complicate the intersection of built space with its human occupation.
- The Wexner Center acquires its distinctive features through interpretation of the metaphysical content of the Christian religion, which gives a unique identity not only to the site but also to the whole urban context.
- Eisenman embraces his own theories about identity, community, culture, and globalisation, and turns them into architecture.
- In architecture, translating metaphysical content through signs and icons is used to give a deeper dimension to the building scenario, which subsequently reflects on urban context.
- Eisenman was influenced by the deconstructivist philosophy of the French philosopher Jacques Derrida, who frequently argues that western philosophy has uncritically allowed metaphorical models to govern the concepts of language and consciousness.

## REFERENCES

Andonian, D.G. (2012). Book review: Derrida and Legal Philosophy. Edited by Peter Goodrich, Florian Hoffmann, Michael Rosenfeld, and Cornelia Vismann. *The European Legacy, 17*(3), 399–400. Retrieved from http://www.tandfonline.com/doi/full/10.1080/10848770.2012.673337.

Ansari, I. (2013, September 23). Eisenman's evolution: Architecture, syntax, and new subjectivity. *ArchDaily*. Retrieved from http://www.archdaily.com/429925/eisenman-s-evolution-architecture-syntax-and-new-subjectivity.

Augros, R.M. & Stanciu, G.N. (1984). *The new story of science* (pp. 11–15). Washington, DC: Regnery.

Baumer, F.L. (1977). *Modern European thought, continuity and change in ideas, 1600–1950* (p. 365). London, UK: Macmillan.

Corbo, S. (2015, January 25). From Formalism to Weak Form: The Architecture and Philosophy of Peter Eisenman. *ArchDaily*. Retrieved from http://www.archdaily.com/591214/from-formalism-to-weak-form-the-architecture-and-philosophy-of-peter-eisenman.

Ghaly, W. (2003). *The end of philosophy* (p. 40). The Public Egyptian Institution for Books.

Glover, R. (2013, March 8). Signs. Retrieved from https://ryanglover92.wordpress.com/2013/03/08/signs.

Hoffmann, M.H.G. (2001). *Digital encyclopedia of Charles S. Peirce: The 1903 classification of triadic sign-relations*. Retrieved from http://www.digitalpeirce.fee.unicamp.br/hoffmann/p-sighof.htm.

Jencks, C. (1988). *Architecture today* (Rev. ed., p. 265), York, NY: Harry N. Abrams.

Kemmer, S. (2009). Guided commentary on de Saussure's Course in Modern Linguistics. Retrieved from http://www.ruf.rice.edu/~kemmer/Found/saussureessay.html.

Kolb, D. (1998, August 20). Collisions and Interactions: A Philosophical Perspective on CATAC 98. Retrieved from http://www.dkolb.org/arch.urb/catac-dk.html.

McEwen, I.K. (2003). *Vitruvius: Writing the body of architecture* (pp. 17, 65). Cambridge, MA: MIT Press.

Rizzatti, M.L. (1967). *The life & times of Michelangelo* (p. 16). Verona, Italy: Hamlyn.

Rogers, C. (1998). Wexner Center for the Arts, Columbus. *Galinsky*. Retrieved from http://www.galinsky.com/buildings/wexner/index.htm.

Russell, F. (Ed.). (1984). *Mies van der Rohe: European works* (p. 20). New York, NY: St. Martin's Press.

*Cities' Identity Through Architecture and Arts – Catalani et al. (Eds)*
*© 2018 Taylor & Francis Group, London, ISBN 978-1-138-05409-7*

# Immersing the landscape in its music: The case of iso-polyphony in South Albania

Michele Porsia & Luciano De Bonis
*University of Molise, Italy*

ABSTRACT: This article reflects on cities and landscape as a complex interweaving of processes that it is not possible to control absolutely. However, it is possible to stimulate the generation of processes that could help an area to develop a sustainable economy, for instance, by highlighting its cultural heritage. In this paper, a project is used as a 'tool for research': it helps to explain how to deal with the fragile, but significant, intangible heritage of a specific place. The case study is the iso-polyphonic traditional music of South Albania, recognised by UNESCO, with a core area around Qeparo village. The paper proposes one project (among many possible ones) that is a process itself and that tries to elaborate a *reductio ad unum* of the mutual relations between landscape research, education, creativity and praxis.

*Keywords*: cultural heritage, intangible heritage, iso-polyphonic traditional music, reductio ad unum

## 1 INTRODUCTION

### 1.1 *Discovering a hidden heritage*

If landscape is, as defined in the European Landscape Convention (Council of Europe, 2000), an area as perceived by people, it is important to consider that the view is just one *door of perception* among the many senses that are involved in a synaesthetic experience. Perception of landscape involves a complex relationship between data processing, cultural filters and subjectivities, where human presence is essential. When we speak about cultural landscapes we often think about particular features, depending on artefacts that represent the relations between a specific cultural community and its space. Included within these is the 'intangible cultural heritage' defined by UNESCO (Francioni, 2001).

The opportunity to take part in the Riviera Lab—a September 2016 workshop on "Art, Architecture and Landscaping as tools for boosting improvement of life and social economic growth", organised by Polis University of Tirana—offered the chance to review the role of cultural heritage in the landscape quality assessment of the South Albania coast (known as the Albanian Riviera) and the village of Qeparo. A traditional Albanian folk music, iso-polyphony, is in the UNESCO Representative List of the Intangible Cultural Heritage of Humanity (UNESCO, 2008).

## 2 BETWEEN THEORETICAL FRAMEWORK AND SURVEYS

### 2.1 *Project as a method*

Consideration of the intangible cultural heritage of a landscape helps to focus on a more general issue that underlines how landscape should be considered as a space-time process, where human beings participate in the same way as other animate and inanimate elements. Just like a nice view, music is not a noumenon: it is a phenomenon (enjoyable for some humans) of an extremely complex overlapping of processes. Given this complexity, it is a means that would generate constructive thoughts for major projects. The project aims could define one way (among many) of approaching and interacting with the processes of landscape. So we could say that here a project is used as a methodological research tool to understand why it is necessary and how it is possible to interact with such a labyrinthine issue.

Different skills are needed in order to approach any project properly. As landscape is a matter of perception, it is a process of the human mind, where the projection is a common aspect—even when it is perceived implicitly—for all human beings. We can call this continuous activity '*mind image* generation' (Lynch, 2016). The *mind images* that we build lead our action. A plan or a project is, in this sense, an image (De Bonis, 2001). A project, like a piece of art, can interact with the *mind images* to shift perception into a different *mind area*. A project can definitely be a tool to discover and to show an idea improving the perception skills of people. The planner often plays a subtle role in changing landscape and, solely through a common project, a community could learn how to learn. A project can deconstruct images and stimulate the creation of new ones (De Bonis, 1998).

In this sense, a project should be self-generated from the place and the planner has an intermediator: as a good teacher, who does not impose ideas, but uses them to establish a peer-based dialogue with their students, being ready to learn as well (Dewey, 2012).

It is inadequate to imagine, as some classical planning methods have suggested the maintenance of certain features belonging to the a system of restrictions. Research, planning practice and education could enrich each other with their mutual relationships by treating the project, or plan, as a medium for communication (De Bonis, 2001). So, to deal with a complex process, a catalyst is required that can, for example, potentially speed up some of the simpler processes, slow down others, and trigger new processes. Indeed, the relevant features of a landscape are just a side effect of the processes that permeate it, and the only way to retain some aesthetic features is to try an interaction with those implicit processes and to monitor the ongoing results.

Returning to Albanian iso-polyphonic music, we could consider that some sounds and some music are deeply linked to, for instance, economic processes: farmers' and shepherds' songs are an example of this. Some sounds are even able to establish communication among species: a cry, a whistle or a scream is a *medium* between human beings and animals. Music is often linked to rituals, performed in special times and places that are associated with myths and legends, including the creation of the world (Schneider, 2007). Music is part of a complex pattern in which cutting just one thread could cause consequences for the handing on of cultural heritage.

The idea of the developed project is a long process that could merge the cultural heritage of iso-polyphony with creativity and innovation in order to enrich artistic heritage.

### 2.2 *A few notes about the place*

At first sight, the relationship between the sea coast and the hinterland of Southern Albania is a peculiar feature: the landscape is characterised by a huge altitude differential within a short distance, which is also the reason for an equally huge biodiversity of flora and fauna. A road following the coast reveals panoramic views.

This area also reflects critical issues associated with recent history. The territory was regarded as strategic for the military control of the country and for this reason entry was barred even for its former inhabitants. We have to mention that only after the end of the civil war in 2007 could this region breathe again after several years of trial. A war-based envisioning of the territory means looking through a strategic filter where, for example, a hill represents a check point, and a bay represents a port to constantly monitor the enemy. Several villages were abandoned during the civil war and are actually in ruins, being too far from the main road, with not enough people living there to sustain viable communities. This is the case for Qeparo, which is a village 450 metres above sea level, on the western slopes of Mount Gjivlash. The name of the village probably comes from the Greek word *kipos*, which means 'garden'. In ancient times, the entire region (Himara) was inhabited by Greeks and a mix of Albanian and Greek cultures is still noticeably present.

Terraced hills with olive groves, dry stone walls, and flocks of sheep, goats, cows and donkeys are all remnants of a culture of constant dialogue between human beings and their environment; a relationship with the place lasting longer than any war. The history left attractive and powerful agro-pastoral features that are a potential basis for future sustainable development. However, there are several military bunkers sprawled on the coast.

It could be interesting to avoid the tendency towards forms of economy that bring benefits for only a few stakeholders. 'Old-fashioned' mass tourism, for instance, could destroy some of the qualities of the Riviera, as has already happened in many stretches of the Adriatic coast. Given that we are speaking of music, it would be more interesting to try singing a different song.

The challenge could be to develop new ideas for tourism, directly connected with cultural heritage. Could the high value of artistic artefacts, including iso-polyphonic music, assure a better use of resources? Perhaps the altitude differential could be addressed with rack or escalator systems to enhance the connection between the beaches and the historic villages that could be used for accommodation. Sometimes, what appears as a weakness can become a strength. The critical point is that depopulation means no communities of inhabitants and, for this reason, the future probabilities cannot be met, but at the same time, if there are no favourable conditions, people will not come back to live in this place. Certainly, it is important to focus on infrastructure, but also on reconstruction of a sense of belonging in the people for the place.

### 2.3 *The issue of intangible heritage*

The project *Safeguarding Albanian Folk Iso-polyphony* started in 2006 and ran until 2010 when a second project was funded, called *Inventory of Albanian Folk Iso-polyphony.* The aim was to preserve Albanian iso-polyphonic music, which is a peculiar form of choral singing, performed mostly by men, often without any instrumental support (Tole, 2014).

It is notable that despite two well-conducted international projects, a visit to the Albanian Riviera does not easily reveal its precious musical heritage. Only by researching was it possible to learn about Albanian iso-polyphony, first proclaimed as Humanity Heritage in 2005 and then incorporated on the Representative List of the Intangible Cultural Heritage of Humanity in 2008.

Music could act as a hub to rebuild attachment for the place, rediscovering its agro-pastoral culture. It could definitely act as a powerful attractor for inhabitants to reconstruct the community, as well as for a new generation of experiential tourism.

## 3 THE PROJECT

### 3.1 *Ideas for an iso-polyphonic roots hub*

The first goal of any project for Qeparo could aim to link Albanian iso-polyphonic music, as an example of the Intangible Cultural Heritage of Humanity, with the rural culture that persists. During surveys, olive trees appeared as silent witnesses of history. Dendrochronology shows how hundreds of years of history are recorded in a tree. On the other hand, music played just by voices are to performed and produced by blue-collar jobs. It is interesting to consider that we routinely use expressions such as 'genealogical tree' or 'cultural roots' or 'rootedness' to express relationships. We could say that the tree is also a human symbol, crossing different cultures, representing a link between the ground and the sky. The symbolic value of the tree is exactly the same as the singing that has the power to connect the highest and the deepest parts of our world (Schneider, 2007). The typical square of a Riviera village—the place where people used to meet—has a tree (often an oak) at its centre. Riviera people need to form a community and, at the same time, the incorporeal heritage of iso-polyphonic music needs a 'body', which could represent the rural culture and rhythms.

Some suggestions come from a contemporary art background. For example, the artworks of Giuseppe Penone often involved a tree, sometimes a living one, sometimes a mould-formed metal one. Pinuccio Sciola also influenced our ideas, suggesting that, because stone usage is common in the Qeparo area, it might be possible to imagine some roads with stones transmuted into musical instruments. In addition, the successful experience of the *morske orgulje* (sea organ) of Zadar, Croatia, designed by Nikola Bašić and opened to the public in 2005, has also inspired the project.

The project method clearly showed the need for a physical and symbolic space for the people to act as a community builder and heritage producer.

Is it possible to imagine that a technological device, an object with features that made it part of the Internet of Things (IoT), could help the generation of an immanent and permanent territorial polyphonic network?

### 3.2 *The question*

But how can an "intangible" (but perceptible) element of a cultural heritage be "territorialised" – made even more perceptible—in a manner neither reifying, fetishistic or static but, on the contrary, aimed at fostering the preservation of that heritage as well as at the creation of new heritage, both intangible and tangible, and at maintaining an active relationship with that specific environment?

This question was the starting point for the development of a territorial project, aimed at transforming the iso-polyphonic music into a local community builder, stimulating an increased population involvement in its creation and fruition, and developing a new kind of cultural and sustainable tourism.

### 3.3 *How the project works*

The project is called "From an Iso-Polyphonic Roots Hub to a Territorial Iso-Polyphonic Network" and it is organised in three phases. The physical intervention is a kind of hi-tech land art installation: reuse of an old house on the border of the town could be suitable to build a form of amphitheatre for a small choir, and it could be possible to dig a cistern to collect rainwater for agricultural use and to provide water for livestock. Having a good acoustic, the vacuum and the water in the ground could help to amplify the sound (see Figure 1). At the same time, the roof could be a kind of hi-tech sculpture in the form of a tree trunk with long stems ending in two lateral supports. In terms of the construction material, it is not specified: recycled plastic modelled with a huge 3D printer could be used, as well as a mixture of local materials and refuse using an assemblage technique, so that the result could look more 'handcrafted'. In this last case, one positive aspect is that the inhabitants could help to realise it. This place could have a name linking the music, the cultural roots, and agriculture: it could be something like "Iso-polyphonic Roots Hub", but it could be a good idea to involve the people and conduct brainstorming exercises. What is important, from the technological side, is that some sensors should be installed on the 'roots' with the ability to detect and record, on a device, singing voices (see Figure 2). A touchscreen could be used to collect the personal data of the 'user-singer' in order to connect him/her with a virtual, but nevertheless real, community. Other visitors could come and choose to sing the same song in the same or a different tone in order to specifically create a polyphonic music, or they could sing a different song.

### 3.4 *Three steps*

The aim of the first step of the project is to collect information about the informal network of locals interested in listening to and singing polyphonic music. A multimedia platform will help to engage young people, and older ones too, interested in their common 'roots', and to organise them into a web community. We cannot say in advance how long it could take to achieve a reasonable result.

The second step will be the formalisation of this informal network into a Non-Governmental Organisation (NGO) for iso-polyphonic music preservation and transmission. This association could find internal or external members for a real choir and start to collect sponsors to fund a polyphonic voice solo international festival in order to establish a network of similar heritages in other places (e.g. Sardinian 'cantores'). At this time, the device could

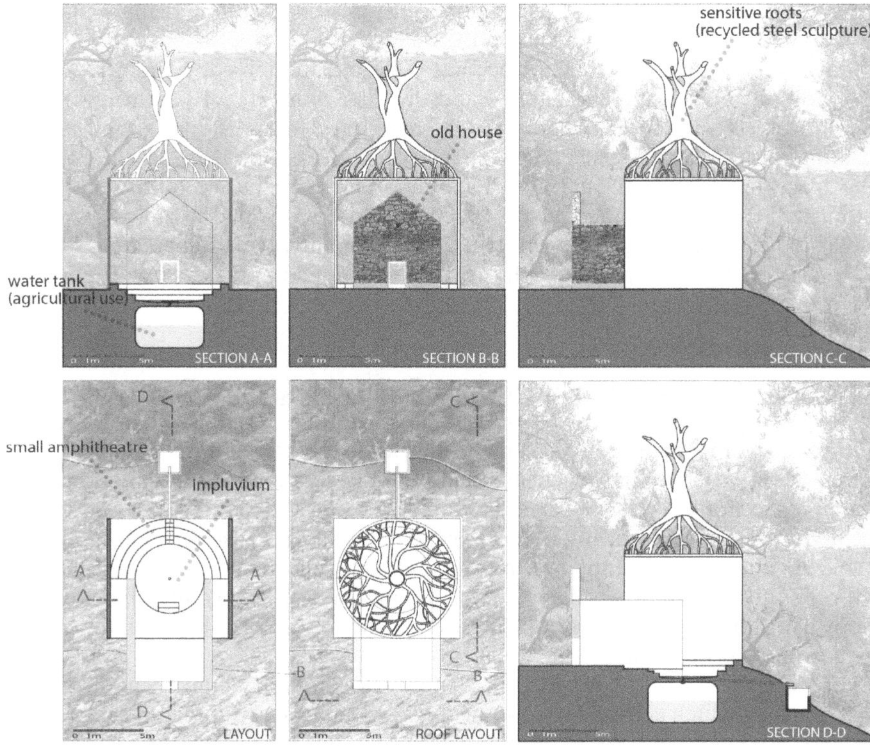


Figure 1. Hypothesis of intervention.

Figure 2. Territorial iso-polyphonic network.

collect materials as documents of the intangible heritage. Songs could be categorized into several themes. Some of which could be labelled as the relaxing one, where people listen to in order to alleviate their pains, other songs could be collected and the hub could continue to attract and record people interested in this topic, as well as simply being a place to meet people. It could become a tourist attraction.

The third step will start when the device has collected enough edited songs and the choir is stable enough to perform. The choir could give live concerts all year round in the Iso-polyphonic Roots Hub and elsewhere. At the same time, long-distance sound relays could broadcast music in strategic places such as the old bunkers that exist on the coast. In this case, the spaces could help to amplify the sound. The association could identify the best places to install the long-distance relays, which should be provided with a proximity sensor and a solar panel to be energy-independent. During and after this step, it would be possible to build a territorial music network with the long-distance sound relays (see Figure 2).

### 3.5 *Conclusions as starting point*

Advertising of the Iso-polyphonic Roots Hub could bring tourists to visit Qeparo and activate the local economy. An international festival could be organised every year; the hub will continue its work of collecting and recording while the association could organise a Call for Composers in order to find musicians that could generate new songs and regenerate the tradition in respect of the re-established heritage.

The core of the project is an Internet of Things (IoT) device, also representing a symbolic object able to gather a community around its iso-polyphonic music, either in real or deferred time. It can, indeed, capture single voices, mixing and transforming them in a virtual polyphony over time, but it can periodically also transform itself into a hall for a live event, actually becoming polyphonic.

The long-distance sound relays, placed in the disused military bunkers and supported by proximity sensors (which activate when someone approaches), can spread the captured voices and the mixed polyphonies throughout the Riviera (along pathways and close to the seaside), discreetly transforming the entire territory into a landscape immersed in its own music.

It is possible to conclude that a symbolic physical object can valorise an item of intangible cultural heritage, becoming an emblem of identity for a city and for a landscape, giving 'identity' to the holistic sum of memberships in a community.

## REFERENCES


Council of Europe. (2000). *European Landscape Convention* (Treaty No. 176). Strasbourg, France: Council of Europe. Retrieved from http://www.coe.int/en/web/conventions/full-list/-/conventions/treaty/176.

De Bonis, L. (1998). *Città e comunicazione. Dalla Pianificazione ordinativa alla pianificazione implicante* [Towns and communication. From Planning to ordinate involving planning] (Doctoral thesis, Department of Architecture and Urban Planning, Sapienza University of Rome, Italy).

De Bonis, L. (2001). Communication technologies and planning 'technologies'. *Plurimondi*, 5, 207–222.

Dewey, J. (2012). *Arte come esperienza* [*Art as experience*] (3rd ed., Original work published 1934). Palermo, Italy: Aesthetica.

Francioni, F. (2001). Intangible cultural heritage—Working definitions. Paris, France: UNESCO. Retrieved from https://ich.unesco.org/doc/src/05359-EN.pdf.

Lynch, K. (2016). *L'immagine della città* [*The image of the city*] (17th ed., Original work published 1960). Milan, Italy: Marsilio.

Schneider, M. (2007). *Il significato della musica* [*The meaning of music*] (Original work published 1970). Milan, Italy: SE.

Tole, V.S. (2014). *Enciklopedia e iso-polifonisë popullore shqiptare* [*Encyclopaedia of Albanian folk iso-polyphony*]. Tirana, Albania: Uegen.

UNESCO. (2008). Albanian folk iso-polyphony. Paris, France: UNESCO. Retrieved from https://ich.unesco.org/en/RL/albanian-folk-iso-polyphony-00155.

*Cities' Identity Through Architecture and Arts – Catalani et al. (Eds)*
*© 2018 Taylor & Francis Group, London, ISBN 978-1-138-05409-7*

# The changing image of Istanbul through its monuments (1923–1973)

Elâ Güngören
*Mimar Sinan Fine Arts University, Istanbul, Turkey*

ABSTRACT: In this paper debates and recent discussions concerning the planning of public spaces/parks and squares will be presented to understand whether an identity for Istanbul had been sought to portray it via the design of its monuments. Blending the characteristics of a transitory non-lieu urban settlement as well as having peculiarities proper to its own culture, societal values, political, religious and economic dynamics, the city is long considered as an intricate set of components. Apart from questioning the persistence of the commonly received idea that the square is an unplanned leftover space from late Ottoman Empire till today in architectural historiography, mainly based on eurocentric and political interpretations of the public space, this paper aims to focus on the aesthetical values of monuments. While main artistic trends will be presented, we will try to explore the kind of modernism proper to their design in post republican era in Turkey.

*Keywords*: public spaces; city identity; architectural historiography; modernism

## 1 INTRODUCTION

The city of Istanbul has been subject to many poems, novels, films, strip cartoons, scientific articles and books. Poets such as Yahya Kemal Beyatlı (1884–1958), Ahmet Hamdi Tanpınar (1901–62), Orhan Veli Kanık (1914–50), novelists as Orhan Pamuk (b. 7 June 1952) or film makers as Joseph Pevney (1911–2008) or Ferzan Özpetek (b. 3 February 1959) have presented their own interpretations of the city. Whether used as a stage for fictional Hollywood films as for example in the film called Istanbul (1957) or considered as database for novels as for example Pierre Loti's semi autobiographic book Aziyade (1879), the city's image changed within time to being confined to its public squares and became an arena wherein struggles and massacres would take place, mainly after the Gezi upheavals which started in May 2013. While it was presented in both popular literature, marketing discourses or academic dissertations as being a bridge and meeting point between the East and the West, it has been observed that after the upheavals especially, emphasis was put on its dichotomic character via its urban components such as its public spaces and squares. The city in line with the political agenda was thus considered as being an arena wherein Muslim religion would conflict with secularism, or different ethnic groups or social layers would not meet but struggle in search of identity, for the right to citizenship or for political recognition. Whereas the city's public spaces and squares are mainly considered as products of hegemonic political relations, during these manifestations surprisingly, the very same spaces turned into arenas wherein political actions acquired an aesthetic dimension and content. Accordingly in this paper it is intended to focus on the aesthetical and semantic values of monuments and the design of its public spaces during the Cold War years. It is aimed to present these criteria concentrating mainly on the twenty monument-sculptures erected to celebrate the fiftieth anniversary of the Republic in 1973, in order to shed light on the artistic currents and themes that were at the root of their creation. While main artistic trends will be presented during the above mentioned years, we will try to explore the kind of modernism or modernisms proper to the design of edifices, sculptures or installations defined as entities and products of common memory and identity.

## 2 THE PUBLIC SPACE IN TURKISH-OTTOMAN URBAN HISTORY

As it can be followed from architectural guides, Istanbul is presented as being geo-graphically divided into four areas namely the Historic Peninsula, Galata, Bosphorus and the Asian Side and the Modern and Contemporary City (Architectural Guide to Istanbul, March 2006, Istanbul: Chamber of Architects of Turkey). In books concerning its architectural and urban history the city is generally seen as a depository and generator of images. The titles of the books such as Çelik's *The Remaking of Istanbul: portrait of an Ottoman city*, dating to 1993 or Gül's *The Emergence of Modern Istanbul, transformation and modernization of a city*, dating to 2009 or Kuban's *Istanbul Yazıları*, dating 1998 all converge to the same idea that the city is an open air museum, a stage wherein buildings, streets, boulevards, squares are components and entities open to change under specific rules and decisions. These narratives changing according to the authors' background, political tendencies or age generally focus on different episodes and dynamics affecting the city's morphological aspect, tracing its peculiarities and its traits over time. These writings are completed by those on Ankara, appearing as an antithesis to Istanbul representing ideologies of the new Republic instaured in 1923 and tightly associated with the modernism of the early republican architecture. For example for Bozdoğan modernism is considered within the framework of a systematic modernization project program and it consists of an imported modernism. Moreover the author argues that Turkey's political and intellectual elite had engaged itself to build up a European, modern and laic nation-state from the ashes of the Ottoman Empire during the proclamation of the Republic in 1923 (Bozdoğan, 2012).

One of the major outcomes of these narratives is that generally the concept of westernisation goes hand in hand with modernization as does the concept of modernity. Architectural historian Kuban (b. 1926) for example situates this phenomenon around the reign of Mahmud I (2 August 1696, Edirne—13 December 1754), whereas for Çelik the years comprised between 1838–1908 are more significant (Çelik, 1993). For Kuban the concept of an inner city square existing in the Western world from the Greek agora onwards does not exist in Ottoman-Turkish urban history (Kuban, 1994) and social life takes place around the mosque situated next to forums which are remnants of the Byzantine city. He points out to morphological changes due to fires ravaging the city during the 18th c. (Cibali 1716, Balat Gate 1727, 1754–1757, 1774–89) as does Çelik who mainly presents projects and proposals emphasizing the urban physical changes due to legislative and administrative decisions. While she deals with the urban transformations the city undergoes basing them on the institutional reforms set in motion by the declaration of the Tanzimat Charter finding their extensions in the built forms-in the urban fabric on a larger scale and in architecture on a smaller scale, she states that the result was the metamorphosis of the classical Ottoman/Islamic urban image into a more cosmopolitan one, penetrated by forms and elements adopted from Western models (Çelik, 1993).

As to public urban spaces the large open areas around mosques or the construction of a covered bazaar can be interpreted as signs for the will to open the city to open air practices. In the Süleymaniye complex (1550–1557) for example, a third courtyard was added outside the inner and outer ones, more commercial, less connected to prayer. Although the public urban space had been of a major concern with the reshaping of former Byzantine squares and the construction of mosques, still the use of sculptures within the square was a missing feature compared to European cities' squares and to Rome specially. The lack of the sculpture as having representative qualities in comformity with the figure is believed to be related to the practices and beliefs of muslim religion. However iconoclastic attitudes have also been experienced during the Byzantine Empire. The absence of edifices from the city's squares is seen after the 8th c. onwards in the Byzantine capital.

## 3 THE PUBLIC SPACE FROM LATE OTTOMAN TO EARLY REPUBLICAN ERA

Çelik who is mainly interested in the nineteeth century because it opens a new era in the city's history, states that a concerted effort was made to transform the Ottoman capital of Istanbul into a Western-style capital in the effort to salvage the Ottoman Empire by reforming its traditional institutions. In the same book the author examines the transformations in the urban form of the Ottoman capital from 1838 to 1908 that is, from the Anglo-Turkish Commercial Treaty, which opened the Empire to foreign capital, to the Young Turk Revolution, which marked the end of Abdülhamit II's autocratic rule. The author believes that although culturally and physically exposed to impositions coming from Europe as many non-Western cities of the nineteenth and early twentieth century, its response to these external influences was unique, being shaped by a complex heritage and a noncolonial status (Çelik, 1993).

Yeşilkaya in line with this rhetoric developed around East/West or Islamic/secular dualities posits that in general apart from monumental architectural constructions the Ottoman urban space does not have sculptural monuments. While the appearance of the sculpture as an independant and indispensable feature within the urban context is said to cover the period from Baroque to the 19th c. among art historians, the appearance of fountains as isolated monumental sculptures and as focal points of a square start with the modernization of the Ottoman Empire considered to take place mainly in the 19th c. (Yeşilkaya, 2002). Examples are multiple for these public fountains within the city. Other components of the Ottoman city are clock towers and fire towers. The fire tower of Bayezid situated in the confines of the former Theodosian Forum is a magnified column and stands in the historic peninsula in the prolongation of the habit to erect obelisks although executed in 1828.

## 4 THE PUBLIC SPACE IN 1973

Art historians point out to the fact that the monument stripped from its ideological character as an aesthetic entity meeting the citizen within the city had a starting point. While Göktaş situates this around 1960s with the construction of the Manifaturacılar (small manufacturer) retail center (1959) by Doğan Tekeli, Sami Sisa and Metin Hepgüler (Göktaş, 1998), considered as the earliest shopping malls and an example of 'mat urbanism situated in Sultanhamam', for Pelvanoğlu the seminal date is 1973, the 50th aniversary of the proclamation of the kemalist Republic extanding till 1993 (url-1).

The retail center is composed of a series of lower-rise small blocks connected by outdoor galleries and courtyards. The courtyards open up to the Süleymaniye Mosque behind, establising urban pedestrian links in between. The site plan and section drawings testify to a respect for the historical silhouette and an intention of infusing the large complex into its environment with small and dynamic steps. This make the building and indisputable milestone marking the departure from the canonic prismatic block in favour of a more fragmented typology and a more public-orientated urban morphology (Bozdoğan, Akcan, 2012). Its inner courts are adorned by sculptural fountains and its blocks facades by mural in ceramic and mosaic. Kuzgun Acar (1928–76), Füreya Koral (1910–97), Bedri Rahmi Eyüpoğlu (1911–75), Yavuz Görey (1912–95), Ali Teoman Germaner (b. 1934), Sadi Diren (b. 1927) and Nedim Günsur (1924–94) are among the artists who have worked to build up the artistic quality of the retail center. It was in the 1950s (together with Turkey's transition into the multi-party political arena) that new stimuli arose in the fine arts with the formation of the group called the turkish Grup Espas in 1955 in the prolongation of the French Groupe Espace founded in 1949, although departing from the French group because of different conceptions concerning the spatial qualities of works of art, their coming together in an exhibition and interdisciplinarity. In the frame of the works executed by the above mentioned artists for at least the next thirty years, the plastic arts revived and reached new limits in both

Figure 1. Kuzgun Acar, "Soyut Kompozisyon" (Abstract Composition), IMÇ, Unkapanı, 1967, 500 × 300 × 70 cm.

representative and abstract expression. Another complex having works of plastic arts under its roof was the Istanbul Trade Center (1963–71) designed by Orhan Şahinler (1928–2016).

A commission called the Board for the celebration of the 50th anniversary of the proclamation of the Turkish Republic having at its head the mayor of Istanbul at a meeting in 1972 decided to implement monuments to adorn the city. But out of fifty monuments only twenty could be erected because of financial problems. In fact this project was part of a broader project concerning other cities as well in the country. For example the memorial for Atatürk, his mother and womens' rights (1972–73) designed by the architect Erkal Güngören (1934–2002) and the sculptor Tamer Başoğlu (b. 1938) in Izmir-Karşıyaka or the park called Anıtpark (1972–73) realized partially in Izmit, covering a total area of 8.000 m$^2$ by the architect Erkal Güngören and the sculptor Ali Teoman Germaner were other examples for such an endeavor. These monuments were commissioned by way of foundations which were financially supported by the government and local municipalities. Although these monuments' names had ideological connotations emphasizing national pride, associated with Atatürk, his companions or his mother as it is the case in Izmir, their designs were done according to modernistic principles. Such was also the case for the monument to workers, in Edirne again by Güngören and Germaner, which can be considered as another early example for the implementation of Worringer's Einfühlung principle paving the way to installations if not landart in Krauss' terms (Krauss, 2002). Güngören and Germaner conceived this utopic park along socialist ideals, intending to spread artistic experience to people, the beholders of the edifice. The monument-installation in Edirne including a museum, coffee shops and reliefs could never see the light of day because of the Cyprus war in 1974.

As to the plastic principles which are at the basis of the design of edifices used for the adornment of the parks of Istanbul, in 1973 for the first time Atatürk or a political leader as representing the power of the state was not chosen as a key figure. Among the twenty sculptures Gürdal Duyar's (1935–2004) 'beautiful Istanbul' portrayed Istanbul as a naked woman. The sculptor had used symbols related to the city. For example the fig represented holiness, the honesuckle the city's fresh air, the bee its population density, its dynamism, and the pomegranate the symbol of the city's legends (Antmen, 2009).

As Ali Teoman Germaner and internationally reknowned sculptor Kuzgun Acar, Duyar had been Rudolph Belling's (1886–1972), Ali Hadi Bara's (1906–71) and Zühtü Müridoğlu's (1906–92) student while studying at the Fine Arts School in Istanbul. Ali Hadi Bara and Zühtü Müridoğlu were among second generation Turkish sculptors who had conceived the edifice (1942) to Barbarossa (1478–1546) across the türbe of the former pirate who became in time Kanuni Süleyman's (1494–1566) admiral in Beşiktaş, Istanbul. Ali Hadi Bara is known to be one of the founders of the Turkish Grup Espas. According to Pelvanoğlu this monument is one of the early monuments to have spatial qualities although not completely freed from its piedestal (url-1).

Another naked woman executed in the classical beaux-art style was the sculpture called 'Nü' (The Nude) by Kâmil Sonad (b. 1914) taking place in the park of Gülhane, the park

Figure 2. Muzaffer Ertoran, "Işçi" (Worker), Tophane, Karabaş Park, 1973, Concrete, 200 × 80 × 80.

Figure 3. Kuzgun Acar, "Soyut Heykel" (Abstract Sculpture), Gülhane Park, 1973, Iron.

situated at the skirts of the Topkapı palace. Contrary to Duyar's expressionist touché, Sonad's is a well balanced figure standing upright as if devoid of feelings. Other sculptors who worked for this project were Muzaffer Ertoran who designed a figurative work called the 'Işçi' (The Worker) in Tophane, Nusret Suman (1905–78) who executed the architect Sinan (1489–1588), Namık Denizhan (1937–2015) who conceived a monument called 'Ikimiz' (The two of us). The city had also been portrayed via the usage of its tombstones. An example for this was Zühtü Müridoğlu's 'Dayanışma' (Solidarity), made out of concrete, installed in Fındıklı Park. The sculpture was a column having a height of 4 m with spiral abstract calligraphic reliefs on its main body, climbing up the column. This column refered to the former Byzantine capital's urban components such as obelisks as well as the Ottoman tombstone tradition.

Among the twenty monuments erected in different parts of the modern city some had figurative features and some had abstract qualities. Accordingly most of these installations had been called Abstract. An example was Haluk Tezonar's sculpture taking place in Maçka, another was Füsun Onur's (b. 1938) 'Soyut Kompozisyon' (Abstract composition) in Fındıklı park or Seyhun Topuz's (b. 1942) 'Soyut heykel' (Abstract sculpture) in the entrance of 4. Levent. Kuzgun Acar's 'Soyut Heykel' (Abstract sculpture) had taken its place in Gülhane Park. This piece made out of wrought iron was an assemblage and its structural features gained importance in the overall design in the prolongation of Acar's inclination to work with iron. Another sculpture again called 'Soyut heykel' (Abstract sculpture) in Bebek was Ali Teoman Germaner's. This sculpture done out of wood and copper had been executed according to the scale of the children because it was installed in a children's playground.

Germaner aimed at avoiding the imposing scale of the classical ideological edifices representing the hegemonic character of the political power in place.

## 5 CONCLUSION

The decade which had spanned from the seventies onwards coincided with syndical movements, political uprisings, the clash among radical groups having controversial political inclinations, repercussions of the petrol crisis on economy. The military coup d'Etat of 1980 followed and ended the period which had already started with the coup of May 1960. Most of the design of monuments and parks had to be interrupted although creative and innovative designs had been prepared. But at large the official monument representing political ideologies had left its place to a more various kind of design and subjects in line with the free will of the artists taking into consideration the human scale.

Some themes that conditioned the arts and architecture in general in the seventies through the implementation of monuments within the city were nationalism/internationalism in the arts, the search for a national culture, for folkloric traditions along with socialist and revolutionist art serving the people of the street and workers.

Multidisciplinarity achieved through competitions organized mainly by the state was another property akin to the conception of monuments and parks. In this decade instead of the art for art's sake, 'l'art pour tout le monde' principle had been sought and applied. Artists had the conviction that the citizen covering almost all social classes and the passerby had the right to experience the Einfühlung principle as the artist and the architect had done while designing. Furthermore, monumentality/scale relations or the architectural space in painting was another topic that was discussed. For example parallels between architecture, painting, and sculpture were discussed in Adnan Çoker's (b. 1927) paintings mainly.

A kind of a land art in the design of monument-parks had been sought but remained as paper architecture. Abstraction in painting and sculpture—abstract expressionism and geometric abstraction was experimented and as architects and sculptors worked and designed monuments together, the monument gained abstract features. Furthermore, if placelessness is a kind of modernism as Krauss mantains concerning landart (Krauss, 2002), it can be asserted that the displacement/migration of edifices from place to place within the city and especially in late modern Istanbul subdued to neglection, still undergoing major transformations in favour of rational solutions for traffic, housing and commerce, or their disappearance because of bad weather conditions can be considered as an unvoluntary, unconscious or better a spontaneous kind of modernism.

## REFERENCES


Antmen, A. 2009. "The Body In Turkish Culture And The 'Beautiful Istanbul' Event". In Electronic Journal of Social Sciences, 367.

*Architectural Guide to Istanbul*, March 2006. İstanbul: Chamber of Architects of Turkey.

Bozdoğan, S. 2012. *Modernizm ve Ulusun İnşası*, Istanbul: Metis.

Bozdoğan, S. & Akcan, E. 2012. *Turkey, Modern Architectures in History*, London, UK: Reaktion.

Çelik, Z. 1993. *The Remaking of Istanbul: portrait of an Ottoman city*. Berkeley and Los Angeles, California: University of California Press, 15.

Germaner, S. 2008. "Türk Sanatının Modernleşme Süreci: 1950–1990". *In Modern ve Ötesi: 1950–2000*, Istanbul: Bilgi Üniversitesi, 1–19.

Göktaş, C. 1998. *Istanbul'da Çağdaş Kent Heykeli Uygulamaları, Master's thesis,* MSÜ, Istanbul.

Krauss, R. 2002. "Mekâna Yayılan Heykel". In YKY (ed.), sanatdünyamız, 82: 103–110.

Kuban, D., 1994. *Istanbul Ansiklopedisi*, vol. 5, 432.

Yeşilkaya, N. 2002. *Osmanlı'da ve Cumhuriyet'te anıt-heykeller ve kentsel mekân*. In YKY (ed.), sanatdünyamız, 82: 147–154.

Web resources: Pelvanoğlu, http://www.sanalmuze.org/sergiler/view.php?type=1&artid=569, 22.1.2017.

Figures resources: Göktaş, C. 1998. *Istanbul'da Çağdaş Kent Heykeli Uygulamaları, Master's thesis*, MSÜ, Istanbul, 26, 52, 44.

*Cities' Identity Through Architecture and Arts – Catalani et al. (Eds)*
 *ISBN 978-1-138-05409-7*

# The cumulative effect of an interdisciplinary project

Alexandros Christophinis
*Architect—Engineer, Cyprus, Greece*

ABSTRACT: This paper is the narration of the implementation process of the project called *Λεμεσός μετά την ανάπτυξη τι;* 'Limassol: The aftermath of development', and, the socio-economic and political framework that led to its development. The project initially aimed to offer insight to citizens around the field of architecture and its local heritage; to document the contemporary city's identity and to bond artists and architects during this project, as well as for future projects. Additionally, through the involvement of authorities and decision makers, it aimed to change the limited ways local society thinks, acts, and perceives its architecture landscape and urban fabric. This paper looks at the role of the architects and artists through the research and experiential practices of a visual social analysis of this specific city's contemporary identity.

*Keywords*: Limassol; architecture; heritage; contemporary city's identity; architecture landscape; urban fabric

## 1 INTRODUCTION

### 1.1 *The measure*

The project assessed that Limassol's population—or at least the decision makers—have a tendency to repeatedly aim towards an ultimate modernity. Consequently, the ideal Lacanian imaginary context of the word development, which is used as a symbolic gesture to transform the urban landscape, that is, our real space, is the model that includes only new structures. This might be the result of a late industrial revolution in Cyprus since the first steam machine arrived as late as 1871 to the city of Limassol (Sergides, 2012). Ultimately, the delay on development has fully come to equilibrium during the past quarter of this century. However, decision makers still have the idea that we are trying to catch up and that we must reach the "ultimate modernity". Consequently, there is a tendency to miss out once again. Although, due to the financial haircut of 2013 and the halting of the building industry, we presently have the opportunity to think things over.

### 1.2 *The contemporary architect*

According to an article published in the newspaper, Βήμα Ιδεών (Vima Ideon) by Yorgos Tzirtzilakis, nowadays the contemporary way to approach architecture is as a Bricoleur. Within the article, the author claims that the industrial model of an architect engineer, like the one of Oscar Niemeyer, which controls everything in the plan came to a cease. Subsequently, a post-industrial model followed, that of a Bricoleur, e.g. the 2017 Pritzker Architecture Prize winners Rafael Aranda, Carme Pigem & Ramón Vilalta, whose work is a combination of remnants, functions and fragments of the pre-existing urban fabric. It does not work in an Aristotelian manner any more but in a Collective one.

### 1.3 *Setting goals*

Having this in mind, the curatorial team believed that this background could only be transforming if we welcome the decision makers into an interdisciplinary project which would mainly try to enlighten the public on architectural matters. Eventually, every participating group would gain knowledge. Furthermore, attracted by the power of City Branding strategies, due to their power to guide the way we perceive the identity, form, content and meaning of a city through guided impressions (Karavatsis, 2008) we used it as a strategic toll.

## 2 THE PROJECT PROGRAM

### 2.1 *Organized crime*

Since architecture is a neglected art in Limassol, and Cyprus in general, we became inspired by the methodology of Edward Bernayz and the field of public relations (Curtis, 2002) and invited some of the most respected art and architecture institutions or affiliated Associations of the island to join forces with us, examples include: Limassol Municipality, Cyprus Architects Association, ICOMOS, Cyprus Architectural Heritage Organization, Visual Artists Association, Limassol Literature Society, and last but not least, the most prestigious cultural institution of the city, the Rialto Theatre. With such a dynamic team, the ministry of Education and Culture and a group of affiliate patron companies were happy to support us.

### 2.2 *The agenda*

The submitted proposal included:

- A series of lectures based on the historic formation of Limassol.
- A series of pedestrian architectural tours.
- A workshop which would gather feedback from local users.
- Research within private archives and old submitted files of the city's planning office in order to gather information and expand our city's architectural archive library.
- An architectural exhibition with images/photographs from the most important urban formation phases.
- A contemporary art exhibition with ten plus one pronounced artists living on the island. Artists who have participated internationally in established forums and exhibitions came to associate architecture as a form of art that society perceives as noble.
- In addition to this, we proposed to the Pritzker Architecture Prize winner, Starchitect Renzo Piano to give a lecture based on the recently inaugurated 566-million-euro Cultural Center Stavros Niarchos Foundation of Athens. However, the offer was turned down since, as he claimed, he would love to join but he would need one more life to achieve so given our strict schedule. Therefore, he addressed/recommended us to his truly helpful personal assistant who suggested we address the invitation to his project leader architect, Giorgio Bianchi. Unfortunately, the event was canceled just a few days before taking place and it is still pending. Hopefully, as the project grows, Renzo Piano will address the lecture after all.

### 2.3 *Getting under the umbrella*

With such a team and strong proposal, the project was placed underneath the umbrella of the celebrations of the European Heritage days and the World Architecture Day. Cumulatively, the preannounced lecture was programed to be part of the Cyprus Architectural Awards events, given by the Urban Planning Office.

### 2.4 *The call for participation*

Having accomplished the aforementioned collaborations, the first lecture was organized and advertised as an open call to the project. Invitations were sent through the architect's association channels and through 50,000 mobile phones registered in the Limassol district with the financial support of our sponsor. The call gathered a crowd that oversized our participating expectations. Therefore, the first lecture was presented within a fully crowded lecture room of Limassol's public university, TEPAK.

## 3 THE PROCESS OF IMPLEMENTATION

### 3.1 *The lectures*

The agenda of the first lecture had three phases. The first one included the introduction of the organizers and the curatorial team, the project's aim, and detailed information about the final eight abovementioned events, (see paragraph 2.2). The second part was a historic narration about the formation of Limassol's urban fabric dating from the late Ottoman era until the contemporary stages of our urban landscape. The third phase featured an explanation of the workshop with a parallel open call for participation.

The second lecture was addressed by the project's research consultant, architect, Christakis Sergides. He is one of Limassol's most pronounced researchers in architectural heritage and the former head of the Urban Planning Office. Sergides elucidated the participants about the contemporary stages of Limassol's urban formation. By using maps, diagrams and images, the different kinds of urban planning strategies, and a layered superimposition that took place were indicated. In addition, 8 different clusters which include the vast majority of repeatable infrastructures were introduced, like the industry zone, the abandoned industrial zone, a low-income housing neighborhood, a high-income housing neighborhood, a refugee settlement area, the city center, the former tourist area and an integrated village within the main urban fabric. A bus tour and a city walk were helpful in approaching the subject, and the participants were asked to take as many pictures as possible in order to proceed to the next stage of the workshop, that of visual sociology. For our records, this lecture was recorded and it has been featured in the homonym publication.

### 3.2 *Engaging the tools of visual sociology*

Personal experiences, local beliefs, myths, and folklores, as part of urban narratives is claimed to be beneficial for a City's Branding strategy (Finnegan, 1988). Therefore, during the next stage the researchers had been asked to submit 20 pictures, sketches, maps, diagrams and a short text no more than two paragraphs for each picture, in the form of PechaKucha. The commentary was based on economic, social, environmental, aesthetic, emotional or associated issues with a sensitivity to keep in mind the researching field which was based upon the Bristol Accord, 2005 and questioned: The design and planning quality, the existence of a safe and active environment and the quality level of connectivity, infrastructure and service. In addition to the maintenance and the management's level.

Inevitably, disqualification of a series of essays which were purely documented was decided. At the end, thirteen well written small essays were selected and included in the homonym publication. The unfortunate disqualification followed since we craved to achieve a series of essays which would work as short literature jewels but at the same time as potential proposals for further academic research.

### 3.3 *Reviewing the 13 proposals*

Nadia Anaxagorou, Head of the Cultural Services of Limassol's Municipality services: Through her essay, she gave a glimpse of the relationship of city and sea by using the example of the regenerated sculpture park. At the same time, she introduced us to some artists of the artworks.

Penelope Vasquez Hadjilyra, architect: with a parallel participation in the art exhibition, using a video camera, she documented the recently inaugurated linear park in the bed of the stream river with a poetic verbal criticism.

Yiannis Kakoullis, General Secretary of the provincial Technical Chamber of Cyprus: claimed that Limassol's new developments and beauty coexist with the timeless problems of the city.

Marina Kassianidou, artist, Ass. Professor at Colorado Boulder University: expressed her inner need to "space" herself by dancing as a ballerina within the urban fabric. She arrived to the conclusion that following the condition of the infrastructures she must drive by car even for buying some basic commodities.

Louis Kilonis, Journalist and Sociologist and the main assistant of the organizing team: Evidenced the coexistence of intercultural spaces within the urban fabric.

Manolis Hadjimanolis, artist & Marinos Panagiotou, undergraduate student: As a team studied a part of the garbage disposal system and the degraded bus stops.

Phidias Pavlides, Researcher, Architect and Quantity Surveyor: Studied five local commercial centers and focused on the refugee settlement area.

Elena Pilakouri, Architect: During the program, she investigated a suburb of high income with its imposing houses and its isolation from all services.

Christina Poourkou & Savina Simillidou, Architects: They focused on the modern architecture era of the city.

Christakis Sergides, Architect and Research Consultant of the project: Studied the irrational phenomenon of the disclosure of the plaster coated stone masonry which creates a fake building identity.

Tasos Stylianou, gallery owner and merchant: Studied the decentralization of the trade centers, the urban sprawl and the lack of pedestrian life in the streets.

Petros Fiakkas, engineer and amateur photographer: proposed a silent photographic tour to explain the relationship between the buildings.

Andria Christou, Architect: while comparing various development programs of the modern city, she studied a low-income neighborhood.

### 3.3 *Initiating the creation of the artworks*

Before presenting these essays at Limassol's City Hall, during the celebratory day's activities, to an audience with a mixed background, the essays were distributed to our pre-selected group of participating artists. Additionally, they were given any necessary material which was requested to start their creative process.

### 3.4 *The urban architectural tour*

At the same time, the curatorial team gathered research material for the architectural tour which took place during the final festivities. The tour connected the two exhibition spaces via a pedestrian route and focused on 30 buildings which represent the different architectural styles that are recognized in Cyprus. At the same time, the tour encouraged the idea of the pedestrian activity in a city that tends to drive even for basic consumables. The tour, which is hosted within the pages of the homonym publication, was also enriched and addressed on some dates by the architect, Phidias Pavlides.

### 3.5 *The architectural exhibition*

The exhibition spaces were carefully selected. Firstly, an abandoned municipal exhibition space was selected due to its vicinity with the Rialto Theater, the intensive involvement of its executive director, George Papageorgiou and the goodwill of the former Mayor, Andreas Christou which led to this action. It is worth mentioning that after the exhibition, the space is available for any event of Rialto. The exhibition space next to the Rialto Theater is located at the only square within the historic center, the Heroes square, and the abandoned municipal exhibition space is located underneath the dominant epicenter monument of the square. Within this space, the architectural exhibition which took place was formed and curated by the organizing team. The exhibits were post cards selected from the archive of the city, placed upon a wooden structure screwed upon a pillar of local plastic baskets. This gesture and the use of plants symbolized the idea that architectural heritage is in reality our commodity (Avraham, 2000). The pictures in the form of post cards contained information about the buildings and their location. Once there is a second publication of the book, this material will be available to the public.

### 3.6 *The construction of a latent new social housing project*

The final destination of the architectural tour led to the contemporary art exhibition which took place in the first municipal social housing project. The abandoned buildings, erected in 1948 by the local municipal authority from the engineer, Nikos Roussos during the British administration, were planned to be demolished by the municipality. The demolition was planned for the completion of a new but latent development project, with similar characteristics to the famous, Pruitt Igoe demolished housing project (Reggio, 1983) and dated approximately three years before our project took place. Originally, the contemporary art show was supposed to take place back then as a protest to the demolition plans of our historic buildings, as an act of what is known as effective art, i.e. artists are committed to work in ways that actually change how the world works in addition to the ways we might perceive the world. However, after the conflict with the municipal authorities and the Mayor, the idea was rejected and the buildings saved since everyone agreed that they are part of our architectural heritage and must, therefore, be protected. Moreover, these abandoned social houses were placed under the umbrella of a European funding program. Thus, now they are eligible for a 2.5 million funding, for their reuse in a different framework that would contribute positively to the new latent numbering social housing project. Therefore, for the purposes of the implemented contemporary art exhibition, four housing units were cleaned, sealed, wired and lighted, with the contribution of the municipality.

### 3.7 *The contemporary art show*

Within the activated building the organizing team curated an art show that gave every participating artist a heavily used empty room. However, the simplicity of lighting, the visible cable wiring and the repetition of the plan managed to present the event as a well-organized art project space.

Helene Black the co-founder of NeMe (www.neme.org). Through a series of photographs and an engraved marble logo questioned the local mentality.

Elina Ioannou by borrowing the forms of architectural material she sculpted what she claimed to be Cyprus's identity.

Helene Black the co-founder of NeMe (www.neme.org). Through a series of photographs and an engraved marble logo questioned the local mentality.

Elina Ioannou by borrowing the forms of architectural material she sculpted what she claimed to be Cyprus's identity.

Stelios Kallinikou, co-founder of the art space, Thkio Ppalies (www.thkioppalies.org). Through a series of photos studied the topography of the city's landscape and thus reviled the complexity of the city's identity.

Marina Kassianidou Ass. Professor of Colorado Boulder University. Through her work indicated that a building could be an architectural landscape even if it is just a minor gesture.

Lia Lapithi, artist and part 2 RIBA architect. She proposed a merge between the city and the citizen.

Marina Xenofontos inspired by the fragments of the existing urban fabric indicated and questioned some failed efforts.

Michalis Papamichael through a series of photographs and by collaborating with two architects studied some unfinished buildings.

Andreas Savva initially suggested an artwork which required 2000 banknotes of €500. Due to the lack of banknotes sponsorship, he proceeded to a second proposal which studied the relationship between structure and habitation.

Constantinos Taliotis, an artist that won a special mention at the 55th Biennale di Venezia with the group participating in Cyprus's pavilion in 2013. He tried to study the modern movement of Limassol within a context of a playful manner.

Evanthia Tselika Ass. Professor & Coordinator of the Fine Art program of the University of Nicosia. With a parallel invitation to one of her undergraduate students, she studied the relationship of the refugee settlement area and the integration with the rest of the urban fabric.

Penelope Vasquez Hadjilyra, by a video installation, questioned the recently inaugurated local linear park in the bed of the stream river.

## 4 CONCLUSIONS

### 4.1 *Bricoleur(ing)*

The interrelated, marketing, branding and public relations strategies we used, arguably, could create the same outcome as a Tourist City Branding program attempts to do. If the population was better informed in the fields of architecture, art and urban planning, then even a simple tourist branding strategy would have included bits and pieces from the architectural heritage of the city without perceiving its heritage as a monumental object. However, this team of artists and architects provided archival material, new awareness and knowledge in the architecture and theoretical field, instigating an ongoing discussion around cities identities within the present urban fabric. Hopefully, this long-term project will give rise to a continuous transformation in consciousness in a way that the city's identity might construct architect bricoleurs to keep pushing the boundaries.

## REFERENCES


Anholt, S. 2010. *Definitions of place branding—Working towards a resolution*. Palgrave Journal of Marketing Analytics. Place Branding and Public Diplomacy.

Avraham, E. 2000. *Cities and their news media images*. Cities, 17(5), pp. 363–370.

Baudrillard, J. 1993. *The Transparency of Evil: Essays on Extreme Phenomena.* New York. Verso.

Curtis A. 2009. *The Century of the Self.* United Kingdom: RDF Television BBC, Film documentary.

Dormans, S. 2006. *A narrative analysis of urban identity: the case of Tilburg*. Governance and Places (GAP) working papers series, 2006/4. Nijmegen: Radboud University Nijmegen.

Embracher, & Buttle. 1989. *A repertory grid analysis of Austria's image as a summer vacation destination.* Journal of Travel Research, 27 (Winter), pp. 3–7.

Finnegan, R. 1997. *Storying the self: personal narratives and identity. In H. Machay (Ed.),* Consumption and everyday life (pp. 65–112). Sage: Thousand Oaks.

Reggio, G. 1982. *Koyaanisqatsi: Life Out of Balance.* United States. Island Alive New Cinema. Film Documentary.

Sergides, C. 2012. *Λεμεσός: Οινοπόλεως Μέστωμα Lemesos inopoleos mestoma*. Limassol. Αφή.

Tzirtzilakis, Y. 2008. *Ο οριστικός κλονισμός της υψηλής νεωτερικότητας.* Βήμα Ιδεών Vima Ideon.

*Cities' Identity Through Architecture and Arts – Catalani et al. (Eds)*
*© 2018 Taylor & Francis Group, London, ISBN 978-1-138-05409-7*

# Unearthing the spatial archetypes of two indigenous South Asian cities: The walled city of Old Delhi and Old Dhaka

Sheikh Itmam Soud
*Department of Architecture, Stamford University Bangladesh, Dhaka, Bangladesh*

Md. Obidul Haque
*Department of Architecture, Premier University, Chittagong, Bangladesh*

ABSTRACT: The cities of Old Delhi and Old Dhaka are frequently depicted as congested, filthy, formless, functionally obsolete, lacking public green space, and possessing narrow twisted streets. All of these criticisms appear true if someone looks from non-indigenous insights. But deep beneath there is an archetypal spatial structure that has functioned since their establishment as cities, where traditional hierarchies and organic spatial structures form a rich mix of indigenous morphology. This study focuses on these spatial dynamics of the urban structure of these two indigenous cities, where a high density of people living, working, and socialising within a self-contained community bondage. Further de-layering of these (so-called) chaotic urban patterns reveals a meaningful spatial structure, which provides valuable clues to the resolution of planning, housing and related community development problems arising from the continuing rapid rate of urbanisation.

*Keywords*: Old Delhi; Old Dhaka; indigenous city; archetypal spatial structure; chaotic urban patterns; rapid urbanisation

## 1 INTRODUCTION

Many scholars have adopted a simplistic scheme in which cities with an orthogonal layout are classified as planned, whereas those that lack the grid principle are considered to be unplanned (Smith, 2007). However, in contemporary times a number of researchers have considered the spatial principles in addition to orthogonal layouts instead of the false dichotomous schemes of planned versus unplanned (or sometimes 'organic'). This paper is an attempt to compile such research output as it has been applied to two cities of India and Bangladesh. Old Dhaka (Bangladesh) and the walled city of Old Delhi (India) both have similar features and backgrounds in the subcontinental history. Both cities had been at the crossroads of international trade routes from ancient times (Figure 1), were established as capitals by Mughal emperors, and were ruled by the British for nearly 150 years at the same time. As a result, a good number of indigenous spatial characteristics innate to their settlements remained unappreciated for a long space of time.

Moreover, rapid urbanisation and uncontrolled growth have been annihilating these indigenous features on a daily basis and planners have imposed Western ideas in their recent projects, ignoring the cities' native cultures, contexts and environments. According to Geddes (1917), "the diagnostic survey ... seeks to unravel the old city's labyrinth and discern how this has grown up. Though, like all organic growth, this may at first seem confused to our modern eyes that have for so long been trained to a mechanical order, gradually a higher form of order can be discerned—the order of life in development".

Figure 1. Map of South Asia showing the locations of Old Delhi and Old Dhaka, which had been at the crossroads of international trade routes from ancient times (adapted from Texas Austin, 2004).

## 2 METHODOLOGY

This study is a successive consideration of three basic questions:

a. Why select Old Delhi and Old Dhaka?
b. How were the cities' spatial patterns established?
c. What were the characteristics they inherited?

Based on the nature of the research, this study will adopt two approaches:

a. Forming an overview of the individual settlement patterns of Old Delhi and Old Dhaka according to the relevant literature;
b. Analysing this information to outline the archetypes of the indigenous spatial planning features of these cities.

At its conclusion, this paper also offers a glimpse of the recent chaotic urban fabrics of these two cities and the alarming condition of the old parts of the cities, which might help architects and urban planners when taking decisions about their conservation and upgrade. The study will also support professionals in future city planning with a more 'indigenous' spirit.

## 3 TRACKING THE ORIGINS OF OLD DELHI AND OLD DHAKA

The walled city of Old Delhi was founded as Shahjahanabad by Mughal Emperor Shah Jahan between 1638 and 1649. It remained the capital of the Mughals until the end of the Mughal dynasty (Nath, 2006). In the 19th century, the British government developed a low-density semi-suburban 'North Delhi' as their garrison (cantonment) and administrative civil lines to the north of the walled city where most of the residences were bungalow-types with huge open spaces to ride horses and horse carriages. Then, in the early 20th century, the British government transferred its administrative capital from Calcutta to Delhi. This new Indian capital was developed on the southern edge of Old Delhi and exhibited a dramatic contrast with its older neighbour by virtue of its street planning and urban design decisions. Two British architects, Edwin Lutyens and Herbert Baker, ignored the ingenuity of Old Delhi in planning this 'New Delhi' and designed it with low-density, wide streets and a complex Renaissance street pattern featuring a diversity of geometric shapes.

The history of Dhaka begins with the existence of urbanised settlements in the area that is now Dhaka that date from the 7th century. The city area was ruled by the Buddhist kingdom of Kamarupa before control passed to the Sena dynasty in the 9th century (Dhaka City Corporation, 2006). After the Sena dynasty, Dhaka was successively ruled by Turkic and Afghan governors, descending from the Delhi Sultanate, before the arrival of the Mughals in 1608. Following the Mughals, the British ruled the region for over 150 years until the independence

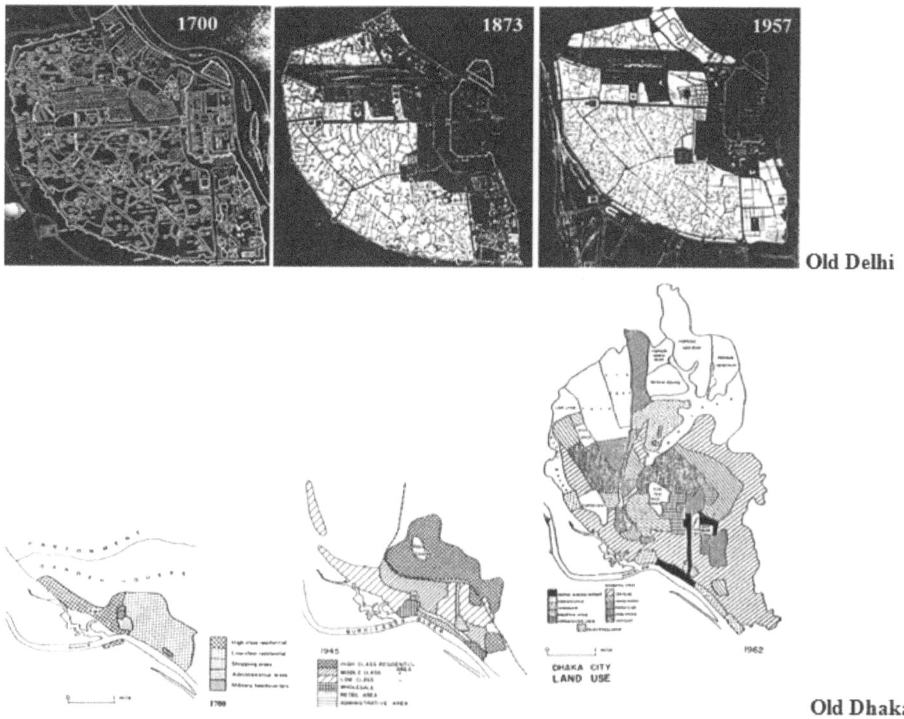


Figure 2. Expansion of two cities (Fonseca, 1969; Ahsan, 1991).

of India. In 1947, Dhaka became the capital of East Bengal province under the dominion of Pakistan, and then became the capital of the new state after the independence of Bangladesh in 1971. Over the passage of 400 years of history, the medieval trading town of Dhaka has extended from one square mile in 1600 to a large conurbation (Dhaka Statistical Metropolitan Area, DSMA) of 522.34 square miles in 1991. Over a similar period, the population has increased from 200,000 (in 1640) to 6,950,920 (in 1991). Dhaka has been experiencing phenomenal growth for the last three decades. By 2001, the city was deemed one of the mega cities of the world, with a population of 10.71 million (Nazem, 2003).

## 4 SPATIAL STRUCTURE OF OLD DELHI

### 4.1 *The settlement pattern of rural communities*

Indian society was traditionally perceived as divided into five castes: Brahmins, Kshatriyas, Vaishyas, Shudras and menials (untouchables). This was the case only in ancient times. Later, the basic subdivision was formed by social units known as 'jatis', each of them having their own rules and regulations. Specific groups of jati clustered in communes, where upper caste groups maintained their superiority over inferior groups. The spatial structure within a rural community sometimes reflects the hierarchical order based on ritual/religious purity and the norms of the community (Figure 3a). Generally, two or more households are grouped together around a common courtyard with separate areas for men and women (Figure 3c). Women's courtyards are located at the rear, behind the men's quarters. The spatial structure of the community is one of enclaves, where unclean occupations cluster on the periphery while clean or dominant groups occupy prime central locations (Fonseca, 1969). The boundaries between the communes or enclaves are mostly open ground, where service facilities such as water-wells, temples and bazaars are located (Figure 3b). For groups of individual houses, the boundaries

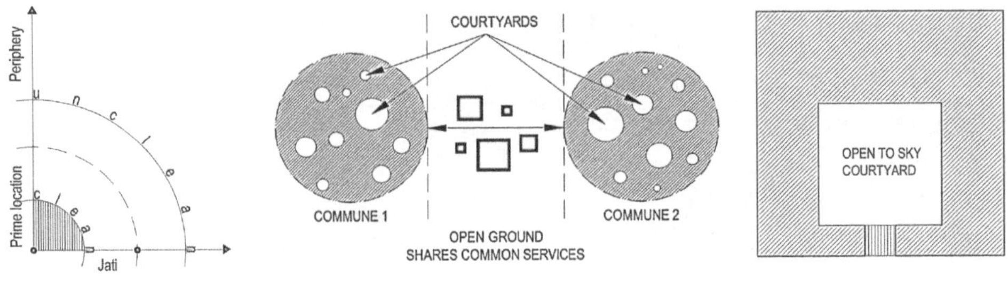


a b c

Figure 3. (a) Enclaves of community; (b) Patchwork pattern of clusters; (c) Multiple households around a common courtyard.

are simply common walls of houses or courtyards or streets. Fonseca (1969, p. 107) describes how Indian settlements thus present a patchwork pattern of clusters, embodying the living principles of social behaviour, rather than the annual growth ring development observed in European settlements, carried over with modifications into the indigenous urban scene.

### 4.2 *The bazaar city*

The streets and alleyways of Old Delhi are among the best known examples of traditional urban environments in India. They are characterised by a great variety of users and activities. Railroads and major road connections, such as the ones connecting Calcutta and Bombay, had importance as principal commercial centres for bazaars and handicrafts. The walled city that is Old Delhi is still a bazaar city, with a tradition of fine craftsmanship in gold, silver, jewellery, ivory, carving, miniature painting and weaving. The entire city may be described as a bazaar, with peaks and dips in activity as one moves from primary to secondary to tertiary streets (Fonseca, 1969). Generally, primary streets formed large clusters with retail and wholesale outlets while bazaars of specialised single items formed smaller clusters among the secondary streets. The interior courtyards and alleyways away from primary streets were the actual production and storage houses of the artefacts and other utilitarian items. This meant that every residential courtyard had an important role in the city economy (Figure 4a), and is why Fonseca (1969, p. 108) argued that production, storage and service centres were either immediately behind the retail outlets or a short distance away, in residential areas.

### 4.3 *An informal reflection of gridiron pattern in access routes*

Looking at the public spaces of Old Delhi, an overcrowded scenario will be a common one where there is little or no formal spatial separation between vehicles and humans. One will also find a complex, congested relationship between street commerce, parking and traffic movements. This has happened because motor vehicles are not banned but the urban design of such neighbourhoods originally envisioned use by pedestrians, carts, bicycles, cattle and draught animals and is still in place: it is not conducive to car use. Nevertheless, Bromley (2016) argued that, despite all this hustle and bustle, there are informal norms and practices that influence the use of space and that keep accidents, property damage and criminal activity to low levels. Closer inspection of this complex road network reveals an essential gridiron pattern of primary and secondary streets within the walled city, although it is not recognisable at first sight (Figure 4b).

### 4.4 *Administrative divisions and social groupings*

Old Delhi is divided into eleven municipal wards or *thanas*, which are separated by bazaars on primary and secondary streets and mostly named after the main streets. At a macro level,

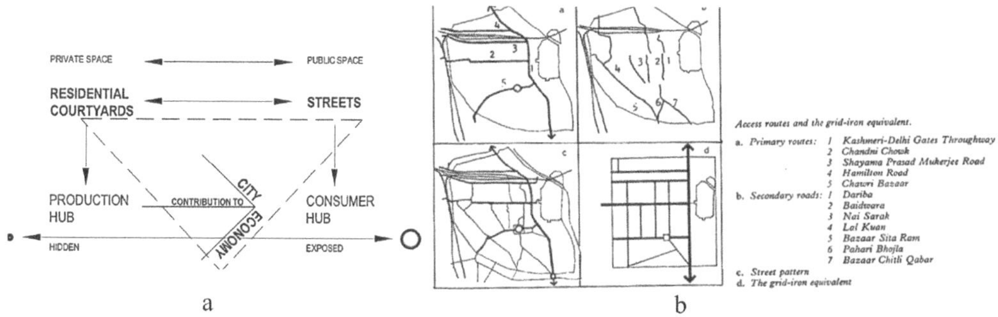


Figure 4. (a) Production vs. consumer hub in *mohalla* enclave; (b) Unseen gridiron pattern in access routes (Fonseca, 1969).

Figure 5. A *katra* and street views of Old Delhi (sources: www.fabhotels.com, www.aroundtheworldl.com).

the city is a combination of five essential elements: *mohallas*, *kuchas*, *galis*, *katras* and *chattas*. The word *gali* is frequently used interchangeably with *kucha*, *katra* and *mohalla*. A linear entity (conventionally, a street) is implied by the use of *kucha* and *gali*, while *mohalla* and *katra* imply zones (Fonseca, 1969). A market with residential quarters and storage facilities is known as a *katra*, which is enclosed by high walls and entered through a gate (see Figure 5). When the upper storey of a residential structure crosses over a street or lane, the term *chatta* is used instead of *kucha* or *gali* (Fonseca, 1969). This is a feature unique in its use of the air rights above public commuting space, and also provides a means of climate protection. Arguably, this architectural feature increases the attractiveness of the streets, as do canals and shading foliage.

### 4.5 *Unit of social identification: Mohalla*

A resident of the walled city traditionally belongs to a particular *mohalla*. A *mohalla* is a geographical area which relates to the occupation, religion or geographical origin of those dwelling within. Fonseca (1969, p. 109) defined this segmented area thus: a *mohalla* is a clearly defined area of residential and commercial activity fronting onto a spine street that connects to a primary or secondary bazaar street. Most often, there is a network of *galis*, *kuchas*, *katras* and *chattas* spreading out from the spine into the interior of the *mohalla*. Some *mohallas* had controlled social hierarchies with their own bazaar, mosque or temple, schools and orphanage, which were governed by a *mohalla* council. However, the isolated aspect of these social units was broken by the need for community co-operation during religious festivals and by the bazaar, which retails the artefacts produced within the unit (Fonseca, 1969).

### 4.6 *Unit of group interaction: Chowk*

The majority of the buildings of Old Delhi are residential and commercial. However, there is a great variety of worship houses, historic monuments and other built forms available within a walkable distance. Old Delhi is a walkable city. The streetscape of Old Delhi is continuous because most of the buildings fully occupy their lots with no setbacks around them. This essentially continuous streetscape is occasionally broken by narrow private alleyways that lead to backyard spaces or by broader public alleyways providing access to properties that do not front onto the streets. Bromley (2016) claimed that, occasionally, an alley may also open out a little to create a pocket plaza around a shade tree or well. Another feature unique to the indigenous urban setting is the *chowk* (see Figure 6). This is often no more than a widening of the street as it turns a corner or at the junction of two or more streets, and more commonly at the termination of a *kucha* or *gali* (Fonseca, 1969). Sometimes a *chowk* is no more than a setback in the street around a precious tree, which served as a market space for vendors, a play area for children and adults, a place of debate of political issues, and so on. Around a *chowk*, the land use type changes from residential to commercial activities. This change may appear in the form of a tea shop or a general merchandise shop where the daily necessities of life can be purchased. Thus, a *chowk* tends to serve the immediate community and becomes the main focal point where the inhabitants of two or more *mohallas* meet (Fonseca, 1969).

### 4.7 *The introverted garden city: Courtyard*

An outside observer may see the streets of the walled city as a pattern of dark voids between buildings and is thus apt to form an erroneous impression. However, behind these narrow dark lanes bounded by blank walls are sunny courtyards where most of the activity takes place (Clinard, 1966). This pattern is reflected in the land use, where only 10–12 per cent of a ward is devoted to streets, whereas nearly 25 per cent is interior courtyards (Fonseca, 1969). This is clearly a case of an introverted 'garden city', where the open space varies between 30 and 40 per cent of the gross area and is public property (Marshall, 1931). At a deeper level, a resident of the walled city had three different tiers of protection in front of his residence: sequentially, the interior courtyard, the lane outside his door and the world outside the *mohalla*. The metamorphosis is completed as he increases his distance from the *mohalla* entrance. Returning, he regains his privacy in increments: first in familiar sights, then faces he recognises, and finally in people, who, greeting him, invite him to join them for a cup of tea (Fonseca, 1969).

Figure 6. *Chowks* of Old Delhi (sources: https://dailyjag.com, https://poraschaudhary.photoshelter.com).

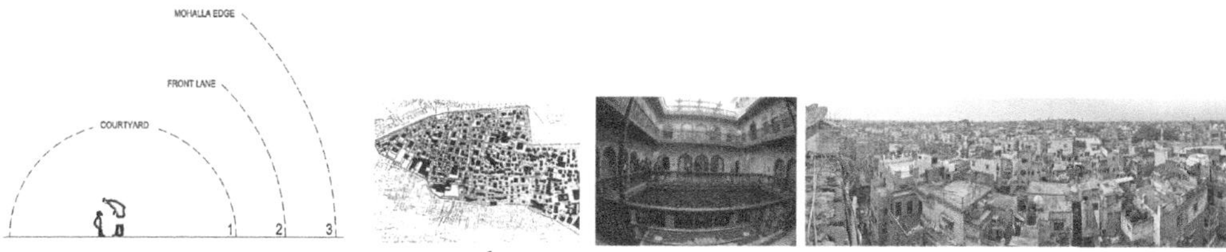


a b

Figure 7. (a) Three different tiers of protection up to the residence; (b) Courtyards of Old Delhi (sources: Fonseca, 1969, www.business-standard.com, https://sujoysen.wordpress.com).

## 5 SPATIAL STRUCTURE OF OLD DHAKA

### 5.1 *Trading hub and the bazaars*

The street and the court are the two types of component that exist in Old Dhaka, which are connected in a continuous network of open and semi-open, public and semi-public spaces. The intersecting points often turn into *chowks* or nodes (Figure 8b). Another important element is a 'bazaar street', which signifies the commercial interface or market centre of the city (Figure 8b). This was the core concept for generating Islamic bazaar streets, because these heterogeneous types of spaces generated human activity at the pedestrian level (Ferdous, 2012). Rappaport (1977, p. 219) depicted the transformation of simple homogenous streets into complex indigenous bazaar streets (Figure 8a). Generally, these types of bazaar streets shadow the residential neighbourhoods, which are morphologically indigenous and sometimes of mixed use in character. All of these localities were confined within the circuit of the old Dholaikhal. The *Tanti* (weavers) and *Sankhari* (shell cutters) are believed to be the oldest inhabitants of the city, and they still live in the area (Dani, 1962). The cottage industries and trading areas of the pre-Mughal period housed the majority of the city's lower-class population, which consisted of artisans, labourers, and traders who were effectively segregated from the higher-class residential areas. During the Mughal period, Chowk Bazaar was developed as the main business centre, which served both upper- and lower-class residential areas, and another commercial centre was located at Bangla Bazaar, which was the main shopping centre before the Mughal period (Taifoor, 1956).

### 5.2 *Access routes and organic street patterns*

There are, in fact, two types of access system in this area: the waterways where *ghats* (landing platforms) are the significant feature, and the roadways where main roads intersect with many secondary and tertiary roads, referred to locally as *gali*. River *ghats*, significant transitory spaces linking the river with the land, had commercial, social and religious roles. They were used as wholesale points for primary produce, embarkation places, and sacrificial and cremation points (Rahman & Imon, 2016). All types of formal and domestic activities and regular and seasonal rituals evolved around the water, giving sense to the particular morphology featured in the area's orientation towards the water (Ashraf, 2012). The city was divided into a number of neighbourhoods, which were clusters of houses webbed with intricate narrow lanes (Islam, 1996). The long narrow shop houses and houses facing inner courtyards generated a dense settlement with intimate social spaces (Figure 9). The winding lanes, often ending at the river bank, created social spaces at the street level. The street junctions and sudden widening of the lanes due to placing of building would form popular hang-out spaces for all ages (Khan, 1985). The *morhs* and *chowks* became the magnets of the settlements, special places where people meet, gossip, and enjoy and exchange products (Mowla, 1997).

### 5.3 *Administrative division and social groupings*

During the Sultanate and Mughal periods, "...each *mohalla* had its *punchayet* which engaged much of the people's attention and drew on their community spirit, since all could participate

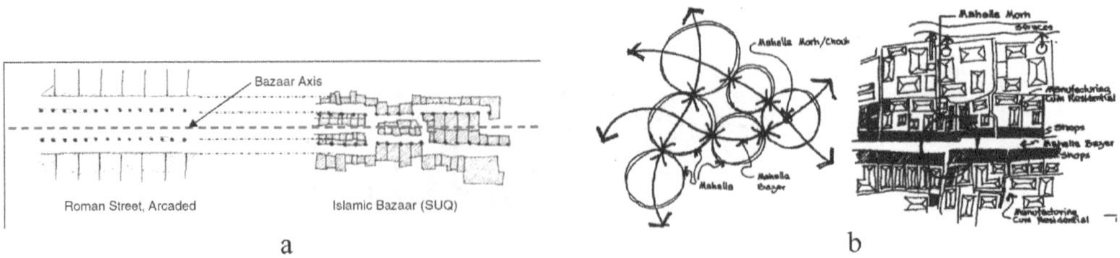


Figure 8. (a) Transformation of Roman street into Islamic bazaar (adapted from Rapoport, 1977); (b) *Mohalla* bazaar in *mohalla* morphology (Mowla, 1997).

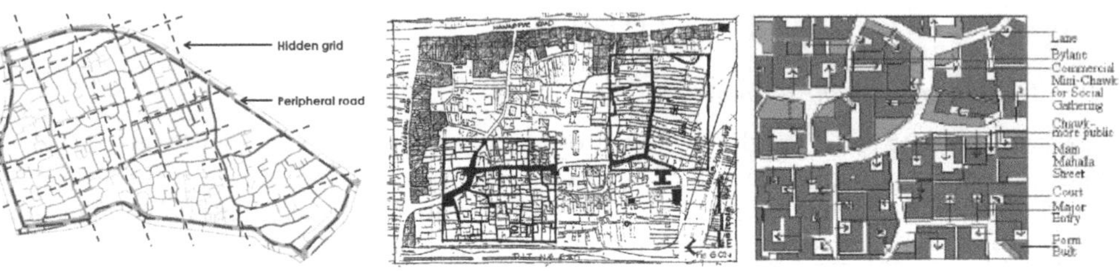


Figure 9. The complex webbed street pattern of Old Dhaka (Author: Khan, 1985; Rahman & Haque, 2001).

Figure 10. Interactive spaces in indigenous Old Dhaka (Sources: www.leeabbamonte.com, http://dhakadailyphoto.blogspot.com, Khan, 1982).

in its deliberations" (Ahmed, 1986, p. 15). The *mohallas* were semi-autonomous socio-spatial units, with very little control in their internal matters from any higher level in the hierarchy of the administration and judiciary. Each *mohalla* had a *panchayat* (village council) office, also used as a community centre, called a *bangla*, situated in a conspicuous part of the *mohalla*. During the day the *bangla* would be used as a *maktab* (elementary school) and at night as a social club for adults of the locality (Mowla, 1997).

The Mughal elites had garden houses for recreation, festivities and receptions (Hossain, 2013). Other favourite pastimes of *zamindars* (aristocrats) included cockfighting and goat fighting. These events would usually be held in streets and *chowks* (Rahman & Haque, 2001). The intersections of narrow lanes formed wide and irregular nodes that acted as a civic space at the local level. Some of the local nodes turned into *chowks*—larger social gathering and festival spaces—for the local *mohallas*, whereas other nodes were rather intimate in nature and held local social gatherings (Nilufar, 2011). The sense of enclosure of these spaces was very intimate in scale. The area is celebrated with lots of colour in different festivals and religious occasions and most of them take place in the streets, *chowks* and sometimes in the inner courtyards and upper roof terraces. The extended plinths and low-height boundary walls are the *mohalla*-level interactive spaces for both the young and adults (Figure 10). The small tea stalls in the *chowks* became an institution for local natives. This indigenous settlement pattern integrates richer and poorer in living and working as a family and neighbourhood, which offers a greater sense of community and the option of a less expensive lifestyle.

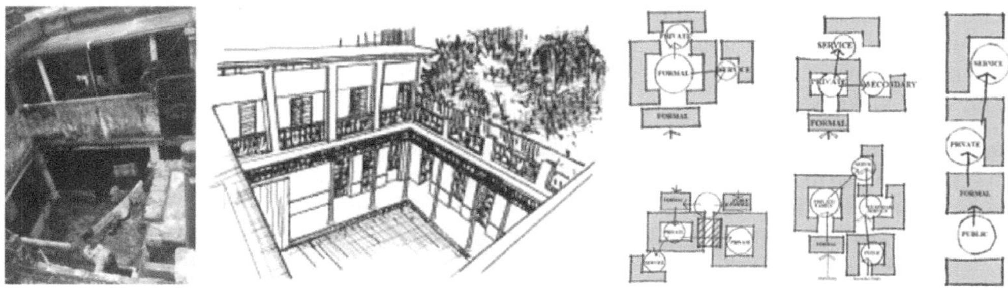

Figure 11. Different types of courtyard of Old Dhaka (Rahman & Haque, 2001).

### 5.4 *Unit of social identification: Mohalla*

According to Conzen (2004), the diversity of morphology in South Asian cities arises from the diversity of historical development, functional types and different combinations of morphological characteristics. Tankel (1963, p. 58) believes that there are two types of urban space of which people are aware, and another of which they are more likely to be unaware but which affects their lives and helps shape the pattern of development. The first categories of space are used, viewed and felt, that is, they provide a range of active and passive recreation activities, circulation and privacy, insulation, and a sense of spaciousness and scale. The indigenous *uthans*, *galis*, *morhs*, *chowks* and bazaars of the *mohallas* are in this category. Nilufar (2004) reveals that neighbourhood (i.e. *mohalla* in Urdu or *para* in Bengali) and locality are the two territorial units in the cognitive imagination of the inhabitants of Old Dhaka. Neighbourhood is primarily a social phenomenon arising from cohabitation in a physical area. The traditional neighbourhoods in Old Dhaka were mostly named after their occupation or caste. Localities have strong agreement as to their named identity and also have robust images in the cognition of the inhabitants. In addition to their symbolic identity with the name, the objective social attributes (like economy, ethnicity) of the localities act as social labels for these physical areas, which have a distinguishable environment and ecological position defined by distinctive spatial characteristics within the city structure.

### 5.5 *Unit of family interaction: Courtyard*

The areas to the south and south-west of the Old Fort up to the river bank grew mainly as commercial areas, whereas the areas to the north and north-east grew as residential areas (Chowdhury & Faruqui, 1991). In urban areas, the traditional rural house was often reproduced in compressed form (Rahman & Haque, 2001). In rural areas a house is composed of multiple rooms around an *uthan* (courtyard). In general, larger families have two courtyards where the inner one is the female or children's domain and the outer one is that of the males; the courtyards are the key socialisation spaces at a family level, as well as being used for multiple household activities. A *mohalla* was formed from a few houses; thus, the attributes of the houses are not only combined within it but imposed on the form of the *mohalla* to support these houses. This is the oldest pattern identified at various locations of indigenous Old Dhaka (Mowla, 1997). The celebration of native festivals and rituals was also still an integral part of Dhaka's social life. Though many of these rituals were religious in origin, they were pan-religious in celebration (Rahman & Haque, 2001) and were celebrated from inner courtyard to the *chowk* of the *mohalla*.

## 6 DISCUSSION

Altman and Wandersman (1987) pointed out that communal life, such as the presence of local institutions, official recognition, the type of housing, the pattern of social interaction and organisation, and the ethnic, socioeconomic, and demographic makeup of residents, provides the communal identity and essential characteristics of an area. The following list attempts to summarise the archetypal characteristics of the two indigenous cities, Old Delhi and Old Dhaka:

a. Demarcation of city fringe: by natural defence (river and canals) and fortification.
b. Access patterns: organic/greatly modified gridiron/webbed.
c. Access hierarchy: primary roads (connected with city gates or *ghats*); secondary roads (town bazaars); tertiary road (main streets of *mohallas*); quaternary and quinary roads (lanes, byways/*galis*).
d. Integrated business hubs: neighbourhood shops; *chowks*; *katras*; *ghats*.
e. Integrated social hubs: mostly *chowks*.
f. Integrated recreational hubs: streets (*gali*), but mostly street nodes (*morhs*).
g. Neighbourhood patterns: localities with clusters of *mohallas* (mostly single-occupation).
h. Settlement texture: introverted garden city with shared walls.
i. Settlement patterns: *havelis* (mansions) and courtyard houses are the most dominant residential units.
j. Location of institutions within neighbourhood: mostly religious structures along with the main access; dominant presence of *panchayat* office (*bangla*) within *mohalla*.
k. Private/family hubs: courtyard(s) (inner, outer) to adjacent *gali* (street) and connected roof terraces.

## 7 CONCLUSION

After taking control of the Indian subcontinent, British rulers developed a lot of infrastructure, buildings, cities and urban areas. In most cases, the British designers ignored the traditional patterns and instead imposed Western ideas of planning and design. In developing both New Delhi and the extension of Old Dhaka, the street patterns and urban design seemed deliberately contradictory and negating of existing parts. Moreover, these urban developments were not uniform throughout the city but were concentrated in the isolated areas where European and native elites resided. In contrast, the old cities were seen as impervious, perilous and a place of communal conflict. On the basis of these preconceived viewpoints, the Western designers saw these cities as unpredictable objects to reform or be remade and they remained fully unembraced by the newer developments. As a result, infrastructure improvements bypassed these areas and led to further decline, thus confirming prevailing discourses. In consequence, places such as Old Delhi and Old Dhaka, once seen as the focus of subcontinental culture, were now seen in negative terms, that is, in terms of what they lacked or the threat they proffered.

In a congested and seemingly chaotic environment, it is very difficult to enforce rules because so many seem to be breaking them. Instead, to a very large extent, the continuing peaceful functioning of public spaces depends on the common sense and mutual 'give and take' of the many people who use them. The concept of *jugaad* (a word taken from Hindi which captures the meaning of finding a low-cost solution to any problem in an intelligent way), the legendary Indian capacity to find solutions to problems through ingenuity, is crucial in understanding how the public realm functions in a highly congested area like Old Delhi (Gandhia, 2015).

The survival and success of old South Asian cities require great ingenuity and adaptability, but the most important factor is the human ecology. Here, in the indigenous context, the complex human ecology provides an astonishing variety of prospects within an inconsequential range. As a result, the distances between the availability of facilities and settlements are shorter than in the countryside or on the metropolitan periphery. At present, it seems that the areas of Old Delhi and Old Dhaka are fully occupied and an unfit place to survive if one looks from non-indigenous insights. But these urban environments remain historic treasures and have yet to reveal the archetypes of existence with religious tolerance and high-density mixed-use environments that continue to be viable and desirable in the 21st century.

However, in recent times these indigenous contexts and ecological prospects have been changing and changing violently. Fonseca (1969) claimed that it was a tragic loss to India when planners rejected all of the traditions and institutions of the indigenous urban scene—institutions uniquely Indian and arguably superior to the Western tradition that replaced them. Mahmud (2007) argued that, in Old Dhaka, the traditional way of life within the neighbourhood is now changing and people are losing the neighbourhood identity they were once proud of, because the physical built environment is transforming. In conclusion, we cite the thoughtful quotation of Doshi (1995) on efficient urban design:

> *The models of indigenous cities have their own problems too. Congestion, pollution and traffic chaos may have been synonymous with old neighbourhoods but it would be wrong to blame the physical form or the built environment of the place. The culprit is the unhealthy densification and an overtaxed infrastructure and not the spatial configuration.*

## REFERENCES


Ahmed, S.U. (1986). *Dacca: A study in urban history and development*. Dhaka, Bangladesh: Curzon Press.

Ahsan, R.M. (1991). *Dhaka: Past Present Future*. Chapter-Changing Pattern of the Commercial Area of Dhaka City (pp. 432–437), Dhaka, Bangladesh: The Asiatic Society of Bangladesh.

Ali, M.M. (1996). Future of the past. Paper presented at the International Seminar on Future of the Past: Architectural Heritage of Dhaka, 13–14 January 1996, Bangladesh University of Engineering and Technology (BUET), Dhaka, Bangladesh.

Altman, I. & Wandersman, A. (1987). *Neighbourhood and community environments: Human behavior and environments*. New York, NY: Plenum Press.

Ashraf, K.K. (2012). *Designing Dhaka: A manifesto for a better city*. Dhaka, Bangladesh: Loka Press.

Bromley, R. (2016), The Streets of Old Delhi: Order in a Seemingly-Chaotic Public Realm, *Governing Megacities in the MENA and Asia*. Retrieved from: http://www.mei.edu/content/map/streets-old-delhi-order-seemingly-chaotic-public-realm.

Chowdhury, A.M. & Faruqui, S. (1991). *Physical growth of Dhaka*. Dhaka, Bangladesh: Asiatic Society.

Clinard, M.B. (1966). *Slums and community development: Experiments in self-help* (pp. 73–74). New York, NY: Free Press.

Conzen, M.R.G. (2004), *Thinking About Urban Form: Papers on Urban Morphology, 1932–1998*. Bern, Switzerland: Peter Lang.

Dani, A.H. (1962). *Dacca: A record of its changing fortune*. Dacca, Pakistan: Asiatic Society.

Dhaka City Corporation. (2006, September 5). Pre-Mughal History of Dhaka.

Doshi, V.B. (1995). Architecture and Society. *EARTH*, *2*(1), 6.

Ferdous, F. (2012). The spatial analysis and morphological evolution of the 'Bazaar Streets' and urban structure of Dhaka city. *URBAN DESIGN International*, *17*(3), 206–220. doi:10.1057/udi.2012.10.

Fisher, T. (1993). Do the suburbs have a future? *Progressive Architecture*, *13*, 37.

Fonseca, R. (1969). The walled city of Old Delhi. In P. Oliver (Ed.), *Shelter and society* (pp. 103–115). London, UK: Barrie & Rockliff.

Gandhia, A. (2015). Delicious Delhi: Nostalgia, consumption and the old city. *Identities*, *23*, 1–17. doi: 10.1080/1070289X.2015.1034130.

Geddes, P. (1917). *Report on town planning: Dacca*. Calcutta, India: Bengal Secretariat Book Depot.

Hossain, M.S. (2013). Strategies to integrate the Mughal settlements in Old Dhaka. *Frontiers of Architectural Research*, *2*, 420–434.

Islam, N. (1996). Dhaka from city to megacity. In G. Clarke & J.R. Stubbs (Eds.), *Megacity management in the Asian and Pacific region* (Vol. 2). Manila, Philippines: Asian Development Bank.

Khan, I.M. (1982). *Alternative approach to the redevelopment of old Dacca* (Doctoral dissertation, Katholieke University of Leuven, Belgium).

Khan, I.M. (1985). Liveability of Old Dhaka: Evolving residential patterns in mahallas. In R. Powell (Ed.), *Regionalism in architecture* (pp. 123–135). Singapore: Concept Media. Retrieved from https://archnet.org/publications/2709.

Mahmud, S. (2007). Identity crisis due to transformation of home environment: The case for two Muslim cities, Dhaka and Hofuf. *Middle East Technical University Journal of the Faculty of Architecture*, *24*(2), 37–56.

Marshall, J. (Ed.). (1931). *Mohenjo-Daro and the Indus civilization*. London, UK: Arthur Probsthain.
Mowla, Q.A. (1997). Settlement texture: Study of mahalla in Dhaka. *Journal of Urban Design*, *2*(3), 259–275.
Muktadir, M.A. & Hasan, D.M. (1985). Traditional house form in rural Bangladesh: A case study for regionalism in architecture. In R. Powell (Ed.), *Regionalism in architecture* (pp. 81–86). Singapore: Concept Media. Retrieved from https://archnet.org/publications/3676.
Nath, R. (2006). *History of Mughal architecture*. New Delhi, India: Abhinav.
Nazem, N.I. (2003), "Urban Growth and Urbanization in Bangladesh: Interpretation of Census 2001" in CUS Bulletin on Urbanization and Development No. 45, July–December, 9–11.
Nilufar, F. (2004). Hidden morphological order in an organic city. *Protibesh, Journal of Department of Architecture, Bangladesh University of Engineering and Technology*, *9*, 34–41.
Nilufar, F. (2011). Urban morphology of Dhaka City: Spatial dynamics of growing city and the urban core. In *400 Years of Capital Dhaka and Beyond, Vol. III, Urbanization and urban development* (pp. 187–210). Dhaka, Bangladesh: Asiatic Society of Bangladesh. Retrieved from http://www.academia.edu/245137/Urban_Morphology_of_Dhaka_City_Spatial_Dynamics_of_Growing_City_and_the_Urban_Core.
Rahman, M. & Haque, F.A. (2001). Multiple courtyard mansions of Dhaka: Form and context. *Traditional Dwellings and Settlements Review*, *12*(2), 57–71.
Rahman, M.M. & Imon, S.S. (2016). How can historic waterfront conservation help to improve the quality of life in Old Dhaka. *American Scientific Research Journal for Engineering, Technology, and Sciences (ASRJETS)*, *26*(2), 200–218.
Rapoport, A. (1977). *Human aspects of urban form—Towards a man environment approach to urban form and design*. Oxford, UK: Pergamon Press.
Roy, S. (2011, September 1). A tale of two cities. *Hindustan Times*. Retrieved from http://www.hindustantimes.com/delhi-news/a-tale-of-two-cities/story-8jkSX5QwS3DVpl3pK4KqZJ.html.
Smith, M.E. (2007). Form and meaning in the earliest cities: A new approach to ancient urban planning. *Journal of Planning History*, *6*, 3–47.
Taifoor, S.M. (1956). *Glimpses of Old Dhaka*. Dacca, Pakistan: S M Perwez.
Tankel, S.B. (1963). The importance of open space in the urban pattern. In I. Wingo (Ed.), *Cities and space*. Baltimore, MD: Johns Hopkins University Press.
Texas Austin, U. (2004). Southern Asia. Retrieved from http://www.lib.utexas.edu/maps/asia.html.

*Cities' Identity Through Architecture and Arts – Catalani et al. (Eds)*
*© 2018 Taylor & Francis Group, London, ISBN 978-1-138-05409-7*

# Author index

For Product Safety Concerns and Information please contact our
EU representative GPSR@taylorandfrancis.com Taylor & Francis
Verlag GmbH, Kaufingerstraße 24, 80331 München, Germany